Physical Chemistry
of Surfaces

Physical Chemistry of Surfaces

Fourth Edition

ARTHUR W. ADAMSON

Department of Chemistry, University of Southern California
Los Angeles, California

A Wiley-Interscience Publication

JOHN WILEY & SONS

New York • Chichester • Brisbane • Toronto • Singapore

Library of Congress Cataloging in Publication Data:

Adamson, Arthur W.
Physical chemistry of surfaces.

"A Wiley-Interscience publication."
Bibliography: p.
Includes index.
1. Surface chemistry. 2. Chemistry, Physical
and theoretical. I. Title.

QD506.A3 1982 541.3′453 82-2711
ISBN 0-471-07877-8 AACR2

Printed in the United States of America

10 9 8 7

To the students
past, present, and future

Preface

The character and the orientation of this fourth edition are much the same as those of the third. I hope that *Physical Chemistry of Surfaces* will continue to serve as a textbook for senior and graduate level courses, both of academic and of industrial venue. As before, it is assumed that students have completed the usual undergraduate year course in physical chemistry.

I hope, too, that professional chemists will continue to find the book useful both for its references and as an entrée into a field that is often neglected in formal training. Surface chemistry remains, of course, as important as ever to large areas of established industrial chemistry and technology. There are, in addition, newer classes of entrants into the field of surface chemistry to whom this book may be useful. Molecular spectroscopists and theoreticians are, for example, finding the phenomenology of high-vacuum surface chemistry to be an enticing challenge. Secondly, a variety of organic and inorganic chemists are realizing the importance of high-surface-area systems to the successful conversion of solar energy.

The book is, however, primarily a textbook—a teaching book. There is a mass of established, fundamental material to be covered—a mass that grows relatively slowly with time. This portion of the material has, correspondingly, changed slowly with successive editions. However, as usual, I have added timely new theory and fundamental material and deleted or skimped older material of diminished interest (the book is long enough).

Some specific other changes are as follows. There is a separate chapter on high-vacuum or "dry" surface chemistry, dealing with diffraction and spectroscopic effects, now multitudinous. The illustrative material has been updated throughout the book; there is a considerable turnover of literature citations, and many of the older figures have been replaced by new ones. It remains my opinion that the SI system of units is not particularly useful or relevant to physical chemistry, but in deference to foreign users and to a certain fraction of U.S. students, there is now some detailed attention to the interconversion of SI and cgs units. This is particularly true of electrical quantities (see Chapter VI). A number of new problems have been written, including some that use SI units.

The drifting apart of the "dry" and the more traditional, "wet," surface chemists was noted in the Preface to the third edition. This separation has been deplored by the Division of Colloid and Surface Chemistry (of the American Chemical Society), leading the Division to establish the Langmuir Award Lectures, in which an annual back-to-back pair of formal lectures is given by a wet and by a dry surface chemist. I do hope that efforts such as the above will multiply and will be successful in keeping the two groups in contact. Each, I am convinced, has much to learn from the other. Where possible, I have tried to introduce or to point out bridges between the two types of fields.

Much of this edition was prepared during a pleasant stay as Visiting Professor at the University of Queensland, Brisbane, Australia. My wife and I gratefully acknowledge the warm hospitality of the Chemistry Department as a whole and of Sally and John Hall and of Beth and Geoffrey Barnes in particular. I acknowledge with much thanks the many colleagues who kindly responded to my request for comments on the third edition and for reprints of their more recent work. Also, B. V. Derjaguin kindly added an appendix on disjoining pressure to the Russian translation of the third edition, and I have borrowed from this appendix in the present edition. My thanks go to L. Dormant, W. A. Steele, G. Somorjai, and particularly to B. Crawford for looking over the sections on units. I am, finally, most grateful to V. Slawson and to my wife, Virginia, for the time they spent in proofreading. The book is distinctly more error-free because of their efforts.

ARTHUR W. ADAMSON

May 1982

Answers to problems are available from the author
c/o Department of Chemistry
University of Southern California
Los Angeles, California 90007

Contents

I. General Introduction **1**

II. Capillarity **4**

1. Surface Tension and Surface Free Energy 4
2. The Equation of Young and Laplace 6
3. Some Experiments with Soap Films 8
4. The Treatment of Capillary Rise 10
   A. Introductory Discussion 10
   B. Exact Solutions to the Capillary Rise Problem 12
   C. Experimental Aspects of the Capillary Rise Method . . . 17
5. The Maximum Bubble Pressure Method 18
6. The Drop Weight Method 20
7. The Ring Method 23
8. Wilhelmy Slide Method 24
9. Methods Based on the Shape of Static Drops or Bubbles . . . 27
   A. Pendant Drop Method 28
   B. Sessile Drop or Bubble Method 29
   C. Deformed Interfaces 34
10. Dynamic Methods of Measuring Surface Tension 36
    A. Flow Methods 37
    B. Capillary Waves 38
11. Surface Tension Values as Obtained by Different Methods . . 39
12. Problems 39
    General References 45
    Textual References 46

III. The Nature and Thermodynamics of Liquid Interfaces **49**

1. One-Component Systems 49
   A. Surface Thermodynamic Quantities for a Pure Substance . 49
   B. The Total Surface Energy, E^s 52
   C. Change in Vapor Pressure for a Curved Surface 54
   D. Effect of Curvature on Surface Tension 55
   E. Effect of Pressure on Surface Tension 56
2. The Structural and Theoretical Treatment of Liquid Interfaces . 56
   A. Further Development of the Thermodynamic Treatment of the Surface Region 59

B. Calculation of the Surface Energy and Free Energy of Liquids . . . 61
3. Orientation at Interfaces . . . 63
4. The Surface Tension of Solutions . . . 65
5. Thermodynamics of Binary Systems—The Gibbs Equation . . 70
A. Definition of Surface Excess . . . 70
B. The Gibbs Equation . . . 72
C. Alternate Methods of Locating the Dividing Surface . . . 73
D. The Thermodynamics of Surfaces Using the Concept of a Surface Phase . . . 76
E. Other Surface Thermodynamic Relationships . . . 77
6. Verification of the Gibbs Equation—Direct Measurements of Surface Excess Quantities . . . 78
A. The Microtome Method for Measuring Surface Excess . . 79
B. The Tracer Method for Measuring Surface Excess 80
C. Ellipsometric Method for Measuring Surface Excess . . . 82
7. Gibbs Monolayers . . . 83
A. The Two-Dimensional Ideal Gas Law . . . 84
B. Nonideal Two-Dimensional Gases . . . 86
C. The Osmotic Pressure Point of View . . . 88
D. Dynamic Surface Properties of Solutions . . . 90
E. Traube's Rule . . . 92
F. Some Further Comments on Gibbs Monolayers . . . 93
8. Problems . . . 93
General References . . . 96
Textual References . . . 97

IV. Surface Films on Liquid Substrates . . . 100
1. Introduction . . . 100
2. The Spreading of One Liquid on Another . . . 103
A. Criteria for Spreading . . . 104
B. Kinetics of Spreading Processes . . . 109
C. Lenses . . . 111
3. Experimental Techniques for the Study of Monomolecular Films . . . 111
A. Measurement of π . . . 112
B. Surface Potentials . . . 114
C. Surface Viscosities . . . 117
D. Optical Properties of Monolayers . . . 121
E. The Ultramicroscope . . . 124
F. Electron Microscopy and Diffraction . . . 124
G. Other Techniques . . . 124
4. States of Monomolecular Films . . . 124
5. Correspondence between π and a Three-Dimensional Pressure . 128
6. Further Discussion of the States of Monomolecular Films . . . 128
A. Gaseous Films . . . 128
B. The L_1–G Transition . . . 129
C. The Liquid Expanded State . . . 130
D. Intermediate and L_2 Films . . . 132

E. The Solid State . . . 133
F. Effect of Changes in the Aqueous Substrate . . . 134
G. Rheology of Monolayers . . . 135
H. General Correlations between Molecular Structure and the Type of Film Formed . . . 136
7. Mixed Films . . . 139
8. Evaporation Rates through Monomolecular Films . . . 143
9. Rate of Dissolving of Monolayers . . . 146
10. Reactions in Monomolecular Films . . . 148
A. Kinetics of Reactions in Films . . . 148
B. Kinetics of Formation and Hydrolysis of Esters . . . 150
C. Other Chemical Reactions . . . 152
11. Films of Biological and Polymeric Materials . . . 153
A. General Properties and Structure . . . 153
B. Reactions of Protein Films . . . 157
C. Films at the Oil–Water Interface . . . 158
D. Polymer Films . . . 160
12. Films at Liquid–Liquid Interfaces and on Liquid Surfaces Other than Water . . . 161
13. Charged Films . . . 162
A. Equation of State of a Charged Film . . . 162
B. Interfacial Potentials . . . 167
14. Capillary Waves . . . 168
15. Films Deposited on Solids . . . 172
A. Built-Up Films . . . 172
B. Monolayers . . . 173
16. Problems . . . 174
General References . . . 176
Textual References . . . 176

V. Electrical Aspects of Surface Chemistry . . . 185
1. Introduction . . . 185
2. The Electrical Double Layer . . . 185
3. Units—the SI System . . . 189
A. Potential . . . 190
B. Coulomb's Law and Equations of Electrostatics . . . 190
4. The Stern Treatment of the Electrical Double Layer . . . 191
5. Further Treatment of the Stern and Diffuse Layers . . . 193
6. The Free Energy of a Diffuse Double Layer . . . 195
7. Repulsion between Two Plane Double Layers . . . 196
8. The Zeta Potential . . . 198
A. Electrophoresis . . . 199
B. Electroosmosis . . . 200
C. Streaming Potential . . . 202
D. Sedimentation Potential . . . 203
E. Further Developments in the Theory of Electrokinetic Phenomena . . . 204
F. General Observation of ζ Potentials—Stability of Colloids . . . 204

9. Electrocapillarity 206
   A. Thermodynamics of the Electrocapillary Effect 208
   B. Experimental Methods 211
   C. Results for the Mercury–Aqueous Solution Interface . . . 212
   D. Effect of Uncharged Solutes and Changes of Solvent . . . 214
   E. Other Electrocapillary Systems 217
10. The Electrified Solid–Liquid Interface 217
   A. Electrode–Solution Interface 218
   B. Silver Halide–Solution Interface 218
11. Types of Potentials and the Meaning of Potential Differences When Two Phases are Involved 219
   A. The Various Types of Potentials 219
   B. Volta Potentials, Surface Potential Differences, and the Thermionic Work Function 221
   C. Electrode Potentials 222
   D. Irreversible Electrode Phenomena 224
12. Problems 227
   General References 229
   Textual References 229

VI. Long-Range Forces **232**

1. Introduction 232
2. Forces between Atoms and Molecules 233
3. The SI System 237
4. Long-Range Forces 238
   A. Attraction Due to the Dispersion Effect 239
   B. The Retarded Dispersion Attraction 240
   C. Experimental Verification 241
5. Long-Range Forces in Solution 244
   A. Dispersion Attraction in a Condensed Medium 244
   B. Electrical Double Layer Repulsion 245
   C. Other Experimental Approaches 247
6. The Disjoining Pressure 248
7. Dipole-Induced Dipole Propagation 249
8. Evidence for Deep Surface Orientation 251
9. Anomalous Water 253
10. Problems 254
   General References 256
   Textual References 257

VII. Surfaces of Solids **260**

1. Introduction 260
   A. The Surface Mobility of Solids 260
   B. Effect of Past History on the Condition of Solid Surfaces . 262
2. Thermodynamics of Crystals 263
   A. Surface Tension and Surface Free Energy 263
   B. The Equilibrium Shape of a Crystal 264
   C. The Kelvin Equation 266

3. Theoretical Estimates of Surface Energies and Free Energies . 266
A. Covalently Bonded Crystals 266
B. Rare Gas Crystals 267
C. Ionic Crystals 270
D. Molecular Crystals 273
E. Metals 273
4. Factors Affecting the Surface Energies and Surface Tensions of Actual Crystals 274
A. State of Subdivision 274
B. Deviations from Ideal Considerations 275
C. Dislocations 277
D. Surface Heterogeneity 280
5. Experimental Estimates of Surface Energies and Free Energies 280
A. Methods Depending on the Direct Manifestation of Surface Tensional Forces 280
B. Surface Energies and Free Energies from Heats of Solution 282
C. Relative Surface Tensions from Equilibrium Shapes of Crystals 283
D. Dependence of Other Physical Properties on Surface Energy Changes at a Solid Interface 284
6. Reactions of Solid Surfaces 285
7. Problems 287
General References 289
Textual References 290

VIII. Surfaces of Solids: Microscopy and Spectroscopy **294**
1. Introduction 294
2. Microscopy of Surfaces 294
A. Optical and Electron Microscopy 294
B. Scanning Electron Microscope 299
C. Field Emission and Field Ion Microscopy 300
3. Low Energy Electron Diffraction (LEED) 305
4. Spectroscopic Methods 310
A. Auger Electron Spectroscopy (AES) 310
B. Photoelectron Spectroscopy (ESCA) 311
C. Ion Scattering (ISS, LEIS) 312
5. Problems 315
General References 316
Textual References 316

IX. The Formation of a New Phase—Nucleation and Crystal Growth . . **319**
1. Introduction 319
2. Classical Nucleation Theory 320
3. Results of Nucleation Studies 325
4. Crystal Growth 327
5. Problems 329
General References 329
Textual References 329

X. The Solid–Liquid Interface—Contact Angle **332**
1. Introduction 332
2. Surface Free Energies from Solubility Changes 332
3. Surface Energy and Free Energy Differences from Immersion, Adsorption, and Engulfment Studies 334
   A. Heat of Immersion 334
   B. Surface Energy and Free Energy Changes from Adsorption Studies 335
   C. Engulfment 337
4. Contact Angle 338
   A. Young's Equation 338
   B. Nonuniform Surfaces 340
5. Experimental Methods and Results of Contact Angle Measurements 341
   A. Measurement of Contact Angle 341
   B. Hysteresis in Contact Angle Measurements 344
   C. Results of Contact Angle Measurements 348
6. Some Theoretical Aspects of Contact Angle Phenomena . . . 354
   A. Thermodynamics of the Young Equation 354
   B. Semiempirical Models—The Girifalco–Good–Fowkes–Young Equation 357
   C. Potential–Distortion Model 359
   D. The Microscopic Meniscus Profile 360
7. Problems 361
   General References 364
   Textual References 364

XI. The Solid–Liquid Interface—Adsorption from Solution **369**
1. Adsorption of Nonelectrolytes from Dilute Solution 369
   A. Adsorption Isotherms 370
   B. Qualitative Results of Adsorption Studies—Traube's Rule . 373
   C. Multilayer Adsorption 377
2. Adsorption of Polymers 378
3. Surface Area Determination 381
4. Adsorption in Binary Liquid Sytems 382
   A. Adsorption at the Solid–Solution Interface 382
   B. Heat of Adsorption at the Solid–Solution Interface 388
5. Adsorption of Electrolytes 388
   A. Stern Layer Adsorption 389
   B. Surface Areas from Negative Adsorption 393
   C. Counterion Adsorption—Ion Exchange 394
6. Problems 396
   General References 398
   Textual References 399

XII. Friction and Lubrication—Adhesion **402**
1. Introduction 402
2. Friction Between Unlubricated Surfaces 402
   A. Amontons' Law 402

B. Nature of the Contact Between Two Solid Surfaces 403
C. Role of Shearing and Plowing—Explanation of Amontons' Law 405
D. Static and "Stick-Slip" Friction 407
E. Rolling Friction 408
3. Two Special Cases of Friction 409
A. Use of Skid Marks to Estimate Vehicle Speeds 409
B. Ice and Snow 410
4. Metallic Friction—Effect of Oxide Films 411
5. Friction Between Nonmetals 412
A. Relatively Isotropic Crystals 412
B. Layer Crystals 412
C. Plastics 413
6. Some Further Aspects of Friction 413
7. Friction Between Lubricated Surfaces 415
A. Boundary Lubrication 415
B. The Mechanism of Boundary Lubrication 417
8. Adhesion 424
A. Ideal Adhesion 424
B. Practical Adhesion 425
9. Problems 428
General References 429
Textual References 430

XIII. Wetting, Flotation, and Detergency **433**
1. Introduction 433
2. Wetting 433
A. Wetting as a Contact Angle Phenomenon 433
B. Wetting as a Capillary Action Phenomenon 435
C. Tertiary Oil Recovery 436
3. Water Repellency 437
4. Flotation 438
A. The Role of Contact Angle in Flotation 439
B. Flotation of Metallic Minerals 443
C. Flotation of Nonmetallic Minerals 444
5. Detergency 446
A. General Aspects of Soil Removal 446
B. Properties of Colloidal Electrolyte Solutions 448
C. Factors in Detergent Action 452
D. Adsorption of Detergents on Fabrics 454
E. Detergents in Commercial Use 455
6. Problems 456
General References 458
Textual References 458

XIV. Emulsions and Foams **461**
1. Introduction 461
2. Emulsions—General Properties 462

3. Factors Determining Emulsion Stability 464
A. Macroscopic Theories of Emulsion Stabilization 464
B. Specific Chemical and Structural Effects 466
C. Long-Range Forces as a Factor in Emulsion Stability . . . 467
D. Stabilization of Emulsions by Solid Particles 470
4. The Aging and Inversion of Emulsions 470
A. Flocculation and Coagulation Kinetics 471
B. Inversion and Breaking of Emulsions 472
5. Spontaneous Emulsification—Micellar or Microemulsions . . . 474
6. The Hydrophile–Lipophile Balance 475
7. Practical Aspects of Emulsions 477
8. Foams—The Structure of Foams 478
9. Foam Drainage 480
A. Drainage of Single Films 481
B. Drainage of Foams 483
10. The Stability of Foams 484
11. Foaming Agents and Foams of Practical Importance 486
12. Problems 486
General References 487
Textual References 488

XV. The Solid–Gas Interface—General Considerations 492
1. Introduction 492
2. The Surface Area of Solids 493
A. The Meaning of Surface Area 493
B. Methods Requiring Knowledge of the Surface Free Energy or Total Energy 495
C. Rate of Dissolving 496
D. The Mercury Porosimeter 498
E. Other Methods of Surface Area Estimation 501
3. The Structural and Chemical Nature of Solid Surfaces 502
4. The Nature of the Solid–Adsorbate Complex 503
A. Effect of Adsorption on Adsorbate Properties 503
B. Effect of the Adsorbate on the Adsorbent 510
C. The Adsorbate–Adsorbent Bond 512
5. Problems 512
General References 513
Textual References 514

XVI. Adsorption of Gases and Vapors on Solids 517
1. Introduction 517
2. The Adsorption Time 519
3. The Langmuir Adsorption Isotherm 521
A. Kinetic Derivation 521
B. Statistical Thermodynamic Derivation 523
C. Adsorption Entropies 526
D. Lateral Interaction 529
E. Experimental Applications of the Langmuir Equation . . . 531
4. Experimental Procedures 531

5. The BET and Related Isotherms 533
A. Derivation of the BET Equation 535
B. Properties of the BET Equation 537
C. Modifications of the BET Equation 538
6. Isotherms Based on the Equation of State of the Adsorbed Film 539
A. Film Pressure–Area Diagrams from Adsorption Isotherms . 539
B. Adsorption Isotherms from Two-Dimensional Equations of State 540
7. The Potential Theory 543
A. The Polanyi Treatment 543
B. Correspondence between the Potential Theory and That of a Two-Dimensional Film 545
C. Isotherms Based on an Assumed Variation of Potential with Distance 546
D. The Polarization Model 547
8. Comparison of the Surface Areas from the Various Multilayer Models 549
9. The Characteristic Isotherm and Related Concepts 551
10. Potential Theory as Applied to Submonolayer Adsorption . . . 553
11. Adsorption Steps and Phase Transformations 556
12. Adsorption in the Case of a Nonwetting System 559
13. Thermodynamics of Adsorption 559
A. Theoretical Considerations 559
B. Experimental Heats and Entropies of Adsorption 565
14. Critical Comparison of the Various Models for Adsorption . . 570
A. The Langmuir–BET Model 570
B. Two-Dimensional Equation of State Treatments 571
C. The Potential Model 572
15. Adsorption on Heterogeneous Surfaces 573
A. Site Energy Distributions 574
B. Thermodynamics of Adsorption on Heterogeneous Surfaces 580
16. Rate of Adsorption 581
17. Adsorption on Porous Solids—Hysteresis 582
A. Molecular Sieves 582
B. Capillary Condensation 584
C. Micropore Analysis 590
18. Problems 591
General References 594
Textual References 595

XVII. Chemisorption and Catalysis 601
1. Introduction 601
2. Chemisorption—The Molecular View 602
A. LEED Structures 602
B. Spectroscopy of Chemisorbed Species 603
C. Work Function and Related Measurements 603
D. Flash Desorption 605

3. Chemisorption Isotherms 607
A. Variable Heat of Adsorption 607
B. Effect of Site and Adsorbate Coordination Number 610
C. Adsorption Thermodynamics 611
4. Kinetics of Chemisorption 611
A. Activation Energies 611
B. Rates of Adsorption 613
C. Rates of Desorption 615
5. Surface Mobility 616
6. The Chemisorption Bond 618
A. The Localized Bond Approach 618
B. Metals 621
C. Semiconductors 623
D. Acid–Base Systems 626
7. Mechanisms of Heterogeneous Catalysis 627
A. Adsorption or Desorption as the Rate-Determining Step . . 628
B. Reaction Within the Adsorbed Film as the Rate-Determining Step 629
8. Influence of the Adsorption Isotherm on the Kinetics of Heterogeneous Catalysis 631
A. Unimolecular Surface Reactions 632
B. Bimolecular Surface Reactions 633
C. Effect of Isotherm Complexities 634
9. Mechanisms of a Few Catalyzed Reactions 634
A. Ammonia Synthesis 634
B. Fischer–Tropsch Type Reactions 636
C. Hydrogenation of Ethylene 637
D. Catalytic Cracking of Hydrocarbons and Related Reactions 639
E. Photochemical and Photoassisted Processes at Surfaces . . 640
10. Problems 641
General References 642
Textual References 643

Index **649**

Physical Chemistry of Surfaces

CHAPTER I

General Introduction

The material of this book, according to its title, deals with the physical chemistry of surfaces. Although an obvious enough point, it is perhaps worth noting that in reality we will always be dealing with the *interface* between two phases and that in general the properties of an interface will be affected by changes in either of the two phases involved.

The interfaces possible can be summarized in a formal way in terms of the three states of matter—solid, liquid, and gas as follows:

Gas–Liquid
Gas–Solid
Liquid–Liquid
Liquid–Solid
Solid–Solid

A general prerequisite for the stable existence of an interface between two phases is that the free energy of formation of the interface be positive; were it negative or zero, the effect of accidental fluctuations would be to expand the surface region continuously and to lead to eventual complete dispersion of one material into the other. Examples of interfaces whose free energy per unit area is such as to offer no opposition to dispersive forces would be those between two dilute gases or between two miscible liquids or solids.

As implied, thermodynamics constitutes an important discipline within the general subject. It is one in which surface area joins the usual extensive quantities of mass and volume, and in which surface tension and surface composition join the usual intensive quantities of pressure, temperature, and bulk composition. The thermodynamic functions of free energy, enthalpy, and entropy can be defined for an interface as well. Chapters II and III are largely thermodynamic in nature, as is Chapter V, in which electrical potential and charge are added as thermodynamic variables.

Systems involving an interface are often *metastable*, that is, essentially equilibrium behavior is exhibited in certain aspects although the system as a whole may be unstable in other aspects. The solid-vapor interface is a common example. We can have adsorption equilibrium and calculate various thermodynamic quantities for the adsorption process; yet the particles of solid are unstable toward a drift to the final equilibrium condition of a single, perfect crystal. Much of Chapters X and XVI are thus thermodynamic in content.

The physical chemist is very interested in kinetics—in the mechanisms of chemical reaction, in the theory of rate processes, and in general in time as a variable. Correspondingly, the dynamics of interfaces is an important topic. As may be imagined, there is a wide spectrum both of types of rate phenomena and in the sophistication achieved in dealing with them. In some cases changes in area or in amounts of phases are involved, as in rates of evaporation, condensation, dissolving, precipitation, flocculation and coagulation, and adsorption and desorption. In other cases surface composition is changing, as with reactions in monolayers. The field of catalysis is largely one of the mechanistic kinetic study of surface reactions.

An important characteristic of an interface is that it is directional. Properties vary differently along an interface and perpendicular to an interface. This aspect provides leverage in the study of forces, especially long-range forces, between molecules. It is possible, for example, to measure *directly* the van der Waals force between two portions of matter. This area is one in which surface physical chemists have made fundamental contributions to physical chemistry as a whole. In addition, potentials for intermolecular attraction and repulsion, specifically discussed in Chapter VI, play a recurring role in this book.

Structure is as important in surface physical chemistry as in chemistry generally. The structure of a crystalline solid can be determined by x-ray diffraction studies, and the surface structure of a solid can, somewhat analogously, be determined by low energy electron diffraction (LEED). Both the structure and the chemical state of adsorbed molecules can be investigated by means of various surface spectroscopic techniques, and Chapter VIII is devoted to this topic. High vacuum surface spectroscopy has in fact become a major field. More recent and less developed is the subject of the photochemistry and excited state characteristics of adsorbed molecules.

We attempt to draw a line between surface physical chemistry and surface chemical physics and solid state physics of surfaces. These last two subjects are largely wave-mechanical in nature, and can be highly mathematical; they properly form a discipline of their own.

We also attempt to draw lines between surface physical chemistry and colloid and polymer physical chemistry. The distinction is not always easy and not always appropriate. The emphasis in surface physical chemistry, however, is on the thermodynamics, structure, and rate processes involving an interface, where the properties of the interfacial region are directly emphasized and more or less directly studied. In colloid and polymer physical chemistry, the emphasis is more on the collective properties of a disperse system. Light scattering by a suspension is not, for example, of central interest to the surface chemist. Nor is he directly concerned with random coil configurations of a long chain polymer in solution, or with polymer elasticity. Both topics do become of interest, however, if the polymer is adsorbed at an interface; the divisions between fields that have been so indicated are thus by no means sharp.

Finally, phenomenological, that is, macroscopically viewed surface physical chemistry finds a host of special situations of great practical importance. Contact angle, representing a balance between surface tensions at a three-phase boundary, is central to the enormous flotation industry. Wetting, adhesion, and detergency depend importantly on the control of interfacial tensions. Emulsions and foams are stabilized or destabilized through the judicious use of surface active agents, and so on. A variety of such topics is included in the book. The emphasis is on those aspects that have received sufficient attention for the subject to be somewhat under control. The surface chemical principles involved are reasonably well understood, and useful physical chemical models have been developed.

Clearly, the "physical chemistry of surfaces" covers a wide range of topics. Most of these topics are sampled in this book, with emphasis on fundamentals and on important theoretical models. With each topic there is some annotation of current literature, the citations often being chosen because they contain bibliographies that will provide detailed source material.

CHAPTER II

Capillarity

1. Surface Tension and Surface Free Energy

The topic of capillarity concerns interfaces that are sufficiently mobile to assume an equilibrium shape. The most common examples are meniscuses and drops formed by liquids in air or in another liquid and thin films such as that forming a soap bubble. Because it deals with equilibrium configurations, capillarity occupies a place in the general framework of thermodynamics—it deals with the macroscopic and statistical behavior of interfaces rather than with the details of their molecular structure.

Although referred to as a free energy per unit area, surface tension may equally well be thought of as a force per unit length. Two examples serve to illustrate these viewpoints. Consider, first, a soap film stretched over a wire frame, one end of which is movable (Fig. II-1). Experimentally one observes that a force is acting on the movable member in the direction opposite to that of the arrow in the diagram. If the value of this force per unit length is denoted by γ, then the work done in extending the movable member a distance dx is

$$\text{work} = \gamma l \, dx \qquad \text{(II-1)}$$

Equation II-1 could be equally well written as

$$\text{work} = \gamma \, d\mathcal{A} \qquad \text{(II-2)}$$

where $d\mathcal{A} = l \, dx$ and thus gives the change in area. In this second formulation, γ appears to be an energy per unit area. Customary units, then, may either be ergs per square centimeter (erg/cm^2) or dynes per centimeter (dyne/cm); these are identical dimensionally. The corresponding SI units are, of course, joules per square meter (J/m^2) or newtons per meter (N/m).

A second illustration involving soap films is that of the soap bubble. We will choose, here, to think of γ in terms of energy per unit area. In the absence of fields, such as gravitational, a soap bubble is spherical, this being the shape of minimum surface area for a given enclosed volume. Consider a soap bubble of radius r, as illustrated in Fig. II-2. Its total surface free energy is $4\pi r^2\gamma$ and, if the radius were to decrease by dr, then the change in surface free energy would be $8\pi r\gamma \, dr$. Since shrinking decreases the surface energy, the tendency to do so must be balanced by a pressure

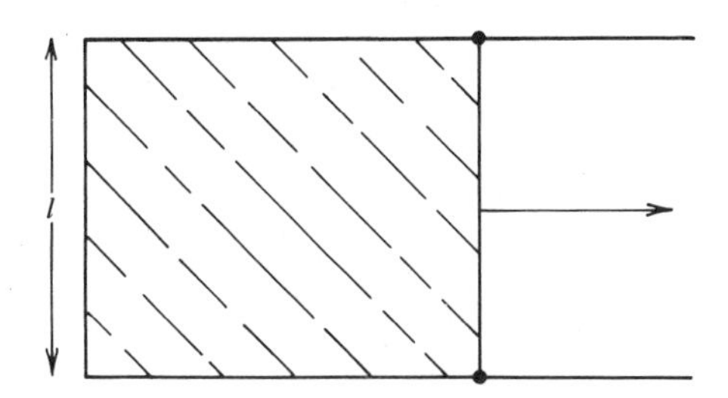

Fig. II-1

difference across the film $\mathbf{\Delta P}$, such that the work against this pressure difference $\mathbf{\Delta P}4\pi r^2\, dr$ is just equal to the decrease in surface free energy. Thus,

$$\mathbf{\Delta P}4\pi r^2\, dr = 8\pi r\gamma\, dr \tag{II-3}$$

or

$$\mathbf{\Delta P} = \frac{2\gamma}{r} \tag{II-4}$$

One thus arrives at the important conclusion that the smaller the bubble, the greater the pressure of the air inside compared to that outside. This conclusion is easily verified experimentally by arranging two bubbles with a common air connection, as illustrated in Fig. II-3. The arrangement is unstable, and the smaller of the two bubbles will shrink while the other enlarges. Note, however, that the smaller bubble does not shrink indefinitely; once its radius is equal to that of the tube, the radius will begin to increase with further shrinkage, and a stage must be reached such that the two radii become equal, as shown by the dotted lines. This final state is now one of mechanical equilibrium.

It might be noted that common usage defines γ as the surface tension for *one* interface. Because of this, it would be better to use the quantity 2γ

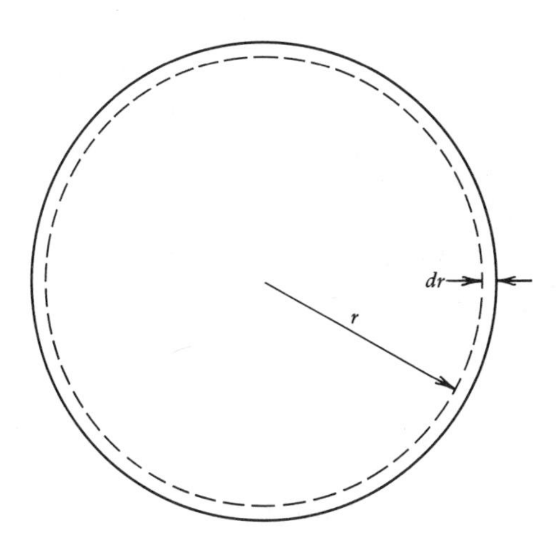

Fig. II-2

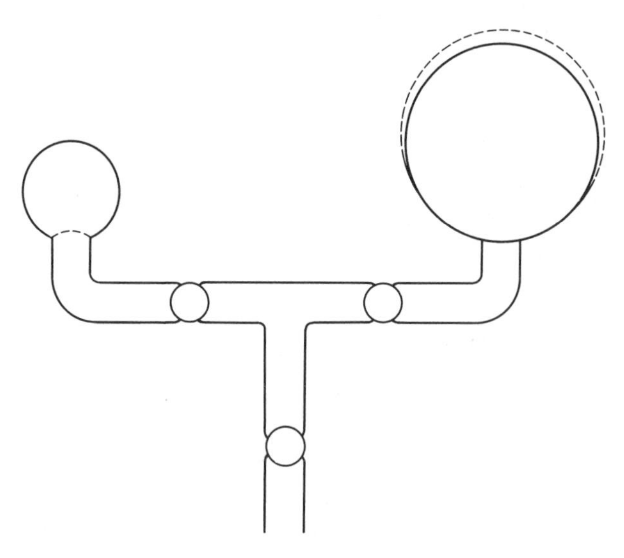

Fig. II-3. Illustration of the equation of Young and Laplace.

instead of γ in the preceding equations when they *are applied to soap or other two-sided films*.

The foregoing examples illustrate the point that equilibrium surfaces may be treated mathematically, using either the concept of surface tension or the (mathematically) equivalent concept of surface free energy. (The derivation of Eq. II-4 from the surface tension point of view is given as an exercise at the end of the chapter.) This mathematical equivalence holds everywhere in capillarity phenomena. As discussed in Section III-2, a similar duality of viewpoint can be argued on a molecular scale so that the decision as to whether *surface tension* or *surface free energy* is the more fundamental concept becomes somewhat a matter of individual taste. The two terms generally are used interchangeably in this book.

The term surface tension is the earlier of the two; it goes back to early ideas that the surface of a liquid had some kind of contractile "skin." More subtly, it can convey the erroneous impression that extending a liquid surface somehow stretches the molecules in it. In contrast, the term surface free energy implies only that work is required to form more surface, that is, to bring molecules from the interior of the phase into the surface region. For this reason, and also because it ties more readily into conventional chemical thermodynamic language, this writer considers the surface free energy concept to be preferable if a choice must be made.

2. The Equation of Young and Laplace

Equation II-4 is a special case of a more general relationship that is the basic equation of capillarity and was given in 1805 by Young (1) and by Laplace (2). In general, it is necessary to invoke two radii of curvature to describe a curved surface; these are equal for a sphere, but not necessarily otherwise. A small section of an arbitrarily curved surface is shown in Fig. II-4. The two radii of curvature, R_1 and R_2,† are indicated in the figure, and the section of surface taken is small enough so that R_1 and R_2 are essentially constant. Now if the surface is displaced a small distance outward, the change in area will be

$$\Delta\mathcal{A} = (x + dx)(y + dy) - xy = x\,dy + y\,dx$$

The work done in forming this additional amount of surface is then

$$\text{Work} = \gamma(x\,dy + y\,dx)$$

There will be a pressure difference $\Delta\mathbf{P}$ across the surface; it acts on the area xy and through a distance dz. The corresponding work is thus

$$\text{Work} = \Delta\mathbf{P}xy\,dz$$

From a comparison of similar triangles, it follows that

$$\frac{x + dx}{R_1 + dz} = \frac{x}{R_1} \quad \text{or} \quad dx = \frac{x dz}{R_1}$$

† It is perhaps worthwhile to digress briefly on the subject of radii of curvature. The two radii of curvature for some arbitrarily curved surface are obtained as follows. One erects a normal to the surface at the point in question and then passes a plane through the surface and containing the normal. The line of intersection in general will be curved, and the radius of curvature is that for a circle tangent to the line at the point involved. The second radius of curvature is obtained by passing a second plane through the surface, also containing the normal, but perpendicular to the first plane. This gives a second line of intersection and a second radius of curvature.

If the first plane is rotated through a full circle, the first radius of curvature will go through a minimum, and its value at this minimum is called the principal radius of curvature. The second principle radius of curvature is then that in the second plane, kept at right angles to the first. Because Fig. II-4 and Eq. II-7 are obtained by quite arbitrary orientation of the first plane, the radii R_1 and R_2 are not necessarily the principal radii of curvature. The pressure difference $\Delta\mathbf{P}$, cannot depend upon the manner in which R_1 and R_2 are chosen, however, and it follows that the sum $(1/R_1 + 1/R_2)$ is independent of how the first plane is oriented (although, of course, the second plane is always at right angles to it).

Most of the situations encountered in capillarity involve figures of revolution, and for these it is possible to write down explicit expressions for R_1 and R_2, by choosing plane 1 so that it passes through the axis of revolution. As shown in Fig. II-10(a), R_1 then swings in the plane of the paper, i.e., it is the curvature of the profile at the point in question. R_1 is therefore given simply by the expression from analytical geometry for the curvature of a line

$$1/R_1 = y''/(1 + y'^2)^{3/2} \tag{II-5}$$

where y' and y'' denote the first and second derivatives with respect to x. The radius R_2 must then be in the plane perpendicular to that of the paper and, for figures of revolution, must be given by prolonging the normal to the profile until it hits the axis of revolution, again as shown in Fig. II-10(a). Turning to Fig. II-10(b), the value of R_2 for the coordinates (x, y) on the profile is given by $1/R_2 = \sin\phi/x$, and since $\tan\phi$ is equal to y', one obtains the following expression for R_2

$$1/R_2 = y'/x(1 + y'^2)^{1/2} \tag{II-6}$$

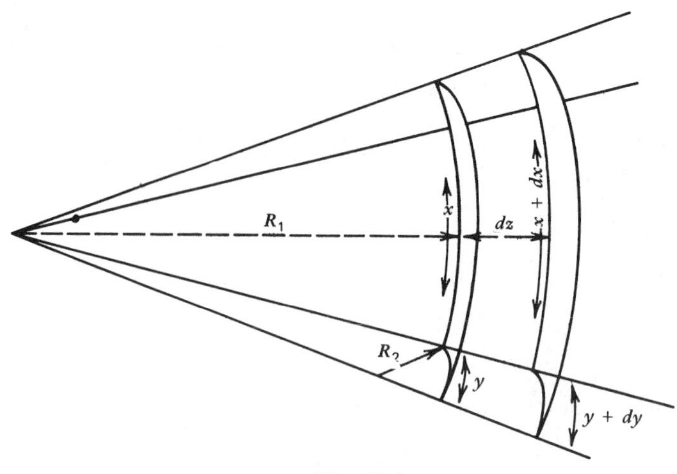

Fig. II-4

and

$$\frac{y + dy}{R_2 + dz} = \frac{y}{R_2} \quad \text{or} \quad dy = \frac{y\,dz}{R_2}$$

If the surface is to be in mechanical equilibrium, the two work terms as given must be equal, and on equating them and substituting in the expressions for dx and dy, the final result obtained is

$$\Delta \mathbf{P} = \gamma\left(\frac{1}{R_1} + \frac{1}{R_2}\right) \tag{II-7}$$

Equation II-7 is the fundamental equation of capillarity and will recur many times in this chapter.

It is apparent that Eq. II-7 reduces to Eq. II-4 for the case of both radii being equal, as is true for a sphere. For a plane surface, the two radii are each infinite and $\Delta\mathbf{P}$ is therefore zero; thus there is no pressure difference across a plane surface.

3. Some Experiments with Soap Films

There are a number of relatively simple experiments with soap films that illustrate beautifully some of the implications of the Young and Laplace equation. Two of these have already been mentioned. Neglecting gravitational effects, a film stretched across a frame as in Fig. II-1 will be planar because the presssure is the same on both sides of the film. The experiment depicted in Fig. II-3 illustrates the relation between the pressure inside a spherical soap bubble and its radius of curvature; by attaching a manometer, $\Delta\mathbf{P}$ could be measured directly.

An interesting set of shapes results if one forms a soap bubble between two cylindrical supports, as shown in Fig. II-5. In Fig. II-5*a*, the upper support is open to the atmosphere so that the pressure is everywhere the same, and **ΔP** must be zero. Although the surface appears to be curved, Eq. II-7 is nonetheless not contradicted. The two radii of curvature are indicated in Fig. II-5*a*, where R_1 swings in the plane of the paper and R_2 swings in the plane at right angles to it. It turns out that R_1 and R_2 are equal in magnitude but are opposite in sign because they originate on opposite sides of the film; they just cancel each other in Eq. II-7.

Instability of Cylindrical Columns. C. V. Boys, in his elegant little monograph (of 1890!) (3), discusses an important further aspect of the properties of quasi-cylindrical films. If, in Fig. II-5*a*, the upper cylindrical support is closed and the lower one is connected to an air-line and manometer, one may, by adjusting the gas pressure, make the soap film essentially a uniform cylinder in shape. An important phenomenon now appears. A uniform cylindrical bubble possesses a critical length beyond which it is unstable toward necking in at one end and bulging at the other, as shown in Fig. II-5*b*. This length is that which equals the circumference of the cylinder. A cylinder of length greater than this critical value thus promptly collapses into a smaller and larger bubble. The same is true of a cylinder of liquid as, for example, in the case of a stream emerging from a circular nozzle. Figure II-6 reproduces a photograph of such a stream (4) and illustrates clearly how the necking

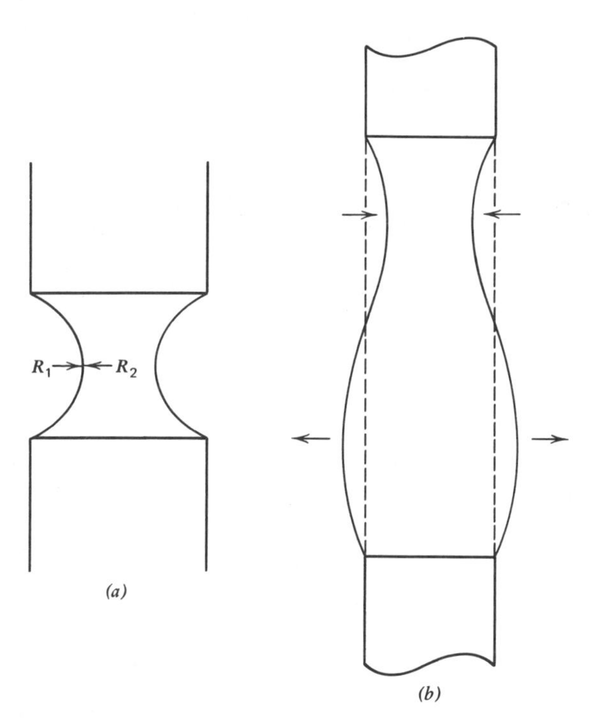

Fig. II-5. (*a*) A cylindrical soap film. (*b*) Manner of collapse of a cylindrical soap film of excessive length.

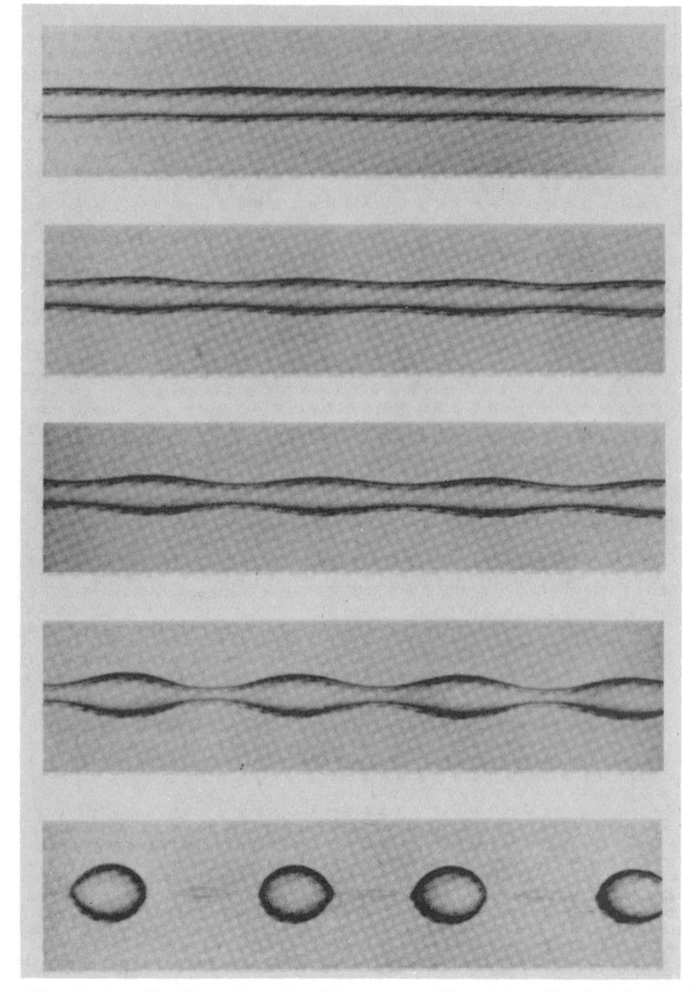

Fig. II-6. Necking in a liquid stream. [Courtesy S. G. Mason (4).]

in process leads to a breakup of the stream into alternate smaller and larger drops. A more recent discussion of the effect is given in Ref. 5.

Returning to equilibrium shapes, the equation of Young and Laplace has been solved for a number of boundary conditions, including cases where gravity cannot be neglected. See Refs. 6 and 7 for some recent reviews.

4. The Treatment of Capillary Rise

A. Introductory Discussion

An approximate treatment of the phenomenon of capillary rise is easily made in terms of the Young and Laplace equation. If the liquid wets the wall of the capillary, the liquid surface is thereby constrained to lie parallel with the wall, and the complete surface must therefore be concave in shape. The pressure difference across the interface is given by Eq. II-7, and its sign is such that the pressure is less in the liquid than in the gas phase. In this connection, it is helpful to remember that the radii of curvature (where both are of the same sign) always lie on that side of the interface having the greater pressure.

If the capillary is circular in cross section and not too large in radius, the meniscus will be approximately hemispherical, as illustrated in Fig. II-7. The two radii of curvature are thus equal to each other and to the radius of the capillary. Equation II-7 then reduces to

$$\Delta \mathbf{P} = \frac{2\gamma}{r} \tag{II-8}$$

where r is the radius of the capillary. If h denotes the height of the meniscus above a *flat* liquid surface (for which $\Delta \mathbf{P}$ must be zero), then $\Delta \mathbf{P}$ in Eq. II-8 must also equal the hydrostatic pressure drop in the column of liquid in the capillary. Thus $\Delta \mathbf{P} = \Delta\rho g h$, where $\Delta\rho$ denotes the difference in density between the liquid and gas phase and g is the acceleration due to gravity. Equation II-8 becomes

$$\Delta\rho g h = \frac{2\gamma}{r} \tag{II-9}$$

or

$$a^2 = \frac{2\gamma}{\Delta\rho g} = rh \tag{II-10}$$

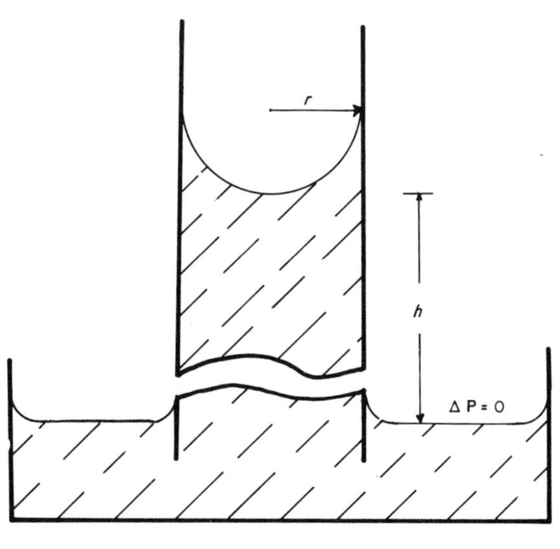

Fig. II-7. Capillary rise (capillary much magnified in relation to dish).

The quantity a, defined by Eq. II-10, is known as the *capillary constant* (as a source of confusion, some authors have defined $a^2 = \gamma/\Delta\rho g$).

Similarly, for a liquid that completely fails to wet the walls of the capillary, that is, one whose contact angle with the wall is 180° instead of 0°, the simple treatment yields the identical equation. There is now a capillary depression, however, because the meniscus is convex, and h is now the depth of this depression, as illustrated in Fig. II-8.

A slightly more general case is that in which the liquid meets the circularly cylindrical capillary wall at some angle θ, as illustrated in Fig. II-9. If the meniscus is still taken to be spherical in shape, it follows from simple geometric consideration that $R_2 = r/\cos\theta$ and, since $R_1 = R_2$, Eq. II-9 then becomes

$$\Delta\rho g h = \frac{2\gamma \cos\theta}{r} \tag{II-11}$$

B. Exact Solutions to the Capillary Rise Problem

The exact treatment of capillary rise must take into account the deviation of the meniscus from sphericity, that is, the curvature must correspond to the $\Delta\mathbf{P} = \Delta\rho g y$ at each point on the meniscus, where y is the elevation of that point above the flat liquid surface. The formal statement of the condition is obtained by writing the Young and Laplace equation for a general point (x, y) on the meniscus, with R_1 and R_2 replaced by the expressions from analytical geometry given in the footnote to Section II-2. We still assume that the capillary is circular in cross section so that the meniscus shape is that of a figure of revolution; as indicated in Fig. II-9, R_1 swings in the plane

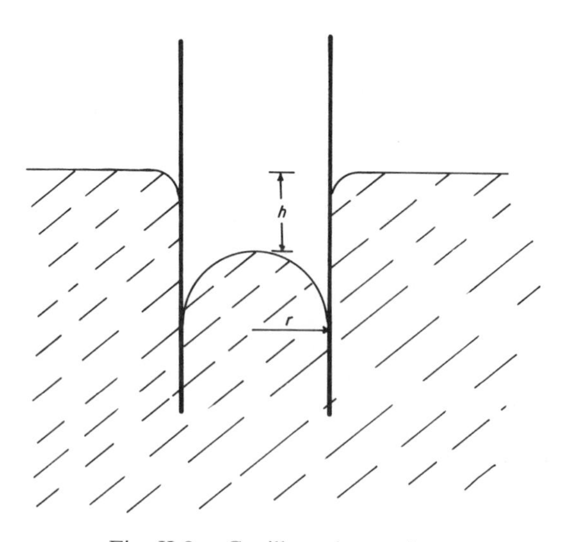

Fig. II-8. Capillary depression.

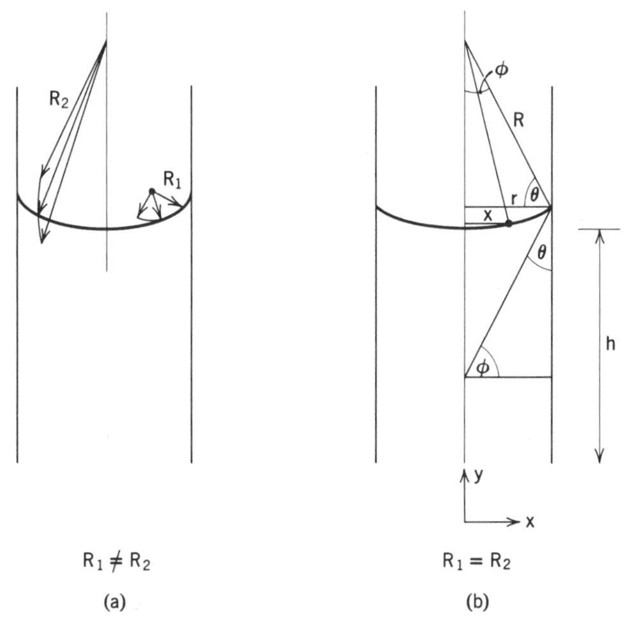

Fig. II-9. The meniscus in a capillary as a figure of revolution.

of paper, and R_2 in the plane perpendicular to the paper. One thus obtains

$$\Delta\rho g h = \gamma\left[\frac{y''}{(1 + y'^2)^{3/2}} + \frac{y'}{x(1 + y'^2)^{1/2}}\right] \tag{II-12}$$

where $y' = dy/dx$ and $y'' = d^2y/dx^2$.

The total weight W of the column of liquid in the capillary may be obtained exactly from Eq. II-12. Let $p = y'$ and, hence, $y'' = p\, dp/dy$. The equation may then be written

$$\frac{2y}{a^2} = \frac{p\, dp/dy}{(1 + p^2)^{3/2}} + \frac{p}{x(1 + p^2)^{1/2}} \tag{II-13}$$

Since W is given by

$$W = 2\Delta\rho g\pi \int_0^r xy\, dx \tag{II-14}$$

elimination of y by the use of Eq. II-13 leads to the equation

$$W = 2\pi\gamma \int \left[\frac{x\, dp}{(1 + p^2)^{3/2}} + \frac{p\, dx}{(1 + p^2)^{1/2}}\right] \tag{II-15}$$

The integrand turns out to be the exact differential of $xp/(1 + p^2)^{1/2}$, so that Eq. II-15 reduces to

$$W = 2\pi\gamma\left[\frac{xp}{(1 + p^2)^{1/2}}\right]_{x=0,\ p=0}^{x=r,\ p=\tan\phi} \tag{II-16}$$

Since $p = dy/dx$, at $x = r$, $p = \tan \phi$, where $\phi = 90° - \theta$. Inserting these limits, one obtains

$$W = 2\pi r\gamma \cos \theta \tag{II-17}$$

Notice that Eq. II-17 is exactly what one would write, assuming the meniscus to be "hanging" from the wall of the capillary and its weight to be supported by the vertical component of the surface tension $\gamma \cos \theta$ multiplied by the circumference of the capillary cross section $2\pi r$. Thus, once again, the mathematical identity of the concepts of "surface tension" and "surface free energy" is observed.

While Eq. II-17 is exact, its experimental use requires the determination of the total weight of liquid in the capillary, and this is not generally convenient to do. More commonly, one measures the height h to the *bottom* of the meniscus.

Unfortunately, it has not been possible to obtain an explicit solution to the general equation (Eq. II-12) in terms of h, the usual experimental parameter. Approximate solutions of entirely adequate accuracy have been obtained, however, for the case of $\theta = 0$. This last restriction is not actually a very limiting one, as the usual observation is that θ is difficult to reproduce experimentally except in the case where it is zero.

These approximate solutions have been obtained in two forms. The first, given by Lord Rayleigh (8), is that of a series approximation. The derivation is not repeated here, but for the case of nearly spherical meniscus, that is, $r \ll h$, expansion around a deviation function led to the equation

$$a^2 = r\left(h + \frac{r}{3} - \frac{0.1288r^2}{h} + \frac{0.1312r^3}{h^2} \cdots\right) \tag{II-18}$$

The first term gives the elementary equation (Eq. II-10). The second term takes into account the weight of the meniscus, assuming it to be spherical (see Problem 3 at the end of the chapter). The succeeding terms provide corrections of deviation from sphericity.

The general case has been solved by Bashforth and Adams (9), using an iterative method, and extended by Sugden (10). See also Refs. 6 and 7. In the case of a figure of revolution, the two radii of curvature must be equal at the apex (i.e., at the bottom of the meniscus in the case of capillary rise). If this radius of curvature is denoted by b, and the elevation of a general point on the surface is denoted by z, where $z = y - h$, then Eq. II-7 can be written

$$\gamma\left(\frac{1}{R_1} + \frac{1}{R_2}\right) = \Delta\rho g z + \frac{2\gamma}{b} \tag{II-19}$$

Thus, at $z = 0$, $\Delta \mathbf{P} = 2\gamma/b$, and at any other value of z, the change in $\Delta \mathbf{P}$ is given by $\Delta\rho g z$. Equation II-19 may be rearranged so as to involve only dimensionless parameters

$$\frac{1}{R_1/b} + \frac{\sin \phi}{x/b} = \beta\left(\frac{z}{b}\right) + 2 \tag{II-20}$$

where R_2 has been replaced by its equivalent, $x/\sin \phi$, and the dimensionless quantity β is given by

$$\beta = \frac{\Delta\rho g b^2}{\gamma} = \frac{2b^2}{a^2} \qquad \text{(II-21)}$$

This parameter is positive for oblate figures of revolution, that is, for a sessile drop, a bubble under a plate, and a meniscus in a capillary. It is negative for prolate figures, that is, for a pendant drop or a clinging bubble.

Bashforth and Adams obtained solutions to Eq. II-20 (with R_1 replaced by the expression in analytical geometry), using a numerical integration procedure (this was before the day of high speed digital computers, and their work required tremendous labor). Their results are reported as tables of values of x/b and z/b for closely spaced values of β and of φ. For a given β value, a plot of z/b versus x/b gives the profile of a particular figure of revolution satisfying Eq. II-20. By way of illustration, their results for β = 80 are reproduced (in abbreviated form) in Table II-1. Observe that x/b reaches a maximum at φ = 90°, so that in the case of zero contact angle the surface is now tangent to the capillary wall and hence $(x/b)_{\max} = r/b$. The corresponding value of r/a is given by $(r/b)\sqrt{\beta/2}$. In this manner, Sugden compiled tables of r/b versus r/a, and his results are given in Tables II-2 and II-3.

The use of these tables is perhaps best illustrated by means of a numerical example. In a measurement of the surface tension of benzene, the following data are obtained:

Radius of capillary: 0.0550 cm

Density of benzene: 0.8785; density of air: 0.0014 (both at 20°C); hence, $\Delta\rho = 0.8771$ g/ml

Height of capillary rise: 1.201 cm

We compute a first approximation to the value of the capillary constant a_1 by means of Eq. II-10 ($a^2 = rh$). The ratio r/a_1 is then obtained and the

TABLE II-1
Solution to Eq. I–20 for β = 80

φ, deg	x/b	z/b	φ, deg	x/b	z/b
5	0.08159	0.00345	100	0.33889	0.17458
10	0.14253	0.01133	110	0.33559	0.18696
20	0.21826	0.03097	120	0.33058	0.19773
30	0.26318	0.05162	130	0.32421	0.20684
40	0.29260	0.07204	140	0.31682	0.21424
50	0.31251	0.09183	150	0.30868	0.21995
60	0.32584	0.11076	160	0.30009	0.22396
70	0.33422	0.12863	170	0.29130	0.22632
80	0.33872	0.14531			
90	0.34009	0.16067			

TABLE II-2
Solution to the Young and Laplace Equation for the Case of Capillary Rise with Zero Contact Angle (Values of r/b for Values of r/a from 0.00 to 2.29)

r/a	0.00	0.01	0.02	0.03	0.04	0.05	0.06	0.07	0.08	0.09
0.00	1.0000	9999	9998	9997	9995	9992	9988	9983	9979	9974
0.10	0.9968	9960	9952	9944	9935	9925	9915	9904	9893	9881
0.20	9869	9856	9842	9827	9812	9796	9780	9763	9746	9728
0.30	9710	9691	9672	9652	9631	9610	9589	9567	9545	9522
0.40	9498	9474	9449	9424	9398	9372	9346	9320	9293	9265
0.50	9236	9208	9179	9150	9120	9090	9060	9030	8999	8968
0.60	8936	8905	8873	8840	8807	8774	8741	8708	8674	8640
0.70	8606	8571	8536	8501	8466	8430	8394	8358	8322	8286
0.80	8249	8212	8175	8138	8101	8064	8026	7988	7950	7913
0.90	7875	7837	7798	7759	7721	7683	7644	7606	7568	7529
1.00	7490	7451	7412	7373	7334	7295	7255	7216	7177	7137
1.10	7098	7059	7020	6980	6941	6901	6862	6823	6783	6744
1.20	6704	6665	6625	6586	6547	6508	6469	6431	6393	6354
1.30	6315	6276	6237	6198	6160	6122	6083	6045	6006	5968
1.40	5929	5890	5851	5812	5774	5736	5697	5659	5621	5583
1.50	5545	5508	5471	5435	5398	5362	5326	5289	5252	5216
1.60	5179	5142	5106	5070	5034	4998	4963	4927	4892	4857
1.70	4822	4787	4753	4719	4686	4652	4618	4584	4549	4514
1.80	4480	4446	4413	4380	4347	4315	4283	4250	4217	4184
1.90	4152	4120	4089	4058	4027	3996	3965	3934	3903	3873
2.00	3843	3813	3783	3753	3723	3683	3663	3633	3603	3574
2.10	3546	3517	3489	3461	3432	3403	3375	3348	3321	3294
2.20	3267	3240	3213	3186	3160	3134	3108	3082	3056	3030

corresponding value of r/b read from Table II-2; in the present case, $a_1^2 = 1.201 \times 0.0550 = 0.0660$; hence, $r/a_1 = 0.0550/0.2569 = 0.2142$. From Table II-2, r/b_1 is then 0.9855. Since b is the value of R_1 and of R_2 at the bottom of the meniscus, the equation

$$a^2 = bh \tag{II-22}$$

TABLE II-3
Values of r/b for Values of r/a Larger than 2.00

r/a	0.0	0.1	0.2	0.3	0.4	0.5	0.6	0.7	0.8	0.9
2.0	0.384	355	327	301	276	252	229	206	185	166
3.0	149	133	119	107	097	088	081	074	067	061
4.0	056	051	047	043	039	035	031	028	025	022
5.0	020	018	017	015	014	012	010	009	008	007
6.0	006	006	005	004	004	003	003	003	002	002

is exact. From the value of r/b_1, we obtain a first approximation to b, that is, $b_1 = 0.0550/0.9855 = 0.05590$. Insertion of this value of b into Eq. II-22 gives a second approximation to a, that is, $a_2^2 = b_1h = 0.05590 \times 1.201 = 0.06710$. A second round of approximation is not needed in this case but would be carried out by computing r/a_2; then, from Table II-2, r/b_2, and so on. The value of 0.06710 for a^2 obtained here leads to 28.86 dyne/cm for the surface tension of benzene (at 20°C).

The calculation may be repeated for those preferring or more familiar with SI units (see, however, Ref. 11). The radius is now 5.50×10^{-4} m, the densities become 878.5 and 1.4 kg/m^3, and h is 1.20×10^{-2} m. We find $a_1^2 = 6.60 \times 10^{-6}$ m^2; the dimensionless ratio r/a_1 remains unchanged. The final approximation gives $a^2 = 6.710 \times 10^{-6}$ m^2, whence

$$\gamma = \frac{877.1 \times 9.807 \times 6.710 \times 10^{-6}}{2} = 2.886 \times 10^{-2} \text{ N/m (or J/m}^2)$$

Note that the SI unit of surface tension is 10^3 times the cgs unit.

Recalculations have been made of the Bashforth and Adams and Sugden tables by Padday and co-workers (12) and by Lane (13). This last author gives quite exact analytical expressions for (b/r) as a function of (r/a) improving, in particular, on the values in Table II-3. Finally, Erikson (14) has published computer calculations on the areas of nonspherical interfaces.

C. Experimental Aspects of the Capillary Rise Method

The capillary rise method is generally considered to be the most accurate of all methods, partly because the theory has been worked out with considerable exactitude and partly because the experimental variables can be closely controlled. This is to some extent an historical accident, and other methods now rival or surpass the capillary rise one in value.

Perhaps the best discussions of the experimental aspects of the capillary rise method are those given by Richards and Carver (15) and Harkins and Brown. (16). *For the most accurate work, it is necessary that the liquid wet the wall of the capillary so that there be no uncertainty as to the contact angle.* Because of its transparency and because it is wet by most liquids, a glass capillary is most commonly used. The glass must be very clean, and even so it is wise to use a receding meniscus. The capillary must be accurately vertical, of accurately known and uniform radius, and should not deviate from circularity in cross section by more than a few percent.

As is evident from the theory of the method, h must be the height of rise above a surface for which $\Delta \mathbf{P}$ is zero, that is, a flat liquid surface. In practice, then, h is measured relative to the surface of the liquid in a wide outer tube or dish, as illustrated in Fig. II-7, and it is important to realize that there may be an appreciable capillary rise in relatively wide tubes. Thus, for water, the rise is 0.04 mm in a dish 1.6 cm in radius, although it is only 0.0009 mm in one of 2.7 cm radius.

The general attributes of the capillary rise method may be summarized as follows. It is considered to be one of the best and most accurate absolute methods, good to a few hundredths of a percent in precision. On the other hand, for practical reasons, a zero contact angle is required, and fairly large volumes of solution are needed. With glass capillaries, there are limitations as to the alkalinity of the solution. For variations in the capillary rise method see Refs. 6, 7, and 17 to 20.

5. The Maximum Bubble Pressure Method

The procedure, as indicated in Fig. II-10, is slowly to blow bubbles of an inert gas in the liquid in question by means of a tube projecting below the surface. As also illustrated in the figure, for *small* tubes, the sequence of shapes assumed by the bubble during its growth is such that, while it is always a section of a sphere, its radius goes through a minimum when it is just hemisphereical. At this point the radius is equal to that of the tube and, since the radius is at a minimum, $\Delta\mathbf{P}$ is at a maximum. The value of $\Delta\mathbf{P}$ is then given by Eq. II-4, where r is the radius of the tube. If the liquid wets the material of the tube, the bubble will form from the inner wall, and r will then be the inner radius of the tube. Experimentally, then, one measures the maximum gas pressure in the tube such that bubbles are unable to grow and break away. Referring again to Fig. II-10, since the tube is some arbitrary distance t below the surface of the liquid, $\Delta\mathbf{P}_{max}$ is given by $(\mathbf{P}_{max} - \mathbf{P}_t)$, where $\mathbf{P}_{max}$ is the measured maximum pressure and $\mathbf{P}_t$ is the pressure corresponding to the hydrostatic head t.

If $\Delta\mathbf{P}_{max}$ is expressed in terms of the corresponding height of a column of the liquid, that is, $\Delta\mathbf{P}_{max} = \Delta\rho gh$, then the relationship becomes identical to that for the simple capillary rise situation as given by Eq. II-10.

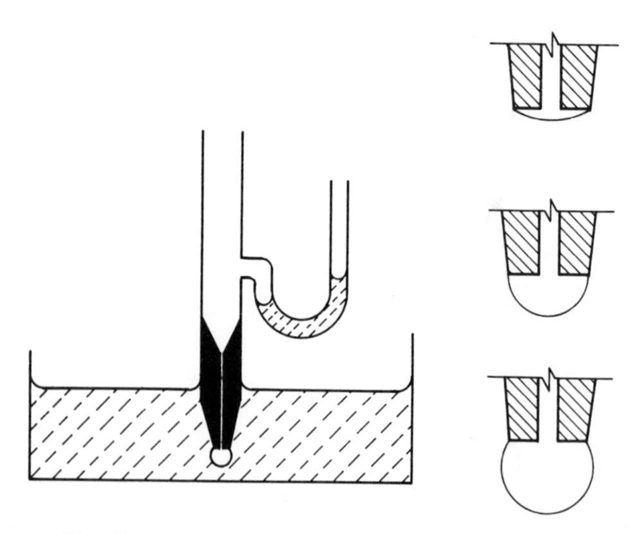

Fig. II-10. Maximum bubble pressure method.

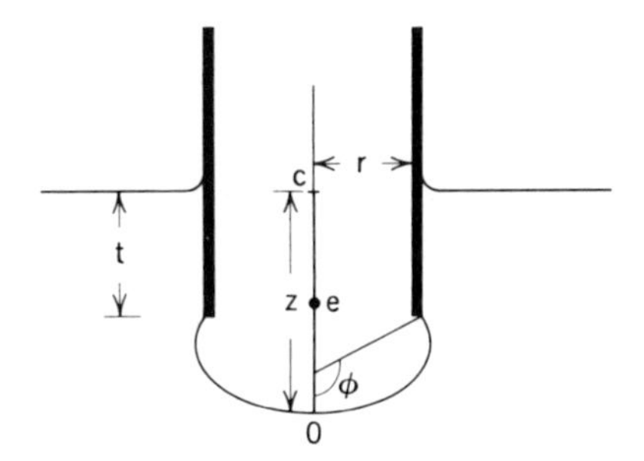

Fig. II-11

It is important to realize that the preceding treatment is the limiting one for sufficiently small tubes and that significant departures from the limiting Eq. II-10 occur for r/a values as small as 0.05. More realistically, the situation is as shown in Fig. II-11, and the maximum pressure may not be reached until ϕ is considerably greater than 90°.

As in the case of capillary rise, Sugden (21) has made use of Bashforth's and Adams' tables to calculate correction factors for this method. Because the figure is again one of revolution, the equation $h = a^2/b + z$ is exact, where b is the value of $R_1 = R_2$ at the origin and z is the distance of OC. The equation simply states that $\mathbf{\Delta P}$, expressed as height of a column of liquid, equals the sum of the hydrostatic head and the pressure change across the interface; by simple manipulation, it may be put in the form

$$\frac{r}{X} = \left(\frac{r}{b}\right) + \left(\frac{r}{a}\right)\left(\frac{z}{b}\right)\left(\frac{\beta}{2}\right)^{1/2} \qquad \text{(II-23)}$$

where β is given by Eq. II-21 and $X = a^2/h$. For any given value of r/a there will be a series of values of r/X corresponding to a series of values of β and of ϕ. For each assumed value of r/a, Sugden computed a series of values of r/b by inserting various values of β in the identity $r/b = (r/a)(2/\beta)^{1/2}$. By means of the Bashforth and Adams Tables (9), for each β value used and corresponding r/b value, a value of z/b and hence of r/X (by Eq. II-23) was obtained. Since r/X is proportional to the pressure in the bubble, the series of values for a given r/a go through a maximum as β is varied. For each assumed value, Sugden then tabulated this maximum value of r/X. His values are given in Table II-4 as X/r versus r/a.

The table is used in much the same manner as are Tables II-2 and II-3 in the case of capillary rise. As a first approximation, one assumes the simple Eq. II-10 to apply, that is, that $X = r$; this gives the first approximation a_1 to the capillary constant. From this, one obtains r/a_1 and reads the corresponding value of X/r from Table II-4. From the derivation of X ($X = a^2/h$), a second approximation a_2 to the capillary constant is obtained, and so on. Some more recent calculations have been made by Johnson and Lane (22).

The maximum bubble pressure method is good to a few tenths percent accuracy, does not depend on contact angle (except insofar as to whether the inner or outer radius of the tube is to be used), and requires only an approximate knowledge of the density of the liquid (if twin tubes are used), and the measurements can be made rapidly. A bubble rate of about 1/sec seems desirable, and the method is therefore a quasi-dynamic one in that freshly formed liquid-air interfaces are involved. It

TABLE II-4
Correction Factors for the Maximum Bubble Pressure Method
(Minimum Values of X/r for Values of r/a from 0 to 1.50)

r/a	0.00	0.01	0.02	0.03	0.04	0.05	0.06	0.07	0.08	0.09
0.0	1.0000	9999	9997	9994	9990	9984	9977	9968	9958	9946
0.1	0.9934	9920	9905	9888	9870	9851	9831	9809	9786	9762
0.2	9737	9710	9682	9653	9623	9592	9560	9527	9492	9456
0.3	9419	9382	9344	9305	9265	9224	9182	9138	9093	9047
0.4	9000	8952	8903	8853	8802	8750	8698	8645	8592	8538
0.5	8484	8429	8374	8319	8263	8207	8151	8094	8037	7979
0.6	7920	7860	7800	7739	7678	7616	7554	7493	7432	7372
0.7	7312	7252	7192	7132	7072	7012	6953	6894	6835	6776
0.8	6718	6660	6603	6547	6492	6438	6385	6333	6281	6230
0.9	6179	6129	6079	6030	5981	5933	5885	5838	5792	5747
1.0	5703	5659	5616	5573	5531	5489	5448	5408	5368	5329
1.1	5290	5251	5213	5176	5139	5103	5067	5032	4997	4962
1.2	4928	4895	4862	4829	4797	4765	4733	4702	4671	4641
1.3	4611	4582	4553	4524	4496	4468	4440	4413	4386	4359
1.4	4333	4307	4281	4256	4231	4206	4181	4157	4133	4109
1.5	4085									

cannot therefore be used very well to study the aging of surfaces, but where pure liquids are involved, the influence of surface active impurities is minimized. The method is also amenable to remote operation and can be used to measure surface tensions of not easily accessible liquids such as molten metals (23). A differential method may be used (24).

6. The Drop Weight Method

This is a fairly accurate method and perhaps the most convenient laboratory one for measuring the surface tension of a liquid–air or a liquid–liquid interface. As illustrated in Fig. II-12, the procedure is to form drops of the liquid at the end of a tube, allowing them to fall into a container until enough have been collected so that the weight per drop can be determined accurately.

The method is a very old one, remarks on it having been made by Tate in 1864 (25), and a simple expression for the weight W of a drop is given by what is known as Tate's law†:

$$W = 2\pi r\gamma \tag{II-24}$$

Here again, the older concept of "surface tension" appears, since Eq. II-

† The actual statement by Tate is: "Other things being equal, the weight of a drop of liquid is proportional to the diameter of the tube in which it is formed." See Refs. 26 and 27 for some discussion.

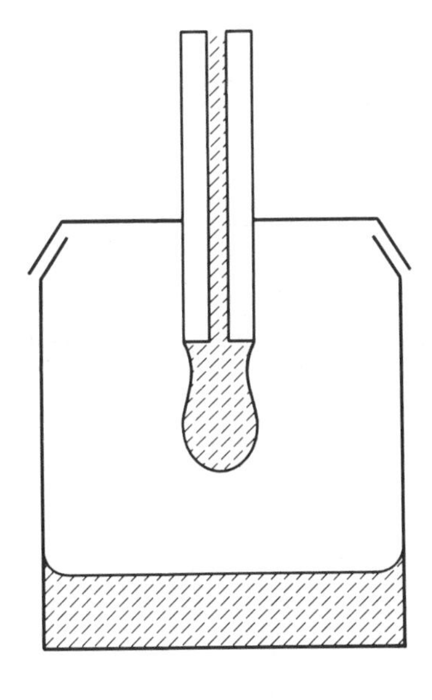

Fig. II-12. Drop weight method (drop and tip enlarged).

24 is best understood in terms of the argument that the maximum force available to support the weight of the drop is given by the surface tension force per centimeter times the circumference of the tip.

In actual practice, a weight W' is obtained, which is less than the "ideal" value W. The reason for this becomes evident when the process of drop formation is observed closely. What actually happens is illustrated in Fig. II-13. The small drops arise from the mechanical instability of the thin cylindrical neck that develops (see Section II-3); in any event, it is clear that only a portion of the drop that has reached the point of instability actually falls—as much as 40% of the liquid may remain attached to the tip.

The usual procedure is to apply a correction factor f to Eq. II-24, so that W' is given by

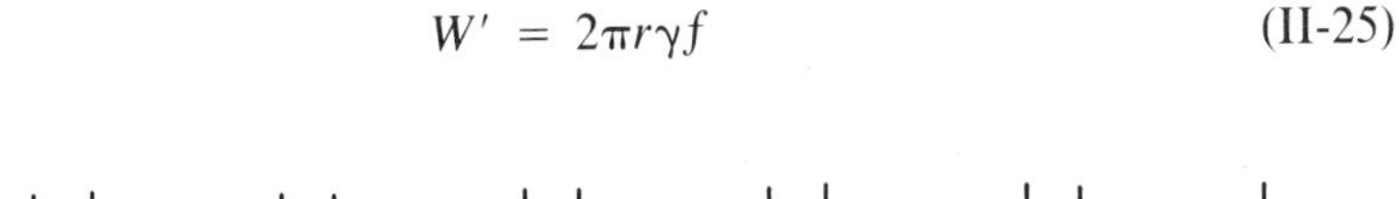

$$W' = 2\pi r \gamma f \tag{II-25}$$

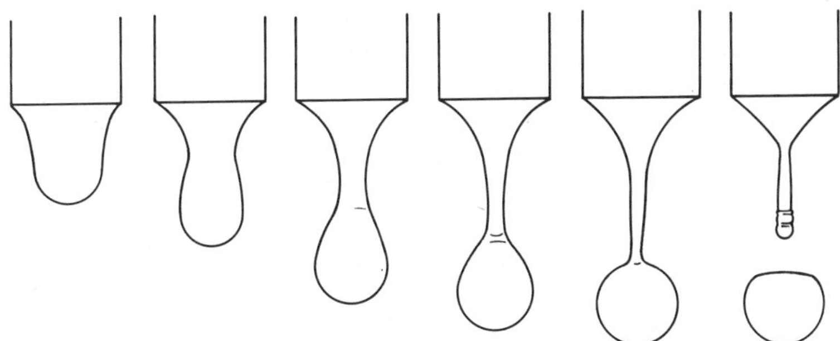

Fig. II-13. High speed photographs of a falling drop.

TABLE II-5
Correction Factors for the Drop Weight Method

$r/V^{1/3}$	f	$r/V^{1/3}$	f	$r/V^{1/3}$	f^a
0.00	(1.0000)	0.75	0.6032	1.225	0.656
0.30	0.7256	0.80	0.6000	1.25	0.652
0.35	0.7011	0.85	0.5992	1.30	0.640
0.40	0.6828	0.90	0.5998	1.35	0.623
0.45	0.6669	0.95	0.6034	1.40	0.603
0.50	0.6515	1.00	0.6098	1.45	0.583
0.55	0.6362	1.05	0.6179	1.50	0.567
0.60	0.6250	1.10	0.6280	1.55	0.551
0.65	0.6171	1.15	0.6407	1.60	0.535
0.70	0.6093	1.20	0.6535		

[a] The values of f in this column are less accurate than the others.

Harkins and Brown (16) concluded that f should be a function of the dimensionless ratio r/a or, alternatively, of $r/V^{1/3}$, where V is the drop volume. (See Refs. 26 and 27 for a more up-to-date discussion.) This they verified experimentally by determining drop weights for water and for benzene, using tips of various radii. Knowing the values of γ from capillary rise measurements, and thence the respective values of a, f could be determined in each case. The resulting variation of f with $r/V^{1/3}$ is tabulated in Table II-5.

It is desirable to use $r/V^{1/3}$ values in the region of 0.6 to 1.2, where f is varying most slowly. The table is used as follows. From the experimental value of m, the mass per drop, the volume per drop V is determined from the density of the liquid, and $r/V^{1/3}$ is evaluated. From the table, the corresponding value of f is determined, and the correct value for the surface tension is then given by

$$\gamma = \frac{mg}{2\pi rf} \tag{II-26}$$

It is to be noted that not only is the correction quite large, but for a given tip radius it depends on the nature of the liquid. It is thus *incorrect* to assume that the drop weights for two liquids are in the ratio of the respective surface tensions when the same size tip is used. Finally, correction factors for $r/V^{1/3} < 0.3$ have been determined, using mercury drops (28).

In employing this method, an important precaution to take is to use a tip that has been ground smooth at the end and which is free from any nicks. In the case of liquids that do not wet the tip, r is the inside radius. For volatile liquids, some sort of closed system should be employed to eliminate evaporation losses, such as is described by Harkins and Brown. The drops should be formed slowly, although actually this is necessary only during the last stages just prior to detachment; even for a drop time of 1 min, only 0.2% error is introduced. The method, however, is good to 0.1%.

The drop method, of course, may be used for the determination of liquid–liquid interfacial tensions. In this case, drops of one liquid are formed within the body of the second. The same equations apply, although it must be remembered that W' and m now denote the weight and mass of the drop minus that of the displaced liquid. Also, the validity of Table II-5 is not fully established as accurate if both fluids are viscous (26).

The method may also be used for solutions, but it is a dynamic one and not well suited to systems that establish their equilibrium surface tension slowly.

7. The Ring Method

A method that has been rather widely used involves the determination of the force to detach a ring or loop of wire from the surface of a liquid. The method belongs in the family of detachment methods, of which the drop weight (Section II-6) and one version of the Wilhelmy slide method (Section II-8) are also examples. It is generally attributed to du Noüy (29). As with all detachment methods, one supposes that a first approximation to the detachment force is given by the surface tension multiplied by the periphery of the surface detached. Thus, for a ring, as illustrated in Fig. II-14,

$$W_{tot} = W_{ring} + 4\pi R\gamma \tag{II-27}$$

Harkins and Jordan (30) found, however, that Eq. II-27 was generally in serious error and worked out an empirical correction factor in much the same way as was done for the drop weight method. Here, however, there is one additional variable so that the correction factor f now depends on two dimensionless ratios. Thus

$$f = \left(\frac{\gamma}{p}\right) = f\left(\frac{R^3}{V}, \frac{R}{r}\right) \tag{II-28}$$

where p denotes the "ideal" surface tension computed from Eq. II-27. and

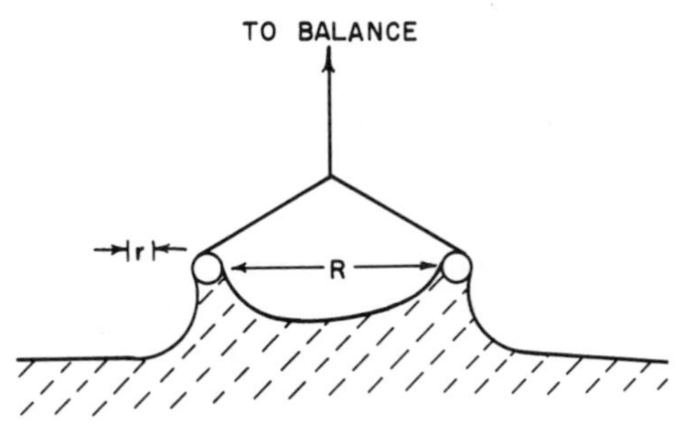

Fig. II-14. Ring method.

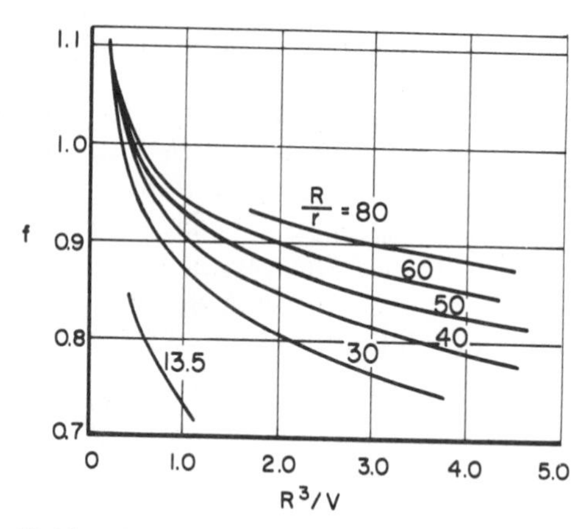

Fig. II-15. Correction factor plots for the ring method.

V is the meniscus volume. The extensive tables of Harkins and Jordan are summarized graphically in Fig. II-15, and it is seen that the simple equation may be in error by as much as 25%. An extension of the tables to cover higher densities and lower γ's is available (31). The detailed theory of the method is quite complicated, but it has been worked out by Freud and Freud (32), and the calculated values of f agree with the empirical ones to within the experimental precision of about 0.25%.

Experimentally, the method is capable of good precision. Harkins and Jordan used a chainomatic balance to determine the maximum pull, but a popular simplified version of the *tensiometer*, as it is sometimes called, makes use of a torsion wire and is quite compact. Among experimental details to mention are that the dry weight of the ring, which is usually constructed of platinum, is to be used, the ring should be kept horizontal (a departure of 1° was found to introduce an error of 0.5%, whereas one of 2.1° introduced an error of 1.6%), and care must be taken to avoid any disturbance of the surface as the critical point of detachment is approached. The ring is usually flamed before use to remove surface contaminants such as grease, and it is desirable to use a container for the liquid that can be overflowed so as to ensure the presence of a clean liquid surface.

A zero or near zero contact angle is necessary; otherwise results will be low. This was found to be the case with surfactant solutions where adsorption on the ring changed its wetting characteristics, and where liquid–liquid interfacial tensions were measured. In such cases a Teflon or polyethylene ring may be used (33). When used to study monolayers, it may be necessary to know the increase in area at detachment, and some calculations of this are available (34).

8. Wilhelmy Slide Method

The methods so far discussed have required more or less tabular solutions, or else correction factors to the respective "ideal" equations. Yet there is one method, attributed to Wilhelmy (35) in 1863, that entails no such corrections and is very simple to use.

The basic observation is that a thin plate, such as a microscope cover glass or piece of platinum foil, will support a meniscus whose weight both as measured statically or by detachment is given very accurately by the "ideal" equation (assuming zero contact angle):

$$W_{\text{tot}} = W_{\text{plate}} + \gamma p \tag{II-29}$$

where p is the perimeter. The experimental arrangement is shown schematically in Fig. II-16. When used as a detachment method, the procedure is essentially the same as with the ring method, but Eq. II-29 holds to within 0.1% so that no corrections are needed (36–38).

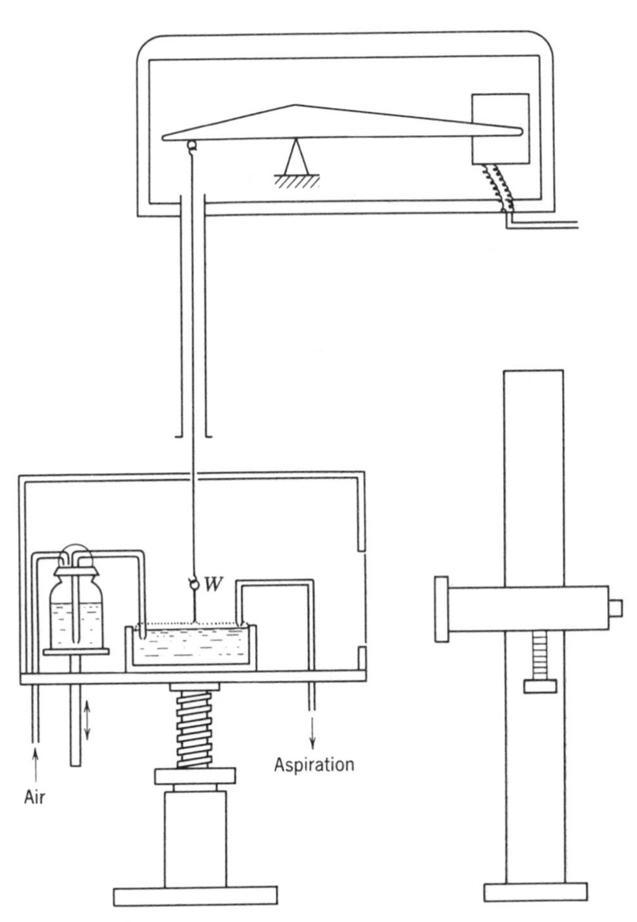

Fig. II-16. Apparatus for measuring the time dependence of interfacial tension (from Ref. 40). The air and aspirator connections allow for establishing the desired level of fresh surface. *W* denotes the Wilhelmy slide, suspended from a Cahn electrobalance with a recorder output.

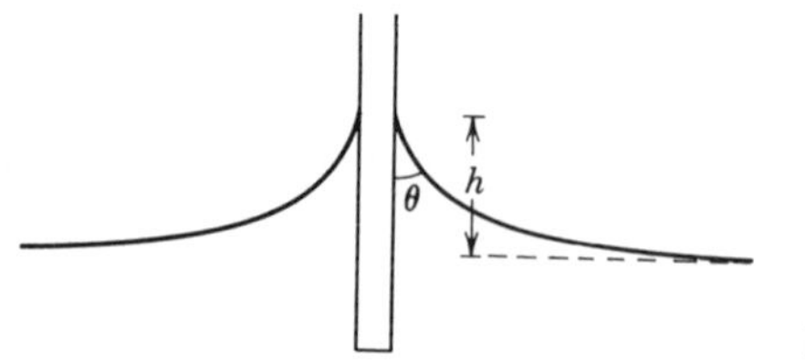

Fig. II-17

It should be noted that here, as with capillary rise, there is an adsorbed film of vapor (see Section X-6D) with which the meniscus merges smoothly. The meniscus is not "hanging" from the plate but rather from a liquidlike film (39). The correction for the weight of such film should be negligible, however.

An alternative and probably now more widely used procedure is to raise the liquid level gradually until it just touches the hanging plate suspended from a balance (often, as indicated in Fig. II-16, a recording electrobalance). The increase in weight is then noted. A general equation is

$$\gamma \cos \theta = \frac{\Delta W}{p} \tag{II-30}$$

where ΔW is the change in weight of (i.e., force exerted by) the plate when it is brought into contact with the liquid, and p is the perimeter of the plate. The contact angle, if finite, may be measured in the same experiment (40). Integration of Eq. II-13 (remembering that R_2 is infinite so that the second

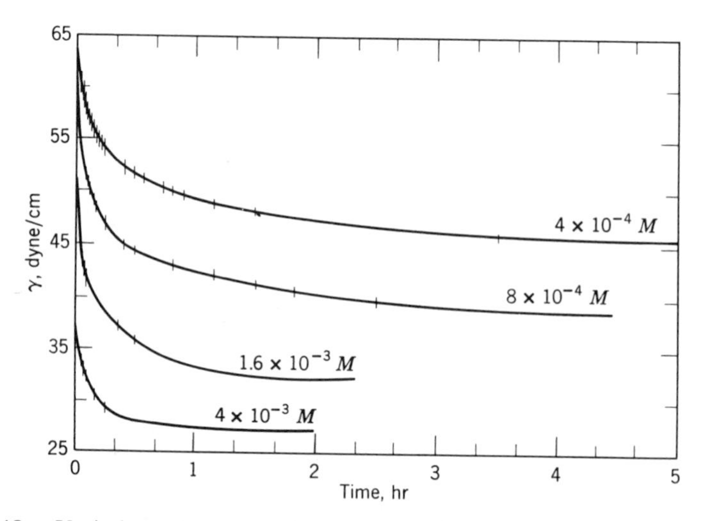

Fig. II-18. Variation with time of aqueous sodium dodecyl sulfate solutions of various concentrations (from Ref. 42). See Ref. 42 for later data with highly purified materials.

term on the right is zero) gives

$$\left(\frac{h}{a}\right)^2 = 1 - \sin\theta \tag{II-31}$$

where, as illustrated in Fig. II-17, h is the height of the top of the meniscus above the level liquid surface. Zero contact angle is preferred, however, if only the liquid surface tension is of interest; it may help to slightly roughen the plate, see Ref. 41.

As an example of the application of the method, Neumann and Tanner (40) followed the variation with time of the surface tension of aqueous sodium dodecyl sulfate solutions. Their results are shown in Fig. II-18, and it is seen that a slow but considerable change occurred.

A modification of the foregoing procedure is to suspend the plate so that it is partly immersed and to determine from the dry and immersed weights the meniscus weight. The procedure is especially useful in the study of surface adsorption or of monolayers, where a change in surface tension is to be measured. This application is discussed in some detail by Gaines (43).

9. Methods Based on the Shape of Static Drops or Bubbles

Small drops or bubbles will tend to be spherical because surface forces depend on the area, which decreases as the square of the linear dimension, whereas distortions due to gravitational effects depend on the volume, which decreases as the cube of the linear dimension. Likewise, too, a drop of liquid in a second liquid of equal density will be spherical. However, when gravitational and surface tensional effects are comparable, then one can determine in principle the surface tension from measurements of the shape of the drop or bubble. The various situations to which Eq. II-19 applies are shown in Fig. II-19.

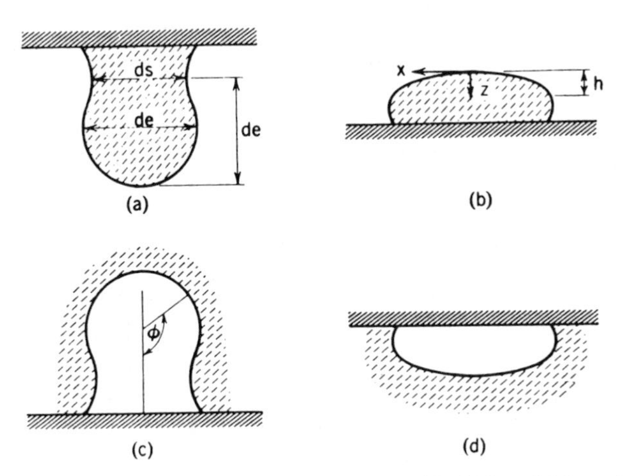

Fig. II-19. Shapes of sessile and hanging drops and bubbles: (*a*) hanging drop, (*b*) sessile drop, (*c*) hanging bubble, (*d*) sessile bubble.

The general procedure is to form the drop or bubble under conditions such that it is not subject to disturbances and then to make certain measurements of its dimensions, for example, from a photograph. Usually, several tenths of a percent accuracy is attainable, and the method is well suited to the observation of long-term changes in surface tension.

A. Pendant Drop Method

A drop hanging from a tip (or a clinging bubble) elongates as it grows larger because the variation in hydrostatic pressure $\Delta \mathbf{P}$ eventually becomes appreciable in comparison with that given by the curvature at the apex. As in the case of a meniscus, it is convenient to write Eq. II-12 in the form of Eq. II-20, where in the present case, the dimensionless parameter β is negative. It is not generally convenient to measure β directly, but as a shape determining variable it is related to others whose determination may be easier.

In the case of the pendant drop, Andreas et al. (44) felt that the most conveniently measurable shape dependent quantity was $S = d_s/d_e$. As indicated in Fig. II-19, d_e is the equatorial diameter and d_s is the diameter measured a distance d_e up from the bottom of the drop. The difficultly measurable size parameter b in Eq. II-20 was combined with β by defining in its stead the quantity $H = -\beta(d_e/b)^2$. Thus

$$\gamma = \frac{-\Delta\rho g b^2}{\beta} = \frac{-\Delta\rho g d_e^2}{\beta(d_e/b)^2} = \frac{\Delta\rho g d_e^2}{H} \tag{II-32}$$

The relationship between the shape dependent quantity H and the experimentally measurable shape dependent quantity S was then determined empirically, using pendant drops of water. A set of quite accurate $1/H$ versus S values has been obtained by Niederhauser and Bartell (45)—see also Refs. 26 and 46. These were obtained by a numerical integration procedure using the Bashforth and Adams tables (9) and based on the fundamental Eq. II-20. Some supplementary tables of the Bashforth and Adams type were also computed, and their results for $\beta = -0.45$ are tabulated in Table II-6. Their values of $1/H$ as a function of $S = d_s/d_e$ are given in Table II-7. The authors point out that, for practical reasons, the size of the tip should be such that r/a is about 0.5 or less. Table II-7 includes S values from 0.3 to 0.67, as published by Stauffer (47).

Andreas et al. measured the variation with time of the surface tension of sodium stearate solutions, and a typical sequence of drop shapes as a given solution aged is shown in Fig. II-20. As in the case of the sodium dodecyl sulphate solutions of Fig. II-18, a considerable change in surface tension took place.

The pendant drop method is a very widely used one, requiring only small quantities of liquid and applicable to the experimentally difficult situation of measurements at high temperature or with reactive materials. With good optical equipment, it is good to a few tenths of a percent.

TABLE II-6
Solutions to Eq. II-20 for $\beta = -0.45$

ϕ	x/b	z/b
0.099944	0.099834	0.004994
0.199551	0.198673	0.019911
0.298488	0.295547	0.044553
0.396430	0.389530	0.078600
0.493058	0.479762	0.121617
0.588070	0.565464	0.173072
0.681175	0.645954	0.232352
0.772100	0.720657	0.298779
0.860590	0.789108	0.371635
0.946403	0.850958	0.450175
1.029319	0.905969	0.533649
1.109130	0.954013	0.621322
1.185644	0.995064	0.712480
1.258681	1.029190	0.806454
1.328069	1.056542	0.902619
1.393643	1.077347	1.000413
1.455242	1.091895	1.099333
1.512702	1.100530	1.198946
1.565856	1.103644	1.298886
1.614526	1.101667	1.398856
1.658523	1.095060	1.498630
1.697641	1.084311	1.598044
1.731653	1.069933	1.697000
1.760310	1.052460	1.795458
1.783338	1.032445	1.893432
1.800443	1.010466	1.990986
1.811310	0.987123	2.088223
1.815618	0.963039	2.185279
1.813050	0.938868	2.282314
1.803321	0.915293	2.379495
1.786207	0.893023	2.476982
1.761593	0.872791	2.574912
1.729517	0.855344	2.673373
1.690226	0.841424	2.772393

B. Sessile Drop or Bubble Method

The cases of the sessile drop and bubble are symmetrical, as illustrated in Fig. II-19, but the use of the former is more common in surface tension determinations, and the present discussion is in terms of it. Porter (48), using the Bashforth and Adams tables (9), has made calculations of the difference Δ between $h^2/2r^2$ and $a^2/2r^2$, where r is the equatorial radius and h denotes the distance from the apex to the equatorial plane (Fig. II19*b*).

TABLE II-7
Numerical Tabulation of $1/H$ versus S Function for Calculation of Boundary Tensions by Pendant Drop Method (Linear Interpolation Warranted)

S	0	1	2	3	4	5	6	7	8	9
0.30	7.09837	7.03966	6.98161	6.92421	6.86746	6.81135	6.75586	6.70099	6.64672	6.59306
0.31	6.53998	6.48748	6.43556	6.38421	6.33341	6.28317	6.23347	6.18431	6.13567	6.08756
0.32	6.03997	5.99288	5.94629	5.90019	5.85459	5.80946	5.76481	5.72063	5.67690	5.63364
0.33	5.59082	5.54845	5.50651	5.46501	5.42393	5.38327	5.34303	5.30320	5.26377	5.22474
0.34	5.18611	5.14786	5.11000	5.07252	5.03542	4.99868	4.96231	4.92629	4.89061	4.85527
0.35	4.82029	4.78564	4.75134	4.71737	4.68374	4.65043	4.61745	4.58479	4.55245	4.52042
0.36	4.48870	4.45729	4.42617	4.39536	4.36484	4.33461	4.30467	4.27501	4.24564	4.21654
0.37	4.18771	4.15916	4.13087	4.10285	4.07509	4.04759	4.02034	3.99334	3.96660	3.94010
0.38	3.91384	3.88786	3.86212	3.83661	3.81133	3.78627	3.76143	3.73682	3.71242	3.68824
0.39	3.66427	3.64051	3.61696	3.59362	3.57047	3.54752	3.52478	3.50223	3.47987	3.45770
0.40	3.43572	3.41393	3.39232	3.37089	3.34965	3.32858	3.30769	3.28698	3.26643	3.24606
0.41	3.22582	3.20576	3.18587	3.16614	3.14657	3.12717	3.10794	3.08886	3.06994	3.05118
0.42	3.03258	3.01413	2.99583	2.97769	2.95969	2.94184	2.92415	2.90659	2.88918	2.87192
0.43	2.85479	2.83781	2.82097	2.80426	2.78769	2.77125	2.75496	2.73880	2.72277	2.70687
0.44	2.69110	2.67545	2.65992	2.64452	2.62924	2.61408	2.59904	2.58412	2.56932	2.55463
0.45	2.54005	2.52559	2.51124	2.49700	2.48287	2.46885	2.45494	2.44114	2.42743	2.41384
0.46	2.40034	2.38695	2.37366	2.36047	2.34738	2.33439	2.32150	2.30870	2.29600	2.28339
0.47	2.27088	2.25846	2.24613	2.23390	2.22176	2.20970	2.19773	2.18586	2.17407	2.16236
0.48	2.15074	2.13921	2.12776	2.11640	2.10511	2.09391	2.08279	2.07175	2.06079	2.04991
0.49	2.03910	2.02838	2.01773	2.00715	1.99666	1.98623	1.97588	1.96561	1.95540	1.94527
0.50	1.93521	1.92522	1.91530	1.90545	1.89567	1.88596	1.87632	1.86674	1.85723	1.84778
0.51	1.83840	1.82909	1.81984	1.81065	1.80153	1.79247	1.78347	1.77453	1.76565	1.75683
0.52	1.74808	1.73938	1.73074	1.72216	1.71364	1.70517	1.69676	1.68841	1.68012	1.67188
0.53	1.66369	1.65556	1.64748	1.63946	1.63149	1.62357	1.61571	1.60790	1.60014	1.59242
0.54	1.58477	1.57716	1.56960	1.56209	1.55462	1.54721	1.53985	1.53253	1.52526	1.51804

0.55	1.51086	1.50373	1.49665	1.48961	1.48262	1.47567	1.46876	1.46190	1.45509	1.44831
0.56	1.44158	1.43489	1.42825	1.42164	1.41508	1.40856	1.40208	1.39564	1.38924	1.38288
0.57	1.37656	1.37028	1.36404	1.35784	1.35168	1.34555	1.33946	1.33341	1.32740	1.32142
0.58	1.31549	1.30958	1.30372	1.29788	1.29209	1.28633	1.28060	1.27491	1.26926	1.26364
0.59	1.25805	1.25250	1.24698	1.24149	1.23603	1.23061	1.22522	1.21987	1.21454	1.20925
0.60	1.20399	1.19875	1.19356	1.18839	1.18325	1.17814	1.17306	1.16801	1.16300	1.15801
0.61	1.15305	1.14812	1.14322	1.13834	1.13350	1.12868	1.13389	1.11913	1.11440	1.10969
0.62	1.10501	1.10036	1.09574	1.09114	1.08656	1.08202	1.07750	1.07300	1.06853	1.06409
0.63	1.05967	1.05528	1.05091	1.04657	1.04225	1.03796	1.03368	1.02944	1.20522	1.02102
0.64	1.01684	1.01269	1.00856	1.00446	1.00037	0.99631	0.99227	0.98826	0.98427	0.98029
0.65	0.97635	0.97242	0.96851	0.96463	0.96077	0.95692	0.95310	0.94930	0.94552	0.94176
0.66	0.93803	0.93431	0.93061	0.92693	0.92327	0.91964	0.91602	0.91242	0.90884	0.90528
0.67	0.90174	89822	89471	89122	88775	88430	88087	87746	87407	87069
0.68	86733	86399	86067	85736	85407	85080	84755	84431	84110	83790
0.69	83471	83154	82839	82525	82213	81903	81594	81287	80981	80677
0.70	80375	80074	79774	79477	79180	78886	78593	78301	78011	77722
0.71	77434	77148	76864	76581	76299	76019	75740	75463	75187	74912
0.72	74639	74367	74097	73828	73560	73293	73028	72764	72502	72241
0.73	71981	71722	71465	71208	70954	70700	70448	70196	69946	69698
0.74	69450	69204	68959	68715	68472	68230	67990	67751	67513	67276
0.75	67040	66805	66571	66338	66107	65876	65647	65419	65192	64966
0.76	64741	64518	64295	64073	63852	63632	63414	63196	62980	62764
0.77	62550	62336	62123	61912	61701	61491	61282	61075	60868	60662
0.78	60458	60254	60051	59849	59648	59447	59248	59050	58852	58656
0.79	58460	58265	58071	57878	57686	57494	57304	57114	56926	56738
0.80	56551	56364	56179	55994	55811	55628	55446	55264	55084	54904
0.81	54725	54547	54370	54193	54017	53842	53668	53494	53322	53150
0.82	52978	52808	52638	52469	52300	52133	51966	51800	51634	51470
0.83	51306	51142	50980	50818	50656	50496	50336	50176	50018	49860
0.84	49702	49546	49390	49234	49080	48926	48772	48620	48468	48316

TABLE II-7 (*Continued*)

S	0	1	2	3	4	5	6	7	8	9
0.85	48165	48015	47865	47716	47568	47420	47272	47126	46980	46834
0.86	46690	46545	46401	46258	46116	45974	45832	45691	45551	45411
0.87	45272	45134	44996	44858	44721	44585	44449	44313	44178	44044
0.88	43910	43777	43644	43512	43380	43249	43118	42988	42858	42729
0.89	42600	42472	42344	42216	42089	41963	41837	41711	41586	41462
0.90	41338	41214	41091	40968	40846	40724	40602	40481	40361	40241
0.91	40121	40001	39882	39764	39646	39528	39411	39294	39178	39062
0.92	38946	38831	38716	38602	38488	38374	38260	38147	38035	37922
0.93	37810	37699	37588	37477	37367	37256	37147	37037	36928	36819
0.94	36711	36603	36495	36387	36280	36173	36067	35960	35854	35749
0.95	35643	35538	35433	35328	35224	35120	35016	34913	34809	34706
0.96	34604	34501	34398	34296	34195	34093	33991	33890	33789	33688
0.97	33587	33487	33386	33286	33186	33086	32986	32887	32787	32688
0.98	32588	32489	32390	32290	32191	32092	31992	31893	31793	31694
0.99	31594	31494	31394	31294	31194	31093	30992	30891	30790	30688
1.00	30586	30483	30379							

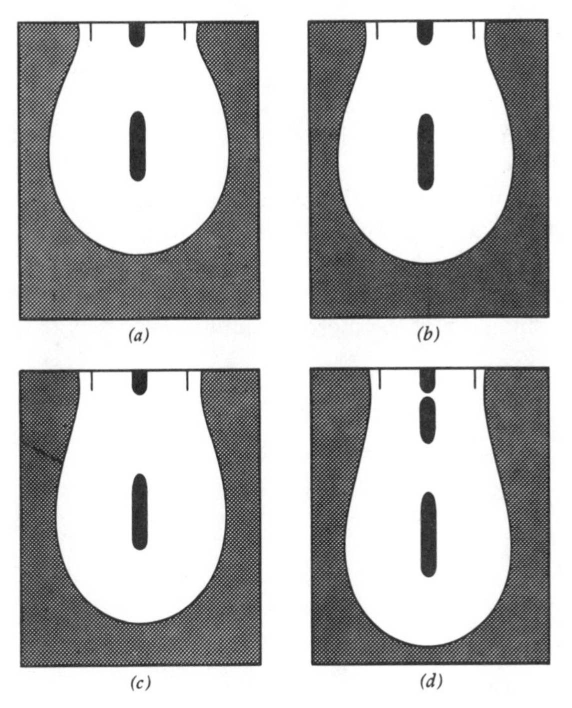

Fig. II-20. Shapes of pendant drops of sodium stearate solutions as a function of time (44); (*a*) age = 10 sec, γ = 71.9; (*b*) age = 60 sec, γ = 58.2; (*c*) age = 120 sec, γ = 54.4; (*d*) age = 1800 sec, γ = 39.2.

The variation of Δ with h/r could be fitted accurately by means of an empirical equation

$$\Delta = 0.3047\left(\frac{h^3}{r^3}\right)\left(1 - \frac{4h^2}{r^2}\right) \tag{II-33}$$

Wheeler and co-workers give some experimental details and further discussion (49). The method is claimed to be good to 0.2%.

The use of Eq. II-33 does require fairly large drops and provides no internal check against possible irregularities in drop contour due to, say, irregular wetting that causes it not to be a figure of revolution. An alternative procedure has been proposed by Smolders and Duyvis (50). From the definition of β, Eq. II-21, it follows that

$$\gamma = \frac{\Delta\rho g b^2}{\beta} \tag{II-34}$$

Now, while b is difficult to determine, the Bashforth and Adams tables (9) give (x_e/b) as a function of β, where x_e is the equatorial radius (Fig. II-19), so that Eq. II-34 may be written

$$\gamma = \frac{\Delta\rho g x_e^2}{[f(\beta)]^2}, \qquad \frac{x_e}{b} = f(\beta) \tag{II-35}$$

Because x_e can be determined quite accurately, the problem reduces to the determination of β, and this was done by comparing the drop profile with the set of profiles for various β values as obtained from the tables. Knowing β, $f(\beta)$ is read from the same tables, and γ is calculated from Eq. II-35. An accuracy of 0,1% was claimed.

As an example of its application, the method was used by Nutting and Long (51) to follow the variation with time of the surface tension of sodium laurate solutions. The method has also been found useful for the measurement of the surface tension of molten metals (52–54); it has been used as well in the determination of the mercury-aqueous electrolyte interfacial tension (55).

The case of very large drops or bubbles is easy because only one radius of curvature (that in the plane of the drawings) is considered. Equation II-12 then becomes

$$\Delta\rho g y = \gamma\left(\frac{y''}{(1 + y'^2)^{3/2}}\right)$$

or

$$\frac{2y}{a^2} = \frac{p\,dp/dy}{(1 + p^2)^{3/2}} \tag{II-36}$$

where $p = dy/dx$. Integration gives

$$\frac{y^2}{a^2} = \frac{-1}{(1 + p^2)^{1/2}} + \text{constant} \tag{II-37}$$

Since h denotes the distance from the apex to the equatorial plane, then at $y = h$, $p = \infty$, and Eq. II-37 becomes

$$\frac{y^2}{a^2} - \frac{h^2}{a^2} = \frac{-1}{(1 + p^2)^{1/2}}$$

Furthermore, at $y = 0$, $p = 0$, from which it follows that $h^2/a^2 = 1$, or $h = a$,

$$\gamma = \frac{\Delta\rho g h^2}{2} \tag{II-38}$$

This very simple result is independent of the value of the contact angle because the configuration involved is only that between the equatorial plane and the apex.

C. Deformed Interfaces

The discussion so far has been of interfaces in a uniform gravitational field. There are several variants from this situation, some of which are useful in the measurement of liquid–liquid interfacial tensions where these are very small. Consider the case of a drop of liquid A suspended in liquid B. If the density of A is less than that of B, on rotating the whole mass, as illustrated in Fig. II-21, liquid A will go to the center, forming a drop astride the axis of revolution. With increasing speed of revolution, the drop of A elongates, since centrifugal force increasingly opposes the surface tensional drive towards minimum interfacial area. In brief, the drop of A deforms from a sphere to a prolate ellipsoid. At a sufficiently high speed of revolution, the drop approximates to an elongated cylinder.

The general analysis, while not difficult, is complicated; however, the limiting case of the very elongated, essentially cylindrical drop is not hard to treat. Consider a section of the elongated cylinder of volume V (Fig. II-21*b*). The centrifugal force on a volume element is $\omega^2 r\ \Delta\rho$ where ω is the speed of revolution and $\Delta\rho$ the difference in density. The potential energy at distance r from the axis of revolution is then $\omega^2 r^2\ \Delta\rho/2$, and the total potential energy for the cylinder of length l is $l \int_0^{r_0} (\omega^2 r^2\ \Delta\rho/2) 2\pi r\ dr = \pi\omega^2\ \Delta\rho\ r_0^4 l/4$. The interfacial free energy is $2\pi r_0 l\gamma$. The total energy is thus

$$E = \frac{\pi\omega^2\ \Delta\rho\ r_0^4 l}{4} + 2\pi r_0 l\gamma = \frac{\omega^2\ \Delta\rho\ r_0^2 V}{4} + \frac{2V\gamma}{r_0}$$

since $V = \pi r_0^2 l$. Setting $dE/dr_0 = 0$, we obtain

$$\gamma = \frac{\omega^2\ \Delta\rho\ r_0^3}{4} \tag{II-39}$$

Equation II-39 has been called Vonnegut's equation (56).

Princen and co-workers have examined the case of the rotating drop (57) (see also Refs. 58 and 59), as well as that of the meniscus profile in a rotating vertical tube (60). Wade and co-workers have applied the rotating drop method (Fig. II-21) to the study of systems of extremely low interfacial tension (61); values around 0.001 dyne/cm (or mN/m) can be measured readily and accurately.

Lucassen (62) describes an alternative procedure whereby one measures the deformation of a small drop suspended in a liquid having a small density gradient. Some typical shapes are shown in Fig. II-22*a*, the peripheral dots giving the matching calculated profiles. Interfacial tensions in the range of 10^{-4} dyne/cm (or mN/m) were measured.

Drops will deform in a shear field (see Ref. 63) and also in an electric field (64), as shown in Fig. II-22*b*. This last effect was noted in 1871 by Lord

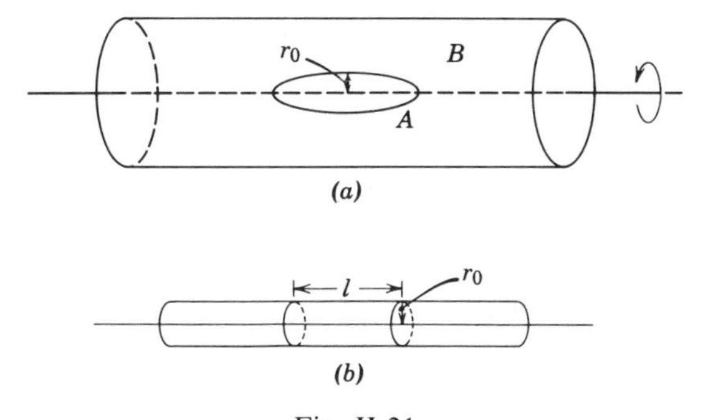

Fig. II-21

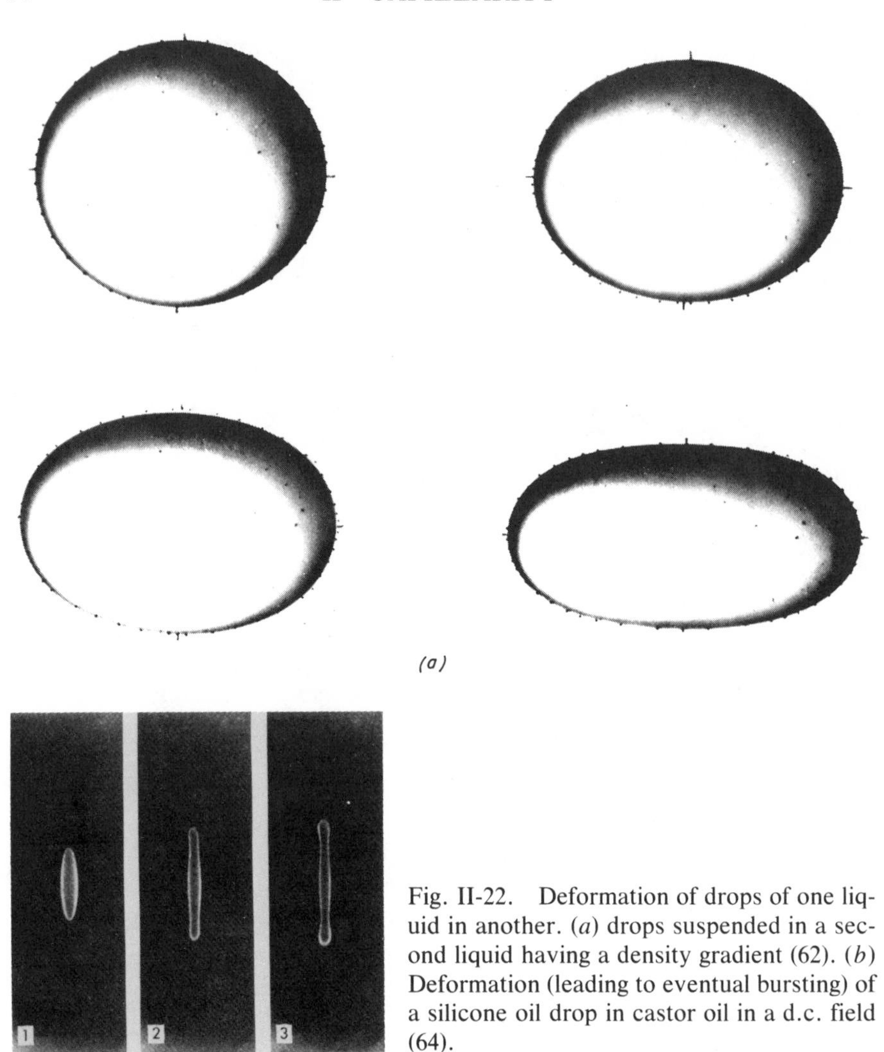

Fig. II-22. Deformation of drops of one liquid in another. (*a*) drops suspended in a second liquid having a density gradient (62). (*b*) Deformation (leading to eventual bursting) of a silicone oil drop in castor oil in a d.c. field (64).

Kelvin (65). The converse situation of systems free of gravity, centrifugal, or other fields has found an interesting and potentially important application. In the absence of such fields, a liquid drop assumes a perfectly spherical shape. In particular, drops of molten metal should do this and, on solidification, spheres of ball bearing quality should result. The idea is being tested in the U.S. space program.

10. Dynamic Methods of Measuring Surface Tension

The capillary rise, Wilhelmy slide, and the pendant or sessile drop or bubble methods are essentially equilibrium ones in the sense that quiescent surfaces are involved. As has been seen, these methods can be used to follow the slow changes in surface tension that solutions sometimes exhibit. It is of interest, however, to be able to study surface aging and relaxation effects on a very small time scale, and for this dynamic methods are needed. The various detachment methods are dynamic in that extension of the surface occurs at the critical point, but it is very difficult to define the exact surface age. This is possible with the methods discussed below.

A. Flow Methods

A jet emerging from a noncircular orifice is mechanically unstable, not only with respect to the eventual breakup into droplets discussed in Section II-3, but, more immediately, also with respect to the initial cross section not being circular. Oscillations develop in the jet, since the momentum of the liquid carries it past the desired circular cross section. This is illustrated in Fig. II-23.

The mathematical treatment was first developed by Lord Rayleigh in 1879, and a more exact one by Bohr has been reviewed by Sutherland (66), who gives the formula

$$\gamma_{\mathrm{app}} = \frac{4\rho v^2(1 + 37b^2/24r^2)}{6r\lambda^2(1 + 5\pi^2r^2/3\lambda^2)}. \tag{II-40}$$

where ρ is the density of the liquid, v is the volume velocity, λ is the wavelength, r is the sum of the minimum and maximum half-diameters, and b is their difference. The required jet dimensions were determined optically, and a typical experiment would make use of jets of about 0.03 cm in size and velocities of about 1 cm^3/sec, giving λ values of around 0.5 cm. To a first approximation, the surface age at a given node is just the distance from the orifice divided by the linear jet velocity and, in the above example, would be about 1 msec per wavelength.

It was determined in a recent study, for example, that the surface tension of water relaxes to its equilibrium value with a relaxation time of 0.6 msec (67).

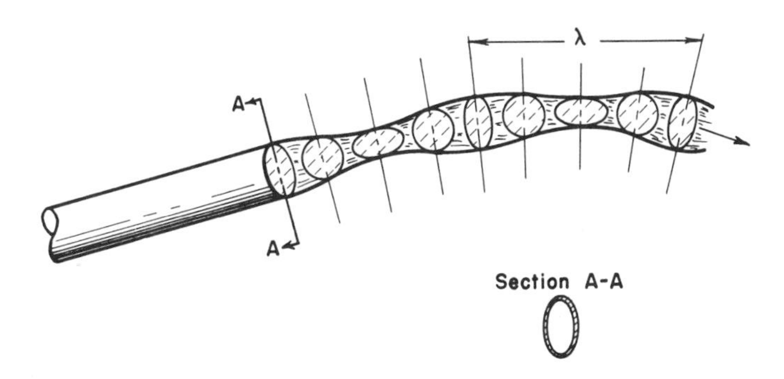

Fig. II-23. Oscillations in an elliptical jet.

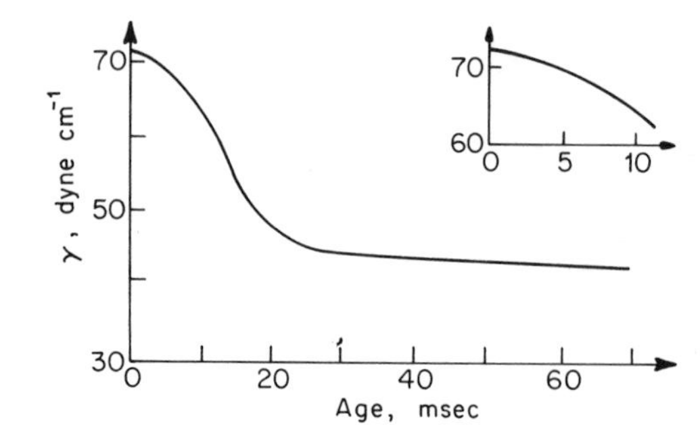

Fig. II-24. Surface tension as a function of age for 0.05 g/100 cm^3 of sodium di-(2-ethylhexyl)sulfosuccinate solution determined with various types of jet orifices (72).

The oscillating jet method has been useful in studying the surface tension of surfactant solutions. Figure II-24 illustrates the usual observation that at small times the jet appears to have the surface tension of pure water. The slowness in attaining the equilibrium value may partly be due to the times required for surfactant to diffuse to the surface and partly due to chemical rate processes at the interface. See Ref. 68 for similar studies with heptanoic acid and Ref. 69 for some anomalous effects.

For times below about 5 msec a correction must be made to allow for the fact that the surface velocity of the liquid *in* the nozzle is zero and takes several wavelengths to increase to the jet velocity after emerging from the nozzle. Correction factors have been tabulated (70, 71); see also Ref. 72.

The oscillating jet method is not suitable for the study of liquid–air interfaces whose ages are in the range of tenths of a second, and an alternative method is based on the dependence of the shape of a falling column of liquid on its surface tension. Since the hydrostatic head, and hence the linear velocity, increases with h, the distance away from the nozzle, the cross-sectional area of the column must correspondingly decrease as a material balance requirement. The effect of surface tension is to oppose this shrinkage in cross section. This method has been developed by Addison and Elliott (73). A more recent discussion is given by Garner and Mina (74).

B. Capillary Waves

The wavelength of ripples on the surface of a deep body of liquid depends on the surface tension. According to a formula given by Lord Kelvin (65),

$$v^2 = \frac{g\lambda}{2\pi} + \frac{2\pi\gamma}{\rho\lambda}$$
$$\gamma = \frac{\lambda^3\rho}{2\pi\tau^2} - \frac{g\lambda^2\rho}{4\pi^2} \qquad \text{(II-41)}$$

where v is the velocity of propagation, λ is the wavelength, and τ is the period of the ripples. For water there is a minimum velocity of about 0.5 mph for $\lambda = 1.7$ cm; for $\lambda = 0.1$ cm, it is 1.5 mph, whereas for $\lambda = 10^5$ cm, it is 89 mph!

Experimentally, the waves are measured as standing waves, and the situation might be thought to be a static one. However, individual elements of liquid in the surface region undergo a roughly circular motion, and the surface is alternately expanded and compressed. As a consequence, damping occurs even with a pure liquid, and much more so with solutions or film-covered surfaces for which transient surface expansions and contractions may be accompanied by considerable local surface tension changes and by material transport between surface layers. A recent review of some current developments has been given by Lucassen and Hansen (75). A more detailed discussion is deferred to Chapter III, but it might be noted here that the dispersion (i.e., variation with frequency) of the damping coefficient provides a way of studying surface relaxation processes.

11. Surface Tension Values as Obtained by Different Methods

The surface tension of a pure liquid should and does come out to be the same irrespective of the method used, although difficulties in the mathematical treatment of complex phenomena can lead to apparent discrepancies. In the case of solutions, however, dynamic methods, including detachment ones, often tend to give high values. Padday and Russell discuss this point in some detail (76). The same may be true of interfacial tensions between partially miscible liquids.

The data given in Table II-8 were selected with the purpose of providing a working stock of data for use in problems as well as a convenient reference to surface tension values for commonly studied interfaces. In addition, a number of values are included for uncommon substances or states of matter (e.g., molten metals) to provide a general picture of how this property ranges and of the extent of the literature on it. While the values have been chosen with some judgment, they are not presented as critically selected best values. Finally, many of the references cited in the table contain a good deal of additional data on surface tensions at other temperatures and for other liquids of the same type as the one selected for entry in the table.

12. Problems

1. Derive Eq. II-4 using the "surface tension" point of view. *Suggestion*. Consider the sphere to be in two halves, with the surface tension along the join balancing the force due to $\Delta \mathbf{P}$, which would tend to separate the two halves.

2. The diagrams in Fig. II-25 represent capillaries of varying construction and arrangement. The diameter of the capillary portion is the same in each case, and all of the capillaries are constructed of glass, unless otherwise indicated. The equilibrium rise for water is shown at the left. Draw meniscuses in each figure to correspond to (*a*) the level reached by water rising up the clean, dry tube and (*b*) the level to which the water would recede after having been sucked up to the end of the capillary. The meniscuses in the capillary may be assumed to be spherical in shape.

TABLE II-8
Surface Tension Values

Liquid	Temp., °C	γ dyne/c, mN/m	Liquid	Temp., °C	γ dyne/cm mN/m
Liquid–Vapor Interfaces					
Water[a]	20	72.88			
	25	72.14			
	30	71.40			
Organic Compounds					
Methylene iodide[b]	20	67.00	Butyl acetate[h]	20	25.09
Iodine[c]	20	54.7	Nonane[a]	20	22.85
Dimethyl sulfoxide[d]	20	43.54	Methanol[a]	20	22.50
Propylene carbonate[e]	20	41.1	Ethanol[a]	20	22.39
Dimethyl aniline[f]	20	36.56		30	21.55
Nitromethane[g]	20	32.66	Octane[a]	20	21.62
Benzene[a]	20	28.88	Heptane[a]	20	20.14
	30	27.56	Ether[a]	25	20.14
Toluene[a]	20	28.52	Perfluoromethylcyclohexane[a]	20	15.70
Chloroform[a]	25	26.67	Perfluoroheptane[a]	20	13.19
Propionic acid[a]	20	26.69	Hydrogen sulfide[i]	20	12.3
Butyric acid[a]	20	26.51	Perfluoropentane[a]	20	9.89
Carbon tetrachloride[a]	25	26.43			
Low-Boiling Substances					
^4He[j]	1°K	0.365	C_2H_6[m]	180.6°K	16.63
H_2[k]	20°K	2.01	Xe[n]	163°K	18.6
D_2[k]	20°K	3.54	N_2O[m]	182.5°K	24.26
N_2[l]	75°K	9.41	Cl_2[a]	−30	25.56
Ar[l]	90°K	11.86	NOCl[a]	−10	13.71
CH_4[a]	110°K	13.71	Br_2[o]	20	31.9
F_2[a]	85°K	14.84			
O_2[a]	77°K	16.48			

Metals					
Hg[a]	20	486.5	Ag[s]	1100	878.5
	25	485.5	Cu[t]	mp	1300
	30	484.5	Ti[u]	1680	1588
Na[p]	130	198	Pt[t]	mp	1800
Ba[q]	720	226	Fe[t]	mp	1880
Sn[r]	332	543.8			
Salts					
NaCl[v]	1073	115	$NaNO_3$[x]	308	116.6
$KClO_3$[w]	368	81	$K_2Cr_2O_7$[w]	397	129
KNCS[w]	175	101.5	$Ba(NO_3)_2$[x]	595	134.8
		Liquid–Liquid Interface			
Liquid 1: Water					
n-Butyl alcohol[y]	20	1.8	Nitrobenzene[y]	20	25.2
Ethyl acetate[y]	20	6.8	Benzene[z]	20	35.0
Heptanoic acid[z]	20	7.0	Carbon tetrachloride[z]	20	45.0
Benzaldehyde[x]	20	15.5	*n*-Heptane[z]	20	50.2
Liquid 1: Mercury					
Water[aa]	20	415	*n*-Heptane[z]	20	378
	25	416	Benzene[aa]	20	357
Ethanol[z]	20	389			
n-Hexane[z]	20	378			
Liquid 1: Fluorocarbon polymer					
Benzene[bb]	25	7.8	Water[bb]	25	57

[a] J. J. Jasper, *J. Phys. Chem. Ref. Data*, **1,** 841 (1972). Somewhat higher values for water are given by W. V. Kayser, *J. Colloid Interface Sci.*, **56,** 622 (1976).

[b] R. Grzeskowiak, G. H. Jeffery, and A. I. Vogel, *J. Chem. Soc.*, **1960,** 4728.

[c] J. E. Fredrickson, *J. Colloid Interface Sci.*, **48,** 506 (1974).

TABLE II-8 (*Continued*)

[d] H. L. Clever and C. C. Snead, *J. Phys. Chem.*, **67,** 918 (1963).
[e] M. K. Bernett, N. L. Jarvis, and W. A. Zisman, *J. Phys. Chem.*, **66,** 328 (1962).
[f] Ref. 15.
[g] C. C. Snead and H. L. Clever, *J. Chem. Eng. Data*, **7,** 393 (1962).
[h] J. B. Griffin and H. L. Clever, *J. Chem. Eng. Data*, **5,** 390 (1960).
[i] C. S. Herrick and G. L. Gaines, Jr., *J. Phys. Chem.*, **77,** 2703 (1973).
[j] K. R. Atkins and Y. Narahara, *Phys. Rev.*, **138,** A437 (1965).
[k] V. N. Grigor'ev and N. S. Rudenko, *Zh. Eksperim. Teor. Fiz.*, **47,** 92 (1964) (through *Chem. Abstr.*, **61,** 12669[e] (1964).
[l] D. Stansfield, *Proc. Phys. Soc.*, **72,** 854 (1958).
[m] A. J. Leadbetter, D. J. Taylor, and B. Vincent, *Can. J. Chem.*, **42,** 2930 (1964).
[n] A. J. Leadbetter and H. E. Thomas, *Trans. Far. Soc.*, **61,** 10 (1965).
[o] M. S. Chao and V. A. Stenger, *Talanta*, **11,** 271 (1964) (through *Chem. Abstr.*, **60,** 4829[e] (1964).
[p] C. C. Addison, W. E. Addison, D. H. Kerridge, and J. Lewis, *J. Chem. Soc.*, **1955,** 2262.
[q] C. C. Addison, J. M. Coldrey, and W. D. Halstead, *J. Chem Soc.*, **1962,** 3868.
[r] J. A. Cahill and A. D. Kirshenbaum, *J. Inorg. Nucl. Chem.*, **26,** 206 (1964).
[s] I. Lauerman, G. Metzger, and F. Sauerwald, *Z. Phys. Chem.*, **216,** 42 (1961).
[t] B. C. Allen, *Trans. Met. Soc. AIME*, **227,** 1175 (1963).
[u] J. Tille and J. C. Kelley, *Brit. J. Appl. Phys.*, **14,** No. 10, 717 (1963).
[v] J. D. Patdey, H. R. Chaturvedi, and R. P. Pandey, *J. Phys. Chem.*, **85,** 1750 (1981).
[w] J. P. Frame, E. Rhodes, and A. R. Ubbelohde, *Trans. Faraday Soc.*, **55,** 2039 (1959).
[x] C. C. Addison and J. M. Coldrey, *J. Chem. Soc.*, **1961,** 468.
[y] D. J. Donahue and F. E. Bartell, *J. Phys. Chem.*, **56,** 480 (1952).
[z] L. A. Girifalco and R. J. Good, *J. Phys. Chem.*, **61,** 904 (1957).
[aa] E. B. Butler, *J. Phys. Chem.*, **67,** 1419 (1963).
[bb] F. M. Fowkes and W. M. Sawyer, *J. Chem. Phys.*, **20,** 1650 (1952).

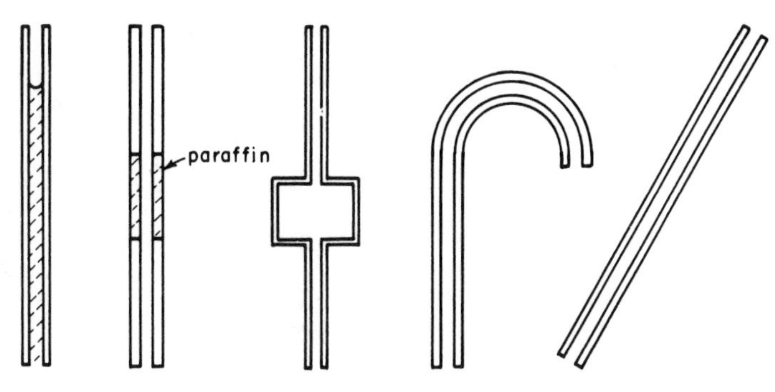

Fig. II-25

3. Show that the second term in Eq. II-18 does indeed correct for the weight of the meniscus. (Assume the meniscus to be hemispherical.)

4. Calculate to 1% accuracy the capillary rise for water at 20°C in a 1.5-cm diameter capillary.

5. Referring to the numerical example following Eq. II-21, what would be the surface tension of a liquid of density 2.500 g/cm^3, the rest of the data being the same?

6. Are the drops shown in Fig. II-20 life size? If not, calculate the reduction or magnification factor.

7. The surface tension of a liquid that wets glass is measured by determining the height Δh between the levels of the two meniscuses in a U-tube having a small radius r_1 on one side and a larger radius r_2 on the other. The following data are known: $h = 1.90 \times 10^{-2}$ m, $r_1 = 1.00 \times 10^{-3}$ m, $r_2 = 1.00 \times 10^{-2}$ m, $\rho = 950$ kg/m^3 at 20°C. Calculate the surface tension of the liquid using (*a*) the simple capillary rise treatment and (*b*) making the appropriate corrections using Tables II-2 and II-3.

8. Derive Eq. II-6.

9. The surface tension of a liquid is determined by the drop weight method. Using a tip whose outside diameter is 6×10^{-4} m and whose inside diameter is 2×10^{-5} m, it is found that the weight of 20 drops is 8×10^{-4} kg. The density of the liquid is 950 kg/m^3, and it wets the tip. Using the appropriate correction factor, calculate the surface tension of this liquid.

10. Derive the equation for the capillary rise between parallel plates, including the correction term for meniscus weight. Assume zero contact angle, a cylindrical meniscus, and neglect end effects.

11. Boucher and Evans (26) object to Eq. II-24 as being incorrect even on the simplest grounds, and give as the first-order equation

$$W = 2\pi r\gamma \sin\phi - \pi r^2\gamma\left(\frac{1}{r_1} + \frac{1}{r_2}\right)$$

Here ϕ is the angle defined in Fig. II-19, and r_1 and r_2 are the radii of curvature, all for the drop surface at the base of the tip. Explain how this equation is obtained.

12. In a rotating drop measurement, what is the interfacial tension if the two liquids differ in density by 0.15 g/cm^3, the speed of rotation is 50 rpm, the volume of the drop is 0.3 cm^3, and the length of the extended drop is 6.1 cm?

13. Derive, from simple considerations, the capillary rise between two parallel

plates of infinite length inclined at an angle of θ to each other, and meeting at the liquid surface, as illustrated in Fig. II-26. Assume zero contact angle and a circular cross section for the meniscus. Remember that the area of the liquid surface changes with its position.

14. The following values for the surface tension of a 10^{-4} *M* solution of sodium oleate at 25°C are reported by various authors: (*a*) by the capillary rise method, $\gamma = 43$ mN/m; (*b*) by the drop weight method, $\gamma = 50$ mN/m; and (*c*) by the sessile drop method, $\gamma = 40$ mN/m. Explain how the above discrepancies might arise. Which value should be the most reliable and why?

15. Drive Eq. II-31.

16. Calculate (r/b) for (r/a) values of 0.50, 1.00, 1.50, 2.00, and 3.00 using the equations of Ref. 13. Compare your answers with the corresponding entries in Tables II-2 and II-3.

17. Molten naphthalene at its melting point of 82°C has the same density as does water at this temperature. Suggest two methods that might be used to determine the naphthalene–water interfacial tension. Discuss your suggestions sufficiently to show that the methods will be reasonably easy to carry out and should give results good to 1% or better.

18. Using Table II-6, calculate S and $1/H$ for $\beta = -0.45$ for a pendant drop. *Hint.* x/b in the table is at a maximum when x is the equatorial radius. Compare your result with the appropriate entry in Table II-7.

19. For a particular drop of a certain liquid of density 0.80, β is -0.45 and d_e is 0.50 cm. (*a*) Calculate the surface tension of the drop and (*b*) calculate the drop profile from apex to the tip, assuming r_t/a to be 0.55, where r_t is the radius of the tip.

20. Using Table II-1, calculate r/b and r/a for the case of a menicus that wets the capillary and whose shape corresponds to $\beta = 80$.

21. The surface tension of mercury is 471 dyne/cm at 24.5°C. In a series of measurements (28) the following drop weight data were obtained, (diameter of tip in centimeters, weight of drop in grams): (0.04293, 0.05443), (0.07651, 0.09093), (0.10872, 0.12331). Calculate the corresponding f and $r/V^{1/3}$ values to fill out Table II-5.

22. Johnson and Lane (22) give the equation for the maximum bubble pressure:

$$h = \frac{a^2}{r} + \frac{2}{3}r + \frac{1}{6}\frac{r^3}{a^2}$$

For a certain liquid, $a^2 = 0.0670$ cm^2 and $r = 0.100$ cm. Calculate, using the equation, the values of X/r and r/a and compare with the X/r value given by Table II-4.

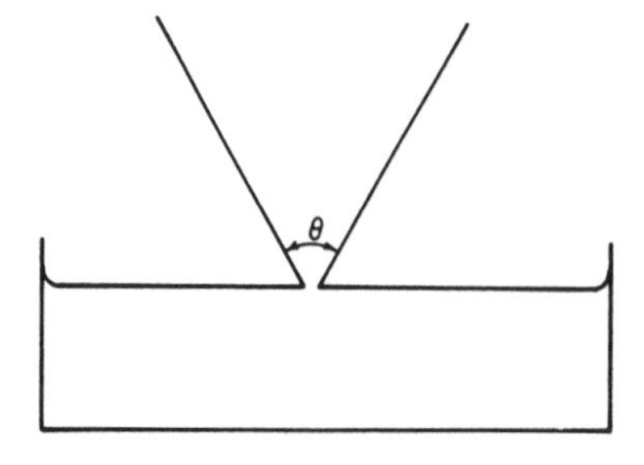

Fig. II-26

23. A drop of liquid rests on a flat surface that it does not wet, but contacts with an angle θ (measured in the liquid phase). The height of the drop above the surface is 0.45 cm, and its equatorial diameter is 1.50 cm; its density is 2.5, and its shape corresponds to $\beta = 80$ (see Table II-1). (*a*) Calculate the surface tension of this liquid. (*b*) Calculate the contact angle θ.

24. According to the simple formula, the maximum bubble pressure is given by $P_{max} = 2\gamma/r$ where r is the radius of the circular cross-section tube, and P has been corrected for the hydrostatic head due to the depth of immersion of the tube. Using the appropriate table, show what maximum radius tube may be used if γ computed by the simple formula is not to be more than 5% in error. Assume a liquid of $\gamma =$ 30 dyne/cm and density 1.10.

25. A liquid of density 2.5 g/cm³ forms a meniscus of shape corresponding to $\beta =$ 80 in a metal capillary tube with which the contact angle is 30°. The capillary rise is 0.074 cm. Calculate the surface tension of the liquid and the radius of the capillary, using Table II-1.

26. Equation II-31 may be integrated to obtain the profile of a meniscus against a vertical plate; the integrated form is given in Ref. 39. Calculate the meniscus profile for water at 20°C for (*a*) the case where water wets the plate and (*b*) the case where the contact angle is 50°. For (*b*) obtain from your plot the value of h, and compare with that calculated from Eq. II-31.

27. Equation II-40 may be written in the form:

$$\gamma = \frac{2 \times 10^3 \rho v^2 (1 + 1.542\, b^2/r^2)}{3r\lambda^2 + 5\pi^2 r^3}$$

Show what the units of γ must be.

General References

N. K. Adam, *The Physics and Chemistry of Surfaces*, 3rd ed., Oxford University Press, London, 1941.

R. Aveyard and D. A. Haydon, *An Introduction to the Principles of Surface Chemistry*, Cambridge University Press, 1973.

J. T. Davies and E. K. Rideal, *Interfacial Phenomena*, 2nd ed., Academic, New York, 1963.

W. D. Harkins, *The Physical Chemistry of Surface Films*, Reinhold, New York, 1952.

S. R. Morrison, *The Chemical Physics of Surfaces*, Plenum, London, 1977.

H. van Olphen and K. J. Mysels, *Physical Chemistry: Enriching Topics from Colloid and Surface Science*, Theorex (8327 La Jolla Scenic Drive), La Jolla, California, 1975.

L. I. Osipow, *Surface Chemistry, Theory and Industrial Applications*, Krieger, New York, 1977.

J. R. Partington, *An Advanced Treatist of Physical Chemistry*, Vol. II, Longmans, Green, New York, 1951.

D. J. Shaw, *Introduction to Colloid and Surface Chemistry*, Butterworths, London, 1966.

Textual References

1. T. Young, *Miscellaneous Works*, G. Peacock, ed., J. Murray, London, 1855, Vol. I, p. 418.
2. P. S. de Laplace, *Mechanique Celeste*, Supplement to Book 10, 1806.
3. C. V. Boys, *Soap Bubbles and the Forces that Mould Them*, Society for Promoting Christian Knowledge, London, 1890; reprint ed., Doubleday Anchor Books, Science Study Series S3, Doubleday, Garden City, New York, 1959.
4. F. D. Rumscheit and S. G. Mason, *J. Colloid Sci.,* **17,** 266 (1962).
5. B. J. Carroll and J. Lucassen, *J. Chem. Soc., Faraday Trans. I,* **70,** 1228 (1974).
6. E. A. Boucher, *Rep. Prog. Phys.,* **43,** 497 (1980).
7. H. M. Princen, *Surface and Colloid Science*, E. Matijevic, ed., Vol. 2, Wiley-Interscience, 1969.
8. Lord Rayleigh (J. W. Strutt), *Proc. Roy. Soc.* (*London*), **A92,** 184 (1915).
9. F. Bashforth and J. C. Adams, *An Attempt to Test the Theories of Capillary Action*, University Press, Cambridge, England, 1883.
10. S. Sugden, *J. Chem. Soc.,* **1921,** 1483.
11. A. W. Adamson, *J. Chem. Ed.,* **55,** 634 (1978).
12. J. F. Padday and A. Pitt, *J. Colloid Interface Sci.,* **38,** 323 (1972).
13. J. E. Lane, *J. Colloid Interface Sci.,* **42,** 145 (1973).
14. T. A. Erikson, *J. Phys. Chem.,* **69,** 1809 (1965).
15. T. W. Richards and E. K. Carver, *J. Am. Chem. Soc.,* **43,** 827 (1921).
16. W. D. Harkins and F. E. Brown, *J. Am. Chem. Soc.,* **41,** 499 (1919).
17. S. S. Urazovskii and P. M. Chetaev, *Kolloidn. Zh.,* **11,** 359 (1949); through *Chem. Abstr.,* **44,** 889 (1950).
18. P. R. Edwards, *J. Chem. Soc.,* **1925,** 744.
19. G. Jones and W. A. Ray, *J. Am. Chem. Soc.,* **59,** 187 (1937).
20. W. Heller, M. Cheng, and B. W. Greene, *J. Colloid Interface Sci.,* **22,** 179 (1966).
21. S. Sugden, *J. Chem. Soc.,* **1922,** 858; **1924,** 27.
22. C. H. J. Johnson and J. E. Lane, *J. Colloid Interface Sci.,* **47,** 117 (1974).
23. Y. Saito, H. Yoshida, T. Yokoyama, and Y. Ogina, *J. Colloid Interface Sci.,* **66,** 440 (1978).
24. R. Razouk and D. Walmsley, *J. Colloid Interface Sci.,* **47,** 415 (1974).
25. T. Tate, *Phil. Mag.,* **27,** 176 (1864).
26. E. A. Boucher and M. J. B. Evans, *Proc. Roy. Soc.* (*London*) **A346,** 349 (1975).
27. E. A. Boucher and H. J. Kent, *J. Colloid Interface Sci.,* **67,** 10 (1978).
28. M. C. Wilkinson and M. P. Aronson, *J. Chem. Soc., Faraday Trans. I,* **69,** 474 (1973).
29. P. Lecomte du Noüy, *J. Gen. Physiol.,* **1,** 521 (1919).
30. W. D. Harkins and H. F. Jordan, *J. Am. Chem. Soc.,* **52,** 1751 (1930).
31. H. W. Fox and C. H. Chrisman, Jr., *J. Phys. Chem.,* **56,** 284 (1952).
32. B. B. Freud and H. Z. Freud, *J. Am. Chem. Soc.,* **52,** 1772 (1930).
33. J. A. Krynitsky and W. D. Garrett, *J. Colloid Sci.,* **18,** 893 (1963).
34. F. van Zeggeren, C. de Courval, and E. D. Goddard, *Can. J. Chem.,* **37,** 1937 (1959).
35. L. Wilhelmy, *Ann. Phys.,* **119,** 177 (1863).
36. R. Ruyssen, *Rec. Trav. Chim.,* **65,** 580 (1946).

37. D. O. Jordan and J. E. Lane, *Australian J. Chem.*, **17,** 7 (1964).
38. J. T. Davies and E. K. Rideal, *Interfacial Phenomena*, Academic Press, New York, 1961.
39. A. W. Adamson and A. Zebib, *J. Phys. Chem.*, **84,** 2619 (1980).
40. A. W. Neumann and W. Tanner, *Tenside*, **4,** 220 (1967).
41. H. M. Princen, *Australian J. Chem.*, **23,** 1789 (1970).
42. J. Kloubek and A. W. Neumann, *ibid.*, **6,** 4 (1969).
43. G. L. Gaines, Jr., *Insoluble Monolayers at Liquid–Gas Interfaces*, Interscience, New York, 1966; *J. Colloid Interface Sci.*, **62,** 191 (1977).
44. J. M. Andreas, E. A. Hauser, and W. B. Tucker, *J. Phys. Chem.*, **42,** 1001 (1938).
45. D. O. Niederhauser and F. E. Bartell, *Report of Progress—Fundamental Research on the Occurrence and Recovery of Petroleum*, Publication of the American Petroleum Institute, The Lord Baltimore Press, Baltimore, 1950, p. 114.
46. S. Fordham, *Proc. Roy. Soc. (London)*, **A194,** 1 (1948).
47. C. E. Stauffer, *J. Phys. Chem.*, **69,** 1933 (1965).
48. A. W. Porter, *Phil. Mag.*, **15,** 163 (1933).
49. O. L. Wheeler, H. V. Tartar, and E. C. Lingafelter, *J. Am. Chem. Soc.* **67,** 2115 (1945).
50. C. S. Smolders and E. M. Duyvis, *Rec. Trav. Chim.*, **80,** 635 (1961).
51. G. C. Nutting and F. A. Long, *J. Am. Chem. Soc.*, **63,** 84 (1941).
52. P. Kosakévitch, S. Chatel, and M. Sage, *CR.*, **236,** 2064 (1953).
53. C. Kemball, *Trans. Faraday Soc.*, **42,** 526 (1946).
54. N. K. Roberts, *J. Chem. Soc.*, **1964,** 1907.
55. H. Vos and J. M. Vos, *J. Colloid Interface Sci.*, **74,** 360 (1980).
56. B. Vonnegut, *Rev. Sci. Inst.*, **13,** 6 (1942).
57. H. M. Princen, I. Y. Z. Zia, and S. G. Mason, *J. Colloid Interface Sci.*, **23,** 99 (1967).
58. J. C. Slattery and J. Chen, *J. Colloid Interface Sci.*, **64,** 371 (1978).
59. G. L. Gaines, Jr., *Polymer Eng.*, **12,** 1 (1972).
60. M. P. Aronson and H. M. Princen, *J. Chem. Soc., Faraday Trans. I*, **74,** 555 (1978).
61. J. L. Cayias, R. S. Schechter, and W. H. Wade, *Adsorption at Interfaces*, K. L. Mittal, ed., ACS Symposium Series, **8,** 234 (1975); L. Cash, J. L. Cayias, G. Fournier, D. MacAllister, T. Schares, R. S. Schechter, and W. H. Wade, *J. Colloid Interface Sci.*, **59,** 39 (1977).
62. J. Lucassen, *J. Colloid Interface Sci.*, **70,** 355 (1979).
63. W. J. Phillips, R. W. Graves, and R. W. Flumerfelt, *J. Colloid Interface Sci.*, **76,** 350 (1980).
64. S. Torza, R. G. Cox, and S. G. Mason, *Phil. Trans. Roy. Soc. (London)*, **269,** 295 (1971).
65. Lord Kelvin (W. Thomson), *Phil. Mag.*, **42,** 368 (1871).
66. K. L. Sutherland, *Australian J. Chem.*, **7,** 319 (1954).
67. N. N. Kochurova and A. I. Rusanov, *J. Colloid Interface Sci.*, **81,** 297 (1981).
68. R. S. Hansen and T. C. Wallace, *J. Phys. Chem.*, **63,** 1085 (1959).
69. W. D. E. Thomas and D. J. Hall, *J. Colloid Interface Sci.*, **51,** 328 (1975).
70. D. A. Netzel, G. Hoch, and T. I. Marx, *J. Colloid Sci.*, **19,** 774 (1964).

71. R. S. Hansen, *J. Phys. Chem.,* **68,** 2012 (1964).
72. W. D. E. Thomas and L. Potter, *J. Colloid Interface Sci.,* **50,** 397 (1975).
73. C. C. Addison and T. A. Elliott, *J. Chem. Soc.,* **1949,** 2789.
74. F. H. Garner and P. Mina, *Trans. Faraday Soc.,* **55,** 1607 (1959).
75. J. Lucassen and R. S. Hansen, *J. Colloid Interface Sci.,* **22,** 32 (1966).
76. J. F. Padday and D. R. Russell, *J. Colloid Sci.,* **15,** 503 (1960).

CHAPTER III

The Nature and Thermodynamics of Liquid Interfaces

1. One-Component Systems

It was made clear in Chapter I that the surface tension is a definite and accurately measurable property of the interface between two fluid phases. Moreover, its value is very rapidly established in the case of pure substances of ordinary viscosity; dynamic methods indicate that a normal surface tension for surfaces is established within a millisecond and probably sooner (1). As is discussed later in this section, calculations indicate that most of the surface free energy is developed within a few molecular diameters of the surface.

A. Surface Thermodynamic Quantities for a Pure Substance

Figure III-1 depicts a hypothetical system consisting of some liquid that fills a box having a sliding cover; the material of the cover is such that the interfacial tension between it and the liquid is zero. If the cover is slid back so as to uncover an amount of surface $d\mathcal{A}$, the work required to do so will be $\gamma\, d\mathcal{A}$. This is reversible work at constant pressure and temperature and thus gives the increase in free energy of the system (see Section XVI-13 for a more detailed discussion of the thermodynamics of surfaces).

$$dG = \gamma\, d\mathcal{A} \qquad \text{(III-1)}$$

The total free energy of the system is then made up of the molar free energy times the total number of moles of the liquid plus G^s, the surface free energy per unit area, times the total surface area. Thus

$$G^s = \gamma = \left(\frac{\partial G}{\partial \mathcal{A}}\right)_{T,P} \qquad \text{(III-2)}$$

Because this process is a reversible one, the heat associated with it gives the *surface entropy*

$$dq = T\, dS = TS^s\, d\mathcal{A} \qquad \text{(III-3)}$$

where S^s is the surface entropy per square centimeter of surface.

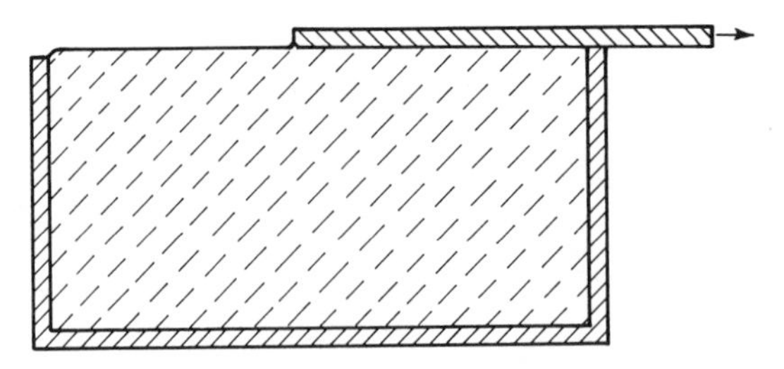

Fig. III-1

Because $(\partial G/\partial T)_P = -S$, it follows that

$$\left(\frac{\partial G^s}{\partial T}\right)_P = -S^s \tag{III-4}$$

or, in conjunction with Eq. III-1,

$$\frac{d\gamma}{dT} = -S^s \tag{III-5}$$

Finally, the total surface enthalpy per square centimeter H^s is

$$H^s = G^s + TS^s \tag{III-6}$$

Often, and as a good approximation, H^s and the surface energy E^s are not distinguished, so Eq. III-6 can be seen in the form

$$E^s = G^s + TS^s \tag{III-7}$$

or

$$E^s = \gamma - \frac{Td\gamma}{dT} \tag{III-8}$$

The total surface energy E^s generally is larger than the surface free energy. It is frequently the more informative of the two quantities, or at least it is more easily related to molecular models.

Other thermodynamic relationships are developed during the course of this chapter. The surface specific heat C^s (the distinction between C_p^s and C_v^s is rarely made), is an additional quantity to be mentioned at this point, however. It is given by

$$C^s = \frac{dE^s}{dT} \tag{III-9}$$

The surface tension of most liquids decreases with increasing temperature in a nearly linear fashion, as illustrated in Fig. III-2. The near linearity has stimulated many suggestions as to algebraic forms that give exact linearity. An old and well-known relationship, attributed to Eötvös (3), is

$$\gamma V^{2/3} = k(T_c - T) \tag{III-10}$$

where V is the molar volume. One does expect the surface tension to go to zero at the critical temperature, but the interface seems to become diffuse at a slightly lower temperature, and Ramsay and Shields (4) replaced T_c in Eq. III-10 by $(T_c - 6)$. In either form, the constant k is about the same for most liquids and has a value of about 2.1 erg/°K. Another form originated by van der Waals in 1894 but developed further by Guggenheim (5) is

$$\gamma = \gamma^\circ \left(1 - \frac{T}{T_c}\right)^n \tag{III-11}$$

where n is 11/9 for many organic liquids but may be closer to unity for metals (5).

There is a point of occasional misunderstanding about dimensions that can be illustrated here. The quantity $\gamma V^{2/3}$ is of the nature of a surface free energy per mole, yet it would appear that its dimensions are energy per mole$^{2/3}$, and that k in Eq. III-10 would be in erg/deg-mole$^{2/3}$. The term "mole$^{2/3}$" is meaningless, however, because

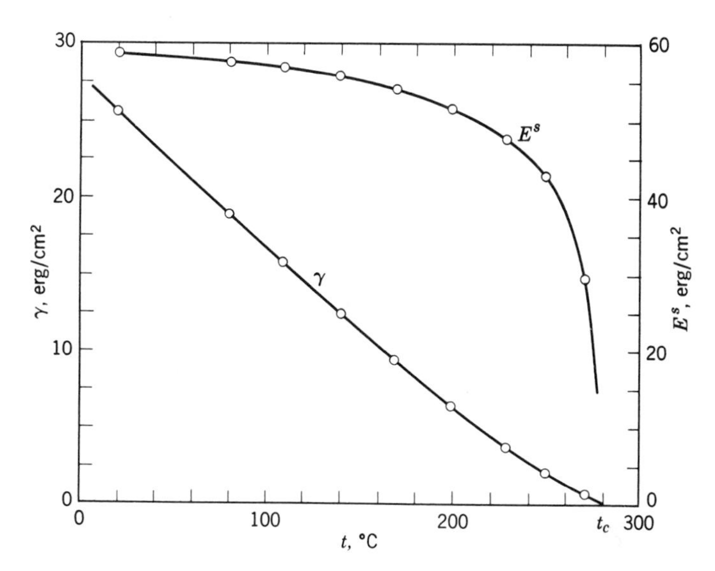

Fig. III-2. Variation of surface tension and total surface energy of CCl_4 with temperature (data from Ref. 2).

"mole" is not a dimension but rather an indication that an Avogadro's number of molecules is involved. Because Avogadro's number is itself arbitrary (depending, for example, on whether a gram or a pound molecular weight system is used), to get at a rational meaning of k one should compute it on the per molecule basis; this gives $k' = 2.9 \times 10^{-16}$ erg/deg-molecule. Lennard-Jones and Corner (7) have pointed out that Eq. III-10 arises rather naturally out of a simple statistical mechanical treatment of a liquid, with k' being the Boltzmann constant (1.37×10^{-16}) times a factor of the order of unity.

Equations III-10 and III-11 are, of course, approximations, and the situation has been examined in some detail by Cahn and Hilliard (8), who find that Eq. III-11 is also approximated by regular solutions not too near their critical temperature.

Another semiempirical relationship is that due to Sugden (9), who defined a quantity P that he called the *parachor*,

$$P = \frac{M\gamma^{1/4}}{\Delta\rho} \qquad \text{(III-12)}$$

where M is the molecular weight and $\Delta\rho$ is the liquid minus vapor density. The parachor can be considered to be a molar volume corrected for the compressive effects of intermolecular forces. In practice, it is a quantity whose value for a given substance is nearly temperature independent.

The parachor turns out to be an approximately additive property of atoms or functional groups and, in this respect, is similar to the molar polarization. One may, for example, try to decide between isomeric formulations for an organic compound by comparing the experimental value of P with various calculated ones. The literature on the parachors of organic compounds has been reviewed by Quayle (10) (see also Ref. 11 and preceding papers). However, it has been pointed out that parachors are not really strictly additive; in fact, today's spectroscopic and crystallographic methods have largely eliminated the need for the use of parachors.

B. The Total Surface Energy, E^s

If the variation of density and hence molar volume with temperature is small, it follows from Eqs. III-10 and III-8 that E^s will be nearly temperature independent. In fact, Eq. III-11 with $n = 1$ may be written in the form

$$\gamma = E^s\left(1 - \frac{T}{T_c}\right) \qquad \text{(III-13)}$$

which illustrates the point that surface tension and energy become equal at 0°K. The temperature independence of E^s generally does hold for liquids not too close to their critical temperature, although, as illustrated in Fig. III-2, E^s eventually drops because it must become zero at the critical temperature.

Inspection of Table III-1 shows that there is a wide range of surface

TABLE III-1
Temperature Dependence of Surface Tension

Liquid	γ, erg/cm^2	Temp., °C	$d\gamma/dT$	E^s, erg/cm^2	$E^{s'}$ cal/mole	Ref.
He	0.308	2.5K	−0.07	0.47	8.7	13
N_2	9.71	75K	−0.23	26.7	585	14
Ethanol	22.75	20	−0.086	46.3	1,340	ICT
Water	72.88	20	−0.138	113	1,590	15
$NaNO_3$[a]	116.6	308	−0.050	146	2,150	16
C_7F_{14}[b]	15.70	20	−0.10	45.0	2,610	17
Benzene	28.88	20	−0.13	67.0	2,680	ICT
n-Octane	21.80	20	−0.10	51.1	2,920	ICT
Sodium	191	98	−0.10	228	3,850	18
Copper	1,550	1,083	−0.176	1,355	10,200	18
Silver	910	961	−0.164	1,234	11,050	18
Iron	1,880	1,535	−0.43	2,657	20,100	19

[a] $E^{s'}$ computed on a per gram ion basis.
[b] Perfluoromethyl cyclohexane.

tension and E^s values. It is more instructive, however, to compare E^s values calculated on an energy per mole basis. For spherical molecules of radius r, the area per mole is

$$A = 4\pi N_0 \left(\frac{3M}{4\pi\rho N_0}\right)^{2/3} \tag{III-14}$$

where N_0 denotes Avogadro's number. Only about one-fourth of the area of the surface spheres will be exposed to the interface and on allowing for this Eq. III-14 becomes

$$A = fN_0^{1/3}V^{2/3} \qquad (\text{cm}^2/\text{mole}) \tag{III-15}$$

where V denotes molar volume and f is a composite geometric factor whose value is around unity. Then,

$$E^{s'} = AE^s \tag{III-16}$$

Values of $E^{s'}$ are given in Table III-1; it is seen that the variation in $E^{s'}$ is much less than that of f or of E^s.

Example. We reproduce the entry for iron in Table III-1 as follows. First, $E^s = 1880 - (1808)(-0.43) = 2657$ erg/cm. Next, estimating V to be about 7.1 cm^3/mole and taking f to be unity, $A = (6.02 \times 10^{23})^{1/3}(7.1)^{2/3} = 3.1 \times 10^8$ cm^2/mole whence $E^{s'} = (2657)(3.1 \times 10^8)/(4.13 \times 10^7) = 20{,}100$ cal/mole.

One may consider a molecule in the surface region as being in a state intermediate between that in the vapor phase and in the interior. Skapski (12) has made the following simple analysis. Considering only nearest neighbor interactions, if n_i and n_s denote the number of nearest neighbors in the interior of the liquid and in the surface region, respectively, then, per molecule:

$$E^{s'} = \left(\frac{N_0 \epsilon}{2}\right)(n_i - n_s) \tag{III-17}$$

where ϵ is the interaction energy. On this basis, the energy of vaporization should be $\epsilon n_i/2$. For close-packed spheres, $n_i = 12$ and $n_s = 9$, so that $E^{s'}$ should be one-fourth of the energy of vaporization (not exactly, because of f). On this basis, the surface energy of metals is somewhat *smaller* than expected. However, over the range of Table III-1 from helium to iron, one sees that the variation in surface energy *per square centimeter* depends almost equally on variation in intermolecular forces and on that of the density of packing or molecular size.

C. Change in Vapor Pressure for a Curved Surface

A very important thermodynamic relationship is that giving the effect of surface curvature on the molar free energy of a substance. The effect is perhaps best understood in terms of the existence of a pressure drop $\Delta \mathbf{P}$ across an interface, as given by Young and Laplace, Eq. II-7. From thermodynamics, the effect of a change in mechanical pressure at constant temperature on the molar free energy of a substance is

$$\Delta G = \int V \, d\mathbf{P} \tag{III-18}$$

or if the molar volume V is considered to be constant and Eq. II-7 is used for $\Delta \mathbf{P}$

$$\Delta G = \gamma V \left(\frac{1}{R_1} + \frac{1}{R_2}\right) \tag{III-19}$$

It is convenient to relate the free energy of a substance to its vapor pressure and, assuming the vapor to be ideal, $G = G^\circ + RT \ln P$. For the case of a spherical surface of radius r, we then have

$$RT \ln \left(\frac{P}{P^0}\right) = \frac{2\gamma V}{r} \tag{III-20}$$

where P^0 is the normal vapor pressure of the liquid and P is that observed over the curved surface. Equation III-20 is frequently called the *Kelvin* equation and, with the Young and Laplace equation, makes the second fundamental relationship of surface chemistry.

For liquid droplets, $\Delta \mathbf{P}$ is positive and there is thus an increased vapor pressure: For water, P/P^0 is about 1.001 if r is 10^{-4} cm, 1.011 if r is 10^{-5} cm, and 1.114 if r is 10^{-6} cm. This increased vapor pressure for small droplets has been verified experimentally for water, dibutyl phthalate, mercury, and other liquids (20), down to radii of the order of 0.1 μ. This phenomenon provides a ready explanation for the ability of vapors to supersaturate. The formation of a new liquid phase proceeds in stages, starting with clusters that may then grow or aggregate to droplets, which in turn grow to macroscopic size. In the absence of dust or other foreign surface on which the foregoing sequence can be bypassed, there will be an activation energy for the formation of these early states corresponding to the increased free energy due to the curvature of the surface (see Section IX-2).

While Eq. III-20 has been verified for small droplets, attempts to do so for liquids in capillaries, for which there should be a vapor pressure reduction, have led to some startling discrepancies. Shereshefsky and co-workers (21) reported that water and organic liquids such as toluene in capillaries of a few microns radius gave vapor pressure lowering *10* to *80* times as great as those calculated by the Kelvin equation. The subject took an amazing twist in the 1960s with confirmation by N. N. Fedyakin and then especially by B. V. Derjaguin and co-workers that water condensed into small capillaries had anomalous properties. It was thought for a while that a new *form* of water had been discovered—anomalous water or polywater. The present consensus is, however, that the various effects were due to one or another kind of impurity (see Section VI-9). Although Everett, Haynes, and McElroy (22) concluded in 1971 that the Kelvin equation lacked experimental verification, the situation has since been partially rectified. Fisher and Israelachvili (22a) report that the concave meniscus of cyclohexane formed between crossed cylinders of molecularly smooth mica shows the expected vapor pressure lowering down to a mean radius of curvature of 4 nm.

D. Effect of Curvature on Surface Tension

Tolman (23), from thermodynamic considerations, finds that with sufficiently curved surfaces the value of the *surface tension itself* should be affected. His result is

$$\frac{\gamma}{\gamma^0} = \frac{1}{(1 + 2\delta/r)} \qquad \text{(III-21)}$$

where δ is expected to be of the order of 10^{-8} cm (see Section III-2B). Thus for $\delta/r = 0.1$, γ/γ^0 is 0.83, and so on.

The effect assumes importance only at very small radii, but it has some application in the treatment of nucleation theory where the excess surface energy of small clusters is involved (see Section IX-2). An intrinsic difficulty with equations such as Eq. III-21 is that the treatment, if not modelistic and hence partly empirical, assumes a continuous medium, yet the effect does not become important until curvature comparable to molecular dimensions is reached. Fisher and Israelachvili (23a) measured the force due to the Laplace pressure for a pendular ring of liquid between crossed mica cylinders and concluded that for several organic liquids the effective surface tension remained unchanged down to radii of curvature as low as 0.5 nm (but only down to 5 nm in the case of water). The usefulness of Eq. III-21 is thus in doubt.

E. Effect of Pressure on Surface Tension

The following relationship holds on thermodynamic grounds (24):

$$\left(\frac{\partial \gamma}{\partial \mathbf{P}}\right)_{\mathcal{A},T} = \left(\frac{\partial V}{\partial \mathcal{A}}\right)_{\mathbf{P},T} \tag{III-22}$$

where $\mathcal{A}$ denotes area. In other words, the pressure effect is related to the change in molar volume when a molecule goes from the bulk to the surface region. This change would be positive, and the effect of pressure should therefore be to increase the surface tension.

Unfortunately, however, one cannot subject a liquid surface to an increased pressure without introducing a second component into the system, such as some inert gas. One thus increases the density of matter in the gas phase and, moreover, there will be some gas adsorbed on the liquid surface, with corresponding volume change V_g. The total change in volume with area is then

$$\frac{\partial \gamma}{\partial P} = -\frac{\Gamma ZRT}{P} + \Delta V^{\sigma} = -\Gamma V_g + \Delta V^{\sigma} \tag{III-23}$$

where Z is the compressibility factor, with ZRT/P reducing to V_g for an ideal gas.

Studies by Eriksson (25) and, more recently, by King and co-workers (26) have shown that the adsorption term dominates. One may also study the effect of pressure of a vapor on its interaction with the interface of a surfactant solution (27). Finally, the reported experimental failure of Eq. III-21 could be due to a compensating change in "γ^0" under the high negative Laplace pressures involved.

2. The Structural and Theoretical Treatment of Liquid Interfaces

It has been pointed out that the surface free energy can be regarded as the work of bringing a molecule from the interior of a liquid to the surface and that this work arises from the fact that, although a molecule experiences no net forces while in the interior of the bulk phase, these forces become unbalanced as it moves toward the surface. As discussed in connection with Eq. III-17 and also in the next sections, a knowledge of the potential function for the interaction between molecules allows a calculation of the total surface energy; if this can be written as a function of temperature, the surface free energy is also calculable.

The unbalanced force on a molecule is directed inward, and it might be asked how this could appear as a surface "tension." A mechanical analogy is shown in Fig. III-3, which illustrates how the work to raise a weight can

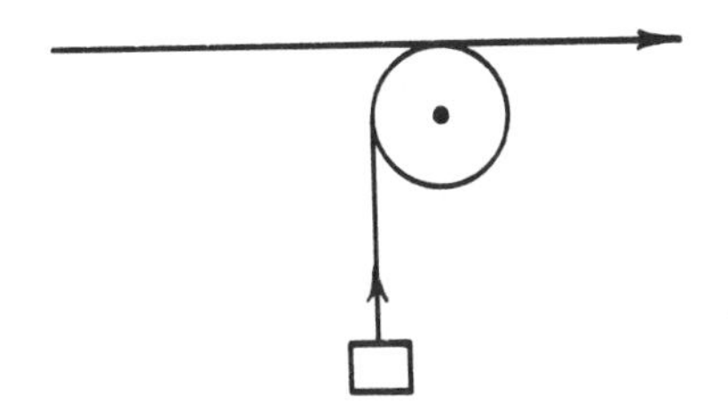

Fig. III-3. Mechanical analogy to surface tension.

appear as a horizontal pull; in the case of a liquid, an extension of the surface results in molecules being brought from the interior into the surface region.

The next point of interest has to do with the question of how deep the surface region or region of appreciably unbalanced forces is. This depends primarily on the range of intermolecular forces and, except where ions are involved, the principal force between molecules is of the so-called van der Waals type (see Section VI-1). This type of force decreases with about the seventh power of the intermolecular distance and, consequently, it is only the first shell or two of nearest neighbors whose interaction with a given molecule is of importance. In other words, a molecule experiences essentially symmetrical forces once it is a few molecular diameters away from the surface, and the thickness of the surface region is of this order of magnitude. The small degree of ellipticity of reflected light when the angle of incidence is the Brewsterian angle (28) confirms this estimate of the thickness of the surface region (see Section IV-3D). (Certain aspects of this statement need modification and are discussed in Section XVI-5.)

It must also be realized that this thin surface region is in a very turbulent state. Since the liquid is in equilibrium with its vapor, then, clearly, there is a two-way and balanced traffic of molecules hitting and condensing on the surface from the vapor phase and of molecules evaporating from the surface into the vapor phase. From the gas kinetic theory, the number of moles striking 1 cm^2 of surface per second is

$$Z = P\left(\frac{1}{2\pi MRT}\right)^{1/2} \tag{III-24}$$

For vapor saturated with respect to liquid water at room temperature Z is about 0.02 mole/cm^2-sec or about 1.2×10^{22} molecules/cm^2-sec. At equilibrium, then, the evaporation rate must equal the condensation rate, which differs from the preceding figure by a factor less than but close to unity (called the condensation coefficient). Thus each square centimeter of liquid water entertains 1.2×10^{22} arrivals and departures per second. The traffic on an area of 10 Å^2, corresponding to the area of a single water molecule, is 1.2×10^7 sec^{-1}, so that the lifetime of a molecule on the surface is of the order of a tenth microsecond.

There is also a traffic between the surface region and the adjacent layers of liquid. For most liquids, diffusion coefficients at room temperature are of the order of 10^{-5} cm^2/sec, and the diffusion coefficient $\mathcal{D}$ is related to the time t for a net displacement x by an equation due to Einstein,

$$\mathcal{D} = \frac{x^2}{2t} \tag{III-25}$$

If x is put equal to a distance of, say 100 Å, then t is about 10^{-6} sec, so that, due to Brownian motion, there is a very rapid interchange of molecules between the surface and the adjacent bulk region.

The picture that emerges is that a "quiescent" liquid surface is actually in a state of violent agitation on the molecular scale with individual molecules passing rapidly back and forth between it and the bulk regions on either side. Under a microscope of suitable magnification, the surface region should appear as a fuzzy blur, with the average density varying in some continuous manner from that of the liquid phase to that of the vapor phase.

In the case of solids, there is no doubt that a lateral tension (which may be anisotropic) can exist between molecules on the surface and can be related to actual stretching or compression of the surface region. This is possible because of the immobility of solid surfaces. Similarly, with thin soap films, whose thickness can be as little as 100 Å, stretching or extension of the film may involve a corresponding variation in intermolecular distances and an actual tension between molecules.

In fairness, however, it must be stated that a case can be made for the usefulness of surface "tension" as a concept even in the case of a normal liquid–vapor interface, and, for its presentation, the reader is referred to papers by Brown (29) and Gurney (30). The matter has its subtleties because the mathematical identity of the concepts of surface "tension" and surface "energy" makes the question of which is the more "real" in the case of liquid surfaces a somewhat philosophical one. The point under discussion is not, of course, the physical reality of surface tension as a measurable force acting parallel to the surface, that is, that work must be done to extend a surface, but rather whether or not it is more fruitful conceptually to consider that the effect arises from the energy requirement of bringing a molecule from the interior to the surface.

The writer believes that a useful choice can be made on the basis of whether or not actual intermolecular distances in the surface region change when the amount of surface is changed. For liquids, where the nature of the surface is independent of its extent, the surface energy concept is then to be preferred. With thin films and with solid surfaces, it may be necessary to recognize the presence of tensions in the surface and, indeed, to consider the surface free energy and the (possibly anisotropic) surface tension as two separate quantities. The reader is referred to Chapter VII for further discussion of this last point. Except where these distinctions are involved, the informal practice of using surface tension and surface free energy interchangeably will be followed.

A. Further Development of the Thermodynamic Treatment of the Surface Region

Consider a liquid in equilibrium with its vapor. The two bulk phases α and β do not change sharply from one to the other at the interface, but rather, as shown in Fig. III-4, there is a region over which the density and local pressure vary. Because the actual interfacial region has no sharply defined boundaries, it is convenient to invent a mathematical dividing surface (31). One then handles the extensive properties (G, E, S, n, etc.) by assigning to the bulk phases the values of these properties that would pertain if the bulk phases continued uniformly up to the dividing surface. The actual values for the system as a whole will then differ from the sum of the values for the two bulk phases by an excess or deficiency assigned to the surface region.

The following relations will then hold:

$$\text{Volume:} \quad V = V^{\alpha} + V^{\beta} \tag{III-26}$$

$$\text{Internal energy:} \quad E = E^{\alpha} + E^{\beta} + E^{\sigma} \tag{III-27}$$

$$\text{Entropy:} \quad S = S^{\alpha} + S^{\beta} + S^{\sigma} \tag{III-28}$$

$$\text{Moles:} \quad n_i = n_i^{\alpha} + n_i^{\beta} + n_i^{\sigma} \tag{III-29}$$

We will use the superscript σ to denote surface quantities calculated on the preceding assumption that the bulk phases continue unchanged to an assumed mathematical dividing surface. For an arbitrary set of variations from equilibrium,

$$dE = T\,dS + \sum_i \mu_i\,dn_i - \mathbf{P}^{\alpha}\,dV^{\alpha} - \mathbf{P}^{\beta}\,dV^{\beta} + \gamma\,d\mathcal{A} + C_1\,dc_1 + C_2\,dc_2 \tag{III-30}$$

where c_1 and c_2 denote the two curvatures (reciprocals of the radii of curvature) and C_1 and C_2 are constants. The last two terms may be written as $\frac{1}{2}(C_1 + C_2)\,d(c_1 + c_2) + \frac{1}{2}(C_1 - C_2)\,d(c_1 - c_2)$, and these plus the term

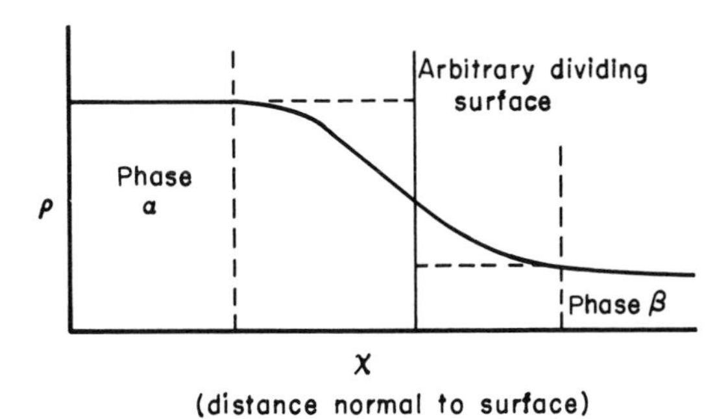

Fig. III-4

$\gamma\, d\mathcal{A}$ give the effect of variations in area and curvature. Because the actual effect must be independent of the location chosen for the dividing surface, a condition may be put on C_1 and C_2, and this may be taken to be that $C_1 + C_2 = 0$. This particular condition gives a particular location of the dividing surface such that it is now called the *surface of tension*.

For the case where the curvature is small compared to the thickness of the surface region, $d(c_1 - c_2) = 0$ (this will be exactly true for a plane or for a spherical surface), and Eq. III-30 reduces to

$$dE = T\,dS + \sum_i \mu_i\,dn_i - \mathbf{P}^\alpha\,dV^\alpha - \mathbf{P}^\beta\,dV^\beta + \gamma\,d\mathcal{A} \qquad \text{(III-31)}$$

Because

$$G = E - TS + \mathbf{P}^\alpha V^\alpha + \mathbf{P}^\beta V^\beta \qquad \text{(III-32)}$$

where G is the Gibbs free energy, it follows that

$$dG = -S\,dT + \sum_i \mu_i\,dn_i + V^\alpha\,d\mathbf{P}^\alpha + V^\beta\,d\mathbf{P}^\beta + \gamma\,d\mathcal{A} \qquad \text{(III-33)}$$

(Equation III-33 is obtained by differentiating Eq. III-32 and comparing with Eq. III-31.) At equilibrium, the energy must be a minimum for a given set of values of S and of n_i, and

$$-\mathbf{P}^\alpha\,dV^\alpha - \mathbf{P}^\beta\,dV^\beta + \gamma\,d\mathcal{A} = 0 \qquad \text{(III-34)}$$

but

$$dV = 0 = dV^\alpha + dV^\beta \qquad \text{(III-35)}$$

so

$$(\mathbf{P}^\alpha - \mathbf{P}^\beta)\,dV^\alpha = \gamma\,d\mathcal{A} \qquad \text{(III-36)}$$

Equation III-36 is the same as would apply to the case of two bulk phases separated by a membrane under tension γ.

If the surface region is displaced by a distance dt

$$d\mathcal{A} = (c_1 + c_2)\mathcal{A}\,dt \qquad \text{(III-37)}$$

and, because

$$dV^\alpha = \mathcal{A}\,dt = -dV^\beta \qquad \text{(III-38)}$$

then

$$(\mathbf{P}^\alpha - \mathbf{P}^\beta)\mathcal{A}\,dt = \gamma(c_1 + c_2)\mathcal{A}\,dt \qquad \text{(III-39)}$$

or

$$\Delta\mathbf{P} = \gamma(c_1 + c_2) \qquad \text{(III-40)}$$

Equation III-40 is the Young and Laplace equation, Eq. II-7.

The foregoing serves as an introduction to the detailed thermodynamics of the surface region; the method is essentially that of Gibbs (31), as reviewed by Tolman (32). An additional relationship is the following

$$\gamma = \int_{-a}^{0} (\mathbf{P}^{\alpha} - \mathbf{p})\, dx + \int_{0}^{a} (\mathbf{P}^{\beta} - \mathbf{p})\, dx \qquad \text{(III-41)}$$

where x denotes distance normal to the surface and the points a and $-a$ lie in the two bulk phases, respectively.

For a plane surface, $\mathbf{P}^{\alpha} = \mathbf{P}^{\beta}$, and

$$\gamma = \int_{-a}^{a} (\mathbf{P} - \mathbf{p})\, dx \qquad \text{(III-42)}$$

Here, **P** is the bulk pressure, which is the same in both phases, and **p** is the local pressure, which varies across the interface. This and other aspects of the Gibbs dividing surface have been reviewed by Lucassen-Reynders (33).

Returning to the matter of how to locate the dividing surface, the location defined by $(C_1 + C_2) = 0$ is in general such that $n^{\alpha} + n^{\beta}$, calculated by assuming the bulk phases to continue unchanged up to the dividing surface, will differ from the actual n. That is, even for a single pure substance, there will be a surface excess Γ that is not zero (and could be positive or negative). This convention, while mathematically convenient, is not pleasing intuitively, and Kirkwood and Buff (34) and Buff (35) have discussed an alternative location of the dividing surface, such that $\Gamma = 0$. From physical considerations, it appears that these two locations of the dividing surface differ in position by a distance of the order of a molecular diameter, with the consequence that the distinction does not become important except for very highly curved surfaces. See Refs. 36 and 37 for some more recent discussions.

B. Calculation of the Surface Energy and Free Energy of Liquids

The function of thermodynamics is to provide phenomenological relationships whose validity has the authority of the laws of thermodynamics themselves. One may proceed further, however, if specific models or additional assumptions are made. For example, use of the van der Waals equation allows a semiempirical analysis of how **P** in Eq. III-42 should vary across an interface, and Tolman (32) was able to calculate that the surface excess of water at 20°C was positive and about 10^{-8} g/cm^2, referred to the surface of tension.

There has been a high degree of development of statistical thermodynamics in this field; the reader may see standard works for details (e.g. Ref. 38); see also Sections XV-4A and XVI-3B.

A great advantage of this approach is that one may define a model system in terms of very fundamental assumptions about the forces between molecules. Fluids are very difficult to treat, but a widely explored particular model is that of a "hard sphere" fluid supposed to consist of rigid spheres of a given size a, which interact with some energy ϵ. Reiss and co-workers (39) have been able to apply the hard sphere treatment to calculations of surface tension by determining how the free energy of the system is affected by introducing a spherical cavity.

The treatment also suggested a relationship between surface tension and the compressibility of the liquid. In a more classical approach, the equation has been obtained (40) as follows:

$$\gamma = \frac{a}{8}\left(\frac{\partial E}{\partial V}\right)_T \tag{III-43}$$

where a is the side of a cube of molecular volume. The coefficient $(\partial E/\partial V)_T$ is just the internal pressure of the liquid, and may be replaced by the expression $(\alpha T/\beta - P)$, where α and β are the coefficients of thermal expansion and of compressibility, and P is the ambient pressure (see Ref. 41).

A problem with the statistical mechanical approach is that it is extremely difficult to obtain from wave mechanics the needed detailed specification of the various energy states. A more practical approach makes use of the radial distribution function $g(r)$, which gives the probability of finding a molecule at distance r from a given one. This function may be obtained experimentally, from x-ray scattering data on the liquid, for example. One needs, in addition, the potential function $\epsilon(r)$ of interaction between two molecules. Kirkwood and Buff (34) showed that

$$\gamma = \left(\frac{\pi}{8}\right)\rho^2 \int_0^\infty g(r)\epsilon'(r)r^4\,dr \tag{III-44}$$

$$E^s = \left(-\frac{\pi}{2}\right)\rho^2 \int_0^\infty g(r)\epsilon(r)r^3\,dr \tag{III-45}$$

where ρ is the average number of molecules per unit volume and $\epsilon'(r)$ denotes $d\epsilon(r)/dr$. A widely used, convenient, and successful form for $\epsilon(r)$ is that due to Lennard-Jones:

$$-\epsilon(r) = 4\epsilon_0\left[\left(\frac{\sigma}{r}\right)^6 - \left(\frac{\sigma}{r}\right)^{12}\right] \tag{III-46}$$

where, as shown in Fig. III-5, ϵ_0 is the potential energy at the minimum and σ is an effective molecular diameter. The internal pressure of a liquid and its internal energy are also functions of $g(r)$ and $\epsilon(r)$, and in a recent approach the two experimental quantities, internal pressure and energy, were used to evaluate the constants ϵ_0 and σ in Eq. III-46. It was then possible to calculate γ and E^s from Eqs. III-44 and III-45 (42). For example, the calculated and experimental values of γ for argon at 84.3 K are 15.1 and 13.2 erg/cm^2 respectively, and for nitrogen at 77 K they are 11.5 and 8.9 erg/cm^2 respectively. Somewhat poorer agreement was obtained for the E^s values.

A similar approach to the treatment of interfacial free energies is discussed in Section X-6B. It is important to note that all of these treatments are at their best for monatomic or at least spherically symmetric molecules since the problem of expanding the functions $g(r)$ and $\epsilon(r)$ to include the angular

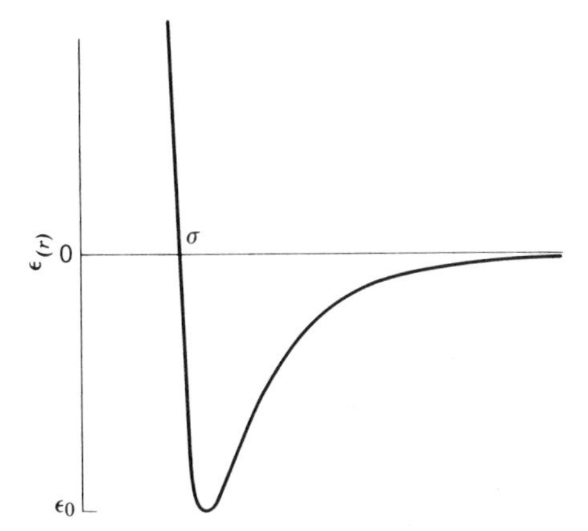

Fig. III-5. The Lennard-Jones potential function.

relationships between molecules is a very difficult one. Furthermore, asymmetric molecules may be highly oriented at an interface, as shown in the next section, so that $g(r)$ angles are very different for the surface and bulk regions. The approach has been applied to molten salt mixtures (43).

3. Orientation at Interfaces

There is one remaining and very significant aspect of liquid–air and liquid–liquid interfaces to be considered before proceeding to a discussion of the behavior and thermodynamics of the interfaces of solutions. This is the matter of molecular orientation at interfaces.

The idea that unsymmetrical molecules will be oriented at an interface is now so well accepted that it hardly needs to be argued, but it is of interest to outline some of the history of the concept. Hardy (44) and Harkins (45) devoted a good deal of attention to the idea of "force fields" around molecules, more or less intense depending on the polarity and specific details of structure. Orientation was treated in terms of a principle of "least abrupt change in force fields," that is, that molecules at an interface should be oriented so as to provide the most gradual possible transition from the one phase to the other. If we read "interaction energy" in place of "force field," the principle could be reworded on the very reasonable basis that molecules will be oriented so that their mutual interaction energy will be a maximum.

A somewhat more quantitative development along these lines was given by Langmuir (46) in what he termed the *principle of independent surface action*. He proposed that, qualitatively, one could suppose each part of a molecule to possess a local surface free energy. Taking ethanol as an example, one can employ this principle to decide whether surface molecules should be oriented according to Fig. III-6*a* or *b*. In the first case, the surface presented would be one of hydroxyl groups whose surface energy should be about 190 erg/cm^2, extrapolating from water. In the second case, a surface energy like that of a hydrocarbon should prevail, that is, about 50 erg/cm^2

(a) (b) Fig. III-6

(see Table III-1). This is a difference of 140 erg/cm^2 or about 30×10^{-14} erg per molecule. Since kT is of the order of 4×10^{-14} erg per molecule, the Boltzmann factor, $\exp(-\epsilon/kT)$, favoring the orientation in Fig. 6*b* should be about 10^5. This conclusion is supported by the observation that the actual surface tension of ethanol is 22 erg/cm^2 or not very different from that of a hydrocarbon. Langmuir's principle may sound rather primitive but, in fact, it is widely used and useful today in one or another (often disguised) form.

Harkins (45) used another argument whose repetition here serves the added useful function of introducing some important definitions. The quantity known as the *work of adhesion* w_{AB} between two phases is given by

$$w_{AB} = \gamma_A + \gamma_B - \gamma_{AB} \tag{III-47}$$

As illustrated in Fig. III-7*a*, w_{AB} gives the work necessary to separate one square centimeter of interface AB into two liquid–vapor interfaces A and B. Similarly, the *work of cohesion* w_{AA} for a single liquid A is

$$w_{AA} = 2\gamma_A \tag{III-48}$$

and corresponds to the reversible work to pull apart a column of liquid, as illustrated in Fig. III-7*b*.

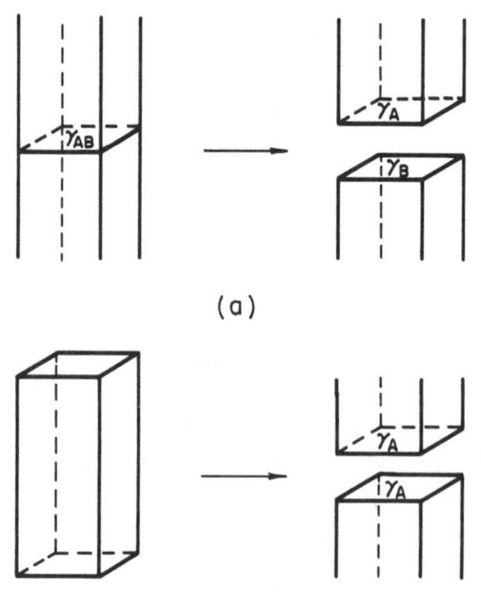

Fig. III-7. Work of adhesion and work of cohesion.

TABLE III-2
Some Values of Works of Adhesion and Cohesion

Liquid–air interface	Work of cohesion	Liquid–liquid interface	Work of adhesion
Octane	44	Octane–water	44
Octyl alcohol	55	Heptane–water	42
Heptanoic acid	57	Octyl alcohol–water	92
Heptane	40	Octylene–water	73
		Heptanoic acid–water	95

Some values of works of cohesion and of adhesion are given in Table III-2. Looking at the data, it is seen that the values of the works of cohesion are about the same for a variety of organic liquids, suggesting that the interfaces are similar, that is, primarily hydrocarbon in nature. The same low values prevail for the works of adhesion of pure hydrocarbons to water. However, the last three values of w_{AB} pertain to the interface between water and a polar–nonpolar material, and these are now much larger. The reasonable conclusion is that at such interfaces the polar end of the organic molecule is oriented toward the water.

There is, of course, a mass of rather direct evidence now available on surface orientation. Much of this will be found in the chapters on surface films and on adsorption. Good (47) calculated the molar surface entropy for a number of liquids, essentially combining Eqs. III-4 and III-15, and found that the values for polar liquids were about 3 cal/(K)(mole) lower than for nonpolar ones. He attributed the difference to the reduced number of orientations available to surface polar molecules as compared to surface nonpolar ones. See also Ref. 48.

Returning to the concept of work of adhesion, there are some complications, taken up in Section IV-2, and still more if a solid is involved, discussed in Section XII-8.

4. The Surface Tension of Solutions

The principal point of interest to be discussed in this section is the manner in which the surface tension of binary systems varies with composition. The effects of other variables such as pressure and temperature are similar to those for pure substances, and the more elaborate treatment for two component systems is not considered here. Also, the case of the interfacial tension of immiscible liquids is taken up in Section IV-2.

A fairly simple treatment, due to Guggenheim (49), is useful for the case of ideal or nearly ideal solutions. An abbreviated derivation follows. The free energy of a species may be written

$$G_i = kT \ln a_i \qquad \text{(III-49)}$$

where a_i is an absolute activity, and may in turn be written as

$$a_i = N_i g_i = g_i \qquad \text{(for a pure liquid } i\text{)} \tag{III-50}$$

where N_i is the mole fraction (if not unity), and g_i derives from the partition function Q_i. For a pure liquid 1 the surface tension may be written as

$$\gamma_1 \sigma_1 = -kT \ln \frac{a_1}{a_1^s} \tag{III-51}$$

or

$$\exp\left(\frac{-\gamma_1 \sigma_1}{kT}\right) = \frac{g_1}{g_1^s} \tag{III-52}$$

where the surface is viewed as a two-dimensional phase of molecular state corresponding to g_1^s, and σ_1 is the molecular area. That is, the work of bringing a molecule into the surface is expressed as a ΔG using Eq. III-49.

The same relations are then applied to each component of a solution

$$\exp\left(\frac{-\gamma \sigma_1}{kT}\right) = \frac{N_1 g_1}{N_1^s g_1^s} \tag{III-53}$$

$$\exp\left(\frac{-\gamma \sigma_2}{kT}\right) = \frac{N_2 g_2}{N_2^s g_2^s} \tag{III-54}$$

where N^s denotes the mole fraction in the surface phase. Equations III-53 and III-54 may be solved for N_1^s and N_2^s, respectively, and substituted into the requirement that $N_1^s + N_2^s = 1$. If it is assumed that $\sigma = \sigma_1 = \sigma_2$, one then obtains

$$\exp\left(\frac{-\gamma \sigma}{kT}\right) = \frac{N_1 g_1}{g_1^s} + \frac{N_2 g_2}{g_2^s} \tag{III-55}$$

and, in combination with Eq. III-52

$$e^{-\gamma\sigma/kT} = N_1 e^{-\gamma_1 \sigma/kT} + N_2 e^{-\gamma_2 \sigma/kT} \tag{III-56}$$

Hildebrand and Scott (50) give an expansion of Eq. III-56 in which it is not assumed that $\sigma_1 = \sigma_2$.

Guggenheim (49) extended his treatment to the case of regular solutions, that is, solutions for which

$$RT \ln f_1 = -\alpha N_2^2, \qquad RT \ln f_2 = -\alpha N_1^2 \tag{III-57}$$

where f denotes activity coefficient. A very simple relationship for such regular solutions comes from Prigogine and Defay (51):

$$\gamma = \gamma_1 N_1 + \gamma_2 N_2 - \beta N_1 N_2 \tag{III-58}$$

where β is a semiempirical constant.

Figure III-8(a) shows some data for fairly ideal solutions (50) where the solid lines 2, 3, and 6 show the attempt to fit the data with Eq. III-56); line

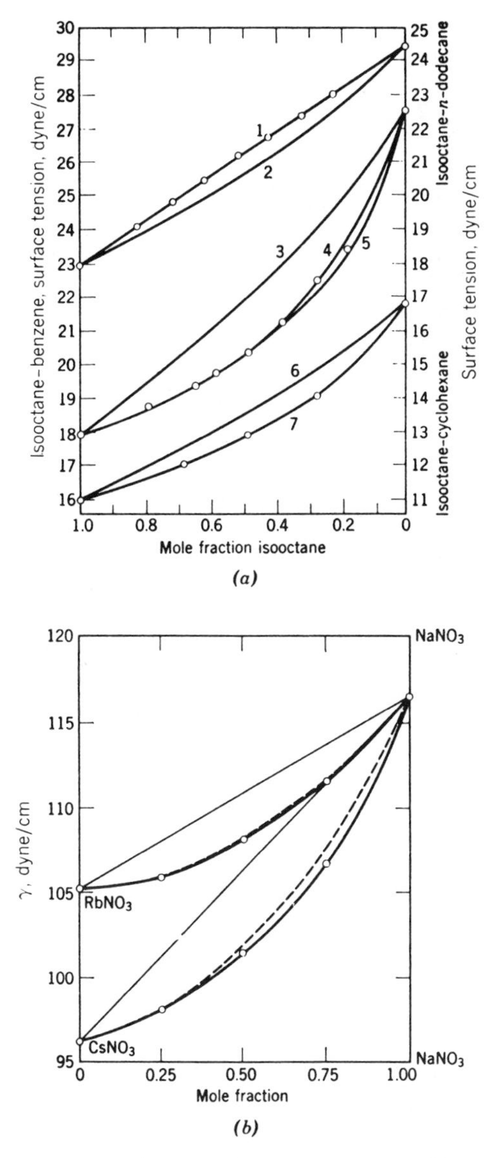

Fig. III-8. Representative surface tension versus composition plots. (*a*) Isooctane-*n*-dodecane at 30°C: 1 linear, 2 ideal, with $\sigma = 48.6$. Isooctane–benzene at 30°C: 3 ideal, with $\sigma = 35.4$, 4 ideal-like, with empirical σ of 112, 5 unsymmetrical, with $\sigma_1 = 136$ and $\sigma_2 = 45$. Isooctane–cyclohexane at 30°C: 6 ideal, with $\sigma = 38.4$, 7 ideal-like, with empirical σ of 109.3 (σ values in Å^2/molecule) (from Ref. 58). (*b*) Surface tension isotherms at 350°C for the systems (Na–Rb) NO_3 and (Na–Cs) NO_3. Dotted lines show the fit to Eq. III-58 (from Ref. 52a). (*c*) Water–ethanol at 25°C. (*d*) Aqueous dodecyldimethylammonium chloride at different NaCl concentrations: $C(M)$: ●, 0; ○, 0.01; ◓, 0.05; ◑, 0.10; ◐, 0.20; ⊖, 0.94 (from Ref. 59). (*e*) Aqueous sodium chloride at 20°C.

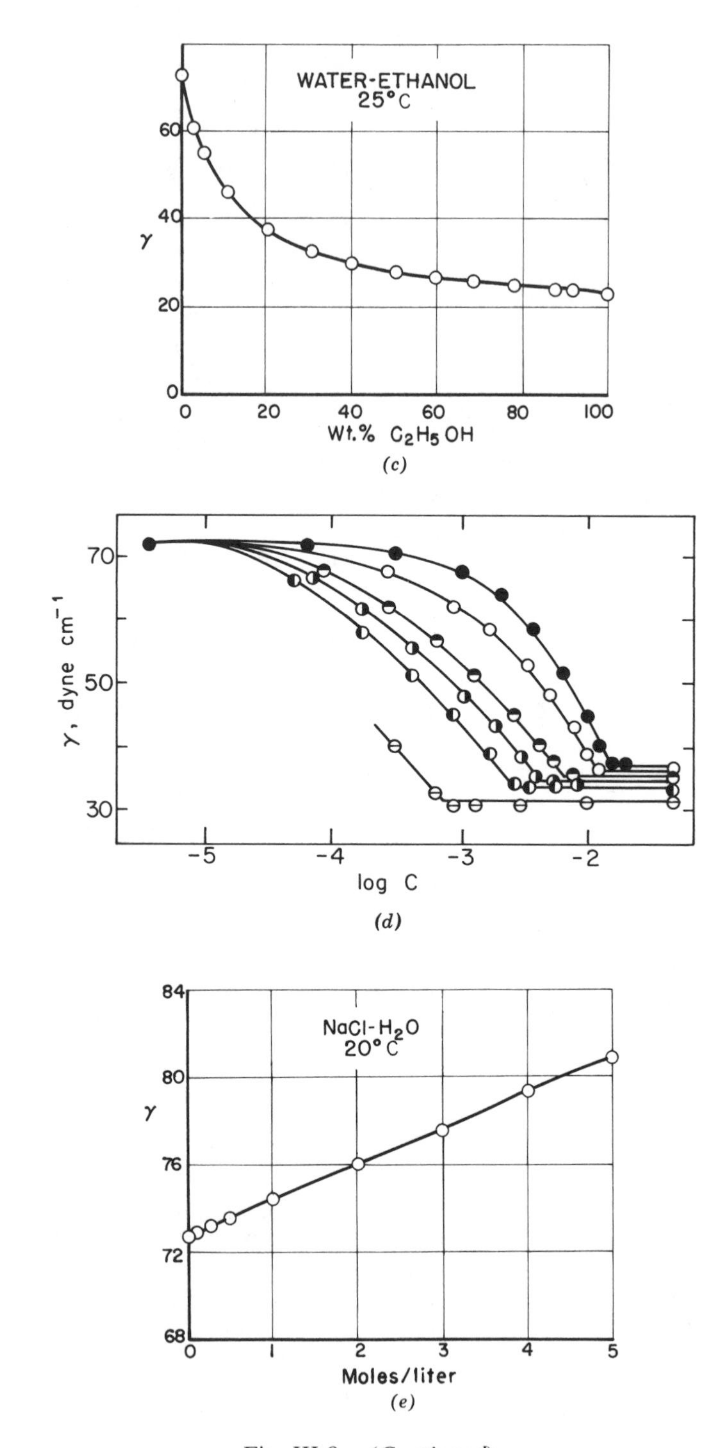

(c)

(d)

(e)

Fig. III-8. (*Continued*)

4, by taking σ as a purely empirical constant; and line 5, by the use of the Hildebrand and Scott equation (50). As a further example of solution behavior, Fig. III-8(*b*) shows some data on fused salt mixtures (52); the dotted lines show the fit to Eq. III-58).

An extensive development has been made for various types of nonideal solutions by Defay, Prigogine, and their co-workers, using a lattice model (see Ref. 51); their treatments allow for interacting molecules of different sizes. Nissen (see Ref. 52) has applied the approach to molten salt mixtures, as has Gaines (53). Gaines and co-workers have also treated the surface tension of polymer solutions (see Ref. 54). Reiss and Mayer (55) developed an expression for the surface tension of a fused salt, using their hard sphere treatment of liquids (Section III-2B), and have extended the approach to solutions. The same is true for a statistical mechanical model developed by Eyring and co-workers (56).

A simple semiempirical treatment due to Eberhart (57) assumes that the surface tension of a binary solution is linear in *surface composition* so that

$$\gamma = N_1^s\gamma_1 + N_2^s\gamma_2 \tag{III-59}$$

and that the two components are distributed between the solution and interfacial phases (see Section III-7C), according to the simple distribution law

$$S_{12} = \frac{K_1}{K_2} = \frac{N_1^s/N_1}{N_2^s/N_2} \tag{III-60}$$

Algebraic manipulation then gives

$$p = 1 - S_{12}\frac{p}{r} \tag{III-61}$$

where

$$p = \frac{\gamma - \gamma_1}{\gamma_2 - \gamma_1}$$

and

$$r = \frac{N_2}{N_1}$$

Equation III-61 was found to fit a variety of two component systems quite well. Goodisman (57a) has commented on various treatments for molten salt mixtures, including Eberhart's.

We have considered the surface tension behavior of some widely different types of systems, and it is now desirable to discuss in slightly more detail the very important case of aqueous mixtures. If the surface tensions of the separate pure liquids differ appreciably, as in the case of alcohol–water mixtures, then the addition of small amounts of the second component generally results in a marked decrease in surface tension from that of the solvent water. The case of ethanol and water is shown in Fig. III-8*c*. As seen in the next section, this effect may be accounted for in terms of a selective adsorption of the alcohol at the interface. For dilute aqueous solutions of organic substances, the semiempirical equation

$$\frac{\gamma}{\gamma_0} = 1 - B \ln\left(1 + \frac{C}{a}\right) \qquad \text{(III-62)}$$

has been used (60), where γ_0 is the surface tension of water, B is a constant characteristic of the homologous series of organic compounds involved, a is a constant characteristic of each compound, and C is its concentration. This equation may be derived on the basis that the surface adsorption follows a Langmuir adsorption equation (see Problem III-8 and Sections XI-1A and XVI-3).

The type of behavior shown by the ethanol-water system reaches an extreme in the case of higher molecular weight solutes of the polar–nonpolar type, as for example the various soaps and detergents. As illustrated in Fig. III-8*d*, the decrease in surface tension now takes place at very low concentrations, and the surface tension then remains essentially constant. The concentration at this point is that of micelle formation (see Section XIII-5B); note the sensitivity to salt concentration.

Finally, aqueous electrolyte solutions behave as exemplified by the case of the NaCl–water system shown in Fig. III-8*e*, that is, the surface tension rises with concentration.

5. Thermodynamics of Binary Systems— The Gibbs Equation

We come now to a very important topic, namely, the thermodynamic treatment of the variation of surface tension with composition. The treatment is due to Gibbs (31), but has been amplified in a more conveniently readable way by Guggenheim and Adam (61).

A. Definition of Surface Excess

As in Section III-2A, it is convenient to suppose the two bulk phases, α and β, to be uniform up to an arbitrary dividing plane S, as illustrated in Fig. III-9. We restrict ourselves to plane surfaces so that c_1 and c_2 are zero, and the condition of equilibrium does not impose any particular location for S. As before, one computes the various extensive quantities on the basis above and compares them with the values for the system as a whole. Any excess or deficiency is then attributed to the surface region.

Taking the section shown in Fig. III-9 to be of unit area in cross section, then, if the phases were uniform up to S, the amount of the ith component present would be

$$xC_i^\alpha + (a - x)C_i^\beta \qquad \text{(III-63)}$$

Here, the distances x and a are relative to planes A and B located far enough

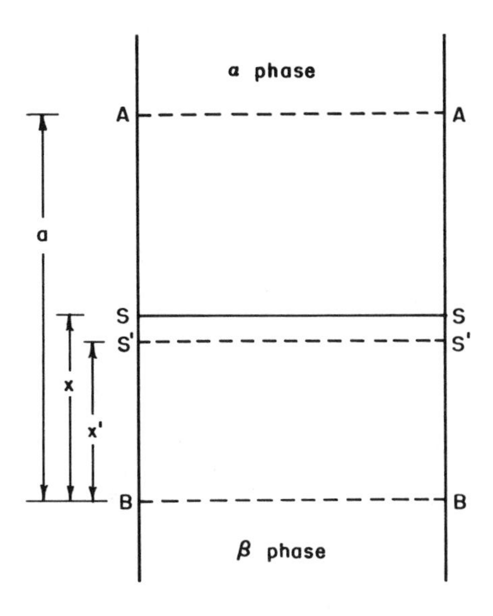

Fig. III-9

from the surface region so that bulk phase properties prevail. The actual amount of component i present in the region between A and B will be

$$xC_i^\alpha + (a - x)C_i^\beta + \Gamma_i^\sigma \tag{III-64}$$

where Γ_i^σ denotes the surface excess per unit area.†

For the case where the phase B is gaseous, C_i^β may be neglected, and quantities III-63 and III-64 become

$$xC_i^\alpha \quad \text{and} \quad xC_i^\alpha + \Gamma_i^\sigma$$

If one now makes a second arbitrary choice for the dividing plane, namely, S' and distance x', it must follow that

$$x'C_i + \Gamma_i^{\sigma\prime} = xC_i + \Gamma_i^\sigma \tag{III-65}$$

(dropping the superscript α as unnecessary), because the same total amount of the ith component must be present between A and B regardless of how the dividing surface is located. One then has

$$\frac{\Gamma_i^{\sigma\prime} - \Gamma_i^\sigma}{C_i} = x - x' \tag{III-66}$$

so that

$$\frac{\Gamma_1^{\sigma\prime} - \Gamma_1^\sigma}{C_1} = \frac{\Gamma_2^{\sigma\prime} - \Gamma_2^\sigma}{C_2} = \text{etc.} \tag{III-67}$$

† The term *surface excess* will be used as an algebraic quantity. If positive, an actual excess of the component is present and, if negative, there is a surface deficiency. An alternative name that has been used is *superficial density*.

or, in general,

$$\frac{(\Gamma_i^{\sigma'} - \Gamma_i^\sigma)}{N_i} = \frac{(\Gamma_j^{\sigma'} - \Gamma_j^\sigma)}{N_j} \tag{III-68}$$

where N denoted mole fraction, or

$$\Gamma_j^\sigma N_i - \Gamma_i^\sigma N_j = \Gamma_j^{\sigma'} N_i - \Gamma_i^{\sigma'} N_j \tag{III-69}$$

Since S and S' are purely arbitrary in location Eq. III-69 can be true only if each side separately equals a constant

$$\Gamma_j^\sigma N_i - \Gamma_i^\sigma N_j = \text{constant} \tag{III-70}$$

B. The Gibbs Equation

With the preceding introduction to the handling of surface excess quantities, we now proceed to the derivation of the third fundamental equation of surface chemistry (the Laplace and Kelvin equations, Eqs. II-7 and III-20, being the other two), known as the *Gibbs* equation.

For a small, reversible change dE in the energy of a system, one has

$$\begin{aligned} dE &= dE^\alpha + dE^\beta + dE^\sigma \\ &= T\,dS^\alpha + \sum \mu_i\,dn_i^\alpha - P^\alpha\,dV^\alpha + T\,dS^\beta \\ &\quad + \sum \mu_i\,dn_i^\beta - P^\beta\,dV^\beta + T\,dS^\sigma + \sum \mu_i\,dn_i^\sigma + \gamma\,d\mathcal{A} \end{aligned} \tag{III-71}$$

Since

$$dE^\alpha = T\,dS^\alpha + \sum \mu_i\,dn_i^\alpha - P^\alpha\,dV \tag{III-72}$$

and similarly for phase β, it follows that

$$dE^\sigma = T\,dS^\sigma + \sum \mu_i\,dn_i^\sigma + \gamma\,d\mathcal{A} \tag{III-73}$$

If one now allows the energy, entropy, and amounts to increase from zero to some finite value, keeping T, $\mathcal{A}$ (area), and the n_i^σ constant, Eq. III-73 becomes

$$E^\sigma = TS^\sigma + \sum \mu_i n_i^\sigma + \gamma\mathcal{A} \tag{III-74}$$

Equation III-74 is generally valid and may now be differentiated in the usual manner to give

$$dE^\sigma = T\,dS^\sigma + S^\sigma\,dT + \sum \mu_i\,dn_i^\sigma + \sum n_i^\sigma\,d\mu_i + \gamma\,d\mathcal{A} + \mathcal{A}\,d\gamma \tag{III-75}$$

Comparison with Eq. III-73 gives

$$0 = S^\sigma\,dT + \sum n_i^\sigma\,d\mu_i + \mathcal{A}\,d\gamma \tag{III-76}$$

or, per unit area

$$d\gamma = -S^\sigma\,dT - \sum \Gamma^\sigma\,d\mu_i \tag{III-77}$$

For a two-component system at constant temperature, Eq. III-77 reduces to

$$d\gamma = -\Gamma_i^\sigma \, d\mu_1 - \Gamma_2^\sigma \, d\mu_2 \qquad \text{(III-78)}$$

Moreover, since Γ_1^σ and Γ_2^σ are defined relative to an arbitrarily chosen dividing surface, it is possible in principle to place that surface so that $\Gamma_1^\sigma = 0$ (this is discussed in more detail below), so that

$$\Gamma_2^1 = -\left(\frac{\partial \gamma}{\partial \mu_2}\right)_T \qquad \text{(III-79)}$$

or

$$\Gamma_2^1 = -\left(\frac{a}{RT}\right)\left(\frac{d\gamma}{da}\right) \qquad \text{(III-80)}$$

where a is the activity of the solute and the superscript 1 on the Γ means that the dividing surface was chosen so that $\Gamma_1^\sigma = 0$. Thus if $d\gamma/da$ is negative, as in Fig. III-8*c*, Γ_2^1 is positive, and there is an actual surface excess of solute. If $d\gamma/da$ is positive, as in Fig. III-8*e*, there is a surface deficiency of solute.

C. Alternate Methods of Locating the Dividing Surface

Returning to Eq. III-78, this may be put in another form by use of the Gibbs-Duhem equation.†

$$N_1 \, d\mu_1 + N_2 \, d\mu_2 = 0 \qquad \text{(III-81)}$$

Using Eq. III-81 to eliminate $d\mu_1$ from Eq. III-78, one obtains

$$-d\gamma = d\mu_2 \left[\Gamma_2^\sigma - \left(\frac{N_2}{N_1}\right)\Gamma_1^\sigma\right] \qquad \text{(III-82)}$$

The term in square brackets must be independent of the choice of location for the dividing surface, so again one arrives at the conclusion stated in Eq. III-70.

A detailed picture of how concentrations might vary across a liquid–vapor interface is given in Fig. III-10. The convention indicated by superscript 1, that is, that $\Gamma_1^\sigma = 0$, is illustrated. The dividing line is drawn so that the two

† Equation III-81 may be obtained by applying to Eq. III-72 the same set of operations as was applied to Eq. III-73.

$$E = TS - PV + \sum \mu_i \, dn_i$$

$$dE = T\,dS + S\,dT - P\,dV - V\,dP + \sum n_i \, d\mu_i + \sum \mu_i \, dn_i$$

$$0 = S\,dT - V\,dP + \sum n_i \, d\mu_i$$

and, for a two-component system at constant temperature and pressure,

$$n_1 \, d\mu_1 = -n_2 \, d\mu_2 \quad \text{or} \quad N_1 \, d\mu_1 = -N_2 \, d\mu_2$$

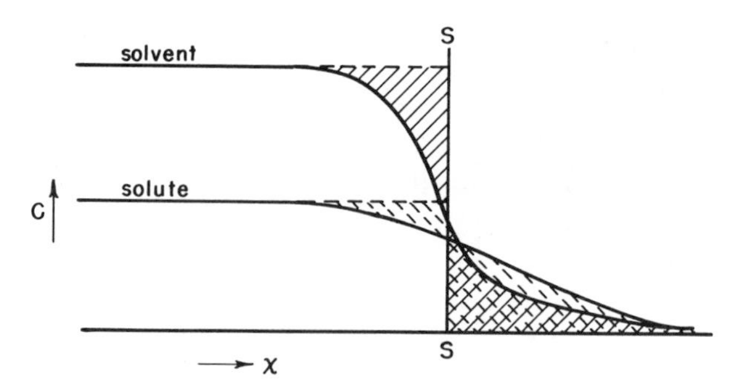

Fig. III-10. Schematic illustration of surface excess.

areas shaded in full strokes are equal, and the surface excess of the solvent is thus zero. The area shaded with dashed strokes, which lies to the right of the dividing surface, minus the smaller similarly shaded area to the left of the dividing surface, corresponds to the (in this case) positive surface excess of solute.

The quantity Γ_2^1 may thus be defined as the (algebraic) excess of component 2 in a 1-cm^2 cross section of surface region over the moles that would be present in a bulk region containing the *same number of moles of solvent as does the section of surface region.*

Obviously, a symmetric definition Γ_1^2 also exists. Here, Γ_2^s is set equal to zero, and Γ_1^2 represents the excess of component *1* in a 1-cm^2 cross section of surface region over the mole that would be present in a bulk region containing the *same number of moles of solute as does the section of surface region.*

Still another way of locating the dividing surface would be such that the algebraic *sum* of the areas in Fig. III-10 to the right of the dividing line was equal to the sum of the areas to the left of the line. The surface excesses so defined are written Γ_i^N. Here Γ_i^N is the excess of the *i* component in a 1-cm^2 cross section of surface region over the moles that would be present in a bulk region containing the *same total number of moles as does the section of surface region.*

Similarly, Γ_i^M is the excess of the *i*th component in the surface region over the moles that would be present in a bulk region of the *same total mass as the surface region*. Finally, Γ_i^V would be the excess of the *i*th component in the surface region over the moles that would be present in a bulk region of the *same total volume as the surface region.*

At this point, a numerical example will undoubtedly be helpful! Let us suppose that we have a solution of alcohol in water, of mole fraction 50%. We take a slice of surface region, that is, a cut deep enough so that some of the bulk solution is included, such as one through the extreme left of the section shown in Fig. III-10. Let us suppose that this sample of the surface region represents a slice of $\mathcal{A}$ cm^2 and is found to contain 10 moles of water and 30 moles of alcohol. Additional data would be that the molecular weights of water and of ethanol are 18 and 46, and that the molar volumes are 18 and 58 cc/mole, respectively.

The variously defined surface excesses are then as follows:

Case 1. Γ_2^1

Surface region:	10 moles water	30 moles alochol
To be compared with a bulk region containing:	10 moles water	10 moles alcohol
Excess:	0	20

Hence, $\Gamma_2^1 = 20/\mathcal{A}$

Case 2. Γ_1^2

Surface region:	10 moles water	30 moles alcohol
To be compared with a bulk region containing:	30 moles water	30 moles alcohol
Excess:	−20	0

Hence, $\Gamma_1^2 = -20/\mathcal{A}$

Case 3. Γ_1^N and Γ_2^N

Surface region:	10 moles water	30 moles alcohol
To be compared with a bulk region containing the same total moles:	20 moles water	20 moles alcohol
Excess:	−10	10

Hence, $\Gamma_1^N = -10/\mathcal{A}$ and $\Gamma_2^N = 10/\mathcal{A}$

Case 4. Γ_1^M and Γ_2^M

Surface region	10 moles water 180 g water	30 moles alcohol 1380 g alcohol
To be compared with a bulk region having the same total number of grams, i.e., 1560 g. This amount of bulk region would have:	24.4 moles water 440 g water	24.4 moles alcohol 1120 g alcohol
Excess:	−14.4 moles −260 g	5.6 moles 260 g

Hence, $\Gamma_1^M = -14.4/\mathcal{A}$ and $\Gamma_2^M = 5.6/\mathcal{A}$

The calculation of Γ_1^V and Γ_2^V is left to the reader.

It will be noticed that the surface excesses obey the relationship

$$P_1\Gamma_1 + P_2\Gamma_2 = 0 \tag{III-83}$$

where P is determined by the specific property invoked in deciding how to choose the location of the dividing surface. Thus, for Γ_1^N, P is unity; for Γ_1^M, P_i is M_i, the molecular weight; and for Γ_i^V, P_i is V_i, the molar volume.

Equation III-83 may be combined with Eq. III-82 to give

$$\begin{aligned} \frac{-d\gamma}{d\mu_2} &= \Gamma_2\left(1 + \frac{P_2N_2}{P_1N_1}\right) \\ \frac{-d\gamma}{d\mu_1} &= \Gamma_1\left(1 + \frac{P_1N_1}{P_2N_2}\right) \end{aligned} \tag{III-84}$$

The entire picture may then be summarized as follows:

$$\frac{-d\gamma}{d\mu_2} = \Gamma_2^1 = \Gamma_2^N\left(1 + \frac{N_2}{N_1}\right) = \Gamma_2^M\left(1 + \frac{M_2N_2}{M_1N_1}\right) \tag{III-85}$$

or

$$\frac{-N_1 d\gamma}{d\mu_2} = N_1\Gamma_2^1 = \Gamma_2^N = \left(\frac{\bar{M}}{M_1}\right)\Gamma_2^M = \left(\frac{\bar{V}}{V_1}\right)\Gamma_2^V \qquad \text{(III-86)}$$

where $\bar{M} = N_1M_1 + N_2M_2$ and $\bar{V} = N_1V_1 + N_2V_2$.

Figure III-11 illustrates how these various types of surface excess quantities vary with composition. Note that Γ_2^1 and Γ_1^2 do not go to zero as N_2 and N_1 approach unity, respectively. Why?

D. The Thermodynamics of Surfaces Using the Concept of a Surface Phase

An approach that has been developed by Guggenheim (62), and which has some use, is to avoid the somewhat artificial concept of the Gibbs dividing surface and to suppose instead that the surface region is actually a bulk region whose upper and lower limits lie somewhere in the α and β phases and not too far from the actual surface, for example, as given by lines *AA* and *BB* in Figure III-9.

The surface region *s*, so defined, is of thickness τ, volume $V^s = \tau\mathcal{A}$ and possesses the normal extrinsic thermodynamic properties. While the pressure is isotropic in the bulk phases α and β, it is not in the surface phase. On the one hand, the force across unit area in a plane parallel to the surface *is* the same as the general pressure, but the force per unit area across a plane perpendicular to the plane of the surface varies as one travels through the surface phase from *AA* to *BB*. It is the integral of this pressure with distance

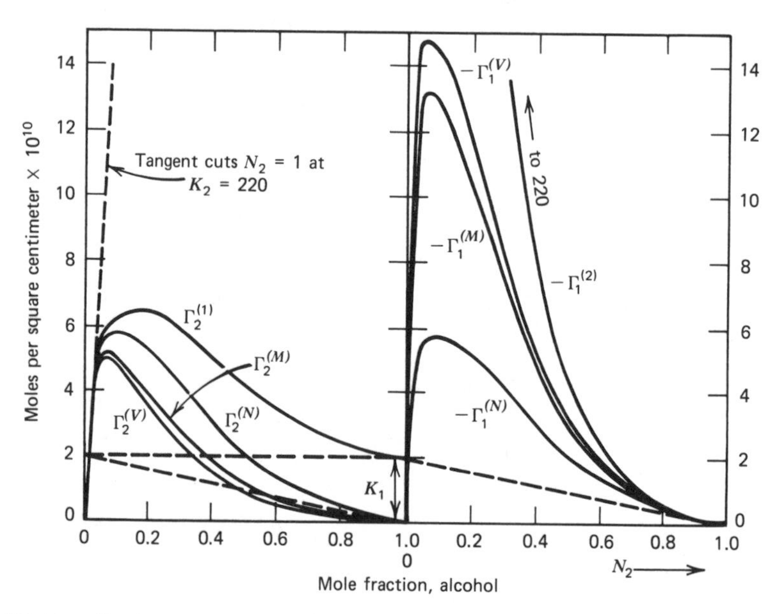

Fig. III-11. Surface excess quantities for the system water–ethanol (from Ref. 61).

that gives the surface tension in Eq. III-42. The thermodynamic development is similar to that in Section III-5B

$$dE^{\mathcal{S}} = T\,dS^{\mathcal{S}} - P\,dV^{\mathcal{S}} + \sum \mu_i\,dn_i^{\mathcal{S}} + \gamma\,d\mathcal{A} \qquad \text{(III-87)}$$

where the superscript $\mathcal{S}$ is used to distinguish this from the Gibbs method of defining surface quantities. On integrating, keeping T, P, and so on constant, and comparing back with Eq. III-87, one obtains

$$d\gamma = -S\,dT + V\,dP - \sum \Gamma_i^{\mathcal{S}}\,d\mu_i \qquad \text{(III-88)}$$

where the extensive quantities are now on a per unit area basis. At constant temperature and pressure, Eq. III-88 reduces to the same form as Eq. III-78 and, with the use of the Gibbs-Duhem relationship, one obtains

$$-d\gamma = d\mu_2\left[\Gamma_2^{\mathcal{S}} - \left(\frac{N_2}{N_1}\right)\Gamma_1^{\mathcal{S}}\right] \qquad \text{(III-89)}$$

which is analogous to Eq. III-82. Comparison with Eq. III-79 completes the correspondence:

$$\Gamma_2^1 = N_2\left(\frac{\Gamma_2^{\mathcal{S}}}{N_2} - \frac{\Gamma_1^{\mathcal{S}}}{N_1}\right) \qquad \text{(III-90)}$$

The quantities $\Gamma_2^{\mathcal{S}}$ and $\Gamma_1^{\mathcal{S}}$ are arbitrary because they will depend on the arbitrary location of the upper and lower boundary, but their combination in Eq. III-90 gives a unique value independent of the location of these boundaries.

E. Other Surface Thermodynamic Relationships

The preceding material of this section has focused on the most important phenomenological equation that thermodynamics gives us for multicomponent systems—the Gibbs equation. Many other, formal thermodynamic relationships have been developed, of course. Many of these are summarized in Ref. 63. The topic is treated further in Section XVI-13, but is worthwhile to give here a few additional relationships especially applicable to solutions.

Using the Gibbs convention for defining surface quantities, we define

$$G^\sigma = E^\sigma - TS^\sigma \qquad \text{(III-91)}$$

so that

$$dG^\sigma = dE^\sigma - T\,dS^\sigma - S^\sigma\,dT \qquad \text{(III-92)}$$

or, in combination with Eq. III-73,

$$dG^\sigma = -S^\sigma\,dT + \sum_i \mu_i\,dn_i^\sigma + \gamma\,d\mathcal{A} \qquad \text{(III-93)}$$

Alternatively, for the *whole* system (i.e., including the bulk phases),

$$dG = -S\,dT - P\,dV + \sum_i \mu_i\,dn_i + \gamma\,d\mathcal{A} \qquad \text{(III-94)}$$

Thus

$$\gamma = \left(\frac{\partial G^\sigma}{\partial \mathcal{A}}\right)_{T,n_i^\sigma} \tag{III-95}$$

$$\gamma = \left(\frac{\partial G}{\partial \mathcal{A}}\right)_{T,V,n_i} \tag{III-96}$$

Integration of Eq. III-93 holding constant the intensive quantities T, μ_i, and γ gives

$$G^\sigma = \sum_i \mu_i n_i^\sigma + \gamma \mathcal{A} \tag{III-97}$$

or

$$G^\sigma = \gamma + \sum_i \mu_i \Gamma_i^\sigma \tag{III-98}$$

where the extensive quantities are now on a per unit area basis. G^σ is the specific surface excess free energy and, unlike the case for a pure liquid (Eq. III-2), it is not in general equal to γ. This last would be true only in the unlikely situation of no surface adsorption, so that the Γ^σ's were zero. (Some authors use the entirely permissible definition $G^\sigma = A^\sigma - \gamma\mathcal{A}$, in which case γ does not appear in the equation corresponding to Eq. III-98—see Ref. 63).

If the surface phase approach of Section III-5D is used, equations analogous to the foregoing are obtained. It must be remembered, however, that $V^\mathcal{S} \neq 0$ and that now

$$G^\mathcal{S} = A^\mathcal{S} + PV^\mathcal{S} \tag{III-99}$$

and

$$H^\mathcal{S} = E^\mathcal{S} + PV^\mathcal{S}$$

where A denotes the Helmholtz free energy.

6. Verification of the Gibbs Equation—Direct Measurements of Surface Excess Quantities

Although Gibbs published his monumental treatise on heterogeneous equilibrium in 1875, his work was not generally appreciated until the turn of the century, and it was not until many years later that the field of surface chemistry developed to the point that experimental applications of the Gibbs equation became important.

It was of interest to many surface chemists to verify the Gibbs equation experimentally. One method, tried by several investigators, was to bubble a gas through the solution and collect the froth in a separate container. The solution resulting from the collapsed froth should differ from the original according to the value of the surface excess of the solute. It was necessary, however, to estimate the surface area of the bubbles and, at any rate, satisfactory results were not obtained.

At this point a brief comment on the justification of testing the Gibbs or any other thermodynamically derived relationship is in order. First, it might be said that such activity is foolish because it amounts to an exhibition of scepticism of the validity of the laws of thermodynamics themselves, and surely they are no longer in doubt! This is justifiable criticism in some specific instances but, in general, this writer feels it is not. The laws of thermodynamics are phenomenological laws about observable or operationally defined quantities, and where one of the more subtle deductions from these laws is involved it may not always be clear just what the operational definition of a given variable really is. This question comes up in connection with contact angles and the meaning of surface tensions of solid interfaces (see Section X-6), and it comes up later with respect to the meaning of "component" in a multicomponent system. Second, thermodynamic derivations can involve the exercise of logic at a very rigorous level, and it is entirely possible for nonsequiturs to creep in, which escape attention until an experimental disagreement forces a reexamination. Often, too, the testing of a thermodynamic relationship reveals unsuspected complexities in the system.

A. The Microtome Method for Measuring Surface Excess

These investigations were carried out by McBain and co-workers (64). Here, an actual slice was taken off the surface of the solutions by means of a device called a microtome. It consisted of a sharp blade mounted on a carriage that rested on rails. The carriage could be propelled at high speed and, with the blade slightly below the surface, a thin layer of solution was scooped up and retained in a reservoir in the blade. A slice of about 0.1 mm would be taken from about 1 m^2 of surface so that a few grams of solution were obtained. If C is the concentration of the solution and ΔC, the difference between it and that of the microtome sample, both expressed as gram solute per gram of water, then is

$$\Gamma_2^1 = \left(\frac{\Delta C}{\mathcal{A}}\right)\left(\frac{w}{1 + C + \Delta C}\right) \qquad \text{(III-100)}$$

Here $\mathcal{A}$ is the area of surface sampled and w is the weight in grams of the sample.

Agreement to within about 10% was found between the Gibbs equation, Eq. III-80, and experiment, using aqueous solutions of *p*-toluidine, phenol, and *n*-hexanoic acid. It was of interest that with increasing concentration Γ_2^1 tended toward the limiting value corresponding to a complete monolayer of solute molecules lying flat on the surface. In the case of aqueous sodium chloride solutions, for which $d\gamma/dC$ is positive, the predicted surface deficiency of salt was found. Alternatively, one can consider that a layer of pure water was present, of depth τ given by

$$\tau = -\frac{1000\Gamma_2^1}{m} \tag{III-101}$$

where m is the molality. These τ values are of the order of a few angstroms, and decrease with increasing salt concentration.

B. The Tracer Method for Measuring Surface Excess

The microtome method was a triumph of equipment construction, and the results answered an important question. It did not, however, represent a very convenient method for obtaining data about surface excesses, and in many cases it is difficult to obtain unambiguous Γ values from surface tension measurements. This is true if highly surface active impurities are unavoidably present, if hydrolytic or other equilibria provide more than one form of existence for the solute, and in general with ionic surfactants where the electrical structure of the surface region may be complex.

An interesting approach was developed by Salley, Dixon, and co-workers (65). The solute to be studied is labeled with a radioisotope that emits weak beta radiation, such as ^{14}C or ^{35}S. One then places a detector close to the surface of the solutions, as illustrated in Fig. III-12, and measures the intensity of radioactivity. Since the range of such beta emitters is small (about 30 mg/cm^2 in the case of ^{14}C, with most of the attenuation in the first two tenths of the range), the measured radioactivity corresponds to that of the surface region plus only a thin layer (about 0.06 mm if ^{14}C) of solution. See Ref. 66 for a more elaborate apparatus description.

Salley et al. obtained data for ^{35}S-labeled Aerosol OTN anionic agent (di-*n*-octylsodium sulfosuccinate, $C_8H_{17}OOCCH_2CH(NaSO_3)\text{-}COOC_8H_{17}$). The results led to a conclusion that a study of the surface tension data alone could not have provided. It was found that the measured surface excesses agreed with those calculated by means of the Gibbs equation only if it was assumed that the surface adsorbed species was the undissociated acid, formed by the hydrolysis of the sodium salt. The two possibilities were as follows:

1. The salt was the surface active species. Its activity would be given by $(a_{Na^+})(a_{X^-})$ and, neglecting activity coefficient effects,

$$\Gamma_2 = -\left(\frac{1}{2RT}\right)\frac{d\gamma}{d}\ln C \tag{III-102}$$

2. The undissociated acid HX was the surface active species; c_{HX} would be proportional to c rather than to C^2 because at the low concentrations employed the pH remained essentially constant and equal to that of the solvent. Hence,

$$\Gamma_2 = -\left(\frac{1}{RT}\right)\frac{d\gamma}{d}\ln C \tag{III-103}$$

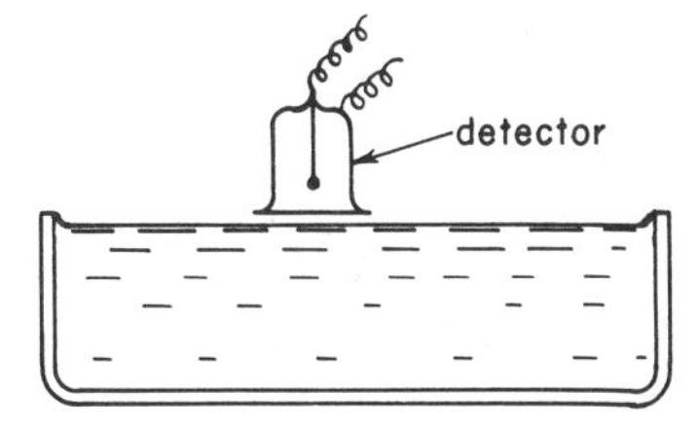

Fig. III-12. Radioactive tracer method for measuring surface excess.

The experimental values of Γ_2 agreed approximately with those given by Eq. III-103 and not at all with the 50% smaller values predicted by Eq. III-102. This illustrates one of the points made in the introduction to this section concerning the value of experimental checking of thermodynamic equations; in this case, three rather than two components are needed to specify the composition of the surface layer (NaX, HX, and H_2O, or NaX, NaOH, and H_2O). In general, when a distribution of species between two phases is involved, if a constituent in one of the phases is formed by a reaction between constituents in the other phase, then there is an increase in the number of components (67).

Other weak beta particle emitters may be used, such as ^{14}C and ^{3}H. The use of tritium (^{3}H) labeling is particularly advantageous, since ^{3}H beta particles are so weak that the correction for the radioactivity coming from bulk solution becomes very small. Tajima and co-workers (66) used this last form of labeling to obtain the adsorption of sodium dodecyl sulfate at the solution–air interface. As illustrated in Fig. III-13, the results agreed very well with those calculated from surface tension data, but now using Eq. III-102 (with an activity coefficient correction). Thus in this case the salt rather than its hydrolysis product is the surface active species. A later study (68) using 0.1*M* sodium chloride as a swamping electrolyte gave results agreed with Eq. III-103, as expected, since (Na^+) was now essentially constant.

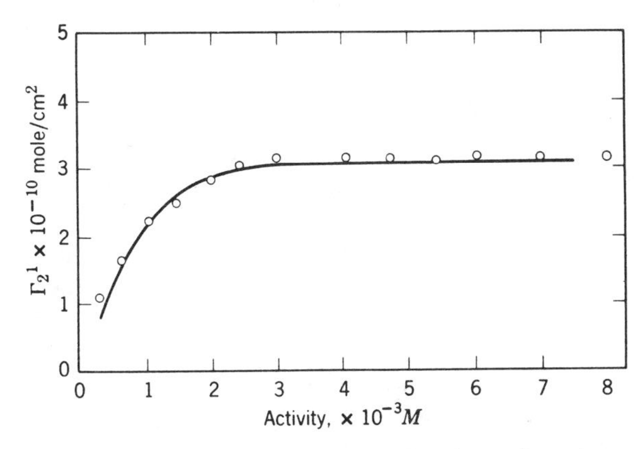

Fig. III-13. Verification of the Gibbs equation by the radioactive tracer method. Observed (○) and calculated (line) values of Γ_2^1 for aqueous sodium dodecyl sulfate solutions. (From Ref. 66.)

Sodium stearate solutions behave like those of Aerosol OTN in that the undissociated acid is the surface adsorbed species (69). The tracer method has also been used successfully with a nonionic surfactant (70).

Steiger and Aniansson (71) made use of the technique of the heavy atom recoil effect to measure surface concentrations. Other examples of the use of the tracer method is that of Shinoda and Ito (72) who used radiocalcium to determine the adsorption of calcium ions at the surface of aqueous sodium dodecyl sulfate solutions, and of Rehfeld (73) who measured the adsorption of tritiated sodium dodecyl sulfate at a polymer–solution interface.

The preceding material illustrates the point that, far from being a dry exercise in "proving" the second law of thermodynamics, the direct measurement of the surface adsorption of surface active materials has yielded information not otherwise obtainable as to the chemical nature of the adsorbed species.

As a concluding example, the tracer method helped finally to resolve a perplexing situation. Early data (74) showed that while the surface tension of lauryl sulfonic acid solutions decreased steadily with increasing concentration if measured immediately, the equilibrium curve of surface tension versus concentration went through a minimum. Since the slope at the minimum is zero, Eq. III-80 gives Γ_2^1 as zero. Yet the low surface tension (of 30 dyne/cm) at the minimum surely meant that surfactant was present at the interface. Later, a similar paradoxical behavior was found for sodium lauryl sulfate solutions. The difficulty was finally traced to the presence of some lauryl alcohol impurity. In an elegant study using tritiated alcohol, Nilsson (75) showed that lauryl alcohol first concentrated in the surface region (see Section IV-7 for a discussion of such "penetration" effects), as the sodium lauryl sulfate concentration was increased, and then went back into the bulk solution at still higher surfactant concentrations. This reversal, which accounted for the surface tension minimum, was probably due to solubilization (see Section XIII-5B) of the alcohol by the aggregates or micelles that form above a certain detergent concentration. The effect, that is, the surface tension minimum, disappears if a highly purified sodium dodecyl sulfate is used (see Ref. 76). The selective surface adsorption of alcohol can be quite large—selectivity factors approaching 60 have been reported (77).

C. Ellipsometric Method for Measuring Surface Excess

The technique of ellipsometry is discussed in Section IV-3D, and it is sufficient to note here that the method allows calculation of the thickness of an adsorbed film from the ellipticity produced in light reflected from the film-covered surface. Knowing this thickness τ, Γ may be calculated from the relationship $\Gamma = \tau/V$ where V is the molecular volume. This last may be estimated either from molecular models or from the bulk liquid density.

Smith (78) studied the adsorption of *n*-pentane on mercury, determining both the surface tension change and the ellipsometric film thickness as a function of the equilibrium pentane pressure. Γ could then be calculated from the Gibbs equation, in the form of Eq. III-117, and from τ. The agreement was excellent.

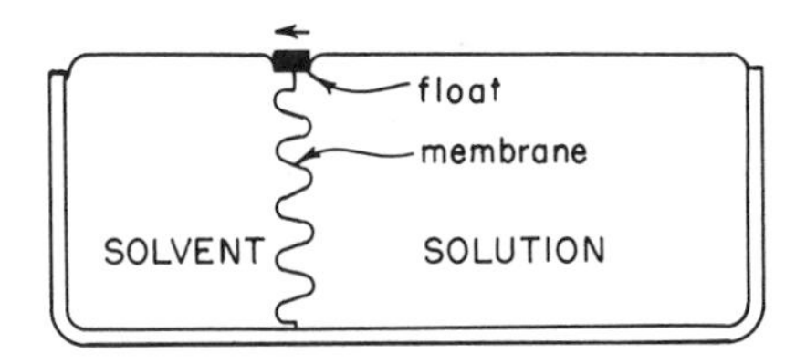

Fig. III-14. The PLAWM trough.

7. Gibbs Monolayers

If the surface tension of a liquid is lowered by the addition of a solute, then, by the Gibbs equation, the solute must be adsorbed at the interface. This adsorption may amount to enough to correspond to a monomolecular layer of solute on the surface. For example, the limiting value of Γ_2^1 in Fig. III-13 gives an area per molecule of 52.0 Å^2, which is about that expected for a close-packed layer of dodecyl sulfate ions. It is thus a physically plausible concept to treat Γ_2^1 as giving the two-dimensional concentration of surfactant in a monomolecular film.

Such a monolayer may be considered to exert a film pressure π, such that

$$\pi = \gamma_{\text{solvent}} - \gamma_{\text{solution}} \tag{III-104}$$

This film pressure (or "two-dimensional" pressure) has the units of dynes per centimeter and can be measured directly. As illustrated in Fig. III-14, if one has a trough divided by a thin rubber membrane into two compartments, one filled with solvent and the other with solution, then a force will be observed to act on a float attached to the upper end of the membrane. In the PLAWM† trough (79), the rubber membrane was very thin, and the portion below the surface was so highly convoluted that it could easily buckle so as to give complete equalization of any hydrostatic differences between the two solutions. The force observed on the float was thus purely surface tensional in origin and resulted from the fact that a displacement in the direction of the surface of higher surface tension would result in a lower overall surface free energy for the system. This force could be measured directly by determining how much opposing force applied by a lever attached to a torsion wire was needed to prevent the float from moving.

In the preceding explanation π arises as a difference between two surface tensions, but it appears physically as a force per unit length on the barrier separating the two surfaces. It is a very fruitful concept to regard the situation as involving two surfaces that would be identical except that on one of them there are molecules of surface-adsorbed solute that can move freely in the plane of the surface but cannot pass the barrier. The molecules of the adsorbed film possess, then, two-dimensional translational energy, and the film pressure π can be regarded as due to the bombardment of the barrier

† Pockels–Langmuir–Adam–Wilson–McBain.

by these molecules. This is analogous to viewing the pressure of a gas as due to the bombardment of molecules against the walls of the container. This interpretation of π allows a number of very pleasing and constructive analogies to be made with three-dimensional systems, and the concept becomes especially plausible, physically, when one is dealing with the quite insoluble monolayers discussed in the next chapter.

It is not the only interpretation, however. Another picture, again particularly useful in the case of insoluble monolayers where the rubber diaphragm of the PLAWM trough is not needed, is to regard the barrier as a semipermeable membrane through which water can pass (i.e., go around actually) but not the surface film. The surface region can then be viewed as a relatively concentrated solution having an osmotic pressure π_{os}, which is exerted against the membrane.

It must be kept in mind that both pictures are modelistic and that in using them extrathermodynamic concepts have been invoked. Except mathematically, there is no such thing as a "two-dimensional" gas, and the "solution" whose osmotic pressure is calculated is not uniform in composition, and its average concentration depends on the depth assumed for the surface layer.

A. The Two-Dimensional Ideal Gas Law

For dilute solutions, solute–solute interactions are unimportant (i.e., Henry's law will hold), and the variation of surface tension with concentration will be linear (at least for nonelectrolytes). Thus

$$\gamma = \gamma_0 - bC \tag{III-105}$$

where γ_0 denotes the surface tension of pure solvent, or

$$\pi = bC \tag{III-106}$$

Then, by the Gibbs equation,

$$-\frac{d\gamma}{dC} = \frac{\Gamma_2^1 RT}{C} \tag{III-107}$$

By Eq. III-105, $-d\gamma/dC$ is equal to b, so that Eq. III-107 becomes

$$\pi = \Gamma_2^1 RT \tag{III-108}$$

or

$$\pi\sigma = kT, \qquad \pi A = RT \tag{III-109}$$

where σ and A denote area per molecule and per mole, respectively. Equation III-109 is analogous to the ideal gas law, and it is seen that in dilute solutions the film of adsorbed solute obeys the equation of state of a two-dimensional ideal gas. Figure III-15*a* shows that for a series of aqueous alcohol solutions π increases linearly with C at low concentrations and, correspondingly, Fig. III-15*c* shows that $\pi A/RT$ approaches unity as π approaches zero.

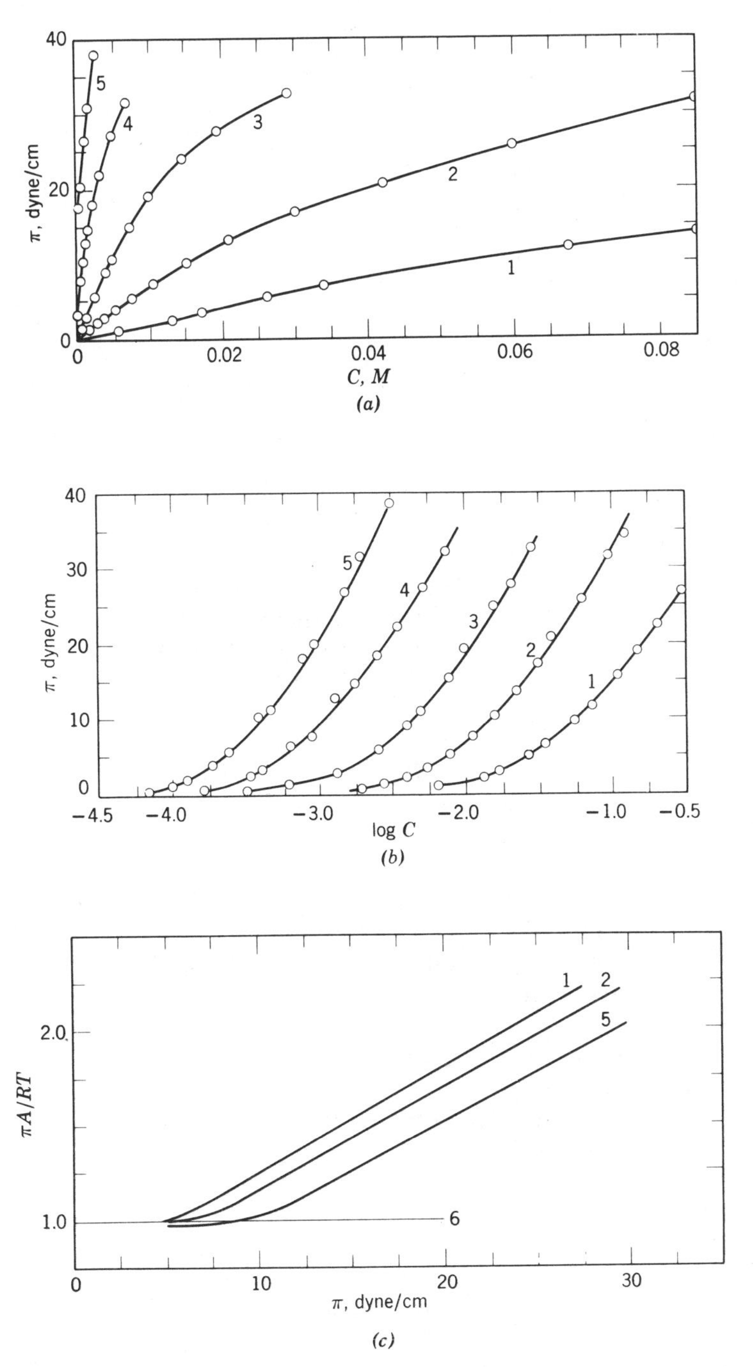

Fig. III-15. Surface tension data for aqueous alcohols; illustration of the use of the Gibbs equation. (1) *n*-butyl; (2) *n*-amyl; (3) *n*-hexyl; (4) *n*-heptyl; (5) *n*-octyl. (Data from Ref. 80.)

A sample calculation shows how Fig. III-15c is computed from the data of Fig. III-15a. Equation III-107 may be put in the form

$$A = -\frac{RT}{d\pi/d \ln C} \tag{III-110}$$

or, at 25°C and with σ in angstrom squared units, $\overset{\circ}{\sigma} = 411.6/[d\pi/d(\ln C)]$. For n-butyl alcohol π is 15.4 dyne/cm for $C = 0.1020$ and is 11.5 dyne/cm for $C = 0.0675$. Taking the slope of the line between these two points, we find

$$\overset{\circ}{\sigma} = \frac{411.6}{(11.5 - 15.4)/[-2.69 - (-2.28)]} = \frac{411.6}{(3.9/0.41)} = 43 \text{ Å}^2 \text{ per molecule}$$

This locates a point at the approximate π value of $(11.5 + 15.4)/2$ or 13.5 dyne/cm. Thus $\pi A/RT = \pi\sigma/kT = 1.41$.

B. Nonideal Two-Dimensional Gases

The deviation of Gibbs monolayers from the ideal two-dimensional gas law may be treated by plotting $\pi A/RT$ versus π, as shown in Fig. III-15c. Here, for a series of straight-chain alcohols, one finds deviations from ideality increasing with increasing film pressure; at low π values, however, the limiting value of unity for $\pi A/RT$ is approached.

This behavior suggests the use of an equation employed by Amagat for gases at high pressure; the two-dimensional form is

$$\pi(A - A^0) = qRT \tag{III-111}$$

where A^0 has the aspect of an excluded area per mole and q gives a measure of the cohesive forces. Rearrangement yields the linear form

$$\frac{\pi A}{RT} = \left(\frac{A^0}{RT}\right)\pi + q \tag{III-112}$$

This form is obeyed fairly well above π values of 5 to 10 dyne/cm in Fig. III-15c. Limiting areas or σ^0 values of about 22 Å^2 per molecule result, nearly independent of chain length, as would be expected if the molecules assume a final orientation that is perpendicular to the surface.

Various other nonideal gas-type two-dimensional equations of state have been proposed, generally by analogy with gases. Volmer and Mahnert (81) added only the covolume correction to the ideal gas law:

$$\pi(A - A^0) = RT \tag{III-13}$$

One may, of course, use a two-dimensional modification of the van der Waals equation:

$$\left(\pi + \frac{a}{A^2}\right)(A - A^0) = RT \tag{III-114}$$

Adsorption may, of course, occur at a liquid–liquid interface. Behavior rather similar to that shown in Fig. III-8d was found, for example, in the case of sodium dodecyl sulfate adsorbed at the n-alkane–water interface

(76). Equation III-102 was then used to calculate Γ_2^1 values. The effect of *pressure* as well as temperature and composition has been studied for 1-octadecanol films at the water–hexane interface (81a).

Adsorption may occur from the vapor phase rather than from the solution phase. Thus Fig. III-16 shows the surface tension lowering when water was exposed for various hydrocarbon vapors; P^0 is the saturation pressure, that is, the vapor pressure of the pure liquid hydrocarbon. The activity of the hydrocarbon is given by its vapor pressure, and the Gibbs equation takes the form

$$-d\gamma = d\pi = \Gamma RT\, d\ln P \qquad \text{(III-115)}$$

(for simplicity we have written just Γ instead of the exact designation Γ_2^1), and Γ may thus be calculated from the analogue of Eq. III-110

$$\Gamma = \left(\frac{1}{RT}\right) d\pi/d\ln P \qquad \text{(III-116)}$$

The data may then be expressed in conventional π versus σ or $\pi\sigma$ versus π plots, as shown in Fig. III-17. The behavior of adsorbed pentane films was that of a nonideal two-dimensional gas, as can be seen from the figure.

The data could be expressed equally well in terms of Γ versus P, or in the form of the conventional adsorption isotherm plot, as shown in Fig. III-18. The appearance of these isotherms is discussed in Section X-6A. The Gibbs equation thus provides a connection between adsorption isotherms and two-dimensional equations of state. For example, Eq. III-62 corresponds to the adsorption isotherm

$$\Gamma = \frac{aC}{1 + bC} \qquad \text{(III-117)}$$

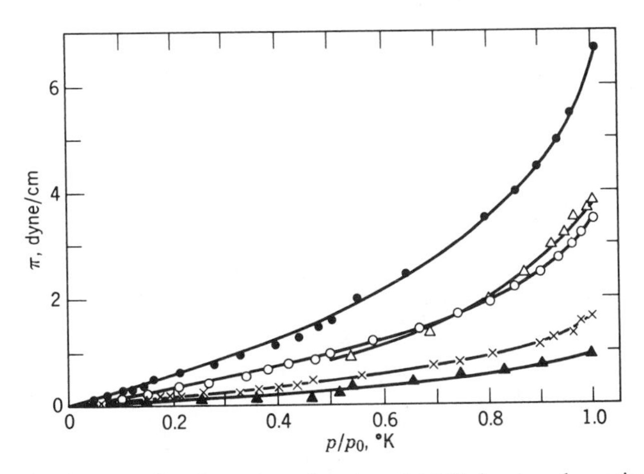

Fig. III-16. Surface tension lowering of water at 15°C due to adsorption of hydrocarbons. ●, *n*-pentane; △, 2,2,4-trimethylpentane; ○, *n*-hexane; ×, *n*-heptane; ▲, *n*-octane (from Ref. 82).

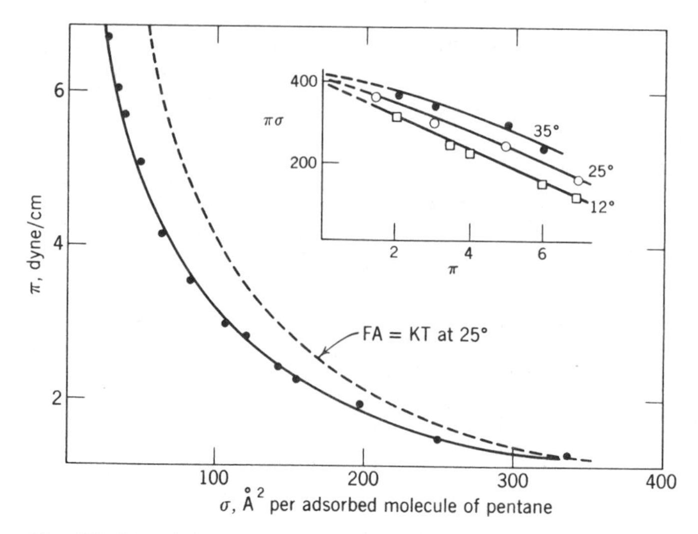

Fig. III-17. Adsorption of pentane on water (from Ref. 83).

(where a and b are constants), which is a form of the Langmuir adsorption equation (see Section XI-1A).

C. The Osmotic Pressure Point of View

It was pointed out at the beginning of this section that π could be viewed as arising from an osmotic pressure difference between a surface region comprising an adsorbed film and that of the pure solvent. It is instructive to develop this point of view somewhat further. The treatment can be made along the line of Eq. III-55, but the following approach will be used instead.

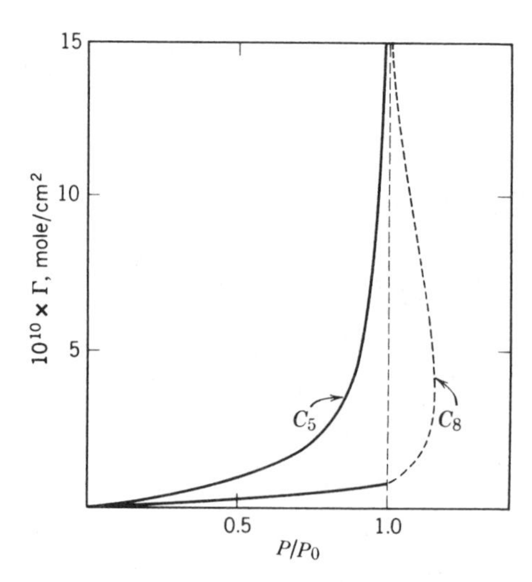

Fig. III-18. Adsorption isotherms for n-pentane and n-octane at 15°C. The dotted curve shows the hypothetical isotherm above $P/P_0 = 1$, and the arrows mark the Γ values corresponding to a monolayer (from Ref. 81).

To review briefly, the osmotic pressure in a three-dimensional situation is that pressure required to raise the vapor pressure of solvent in a solution to that of pure solvent. Thus, remembering Eq. III-18,

$$RT \ln \frac{\overset{\circ}{a}_1}{a_1} = \int V_1 \, dP = \pi_{os} V_1 \tag{III-118}$$

where a_1 denotes the activity of the solvent; it is usually assumed that its compressibility can be neglected, so that the integral may be replaced by $\pi_{os} V_1$, where V_1 is the molar volume. For *ideal* solutions, the ratio $a_1/\overset{\circ}{a}_1$ is given by N_1, the solvent mole fraction, and for *dilute* solutions, $-\ln N_1$ is approximated by N_2, which in turn is approximately n_2/n_1, the mole ratio of solute to solvent. Insertion of these approximations into Eq. III-18 leads to the limiting form

$$\pi_{os} V_1 = \frac{n_2}{n_1} RT \qquad \text{or} \qquad \pi_{os} V = n_2 RT \tag{III-119}$$

Let us now suppose that the surface region can be regarded as having a depth τ and an area $\mathcal{A}$ and hence volume V^s. A volume V^s of surface region, if made up of pure solvent, will be

$$V^s = n_1^{s0} V_1 \tag{III-120}$$

and, if made up of a mixture of solvent and surface adsorbed solute, will be

$$V^s = n_1^s V_1 + n_2^s V_2 \tag{III-121}$$

assuming the molar volumes V_1 and V_2 to be constant. The solute mole fraction in this surface region is then

$$N_2^s = \frac{n_2^s}{n_1^s + n_2^s} = \frac{n_2^s V_1}{n_1^{s0} V_1 - n_2^s (V_2 - V_1)} \tag{III-122}$$

using Eqs. III-120 and III-121 to eliminate n_1. There will be an osmotic pressure given approximately by

$$\pi_{os} V_1 = RTN_2^s \tag{III-123}$$

and if this is viewed as acting against a semipermeable barrier, the film pressure will be the osmotic pressure times the depth of the surface region in which it is exerted, that is,

$$\pi = \pi_{os} \tau \tag{III-124}$$

Equations III-122 and III-123 may now be combined with III-124 to give

$$\pi = \frac{\tau RT n_2^s}{n_1^{s0} V_1 - n_2^s (V_2 - V_1)} \tag{III-125}$$

Now, $n_1^{s0} V_1/\tau$ is just the surface area $\mathcal{A}$, and, moreover, V_1/τ and V_2/τ have the dimensions of molar area. *If* the surface region is considered to be just *one* molecule thick, V_1/τ and V_2/τ become A_1^0 and A_2^0, the actual molar areas, so that Eq. III-125 takes on the form

$$\pi = \frac{RTn_2^s}{\mathcal{A} - n_2^s(A_2^0 - A_1^0)} \qquad \text{(III-126)}$$

or, on rearranging and remembering that $A = \mathcal{A}/n_2^s$,

$$\pi[A - (A_2^0 - A_1^0)] = RT \qquad \text{(III-127)}$$

If further, A_1^0 is neglected in comparison with A_2^0, then Eq. III-127 becomes the same as the nonideal gas law, Eq. III-113.

This derivation has been made in a form calculated best to bring out the very considerable and sometimes inconsistent approximations made. However, by treating the surface region as a kind of solution, an avenue is opened for employing our considerable knowledge of solution physical chemistry in estimating association, interionic attraction, and other nonideality effects. Another advantage, from the writer's point of view, is the emphasis on the role of the solvent as part of the surface region, which helps to correct the tendency, latent in the two-dimensional equation of state treatment, to regard the substrate as merely providing an inert plane surface on which molecules of the adsorbed species may move freely. The approach is not really any more empirical than that using the two-dimensional nonideal gas, and considerable use has been made of it by Fowkes (84).

It has been pointed out (85, 86) that algebraically equivalent expressions can be derived without invoking a surface solution model. Instead, surface excess as defined by the procedure of Gibbs is used, the dividing surface always being located so that the sum of the surface excess quantities equals a given constant value. This last is conveniently taken to be the maximum value of Γ_2^1.

D. Dynamic Surface Properties of Solutions

It was noted in Chapter II that it may take some time for a surfactant solution to establish its equilibrium surface tension (note Figs. II-18 and II-24). Conversely, if the surface area of a solution is changed, the surface adsorbed film or Gibbs monolayer may not immediately adjust in the sense of discharging or acquiring solute from the bulk solution. Instead, the surface film itself is compressed or expanded, that is, if the time scale of the surface area change is short enough, the film acts as an insoluble monolayer (Chapter IV).

The *elasticity* of a film-covered surface (or the *surface dilational modulus*) E is defined as:

$$E = -\frac{d\gamma}{d\ln\mathcal{A}} \qquad \text{(III-128)}$$

where $\mathcal{A}$ is the geometric area of the surface. If there is no molecular traffic with bulk solution, the amount of monolayer is constant, and $\mathcal{A}$ in Eq. III-128 may be replaced by A, the area per mole. In this case, an alternative form of Eq. III-128 is as follows:

$$E = \frac{d\pi}{d \ln \Gamma} \tag{III-129}$$

An experimental arrangement for measuring E is shown in Fig. III-19. The oscillating Teflon bars E,E sit in the surface, riding smoothly on the fixed Teflon support bars D,D which also touched the surface. The area enclosed between the two sets of bars was thus subjected to periodic variation, and the surface tension was followed by means of a Wilhelmy slide (see Fig. II-16) positioned at G. Being in the center, the slide was not disturbed by the oscillations of the bars.

Figure III-20 illustrates the kind of behavior observed, for a surfactant of the $RO[CH_2CH_2O]_nH$ type (87). For the freshly formed surface, π was zero, that is, no appreciable adsorption of surfactant had occurred. The film pressure gradually rose, and, as it did, it also responded to the periodic changes in area. An E_{app} could be calculated from the amplitude and time lag of the oscillation of π as a function of the ageing and hence of the film pressure of the solution. E_{app} increased with increasing frequency of the area oscillation, reaching a limiting value E to which Eq. III-129 should now apply. At lower frequencies, the ratio E_{app}/E could be realted to the bulk-to-surface diffusion rate. For a series of surfactants with R a C_{12} to C_{14} hydrocarbon and n = 3–6, the plot of E versus π was linear with a slope of two up to an E value of about 70 dyne/cm mN/m) and corresponding π value of 40 dyne/cm (note Problem 28 of this chapter).

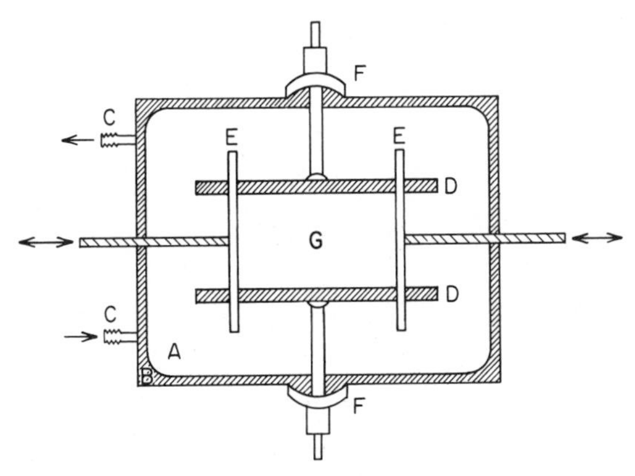

Fig. III-19. Trough for dynamic surface measurements: A, stainless steel dish; B, aluminum mantle; C, inlet thermostatting water; D, lower PTFE bars; E, oscillating bars; F, attachment lower bars; G, Wilhelmy plate. (From Ref. 87a.)

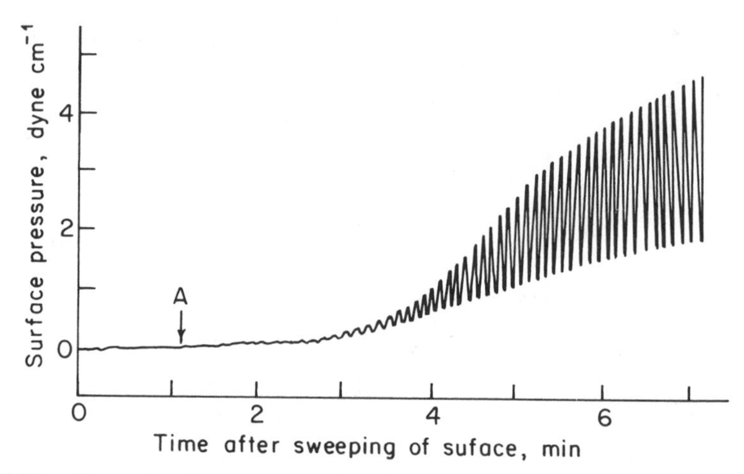

Fig. III-20. Example of surface tensio-elastogram, giving change of surface tension and dilational modulus with time. Frequency 10 cpm. *A* markes the start of periodic surface deformation (from Ref. 87).

E. Traube's Rule

The surface tensions for solutions of organic compounds belonging to a homologous series, for example, $R(CH_2)_nX$, show certain regularities. Roughly, Traube (88) found that for each additional CH_2 group, the concentration required to give a certain surface tension was reduced by a factor of 3. This rule is manifest in Fig. III-15*b*; the successive curves are displaced by nearly equal intervals of 0.5 on the log C scale.

Langmuir (89) gave an instructive interpretation to this rule. The work W to transfer one mole of solute from bulk solution to surface solution should be

$$W = RT \ln \frac{C^s}{C} = RT \ln \frac{\Gamma}{\tau C} \tag{III-130}$$

where C^s is the surface concentration and is given by Γ/τ, where τ is the thickness of the surface region. For solutes of chain length n and $(n - 1)$, the difference in work is then

$$W_n - W_{n-1} = RT \ln \frac{\Gamma_n/\tau C_n}{\Gamma_{n-1}/\tau C_{n-1}} \tag{III-131}$$

By Traube's rule, if $C_{n-1}/C_n = 3$, then $\gamma_n = \gamma_{n-1}$, and, as an approximation, it is assumed that the two surface concentrations are also the same. If so, then

$$W_n - W_{n-1} = RT \ln 3 = 640 \text{ cal/mole} \tag{III-132}$$

The figure 640 cal/mole may be regarded as the work to bring one CH_2 group from the body of the solution into the surface region. Since the value per CH_2 group appears to be independent of chain length, it is reasonable to suppose that all the CH_2 groups are similarly situated in the surface, that is, that the chains are lying flat.

If the dependence on temperature as well as on composition is known for a solution, enthalpies and entropies of adsorption may be calculated from the appropriate thermodynamic relationships (see Ref. 51). Nearn and Spaull (90) have, for example, calculated the enthalpies of surface adsorption for a series of straight chain alcohols. They find an increment in enthalpy of about 470 cal/mole per CH_2 group.

F. Some Further Comments on Gibbs Monolayers

It is important to realize that there is, in principle, no necessary difference between the nature of the adsorbed films so far discussed and those formed by spreading monolayers of insoluble substances on a liquid substrate or by adsorption from either a gas or a liquid phase onto a solid (or a liquid) surface. The distinction that *does* exist has to do with the nature of the accessible experimental data. In the case of Gibbs monolayers, one is dealing with fairly soluble solutes, and the direct measurement of Γ is not easy to carry out. Instead, one measures the changes in surface tension and obtains Γ through the use of the Gibbs equation. With spread monolayers, the solubility of the material is generally so low that its concentration in solution is not easily measurable, but Γ is known directly, as the amount per unit area that was spread onto the surface, and the surface tension also can be measured directly. In the case of adsorption, Γ is known from the decrease in concentration (or pressure) of the adsorbate material, so that both Γ and the concentration or pressure (if it is gas adsorption) are known. It is not generally possible, however, to measure the surface tension of a solid surface. Thus, it is usually possible to measure only two out of the three quantities γ, Γ, and C or P.

The succeeding material is broadly organized according to the types of experimental quantities measured because much of the literature is so grouped. In the next chapter spread monolayers are discussed, and in later chapters the topics of adsorption from solution and of gas adsorption are considered. Irrespective of the experimental compartmentation, the conclusions as to the nature of mobile adsorbed films, i.e., their structure and equations of state, will tend to be of a general validity. Thus, only a limited discussion of Gibbs monolayers has been given here, and none of such related aspects as the contact potentials of solutions or of adsorption at liquid–liquid interfaces, as it is more efficient to treat these topics later.

8. Problems

1. A 2% by weight aqueous surfactant solution has a surface tension of 69.0 dyne/cm (or mN/m) at 20°C. (*a*) Calculate σ, the area of surface containing one molecule. State any assumptions that must be made to make the calculation from the preceding data. (*b*) The additional information is now supplied that a 2.2% solution has a surface tension of 68.8 dyne/cm. If the surface adsorbed film obeys the equation of state $\pi(\sigma - \sigma_0) = kT$, calculate from the combined data a value of σ_0, the actual area of a molecule.

2. Given that $d\gamma/dT$ is -1×10^{-4} Jm^2/K for methanol at 20°C, calculate E^s and $E^{s\prime}$. Look up other data on the physical properties of methanol as needed.

3. Referring to Problem 2, calculate S^s and $S^{s\prime}$ for methanol at 20°C. Do the same for n-octane (data from Table III-1) and compare and discuss your results.

4. Calculate the vapor pressure of water when present in a capillary of 1 μm radius (assume zero contact angle). Express your result as percent change from the normal value at 25°C.

5. Complete the calculation of Γ_1^V and Γ_2^V in Section III-5C.

6. Show that Eq. III-100 does indeed define a Γ_2^1 quantity.

7. The thickness of the equivalent layer of pure water τ on the surface of a 3 m sodium chloride solution is about 1 Å. Calculate the surface tension of this solution assuming that the surface tension of salt solutions varies linearly with concentration. Neglect activity coefficient effects.

8. There are three forms of the Langmuir–Szyszkowski equation, Eq. III-62, Eq. III-117, and a third form which expresses π as a function of Γ. (a) derive Eq. III-62 from Eq. III-117 and (b) derive the third form.

9. The following data have been obtained for aqueous phenol solutions at 20°C (*Handbook of Chemistry and Physics,* U.S. Rubber Co.):

Wt % phenol	0.024	0.047	0.118	0.417	0.941	3.76	5.62
γ, dyne/cm	72.60	72.20	71.30	66.50	61.10	46.00	42.30

Make a plot of γ versus concentration. Using a smoothed curve, convert the data to plots analogous to Figs. 15b and 15c. Neglect activity coefficient corrections.

10. The following data have been reported for methanol–water mixtures at 20°C (*Handbook of Chemistry and Physics,* U.S. Rubber Co.):

Wt % methanol	7.5	10.0	25.0	50.0	60.0	80.0	90.0	100
γ, dyne cm	60.90	59.04	46.38	35.31	32.95	27.26	25.36	22.65

Make a theoretical plot of surface tension versus composition according to Eq. III-56, and compare with experiment. (Calculate the equivalent spherical diameter for water and methanol molecules and take σ as the average of these.)

11. Calculate S_{12} for methanol–water from a fit of Eq. III-61 to the data of Problem 10.

12. The surface tension of water at 25°C exposed to varying relative pressures of a hydrocarbon vapor changes as follows:

P/P^0	0.10	0.20	0.30	0.40	0.50	0.60	0.70	0.80	0.90
π, dyne/cm	0.22	0.55	0.91	1.35	1.85	2.45	3.15	4.05	5.35

Calculate and plot π versus σ (in Å^2 per molecule) and Γ versus P/P^0. Does it appear that this hydrocarbon wets water (note Ref. 82)?

13. Derive the equation of state, that is, the relationship between π and σ, of the adsorbed film for the case of a surface active electrolyte. Assume that the activity coefficient for the electrolyte is unity, that the solution is dilute enough so that surface tension is a linear function of the concentration of the electrolyte, and that the electrolyte itself (and not some hydrolyzed form) is the surface adsorbed species. Do this for the case of a strong 1 : 1 electrolyte and a strong 1 : 3 electrolyte.

14. The surface tension of an aqueous solution varies with the concentration of solute according to the equation $\gamma = 72 - 350C$ (provided that C is less than 0.05 M). Calculate the value of the constant k for the variation of surface excess of solute with concentration, where k is defined by the equation $\Gamma_2^1 = kC$. The temperature is 25°C.

15. The data in Table III-3 have been determined for the surface tension of isooctane–benzene solutions at 30°C. Calculate Γ_1^2, Γ_2^1, Γ_1^N, and Γ_2^N for various concentrations and plot these quantities versus the mole fraction of the solution. Assume ideal solutions.

16. An adsorption isotherm known as the Temkin equation (91) has the form: $\pi = \alpha\Gamma^2/\Gamma^\infty$ where α is a constant and Γ^∞ is the limiting surface excess for a close-packed monolayer of surfactant. Using the Gibbs equation find π as a function of C and Γ as a function of C.

17. Estimate, by means of Eq. III-43, the surface tensions of benzene and of water at 20°C. Look up the necessary data on thermal expansion and compressibility.

18. Tajima and co-workers (66) determined the surface excess of sodium dodecyl sulfate by means of the radioactivity method, using tritiated surfactant of specific activity 9.16 Ci/mole. The area of solution exposed to the detector was 37.50 cm^2. In a particular experiment, it was found that with 1.0×10^{-2} M surfactant the surface count rate was 17.0×10^3 counts per minute. Separate calibration showed that of this count rate 14.5×10^3 came from underlying solution, the rest being surface excess. It was also determined that the counting efficiency for surface material was 1.1%. Calculate Γ for this solution.

19. Derive Eq. III-22 from the first and second laws of thermodynamics and related definitions.

20. The following two statements seem mutually contradictory and seem to describe a paradoxical situation. (*a*) The chemical potential of a species must be everywhere the same for an equilibrium system at constant temperature and pressure; therefore, if we have a liquid in equilibrium with its vapor (the interface is planar), the chemical potential of the species must be the same in the surface region as it is in the bulk liquid, and no work is required to move a molecule from the bulk region to the surface region. (*b*) It must require work to move a molecule from the bulk region to the surface region because to do so means increasing the surface area and hence the surface free energy of the system.

Discuss these statements and reconcile the apparent contradiction. (Note Ref. 85.)

21. Some data obtained by Nicholas et al. (92) are given in Table III-4, for the

TABLE III-3
Data for the System Isooctane–Benzene

Mole fraction isooctane	Surface tension, dyne/cm	Mole fraction isooctane	Surface tension, dyne/cm
0.000	27.53	0.583	19.70
0.186	23.40	0.645	19.32
0.378	21.21	0.794	18.74
0.483	20.29	1.000	17.89

TABLE III-4
Adsorption of Water Vapor on Mercury at 25°C

Water vapor pressure, atm	Surface tension, dyne/cm	Water vapor pressure, atm	Surface tension, dyne/cm
0	483.5	7.5	459
1×10^{-3}	483	10	450
3	482	15	442
4	481	20	438
5	477	25	436
6	467		

surface tension of mercury at 25°C in contact with various pressures of water vapor. Calculate the adsorption isotherm for water on mercury, and plot it as Γ versus P.

22. The surface tension of liquid cadmium is 570 mJ/m^2 at its melting point, and its critical temperature is 2290 K. Calculate $E^{s\prime}$ (of Eq. III-16) for cadmium. Look up any necessary additional data; explain any assumptions that are made.

23. Calculate, using the data of Fig. III-8*a* and Eq. III-56, the surface tension versus mole fraction plot for mixtures of cyclohexane and benzene.

24. Derive an equation for the heat of vaporization of a liquid as a function of drop radius r.

25. Using Langmuir's principle of independent surface action, make qualitative calculations and decide whether the polar or the nonpolar end of ethanol should be oriented toward the mercury phase at the ethanol–mercury interface.

26. S values for the following fused salt mixtures are: $S_{RbNO_3,KNO_3} = 1.43$, $S_{RbNO_3,NaNO_3} = 2.92$. The surface tensions are 112, 117, 103 dyne/cm for KNO_3, $NaNO_3$, and $RbNO_3$, respectively. All data are for 350°C. Referring to Eq. III-61, calculate the surface tension for (*a*) a 25 mole percent solution of $RbNO_3$ in $NaNO_3$ and (*b*) a 25 mole percent solution of KNO_3 in $NaNO_3$.

27. S (of Eq. III-61) is 1.77 for the copper–nickel system at 1550°C. Calculate the surface mole fraction of Cu for a system whose overall composition is 25 mole percent copper.

28. In Section III-7D, it was reported that the plot of surface elasticity, E versus π was linear with a slope of two for a series of surfactants. To what equation of state does this correspond, that is, what is the function relating π and σ?

General References

R. Defay, I. Prigogine, A. Bellemans, and D. H. Everett, *Surface Tension and Adsorption,* Longmans, Green and Co., London, 1966.

R. H. Fowler and E. A. Guggenheim, *Statistical Thermodynamics,* Cambridge University Press, London, 1939.

W. D. Harkins, *The Physical Chemistry of Surfaces,* Reinhold, New York, 1952.

J. H. Hildebrand and R. L. Scott, *The Solubility of Nonelectrolytes,* 3rd ed., Van Nostrand-Reinhold, Princeton, New Jersey, 1950.

G. N. Lewis and M. Randall, *Thermodynamics,* 2nd ed. (revised by K. S. Pitzer and L. Brewer), McGraw-Hill, New York, 1961.

H. van Olphen and K. J. Mysels, *Physical Chemistry: Enriching Topics from Colloid and Surface Science,* Theorex (8327 Ja Jolla Scenic Drive), California, 1975.

Textual References

1. D. A. Netzel, G. Hoch, and T. I. Marx, *J. Colloid Sci.,* **19,** 774 (1964).
2. K. L. Wolf, *Physik und Chemie der Grenzfläschen,* Springer-Verlag, Berlin, 1957.
3. R. Eötvös, *Wied Ann.,* **27,** 456 (1886).
4. W. Ramsey and J. Shields, *J. Chem. Soc.,* **1893,** 1089.
5. E. A. Guggenheim, *J. Chem. Phys.,* **13,** 253 (1945).
6. A. V. Grosse, *J. Inorg. Nucl. Chem.,* **24,** 147 (1962).
7. J. E. Lennard-Jones and J. Corner, *Trans. Faraday Soc.,* **36,** 1156 (1940).
8. J. W. Cahn and J. E. Hilliard, *J. Chem. Phys.,* **28,** 258 (1958).
9. F. B. Garner and S. Sugden, *J. Chem. Soc.,* **1929,** 1298.
10. O. R. Quayle, *Chem. Rev.,* **53,** 439 (1953).
11. R. Grzeskowiak, G. H. Jeffery, and A. I. Vogel, *J. Chem. Soc.,* **1960,** 4719; O. Exner, *Nature,* **196,** 890 (1962).
12. A. S. Skapski, *J. Chem. Phys.,* **16,** 386 (1948).
13. K. R. Atkins and Y. Narahara, *Phys. Rev.,* **138,** A437 (1965).
14. D. Stansfield, *Proc. Phys. Soc.,* **72,** 854 (1958).
15. W. V. Kayser, *J. Colloid Interface Sci.,* **56,** 622 (1976).
16. C. C. Addison and J. M. Coldrey, *J. Chem. Soc.,* **1961,** 468.
17. J. B. Griffin and H. L. Clever, *J. Chem. Eng. Data,* **5,** 390 (1960).
18. S. Blairs, *J. Colloid Interface Sci.,* **67,** 548 (1978).
19. B. C. Allen, *AIME Trans.,* **227,** 1175 (1963).
20. V. K. LaMer and R. Gruen, *Trans. Faraday Soc.,* **48,** 410 (1952).
21. See M. Folman and J. L. Shereshefsky, *J. Phys. Chem.,* **59,** 607 (1955).
22. D. H. Everett, J. M. Haynes, and P. J. McElroy, *Sci. Prog. Oxf.* **59,** 279 (1971).

22a. L. R. Fisher and J. N. Israelachvili, *Nature,* **277,** 548 (1979).

23. R. C. Tolman, *J. Chem. Phys.,* **17,** 333 (1949).

23a. L. R. Fisher and J. N. Israelachvili, *Chem. Phys. Lett.* (in press).

24. G. N. Lewis and M. Randall, *Thermodynamics and the Free Energy of Chemical Substances,* McGraw-Hill, New York, 1923, p. 248.
25. J. C. Eriksson, *Acta Chem. Scand.,* **16,** 2199 (1962).
26. C. Jho, D. Nealon, S. Shogbola, and A. D. King, Jr., *J. Colloid Interface Sci.,* **65,** 141 (1978).
27. C. Jho and A. D. King, Jr., *J. Colloid Interface Sci.,* **69,** 529 (1979).
28. C. V. Raman and L. A. Ramdas, *Phil. Mag.,* **3,** No. 7, 220 (1927).
29. R. C. Brown, *Proc. Phys. Soc.,* **59,** 429 (1947).
30. C. Gurney, *Proc. Phys. Soc.,* **A62,** 639 (1947).
31. J. W. Gibbs, *The Collected Works of J. W. Gibbs,* Longmans, Green, New York, 1931, Vol. I, p. 219.
32. R. C. Tolman, *J. Chem. Phys.,* **16,** 758 (1948); **17,** 118 (1949).
33. E. H. Lucassen-Reynders, *Prog. Surf. Membrane Sci.,* **10,** 253 (1976).
34. J. G. Kirkwood and F. P. Buff, *J. Chem. Phys.,* **17,** 338 (1949).
35. F. Buff, "The Theory of Capillarity," in *Handbuch der Physik,* Vol. 10, S. Flügge and Marburg, Eds., Springer-Verlag, Berlin, 1960.
36. Boruvka and A. W. Neumann, *J. Chem. Phys.,* **66,** 5464 (1977).

37. M. J. Mandell and H. Reiss, *J. Stat. Phys.*, **13**, 107 (1975).
38. T. L. Hill, *Statistical Mechanics*, McGraw-Hill, New York, 1956.
39. H. Reiss, H. L. Frisch, and E. Hefland, *J. Chem. Phys.*, **32**, 119 (1960).
40. H. T. Davis and L. E. Scriven, *J. Phys. Chem.*, **80**, 2805 (1976).
41. A. W. Adamson, *Textbook of Physical Chemistry*, 2nd ed., Academic, 1979.
42. P. D. Shoemaker, G. W. Paul, and L. E. Marc de Chazal, *J. Chem. Phys.*, **52**, 491 (1970).
43. J. Goodisman and R. W. Pastor, *J. Phys. Chem.*, **82**, 2078 (1978).
44. W. B. Hardy, *Proc. Roy. Soc.* (*London*), **A88**, 303 (1913).
45. W. D. Harkins, *Z. Phys. Chem.*, **139**, 647 (1928); *The Physical Chemistry of Surface Films*, Reinhold, New York, 1952.
46. I. Langmuir, *Colloid Symposium Monograph*, The Chemical Catalog Company, New York, 1925, p. 48.
47. R. J. Good, *J. Phys. Chem.*, **61**, 810 (1957).
48. R. J. Good, *J. Colloid Interface Sci.*, **59**, 398 (1977).
49. E. A. Guggenheim, *Trans. Faraday Soc.*, **41**, 150 (1945).
50. J. H. Hildebrand and R. L. Scott, *Solubility of Nonelectrolytes*, Reinhold, New York, 1950, Chap. 21.
51. R. Defay, I. Prigogine, A. Bellemans, and D. H. Everett, *Surface Tension and Adsorption*, Longmans, Green and Co., London, 1966.
52. D. A. Nissen, *J. Phys. Chem.*, **82**, 429 (1978).
52a. G. Bertozzi and G. Sternheim, *J. Phys. Chem.*, **68**, 2908 (1964).
53. G. L. Gaines, Jr., *Trans. Faraday Soc.*, **65**, 2320 (1969).
54. G. L. Gaines, Jr., *J. Polym. Sci.*, **10**, 1529 (1972).
55. H. Reiss and S. W. Mayer, *J. Chem. Phys.*, **34**, 2001 (1961).
56. T. S. Ree, T. Ree, and H. Eyring, *J. Chem. Phys.*, **41**, 524 (1964).
57. J. G. Eberhart, *J. Phys. Chem.*, **70**, 1183 (1966).
57a. J. Goodisman, *J. Colloid Interface Sci.*, **73**, 115 (1980).
58. H. B. Evans, Jr., and H. L. Clever, *J. Phys. Chem.*, **68**, 3433 (1964).
59. S. Ozeki, M. Tsunoda, and S. Ikeda, *J. Colloid Interface Sci.*, **64**, 28 (1978).
60. B. von Szyszkowski, *Z. Phys. Chem.*, **64**, 385 (1908); H. P. Meissner and A. S. Michaels, *Ind. Eng. Chem.*, **41**, 2782 (1949).
61. E. A. Guggenheim and N. K. Adam, *Proc. Roy. Soc.* (*London*), **A139**, 218 (1933).
62. E. A. Guggenheim, *Trans. Faraday Soc.*, **36**, 397 (1940).
63. D. H. Everett, *Pure Appl. Chem.*, **31**, 579 (1972).
64. J. W. McBain and C. W. Humphreys, *J. Phys. Chem.*, **36**, 300 (1932); J. W. McBain and R. C. Swain, *Proc. Roy. Soc.* (*London*), **A154**, 608 (1936).
65. D. J. Salley, A. J. Weith, Jr., A. A. Argyle, and J. K. Dixon, *Proc. Roy. Soc.* (*London*), **A203**, 42 (1950); J. K. Dixon, C. M. Judson, and D. J. Salley, *Monomolecular Layers*, Publication of the American Association for the Advancement of Science, Washington, D.C., 1954, p. 63.
66. K. Tajima, M. Muramatsu, and T. Sasaki, *Bull. Chem. Soc. Japan*, **43**, 1991 (1970).
67. J. W. Gibbs, *The Collected Works of J. W. Gibbs*, Longmans, Green, New York, 1931, Vol. I, p. 63.
68. K. Tajima, *Bull. Chem. Soc. Japan*, **43**, 3063 (1970).
69. K. Sekine, T. Seimiya, and T. Sasaki, *Bull. Chem. Soc. Japan*, **43**, 629 (1970).

70. K. Tajima, M. Iwahashi, and T. Sasaki, *Bull. Chem. Soc. Japan,* **44,** 3251 (1971).
71. N. H. Steiger and G. Aniansson, *J. Phys. Chem.,* **58,** 228 (1954).
72. K. Shinoda and K. Ito, *J. Phys. Chem.,* **65,** 1499 (1961).
73. S. J. Rehfeld, *J. Colloid Interface Sci.,* **31,** 46 (1969).
74. J. W. McBain and L. A. Wood, *Proc. Roy. Soc. (London),* **A174,** 286 (1940).
75. G. Nilsson, *J. Phys. Chem.,* **61,** 1135 (1957).
76. S. J. Rehfeld, *J. Phys. Chem.,* **71,** 738 (1967).
77. K. Shinoda and J. Nakanishi, *J. Phys. Chem.,* **67,** 2547 (1963).
78. T. Smith, *J. Colloid Interface Sci.,* **28,** 531 (1968).
79. J. W. McBain, J. R. Vinograd, and D. A. Wilson, *J. Am. Chem. Soc.,* **62,** 244 (1940); J. W. McBain, T. F. Ford, and D. A. Wilson, *Kolloid-Z.,* **78,** 1 (1937).
80. A. M. Posner, J. R. Anderson, and A. E. Alexander, *J. Colloid Sci.,* **7,** 623 (1952).
81. M. Volmer and P. Mahnert, *Z. Phys. Chem.,* **115,** 239 (1925); M. Volmer, *ibid.,* **115,** 253 (1925).
81a. N. Matubayasi, K. Motomura, M. Aratono, and R. Matuura, *Bull. Chem. Soc. Japan,* **51,** 2800 (1978).
82. F. Hauxwell and R. H. Ottewill, *J. Colloid Interface Sci.,* **34,** 473 (1970).
83. C. L. Cutting and D. C. Jones, *J. Chem. Soc.,* **1955,** 4067; M. Blank and R. H. Ottewill, *J. Phys. Chem.,* **68,** 2206 (1964).
84. See F. M. Fowkes, *J. Phys. Chem.,* **68,** 3515 (1964), and preceding papers.
85. E. H. Lucassen-Reynders and M. van den Tempel, *Proc. IV Int. Cong. Surface Active Substances, Brussels, 1964,* Vol. 2, J. Th. G. Overbeek, ed., Gordon and Breach, New York, 1967, p. 779.
86. E. H. Lucassen-Reynders, *J. Colloid Interface Sci.,* **41,** 156 (1972).
87. J. Lucassen and D. Giles, *J. Chem. Soc., Faraday Trans. I,* **71,** 217 (1975); E. H. Lucassen-Reynders, J. Lucassen, P. R. Garrett, D. Giles, and F. Hollway, *Adv. Chem.,* **144,** 272 (1975).
88. I. Traube, *Annalen,* **265,** 27 (1891).
89. I. Langmuir, *J. Am. Chem. Soc.,* **39,** 1848 (1917).
90. M. R. Nearn and A. J. B. Spaull, *Trans. Far. Soc.,* **65,** 1785 (1969).
91. M. I. Temkin, *Zh. Fiz. Khim.,* **15,** 296 (1941).
92. M. E. Nicholas, P. A. Joyner, B. M. Tessem, and M. D. Olson, *J. Phys. Chem.,* **65,** 1373 (1961).

CHAPTER IV

Surface Films on Liquid Substrates

1. Introduction

When a slightly soluble substance is placed at a liquid–air interface it may spread out to a thin and in most cases monomolecular film. Although the thermodynamics for such a system are in principle the same as for Gibbs monolayers, the concentration of the substance in solution is no longer an experimentally convenient quantity to measure. The solution concentration, in fact, is not usually of much interest, so little use is made even of the ability to compute changes in it through the Gibbs equation. The emphasis shifts to more direct measurements of the interfacial properties themselves, and these are discussed in some detail, along with some of the observations and conclusions.

First, it is of interest to review briefly the historical development of the subject. Gaines in his monograph (1) reminds us that the calming effect of oil on a rough sea was noted by Pliny the Elder and by Plutarch, and that Benjamin Franklin in 1774 characteristically made the observation more quantitative by remarking that a teaspoon of oil sufficed to calm a half-acre surface of a pond. Later, in 1890, Rayleigh (2) noted that the erratic movements of camphor on a water surface were stopped by spreading an amount of oleic acid sufficient to give a film only about 16 Å thick. This, incidentally, gave an upper limit to the molecular size and hence to the molecular weight of oleic acid so that a minimum value of Avogadro's number could be estimated. This comes out to be about the right order of magnitude.

About this time Miss Pockels (3) showed how films could be confined by means of barriers; thus she found little change in the surface tension of fatty acid films until they were confined to an area corresponding to about 20 $Å^2$ per molecule (the Pockels point). In 1899, Rayleigh (4) commented that a reasonable interpretation of the Pockels point was that at this area the molecules of the surface material were just touching each other. The picture of a surface film that was developing was one of molecules "floating" on the surface, with little interaction until they actually came into contact with each other. Squeezing a film at the Pockels point put compressive energy into the film that was available to reduce the total free energy to form more surface; that is, the surface tension was reduced.

Also, these early experiments made it clear that a monomolecular film could exert a physical force on a floating barrier. A loosely floating circle of thread would stretch taut to a circular shape when some surface active material was spread inside its confines. Physically, this could be visualized as being due to the molecules of the film pushing against the confining barrier. Devaux (5) found that light talcum powder would be pushed aside by a film spreading on a liquid surface, and that some films were easily distorted by air currents, whereas others appeared to be quite rigid.

Langmuir (6) in 1917 gave a great impetus to the study of monomolecular films by developing a new experimental technique. As illustrated in Fig. IV-1, he confined the film with a rigid but adjustable barrier on one side and with a floating one on the other. The film was prevented from leaking past the ends of floating barrier by means of small air jets. The actual force on the barrier was then measured directly to give π, the film pressure (see Section III-7). As had been observed by Miss Pockels, he found that one could sweep a film off the surface quite cleanly simply by moving the sliding barrier, always keeping it in contact with the surface. As it was moved along, a fresh surface of clean water would form behind it. The floating barrier was connected to a knife-edge suspension by means of which the force on the barrier could be determined. The barriers were constructed of paper coated with paraffin so as not to be wet by the water.

Langmuir also gave needed emphasis to the importance of employing pure substances rather than the various natural oils previously used. He thus found that the limiting area (at the Pockels point) was the same for palmitic, stearic, and cerotic acids, namely, 21 $Å^2$ per molecule. (For convenience to the reader, the common names associated with the various hydrocarbon derivatives most frequently mentioned in this chapter are given in Table IV-1.)

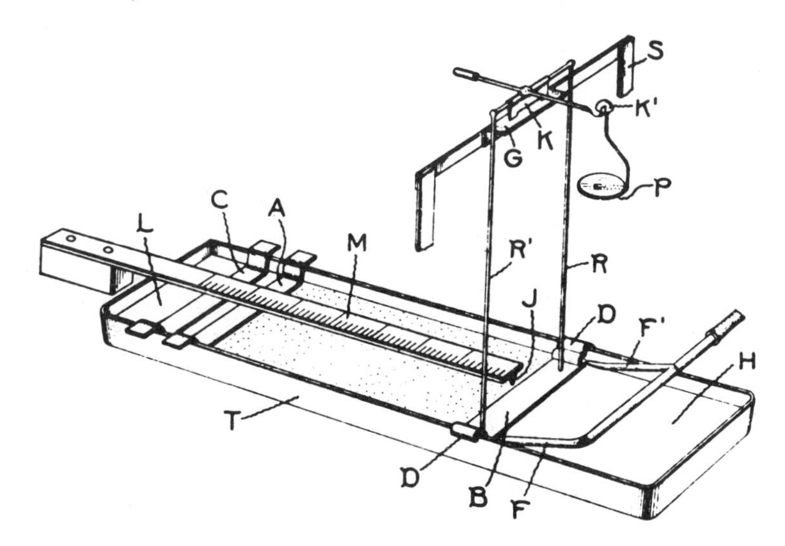

Fig. IV-1. Langmuir's film balance (from Ref. 6).

TABLE IV-1
Common Names of Long Chain Compounds

Formula	Name	Geneva name
$C_{10}H_{21}COOH$	Undecoic acid	Undecanoic acid
$C_{11}H_{23}OH$	Undecanol	1-Hendecanol
$C_{11}H_{23}COOH$	Lauric acid	Dodecanoic acid
$C_{12}H_{25}OH$	Lauryl alcohol, dodecyl alcohol	1-Dodecanol
$C_{12}H_{25}COOH$	Tridecylic acid	Tridecanoic acid
$C_{13}H_{27}OH$	Tridecyl alcohol	1-Tridecanol
$C_{13}H_{27}COOH$	Myristic acid	Tetradecanoic acid
$C_{14}H_{29}OH$	Tetradecyl alcohol	1-Tetradecanol
$C_{15}H_{31}COOH$	Palmitic acid	Hexadecanoic acid
$C_{16}H_{33}OH$	Cetyl alcohol	1-Hexadecanol
$C_{16}H_{33}COOH$	Margaric acid	Heptadecanoic acid
$C_{17}H_{35}OH$	Heptadecyl alcohol	1-Heptadecanol
$C_{17}H_{35}COOH$	Stearic acid	Octadecanoic acid
$C_{18}H_{37}OH$	Octadecyl alcohol	1-Octadecanol
$C_8H_{17}CH{=}CH(CH_2)_7COOH$	Elaidic acid	*trans*-9-Octadecenoic acid
$C_8H_{17}CH{=}CH(CH_2)_7COOH$	Oleic acid	*cis*-9-Octadecenoic acid
$CH_3(CH_2)_7CH{=}CH(CH_2)_8OH$	Oleyl alcohol	*cis*-9-Octadecenyl alcohol
$CH_3(CH_2)_7CH{=}CH(CH_2)_8OH$	Elaidyl alcohol	*trans*-9-Octadecenyl alcohol
$C_{18}H_{37}COOH$	Nonadecylic acid	Nonadecanoic acid
$C_{19}H_{39}OH$	Nonadecyl alcohol	1-Nonadecanol
$C_{19}H_{39}COOH$	Archidic acid	Eicosanoic acid
$C_{20}H_{41}OH$	Eicosyl alcohol, arachic alcohol	1-Eicosanol
$C_{21}H_{45}COOH$	Behenic acid	Docosanoic acid
$CH_3(CH_2)_7CH{=}CH(CH_2)_{11}COOH$	Erucic acid	*cis*-13-Docosenoic acid
$CH_3(CH_2)_7CH{=}CH(CH_2)_{11}COOH$	Brassidic acid	*trans*-13-Docosenoic acid
$CH_{25}H_{51}COOH$	Cerotic acid	Hexacosanoic acid

Note: Notice that is generally those acids containing an even number of carbon atoms that have special common names. This is because these are the naturally occurring ones in vegetable and animal fats.

This observation that the length of the hydrocarbon chain could be varied from 16 to 26 carbon atoms without affecting the limiting area could only mean that at this point the molecules were oriented vertically. From the molecular weight and density of palmitic acid, one computes a molecular volume of 495 Å^3; a molecule occupying only 21 Å^2 on the surface could then be about 4.5 Å on the side but must be about 23 Å long. In this way one begins to obtain information about the shape and orientation as well as the size of molecules.

The preceding evidence for orientation at the interface, plus the considerations given in Section III-3, make it clear that the polar end is directed toward the water and the hydrocarbon tails toward the air. On the other hand, the evidence from the study of the Gibbs monolayers (Section III-7) was that the smaller molecules tended to lie flat on the surface. It will be seen that the orientation depends not only on the chemical constitution but also on other variables, such as the film pressure π.

To resume the brief historical sketch, the subject of monolayers developed rapidly during the interwar years, with the names of Langmuir, Adam, Harkins, and Rideal perhaps the most prominent; the subject became one of precise and mature scientific study. The post-World War II period has been one of even greater quantitative activity, although it has lagged behind the enormous expansion of scientific work generally and has been less associated with a few great centers of study. A belief that solid interfaces are easier to understand than liquid ones has shifted emphasis to the former; but the subjects are not really separable, and the advances in the one are giving impetus to the other. There is increasing interest in films of biological and of liquid crystalline materials, and the current preoccupation with solar energy conversion has stimulated interest in photochemically active films.

On the environmental side, it turns out that the surface of oceans and lakes are usually coated with natural films, mainly glycoproteins (7). As they are biological in origin, the extent of such films seems to be seasonal. Pollutant slicks, especially from oil spills, are of increasing importance, and their clean-up can present interesting surface chemical problems.

A final comment on definitions of terms should be made. The terms *film* and *monomolecular film* have been employed somewhat interchangeably in the preceding discussion. Strictly speaking, a film is a layer of substance, spread over a surface, whose thickness is small enough that gravitational effects are negligible. A molecular film or, briefly, monolayer, is a film considered to be only one molecule thick. A *duplex film* is a film thick enough so that the two interfaces (e.g., liquid–film and film–air) are independent and possess their separate characteristic surface tensions. In addition, material placed at an interface may form a *lens*, that is, a thick layer of finite extent whose shape is constrained by the force of gravity. Combinations of these are possible; thus material placed on a water surface may spread to give a monolayer, the remaining excess collecting as a lens.

2. The Spreading of One Liquid on Another

Before proceeding to the main subject of this chapter—namely, the behavior and properties of spread films on liquid substrates—it is of interest to consider the somewhat wider topic of the spreading of a substance on a liquid surface. Certain general statements can be made as to whether spreading will occur, and the phenomenon itself is of some interest.

A. *Criteria for Spreading*

If a mass of some substance be placed on a liquid surface so that initially it is present in a layer of appreciable thickness, as illustrated in Fig. IV-2, then two possibilities exist as to what may happen. These are best treated in terms of what is called the spreading coefficient.

At constant temperature and pressure a small change in the surface free energy of the system shown in Fig. IV-2 is given by the total differential

$$dG = \left(\frac{\partial G}{\partial \mathscr{A}_{\mathrm{A}}}\right) d\mathscr{A}_{\mathrm{A}} + \left(\frac{\partial G}{\partial \mathscr{A}_{\mathrm{AB}}}\right) d\mathscr{A}_{\mathrm{AB}} + \left(\frac{\partial G}{\partial \mathscr{A}_{\mathrm{B}}}\right) d\mathscr{A}_{\mathrm{B}} \qquad \text{(IV-1)}$$

but

$$d\mathscr{A}_{\mathrm{B}} = -d\mathscr{A}_{\mathrm{A}} = d\mathscr{A}_{\mathrm{AB}}$$

where liquid A constitutes the substrate, and

$$\left(\frac{\partial G}{\partial \mathscr{A}_{\mathrm{A}}}\right) = \gamma_{\mathrm{A}}, \text{ and so on.}$$

The coefficient $-(\partial G/\partial \mathscr{A}_{\mathrm{B}})_{\mathrm{area}}$ gives the free energy change for the spreading of a film of liquid B over liquid A and is called the *spreading coefficient* of B on A. Thus

$$S_{\mathrm{B/A}} = \gamma_{\mathrm{A}} - \gamma_{\mathrm{B}} - \gamma_{\mathrm{AB}} \qquad \text{(IV-2)}$$

$S_{\mathrm{B/A}}$ is positive if spreading is accompanied by a decrease in free energy, that is, is spontaneous. From the definitions of work of adhesion and cohesion (Eqs. III-47 and III-48), it is seen that the spreading coefficient is the difference between the work of adhesion of A to B and the work of cohesion of B

$$S_{\mathrm{B/A}} = w_{\mathrm{AB}} - w_{\mathrm{BB}} \qquad \text{(IV-3)}$$

The process described by Eqs. IV-2 and IV-3 is that depicted in Fig. IV-2, in which a thick or duplex film of B forms and is the first thing that actually occurs when one liquid is placed on the surface of another if S is positive. Generally speaking, this happens when a liquid of low surface tension is placed on one of high surface tension. Some illustrative data are shown in Tables IV-2a through IV-2d; it is seen that benzene and long chain

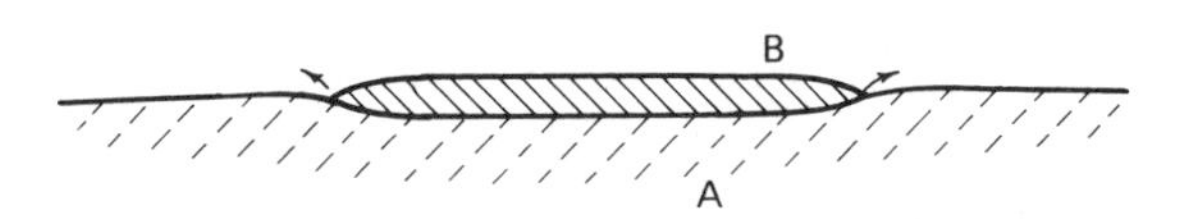

Fig. IV-2

TABLE IV-2a
Spreading Coefficients at 20°C of Liquids on Water

Liquid, B	$S_{B/A}$	Liquid, B	$S_{B/A}$
Isoamyl alcohol	44.0	Nitrobenzene	3.8
n-Octyl alcohol	35.7	Hexane	3.4
Heptaldehyde	32.2	Heptane (30°C)	0.2
Oleic acid	24.6	Ethylene dibromide	−3.2
Ethyl nonanoate	20.9	*o*-Monobromotoluene	−3.3
p-Cymene	10.1	Carbon disulfide	−8.2
Benzene	8.8	Iodobenzene	−8.7
Toluene	6.8	Bromoform	−9.6
Isopentane	9.4	Methylene iodide	−26.5

TABLE IV-2b
Liquids on Mercury (8)

Liquid, B	$S_{B/A}$	Liquid, B	$S_{B/A}$
Ethyl iodide	135	Benzene	99
Oleic acid	122	Hexane	79
Carbon disulfide	108	Acetone	60
n-Octyl alcohol	102	Water	−3

TABLE IV-2c
Initial versus Final Spreading Coefficients on Water (8, 9)

Liquid, B	γ_B	$\gamma_{B(A)}$	$\gamma_{A(B)}$	γ_{AB}	$S_{B/A}$	$S_{B(A)/A(B)}$	$S_{A/B}$	$S_{A(B)/B(A)}$
Isoamyl alcohol	23.7	23.6	25.9	5	44	−2.7	−54	−1.3
Benzene	28.9	28.8	62.2	35	8.9	−1.6	−78.9	−68.4
CS_2	32.4	31.8		48.4	−7	−9.9	−89	
n-Heptyl alcohol	27.5			7.7	40	−5.9	−56	
CH_2I_2	50.7			41.5	−27	−24	−73	

TABLE IV-2d
Initial versus Final Spreading Coefficients on Mercury[a]

Liquid, B	γ_B	$\gamma_{B(A)}$	$\gamma_{A(B)}$	γ_{AB}	$S_{B/A}$	$S_{B(A)/A(B)}$	$S_{A/B}$	$S_{A(B)/B(A)}$
Water	72.8	(72.8)	448	415	−3	−40	−817	−790
Benzene	28.8	(28.8)	393	357	99	7	−813	−721
n-Octane	21.8	(21.8)	400	378	85	0	−841	−756

[a] Data for equilibrium film pressures on mercury are from Ref. 10.

alcohols would be expected to spread on water, whereas CS_2 and CH_2I_2 should remain as a lens. As an extreme example, almost any liquid will spread to give a film on a mercury surface, as examination of the data in the table indicates. Conversely, a liquid of high surface tension would not be expected to spread on one of much lower surface tension; thus $S_{A/B}$ is negative in all of the cases given in Table IV-2d.

A complication now arises. The surface tensions of A and B in Eq. IV-2 are those for the pure liquids. However, when two substances are in contact, they will become mutually saturated, so that γ_A will change to $\gamma_{A(B)}$, and γ_B to $\gamma_{B(A)}$. That is, the convention will be used that a given phase is saturated with respect to that substance or phase whose symbol follows in parentheses. The corresponding spreading coefficient is then written $S_{B(A)/A(B)}$.

For the case of benzene on water,

$$S_{B/A} = 72.8 - (28.9 + 35.0) = 8.9 \qquad \text{(IV-4)}$$

$$S_{B(A)/A} = 72.8 - (28.8 + 35.0) = 9.0 \qquad \text{(IV-5)}$$

$$S_{B(A)/A(B)} = 62.2 - (28.8 + 35.0) = -1.6 \qquad \text{(IV-6)}$$

The final spreading coefficient is therefore negative; thus if benzene is added to a water surface, a rapid initial spreading occurs, and then, as mutual saturation takes place, the benzene retracts to a lens. The water surface left behind is not pure, however; its surface tension is 62.2, corresponding to the Gibbs monolayer for a saturated solution of benzene in water (or, also, corresponding to the film of benzene that is in equilibrium with saturated benzene vapor).

The situation illustrated by the case of benzene appears to be quite common for water substrates. Low surface tension liquids will have a positive initial spreading coefficient, but a near zero or negative final one; this comes about because the film pressure π of the Gibbs monolayer is large enough to reduce the surface tension of the water–air interface to a value below the sum of the other two. Thus the equilibrium situation in the case of organic liquids on water generally seems to be that of a monolayer with any excess liquid collected as a lens.

The spreading coefficient $S_{B(A)/A}$ can be determined directly, and Zisman and co-workers report a number of such values (11). (Note Problem 6 of this chapter.)

There has been considerable theoretical development in the treatment of interfacial tension. The approach used is similar to that of Eq. III-44 but using average densities rather than the actual radial distribution functions—these are generally not available for systems such as those of Table IV-2. As illustrated in Fig. IV-3, the work or free energy of extending the interface between liquids A and B may be regarded as the sum of the works of bringing molecules A and B to their respective liquid–vapor interfaces, or $\gamma_A + \gamma_B$,

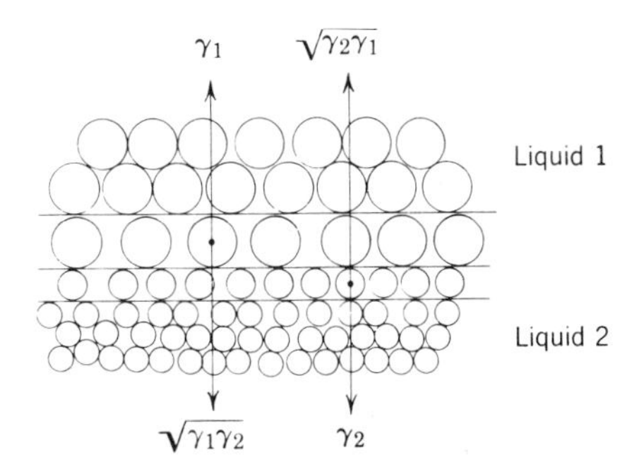

Fig. IV-3. The Good–Fowkes model for calculating interfacial tension (from Ref. 12).

less the free energy of interaction across the interface. From Eq. III-47

$$\gamma_{AB} = \gamma_A + \gamma_B - w_{AB} \tag{III-47}$$

The required theoretical calculation is that of the work of separating the two interfaces w_{AB}, and this can be carried out if the potential function for A–B interactions is known. Girifalco and Good (9) assumed the geometric mean rule ($\epsilon_{AB}(r) = [\epsilon_A(r)\epsilon_B(r)]^{1/2}$, referring to Eq. III-44, to obtain

$$\gamma_{AB} = \gamma_A + \gamma_B - 2\Phi(\gamma_A\gamma_B)^{1/2} \tag{IV-7}$$

where Φ is a function of the molar volumes of the two liquids; empirically, its values ranged from 0.5 to 1.15. Equation IV-7 also follows if the Skapski type of approach is used of neglecting all but nearest-neighbor interactions (see Eq. III-17) (13).

There are different kinds of intermolecular forces (dispersion, dipole–dipole, hydrogen bonding, etc., see Section VI-1), however, and those important in A–A or B–B interactions may not equally contribute to A–B interactions. Consider, for example, the water–hydrocarbon system. The potential function $\epsilon_W(r)$ for water contains important hydrogen bonding contributions—contributions absent in $\epsilon_{WH}(r)$, the function for water–hydrocarbon interaction. In recognition of this point, the more general form of Eq. IV-7 is

$$\gamma_{AB} = \gamma_A + \gamma_B - 2\Phi(\gamma'_A\gamma'_B)^{1/2} \tag{IV-8}$$

where the primes denote the surface tension that liquids A and B would have if their intermolecular potentials contained only the same kinds of interactions as those involved between A and B (see Refs. 12, 14–16). In the case of hydrocarbons, Fowkes (12) assumed that $\epsilon_H(r)$ derived only from dispersion interactions, and likewise $\epsilon_{WH}(r)$ so that for the water–hydrocarbon system, Eq. IV-8 became

$$\gamma_{WH} = \gamma_W + \gamma_H - 2(\gamma_W^d\gamma_H)^{1/2} \tag{IV-9}$$

where Φ has been approximated as unity and γ_W^d is the contribution to the surface tension of water from the dispersion effect only. A number of systems obey Eq. IV-9 very well, with $\gamma_W^d = 21.8$ erg/cm^2 at 20°C (17), as illustrated in Table IV-3. Variations from the 21.8 dyne/cm figure have been discussed by Fowkes (18), and attributed to anisotropic dispersion interaction in hydrocarbons.

For the case of both liquids polar, Fowkes (see Ref. 18) has written

$$w_{AB} = w_{AB}^d + w_{AB}^p + w_{AB}^h \tag{IV-10}$$

where superscripts p and h denote dipolar and hydrogen bonding interaction, respectively. That is, the approximation is made of treating different kinds of van der Waals forces (see Chapter VI) as independently acting. This type of approach has also been taken by Tamai and co-workers (19) and Panzer (20).

The Good-Fowkes approach calculates the work of separating an interface to give the two pure liquid surfaces and it is therefore inherent in the treatment that γ_A and γ_B rather than $\gamma_{A(B)}$ and $\gamma_{B(A)}$ appear in Eqs. IV-7 and IV-8. If liquids A and B are themselves solutions, then the separation process is taken to leave hypothetical surfaces A and B whose compositions are the same as in the respective bulk phases. Good (15) then writes

$$\gamma_{AB} = g_A + g_B - 2(g_A g_B)^{1/2} \tag{IV-11}$$

For two-component solutions the g's are approximated by

$$\begin{aligned} g_A &= N_{1,A}\gamma_1 + N_{2,A}\gamma_2 \\ g_B &= N_{1,B}\gamma_1 + N_{2,B}\gamma_2 \end{aligned} \tag{IV-12}$$

where γ_1 and γ_2 are the surface tensions of the pure liquids, and N denotes solution mole fraction.

These treatments are discussed further in Section X-6B, but some reservations might be noted here. The question has been raised (21a,b,c) whether the calculational procedure for obtaining w_{AB} in Eq. III-47 is correct in assuming that the bulk structures and compositions of the two liquid phases are maintained right up to the surface and are retained at all stages of the separation process. The reversible separation process would recognize surface density and composition changes as the separation occurred, and the analog of Eq. IV-7 would now have $\gamma_{A(B)}$ and $\gamma_{B(A)}$ in place of γ_A

TABLE IV-3
Calculation of γ_W^d at 20°C (17)

Hydrocarbon	γ_H	γ_{WH}, erg/cm^2	γ_W^d	
n-Hexane	18.4	51.1	21.8	
n-Heptane	20.4	50.2	22.6	
n-Octane	21.8	50.8	22.0	
n-Decane	23.9	51.2	21.6	
n-Tetradecane	25.6	52.2	20.8	21.8 ± 0.7
Cyclohexane	25.5	50.2	22.7	
Decalin	29.9	51.4	22.0	
White Oil	28.9	51.3	21.3	

and γ_B. The fact that the Girifalco-Good equation (Eq. IV-7) and its progeny work so amazingly well may be due to certain cancellations of errors that occur in the derivations.

The assumption made in writing Eq. IV-9 has been questioned (13), namely that the water–hydrocarbon interaction is one of dispersion-only forces. There is good evidence that strong polar forces are involved in the interaction of the first hydrocarbon layer with water (21–23). The high water–hydrocarbon interfacial tension and hence (by Eq. III-47) the low work of adhesion may be explained in terms of a modification of the γ_W^d concept. In the reversible process of separating the two phases, a film of hydrocarbon is left on the water surface that approaches a monolayer in thickness. Much of the interaction against which work is done is therefore film–hydrocarbon rather than water–hydrocarbon in nature. The effective γ_W, that is, γ'_W, is that of the film–air interface. This last approximates that of a hydrocarbon–air interface, so we expect $\gamma'_W(=$ "γ_W^d") to be about 20 erg/cm^2.

A relationship that antecedes Eq. IV-7 is an empirical one known as *Antonow's* rule (24), which states that

$$\gamma_{AB} = |\gamma_{A(B)} - \gamma_{B(A)}| \tag{IV-13}$$

This rule is approximately obeyed by a large number of systems but many exceptions are now known. See Refs. 25–28 for additional discussion. The rule can be understood in terms of a simple physical picture. There should be an adsorbed film or Gibbs monolayer of substance B (the one of lower surface tension) on the surface of liquid A. If we regard this film as having the properties of bulk liquid B, then $\gamma_{A(B)}$ is effectively the interfacial tension of a duplex surface and would be equal to $[\gamma_{AB} + \gamma_{B(A)}]$. Note the similarity of this picture to that invoked in the preceding discussion of γ_W^d.

B. Kinetics of Spreading Processes

The spreading process itself has been the object of some study. It was noted very early (29) that the disturbance due to spreading was confined to the region immediately adjacent to the expanding perimeter of the spreading substance. Thus if talc or some other inert powder is sprinkled on a water surface and a drop of oil is added, spreading oil sweeps the talc back as an accumulating ridge at the periphery of the film, but the talc further away is entirely undisturbed. Thus the "driving force" for spreading is localized at the linear interface between the oil and the water and is probably best regarded as a steady bias in molecular agitations at the interface, giving rise to a fairly rapid net motion.

Spreading velocities v are generally of the order of 15–30 cm/sec; a number of these has been tabulated by O'Brien, Feher, and Leja (30). The velocities for a homologous series tend to vary linearly with the film pressure, although in the case of alcohols a minimum π seemed to be required for v to be appreciable. The surface and underlying substrate showed cooling, possibly due to a latent heat of spreading; thus for fatty acids, v was linear in $d\pi/T$. Also, as illustrated in Fig. IV-4, it is interesting that the substrate water may be entrained to some depth (0.5 mm in the case of oleic acid), a compensating counterflow being present at greater depths (31). Related to this is the observation that v tends to vary inversely with substrate viscosity (32–34).

The foregoing discussion has dealt with the rate at which the film front advances.

If the spreading is into a limited surface area, as in a laboratory experiment, this front rather quickly reaches the boundaries of the trough. The film pressure at this stage is low, and the now essentially uniform film more slowly increases in π to the final equilibrium value. The rate of this second-stage process is mainly determined by dn/dt, the rate of release of molecules from the source (e.g. a crystal). Saylor and Barnes (35) find that surface concentration is now a better driving force variable than π, especially if the film undergoes phase changes in its approach to equilibrium (see Section IV-4), and propose the equation

$$\frac{dn}{dt} = k_{sp}\ell(\Gamma_e - \Gamma) \tag{IV-14}$$

where k_{sp} is the spreading rate constant, ℓ is the width of the source, Γ_e is the equilibrium surface concentration, and Γ, the value at time t.

The topic of spreading rates is of importance in the technology of the use of monolayers for evaporation control (see Section IV-8); it is also important, in the opposite sense, in the lubrication of fine bearings, as in watches where it is necessary that the small drop of oil remain in place and not be dissipated by spreading. Zisman and co-workers have found that spreading rates can be enhanced or reduced by the presence of small amounts of impurities; in particular, strongly adsorbed surfactants can form a film over which the oil will not spread (36).

The dependence of spreading rates on substrate viscosity, mentioned previously, indicates that films indeed interact strongly with the bulk liquid phase and cannot be regarded as merely consisting of molecules moving freely in a two-dimensional realm. This effect complicates the interpretation of monolayer viscosities (Section IV-3C). It is also an aspect of what is known as the *Marangoni effect*, namely, the carrying of bulk material through motions energized by surface tension gradients. A familiar (and Biblical) example is the formation of tears of wine in a glass. Here, the evaporation of alcohol from the meniscus film of the glass leads to a local raising of surface tension, which, in turn, induces a surface and accompanying bulk flow upward, the accumulating liquid returning in the form of drops.

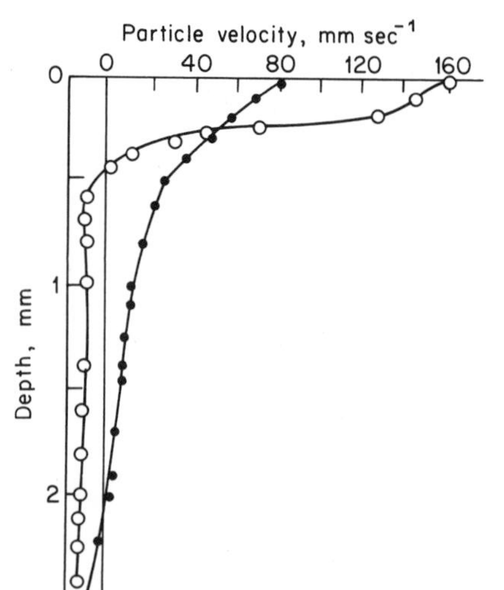

Fig. IV-4. Velocity profiles for particles suspended in water with elapsed time, due to spreading of oleic acid. Time after onset of spreading: ○, 1/8 s, ●, 1/2 s. (From Ref. 31.)

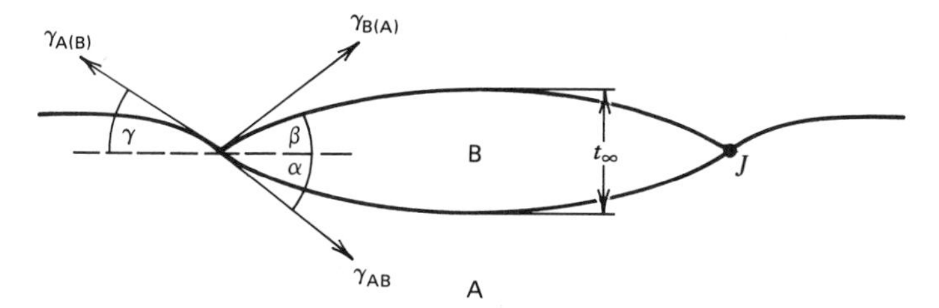

Fig. IV-5. Profile of a lens on water.

A drop of oil on a surfactant solution may send out filamental streamers as it spreads (37). The speed of spreading from a crystal may vary with direction, depending on the contour and, perhaps, on the crystal fact. In such a case, there may be sufficient unbalance to cause the solid particle to move around rapidly, as does camphor when placed on a clean water surface. These effects, and there are many of them, have been reviewed by Sternling and Scriven (38).

C. Lenses

The equilibrium shape of lenses floating on a liquid surface was considered by Langmuir (39), by Miller (40), and by Donahue and Bartell (41). The profile of an oil lens floating on water is shown in Fig. IV-5. The three surface tensions may be represented by forces whose balance may be treated by the principles of Newman's triangle:

$$\gamma_{A(B)} \cos \gamma = \gamma_{B(A)} \cos \beta + \gamma_{AB} \cos \alpha \tag{IV-15}$$

Donahue and Bartell verified Eq. IV-15 for several organic alcohol–water systems.

For very large lenses, a limiting thickness t_∞ is reached. Langmuir (39) gave the equation

$$t_\infty^2 = -\frac{2S\rho_A}{g\rho_B\,\Delta\rho} \tag{IV-16}$$

relating this thickness to the spreading coefficient and the liquid densities ρ_A and ρ_B. Actually, the observed thickness t may be appreciably greater than t_∞ because of the compressive effect of the *linear tension* f at point J in the figure. For paraffin oil, f was about 6 dyne.

Princen and Mason (42) have extended and generalized the analysis of possible contour shapes for lenses. They include the case of a lens at the interface between two immiscible liquids.

3. Experimental Techniques for the Study of Monomolecular Films

The experimental quantities that are conveniently accessible when one is dealing with films of insoluble materials no longer include the concentration of the material in solution. Instead, one knows directly the value of Γ, since this is simply the amount of material put on the surface per unit area. Also, instead of surface tension, one frequently measures the film pressure

π directly, by means of a film balance. Contact potentials, surface viscosities, and so on may also be determined. In the following, some of these quantities are operationally defined and some of the art of their measurement is discussed.

A. Measurement of π

The film pressure π is defined as the difference between the surface tension of the pure solvent and that of the film-covered surface. Therefore, any method of surface tension measurement in principle can be used. However, most of the methods of capillarity are, for one reason or another, ill suited for work with film-covered surfaces, with the principal exception of the *Wilhelmy slide method* (Section II-8), which works very well with fluid films (43) and is capable of measuring low values with a precision of 0.01 dyne/cm.

It seems best to use the procedure whereby the slide is partly immersed and to determine either the change in height with constant upward pull or the change in pull at constant position of the slide. If the latter procedure is used, then

$$\pi = \Delta w p$$

where Δw is the change in pull, for example, as measured by means of a balance (see Fig. II-16), and p is the perimeter of the slide. Thin glass, platinum, mica, and filter paper (44) have been used as slide materials.

The method suffers from two disadvantages. First, it measures γ or changes in γ rather than π directly. As a consequence, any temperature drifts or adventitious impurities can cause changes in γ that can be attributed mistakenly to changes in film pressure. Second, while ensuring zero contact angle usually is not a problem in the case of pure liquids, it may be with film-covered surfaces, as film material may adsorb on the slide. This problem can be a serious one; roughening the plates may help, and some of the more recent literature on techniques is summarized by Gaines (45). On the other hand, the equipment for the Wilhelmy slide method is relatively simple and cheap to construct and use, and can be fully as accurate as the film balance described in the following paragraphs.

Film pressure π is very often measured directly by means of a *film balance*. As indicated schematically in Fig. IV-1, the principle of the method involves the direct measurement of the horizontal force on a float separating the film from clean solvent surface. The film balance has been considerably refined since the crude model used by Langmuir and, in many laboratories, has been made into a precision instrument capable of measuring film pressures with an accuracy of hundredths of a dyne per centimeter. The various method of measuring film pressure have been reviewed in detail by Gaines (1, 45); see also Costin and Barnes (46). Kim and Cannell (46a) describe a film balance for measuring very small film pressures.

A general view of a modern film balance is shown in Fig. IV-6. The trough is a shallow rectangular tray 2 or 3 ft long and perhaps 1 ft wide, made of brass, stainless

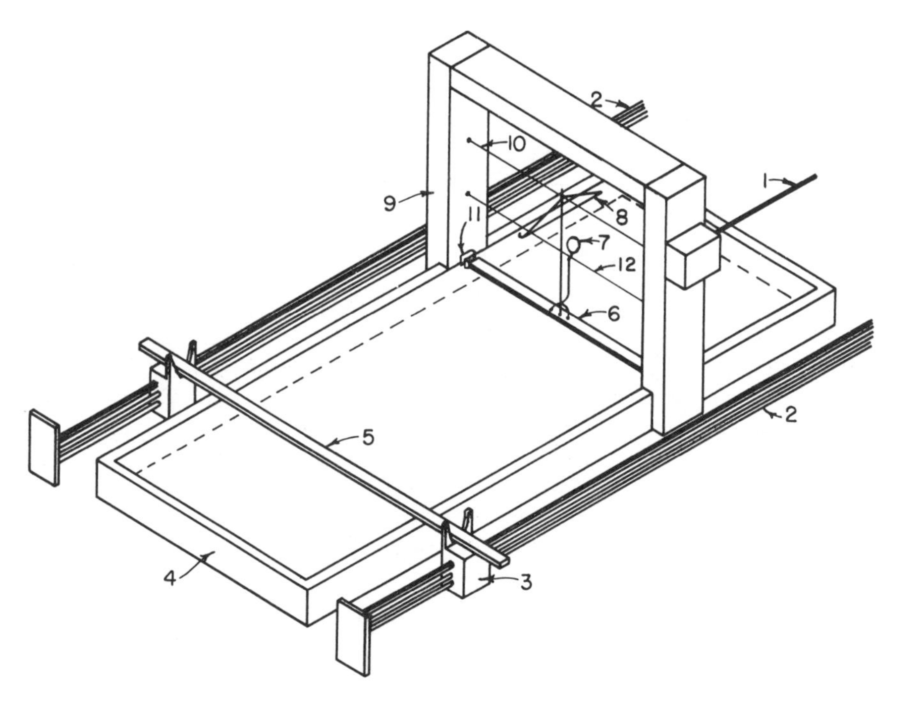

Fig. IV-6. A modern film balance; 1, torsion wire control; 2, sweep control; 3, sweep holder; 4, trough; 5, sweep; 6, float; 7, mirror; 8, calibration arm; 9, head; 10, main torsion wire; 11, gold foil barriers; 12, wire for mirror.

steel, glass, or Teflon. It is necessary to sweep the surfaces (see below) and, to contain the film, that the water level stand above the rim of the trough so that the tray, if made of metal or glass, must be coated with paraffin, lacquer, and so on, to ensure a high contact angle. Scrupulous purity with respect to surfactant contaminants is essential, of course (see Ref. 47).

The "head" is located about one-third of the way from one end, and consists of posts that support the torsion wire, mirror, and float suspension system. The float itself consists of a thin, paraffined, or plastic-coated strip of mica, copper, or platinum, or of a Teflon strip, and floats freely on the surface. The gaps at the ends are sealed off by thin gold or plastic foil barriers that allow movement of the float but prevent escape of the film. A stiff wire connects the float to the torsion wire, through which force is applied to the float, and the float position is monitored by means of a second stirrup attached to a second wire having a mirror attached to it. Any motion of the float rotates the mirror and can be measured accurately by following the displacement of a reflected spot of light.

For precise work, the film balance is placed inside an air thermostat, and all of the controls are operated from the outside. The general procedure for making measurements is as follows. The trough is filled with the substrate liquid, and the assembly is allowed to come to thermal equilibrium. A number of long sweep rods are placed at either end of the trough. The sweep controls are then operated so as to clean the liquid surface in front of and behind the float. This is done by carrying a sweep down to the head, lowering it so that it rests on the rim of the trough, and then moving it back toward the end. The sweep thus carries ahead of it any dust or other contamination that may have been on the surface. This is done several times, using a fresh sweep each time. When the surfaces are clean, the rest point of the float is checked and the film is spread.

It is not convenient to weigh the film-forming material onto the surface, as only a few tenths of a microgram are involved, although this has been done. Usually, the material to be studied is dissolved in some volatile solvent such as benzene or petroleum ether, and a known amount of such solution is then carefully pipetted onto the surface in front of the float. After the solvent has evaporated off, the film of substance to be studied is left on the surface, between the float and a sweep, which has been positioned well back toward the end of the trough. One then moves the sweep forward by stages, noting the film pressure at each point as determined by the force needed to retain the float in its null position. The data are usually reported as π versus σ. It might be mentioned that there has been concern and even controversy over whether evaporation of solvent is always complete and over whether the choice of solvent affects the π–σ plots (1, 48, 49; see also 46 and 53).

Instead of compressing a fixed amount of film material, one may add successive increments at constant area. The resulting π–σ data may differ somewhat, as in the case of stearic acid monolayers (50).

The limiting compression (or maximum π value) is, theoretically, that which places the film in equilibrium with bulk material. Compression beyond this point should force film material into patches of bulk solid or liquid, but in practice one may sometimes compress past this point. Thus in the case of stearic acid, with slow compression collapse occurred at about 15 dyne/cm (51), that is, film material began to go over to a three-dimensional state. With faster rates of compression, the π–σ isotherm could be followed up to 50 dyne/cm, or well into a metastable region. The mechanism of collapse may involve folding of the film into a bilayer (note Fig. IV-22).

There has been some interest in automating both Wilhelmy slide and film balance equipment, not only for saving labor, but also to make possible smooth, steady rates of compression or the measurement of film area versus time at constant pressure, where kinetic studies are involved. Descriptions of such equipment have been given by Gaines (1, 45) in his monographs. Transducers and optical position sensing devices may replace the traditional torsion wire; a sensitivity of 10^{-4} dyne/cm is claimed for one such sensor (52).

B. Surface Potentials

A second type of measurement that may be made on films, usually in conjunction with force–area measurements, is that of the contact or surface potential. One measures, essentially, the Volta potential between the surface of the liquid and that of a metal probe.

There are two procedures for doing this. The first, diagrammed in Fig. IV-7, makes use of a metal probe coated with an α emitter such as polonium or ^{241}Am (around 1 mCi). The resulting air ionization makes the gap between the probe and the liquid sufficiently conducting that the potential difference can be measured by means of a high impedence d-c voltmeter that serves as a null indicator in a standard potentiometer circuit. Electrode E may be a silver–silver chloride electrode. One

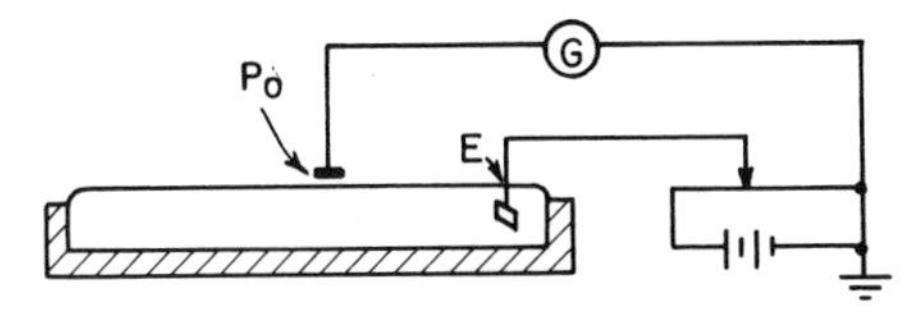

Fig. IV-7. Polonium electrode method for measuring surface potentials; E, reference electrode; G, galvanometer.

generally wishes to compare the potential of the film covered surface with that of the film-free one, and procedures for doing this sequentially or simultaneously have been reported (54, 55).

An alternative method not requiring the use of an ultrahigh resistance voltmeter, and usable at liquid–liquid interfaces, is that known as the vibrating electrode method (1, 56). The block diagram is shown in Fig. IV-8. Here, an audiofrequency current drives a Rochelle salt or a loudspeaker magnet, and the vibrations are transmitted mechanically to a small disk mounted parallel with the surface and about 0.5 mm above it. The vibration of the disk causes a corresponding variation in the capacity across the air gap so that an alternating current is set up in the second circuit, whose magnitude depends on the voltage difference across the gap. This current is amplified by means of an a-c amplifier and heard as a hum in earphones. The potentiometer is adjusted until no noise is heard. The method is capable of giving potentials to about 0.1 mV, and is somewhat more precise than the polonium electrode method, although it is more susceptible to malfunctionings.

One does not ordinarily attempt to interpret the single surface potential values but deals instead with the difference in surface potentials between that for the substrate and that for the film-covered substrate. This difference ΔV is then attributed to the film and is generally given a qualitative interpretation in terms of the analogy with a condenser. Two conducting plates separated by a distance d and enclosing a charge density σ will have a potential difference ΔV given by a formula attributed to Helmholtz

$$\Delta V = \frac{4\pi\sigma d}{D} \qquad \text{(IV-17)}$$

where D is the dielectric constant. Actually, one supposes the situation to

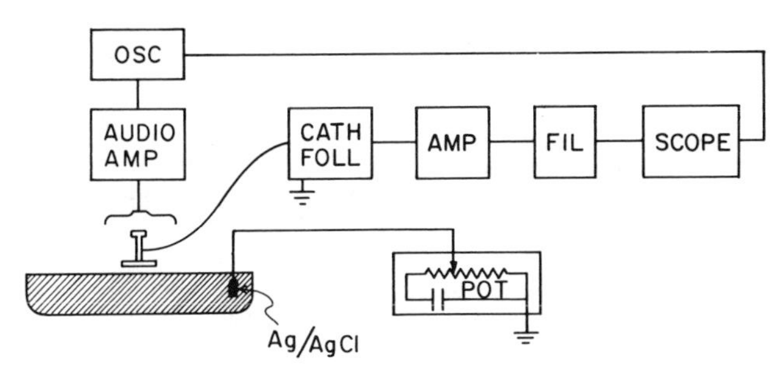

Fig. IV-8. Vibrating electrode method for measuring surface potentials. (From Ref. 1.)

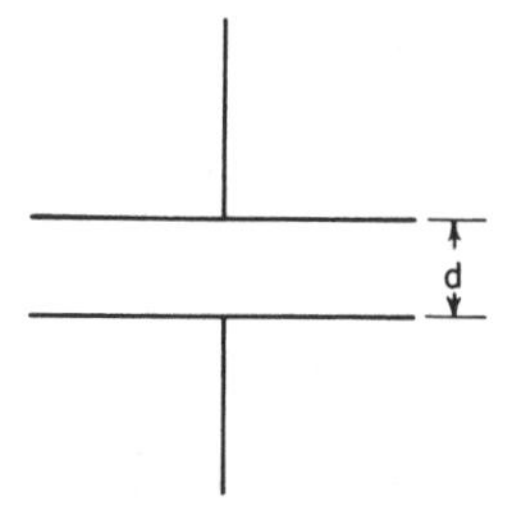

Fig. IV-9

be that illustrated in Fig. IV-9, in which charge separation due to the presence of an effective dipole moment $\bar{\mu}$ simulates a parallel plate condenser. If there are $\mathbf{n}$/cm^2 (perhaps corresponding to Γ molecules of polar film substance per square centimeter), then $\boldsymbol{\sigma} d = \mathbf{n}ed = \mathbf{n}\bar{\mu}$ and Eq. IV-17 becomes

$$\Delta V = \frac{4\pi\bar{\mu}\mathbf{n}}{D} = 4\pi\mathbf{n}\mu \cos\theta \dagger \qquad \text{(IV-18)}$$

Customarily, it is assumed that D is unity and that $\bar{\mu} = \mu \cos\theta$, where θ is the angle of inclination of the actual dipoles to the normal. Harkins and Fischer (57) point out the empirical nature of the preceding interpretation and prefer to consider only that ΔV is proportional to the surface concentration Γ and that the proportionality constant is some quantity characteristic of the film. This was properly cautious as there are many indications that the surface of water is structured and that the structure is altered by film-forming material (see Ref. 23). Accompanying any such structural rearrangement of the substrate at the surface should be a change in its contribution to the surface potential so that ΔV values should not be assigned entirely and too literally to the film molecules.

Nonetheless, Eq. IV-16 allows a number of pleasing interpretations to be made of the effect of film pressure and substrate composition on ΔV values, and as a consequence it is widely used. While there is some question about the interpretation of absolute ΔV values, such measurements are very useful as an alternative means of determining the concentration of molecules in a film (as in following rates of reaction or of dissolving) and in ascertaining whether a film is homogeneous. If decided fluctuations in ΔV occur on probing various regions of a film, there may be two phases present.

There are some theoretical complications, see Refs. 58 and 59 for example. Experimental complications include adsorption of solvent or of film on the electrode (60, 61); the effect may be used to detect atmospheric contaminants (62).

† Equations IV-17 and IV-18 are for the cgs/esu system of units; σ is in esu per square centimeter and ΔV will be in volts esu (1 V_{esu} = 300 ordinary V). In the SI system, the equations become $\Delta V = \sigma d/\epsilon_0 D$ and $\Delta V = \mu\mathbf{n}/\epsilon_0 D = \mathbf{n}\mu \cos\theta/\epsilon_0 D$, where $\epsilon_0 = 1 \times 10^7/4\pi c^2 = 8.85 \times 10^{-12}$. Charge density is now in coulombs per square meter and ΔV will be in ordinary volts. See Section V-3.

C. Surface Viscosities

The title of this section indicates that the material is somewhat limited in scope and does not cover the more general subject of surface rheology. This is partly because of space limitations and partly because the full spectrum of rheological properties, which would include viscoelastic effects with their time and shear rate dependencies, is yet somewhat undeveloped. Van den Tempel (63) has reviewed a number of aspects of surface rheology. The topic of wave damping is taken up in Section IV-15.

We limit ourselves here to two types of surface viscosity, *dilational* and *shear*. In the case of three-dimensional systems, the former is also known as the bulk viscosity and relates to the rate of yielding of a liquid or a solid to applied isotropic pressure. For films, the analogous definition for surface dilational viscosity κ is given by

$$\Delta\gamma = \kappa \frac{1}{\mathcal{A}} \frac{d\mathcal{A}}{dt} \qquad \text{(IV-19)}$$

that is, $1/\kappa$ is the fractional change in area per unit time per unit applied surface pressure.

A method for measuring κ is described by van den Tempel and co-workers (64). The surface is extended by means of two barriers, moving apart at such a velocity that the relative rate of extension, $d \ln \mathcal{A}/dt$, is constant. The dilation or depletion of the film results in a higher surface tension, measured by means of a Wilhelmy slide located at the center between the two barriers, where no liquid motion occurs. The procedure was applied to surfactant solutions but should be applicable to insoluble monolayers.

A related measurement is that of the rate of change of surface dipole orientation, as measured by contact potential change following a change in surface area (65).

The equilibrium quantity corresponding to κ is the modulus of surface elasticity E, defined as

$$E = \frac{d\gamma}{d \ln \mathcal{A}} \qquad \text{(III-128)}$$

The compressibility of a film K is just $1/E$. The measurement of E for surfactant solutions was discussed in Section III-7D. While bulk viscosity is a rather obscure property of a liquid or a solid, mainly because its effect is of little practical importance, K and E are probably the most important rheological properties of interfaces and films. They are discussed further in connection with capillary waves (Section IV-14) and foams (Section XIV-8), and their importance lies in the fact that interfaces are more often subjected to dilational than to shear stresses.

The shear viscosity of a film η^s, although of less actual importance in surface chemical situations, has been studied much more than has the dilational viscosity; perhaps this has been the result of the subjective influence of the known importance of shear viscosity in the case of liquids. The Newtonian definition of a two-dimensional shear viscosity is analogous to that for three dimensions. If two line elements in a surface (as opposed to two area elements in three dimensions) are to be moved relative to each other with a velocity gradient dv/dx, as illustrated in Fig. IV-10, then the required force is

$$f = \frac{\eta^s l \, dv}{dx} \tag{IV-20}$$

where l is the length of the element.

The measurement of η^s can be made in a manner entirely analogous to the Poiseuille method for liquids by determining the rate of flow of a film through a narrow canal, under a two-dimensional pressure head $\Delta\gamma$. The apparatus is illustrated diagrammatically in Fig. IV-11*a*. The equation for viscous flow in a canal may be derived as follows. The viscous force acting on a line located a distance x from the center of a canal of length l, as shown in Fig. IV-11*b*, will be

$$f_\eta = \frac{2\eta^s l \, dv}{dx} \tag{IV-21}$$

This is balanced by the force arising from the two-dimensional pressure gradient, acting on area $2x$:

$$f = 2x \, \Delta\gamma \tag{IV-22}$$

Setting the sum equal to zero and solving for dv:

$$dv = \frac{-\Delta\gamma}{l\eta^s} x \, dx \tag{IV-23}$$

On integrating and evaluating the constant of integration b by the condition that at $x = d$, $v = 0$, one obtains

$$v = \frac{\Delta\gamma}{2\eta^s l}(d^2 - x^2) \tag{IV-24}$$

where $d = a/2$ is the half-width of the canal. The area flow $d\mathscr{A}/dt$ is just

$$2 \int_0^d v \, dx$$

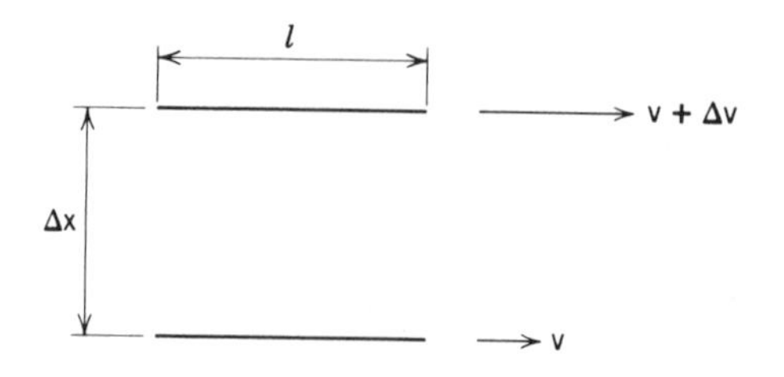

Fig. IV-10

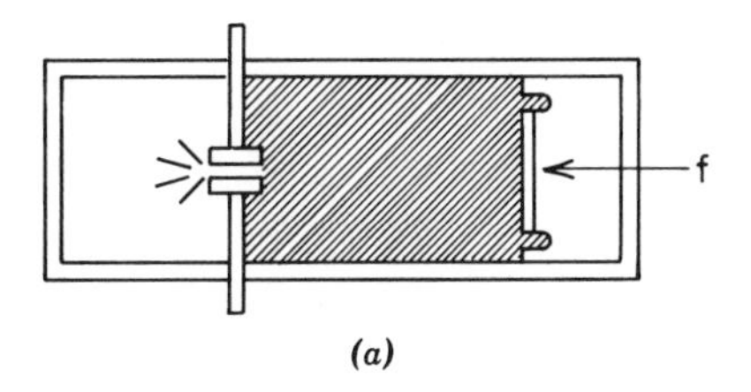

(a)

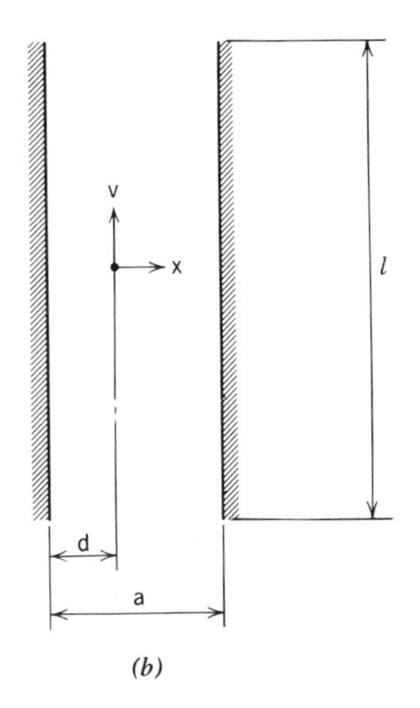

(b)

Fig. IV-11. Canal-type viscometer. (*a*) Top view (from Ref. 1; see also Ref. 68). (*b*) Enlarged view of the channel.

and, on substituting v from Eq. IV-24 and integrating, one obtains

$$\frac{d\mathcal{A}}{dt} = \frac{2}{3}\left(\frac{\Delta\gamma d^3}{\eta^s l}\right) \tag{IV-25}$$

or

$$\eta^s = \frac{\Delta\gamma a^3}{12l(d\mathcal{A}/dt)} \tag{IV-26}$$

Equation IV-26 was given by Myers and Harkins (66) and later, in a corrected form, by Harkins and Kirkwood (67):

$$\eta^s = \frac{\Delta\gamma a^3}{12l(d\mathcal{A}/dt)} - \frac{a\eta}{\pi} \tag{IV-27}$$

where the added term allows for the drag of underlying solvent of viscosity η. For applications of the method, see Nutting and Harkins (69), Ewers and Sack (70), and Hansen (71). Film viscosities are of the order of 10^{-2} to 10^{-4} g/sec or surface poises (sp), and by way of illustration, a film of 0.01 sp will flow through a 0.1-cm side slit 5 cm long at about 2 cm^2/sec if the surface pressure difference is 10 dyne/cm. The correction term in Eq. IV-27 is negligible in this case.

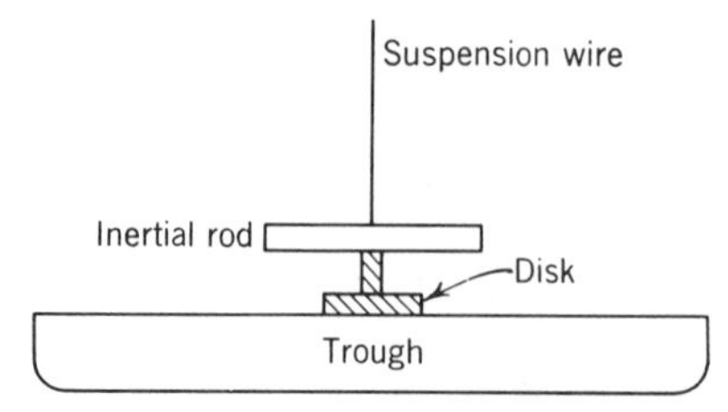

Fig. IV-12. Torsion pendulum surface viscometer (from Ref. 1).

The canal viscometer method gives absolute viscosities, and the effect of substrate drag can be analyzed theoretically, but the measurement cannot be made at a single film pressure, as a gradient is needed, nor is the shear rate constant. A second basic method, more advantageous in these respects, is one that goes back to Plateau (72). This involves the determination of the damping of the oscillations of a torsion pendulum, disk, or ring such as illustrated in Fig. IV-12. Gaines (1) gives the equation

$$\eta^s = \left(\frac{\tau I}{4\pi^2}\right)^{1/2}\left(\frac{1}{a^2} - \frac{1}{b^2}\right)\left[\frac{\lambda}{4\pi^2 + \lambda^2} - \frac{\lambda_0}{4\pi^2 + \lambda_0^2}\right] \tag{IV-28}$$

where a is the radius of the disk or ring, b is the radius of the (circular) film covered area, λ and λ_0 are the natural logarithms of the ratio of successive amplitudes in the presence and absence of film, respectively, and I is the moment of inertia of the pendulum. The torsion constant τ is given by

$$\tau = \frac{4\pi^2 I}{P_a^2} \tag{IV-29}$$

where P_a is the period of the pendulum in air. Tschoegl (73) has made a detailed analysis of torsion pendulum methods, including the difficult theoretical treatment of substrate drag.

In the *viscous traction viscometer* the film is spread in a circular annular canal formed by concentric cylinders (see Ref. 74). The problem of substrate drag is greatly reduced in the case of a rotating ring making a knife-edge contact with the liquid–air or liquid–liquid interface (75, 76); see Fig. IV-13. In first order, it turns out that the

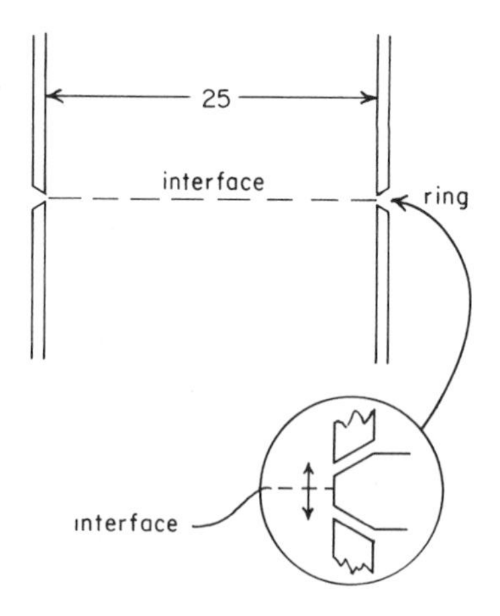

Fig. IV-13. Knife-edge viscometer. A cylindrical vessel with a depth at least equal to its radius a is equipped with a knife-edge ring inserted into its wall. The ring rotates, its knife-edge being flush with the cylinder wall and at the interface (from Ref. 75).

shear gradient is mainly confined to the outer portion of the circularly confined film, the inner region simply rotating as a uniform island. A simple equation that works well is:

$$\eta^s = 0.5631 \left(\frac{\Omega}{\omega}\right) \eta a \qquad \text{(IV-30)}$$

where Ω is the rotational speed of the central island and ω is that of the knife-edge ring; a is the radius of the outer cylindrical contained (and of the knife-edge ring).

An interesting comparison may be made between film and ordinary fluid viscosities. Turning to Fig. IV-10, the corresponding derivation in three dimensions would replace the line l by a plane perpendicular to the page, and l in Eq. IV-20 would be replaced by the area of this plane. If the viscosity of a film actually develops over a film thickness τ, then one would derive an ordinary three-dimensional viscosity where the area of the plane would be written as $l\tau$. Thus

$$\eta^s = \eta_f \tau \qquad \text{(IV-31)}$$

where η_f is the film viscosity treated as a three-dimensional phenomenon. A typical film thickness is 20 Å so that the typical η^s of about 10^{-3} surface poise corresponds to an η_f of $10^{-3}/2 \times 10^{-7}$, or 5000 P. Such films thus appear to have the consistency of a heavy grease.

The calculation above illuminates a current problem. Theoretical modeling of film viscosity leads to values about 10^6 times smaller than those observed (77, 78). It may be that the experimental phenomenology is not that supposed in derivations such as those of Eqs. IV-27 and IV-28; that is, the experiments may not be measuring what they are thought to. Alternatively, it may be that virtually all of the measured surface viscosity is developed in the substrate through its interactions with the polar portion of the film-forming molecule. Note Fig. IV-4. The matter is discussed further in Section IV-6G.

D. Optical Properties of Monolayers

The detailed examination of the behavior of light passing through or reflected by an interface can, in principle, allow the determination of the thickness of the inhomogeneous region, its index of refraction, and its absorption coefficient, the latter two as a function of wavelength. The subject of *ellipsometry* deals with this detailed approach; we only sketch this subject here, and the reader is referred to various monographs for more complete treatments (79–81).

Plane polarized light impinging onto a surface from some angle can be resolved into components parallel to and perpendicular to the plane of incidence. These two components are reflected differently, and, as a consequence, the reflected beam becomes elliptically polarized. As a special case, for example, if the angle of incidence is the Brewsterian angle $\tan^{-1} n$ where

n is the index of refraction of the substrate medium, only the perpendicular component will be reflected, assuming that the surface between the medium and air is absolutely sharp. Reflected ordinary (unpolarized) light will under these conditions be completely plane-polarized. Raman and Ramdas found that for clean water surfaces this was almost true. The slight degree of ellipticity found indicated the presence of a transition region about one molecule thick—another indication of the thinness of the surface region.

The typical ellipsometric experiment is that sketched in Fig. IV-14. A monochromatic light source (nowadays sometimes a small laser) is first plane polarized and then impinges on the interface to be studied (the optimum angle not in general being the Brewsterian angle for the clean surface). The reflected beam will be elliptically polarized and is returned to plane polarization by the compensator; the angle of this polarization is determined by the setting of the analyzer so that the detector shows complete extinction. The experimentally measured quantities are thus the polarizer angle p and the analyzer angle a. These give the fundamentally desired quantities, namely, the phase shift between the parallel and perpendicular components Δ and the change in the ratio of the amplitudes of these components $\tan \psi$. Thus $\Delta = 2p + \pi/2$ and $\psi = a$.

If the surface is clean, Δ and ψ are directly related to the complex index of refraction of the surface region $\hat{n}^s = n^s(1 - ik^s)$, where n^s is the ordinary index of refraction and k^s is the *attenuation index* or rational *extinction coefficient*. If a film of different index n_f is present, Δ and ψ are related to $\hat{n}^s$, $\hat{n}_f$, and to the film thickness. The technique can be very sensitive, detecting films of average thickness about 0.2 Å or about 0.05 of a monolayer.

Archer and co-workers (81) have given the very useful relationships for films less than a monolayer:

$$\Delta = \bar{\Delta} - ad \qquad \text{(IV-32)}$$

$$\psi = \bar{\psi} + \beta d \qquad \text{(IV-33)}$$

The quantities Δ and ψ refer to the clean surface, and $\bar{\Delta}$ and $\bar{\psi}$, to the film covered surface; d is the film thickness, and α and β are constants determined by the angle of incidence of the beam n_f, n^s and the wavelength of the light.

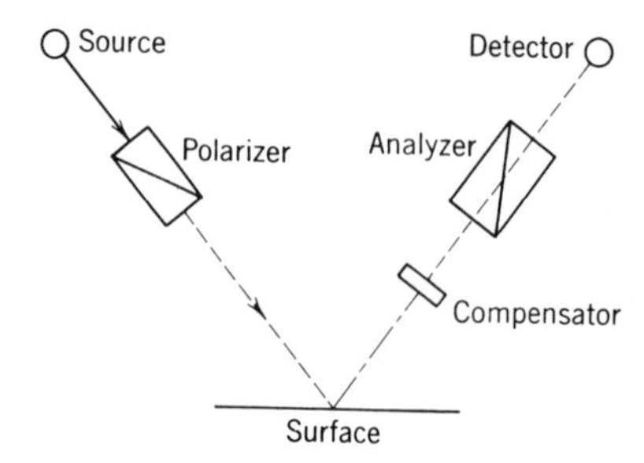

Fig. IV-14. Schematic apparatus for ellipsometry; ———, unpolarized light; - - -, plane-polarized; — — —, elliptically polarized. (From Ref. 1.)

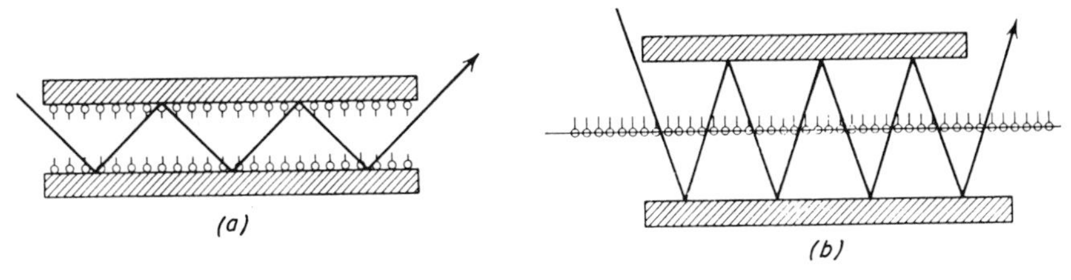

Fig. IV-15. Arrangements to obtain enough interaction of a light beam with a monolayer. (*a*) The monolayer is deposited on reflecting plates and examined by reflection. (*b*) Mirrors above and below the interface reflect the light beam back and forth through the monolayer on the liquid subphase. (From Ref. 1.)

In the case of spread monolayers, film thickness and index of refraction have not been given much attention. An interesting example, however, is a study by Smith of stearic acid monolayers on mercury (82); film thicknesses ranged from 3 to 6 Å. There has been more study of films on solid surfaces, see Section IV-15 and X-6. See Ref. 83 for some bibliography on apparatus and details on a scanning ellipsometer.

Interferometry is based on the fact that light reflected from the front and back interfaces of a film travels different distances, giving rise to interference effects. Light reflected from the front surface suffers a phase reversal because the value of n is greater than 1, whereas that from the back surface does not, in the case of a soap film or where the index of refraction of the backing material is less than that of the film. See Ref. 84 for a general review. The method has been applied to films deposited on a solid backing (85, 86), to soap films (Section XIV-8), and, interestingly, to the measurement of the temperature gradient under a surface that is evaporating (Section IV-8).

Of more interest to the study of insoluble monolayers is absorption spectroscopy, at the moment limited mainly to visible light. A major problem is that the film, being thin, absorbs little, so that some means of enhancing the effect is needed. Gaines (1) describes several procedures, two of which are illustrated in Fig. IV-15. One method, suitable for examination of spread monolayers *in situ*, is that shown in Fig. IV-15*b*, in which light is multiply reflected by mirrors placed above and below the interface (87). Infrared spectroscopy, while well developed for films on solids (see Section XV-4A), has been very difficult to apply to monolayers on liquid substrates.

It has been possible to measure fluorescent emission from spread monolayers, and an apparatus for this is described by Tweet and co-workers (88) and by Trosper and co-workers (89), along with its application to the study of emission from chlorophyll monolayers. More recently, Gaines and co-workers have reported on the emission from monolayers of Ru(bipyridine)$_3^{2+}$ one of the bipyridine ligands having attached C_{18} hydrocarbon chains (90). For studies of emission from films adsorbed on solids see Refs. 91 and 92.

E. The Ultramicroscope

In 1930 Zocher and Stiebel (93) reported that, under divergent light illumination, microscopic examination of a film would show up heterogeneity with great sensitivity. Patches of unspread material or of aggregates appear as more brilliantly illuminated than a monolayer, which appears dark. A particularly dramatic effect occurs on compressing a gaseous film through the region of condensation to a liquid; a two-dimensional colloidal mist of liquid film forms that shows up brilliantly. A more recent study using the technique is that of Bruun (94), on collapsed films.

F. Electron Microscopy and Diffraction

Ries and co-workers have been able to transfer spread monolayers of *n*-hexatriacontanoic (C_{36}) acid (95a), 2-hydroxytetracosanoic acid (95b), lecithin (95c), and cholesterol (95d) onto a collodion film supported by a screen, and then, by shadow casting, obtain very interesting electron micrographs. (See Fig. IV-22 and Section IV-6E.)

Electron diffraction studies have also been made, but also only with films transferred to a support (96). The technique of low energy electron diffraction is discussed in connection with gas adsorption (Section VIII-3).

G. Other Techniques

Just as radioactive tracers have been used in the study of Gibbs monolayers (Section III-6B), so may they be applied to adsorbed films in general. A few measurements have been made with spread monolayers *in situ*; Gaines (97) found that for radiostearic acid this procedure gave about the same results as after transferring the film to a mica plate (see Section IV-15). Spink (98) reports an extensive study of the transfer efficiency or *transfer ratio* for transferring ^{14}C labeled stearic acid monolayers to a variety of solids. Cook and Ries (99), again using radiostearic acid, found evidence for heterogeneity at areas greater than 35 $Å^2$ per molecule.

Esr measurements are possible on monolayers of spin-labeled molecules (99a); indications are that the π–σ behavior is not a good indication of the type of molecular motion present. No nuclear magnetic resonance studies seem to have been reported for spread monolayers.

4. States of Monomolecular Films

The very title of this section begs a question: Are there "states" of monolayers and how are they defined? What is meant, first, is that some gross differences are found in the behavior of various types of films, which, pursuing an analogy to bulk matter, allow them to be characterized as gaslike, liquidlike, and solidlike. Second, and with less excuse, subcategories of behavior have been named, often differently by different investigators, as one or another kind of state. The complexity of behavior undoubtedly arises from the fact that film-forming molecules are all polar–nonpolar in type; the polar end (or head) interacts strongly with the substrate water, and this may be in a structured way with much hydrogen bonding, ion atmosphere, and such effects, while the nonpolar sections (or tails) also interact, but with each other and in some different way. As will be seen, much of the interpretation of behavior of monolayers has been on the basis of combining separate analyses for the two sets of behavior, treated as being independent.

A schematic diagram illustrating most of these types of behavior is shown in Fig. IV-16. They may be described as follows:

1. Gaseous films (G). The film obeys the equation of state of a more or less perfect gas; the area per molecule is large compared to actual molecular areas, and the film may be expanded indefinitely without phase change. As with bulk matter, this state should always be reached at sufficiently large areas, although in practice film pressures even as low as 0.001 dyne/cm may not suffice in some cases.

2. Liquid films (L). Such films are coherent in that some degree of cooperative interaction is present; they appear to be fluid (as opposed to rigid or showing a yield point) and their π–σ plots extrapolate to zero π at areas larger (up to several times larger) than corresponding to a molecular cross section so that some looseness or disorganization in the structure is indicated.

(*a*) *Liquid expanded films* (L_1). There are at least two distinguishable types of L films. The first was called *liquid expanded* and designated L_1 by both Adam (100) and Harkins (8). Such films tend to extrapolate to a limiting

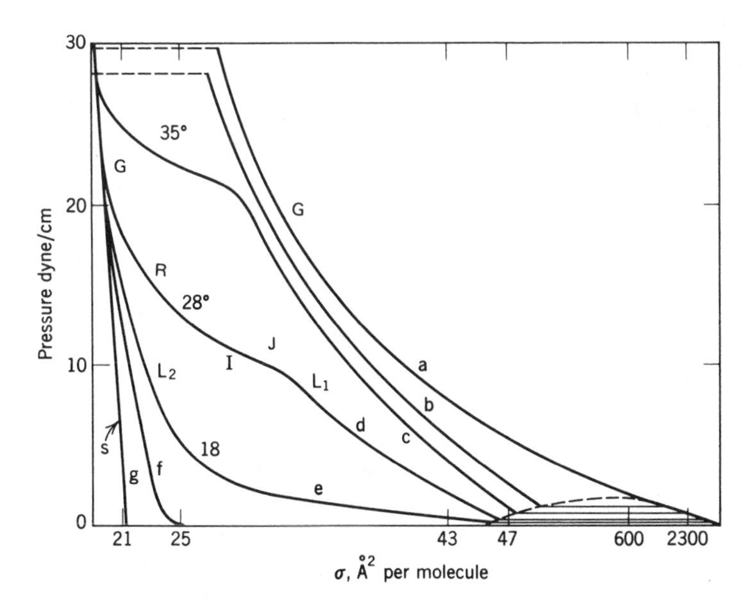

Fig. IV-16. States of monolayers (schematic).

or zero π area of about 50 Å^2, in the case of single chain molecules, as illustrated by curve *d* in Fig. IV-16. This type of film has a rather high compressibility if compared to bulk liquid, but appears to be single phase in that no islands or patches are discernible, for example, by surface potential probing. Characteristically, L_1 films may show a first-order transition to gaseous films at low pressures. Also, on compression, a point of fairly sharp but not abrupt change to a film of much higher compressibility occurs. This high compressibility region, called intermediate (I), is discussed in Section. IV-6.

(*b*) *Liquid condensed films* (L_2). Compression of an intermediate-type film leads to a region of gradual transition to a linear π–σ behavior, but of relatively low compressibility. The film in this region has been called a liquid condensed or L_2 type of film (8) or a film with "close-packed heads, rearranged on compression" (80). At lower temperatures, as indicated in Fig. IV-16, curve *e*, the L_2–I–L_1 sequence may give way to an L_2 type film that may either undergo a phase transition to a gaseous film or may appear to expand indefinitely without discontinuity, as though it were above its critical temperature. Adam has called such films "vapor expanded" because the terminus of their expansion is a G-type rather than an L_1-type film.

As perhaps a limiting case of L_1-type films the compressibility may be as small as that for solid films, with the linear portion of the π–σ plot extrapolating to about 22 Å^2 (about 20% larger than the cross section of a hydrocarbon chain). Cetyl alcohol shows this behavior, although rheologically it appears liquid.

3. Solid films (S). Some films, as for example those of fatty acids on water, may show quite linear π–σ plots, of a low compressibility about equal to that for bulk matter, and which extrapolate to an area at zero pressure of 20.5 Å^2. This area is probably that of close-packed hydrocarbon chains, and such films appear to be quite rigid; if talc is sprinkled over the surface, it is observed that whole sections of the film will move together. Alternatively, as in curve *f* of Fig. IV-16, there may be a break at lower pressures to an L_2 type of behavior.

The foregoing classification is much simplified and also presumes a better rheological knowledge of the various types of films than really exists, but it does serve to introduce characteristic behavior and nomenclature. Substances not of the simple straight chain structure tend to give much less sharply defined behavior. At very low pressures they can be gaseous, and at higher ones, in some coherent state that is liquid- or plastic-like.

The various states are discussed in more detail in Section IV-6, but a few additional comments are made here about their nature. As stated the L_1–G transition clearly is first order; not only is there a discontinuity in the π–σ plot corresponding to the straight lines shown in the dotted region of Fig. IV-16, but a latent heat of vaporization is associated with the transition, and there is evidence that, during it, the film consists of patches of two different phases. The surface potential is everywhere uniform for L_1-type or G-type films but fluctuates wildly as the probe is moved around over a film in the middle of an L_1–G transition (101). Similarly, the ellipticity of reflected light fluctuates from one region to another, corresponding to islands of condensed film surrounded by vapor film (62) and, as mentioned in the preceding section, ultramicroscopic examination also shows the film in the transition region to be inhomogeneous.

The general variation of surface potential ΔV and of calculated effective dipole moment $\bar{\mu}$ with σ, the area per molecule, is shown in Fig. IV-17. It is seen that while ΔV decreases continuously through the I and L_1 states, $\bar{\mu}$ is more nearly constant, suggesting that, while the surface density is changing greatly, the average orientation of the polar portion of the molecules is remaining more nearly constant. There is thus no indication of any first-order phase change in the I region; the nature of the I "phase" is discussed further in Section IV-6.

In addition to L–G transitions another resemblance to bulk systems is that it appears possible to apply the phase rule concept. Crisp (102) gives the following form:

$$F = C - P^b - (P^s - 1) \qquad \text{(IV-34)}$$

where F denotes the number of degrees of freedom; $C = C^b + C^s$, where C^b is the number of components equilibrated throughout the system and C^s is the number confined to the surface; P^s gives the number of surface phases; and P^b gives the number of bulk phases. It is assumed that the general pressure and temperature are constant, so that F is the additional number of degrees of freedom over these two. Thus for an insoluble monolayer with L_1 and gas phases in equilibrium, $C = 1 + 1 = 2$, $P^S = 2$, and $F = 0$.

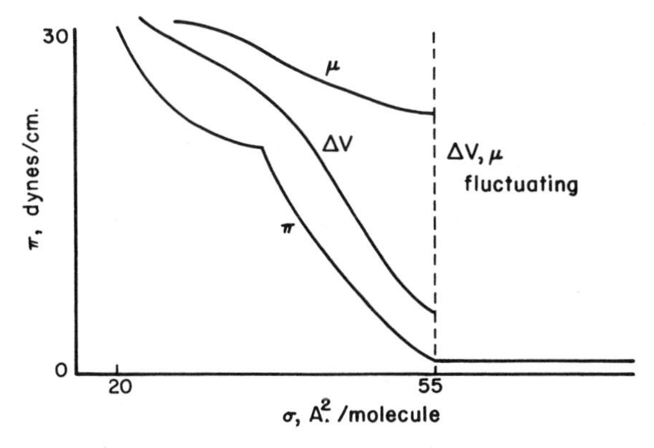

Fig. IV-17. Variation of V and $\bar{\mu}$ with monolayer state. Myristic acid on 0.01 N hydrocloric acid.

5. Correspondence between π and a Three-Dimensional Pressure

Before proceeding to the more detailed discussion of the behavior of monolayers, it is of interest to orient the reader to what sort of correspondence can be made between π values and ordinary pressure in three dimensions. For example, one may consider that although π has the dimensions of force per unit length, it is in reality a pressure distributed over the thickness of the film. By this interpretation

$$P = \frac{\pi}{\tau} \tag{IV-35}$$

where τ is the film thickness. Since τ is of the order of 10 Å for monolayers, $P = 10^7$ dyne/cm^2 for $\pi = 1$ dyne/cm; that is, π multiplied by 10 gives the equivalent three-dimensional pressure in atmospheres. This is not unreasonable; for example, the two-dimensional vapor pressures corresponding to the L_1–G transition are of the order of a few tenths of a dyne per cm and thus correspond to a few atmospheres bulk pressure. The compressibility of an S film is about 5×10^{-4} (dyne/cm)$^{-1}$ or about 5×10^{-5} atm^{-1}, using the above conversion factor; this is the same magnitude as for bulk liquids or solids.

As far as equivalent P values, the alternative osmotic pressure approach in Section III-7C gives the same result. Also, of course, Eq. IV-35 follows from Eq. III-42 if one assumes that a constant pressure p exists in the film, over a region $2a$ equal to τ.

The purpose of introducing at this time a qualitative discussion on the correspondence between π and P is that it is helpful in considering the physical nature of the various states of monolayers to realize that the rather feeble forces of a few dynes or tens of dynes measured on a film balance may correspond to hundreds or thousands of atmospheres pressure in terms of the compressive effect at the molecular level.

According to the preceding analysis the film pressures and compressibilities of gaseous, liquid, and solid monolayers are about what would be expected in terms of the values for the corresponding bulk phases. It is curious that viscosities of monolayers fail in this correspondence test (see Section IV-3C), being 10^6-fold higher than expected.

6. Further Discussion of the States of Monomolecular Films

A. Gaseous Films

Gaseous films have been considered earlier, in connection with Gibbs monolayers (Section III-7), but some further discussion of the two-dimensional surface solution model is appropriate here. It will be recalled that this model yields a film pressure arising from the osmotic pressure difference between an interface containing a second component and a pure solvent interface, assumed to be acting over a depth τ. Some of the original ideas are due to Ter Minassian-Saraga and Prigogine (103), with some developments of them by Fowkes (104) and by Gaines (105).

In this model, a gaseous film is considered to be a dilute surface solution of surfactant in water, and one may use Eq. III-119 in the precursor form wherein the osmotic pressure π_{os} is written as

$$\pi_{os} = -\left(\frac{RT}{V_1}\right) \ln N_1^s \tag{IV-36}$$

where V_1 is the molar volume of the solvent and N_1^s is its mole fraction in the surface region. If the region is of thickness τ, then V/τ has the dimensions of molar area and Eq. IV-36 can be put in the form

$$\pi = -\left(\frac{RT}{A_1}\right) \ln f_1^s N_1^s \tag{IV-37}$$

where A_1 is the molar area of the solvent (strictly, $\bar{A}_1 = (\partial \mathscr{A}/\partial n_1^s)_T$), and f_1^s is the activity coefficient of solvent in the film, taken to be unity for gaseous films [Gaines (105) computes values of 1.01 to 1.02 for pentadecanoic acid films]. We add the condition

$$\mathscr{A} = A_1 n_1^s + A_2 n_2^s \tag{IV-38}$$

so that Eq. IV-37 can be written

$$\pi = \frac{RT}{A_1}\left[\ln\left(1 + \frac{A_1\Gamma_2}{1 - A_2\Gamma_2}\right) - \ln f_1^s\right] \tag{IV-39}$$

where Γ_2 is the moles of film material per unit area. A_1 is known from the estimated molecular area of water, 9.7 Å^2, and A_2 can either be estimated, thus allowing calculation of f_1^s, or left as an empirical parameter.

The alternative approach is to treat the film as a nonideal two-dimensional gas. One may use an appropriate equation of state, such as Eq. III-114. Alternatively, the formal thermodynamics has been developed for calculating film activity coefficients as a function of film pressure (105a).

B. The L_1–G Transition

As noted in Section IV-4, on compression of a gaseous film, a first-order transition to the L_1 state may occur. This transition has been well studied; data for pentadecanoic acid are shown in Fig. IV-18, the curved envelope marking the boundaries of the two-phase region. The situation is reminescent of three-dimensional vapor–liquid condensation and can be treated by the two-dimensional van der Waals equation (Eq. III-114). The π–σ isotherms around the critical temperature fit very poorly in the case of pentadecanoic acid, however, and Stoeckly (106) proposed that considerable clustering was present in the gas phase.

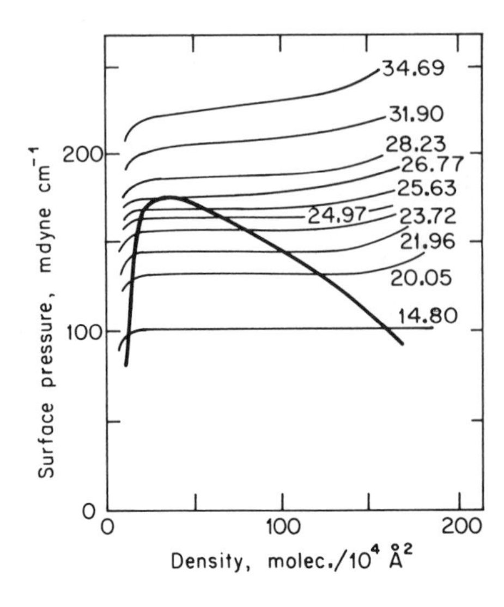

Fig. IV-18. Surface pressure versus density for a film of pentadecylic acid spread on a pH 2 water substrate. The curved envelope marks the L_1–G two-phase region; the isotherms are labeled in degrees Celsius. (From Ref. 46a.)

One may apply the two-dimensional analogue of the Clapeyron equation,

$$\frac{d\pi}{dT} = \frac{\Delta H}{T\Delta A} \tag{IV-40}$$

where ΔH is now the latent heat of vaporization of the L_1 state. Representative values are 2, 3.2, and 9.5 kcal/mole for tridecylic, myristic, and pentadecylic acid, respectively (100). Since the polar carboxyl groups are solvated both in the G and L_1 phases, the latent heat of vaporization can be viewed as arising mainly from the attraction between the hydrocarbon tails. It is thus understandable that these latent heats are much less than the ones for the bulk materials.

C. The Liquid Expanded State

The L_1 state generally may be observed with long chain compounds having highly polar groups, such as acids, alcohols, amides, and nitriles. The π–σ plots tend to extrapolate to a value in the range from 40 to 70 Å^2 at zero π, depending on the compound. Some representative data for pentadecylic acid are shown in Fig. IV-19 (107).

There is a continuing debate on how best to formulate an equation of state of the liquid expanded type of monolayer. Such monolayers are fluid and coherent (and thus liquidlike), yet the average intermolecular distance is much greater than for bulk liquids. A typical bulk liquid is perhaps 10% less dense than its corresponding solid state, yet a liquid expanded monolayer may exist at molecular areas twice that for the solid state monolayer.

As noted by Gershfeld (108), various modifications of the two-dimensional van der Waals equation can be successful in fitting data. This writer prefers the more structurally explicit model of Mittelmann and Palmer (109). As indicated in Fig. IV-20, the liquid expanded state is regarded as being in

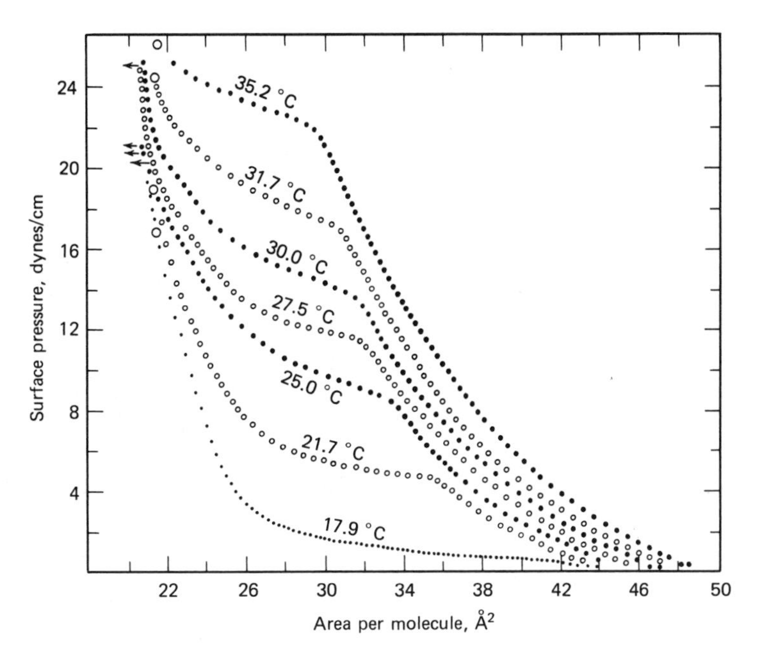

Fig. IV-19. Pressure-area relations for monolayers of pentadeclylic acid at pH 2 (from Ref. 107).

transition from the gaseous one in which all molecules lie flat on the surface and the condensed one in which the molecules are all oriented perpendicular to the surface. It is assumed that there is a statistical distribution among the various possible configurations, the energy for each being given by

$$\epsilon = (\pi + w)al' + \epsilon' \tag{IV-41}$$

where $\pi al'$ is the work against the film pressure, a being the width of the chain and l' being the length of the chain in the surface; w is the work of adhesion of a CH_2 group to water (small), and ϵ' is the work to rotate a C–C bond (about 800 cal/mole). A Boltzmann expression is then used to give the probability of each configuration:

$$p = k \exp\left(-\frac{\pi al' + \epsilon'}{kT}\right) \tag{IV-42}$$

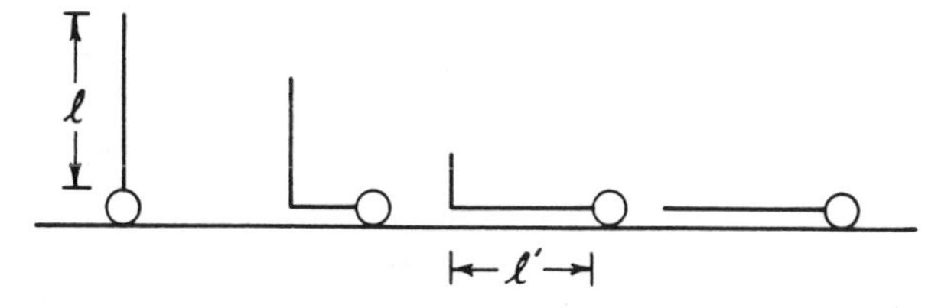

Fig. IV-20

The average area per molecule is then

$$\sigma = \sigma_n + \frac{\alpha \sum_{0}^{l} l' \exp[-(\pi a l' + \epsilon')/kT]}{\sum_{0}^{l} \exp[-(\pi a l' + \epsilon')/kT]} \tag{IV-43}$$

where σ_n denotes the area of the head group. The treatment gave fair agreement with the observed π versus σ values for the L_1 state of oleic acid and also was able to predict the effect of film pressure on the chemical reactivity of the double bond (Section IV-10). Although elaborated on since (82), the approach suffers in not properly accounting for entropy aspects and in not suggesting how there could be a first-order phase transition between the L_1 and G states.

Turning to the two-dimensional surface solution model, Eq. IV-39 may be applied to the L_1 state. In the case of pentadecanoic acid, values of f_1^s are now 1.1 to 1.2 (105). The approach, however, does not seem to lend itself to treatment of the L_1–G phase transition.

D. Intermediate and L_2 Films

The L_2-type film was viewed by Adam (100) and by Langmuir (39) as a semisolid film having more or less water between the polar heads. On compression, the water is squeezed out until a solid film is obtained, or in some cases where the polar heads are large they may gradually assume a staggered arrangement, that is, may "rearrange on compression," Harkins (8) considered that the degree of hydrogen bonding with solvent may determine the degree of compressibility of an L_2-type film.

For L_2 films, the π–σ plots are nearly linear, and can thus be represented by an equation of the form

$$\sigma = b - a\pi \tag{IV-44}$$

If a is small, the compressibility will be nearly constant; values are of the order of 10^{-3} to 10^{-2} for such films, in cgs units.

A film that obeys Eq. IV-44 down to low pressures will generally go through an L_2–I transition at intermediate pressures, if a somewhat higher temperature is used or if a shorter chain length species is substituted. Roughly, a difference of one CH_2 group is equivalent to a 5° change in temperature. The L_2–I transition was considered by Langmuir (39) to involve a breaking up to the L_2 film into clusters of molecules or "micelles." These micelles were small enough to act like an imperfect gas and exert an appreciable vapor pressure. In turn, however, they could dissociate into the more random L_1 state. Adapting Mittelmann and Palmer's picture of the L_1 state (109), the general sequence of structures would be that indicated in Fig. IV-21.

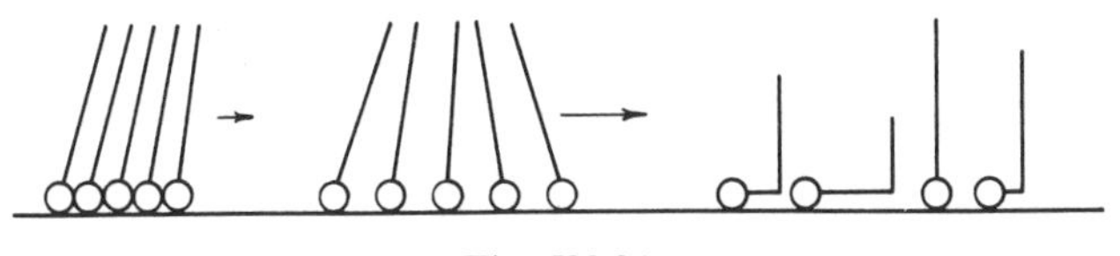

Fig. IV-21

The micelle model was criticized by Harkins and Boyd (107) on the grounds that, if micelles do exist, they must vary in size in order to explain the behavior of the energy of expansion in this region. An alternative explanation for the I state would be that at point J in Fig. IV-16 the system begins to lose rotational degrees of freedom progressively until, at R, the molecules are no longer free to rotate (see also Refs. 110, 111).

A more sweeping question that has been raised is whether the intermediate state may be a nonequilibrium, metastable one (105, 108). The L_1–I transition appears to be a second-order one, and it is difficult on theoretical grounds to accept that the compressibility of a system could *increase* (as it appears to) on going from the lower to the higher density state. The experimental isotherms seem to be reversible, although this has been questioned (112). The whole matter remains unsettled.

E. The Solid State

The general appearance of S-type films is that of a high density and either rigid or plastic phase. Most fatty acids and alcohols exhibit this type of film at sufficiently low temperatures or with sufficiently long chain lengths. The equation of state is simply Eq. IV-44, a typical a value being 0.02 in cgs units, with a limiting value of 20.5 Å^2 for the fatty acids. This is greater than the value of 18.5 Å^2 obtained from the structure of the three-dimensional crystals, and an early suggestion was that the chains were uniformly tilted by about 26° from the vertical; this particular angle provides the added feature that interlocking of the zigzag hydrocarbon chains could then occur. It has been pointed out, however, that the preferred arrangement of carboxyl groups in the surface leads naturally to an area per molecule of 20.5 Å^2 without the necessity of assuming the chains to be tilted (113, 114). A calculation of the electrostatic interaction energy of close-packed dipoles shows that this may be either positive or negative, depending on the geometry of the array (115). It was concluded, however, that the electrostatic contribution was always small compared to the attractive forces between the hydrocarbon tails.

Consideration of the solvent structure in layers adjacent to that of the film have been lacking in all of the foregoing modeling of the various states. Garfias (116) has emphasized this omission, suggesting that in solid monolayers of long chain alcohols, half of the water molecules in the surface layer are replaced by film molecules, the whole forming a very ordered structure.

A very different structural question is the following. Films have so far been presented as monolayers, perhaps with water included in the interfacial region. Ries and co-workers have obtained electron micrographs of films transferred to collodion and shadow cast (95). In the case of *n*-hexatriacontanoic (C_{36}) acid, the pictures show well separated islands of film, raising the question of whether the original spread film was so structured. The effect could be artifactual, due to evaporation of water and even of some fatty acid during the vacuum drying of the film in its preparation for electron microscopy (117, 118). Even if this is the case, however, there is a sequence that needs to be understood of changing appearance of the microphotographs with increasing film pressure of the (always "solid") film. 2-Hydroxytetracosanoic acid collapses at 68 dyne/cm (without falloff in pressure), and the microphotographs clearly show that ridges ranging up to 2000 Å in height have formed, as shown in Fig. IV-22*a*. Possible collapse sequences are illustrated in Fig. IV-22*b*.

F. Effect of Changes in the Aqueous Substrate

Fairly marked effects of changing the pH of the substrate are frequently observed. An obvious case is that of the fatty acid monolayers; these will be ionized on alkaline substrates and, as a result of the repulsion between the charged polar groups, the film becomes gaseous or liquid expanded in type at a much lower temperature than does the acid (119). Also, the surface

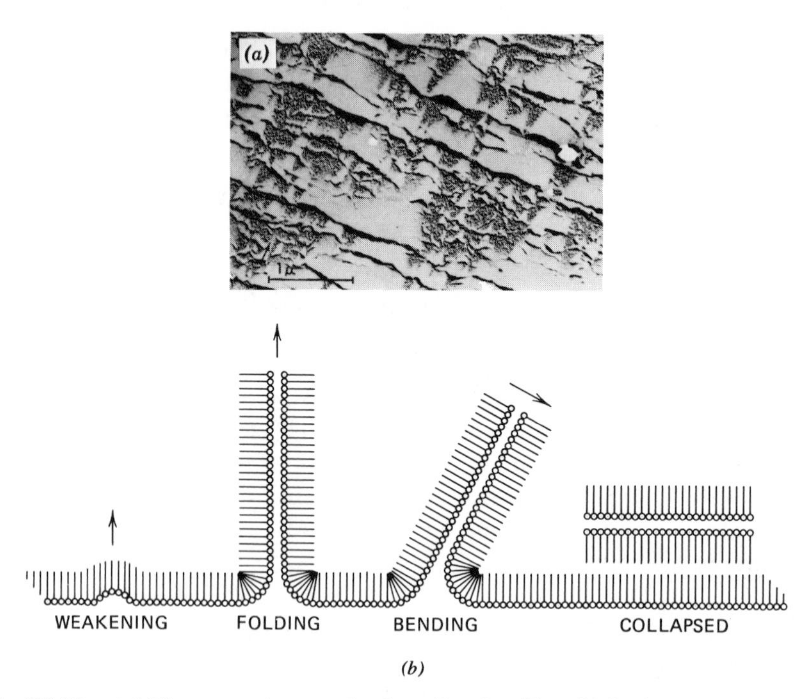

Fig. IV-22. (*a*) Electron micrograph of a collapsing film of 2-hydroxytetraconsanoic acid. Scale bar: 1 μm. (*b*) Mechanism of monolayer collapse. (From Ref. 95b)

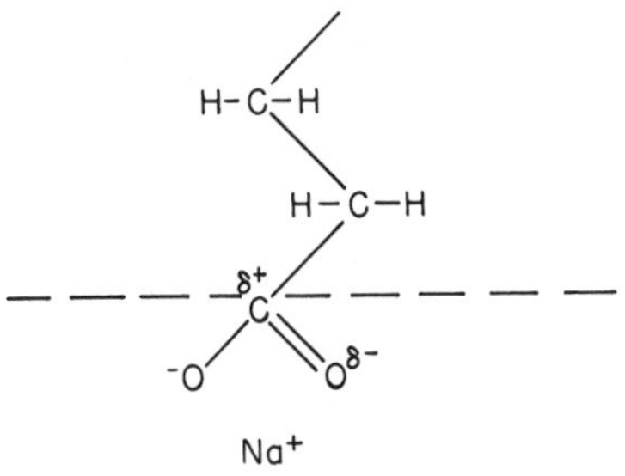

H-C-H

H-C-H

$\delta+$ C

$^{-}$O $\quad$ O$^{\delta-}$

Na^{+}

Fig. IV-23

potential drops since, as illustrated in Fig. IV-23, the presence of nearby counterions introduces a dipole opposite in orientation to that previously present. A similar situation is found with long chain amines on acid substrates (120). The effect is more than just a matter of pH. As shown in Fig. IV-24, stearic acid monolayers are highly expanded on 0.01*M* tetramethyl ammonium hydroxide but hardly at all on 0.01*M* LiOH, with KOH and NaOH having intermediate effects. Clearly, the nature of the counterion is very important. (See also Ref. 120a.) It is dramatically so if there is a tendency to form an insoluble salt with the film ion. Thus the presence of quite low concentrations (10^{-4} *M*) of divalent ions leads to formation of the metal soap of a fatty acid film, unless the pH is quite low. Such films are much more condensed than are the fatty acid monolayers themselves (121, 122; see also Ref. 123). See also Section IV-13 for a discussion of charged monolayers.

Monolayers of $\mathrm{Ru}L_2L'^{2+}$, L being bipyridine and L' being

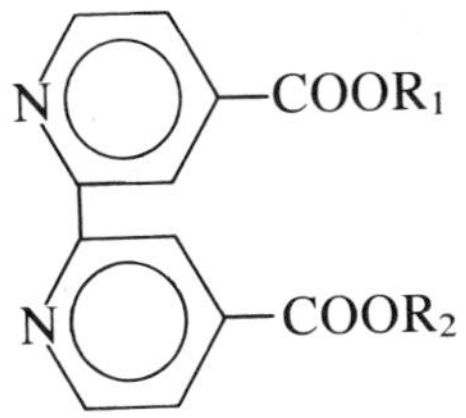

N —COOR$_1$

N —COOR$_2$

where R_1 and R_2 are various straight-chain hydrocarbons, show both π–σ and *emission* characteristics that vary markedly with the nature of the anion present (123a).

G. Rheology of Monolayers

It is not practical to do justice here to the mass of data and of theoretical analysis now extant; the reader is referred to recent monographs by Joly (125) and Barnes (125a). Some illustrative data are shown in Fig. IV-25. Notice the 100-fold increase on passing from the liquid expanded to the condensed state. A complication is that the viscosities are strongly shear rate dependent, in the case of the alcohols [although not in the case of the

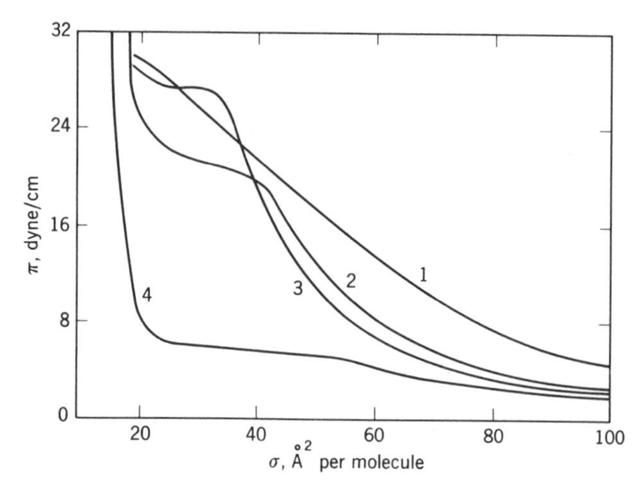

Fig. IV-24. π–σ isotherms of stearic acid at 26°C on subsolutions 0.01 *M* in XOH and 0.09 *M* in XCl. Curve 1: X = tetramethyl ammonium ion; curve 2: $X = Na^+$; curve 3: $X = K^+$; curve 4: $X = Li^+$. (From Ref. 124.)

long chain fatty acids (126)]. Surface viscosities are temperature dependent; apparent activation energies may range from 10 to 15 kcal/mole.

The viscosities of long chain aliphatic compounds are quite sensitive to the nature of the polar group; those for the fatty acids tend to be smaller than those for the corresponding alcohols, which in turn may be smaller than for the amines. The pH of the substrate is important. Long chain amines are ionized if spread on an acidic substrate, the films are expanded and much reduced in viscosity.

A large number of studies have been reported for films of polymeric and of biological materials. Surface viscosities of phospholipid monolayers vary dramatically with small changes in surface area, for example (128; see also 108). Films of polymeric materials may show quite viscoelastic behavior, resembling three-dimensional gels.

The general subject of surface rheology is somewhat in a "fluid" state. It is still not fully resolved how to separate the contribution of substrate drag from that of the film itself. It is very interesting that surface viscosities are about 10^6-fold larger than expected from the bulk properties of the film materials (note Section IV-3C). The implication is that many layers of substrate molecules are involved in an interlocking structure with the film-forming material.

H. General Correlations between Molecular Structure and the Type of Film Formed

The straight chain hydrocarbon derivatives, such as alcohols and acids, can be made to exhibit the various monolayer states described in Section

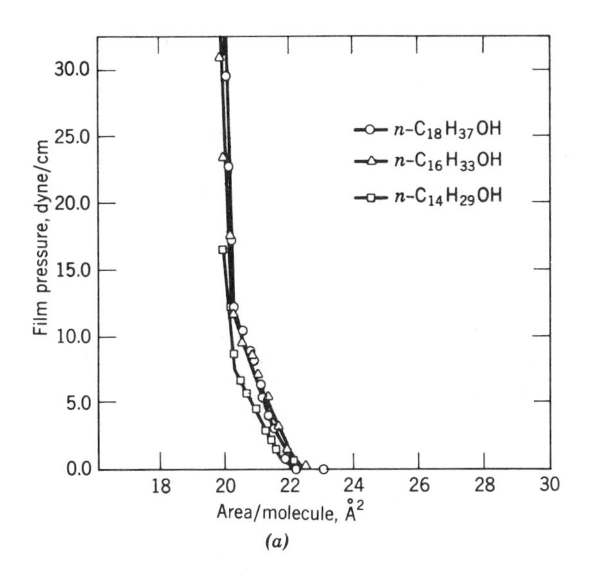

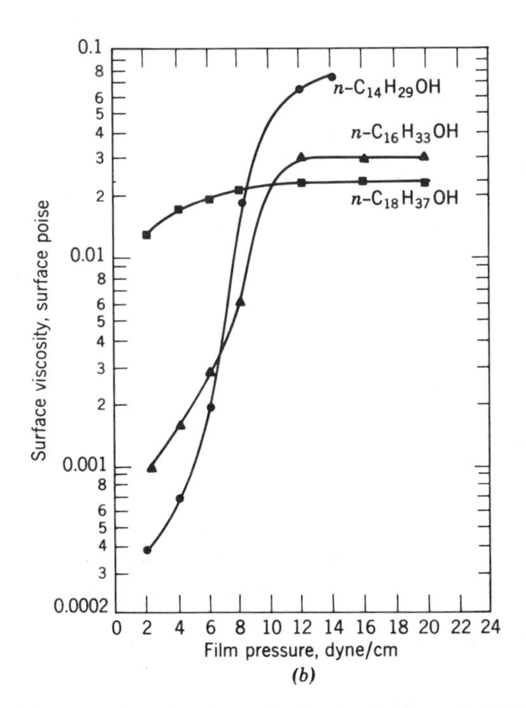

Fig. IV-25. (*a*) π–σ plots for long chain alcohols on 0.01 *N* H_2SO_4 at 20°C. (*b*) Corresponding surface viscosities at the relatively low film flow rate of 0.02 cm^2/sec. Reprinted with permission from Ref. 127. Copyright by the American Chemical Society.

IV-4 by a suitable choice of chain length and temperature. Solid or L_2-type films result with large chain lengths or low temperatures, and I- or L_1- or G-type films result with small chain lengths or high temperatures. As previously remarked, one CH_2 group is roughly equivalent to a 5° change in temperature. As discussed, ionized films have lower expansion temperatures, that is, temperatures below which S-type or L_2-type films result and above which expanded films result, and slightly dissociated salts with polyvalent ions show higher expansion temperatures.

With large, bulky end groups, one tends to get L_2-type rather than S-type films in that the extrapolated areas and the compressibilities are higher than for the standard S type. There may be more than one polar group in the molecule as is the case with oleic acid and other unsaturated compounds, hydroxy acids, lactones, and so on. If the secondary polar group is close to the primary or terminal one, the film may behave as a *monopolar* one, the intervening methylene groups remaining in or lying below the substrate. If the polar centers are well separated, however, as in 9- or 16-hydroxyhexadecanoic acid, the film becomes *bipolar* in type, an inflection in the π–σ isotherms marking the pressure at which the secondary polar group is forced out of the interface (129). Esters give more expanded films than the corresponding acid (see Ref. 130 for an exception).

Alternatively, the molecule may contain more than one hydrocarbon chain, as with esters and glycerides such as tristearin and pentaerythritol tetra esters. These behave somewhat similarly to the acids, giving either condensed or expanded films, depending on chain lengths and temperature. The importance of the *nature* of the hydrocarbon portion is well illustrated by the observation that the straight chain brassidic acid (*trans*-12-docosenoic acid) gives a condensed film whereas the bent chain erucic acid (*cis*-docosenoic acid) gives a very expanded film (131). Films of cholesterol and related compounds may differ noticeably in their behavior according to substituents present and the isomeric configuration (132). Protein films are discussed in Section IV-11.

Monolayers of porphyrin esters tend to be rigid, with the porphyrin planes oriented vertically at an aqueous interface. The absorption spectra are considerably shifted from those in solution (133).

The behavior of monolayers of progressively fluorinated fatty acids has been studied by Bernett and Zisman (134). Although containing 17 to 23 carbon atoms, the films were of the L_2 type rather than S type, presumably because of packing difficulties involving the bulky fluorinated portions. While ΔV values are positive for fatty acids, they were negative and quite large (up to 0.9 V) for the fluorinated ones, explainable in terms of the large negative-outward dipoles that should be present.

As a somewhat specialized topic, thin films and lenses of liquid crystalline materials present ordered structures visible under polarized light (135). It appears that *line tensions* may be important, that is, the tension at the substrate–vapor–liquid crystal junction.

7. Mixed Films

The study of mixed films has become of considerable interest. From the theoretical side, there are pleasing extensions of the various models for single component films; and from the more empirical side, one moves closer to modeling biological membranes. Following Gershfeld (108), we categorize systems as follows:

1. Both components form insoluble monolayers.
   (*a*) Equilibrium between mixed condensed and mixed vapor phases can be observed.
   (*b*) Only condensed phases are observed.
2. One component forms an insoluble monolayer while the other is soluble. Historically, the phenomenon is known as *penetration*.
3. Both components are soluble.

Category 3 was covered in Chapter III, and category 2 is treated later in this section.

Condensed phases of systems of category 1 may exhibit essentially ideal solution behavior, very nonideal behavior, or nearly complete immiscility. It is convenient at this point to define

$$A_{\text{av}} = N_1 A_1 + N_2 A_2 \tag{IV-45}$$

where A_1 and A_2 are the molar areas at a given π for the pure components. An *excess* area, A_{ex} is then given by

$$A_{\text{ex}} = A - A_{\text{av}} \tag{IV-46}$$

If an ideal solution is formed, then the actual molar area A is just A_{av} (and $A_{\text{ex}} = 0$). Unfortunately, the *same* result obtains if the components are completely immiscible! The distinction can be made, however, if category 1*a* applies, that is, if the equilibrium between condensed film and gaseous film can be observed, so that surface vapor pressures can be measured. The system myristic acid–oleic acid, shown in Fig. IV-26*a*, forms essentially ideal solutions in the condensed L_1 phase.

We may also define a free energy of mixing (136). The alternative (and equally acceptable) definition of G^σ given in Eq. III-91 is:

$$G^{*\sigma} = E^\sigma - TS^\sigma - \gamma\mathcal{A} \tag{IV-47}$$

Differentiation and combination with Eq. III-73 yields

$$dG^{*\sigma} = -S^\sigma\, dT + \sum \mu_i\, dn_i - \mathcal{A}\, d\gamma \tag{IV-48}$$

At constant temperature and mole numbers,

$$dG^{*\sigma} = -\mathcal{A}\, d\gamma \tag{IV-49}$$

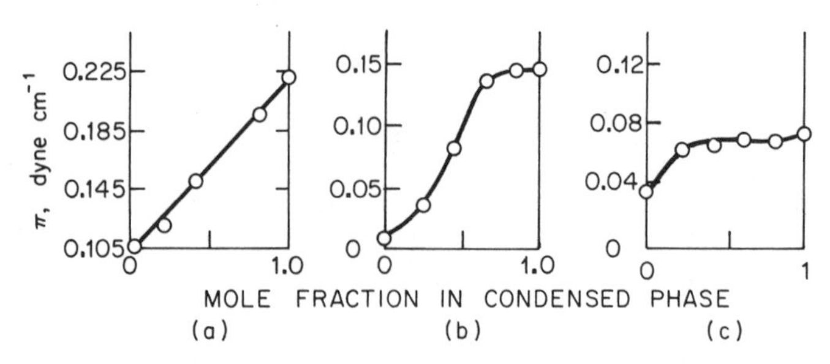

Fig. IV-26. Plots of surface vapor pressure versus condensed phase composition. (*a*) Myristic acid + oleic acid, 27.5°C, pH 2.0 substrate. (*b*) Palmitoleic acid + cholesterol, 27.5°C, pH 2.0 substrate. (*c*) Oleic acid + palmitic acid, 15°C, pH 2.0 substrate. (From Ref. 136).

Consider the mixing process

$$[\mu_1 \text{ moles of film (1) at } \pi] + [\mu_2 \text{ moles of film (2) at } \pi] = (\text{mixed film at } \pi)$$

First, the films separately are allowed to expand to some low pressure, π^*, and by Eq. IV-49 the free energy change is

$$\Delta G^{*\sigma}_{1,2} = -N \int_{\pi^*}^{\pi} A_1 \, d\pi - N_2 \int_{\pi^*}^{\pi} A_2 \, d\pi$$

The pressure π^* is sufficiently low that the films behave ideally, so that on mixing

$$\Delta G^{*\sigma}_{\text{mix}} = RT(N_1 \ln N_1 + N_2 \ln N_2)$$

The mixed film is now compressed back to π:

$$\Delta G^{*\sigma}_{12} = \int_{\pi^*}^{\pi} A_{12} \, d\pi$$

$\Delta G^{*\sigma}$ for the overall process is then

$$\Delta G^{*\sigma} = \int_{\pi^*}^{\pi} (A_{12} - N_1 A_1 - N_2 A_2) \, d\pi + RT(N_1 \ln N_1 + N_2 \ln N_2) \quad \text{(IV-50)}$$

and the excess free energy of mixing is thus

$$\Delta G^{*\sigma}_{\text{ex}} = \int_{\pi^*}^{\pi} (A_{12} - N_1 A_1 - N_2 A_2) \, d\pi \quad \text{(IV-51)}$$

A plot of G^*_{ex} versus composition is shown in Fig. IV-27 for condensed films of octadecanol + docosyl sulfate. Gaines (138) and Cadenhead and Demchak (139) have extended the above approach and the subject has been extended and reviewed by Barnes and co-workers (see Ref. 140).

For additional studies on mixed films see Ries and Swift (141), Cadenhead and Müller-Landau (143), and Tajima and Gershfeld (144). A case of immiscibility at low π is discussed by Cadenhead and co-workers (145). Mixed films of a phospholipid and a polysoap have been treated by Ter-Minassian-Sarago in terms of adsorption on a linear adsorbent (146). Motomura and co-workers (see Ref. 147) have developed a related thermodynamic approach in their treatment of mixed films. Hendrikx (148) reports on the three component system of anionic soap + cationic soap + cetyl alcohol. Shah and Shiao (149) discuss chain length compatibility of long chain alcohols.

In the alternative surface phase approach, Eq. IV-39 may be expanded for mixed films to give (150):

$$\pi = \frac{RT}{A_1}\left\{ \ln\left[1 + \frac{A_1(\Gamma_2 + \Gamma_3)}{1 - A_2\Gamma_2 - A_3\Gamma_3}\right] - \ln f_1^s \right\} \tag{IV-52}$$

Category 2 mixed films, or those formed by penetration, have also been of some interest. Here, a more or less surface active consituent of the substrate enters into a spread monolayer, in some cases to the point of diluting it extensively. Thus monolayers of long chain amines and of sterols are considerably expanded if the substrate contains dissolved low molecular weight acids or alcohols, such as acetic acid. Rideal (151) has summarized much of the earlier work on penetration phenomena. A great deal of work has been carried out, for example, on the penetration of sodium cetyl sulfate and similar detergent species into films of biological materials. Goddard and Schulman (152) consider that sodium cetyl sulfate forms 1:1 complexes with films of digitonin and eicosylamine. Pethica and co-workers (153) and Fowkes (154) studied the penetration of cetyl alcohol films by sodium do-

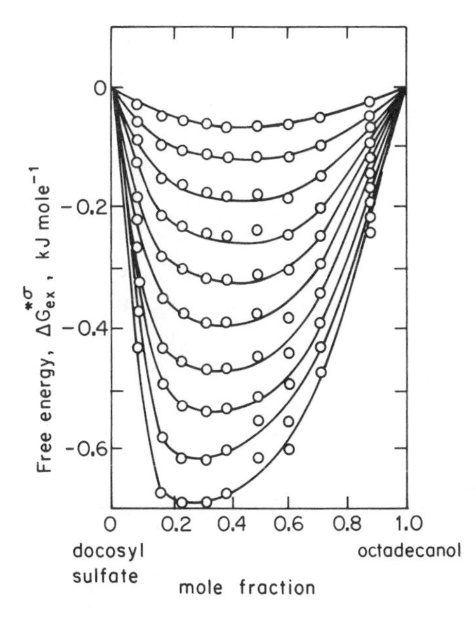

Fig. IV-27. Excess free energy of mixing of condensed films of octadecanol–docosyl sulfate at 25°C, at various film pressures. Top curve: π = 5 dyne/cm; bottom curve: π = 50 dyne/cm; intermediate curves at 5 dyne/cm intervals. The curves are uncorrected for the mixing term at low film pressure. (From Ref. 142.)

decyl sulfate, and Fowkes (155), mixed monolayers of cetyl alcohol and sodium cetyl sulfate, using an aqueous sodium chloride substrate to reduce the solubility of the detergent. Lucassen-Reynders has applied equations of the type of Eq. IV-52 to systems such as sodium laurate–lauric acid (156, 157).

In actual practice the soluble component usually is injected into the substrate solution *after* the insoluble monolayer has been spread. The reason is that if one starts with the solution, the surface tension may be low enough that the monolayer will not spread easily. McGregor and Barnes have described a useful injection technique (158).

A difficulty in the physical chemical study of penetration is that the amount of soluble component present in the monolayer is not an easily accessible quantity. It may be measured directly, through the use of radioactive labeling (Section III-6) (157, 159), but the technique has so far only been used to a limited extent.

Two alternative means around the difficulty have been used. One, due to Pethica (160) [but see also Alexander and Barnes (161)], is as follows. The Gibbs equation, Eq. III-77, becomes for a three-component system at constant temperature and locating the dividing surface so that Γ_1 is zero:

$$d\pi = RT\Gamma_f^1 \, d \ln a_f + RT\Gamma_s^1 \, d \ln a_s \tag{IV-53}$$

where subscripts f and s denote insoluble monolayer and surfactant, respectively, and a is the rational activity. We can eliminate the experimentally inaccessible $\ln a_f$ quantity from Eq. IV-53 as follows. The definition of partial molal area is $(\partial \mathcal{A}/\partial n_i)_{T,n_j} = \bar{A}_i$ and we obtain from Eq. IV-49 (by differentiating with respect to n_f):

$$RT\left(\frac{\partial \ln a_f}{\partial \pi}\right)_{T,n_j} = \bar{A}_f \tag{IV-54}$$

Combination with Eq. IV-53 gives

$$d\pi = \left(\frac{A_f}{A_f - \bar{A}_f}\right) RT\,\Gamma_s^1 \, d \ln m_s \tag{IV-55}$$

where $A_f = 1/\Gamma_f^1$ and, since the surfactant solution is usually dilute, a_s is approximated by the molality m_s. (A factor of 2 may appear in Eq. IV-55 if the surfactant is ionic and no excess electrolyte is present—see Section III-6B.) Pethica assumed that at a given film pressure $\bar{A}_f$ would be the same in the mixed film as in the pure film. A_f is just the experimental area per mole of film material in the mixed film. The coefficient $(\partial\pi/\partial \ln m_s)_{T,n_f}$ could be obtained from the experimental data, and Γ_s^1 follows from Eq. IV-55. Figure IV-28 shows his results for the penetration of cholesterol monolayers by sodium dodecyl sulfate.

Alexander and Barnes (161) have pointed out an approximation in the

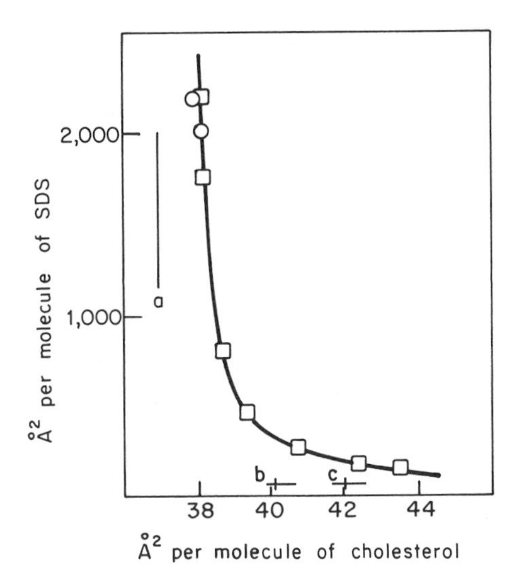

Fig. IV-28. Penetration of cholesterol monolayers by sodium dodecyl sulfate. Film pressure is 30 dyne/cm; 20°C. No added electrolyte, ⊡ 0.145 *M* NaCl. Markers *a*, *b*, and *c* give the molecular areas for cholesterol alone, and sodium dodecyl sulfate alone with and without added electrolyte, respectively (all for π = 30 dyne/cm). (From Ref. 160.)

preceding thermodynamics that renders the procedure suspect except under certain conditions. An alternative means of estimating Γ_s^1 makes the assumption that penetrant molecules have accessible that area not actually occupied by film molecules, that is, the area $a = \mathcal{A} - n_f A_f^0$, and to write

$$\Gamma_s^1 = \frac{\mathcal{A} - n_f A_f^0}{\mathcal{A}} \Gamma_{s,w}^1 = \frac{A_f - A_f^0}{A_f} \Gamma_s^1 \qquad \text{(IV-56)}$$

where A_f^0 is the actual area of a mole of film molecules, and $\Gamma_{s,w}^1$ is the surface excess of surfactant at pressure π but in the absence of film material. In effect, a geometric additivity is assumed. See Ref. 162.

Studies have been made on the rheology of mixed films. For completely immiscible films one expects the reciprocal of viscosity, that is, the fluidity, to be averaged, and probably likewise for ideal mixtures. The subject has been reviewed by Joly (125).

Still another manifestation of mixed film formation is the absorption of organic vapors by films. Stearic acid monolayers strongly absorb hexane up to a limiting ratio of 1 : 1 (163), and data reminiscent of adsorption isotherms for gases on solids are obtained, with the surface density of the monolayer constituting an added variable.

8. Evaporation Rates through Monomolecular Films

An interesting consequence of covering a surface with a film is that the rate of evaporation of the substrate is reduced. Most of these studies have been carried out with films spread on aqueous substrates; in such cases the activity of the water is practically unaffected because of the low solubility of the film material, and it is only the rate of evaporation and not the equilibrium vapor pressure that is affected.

One procedure makes use of a box on whose silk screen bottom powdered desiccant has been placed, usually lithium chloride. The box is positioned 1 to 2 mm above the surface and the rate of gain in weight is measured for the film-free and the film-covered surface. The rate of water uptake is reported as $v = m/t\mathcal{A}$, or in g/sec $-$ cm^2. This is taken to be proportional to $(C_w - C_d)/R$ where C_w and C_d are the concentrations of water vapor in equilibrium with water and with the desiccant, respectively, and R is the diffusional resistance across the gap between the surface and the screen. Qualitatively, R can be regarded as actually being the sum of a series of resistances corresponding to the various diffusion gradients present:

$$R_{\text{total}} = R_{\text{surface}} + R_{\text{film}} + R_{\text{desiccant}} = R_0 + R_{\text{film}} \qquad \text{(IV-57)}$$

Here R_0 represents the resistance found with no film present. We can write:

$$r = (C_w - C_d)\left(\frac{1}{v_f} - \frac{1}{v_w}\right) \simeq C_w\left(\frac{1}{v_f} - \frac{1}{v_w}\right) \qquad \text{(IV-58)}$$

where r is the specific evaporation resistance (sec/cm), and the subscripts f and w refer to the surface with and without film, respectively.

Some fairly typical results, obtained by LaMer and co-workers (164) are shown in Fig. IV-29. At the higher film pressures, the reduction in evapo-

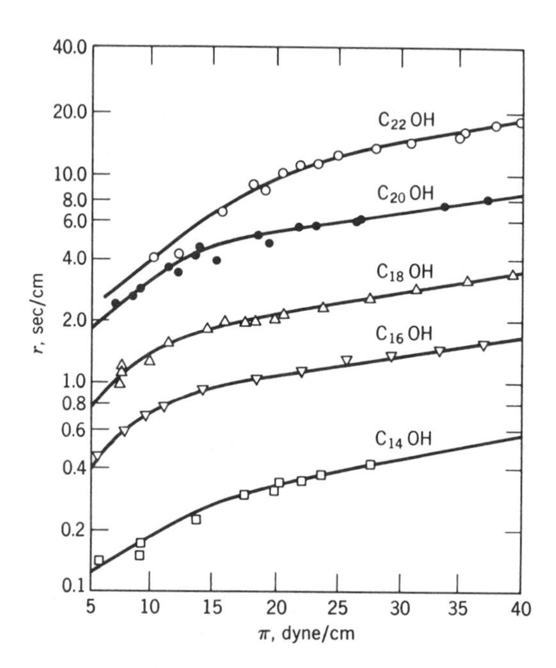

Fig. IV-29. Effect of alkyl chain length of n-alcohols on the resistance of water evaporation at 25°C. (From Ref. 164.)

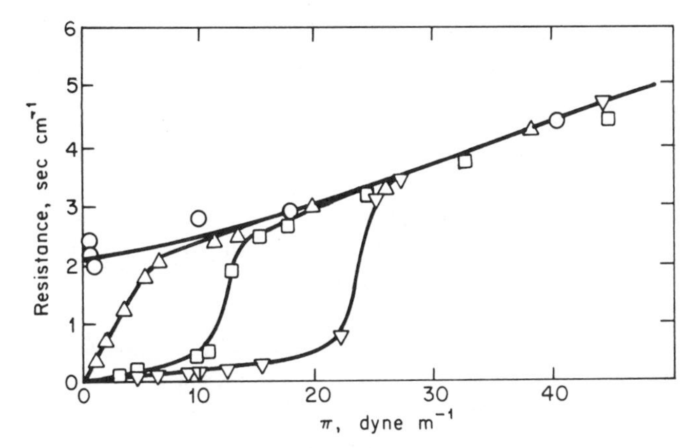

Fig. IV-30. Evaporation resistance of octadecanol films on water: □, original commercial sample; ▽, "purified" by preparative gas–liquid chromatography; △, purified by column chromatography; ○, purified by urea clathration. (From Ref. 176.)

ration rate may be 60 to 90%—a very substantial effect. Similar results have been reported for the various fatty acids and their esters (165, 166). Barnes and co-workers have reported on mixed monolayer systems (167, 168). The "accessible" area, $a = \mathcal{A} - n_f A_f^0$ (see Eq. IV-56) is a good correlating parameter, suggesting that $v_f = \frac{a}{a} v_w$ or

$$r = \left(\frac{\mathcal{A} C_w}{v_w}\right)\left(\frac{\mathcal{A}}{a} - 1\right) \tag{IV-59}$$

Qualitatively, then, it is more significant to compare systems at constant A (or Γ) than at constant π, as has been traditional.

The temperature dependence of r gives an apparent activation energy, and the increment per CH_2 group is about 200 cal/mole in the case of the long chain alcohols, but it is pressure dependent and in a way that suggests that the energy requirement is one of forming a hole in a close-packed monolayer (169). Barnes and co-workers (see Ref. 170) have treated evaporation resistance in terms of a steady state kinetic model which pictures a water molecule as jumping from cavity to cavity along a hydrocarbon chain. A potentially serious problem to quantitative analysis is that evaporation cools the water layers immediately below the interface, the cooling being less if there is retardation (171, 172). There is also a problem in determining true rather than net evaporation rates, that is, the evaporation rate into vaccum. Ideally this is given by Eq. III-24, but with the assumption that molecules hitting the surface from the gas phase stick with unitary efficiency or, alternatively, that the *evaporation coefficient* α is unity (173, 174). Cammenga (175) has reviewed various aspects of the evaporation mechanism of liquids.

Finally, there have been problems in obtaining reproducible values for r (176). These have been traced to impurities either in the monolayer material or in the spreading solvent used, and the effect can be serious, as illustrated in Fig. IV-30.

Note that quite elaborate purification is needed. Also, the effect is most important at low π; at high compression impurities are evidently forced out of the interface.

The importance of these investigations on evaporation retardation is fairly obvious. LaMer (169) estimated that 16 million acre-feet of water are lost by evaporation annually from the western United States impoundment reservoirs alone. The first attempts to use monolayers to reduce reservoir evaporation were made by Mansfield (177) in Australia, and since then moderately successful tests have been made in a number of locations.

In all of these tests, cetyl alcohol was used as a commercially accessible surfactant that also offered a good compromise between specific resistance and rate of spreading. High spreading rates are extremely important; not only must the film form easily on application but also it must be able to do so against wind friction and be able quickly to heal ruptures caused by waves or boat wakes. Other concerns, of course, are that the film-forming material not be rapidly biodegraded and that it not interfere with aquatic life, for example, through prevention of adequate aeration of the water.

9. Rate of Dissolving of Monolayers

The rate of dissolving of monolayers constitutes an interesting and often practically important topic. It affects, for example, the rate of loss of monolayer material used in evaporation control. From the physical chemical viewpoint, the topic represents a probe into the question of whether insoluble monolayers are in equilibrium with the underlying bulk solution. Film dissolution also represents the reverse of the process that gives rise to the slow aging or establishment of equilibrium surface tension in the cases of surfactant solutions discussed in Chapter II (see Fig. II-18).

The usual situation appears to be that after a short initial period the system obeys the equation

$$\mathcal{A} = \mathcal{A}^0 e^{-kt} \tag{IV-60}$$

if the film is kept at a constant film pressure. The differential form of Eq. IV-60 may be written as

$$\left(\frac{dn}{dt}\right)_\pi = -kn \tag{IV-61}$$

where n is the number of moles of film present. As discussed by Ter Minassian-Saraga (178), this form can be accounted for in terms of a steady state diffusion process. As illustrated in Fig. IV-31, we suppose the film to be in equilibrium with the immediately underlying solution, to give a concentration C_f. The rate limiting process is taken to be the rate of diffusion across a thin stagnant layer of solution, of thickness δ. According to Fick's law, and remembering that $n = \mathcal{A}\Gamma$

$$\frac{dn}{dt} = -\mathcal{A}\mathcal{D}\frac{dC}{dx} = -\left(\frac{n}{\Gamma}\right)\left(\frac{\mathcal{D}}{\delta}\right)(C_f - C) \tag{IV-62}$$

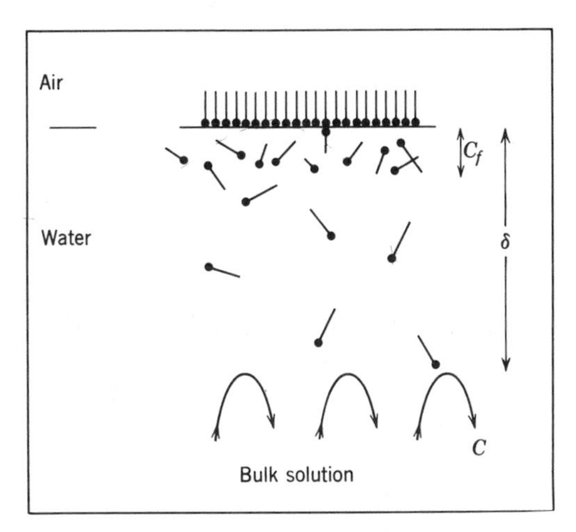

Fig. IV-31. Steady state diffusion model for film dissolution. (From Ref. 179.)

Here $\mathscr{D}$ is the diffusion coefficient and C is the concentration in the general bulk solution. For initial rates C can be neglected in comparison to C_f so that from Eqs. IV-61 and IV-62 we have

$$k = \frac{\mathscr{D}C_f}{\delta\Gamma} = \frac{\mathscr{D}}{\delta K} \tag{IV-63}$$

where

$$K = \frac{\Gamma}{C_f} \tag{IV-64}$$

Allowance may be made for the variation of the activity coefficient of the film material and, on doing so, Gershfeld and Patlak (105a) obtained Eq. IV-65, which was obeyed well in the cases of lauric acid and cetyl alcohol monolayers.

$$\frac{d \ln k}{d\pi} = \frac{\mathscr{A}}{RT} - \kappa \tag{IV-65}$$

The diffusion mechanism gives the upper limit to the rate of dissolving of a monolayer. There may be a chemical activation barrier to the transfer of material from film to underlying solution, and while Eq. IV-61 would still give the rate law, the value of k would now be smaller than that predicted by the diffusion mechanism. Hansen (180) has treated the case of coupled chemical and diffusional processes.

It is fortunate for the study of monolayers that dissolving processes are generally slow enough to permit the relatively unperturbed study of equilibrium film properties, because many films are inherently unstable in this respect. Gaines (1) notes that the equilibrium solubility of stearic acid is 3 mg/l, and that films containing only a tenth of the amount that should dissolve in the substrate can be studied with no evidence of solution occurring. Similarly, there appears to be a considerable barrier to the dissolving of protein films. However, dissolving can often be a problem, especially with lower molecular weight or with charged monolayers.

10. Reactions in Monomolecular Films

The study of reactions in monomolecular films is rather interesting. Not only can many of the usual types of chemical reactions be studied (subject to the limitation, of course, that one of the species forms a monolayer), but also there is the special feature of being able to control the orientation of molecules in space by varying the film pressure. Furthermore, a number of processes that occur in films are of special interest because of their resemblance to biological systems. An early review is that of Davies (181).

A. Kinetics of Reactions in Films

In general it is not convenient and sometimes not possible to follow reactions in films by the same types of measurements as employed in bulk systems. It is awkward and inconvenient to try to make chemical analyses to determine the course of a reaction and, even if one of the reactants is in solution in the substrate, the amounts involved are rather small (a micromole at best). Such analyses are greatly facilitated, of course, if radioactivity labeling is used. Also, film collapsed and collected off the surface has been analyzed by its infrared spectrum (182). *In situ* measurements can be made if a labeled fragment enters or leaves the interface as part of the reaction, or if the film material has a strong absorption spectrum that is altered by the reaction, as in the case of chlorophyll (183). Also, analysis for reactant and reaction products has been possible by means of high performance liquid chromatography (184).

The most common situation studied is that of a film reacting with some species in solution in the substrate, such as in the case of the hydrolysis of ester monolayers and of the oxidation of an unsaturated long chain acid by aqueous permanganate. As a result of the reaction, the film species may be altered to the extent that its area per molecule is different, or it may be fragmented so that the products are soluble. One may thus follow the change in area at constant film pressure, or the change in film pressure at constant area (much as with homogeneous gas reactions); in either case concomitant measurements may be made of the surface potential.

Case 1. A chemical reaction occurs at constant film pressure, where the product is soluble. Here $\mathcal{A}^\infty = 0$, and

$$\frac{n_A}{n_A^0} = \frac{\mathcal{A}}{\mathcal{A}^0} \tag{IV-66}$$

In the usual circumstance of a reaction with a species in the substrate, whose concentration remains essentially constant, a first-order rate law will apply:

$$n_A = n_A^0 e^{-kt} \tag{IV-67}$$

Therefore,

$$\mathcal{A} = \mathcal{A}^0 e^{-kt} \tag{IV-68}$$

Case 2. A chemical reaction occurs at constant film pressure where the product(s) are insoluble and remain in monolayer form. Here it is usually assumed that area is an additive property, that is, that the total area will be given by the sum of the areas that would be occupied by the reactant and products separately. Where the change due to the chemical reaction is not too great, this assumption may be satisfactory, although the discussion on mixed monolayers in Section IV-7 indicates that serious errors may be involved. On the basis of this assumption, however,

$$\frac{n_A}{n_A^0} = \frac{\mathcal{A} - \mathcal{A}^\infty}{\mathcal{A}^0 - \mathcal{A}^\infty} \tag{IV-69}$$

Again, if the first step is rate determining and governed by a constant concentration of substrate reactant, a first-order rate law will be observed:

$$\frac{\mathcal{A} - \mathcal{A}^\infty}{\mathcal{A}^0 - \mathcal{A}^\infty} = e^{-kt} \tag{IV-70}$$

Case 3. A chemical reaction occurs at constant total area, with π varying. In a formal way, an equation similar to Eq. IV-70 can be written

$$\frac{\pi - \pi^\infty}{\pi^0 - \pi^\infty} = e^{-kt} \tag{IV-71}$$

However, this implies that film pressure is an additive function of composition at constant area. Such a restriction would be valid for a mixture of ideal two-dimensional gases or for mixed films obeying Eq. IV-44 with a constant and b linear in composition, but the conditions are rather special.

Case 4. The surface potential is measured as a function of time. Here, since by Eq. IV-19 $\Delta V = 4\pi \mathbf{n}\bar{\mu}/D$, then

$$\mathcal{A}\,\Delta V = (4\pi) n_A \bar{\mu}_A = \alpha_A n_A \tag{IV-72}$$

Because mole numbers are additive, it follows that the product $\mathcal{A}\,\Delta V$ will be an additive quantity provided that α for each species remains constant during the course of the reaction. This last condition implies, essentially, that the effective dipole moments and hence the orientation of each species remain constant, which is most likely to be the case at constant film pressure. Then

$$\frac{\mathcal{A}\,\Delta V - \mathcal{A}^\infty\,\Delta V^\infty}{\mathcal{A}^\circ\,\Delta V^\circ - \mathcal{A}^\infty\,\Delta V^\infty} = e^{-kt} \tag{IV-73}$$

if, again the reaction is first order.

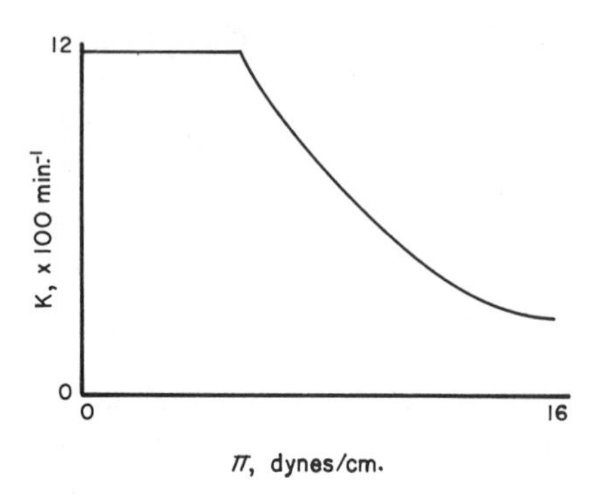

Fig. IV-32. Rate of lactonization of γ-hydroxystearic acid as a function of film pressure.

B. Kinetics of Formation and Hydrolysis of Esters

An example of an alkaline hydrolysis is that of the saponification of monolayers of α-monostearin (185); the resulting glycerine dissolved while the stearic acid anion remained a mixed film with the reactant. Equation IV-70 was obeyed, with $k = k'(\mathrm{OH}^-)$ and showing an apparent activation energy of 10.8 kcal/mole. Davies (186) studied the reverse type of process, the lactonization of γ-hydroxystearic acid (on acid substrates). Separate tests showed that ΔV for mixed films of lactone and acid was a linear function of composition at constant π, which allowed a modification of Eq. IV-73 to be used. The pseudo first-order rate constant was proportional to the hydrogen ion concentration and varied with film pressure, as shown in Fig. IV-32. This variation of k with π could be accounted for by supposing that the γ-hydroxystearic acid could assume various configurations, as illustrated in Fig. IV-33, each of which was weighted by a Boltzmann factor in the manner employed by Mittelmann and Palmer (Section IV-6B). The steric factor for the reaction was then computed as a function of π, assuming that the only configurations capable of reacting were those having the hydroxy group in the surface. Likewise, the change in activation energy with π was explainable in terms of the estimated temperature coefficient of the steric factors; thus

$$k = \phi z p e^{-E/RT} \tag{IV-74}$$

where ϕ is the variable steric factor, so that

$$-\frac{d \ln k}{d(1/T)} = \frac{E}{R} - \frac{d \ln \phi}{d(1/T)}$$

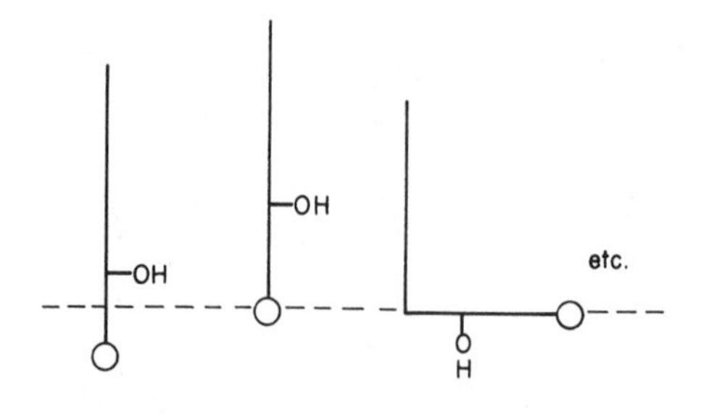

Fig. IV-33. Possible orientations for γ-hydroxystearic acid.

and

$$E_{\text{apparent}} = E + E_{\text{steric}} \tag{IV-75}$$

A more elaborate treatment of ester hydrolysis was attempted by Davies and Rideal (187) in the case of the alkaline hydrolysis of monolayers of monocetylsuccinate ions. The point in mind was that since the interface was charged, the local concentration of hydroxide ions would not be the same as in the bulk substrate. The surface region was treated as a bulk phase 10 Å thick and, using the Donnan equation, actual concentrations of ester and hydroxide ions were calculated, along with an estimate of their activity coefficients. Similarly, the Donnan effect of added sodium chloride on the hydrolysis rates was measured and compared with the theoretical estimate. The computed concentrations in the surface region were rather high (1–3 M), and, since the region is definitely not isotropic because of orientation effects, this type of approach would seem to be semiempirical in nature. On the other hand, there was quite evidently an electrostatic exclusion of hydroxide ions from the charged monocetylsuccinate film, which could be predicted approximately by the Donnan relationship. A more elegant approach is taken in Section IV-13, dealing with charged monolayers.

However, by way of example of the Donnan effect, at $\pi = 10$, the film occupied 55 Å^2 per molecule, and its nominal surface concentration was then 3.0 M. The Donnan condition arises from equating sodium hydroxide activities, or roughly $(Na^+)(OH^-)$, in the bulk and surface phases

$$(Na^+)_b(OH^-)_b = (Na^+)_s(OH^-)_s \tag{IV-76}$$

or

$$C^2 = (3.0 + x)(x) \tag{IV-77}$$

where C is the bulk sodium hydroxide concentration and x is its surface concentration (there is 3.0 M Na^+ associated with the monocetylsuccinate ion, assuming electroneutrality in the surface phase). Thus if $C = 0.1$, $x = 0.0033$, and if $C = 1.0$, $x = 0.3$, and a tenfold increase in C should produce an almost 100-fold increase in x and hence in reaction rate. This was essentially the effect observed.

A recent example of a two-stage hydrolysis is that of the sequence shown in Eq. IV-78 (note Section IV-6F also). The kinetics, illustrated in Fig. IV-34, is approximately that of successive first-order reactions, but complicated by the fact that the intermediate II is ionic (184).

$[(bpy)_2Ru(\text{bipyridine}-(COOR)_2)]^{2+} \xrightarrow[-ROH]{OH^-} [(bpy)_2Ru(\text{bipyridine}-(COOR)(COO^-))]^{1+} \xrightarrow[-ROH]{OH^-} [(bpy)_2Ru(\text{bipyridine}-(COO)_2)]^{0}$

I. $R = R = C_{18}H_{37}$ II. $R = C_{18}H_{37}$ III

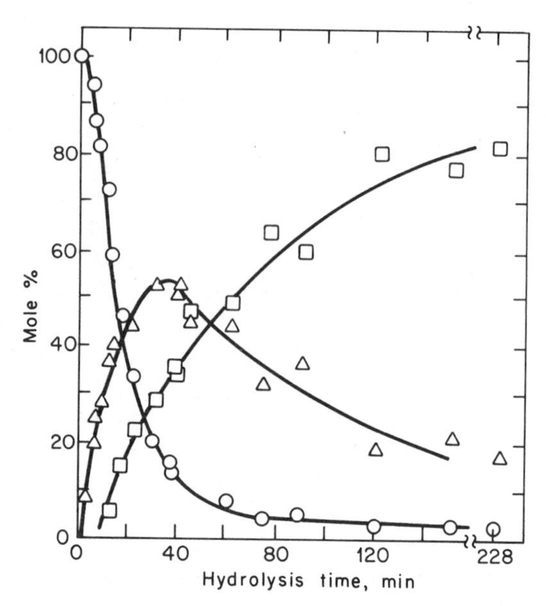

Fig. IV-34. Relative concentrations of reactant I, ○, and products II, △, and III, □ (Eq. IV-78) during the hydrolysis on 0.1 *M* $NaHCO_3/NaCl$ substrate at 10 dyne/cm film pressure and 22°C. (Reproduced with permission from Ref. 184.)

C. Other Chemical Reactions

Another type of reaction that has been studied is that of the oxidation of a double bond. In the case of triolein, Mittelmann and Palmer (109) found that, on a dilute permanganate substrate, the area at constant film pressure first increased and then decreased. The increase was attributed to the reaction:

$$—CH_2—CH{=}CH—CH_2— \longrightarrow —CH_2—CHOH—CH_2—$$

with a consequent greater degree of anchoring of the middle of the molecule in the surface region. The subsequent decrease in area seemed to be due to fragmentation into two relatively soluble species. The initial reaction obeyed Eq. IV-69 and the pseudo-first-order rate constant was proportional to the permanganate concentration. The rate constant also fell off with increasing film pressure, and this could be accounted for semiquantitatively by calculating the varying probability of the double bond being on the surface (see Section IV-6B for the general procedure).

Reactions in which a product remains in the film (as above) are complicated by the fact that the areas of reactant and product are not additive, that is, a nonideal mixed film is formed. Thus Gilby and Alexander (188), in some further studies of the oxidation of unsaturated acids on permanganate substrates, found that mixed films of unsaturated acid and dihydroxy acid (the immediate oxidation product) were indeed far from ideal. They were, however, able to fit their data for oleic and erucic acids fairly well by taking into account the separately determined departures from ideality in the mixed films.

Other examples include polymerization reactions, such as that of the stearic aldehyde (184), photochemical processes, and various biological reactions. Examples of photochemical reactions include various photochemical processes with monolayers of proteins (190), and the photodegradation as well as fluorescence quenching in chlorophyll monolayers (183, 191). In a very interesting recent study, Whitten (190) observed a substantial decrease in the area of mixed films of tripalmitin and a *cis*-thioindigo dye as isomerization to the *trans* occurred on irradiation with ultraviolet light. Submonolayer films of unsaturated fatty acids or their esters on silica gel undergo autoxidation at a reduced rate relative to homogeneous solution (and the oxidation is chemiluminescent) (192).

11. Films of Biological and Polymeric Materials

A. General Properties and Structure

The subject of protein and other polymer films is complex, both experimentally and theoretically. An experimental problem in the case of both natural and synthetic polypeptides is, that while they are generally soluble in the aqueous substrate used, the formation of an equilibrium surface film is very slow, and in practice it is necessary to use the spread monolayer technique. A solution of the protein may be allowed to flow slowly down a rod that touches the surface (193) or may be applied dropwise and very slowly from a syringe (194–196). Some protein appears inevitably to be lost by mixing into the bulk substrate, but the rest forms a film on which conventional π–σ measurements can be made. These isotherms are reasonably reversible provided the film is not compressed past about 20 dyne/cm; with too much compression irreversible formation of an insoluble curd may occur (see Ref. 197 for example).

Some data for gliadin and for egg albumin are shown in Fig. IV-35 and for various synthetic amino acid polymers, in Fig. IV-36. At very low film pressures two-dimensional gas law behavior is approached; Bull (198) found Eq. III-127 to represent his data very well. In the limit of ideal gas behavior, the protein molecular weight can be estimated from the equation

$$M = \frac{wRT}{\pi \mathcal{A}} \tag{IV-79}$$

where w is the mass of protein spread. For example, Gaustalla (199) estimated the molecular weight of egg albumin to be 40,000, in agreement with ultracentrifugation values. Beginning at around 1 m^2/mg or about 17 Å^2 per amino acid residue, π begins to rise rapidly, and collapse occurs at around 5 to 10 Å^2 per residue; the compressibility (see Eq. III-128) may actually be rising with increasing film pressure, however (195, 200). Surface potentials

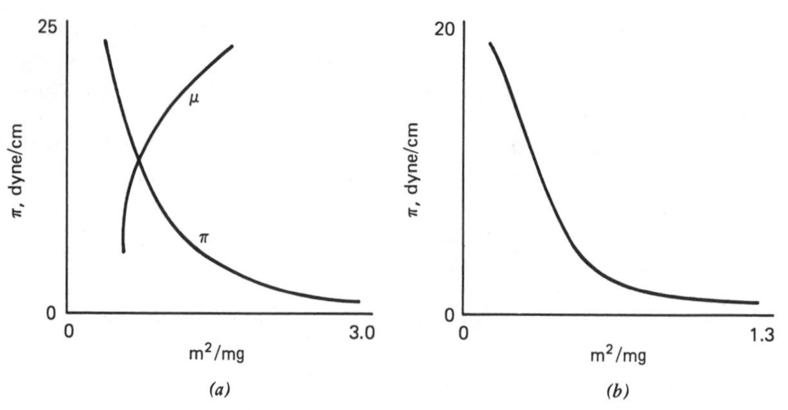

Fig. IV-35. Protein monolayers: (*a*) gliadin, pH 5.9; (*b*) egg albumin on 35% ammonium sulfate. (From Ref. 201.)

often do not vary much with film pressure (see Fig. IV-36), which means a rising vertical component of the film dipole moment with decreasing π (Fig. IV-35*a* and IV-36). Depending on the type of polypeptide, the isotherm may show a nearly flat region, indicating that some kind of phase transition occurs, as illustrated in Fig. IV-36.

The theoretical treatment is greatly complicated by uncertainties as to the structure of the films. At one point it was felt that spread protein films were largely unfolded to an extended linear polymer or β configuration (203), with the polar groups hydrogen bonded to the substrate water and the side chains projecting either up or down, depending on their nature (see Fig. IV-

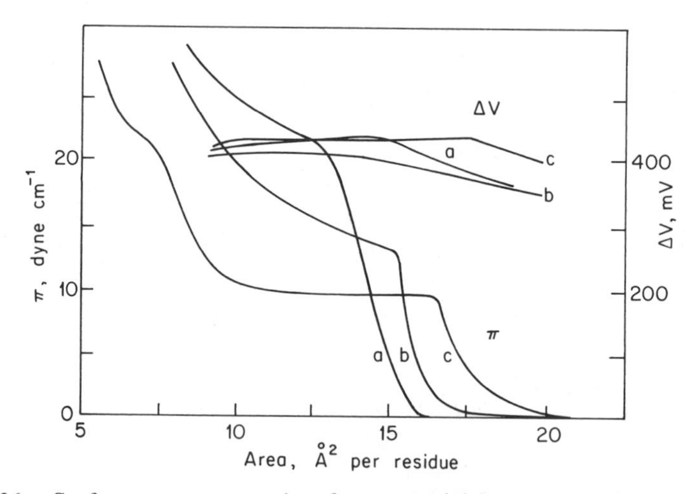

Fig. IV-36. Surface pressure π and surface potential ΔV versus area for (*a*) poly(L-leucine), (*b*) poly(L-methione), and (*c*) poly(L-norleucine) on distilled water. (From Ref. 202.)

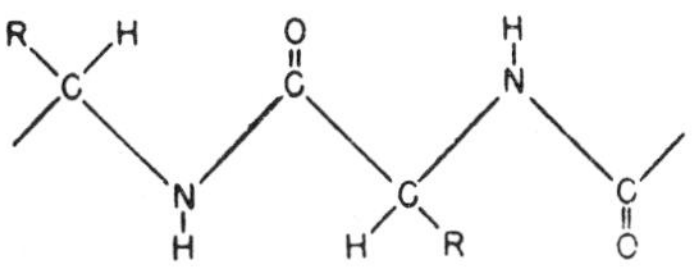

Fig. IV-37

37). Depending on the conditions of spreading, some of the original α-helix structure may also be present, as illustrated in Fig. IV-38. The presence of this last type of structure is indicated by certain characteristics of the infrared absorption spectra of collapsed films (204). Since it can be argued that the helical configuration might be reformed in the collapse process, an important additional piece of evidence is that the rate of deuterium exchange with substrate is much too slow to be consistent with an unfolded film structure (205). These observations, plus the finding of crystallinity in collapsed films (by electron diffraction) (205) led to the rather different picture shown in Fig. IV-39. The film is now viewed as consisting of intact α-helices with second (and further) layers forming on compression. See also Ref. 206.

These structural questions are not yet settled. A point to remember is that protein films cannot be in equilibrium with dissolved protein (note Problem 24) so that some barrier, presumably structural, must be present that prevents film molecules from dissolving on compression.

Films at very low surface pressures may well exist in the unfolded form, and Singer (207) has extended the Flory-Huggins treatment of polymer solutions to this situation. One assumes a chain of n links (amino acid residues

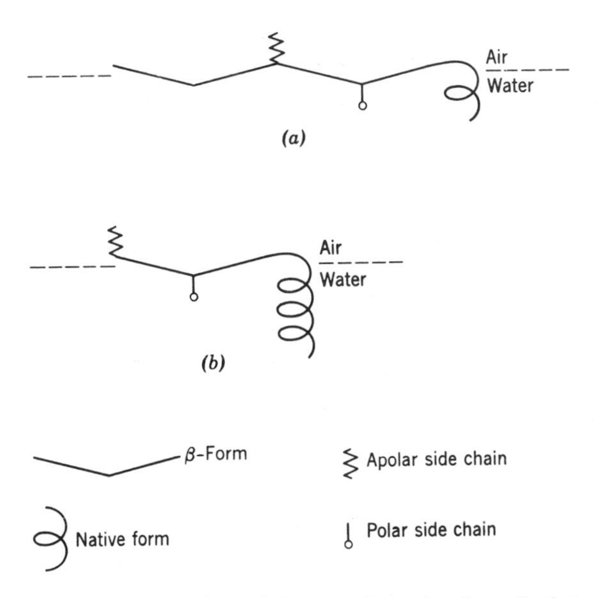

Fig. IV-38. Schematic illustration of the protein molecule at the interface. (*a*) Completely unfolded; (*b*) incompletely unfolded. (From Ref. 196.)

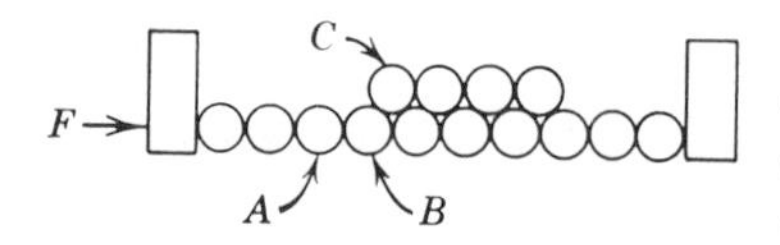

Fig. IV-39. Polypeptide monolayers as arrays of helices, with bilayer formation on compression (from Ref. 205). By permission of The Royal Society.

in the case of proteins), each link having a degree of flexibility given by z, where z is 2 for a rigid chain and may be as high as 4 for a flexible one. The close-packed area of a segment is b, and the average per segment is σ. Left to itself, a long flexible chain will be neither completely unfolded nor completely folded but will have some intermediate most probable spread; on compression, it is compacted, but at the expense of a decrease in configurational entropy. Thus both n and z will affect the film pressure.

The equation given by Singer is (in simplified form)

$$\pi = -\frac{kT}{b}\left[\ln\left(1-\frac{b}{\sigma}\right) - \left(\frac{n-1}{n}\right)\left(\frac{z}{2}\right)\ln\left(1-\frac{2b}{z\sigma}\right)\right] \qquad \text{(IV-80)}$$

Data for ovalbumin are shown in Fig. IV-40, the solid line being Eq. IV-80 with $z = 2.015$. At low π the equation reduces to the ideal gas law, and for $z = 2$, to Eq. IV-37 (with $f_1^s = 1$).

Frisch and Simha (208) have considered a more general treatment that takes into consideration the probability that portions of a polymer chain may dip into the solution phase; their equation reduces to Singer's in the absence of this effect. The procedures used in these theories are essentially stochastic, in that one counts step by step the number of kinds of configurations possible. To do this, it is decidedly helpful to simplify matters by assuming that adsorption (and adjacent solution) positions lie in some simple geometric array or lattice. Figure IV-41 gives a view looking *upward* toward the underside of a surface assumed to have a square lattice of sites, and it illustrates

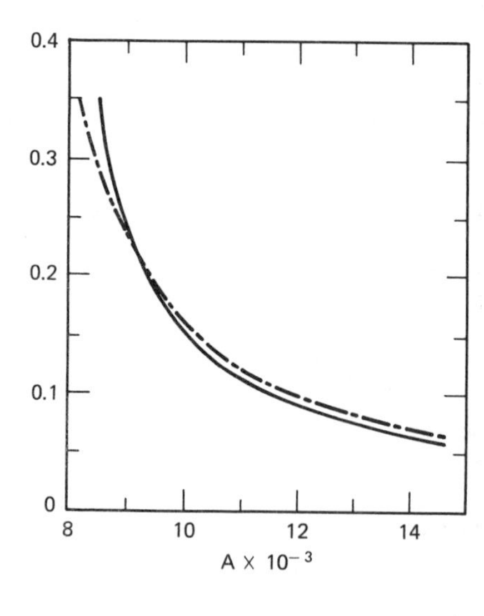

Fig. IV-40. Test of Singer's equation; the egg albumin is on 35% ammonium sulfate (low pressure region). The solid line represents the experimental curve. (From Ref. 206.)

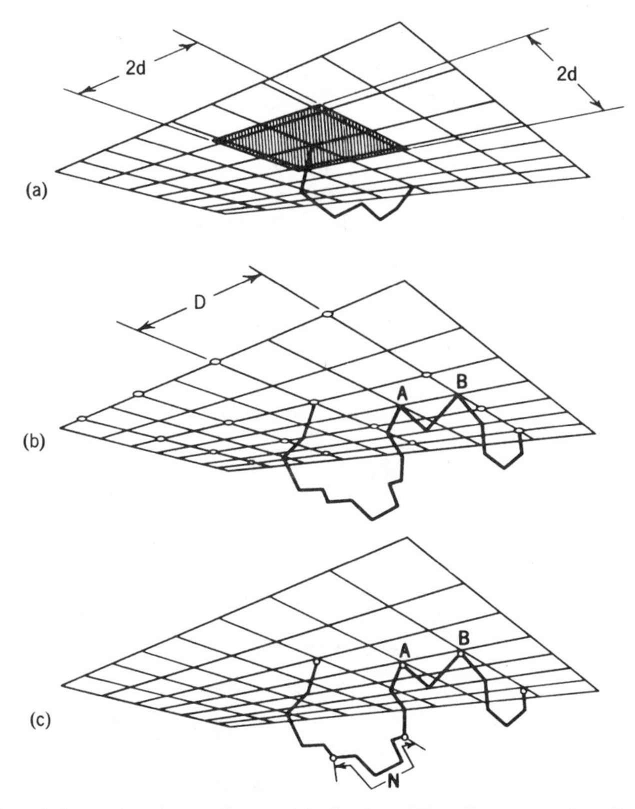

Fig. IV-41. Adsorption loops in a cubic lattice. The diagrams visualize the surface sites as seen from below. (From Ref. 209.)

types of configurations having loops that dip into bulk solution. The figure is attributed to Silberberg (209), who considered the impact of various structural restrictions on the adsorption equilibrium, for example, when not all of the surfaces sites or not all of the polymer segments are adsorbing.

B. Reactions of Protein Films

A number of interactions in mixed films have been reported. Dervichian (210) has, for example, discussed a number of instances of stoichiometric interactions in mixtures of biological materials (although see Section III-8 concerning possible difficulties in such interpretations). An interesting series of studies showed that the Wasserman antibody penetrated considerably more rapidly than did human globulin into a 1:1 mixed monolayer of cardiolipin and cholesterol (211). Vitamin K_1, a component of chloroplasts, did not interact strongly in mixed films with chlorophyll *a* but was an effective fluorescence quenching agent (212). Molecular packing in steroid–lecithin monolayers has been examined by Müller-Landau and Cadenhead (213) and electrostatic interactions involving the Folch-Lees apoprotein by Ter-Minassian-Saraga and co-workers (214), as well as myelin–phospholipid mixed films (214a).

Much interest has centered on the question of whether monolayers of enzymes retain their catalytic activity, and here the evidence is somewhat conflicting. Kaplan (215) found that catalase spread to a monolayer and then compressed to a fiber did retain ability to catalyze the decomposition of hydrogen peroxide, while Cheesman and Schuller (216) questioned whether fully unfolded films had been used and found that pepsin monolayers deposited on paper by pulling the substrate through filter paper did not retain activity. Hayashi (217) has reported that pepsin, in a mixed film with albumin, could catalyze the hydrolysis of the latter. Recovered trypsin films retained enzymatic activity in degrees consistent with other estimates of the proportion of unfolded structure present (195).

An example of chemical reaction is that of the surface pressure dependence of the enzymatic hydrolysis of lipid monolayers (218). Also, some photochemical and radiation decomposition studies have been made. Kaplan and co-workers (215, 219) comment that the expansion of ovalbumin monolayers under ultraviolet irradiation is probably due to destruction of bonds other than the relatively weak ones (probably mostly hydrogen bonds) broken during spreading.

C. Films at the Oil–Water Interface

Much of the interest in films of proteins, steroids, lipids, and so on, has a biological background. While studies at the air–water interface have been instructive, the natural systems approximate more closely to an water–oil interface. A fair amount of work has therefore been reported for such interfaces, in spite of the greater experimental difficulties.

Protein monolayers tend to be expanded relative to those at the air–water interface (220). Davies (221) studied hemoglobin, serum albumin, gliadin, and synthetic polypeptide polymers at the water–petroleum ether interface with the view of determining the behavior of the biologically important ϵ-NH_2 groups, and concluded that on compression they were forced into the oil phase.

The π–σ data for such films still fit Eq. IV-80, but with a larger z value, indicating more flexibility in the chain. Presumably this is because the presence of the oil phase reduces cohesion between the hydrophobic side chains of the protein molecule.

Pethica and co-workers have studied phospholipid monolayers and, as illustrated in Fig. IV-42, a quasi-first-order phase transition can occur. This is surprising because one expects chain–chain interactions to be much reduced at the water–oil interface relative to the water–air interface.

The behavior could be described with the use of a lattice model for the interface, much in the manner whereby Eq. IV-80 was obtained (222). A special assump-

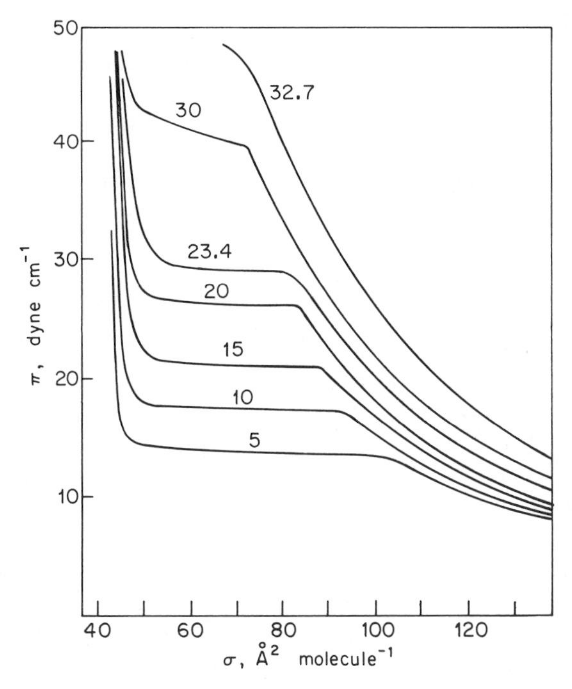

Fig. IV-42. π–σ isotherms for 1,2-distearoyl-lecithin at the *n*-heptane/aqueous NaCl interface. Numbers give the temperature in degrees Celsius. (From Ref. 227.)

tion is that of orientational order-disorder. Orientational states 1 and 2 are taken to have an interaction energy $-(\epsilon + \epsilon')$ for adjacent 1–1 and 2–2 pairs but energy $-(\epsilon - \epsilon')$ for adjacent 1–2 pairs. The resulting equation is:

$$\frac{\pi A_0}{kT} = \tfrac{1}{2}z \ln\left[\frac{u + \beta(1 + s^{z-1})}{u}\right] + (\tfrac{1}{2}z - 1)\ln(1 - \theta) \qquad \text{(IV-81)}$$

where A_0 is the area of a lattice site, z is the nearest neighbor number, and s is an orientational parameter, the system being disordered when $s = 1$ and showing long range order if $s > 1$. The parameter u is defined by

$$u = \frac{s^{z-1} - s - (s^z - 1)\alpha}{\beta(s - 1)} \qquad \text{(IV-82)}$$

where α and β are the Boltzmann factors $\exp(-2\epsilon'/kT)$ and $\exp[-\frac{1}{2}(\epsilon + \epsilon')/kT]$, respectively; θ is the fraction of sites occupied. In this model, the high area side of the transition in Fig. IV-42 corresponds to a disordered fluid phase while the low area side corresponds to a state with long range orientational order. The application of Eq. IV-81 is complicated because s is not a constant of the system, but varies with the film density.

As a point of interest, it is possible to form very thin films or membranes in water, that is, to have the water–film–water system. Thus a solution of lipid can be stretched on an underwater wire frame and, on thinning, the film goes through a succession of interference colors and may end up as a black film of 60 to 90 Å thickness (224). The situation is reminiscent of soap films in air (see Section XIV-9); it also represents a potentially important modeling of biological membranes. A theoretical model has been discussed by Good (225).

Surfaces can be active in inducing blood clotting, and there is much current searching for thromboresistant synthetic materials for use in surgical repair of blood vessels (see Ref. 226). It may be important that a protective protein film be strongly adsorbed (227). The role of water structure in cell–wall interactions may be quite important as well (228).

D. Polymer Films

As has been pointed out, films of high molecular weight polymers frequently resemble those of proteins in physical properties. The π–σ plots are similar in general appearance, and the theoretical analyses discussed in Section IV-11A apply to polymer as well as to protein films. The plots tend to be moderately linear or L_2 in type (Section IV-4) at intermediate pressures, and extrapolation of this region to zero pressure gives an area figure useful in characterizing the film. This area (often reported on a per segment basis, along with the collapse pressure), the value of b from fitting Eq. IV-80 or related forms, and perhaps the compressibility comprise the usual information in tabulated summaries of polymer film behavior. Reviews of earlier literature are by Crisp (229) and by Eirich and co-workers (230).

A great many polymers appear to form films having a flat molecular configuration. Thus various polyesters (231) gave extrapolated areas of about 2.5 m^2/mg corresponding to about the calculated 60 to 70 Å^2 area per segment, or monolayer thickness of 3 to 5 Å. A similar behavior was noted for poly(vinyl acetate) monolayers yet, by contrast, it was found that the behavior of poly(vinyl benzoate) was quite different (232). The poly(vinyl benzoate) gave a very compact monolayer of extrapolated area 9 Å^2 per monomer unit, corresponding to a film thickness of about 20 Å; its compressibility was more like that of stearic acid, that is, about 0.006 cm/dyne, rather than the usual polymer film value of about 0.02 to 0.1 cm/dyne. Apparently, in this case, close packing of the benzene rings occurred. Ries and co-workers (233) have also studied stereoregular poly(methyl methacrylates) and found quite different π–σ plots for the isotactic and the syndiotactic forms.†

Gaines (234) has reported on dimethylsiloxane-containing block copolymers. Interestingly, if the organic block would not in itself spread, the area of the block polymer was simply proportional to the siloxane content, indicating that the organic blocks did not occupy any surface area. If the organic block was separately spreadable, then it contributed, but nonadditively, to the surface area of the block co-polymer.

† If a polymer of the type $\text{-(}CH_2\text{—}CHR\text{)-}_n$ is imagined as stretched out so that the carbon skeleton forms a planar zigzag chain, the H and R atoms will lie on one side or the other of the plane. An *isotactic* polymer is one in which all the R groups are on the same side of the plane; a *syndiotactic* polymer is one in which the R groups alternate as to side, and an *atactic* polymer has a random arrangement.

12. Films at Liquid–Liquid Interfaces and on Liquid Surfaces Other than Water

An interesting field that has received increasing attention in recent years is that of films at liquid–liquid interfaces, where usually one liquid is water. Some work of this type was described in the preceding section on protein and polymer films. Hutchinson (235) describes a number of measurements on films of straight chain acids and alcohols at the interfaces of water–benzene, water–cyclohexane, and water–$C_{14}H_{30}$. The alcohols (hexanol, octanol, decanol, and dodecanol) were more condensed than the acids (butyric, caproic, caprylic, and lauric) and gave close-packed areas of 20 Å^2. In general, areas for alcohols and acids are greater at a given film pressure as compared to those at the water–air interface, as might be expected since the nonpolar solvent should tend to reduce the cohesion between hydrocarbon chains. However, the pressure at which the transition to a condensed phase occurs is lower, possibly because of a greater lateral adhesion between polar groups at high compressions coming in to overbalance the decreased cohesion between the tails. Some work has been done with charged monolayers at oil–water interfaces, and this is discussed briefly in the next section.

As to experimental techniques, several methods have been employed. First, a modification of the film balance can be employed so that direct measurements of film pressures can be made (236, 237); barrier leakage is a problem, however. Since in many cases the film-forming substance is soluble in the oil phase, interfacial tension measurements at various concentrations may suffice. Hutchinson (238) employed a sessile bubble method, while others have used the ring or Wilhelmy slide methods (239) (see Section II-8). Brooks and Pethica (240) designed a film balance type of trough such that the oil–water interface could be swept and interfacial films could be directly compressed, but they used a hydrophobic Wilhelmy slide for measuring γ and hence π values. The procedure was claimed to be superior to the fixed area interface one, where film pressure is built up by successive addition of film-forming material, and spreading against an existing high surface pressure may not always be complete. Interfacial potentials have been measured by means of the vibrating electrode (241); with polar oils direct measurements with a high impedance voltmeter are possible (242). For film viscosity, a torsion pendulum may be used (243).

There is now a fair amount of work reported with films at the mercury–air interface; Smith (82) summarized much of this in a review that covers experimental techniques as well as results.

Ellison and Zisman (237) have reported studies of monolayers of polymethylsiloxane polymer and of the protein zein on such substrates as white mineral oil, *n*-hexadecane, and tricresyl phosphate. Jarvis and Zisman (244) report qualitative spreading studies with a number of fluorinated organic compounds at a variety of organic liquid–air interfaces and π–σ data for some of the Gibbs monolayer systems.

13. Charged Films

A. *Equation of State of a Charged Film*

An important area of development is that of the behavior and theory of charged films, such as those of a fatty acid on an alkaline substrate, or of quaternary amine salts. As pointed out in Section IV-10B, one approach has been to consider the surface region as a thin bulk region and apply the Donnan relationship to determine its ionic makeup. A picture of electrical lines of force is given in Fig. IV-43 (Davies, 245); in the plane *CD* of the ionic groups, it will be a periodic field, whereas a little further into the solution the effect will be more that of a uniformly charged surface. The Donnan treatment is probably best justified if it is supposed that ions from solution penetrate into the region of *CD* itself and might in fact lie between *CD* and *AB*.

It will be recalled that the Donnan effect acts to exclude like charged substrate ions from a charged surface region (Eq. IV-76), and this exclusion, as well as the concentration of oppositely charged ions, can be expressed in terms of a Donnan potential ψ_D. Thus for a film of positively charged surfactant ions S^+ one can write

$$\frac{c_s^+}{c^+} = e^{-e\psi_D/kT} \qquad \text{(IV-83)}$$

and

$$\frac{c_s^-}{c^-} = e^{e\psi_D/kT} \qquad \text{(IV-84)}$$

where c^+ and c^- are the concentrations of nonsurfactant ions and the subscript s denotes that the concentration is for an interfacial region of thickness τ. On multiplying the two equations, one obtains the required condition that, neglecting activity coefficients, $(c^+)(c^-) = (c_s^+)(c_s^-)$ or, in the case of an insoluble surfactant ion, since $c^+ = c^- = c$, the bulk electrolyte concentration, the condition becomes

$$c^2 = (c_s^+)(c_s^-) \qquad \text{(IV-85)}$$

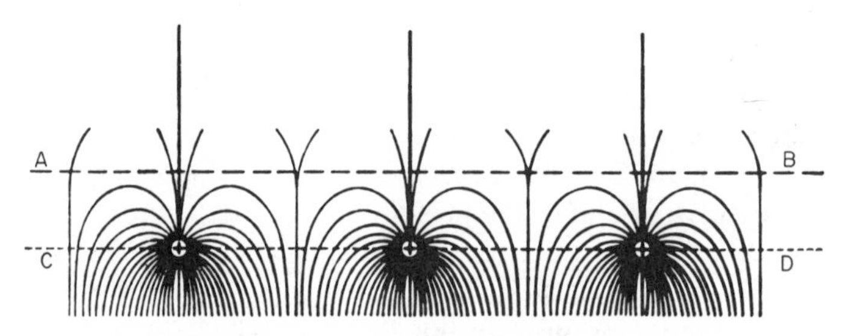

Fig. IV-43. Electrical lines of force for a charged monolayer. (From Ref. 245.)

In the interfacial region, electroneutrality requires that $c_s^- = (S^+) + c_s^+$, so that Eq. IV-85 becomes

$$(S^+) = \frac{c^2}{c_s^+} - c_s^+ \tag{IV-86}$$

or, using Eq. IV-83,

$$(S^+) = (e^{e\psi_D/kT} - e^{-e\psi_D/kT})c \tag{IV-87}$$

$$= 2c \sinh\left(\frac{e\psi_0}{DkT}\right)^{\dagger}$$

Now, $(S^+) = 1000\Gamma/\tau$, where Γ is the surface excess in moles per square centimeter or $(S^+) = 1000 \times 10^{16}/N\tau\sigma$ where σ is in Å^2 per molecule. With these substitutions, Eq. IV-87 may be solved for ψ_D to give

$$\psi_D = \left(\frac{kT}{e}\right) \sinh^{-1}\left(\frac{1000 \times 10^{16}}{2N\tau\sigma c}\right) \tag{IV-88}$$

At this point an interesting simplification can be made if it is assumed that τ, as representing the depth in which the ion discrimination occurs, is taken to be just equal to $1/\kappa$, the ion atmosphere thickness given by Debye-Hückel theory (see Section V-2). In the present case of a 1:1 electrolyte, $\kappa = (8\pi e^2 N/1000DkT)^{1/2} c^{1/2}$, and on making the substitution into Eq. IV-88 and inserting numbers (for the case of water at 20°C), one obtains for ψ_D in millivolts:

$$\psi_D = 25.2 \sinh^{-1}\left(\frac{2 \times 134}{\sigma c^{1/2}}\right) \tag{IV-89}$$

We can now calculate the Donnan contribution to film pressure through the use of Eq. III-123 in the approximate form:

$$\pi_{os} V_1 = RTN_2^s = \frac{RTn_2^s}{n_1^0} \tag{IV-90}$$

or,

$$\pi_{os} = RT\left(\frac{n_2^s}{\mathcal{A}}\right) \tag{IV-91}$$

The total moles n_2 in the surface region is given by $C\mathcal{A}\tau/1000$, where C is the sum of the concentrations of the ionic species present

$$C = (S^+) + c_s^+ + c_s^- = (S^+) + 2c \cosh\left(\frac{e\psi_D}{kT}\right) \tag{IV-92}$$

Actually, it is the *net* concentration that is needed,

$$C_{\text{net}} = (S^+) + 2c\left[\cosh\left(\frac{e\psi_D}{kT}\right) - 1\right] \tag{IV-93}$$

† The reader is reminded that $\sinh(x) = \frac{1}{2}(e^x - e^{-x})$ and that $\cosh x = \frac{1}{2}(e^x + e^{-x})$.

On combining these results with Eq. IV-91,

$$\pi = RT\Gamma_{S^+} + \left(\frac{2c\tau RT}{1000}\right)\left[\cosh\left(\frac{e\psi_D}{kT}\right) - 1\right] \tag{IV-94}$$

The contribution of the Donnan effect is that of the second term in Eq. IV-94, that is,

$$\pi_D = \left(\frac{2c\tau RT}{1000}\right)\left[\cosh\left(\frac{e\psi_D}{kT}\right) - 1\right] \tag{IV-95}$$

and, in combination with Eq. IV-89 for water at 20°C,

$$\pi_D = 1.52c^{1/2}\left[\cosh\sinh^{-1}\left(\frac{2 \times 134}{\sigma c^{1/2}}\right) - 1\right] \tag{IV-96}$$

again taking τ to be $1/\kappa$.

If $(2 \times 134/\sigma c^{1/2})$ is large enough (about 4), Eq. IV-96 reduces to

$$\pi_D = \frac{kT}{\sigma} - 1.52c^{1/2}$$

or, including the contribution from (S^+),

$$\pi_{\text{ions}} = \frac{2kT}{\sigma} - 1.52c^{1/2} \tag{IV-97}$$

This treatment assumes the surface charge to be diffused over a thickness $\tau = 1/\kappa$. As an alternative, the charge may be taken to be spread uniformly on the plane *CD*, in which case the counter-ions are treated as a diffuse double layer (see Section V-2), and the potential, now called the Gouy potential, is

$$\psi_G = 50.4\sinh^{-1}\left(\frac{134}{\sigma c^{1/2}}\right) \tag{IV-98}$$

again, for water at 20° and assuming a 1:1 electrolyte and expressing ψ_G in millivolts. This is not very different from ψ_D as given by Eq. IV-89. Thus, if $(134/\sigma c_{1/2})$ is 2 (e.g., $\sigma = 67$ Å^2 per molecule and substrate concentration 1 *M*), ψ_G is 73 mV and ψ_D is 53 mV. ψ_D should be smaller because the interfacial charge is diffused in a thickness τ rather than concentrated on a plane.

Again referring to Section V-2, the double layer system associated with a surface whose potential is some value ψ_0 requires for its formation a free energy per unit area or a π of

$$\pi_e = 6.10c^{1/2}\left[\cosh\left(\frac{e\psi_0}{2kT}\right) - 1\right] \tag{IV-99}$$

(again water at 20°C and a 1:1 electrolyte). Equation IV-99 and IV-98 may be combined to give

$$\pi_e = 6.10c^{1/2}\left[\cosh\sinh^{-1}\left(\frac{134}{\sigma c^{1/2}}\right) - 1\right] \tag{IV-100}$$

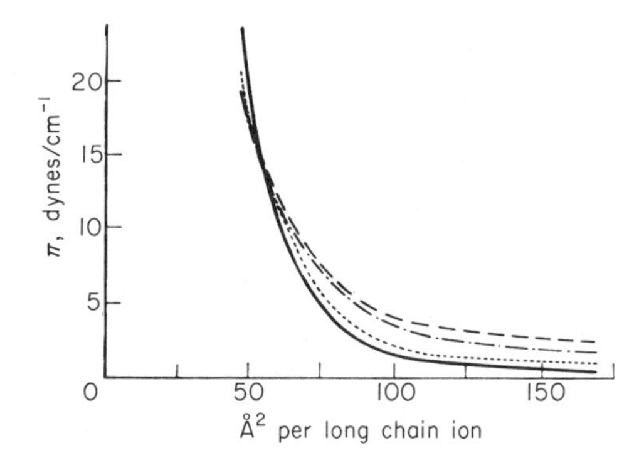

Fig. IV-44. Force–area curves for $C_{18}H_{37}N(CH_3)_3^+$ on aqueous sodium chloride at 21°C. (From Ref. 245.)

Equation IV-100 was given by Davies (245), assuming that ψ_G and ψ_0 were the same.

The assumption was tested as follows, using data on monolayers of $C_{18}H_{37}N(CH_3)_3^+$ on aqueous sodium chloride substrates. The force–area results are illustrated in Fig. IV-44. At 85 Å^2 per molecule and $c = 0.01$, ψ_G may be found from Eq. IV-98 to be 177 mV. This was set equal to ψ_0 in Eq. IV-99, giving a π_e value of about 9. Since the measured film pressure was 5.5 dyne/cm, that of the hypothetical unionized film π_0 was then -3.5 dyne/cm, that is, the assumption was made that

$$\pi_{\text{film}} = \pi_e + \pi_0$$

Knowing the value of π_0, values of ψ_0 could then be calculated for various salt concentrations, with results as shown in Fig. IV-45. The solid line represents ψ_G calculated from the Gouy equation (Eq. IV-98), and it is seen that ψ_0 and ψ_G are essentially the same over the range of data involved.

It is difficult to know how much to read into the fitting of data by the preceding equations. The use of a π_0 term is very empirical, being based on

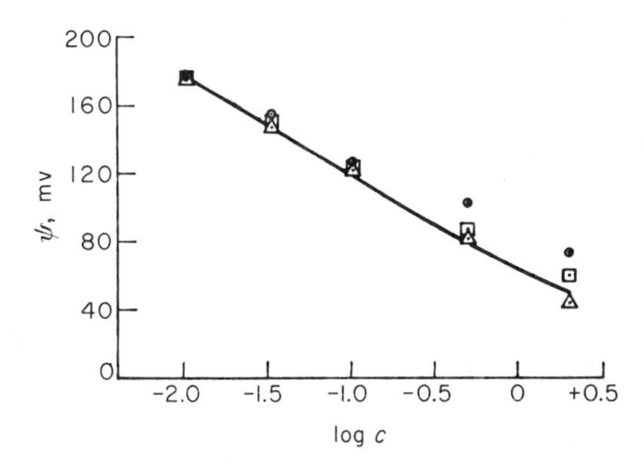

Fig. IV-45. Values of ψ as a function of sodium chloride concentrations for monolayers of $C_{18}H_{37}N(CH_3)_3^+$. The solid line is calculated by Gouy theory: △, ψ_{AB} from ΔV; ○, ψ_0 from $(\delta\pi/\delta t)$ 85 Å^2; □, from π–σ. (From Ref. 245.)

a supposition that the film is duplex (see Section IV-6B); the actual film pressure (as opposed to the spreading pressure of a thick film) for a hydrocarbon is positive, not negative. If slightly positive values for π_0 are assumed, then the data of Fig. IV-44 can be fit by the Donnan-based Eq. IV-96. Both treatments neglect activity coefficient (i.e., interionic attraction) effects and discrete ion effects. Thus Sears and Schulman (246) observed that the force–area plots for stearic acid films on aqueous M OH substrates depended considerably on whether M was Li^+, Na^+ or K^+.

If $(134/\sigma c^{1/2})$ is large enough, Eq. IV-100 reduces to the form

$$\pi_e = \frac{2kT}{\sigma} - 6.1c^{1/2} \qquad \text{(IV-101)}$$

which is similar to Eq. IV-97, based on the Donnan treatment. If the un-ionized part of the film contributes another ideal gas term, then one obtains

$$\pi_{\text{film}} = \frac{3kT}{\sigma} - 6.1c^{1/2} \qquad \text{(IV-102)}$$

Equation IV-102 was approximately obeyed by another quaternary ammonium salt if the spreading was now at an oil–water interface (247). However, other work gives $\pi\sigma = 2kT$ as the limiting form at low c (248).

Davies (249) later proposed a corrected equation:

$$\left[\pi + \frac{Kn}{\sigma^{3/2}} - 6.1c^{1/2}\left(\cosh \sinh^{-1}\frac{134}{\sigma c^{1/2}} - 1\right)\right](\sigma - \sigma_0) = kT \qquad \text{(IV-103)}$$

where σ_0 is the actual molecular area and K has the value 400 if σ is in Å^2, with n being the number of CH_2 groups. The term in K represents cohesion of the chains, following a proposal by Guastella (250). Mingins et al. (251) comment, however, that Eq. IV-103 is not well obeyed even at low π values and suggest that chain cohesion is not correctly accounted for.

Gaines (252) has applied the surface phase approach to ionized films, obtaining:

$$\pi = \frac{RT}{\bar{A}_1}\left[\ln\left(1 + \frac{\nu^s \bar{A}_1}{A - \bar{A}_3}\right) - \ln f_1^s + \ln f_1 \nu_1\right] \qquad \text{(IV-104)}$$

Eq. IV-104 is analogous to Eq. IV-52, with component 1 the solvent, component 2 the electrolyte, and component 3 the film substance. It is assumed that the film substance dissociates into ν^s ions, the counterion being common with the electrolyte. $\bar{A}_1$ and $\bar{A}_3$ are the partial molal areas, and the dividing surface is located so that Γ_2 is zero. Finally the term $\ln f_1 \nu_1$ allows for nonzero electrolyte concentration in the substrate. Eq. IV-104 represented fairly well the π–σ behavior of docosyltrimethyl ammonium bromide on 0.01 M subphases of various 1:1 electrolytes. As before, we see that two rather different appearing models can represent experimental data in a comparable way. The general subject of the interaction of ions with monolayers has been reviewed by Ter-Minassian-Saraga (253).

B. *Interfacial Potentials*

For an ionized film, Cassie and Palmer (254) suggested the equation

$$\Delta V = \frac{12\pi\bar{\mu}}{\sigma} = \frac{12\pi\bar{\mu}_0}{\sigma} + \psi_{AB} \tag{IV-105}$$

(the 12 arises by taking $\bar{\mu}$ to be in millidebyes, σ in Å^2 per charge, and ΔV in millivolts). Then,

$$\left(\frac{\partial\bar{\mu}}{\partial\sigma}\right)_c = \left(\frac{\partial\bar{\mu}_0}{\partial\sigma}\right)_c + \frac{1}{12\pi}\left[\frac{\partial(\sigma\psi_{AB})}{(\partial\sigma)_c}\right]$$

and

$$12\pi\frac{\partial^2\bar{\mu}}{\partial\sigma\partial\ln c} = \frac{\partial^2(\sigma\psi_{AB})}{\partial\sigma\partial\ln c} \tag{IV-106}$$

From Eq. IV-98,

$$12\pi\frac{\partial^2\bar{\mu}}{\partial\sigma\partial\ln c} = -\frac{kT}{e} \tag{IV-107}$$

Davies' results agreed well with Eq. IV-107, provided that the salt concentration was less than about 0.1 *M*; above this a considerable deviation set in (255) which, moreover, was specific as to the nature of the anion of the dissolved electrolyte. The interpretation was that at these higher concentrations counterions enter into the surface region, and the assumption of a charged plane surface is grossly violated; under these circumstances the Donnan treatment may give better results.

Davies and Rideal (256) discuss the matter of interfacial potentials in some detail and draw the diagram shown in Fig. IV-46 for the variation of

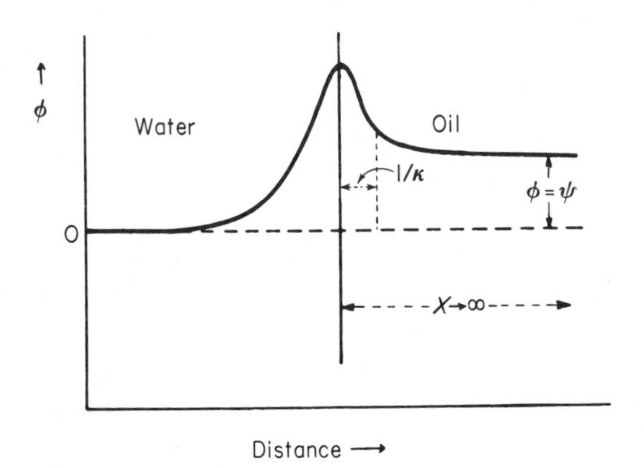

Fig. IV-46. Variation of the Galvani potential across an oil–water interface. (From Ref. 255.)

ϕ, the Galvani potential (see Section V-10), across a water–oil interface. In the case of a surface-adsorbed positive ion, ϕ rises to a maximum across the phase boundary, but then decreases again due to the negative double layer that builds up in the oil phase. If the electrolytes involved are very slightly soluble in the oil phase, the thickness of this double layer becomes large and its buildup slow. In this case, which is also that of the water–air interface, the vibrating electrode lies well within the double layer region, and ΔV gives essentially the change in the phase boundary potential $\Delta\psi$ and thus is directly responsive to the nature of the adsorbed film. However, if there is an appreciable solubility of electrolyte in the oil phase, the double layer is thin and rapidly formed so that the vibrating electrode lies in a region having the properties of the bulk oil phase. The potential change measured at the electrode is now $\Delta\phi$, and if the solubility of the film material is not great, the nature of the bulk phases is little affected by it, and ΔV is small and dependent more on the nature of the electrolytes present than on the film material, that is, the measurement is now that of an electrochemical cell.

14. Capillary Waves

The phenomenon of capillary waves or ripples, that is, waves whose properties are more surface than gravity determined, was mentioned briefly in Section II-10B. The study of such waves has recently received some impetus through the realization that it can provide useful information about time-dependent properties of adsorbed films at liquid interfaces. The field is not a mature one, and space limitations prevent more than a summary presentation here.

The mathematical theory, at least in present form, is rather complex because it involves subjecting the basic equations of motion to the special boundary conditions of a surface that may possess viscoelasticity. An element of fluid can generally be held to satisfy two kinds of conservation equations. First, by conservation of mass,

$$\frac{\partial u}{\partial x} + \frac{\partial v}{\partial y} = 0 \qquad \text{(IV-108)}$$

where x and y denote the horizontal and vertical coordinates for an element of fluid, as illustrated in Fig. IV-47, and u and v are their time derivatives, that is, the velocities of the element. Equation IV-108 can be derived by considering a unit cube and requiring that the sum of the net flows in the x and y directions be zero. The second conservation relationship is that of a force balance or energy conservation and may be written as a pair of equations known as the Navier–Stokes equations. These are

$$\rho \frac{\partial u}{\partial t} + \rho u \frac{\partial u}{\partial x} + \rho v \frac{\partial u}{\partial y} = -\frac{\partial P}{\partial x} + \eta \, \Delta u$$

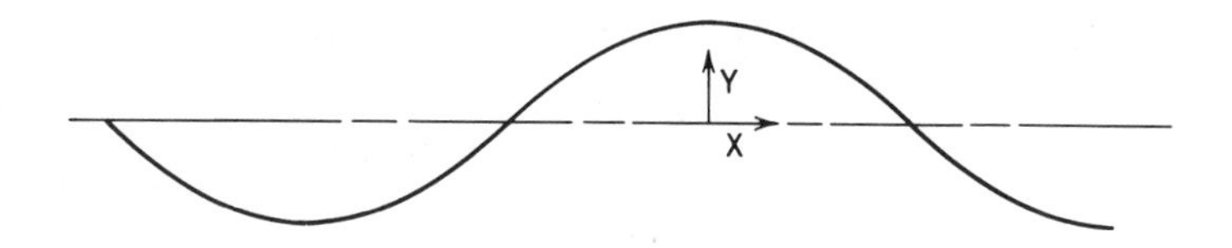

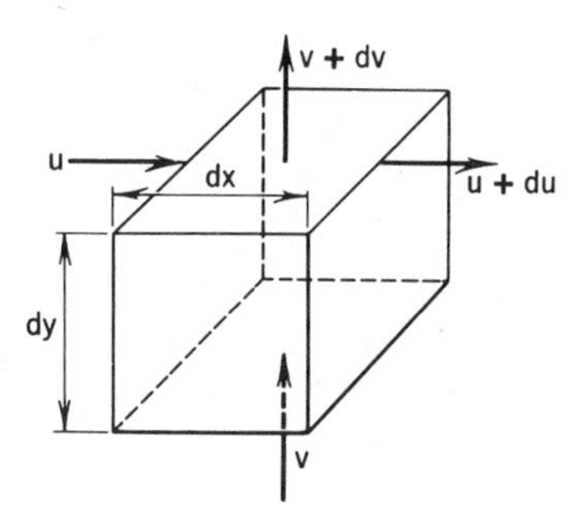

Fig. IV-47

and (IV-109)

$$\rho\frac{\partial v}{\partial t} + \rho u\frac{\partial v}{\partial x} + \rho v\frac{\partial v}{\partial y} = -\frac{\partial P}{\partial y} + \eta\,\Delta v - \rho g$$

(1) (2) (3) (4) (5) (6)

where ρ denotes the density. The three terms on the left are inertial terms, that is, force $= d(mv)/dt = m\,dv/dt + v\,dm/dt$, where mass is changing; term (1) corresponds to the $m\,dv/dt$ component and terms (2) and (3), to the $v\,dm/dt$ component. Term (4) gives the balancing force component due to any pressure gradient; term (5) takes account of viscous friction; and term (6) gives the force due to gravity. The boundary conditions are that at the surface: the vertical pressure component is that of the gas phase plus the Laplace pressure (Eq. II-7 or γ/r for a plane wave), while the horizontal component is given by the gradient of surface tension with area $\partial\gamma/\partial\mathcal{A}$, which is taken to be zero for a pure liquid but involves the surface elasticity of a film-covered surface and any related time-dependent aspects, whereby phase lags may enter. It is customary to assume that no slippage or viscosity anomaly occurs between the surface layer and the substrate.

In the experimental procedure, essentially sinusoidal waves are generated by an oscillating bar, and one therefore writes the solutions to Eqs. IV-108 and IV-109 in the form

$$u = U_1 \exp(iwt) + U_2 \exp(2iwt) \ldots \qquad \text{(IV-110)}$$

$$v = V_1 \exp(iwt) + V_2 \exp(2iwt) \ldots$$

These expressions are inserted in the conservation equations, and the boundary conditions provide a set of relationships defining the U and V coefficients. The outline so far is taken from van den Tempel and van de Riet (257); the subject was treated earlier by Goodrich (258) and, concurrently with van den Tempel, by Hansen (259). Solutions are not obtained in closed form but in terms of successive approximations, as, for example, by Mann and Hansen (260). The entire theoretical situation has been reviewed by Stone and Rice (261).

The detailed mathematical developments are difficult to penetrate, and a simple but useful approach is that outlined by Garrett and Zisman (262). If gravity is not important, Eq. II-41 reduces to

$$v^2 = \frac{2\pi\gamma}{\rho\lambda} \tag{IV-111}$$

The amplitude of a train of waves originating from an infinitely long linear source decays exponentially with the distance x from the source,

$$A = A_0 e^{-kx} \tag{IV-112}$$

and a relationship due to Goodrich (258) gives

$$k = \frac{8\pi\eta\omega}{3\gamma} \tag{IV-113}$$

where ω is the wave frequency. A recent experimental study showed that Eq. IV-113 fits the data for a clean water surface quite well (263). A study of how the damping coefficient k varies with film pressure provides a sensitive indication of structural changes even though it is very difficult to analyze the data into detailed rheological parameters. (It should be remembered that it is as yet very difficult to explain experimental values of film viscosities—a much simpler quantity than k.)

Figure IV-48 illustrates how k may vary with film pressure in a very complicated way although the π–σ plots are relatively unstructured. The results correlated more with variations in film elasticity than with its viscosity and were explained qualitatively in terms of successive film structures with varying degrees of hydrogen bonding to the water substrate and varying degrees of structural regularity. Note the sensitivity of k to frequency; a detailed study of the dispersion of k should give information about the characteristic relaxation times of various film structures.

The experimental procedure used by Hansen and co-workers involved the use of a loudspeaker magnet to drive a rod touching the surface (in an up and down motion) and a detector whose sensitive element was a phonograph crystal cartridge. Zisman and co-workers used an electromechanical transducer to drive a linear knife-edge so that linear waves were produced (262). The standing wave pattern was determined visually by means of stroboscopic illumination. More recently, wave damping has been followed by using a reflected laser beam (263).

The elasticity of monolayer-covered surfaces may be studied in the man-

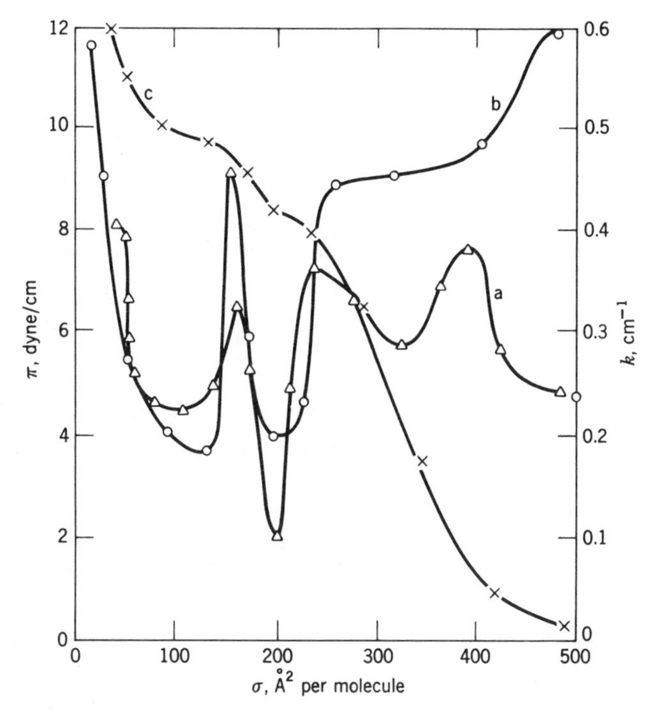

Fig. IV-48. Wave damping behavior of polydimethylsiloxane heptadecamer on water at 25°C at (*a*) 60 cps and (*b*) 150 cps. Curve (*c*) gives the π–σ behavior. (From Ref. 262.)

ner described in Section III-7D. Figure IV-49 shows some results obtained for a poly-L-lysine monolayer with sodium dodecyl sulfate as penetrant (264). The frequency dependence is presumably due to a finite rate at which the penetrant can move in and out of the interfacial region. Ter-Minassian-Saraga and co-workers (265) and van den Tempel and co-workers (266) have discussed some of the theory, the latter introducing the use of "telegraph" equations.

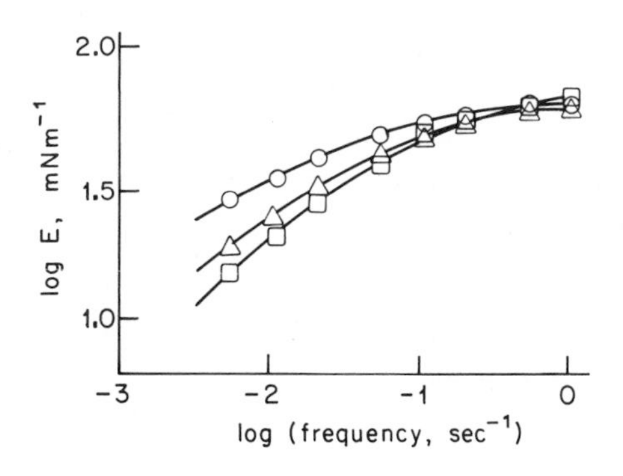

Fig. IV-49. Surface dilational modulus for mixtures of poly-L-lysine (mol. wt. 4800) and sodium dodecyl sufate as a function of frequency. Mixing ratios of SDS/poly-L-lysine: ○, 260; □, 32; △, 5. The solid lines give the theoretical behavior for a single component solution. (From Ref. 264.)

15. Films Deposited on Solids

A. Built-Up Films

There is an interesting but rather complex body of investigations of films that have been deposited on solid supporting surfaces, such as glass or metal plates. For example, if a glass plate is raised up through a barium stearate monolayer spread on water, then, as illustrated in Fig. IV-50, the film that clings to the plate will be oriented with the hydrocarbon surface outward. The surface of such a film-coated plate is hydrophobic and much more so than the surface of solid barium stearate itself. One may then dip the plate *into* the film-covered surface, depositing a second layer "back-to-back." The successive layers that can be built up by this process were termed *Y* films by Miss Blodgett (267). Such films had either hydrophobic or hydrophilic surfaces, depending upon which direction the plate was last moved through the surface. Similarly, *X* films of like oriented monolayers could be built up. These built-up films could be made to consist of as many as a hundred layers, as measured by means of interference fringes.

It appears that successive layers are not necessarily anchored in orientation; Ehlert (268) comments on and adds to a history of x-ray studies on barium stearate built-up films of the *X* and *Y* type and concludes that the internal orientation is the *same* in both. It is presumably the *Y* structure that is stable.

The subject of built-up films is discussed in some detail by Gaines (1). As a more recent development, Kuhn and co-workers have used the built-up film technique to obtain some fascinating results on excited state processes (92 and Van Olphen and Mysels under General References). Swalen (268a) discusses the interesting topic of optical (surface) wave spectroscopy of molecules in thin films.

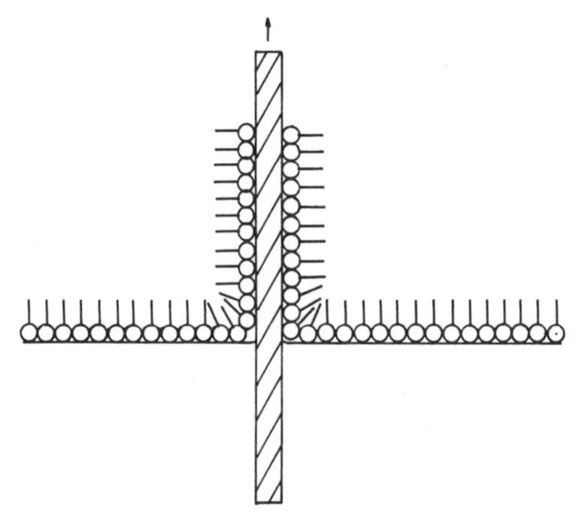

Fig. IV-50

B. Monolayers

Gaines (1) distinguishes between "reactive" and "nonreactive" deposited films. In the former the layer of water initially present between the film and the slide is rapidly expelled, while in the latter it is not. Reactive films are tenaciously held, while nonreactive ones may often be transferred back to a water surface. Spink (98) has confirmed that stearic acid films generally deposit to cover the same area as they did on the water substrate, that is, that the transfer ratio is about unity. Films on mica and on various metal surfaces were stable, but he found that those deposited on silica or glass tended to aggregate into crystalline domains, perhaps because a water film was retained which prevented close interaction of the polar heads with the solid. Gaines and Ward (269) have reported that even with cadmium arachidate, judged to be the best film-forming material, there are fine cracks in films as thin as a double layer. Considerable caution is thus indicated in regarding transferred films as absolutely intact.

It was mentioned that deposited barium stearate monolayers can be extremely hydrophobic. Zisman and co-workers found that "retracted" films, that is, films formed by withdrawing the plate from solution or a melt of a long chain RX-type compound, would often not be wet by the bulk parent compound (see Ref. 270). Such films were called *auto-phobic*. The explanation was similar to Langmuir's, namely, that the outer surface consisted of close-packed methyl groups forming a layer that acted as though it had a very much lower surface tension than did the bulk liquid compound. Zisman's explanation is couched in terms of "critical surface tensions," a concept discussed in Section X-5C.

Deposited monolayers of such RX-type compounds as fatty acid and amines can be extremely tenaciously held; this is evidenced, for example, in frictional wear experiments (271) (see also Section XII-7) and in their stability against evaporation under high vacuum (272). The nature of the substrate is, of course, important; fatty acid films deposited on a NiO surface (with which reaction undoubtedly occurs) showed an alternation in surface potential between odd and even chain lengths, but no such effect if an inert substrate such as Pt was used (273). Even hydrocarbons, such as hexane, form tenacious monolayers on metal surfaces (61), illustrating the hazard in assuming that films deposited from a solution will be free of contamination if the solvent is very volatile.

It has been shown, however, that deposited monolayers on mica and on various metal plates can transfer from the original solid substrate to a second one (274). The suggested mechanism was one of surface diffusion across bridges formed by points of contact between the two solids. Later work, however, gives evidence that the mechanism is one of vapor transport (275, 118).

Films of liquid crystalline material may be strongly oriented and now find important application in "LC" calculator and watch displays since the orientation can be changed by applied electric fields. See Refs. 276–278 for some surface chemical studies of such systems.

16. Problems

1. Benjamin Franklin's experiment is mentioned in the opening paragraphs of this chapter. Estimate, from his results, an approximate value for Avogadro's number; make your calculation clear. The answer is a little off; explain whether more accurate measurements on Franklin's part would have helped.

2. According to Eq. IV-7 (with $\Phi = 1$), the spreading coefficient for a liquid of lower surface tension to spread on one of higher surface tension is always negative. Demonstrate whether this statement is true or not.

3. In detergency, for separation of an oily soil O from a solid fabric S just to occur in an aqueous surfactant solution W, the desired condition is: $\gamma_{SO} = \gamma_{WO} + \gamma_{SW}$. Use simple empirical surface tension relationships to infer whether the above condition might be met if (*a*) $\gamma_S = \gamma_W$, (*b*) $\gamma_O = \gamma_W$, or (*c*) $\gamma_S = \gamma_O$.

4. If the initial spreading coefficient for liquid b spreading on liquid a is to be just zero, what relationship between γ_a and γ_b is implied by the simple Good and Girifalco equation?

5. $S_{B/A}$ is reported to be 36.0 dyne/cm for heptanol on water at 20°C; γ_B is 26.0 dyne/cm. Calculate γ_{AB}.

6. A direct method for determining $S_{B/A}$ is as follows. An inert or "piston" monolayer is spread in a film balance trough and a drop of liquid B is placed on the surface and the film pressure of the system $F_{B/A}$, is measured. Show that $F_{B/A} = S_{B/A}$. Liquid A is the bulk liquid in the trough; the piston monolayer is inert in the sense that it does not mix with B and merely acts as the transmitting medium for the surface pressure generated by the drop of B.

7. Calculate γ_{AB} of Problem 5 using the Girifalco and Good equation, assuming Φ to be 0.85.

8. Derive the expression (in terms of the appropriate works of adhesion and cohesion) for the spreading coefficient for a substance C at the interface between two liquids A and B.

9. Using the Good–Fowkes approach, calculate γ_{WH} for a hydrocarbon whose surface tension at 20°C is 29 dyne/cm. The hydrocarbon might be benzene; compare your result with the experimental γ_{WH} and comment on any discrepancy.

10. Referring to Fig. IV-5, the angles α and β for a lens of n-heptyl alcohol on water are 67° and 16°, respectively. The surface tension of water saturated with the alcohol is 28.8 dyne/cm; the interfacial tension between the two liquids is 7.7 dyne/cm, and the surface tension of n-heptyl alcohol is 26.8 dyne/cm. Calculate the value of the angle γ in the figure. Which equation, IV-7 or IV-13, represents these data better? Calculate the thickness of an infinite lens of n-heptyl alcohol on water.

11. Show what $S_{B(A)/A(B)}$ should be according to Antonow's rule.

12. Obtain Eq. IV-14 in integrated form. Calculate k_{sp} if a spreading source of 2.0-cm width spreads at a rate such that in 15 sec half of the equilibrium surface concentration is reached. The area over which spreading occurs is 2.5 m^2.

13. For a particular monolayer, ΔV is 300 mV at 30 Å^2 per molecule and 175 mV at 45 Å^2 per molecule. Calculate the effective dipole moment; do this both in cgs/

esu units and in SI units. Draw some conclusions from a comparison of the two values of effective dipole moment.

14. Derive the two-dimensional ideal gas law by an analog of the simple gas kinetic theory derivation for a three-dimensional gas.

15. Calculate the theoretical plot of k/k^0 for oleic acid versus π, using Mittelmann and Palmer's method (Section IV-6B). Here, k^0 is the limiting (maximum) rate constant for the oxidation of oleic acid on an aqueous permanganate substrate. Assume the double bond reacts at the maximum rate if it is in the surface and not at all if it is out of the surface.

16. According to the Gibbs equation, how should surface tension vary with concentration if the equation of state of the monolayer is $\pi\sigma = 3kT$?

17. Derive Eq. IV-39 from Eq. IV-37. A particular gaseous film has a film pressure of 0.18 dyne/cm, a value 10% smaller than the ideal gas law value. Taking A_2 to be 25 Å² per molecule, calculate f_1^s. Assume 20°C.

18. Derive Eq. IV-39 from Eq. IV-80, making and stating suitable approximations and assumptions.

19. Assume that an aqueous solute adsorbs at the mercury–water interface according to the Langmuir equation:

$$\frac{x}{x_m} = \frac{bC}{1 + bC}$$

where x_m is the maximum possible amount and $x/x_m = 0.5$ at $C = 0.2\ M$. Neglecting activity coefficient effects, estimate the value of the mercury–solution interfacial tension when C is 0.1 M. The limiting molecular area of the solute is 20 Å per molecule. The temperature is 25°C.

20. The film pressure of a myristic acid film at 20°C is 10 dyne/cm at an area of 23 Å² per molecule; the limiting area at high pressures can be taken as 20 Å² per molecule. Calculate what the film pressure should be, using Eq. IV-39 with $f = 1$, and what the activity coefficient of water in the interfacial solution is in terms of that model.

21. A monolayer of protein containing 0.80 mg of protein per 1.50 m² gives a film pressure of 0.035 erg/cm² at 25°C. Calculate the molecular weight of the protein assuming ideal gas behavior.

22. Figure IV-26*c* appears to have a horizontal portion, suggesting that two surface phases are present. Discuss whether this is consistent with the phase rule, Eq. IV-34.

23. Estimate the heat of two-dimensional vaporization for pentadecylic acid, using the data of Fig. IV-18. Discuss some likely sources of error in your estimate.

24. Taking a typical protein molecular weight as 35,000, estimate the expected increase in solubility of a protein film for a 1 dyne/cm increase in π. The answer provides another indication that protein films are *not* in equilibrium with dissolved protein in the substrate.

25. Calculate the latent heat for the transition shown in Fig. IV-42. Do this by taking successive pairs of isotherms. Comment on your results.

26. Calculate the rate of dissolving of a film of pentadecylic acid at 25°C and a film pressure of 12 dyne/cm. Use Eq. IV-63 and data from Fig. IV-19. The solubility of pentadecylic acid is 10 mg/liter, its diffusion coefficient can be taken to be 1×10^{-6} cm²/sec, and the thickness of the diffusion layer, 1×10^{-3} cm. Repeat the

calculation in SI units. What conclusion can you draw from the result of the calculation?

27. Show under what limiting condition Eqs. IV-89 and IV-98 become the same.

28. At what molecular area should a fatty acid film spread on 0.02 *M* NaOH have a film pressure of 5 dyne/cm, according to the Donnan treatment and to the Gouy treatment? Assume the hydrocarbon part of the film to behave as an ideal gas. Assume 20°C.

29. Davies (245) found that the rates of desorption of sodium laurate and of lauric acid films were in the ratio 6.70:1 at 21.5°C at molecular areas of 70 and 50 Å^2 per molecule, respectively. Calculate ψ_0, the potential at the plane *CD* in Fig. IV-43.

30. Under what physical constraints or assumptions would the rate law

$$\frac{\Delta V - \Delta V^{\infty}}{\Delta V^0 - \Delta V^{\infty}} = e^{-kt}$$

be expected to hold?

31. Derive Eq. IV-52.

32. A film of 0.02 surface poise is flowing under a pressure drop of 8 dyne/cm through a 0.5-cm slit at the rate of 0.3 cm^2/sec. Calculate the linear velocity with which the film is moving at the center of the slit.

General References

N. K. Adam, *The Physics and Chemistry of Surfaces*, 3rd ed., Oxford University Press, London, 1941.

J. T. Davies and E. K. Rideal, *Interfacial Phenomena*, Academic, New York, 1963.

D. H. Everett, Ed., *Specialist Periodical Reports*, Vols. 2 and 3, The Chemical Society London, 1975, 1979.

G. L. Gaines, Jr., *Insoluble Monolayers at Liquid–Gas Interfaces*, Interscience, New York, 1966.

Techniques of Surface and Colloid Chemistry, R. J. Good, R. L. Patrick, and R. R. Stromberg, Eds., Marcel Dekker, 1972.

V. K. LaMer, Ed., *Retardation of Evaporation by Monolayers: Transport Processes*, Academic Press, New York, 1962.

Surface and Colloid Science, E. Matijevic, Ed., Vol. 5, Wiley-Interscience, 1972.

S. R. Morrison, *The Chemical Physics of Surfaces*, Plenum, London, 1977.

H. Van Olphen and K. J. Mysels, *Physical Chemistry: Enriching Topics From Colloid and Surface Chemistry*, Theorex (8327 La Jolla Scenic Drive), La Jolla, California, 1975.

L. I. Osipow, *Surface Chemistry, Theory and Industrial Applications*, Krieger, New York, 1977.

D. J. Shaw, *Introduction to Colloid and Surface Chemistry*, Butterworths, London, 1966.

E. J. W. Verey and J. Th. G. Overbeek, *Theory of the Stability of Lyophobic Colloids*, Elsevier, New York, 1948.

Textual References

1. G. L. Gaines, Jr., *Insoluble Monolayers at Liquid–Gas Interfaces*, Interscience, New York, 1966.

2. Lord Rayleigh, *Proc. Roy. Soc.* (*London*), **47**, 364 (1890).
3. A. Pockels, *Nature*, **43**, 437 (1891).
4. Lord Rayleigh, *Phil. Mag.*, **48**, 337 (1899).
5. H. Devaux, *J. Phys. Radium*, **699**, No. 2, 891 (1912).
6. I. Langmuir, *J. Am. Chem. Soc.*, **39**, 1848 (1917).
7. R. E. Baier, D. W. Goupil, S. Perlmutter, and R. King, *J. Rech. Atmos.*, 571 (1974).
8. W. D. Harkins, *The Physical Chemistry of Surface Films*, Reinhold, New York, 1952, Chap. 2.
9. L. A. Girifalco and R. J. Good, *J. Phys. Chem.*, **61**, 904 (1957).
10. F. E. Bartell, L. O. Case, and H. Brown, *J. Am. Chem. Soc.*, **55**, 2769 (1933).
11. P. Pomerantz, W. C. Clinton, and W. A. Zisman, *J. Colloid Interface Sci.*, **24**, 16 (1967); C. O. Timons and W. A. Zisman, *ibid*, **28**, 106 (1968).
12. F. M. Fowkes, *Adv. Chem.*, **43**, 99 (1964).
13. A. W. Adamson, *Adv. Chem.*, **43**, 57 (1964).
14. F. M. Fowkes, *J. Colloid Interface Sci.*, **28**, 493 (1968).
15. R. J. Good, *Ind. Eng. Chem.*, **62**, 54 (1970).
16. J. C. Melrose, *J. Colloid Interface Sci.*, **28**, 403 (1968).
17. F. M. Fowkes, *J. Phys. Chem.*, **67**, 2538 (1963).
18. F. M. Fowkes, *J. Phys. Chem.*, **84**, 510 (1980).
19. Y. Tamai, T. Matsunaga, and K. Horiuchi, *J. Colloid Interface Sci.*, **60**, 112 (1977).
20. J. Panzer, *J. Colloid Interface Sci.*, **44**, 142 (1973).
21a. A. W. Adamson, *J. Phys. Chem.*, **72**, 2284 (1968).
21b. J. F. Padday and N. D. Uffindell, *J. Phys. Chem.*, **72**, 1407 (1968); see also Ref. 21c.
21c. F. M. Fowkes, *J. Phys. Chem.*, **72**, 3700 (1968).
22a. R. H. Ottewill, Thesis, University of London, Queen Mary College, 1951; see also Ref. 22b.
22b. F. Hauxwell and R. H. Ottewill, *J. Colloid Interface Sci.*, **34**, 473 (1970).
23. M. W. Orem and A. W. Adamson, *J. Colloid Interface Sci.*, **31**, 278 (1969).
24. G. Antonow, *J. Chim. Phys.*, **5**, 372 (1907).
25. W. D. Harkins, *Colloid Symposium Monograph*, Vol. VI, The Chemical Catalog Company, New York, 1928, p. 24.
26. G. L. Gaines, Jr., *J. Colloid Interface Sci.*, **66**, 593 (1978).
26a. G. L. Gaines, Jr., and G. L. Gaines III, *J. Colloid Interface Sci.*, **63**, 394 (1978).
27. M. P. Khosla and B. Widom, *J. Colloid Interface Sci.*, **76**, 375 (1980).
28. J. C. Lang, Jr., P. K. Lim, and B. Widom, *J. Phys. Chem.*, **80**, 1719 (1976).
29. O. Reynolds, *Works*, **1**, 410; *Brit. Assoc. Rept.*, 1881.
30. R. N. O'Brien, A. I. Feher, and J. Leja, *J. Colloid Interface Sci.*, **56**, 474 (1976).
31. R. N. O'Brien, A. I. Feher, and J. Leja, *J. Colloid Interface Sci.*, **51**, 366 (1975).
32. A. Cary and E. K. Rideal, *Proc. Roy. Soc.* (*London*), **A109**, 301 (1925).
33. J. Ahmad and R. S. Hansen, *J. Colloid Interface Sci.*, **38**, 601 (1972).
34. D. G. Sucio, O. Smigelschi, and E. Ruckenstein, *J. Colloid Interface Sci.*, **33**, 520 (1970).
35. J. E. Saylor and G. T. Barnes, *J. Colloid Interface Sci.*, **35**, 143 (1971).

36. M. K. Bernett and W. A. Zisman, *Adv. Chem., Ser.* **43,** 332 (1964); W. D. Bascom, R. L. Cottington, and C. R. Singleterry, *ibid.,* p. 355.
37. F. Sebba, *J. Colloid Interface Sci.,* **73,** 278 (1980).
38. C. V. Sternling and L. E. Scriven, *A.I.Ch.E. J.,* December 1959, 514.
39. I. Langmuir, *J. Chem. Phys.,* **1,** 756 (1933).
40. N. F. Miller, *J. Phys. Chem.,* **45,** 1025 (1941).
41. D. J. Donahue and F. E. Bartell, *J. Phys. Chem.,* **56,** 480 (1952).
42. H. M. Princen and S. G. Mason, *J. Colloid Sci.* **20,** 246 (1965).
43. H. M. Princen, *J. Colloid Sci.,* **18,** 178 (1963).
44. G. L. Gaines, Jr., *J. Colloid Interface Sci.,* **62,** 191 (1977).
45. G. L. Gaines, Jr., *Surface Chemistry and Colloids,* (MTP International Review of Science), M. Kerker, Ed., Vol. 7, University Park Press, Baltimore, 1972.
46. I. S. Costin and G. T. Barnes, *J. Colloid Interface Sci.,* **51,** 94 (1975).
46a. M. W. Kim and D. S. Cannell, *Phys. Rev. A,* **13,** 411 (1976).
47. K. J. Mysels and A. T. Florence, *Clean Surfaces: Their Preparation and Characterization for Interfacial Studies,* G. Goldfinger, Ed., Marcel Dekker, New York, 1970.
48. D. C. Walker and H. E. Ries, Jr., *Nature,* **203,** 292 (1964).
49. G. L. Gaines, Jr., *J. Phys. Chem.,* **65,** 382 (1961).
50. Y. Hendrikx and L. Ter-Minassian-Saraga, *CR,* **276,** Ser. C, 1065 (1973).
51. W. Rabinovitch, R. F. Robertson, and S. G. Mason, *Can. J. Chem.,* **38,** 1881 (1960).
52. N. L. Gershfeld, R. E. Pagano, W. S. Friauf, and J. Fuher, *Rev. Sci. Inst.,* **41,** 1356 (1970).
53. H. E. Ries, Jr., *Nat. Phys. Sci.,* **243,** 14 (1973).
54. J. A. Bergeron and G. L. Gaines, Jr., *J. Colloid Interface Sci.,* **23,** 292 (1967).
55. M. Plaisance and L. Ter-Minassian-Saraga, *CR,* **270,** 1269 (1970).
56. C. D. Kinloch and A. I. McMullen, *J. Sci. Inst.,* **36,** 347 (1959).
57. W. D. Harkins and E. K. Fischer, *J. Chem. Phys.,* **1,** 852 (1933).
58. J. R. MacDonald and C. A. Barlow, *J. Chem. Phys.,* **43,** 2575 (1965) and preceding papers.
59. B. A. Pethica, M. M. Standish, J. Mingins, C. Smart, D. H. Iles, M. E. Feinstein, S. A. Hossain, and J. B. Pethica, *Adv. Chem.,* **144,** 123 (1975).
60. M. Blank and R. H. Ottewill, *J. Phys. Chem.,* **68,** 2206 (1964).
61. K. W. Bewig and W. A. Zisman, *J. Phys. Chem.,* **68,** 1804 (1964).
62. B. Kamiénski, *Bull. Acad. Polon. Sci. Ser.,* **13,** 231 (1965).
63. M. van den Tempel, *J. Non-Newtonian Fluid Mech.,* **2,** 205 (1977).
64. F. van Voorst Vader, Th. F. Erkens, and M. van den Tempel, *Trans. Faraday Soc.,* **60,** 1170 (1964).
65. L. Ter-Minassian-Saraga, I. Panaiotov, and J. S. Abitboul, *J. Colloid Interface Sci.,* **72,** 54 (1979).
66. R. J. Myers and W. D. Harkins, *J. Chem. Phys.,* **5,** 601 (1937).
67. W. D. Harkins and J. G. Kirkwood, *J. Chem. Phys.,* **6,** 53 (1938).
68. D. W. Criddle, *Rheology,* F. R. Eirich, Ed., Vol. 3, Academic, New York, 1960.
69. G. C. Nutting and W. D. Harkins, *J. Am. Chem. Soc.,* **62,** 3155 (1940).
70. W. E. Ewers and R. A. Sack, *Australian J. Chem.,* **7,** 40 (1954).
71. R. S. Hansen, *J. Phys. Chem.,* **63,** 637 (1959).
72. J. Plateau, *Phil. Mag.,* **38,** No. 4, 445 (1869).

73. N. W. Tschoegl, *Kolloid Z.*, **181,** 19 (1962).
74. R. J. Mannheimer, and R. S. Schechter, *J. Colloid Interface Sci.*, **32,** 225 (1970) and preceding papers.
75. F. C. Goodrich, L. H. Allen, and A. Poskanzer, *J. Colloid Interface Sci.*, **52,** 201 (1975).
76. F. C. Goodrich and D. W. Goupil, *J. Colloid Interface Sci.*, **75,** 590 (1980).
77. M. Blank and J. S. Britten, *J. Colloid Sci.*, **20,** 789 (1965).
78. E. R. Cooper and J. A. Mann, *J. Phys. Chem.*, **77,** 3024 (1973).
79. C. Bouhet, *Ann. Phys.*, **15,** 5 (1931).
80. F. L. McCrackin, E. Passaglia, R. R. Stromberg, and H. L. Steinberg, *J. Res. Natl. Bur. Stand.*, **A67,** 363 (1963).
81. R. J. Archer, *Ellipsometry in the Measurement of Surfaces and Thin Films*, E. Passaglia, R. R. Stromberg, and J. Kruger eds.; *Natl. Bur. Stand.*, Misc. Publ. No. 256, 1964, p. 255.
82. T. Smith, *Adv. Colloid Interface Sci.*, **3,** 161 (1972).
83. T. Smith, *Surf. Sci.*, **56,** 212 (1976).
84. R. N. O'Brien, *Physical Methods of Chemistry*, A. Weissberger and B. W. Rossiter, Eds., Part 3 A, Wiley, New York, 1972.
85. K. B. Blodgett and I. Langmuir, *Phys. Rev.*, **51,** 964 (1937).
86. R. E. Hartman, *J. Opt. Soc. Am.*, **44,** 192 (1954).
87. A. G. Tweet, *Rev. Sci. Inst.*, **34,** 1412 (1963).
88. A. G. Tweet, G. L. Gaines, Jr., and W. D. Bellamy, *J. Chem. Phys.*, **40,** 2596 (1964).
89. T. Trosper, R. B. Park, and K. Sauer, *Photochem. Photobiol.* **7,** 451 (1968).
90. S. J. Valenty, D. E. Behnken, and G. L. Gaines, Jr., *Inorg. Chem.*, **13,** 2160 (1979).
91. L. W. Weiss, T. R. Evans, and P. A. Leermakers, *J. Amer. Chem. Soc.*, **90,** 6109 (1968), and references therein.
92. H. Kuhn, D. Mobius, and H. Bucher, *Physical Methods of Chemistry*, A. Weissberger and B. Rossiter, Eds., Vol. 7, part 3B, Chapter 7, p. 577, Wiley-Interscience, New York, 1972.
93. H. Zocher and F. Stiebel, *Z. Phys. Chem.*, **147,** 401 (1930).
94. H. Bruun, *Arkiv. Kemi,* **8,** 411 (1955).
95. (a) H. E. Ries, Jr. and W. A. Kimball, *Proceedings of the Second International Congress of Surface Activity*, Vol. 1, 1957; H. E. Ries, Jr. and D. C. Walker, *J. Colloid Sci.*, **16,** 361 (1961). (*b*) H. E. Ries, Jr., *Nature,* **281,** 287 (1979). (*c*) H. E. Ries, Jr., M. Matsumoto, N. Uyeda, and E. Suito, *Adv. Chem.*, **144,** 286 (1975). (*d*) H. E. Ries, Jr., M. Matsumoto, N. Uyeda, and E. Suito, *J. Colloid Interface Sci.*, **57,** 396 (1976).
96. L. H. Germer and K. H. Storks, *J. Chem. Phys.*, **6,** 280 (1938).
97. G. L. Gaines, Jr., *J. Colloid Sci.*, **15,** 321 (1960).
98. J. A. Spink, *J. Colloid Interface Sci.*, **28,** 9 (1967).
99. H. D. Cook and H. E. Ries, Jr., *J. Phys. Chem.*, **60,** 1533 (1956).
99a. T. R. McGregor, W. Cruz, C. I. Fenander, and J. A. Mann, Jr., *Adv. Chem.*, **144,** 308 (1975).
100. N. K. Adam, *The Physics and Chemistry of Surfaces*, 3rd ed., Oxford University Press, London, 1941.
101. M. W. Kim and D. S. Cannell, *Phys. Rev. A,* **14,** 1299 (1976).
102. D. J. Crisp, *Surface Chemistry*, Butterworths, London, 1949, p. 17.

103. L. Ter-Minassian-Saraga and I. Prigogine, *Mem. Serv. Chim. Etat,* **38,** 109 (1953).
104. F. M. Fowkes, *J. Phys. Chem.,* **66,** 385 (1962).
105. G. L. Gaines, Jr., *J. Chem. Phys.,* **69**(2), 924 (1978).
105a. N. L. Gershfeld and C. S. Patlak, *J. Phys. Chem.,* **70,** 286 (1966).
106. B. Stoeckly, *Phys. Rev. A,* **15,** 2558 (1977).
107. W. D. Harkins and E. Boyd, *J. Phys. Chem.,* **45,** 20 (1941); G. E. Boyd, *J. Phys. Chem.,* **62,** 536 (1958).
108. N. L. Gershfeld, *Ann. Rev. Phys. Chem.,* **27,** 350 (1976).
109. R. Mittelmann and R. C. Palmer, *Trans. Faraday Soc.,* **38,** 506 (1942).
110. J. G. Kirkwood, *Surface Chemistry*, Publication of the American Association for the Advancement of Science, No. 21, 1943, p. 157.
111. D. A. Cadenhead and R. J. Demchak, *J. Chem. Phys.,* **49,** 1372 (1968).
112. F. Müller-Landau and D. A. Cadenhead, *J. Colloid Interface Sci.,* **73,** 264 (1980).
113. M. J. Vold, *J. Colloid Sci.,* **7,** 196 (1952).
114. J. J. Kipling and A. D. Norris, *J. Colloid Sci.,* **8,** 547 (1953).
115. M. C. Phillips, D. A. Cadenhead, R. J. Good, and H. F. King, *J. Colloid Interface Sci.,* **37,** 437 (1971).
116. F. J. Garfias, *J. Phys. Chem.,* **83,** 3126 (1979).
117. G. T. Barnes, private communication.
118. A. W. Adamson and V. Slawson, *J. Phys. Chem.,* **85,** 116 (1981).
119. N. K. Adam and J. G. F. Miller, *Proc. Roy. Soc. (London),* **142,** 401 (1933).
120. J. J. Betts and B. A. Pethica, *Trans. Faraday Soc.,* **52,** 1581 (1956).
120a. G. T. Barnes, *Specialist Periodical Reports*, D. H. Everett, Ed., Vol. 2, The Chemical Society, London, 1975.
121. E. D. Goddard and J. A. Ackilli, *J. Colloid Chem.,* **18,** 585 (1963).
122. J. A. Spink, *J. Colloid Sci.,* **18,** 512 (1963).
123. N. W. Rice and F. Sebba, *J. Appl. Chem. (London),* **15,** 105 (1965); F. Sebba, *Nature,* **184,** 1062 (1959).
123a. G. L. Gaines, Jr., P. E. Behnken, and S. J. Valenty, *J. Am. Chem. Soc.,* **100,** 6549 (1978).
124. E. D. Goddard, O. Kao, and H. C. Kung, *J. Colloid Interface Sci.,* **24,** 297 (1967).
125. M. Joly, *Surface and Colloid Science*, E. Matijevic, Ed., Vol. 5, Wiley-Interscience, 1972.
125a. G. T. Barnes, *Specialist Periodical Reports*, D. H. Everett, Ed., Vol. 3, The Chemical Society, London, 1979.
126. A. Poskanzer and F. C. Goodrich, *J. Colloid Interface Sci.,* **52,** 213 (1975).
127. N. L. Jarvis, *J. Phys. Chem.,* **69,** 1789 (1965).
128. R. W. Evans, M. A. Williams, and J. Tinoco, *Lipids,* **15,** 524 (1980).
129. B. M. J. Kellner and D. A. Cadenhead, *J. Colloid Interface Sci.,* **63,** 452 (1978).
130. B. M. J. Kellner and D. A. Cadenhead, *Chem. Phys. Lipids,* **23,** 41 (1979).
131. J. Marsden and E. K. Rideal, *J. Chem. Soc.,* **1938,** 1163; N. K. Adam, *Proc. Roy. Soc., (London),* **A101,** 516 (1922).
132. F. Müller-Landau and D. A. Cadenhead, *Chem. Phys. Lipids,* **25,** 299 (1979).
133. J. A. Bergeron, G. L. Gaines, Jr., and W. D. Bellamy, *J. Colloid Interface Sci.,* **25,** 97 (1967).

134. M. K. Bernett and W. A. Zisman, *J. Phys. Chem.*, **67**, 1534 (1963).
135. J. E. Proust, E. Perez, and L. Ter-Minassian-Saraga, *Colloid Polym. Sci.*, **256**, 666 (1978).
136. R. E. Pagano and N. L. Gershfeld, *J. Phys. Chem.*, **76**, 1238 (1972).
137. F. C. Goodrich, *Proc. Inter. Congr. Surf. Act. 2nd, London, 1957,* **I**, 85 (1957).
138. G. L. Gaines, Jr., *J. Colloid Interface Sci.*, **21**, 315 (1966).
139. D. A. Cadenhead and R. J. Demchak, *ibid.*, **30**, 76 (1969).
140. K. J. Bacon and G. T. Barnes, *J. Colloid Interface Sci.*, **67**, 70 (1978), and preceeding papers.
141. H. E. Ries, Jr. and H. Swift, *J. Colloid Interface Sci.*, **64**, 111 (1978).
142. I. S. Costin and G. T. Barnes, *J. Colloid Interface Sci.*, **51**, 106 (1975).
143. D. A. Cadenhead and F. Müller-Landau, *Chem. Phys. Lipids*, **25**, 329 (1979).
144. K. Tajima and N. L. Gershfeld, *Adv. Chem.*, **144**, 165 (1975).
145. D. A. Cadenhead, B. M. J. Kellner, and M. C. Phillips, *J. Colloid Interface Sci.*, **57**, 1 (1976).
146. L. Ter-Minassian-Saraga, *J. Colloid Interface Sci.*, **70**, 245 (1979).
147. H. Matuo, N. Yosida, K. Motomura, and R. Matuura, *Bull. Chem. Soc. Jap.*, **52**, 667 (1979), and preceeding papers.
148. Y. Hendrikx, *J. Colloid Interface Sci.*, **69**, 493 (1979).
149. D. O. Shah and S. Y. Shiao, *Adv. Chem.*, **144**, 153 (1975).
150. E. H. Lucassen-Reynders and M. van den Tempel, *Proceedings of the IVth International Congress on Surface Active Substances, Brussels, 1964,* J. Th. G. Overbeek, Ed., Vol. 2, Gordon and Breach, New York, 1967; E. H. Lucassen-Reynders, *J. Colloid Interface Sci.* **42**, 554 (1973); **41**, 156 (1972).
151. E. K. Rideal, *J. Chem. Soc.*, **1945**, 423.
152. E. D. Goddard and J. H. Schulman, *J. Colloid Sci.*, **8**, 309 (1953).
153. P. J. Anderson and B. A. Pethica, *Trans. Faraday Soc.*, **52**, 1080 (1956).
154. F. M. Fowkes, *J. Phys. Chem.*, **66**, 385 (1962).
155. F. M. Fowkes, *J. Phys. Chem.*, **67**, 1982 (1963).
156. E. H. Lucassen-Reynders, *J. Colloid Interface Sci.*, **42**, 563 (1973).
157. Y. Hendrikx and L. Ter-Minassian-Saraga, *Adv. Chem.*, **144**, 177 (1975).
158. M. A. McGregog and G. T. Barnes, *J. Colloid Interface Sci.*, **60**, 408 (1977).
159. Y. Hendrikx and L. Ter-Minassian-Saraga, *CR*, **269**, 880 (1969).
160. B. Pethica, *Trans. Faraday Soc.*, **51**, 1402 (1955).
161. D. M. Alexander and G. T. Barnes, *J. Chem. Soc. Faraday I*, **76**, 118 (1980).
162. M. A. McGregor and G. T. Barnes, *J. Pharma. Sci.*, **67**, 1054 (1978); *J. Colloid Interface Sci.*, **65**, 291 (1978) and preceeding papers.
163. R. B. Dean and K. E. Hayes, *J. Am. Chem. Soc.*, **73**, 5583 (1954).
164. V. K. LaMer, T. W. Healy, and L. A. G. Aylmore, *J. Colloid Sci.*, **19**, 676 (1964).
165. H. L. Rosano and V. K. LaMer, *J. Phys. Chem.*, **60**, 348 (1956).
166. R. J. Archer and V. K. La Mer, *J. Phys. Chem.*, **59**, 200 (1955).
167. I. S. Costin and G. T. Barnes, *J. Colloid Interface Sci.*, **51**, 122 (1975).
168. G. T. Barnes, K. J. Bacon, and J. M. Ash, *J. Colloid Interface Sci.*, **76**, 263 (1980).
169. V. K. LaMer and T. W. Healy, *Science*, **148**, 36 (1965).
170. T. I. Quickenden and G. T. Barnes, *J. Colloid Interface Sci.*, **67**, 415 (1978).
171. R. N. O'Brien, A. I. Feher, K. L. Li, and W. C. Tan, *Can. J. Chem.*, **54**, 2739 (1976).

172. G. T. Barnes and A. I. Feher, *J. Colloid Interface Sci.,* **75,** 584 (1980).
173. G. T. Barnes, *J. Colloid Interface Sci.,* **65,** 566 (1978); see also pp. 574, 576.
174. G. T. Barnes and H. K. Cammenga, *J. Colloid Interface Sci.,* **72,** 140 (1979).
175. H. K. Cammenga, *Current Topics in Materials Science,* in press.
176. I. S. Costin and G. T. Barnes, Proceedings, Fourth International Congress on Surface Active Materials, Carl Hanser Verlag, Munich, 1973, p. 441.
177. W. W. Mansfield, *Nature,* **175,** 247 (1955).
178. L. Ter-Minassian-Saraga, *J. Chim. Phys.,* **52,** 181 (1955).
179. N. L. Gershfeld, *Techniques of Surface and Colloid Chemistry,* R. J. Good, R. L. Patrick, and R. R. Stromberg, Eds., Marcel Dekker (1972).
180. R. S. Hansen, *J. Colloid Sci.,* **16,** 549 (1961).
181. J. T. Davies, *Adv. Catal.,* **6,** 1 (1954).
182. J. Bagg, M. B. Abramson, M. Fishman, M. D. Haber, and H. P. Gregor, *J. Am. Chem. Soc.,* **86,** 2759 (1964).
183. W. D. Bellamy, G. L. Gaines, Jr., and A. G. Tweet, *J. Chem. Phys.,* **39,** 2528 (1963).
184. S. J. Valenty, *J. Am. Chem. Soc.,* **101,** 1 (1979).
185. H. H. G. Jellinek and M. H. Roberts, *J. Sci. Food Agric.,* **2,** 391 (1951).
186. J. T. Davies, *Trans. Faraday Soc.,* **45,** 448 (1949).
187. J. T. Davies and E. K. Rideal, *Proc. Roy. Soc.* (*London*), **A194,** 417 (1948).
188. A. R. Gilby and A. E. Alexander, *Australian J. Chem.,* **9,** 347 (1956).
189. J. E. Bresler, D. L. Talmud, and M. F. Yudin, *J. Phys. Chem.* (*USSR*), **14,** 801 (1940).
190. D. G. Whitten, *J. Am. Chem. Soc.,* **96,** 594 (1974).
191. A. G. Tweet, G. L. Gaines, Jr., and W. D. Bellamy, *J. Chem. Phys.,* **41,** 1008 (1964).
192. V. Slawson, A. W. Adamson, and J. Mead, *Lipids,* **8,** 129 (1973); V. Slawson and A. W. Adamson, *Lipids,* **11,** 471 (1976).
193. H. J. Trurnit, *J. Colloid Sci.,* **15,** 1 (1960).
194. G. I. Loeb and R. E. Baier, *J. Colloid Interface Sci.,* **27,** 38 (1968).
195. L. G. Augenstine, C. A. Ghiron, and L. F. Nims, *J. Phys. Chem.* **62,** 1231 (1958).
196. K. S. Birdi, *Kolloid-Z Z. Polym.,* **250,** 222 (1972).
197. F. MacRitchie, *J. Colloid Sci.,* **18,** 555 (1963).
198. H. B. Bull, *J. Biol. Chem.,* **185,** 27 (1950).
199. J. Guastalla, *CR,* **208,** 1078 (1939).
200. M. Blank, J. Lucassen, and Max van den Tempel, *J. Colloid Interface Sci.,* **33,** 94 (1970).
201. H. B. Bull, *J. Am. Chem. Soc.,* **67,** 4 (1945).
202. B. R. Malcolm, *Applied Chemistry at Protein Interfaces*, R. E. Baier, Ed., *Adv. Chem.,* **145,** 338 (1975).
203. D. F. Cheesman and J. T. Davies, *Adv. Protein Chem.,* **9,** 439 (1954).
204. G. I. Loeb and R. E. Baier, *J. Colloid Interface Sci.,* **27,** 33 (1968).
205. B. R. Malcolm, *Proc. Roy. Soc.,* **A305,** 363 (1968).
206. B. R. Malcolm, *Biopolymers,* **16,** 2591 (1977).
207. S. J. Singer, *J. Chem. Phys.,* **16,** 872 (1948).
208. H. L. Frisch and R. Simha, *J. Chem, Phys.,* **27,** 702 (1957).
209. A. Silberberg, *J. Phys. Chem.,* **66,** 1872 (1962).
210. D. G. Dervichian, in *Surface Phenomena in Chemistry and Biology,* J. F.

Danielli, K. G. A. Pankhurst, and A. C. Riddiford, Eds., Pergamon, New York, 1958.
211. P. Geiduschek and P. Doty, *J. Am. Chem. Soc.*, **74,** 3110 (1952).
212. G. L. Gaines, Jr., A. G. Tweet, and W. D. Bellamy, *J. Chem. Phys.*, **42,** 2193 (1965).
213. F. Müller-Landau and D. A. Cadenhead, *Chem. Phys. Lipids,* **25,** 315 (1979).
214. L. Ter-Minassian-Saraga, G. Albrecht, C. Nicot, T. Nguyen Le, and A. Alpsen, *J. Colloid Interface Sci.,* **44,** 542 (1973).
214a. C. Thomas and L. Ter-Minassian-Saraga, *Bioelectrochem. Bioenerg.*, **8,** 357, 369 (1978).
215. J. G. Kaplan, *J. Colloid Sci.,* **7,** 382 (1952).
216. D. F. Cheesman and H. Schuller, *J. Colloid Sci.,* **9,** 113 (1954).
217. H. Sobotka and S. Rosenberg, *Monomolecular Layers,* Publication of the American Association for the Advancement of Science, Washington, D.C., 1954, p. 175.
218. H. Cohen, B. W. Shen, W. R. Synder, J. H. Law, and F. J. Kézdy, *J. Colloid Interface Sci.,* **56,** 240 (1976).
219. J. G. Kaplan, D. H. Andrews, and M. J. Fraser, *J. Colloid Sci.,* **9,** 203 (1954).
220. C. H. Bamford, A. Elliott, and W. E. Hanby, *Synthetic Polypeptides,* Academic, New York, 1956.
221. J. T. Davies, *Biochem. J.*, **56,** 509 (1954).
222. G. M. Bell, J. Mingins, and J. A. G. Taylor, *J. Chem. Soc., Faraday Trans. II,* **74,** 223 (1978).
223. B. Y. Yue, C. M. Jackson, J. A. G. Taylor, J. Mingins, and B. A. Pethica, *J. Chem. Soc., Faraday Trans. I,* **72,** 2685 (1976).
224. P. Mueller, D. O. Rudin, H. T. Tien, and W. C. Wescott, *J. Phys. Chem.,* **67,** 535 (1963).
225. R. J. Good, *J. Colloid Interface Sci.,* **31,** 540 (1969).
226. R. E. Baier, *Surface Chemistry of Biological Systems,* Plenum Press, p. 235, 1970; see also *J. Biomed. Res.*, **9,** 327 (1975).
227. L. Vroman and A. L. Adams, *J. Biomed. Mater. Res.,* **3,** 43 (1969).
228. W. Drost-Hansen, *Fed. Proc.,* **30,** 1539 (1971).
229. See *Surface Phenomena in Chemistry and Biology,* J. F. Danielli, K. G. A. Pankhurst, and A. C. Riddiford, Eds., Pergamon, New York, 1958.
230. F. Rowland, R. Bulas, E. Rothstein, and F. R. Eirich, *Ind. Eng. Chem.,* September 1965, p. 46.
231. W. M. Lee, R. R. Stromberg, and J. L. Shereshefsky, *J. Res. Natl. Bur. Stand.* **66A,** 439 (1962).
232. H. E. Ries, Jr., N. Beredjick, and J. Gabor, *Nature,* **186,** 883 (1960).
233. N. Beredjick, R. A. Ahlbeck, T. K. Kwei, and H. E. Ries, Jr., *J. Polymer Sci.,* **46,** 268 (1960).
234. G. L. Gaines, Jr., *Adv. Chem.,* **144,** 338 (1975).
235. E. Hutchinson, *J. Colloid Sci.,* **3,** 219 (1948); H. Sobotka, Ed., *Monomolecular Layers*, publication of the American Association for the Advancement of Science, Washington, D. C., 1954, p. 161.
236. F. A. Askew and J. F. Danielli, *Trans. Faraday Soc.,* **36,** 785 (1940).
237. A. H. Ellison and W. A. Zisman, *J. Phys. Chem.,* **60,** 416 (1956).
238. E. Hutchinson, *J. Colloid Sci.,* **3,** 219 (1948).
239. D. J. Cheesman, *Arkiv Kemi, Mineral. Geol.,* **B22,** No. 1, 8 (1946).

240. J. H. Brooks and B. A. Pethica, *Trans. Faraday Soc.*, **60,** 208 (1964).
241. J. T. Davies, *Trans. Faraday Soc.*, **48,** 1052 (1952).
242. J. Mingins, F. G. R. Zobel, B. A. Pethica, and C. Smart, *Proc. Roy. Soc. (London)*, **A324,** 99 (1971).
243. J. T. Davies and G. R. A. Mayers, *Trans. Faraday Soc.*, **56,** 691 (1960).
244. N. L. Jarvis and W. A. Zisman, *J. Phys. Chem.*, **63,** 727 (1959).
245. J. T. Davies, *Proc. Roy. Soc. (London)*, **A208,** 224 (1951).
246. D. F. Sears and J. H. Schulman, *J. Phys. Chem.*, **68,** 3529 (1964).
247. J. T. Davies and E. K. Rideal, *Interfacial Phenomena*, Academic Press, New York, 1963.
248. I. D. Robb and A. E. Alexander, *J. Colloid Interface Sci.*, **28,** 1 (1968).
249. J. T. Davies, *J. Colloid Sci.*, **11,** 377 (1956).
250. J. Guastella, *CR*, **228,** 820 (1949).
251. J. Mingins, J. A. G. Taylor, N. F. Owens, and J. H. Brooks, *Adv. Chem.*, **144,** 28 (1975).
252. G. L. Gaines, Jr., *J. Chem. Phys.*, **69,** 2627 (1978).
253. L. Ter-Minassian-Saraga, *Prog. Surf. Membrane Sci.*, **9,** 223 (1975).
254. A. B. D. Cassie and R. C. Palmer, *Trans. Faraday Soc.*, **37,** 156 (1941).
255. J. T. Davies and E. K. Rideal, *J. Colloid Sci.*, Suppl. No. 1, 1954, p. 1.
256. J. T. Davies and E. K. Rideal, *Can. J. Chem.*, **33,** 947 (1955).
257. M. van den Tempel and R. P. van de Riet, *J. Chem. Phys.*, **42,** 2769 (1965).
258. F. C. Goodrich, *Proc. Roy. Soc. (London)*, **A260,** 503 (1961).
259. R. S. Hansen and J. A. Mann, Jr., *J. Appl. Phys.*, **35,** 152 (1964).
260. J. A. Mann, Jr., and R. S. Hansen, *J. Colloid Sci.*, **18,** 805 (1963).
261. J. A. Stone and W. J. Rice, *J. Colloid Interface Sci.*, **61,** 160 (1977).
262. W. D. Garrett and W. A. Zisman, *J. Phys. Chem.*, **74,** 1796 (1970).
263. D. Byne and J. C. Earnshaw, *J. Colloid Interface Sci.*, **74,** 467 (1980).
264. J. Lucassen, F. Hollway, and J. H. Buckingham, *J. Colloid Interface Sci.*, **67,** 432 (1978).
265. D. S. Dimitrov, I. Panaiotov, P. Richmond, and L. Ter-Minassian-Saraga, *J. Colloid Interface Sci.*, **75,** 483 (1978).
266. A. H. M. Crone, A. F. M. Snik, J. A. Poulis, A. J. Kruger, and Van den Tempel, *J. Colloid Interface Sci.*, **74,** 1 (1980).
267. K. B. Blodgett, *J. Am. Chem. Soc.*, **57,** 1007 (1935).
268. R. C. Ehlert, *J. Colloid Sci.*, **20,** 387 (1965).
268a. J. D. Swalen, *J. Phys. Chem.*, **83,** 1438 (1979).
269. G. L. Gaines, Jr. and W. J. Ward, *J. Colloid Interface Sci.*, **60,** 210 (1967).
270. E. G. Shafrin and W. A. Zisman, *J. Phys. Chem.*, **64,** 519 (1960).
271. O. Levine and W. A. Zisman, *J. Phys. Chem.*, **61,** 1188 (1957).
272. G. L. Gaines, Jr., and R. W. Roberts, *Nature*, **197,** 787 (1963).
273. C. O. Timmons and W. A. Zisman, *J. Phys. Chem.*, **69,** 984 (1965).
274. E. K. Rideal and J. Tadayon, *Proc. Roy. Soc. (London)*, **A225,** 346, 357 (1954).
275. Young, J. E., *Aust. J. Chem.*, **8,** 173 (1955).
276. F. J. Kahn, G. N. Taylor, and H. Schonhorn, *Proc. IEEE*, **61,** 823 (1973).
277. J. E. Proust and L. Ter-Minassian-Saraga, *Colloid Polym. Sci.*, **254,** 492 (1976).
278. E. Perez and J. E. Proust, *J. Colloid Interface Sci.*, **68,** 48 (1979).

CHAPTER V

Electrical Aspects of Surface Chemistry

1. Introduction

It is true of many of the topics treated in this book that they may be extended almost indefinitely into large areas of practical application, on the one hand, and into equally large adjacent areas of chemistry and physics, on the other. This was the case in the last chapter, in which a somewhat arbitrary limit was placed on the extent of excursion into the manifold biological aspects of surface films, and it will be so later when the topics of lubrication, detergency, adsorption, and so on are taken up. Again, in this chapter, somewhat arbitrary limitations are placed on content, since it is not intended to cover large areas of colloid chemistry or electrochemistry or to venture into the fields of metallic and semimetallic conduction.

The discussion is limited to two broad aspects of electrical phenomena at interfaces: the first is that of the consequences of imposing or of having electrical charges at an interface involving an electrolyte solution, and the second is that of the nature of the potentials that occur at phase boundaries. Even with these limitations, frequent reference will be made to various specialized treatises dealing with such subjects rather than attempting any general coverage of the literature directly. One important application, namely to the treatment of long range forces between a molecule and a surface and between particles, is developed in the next chapter.

A now present complication is the matter of units. The introductory derivations are made in the conventional cgs/esu system, but alternative forms appropriate to the Système International d'Unités (SI), are given in Section V-3, along with numerical illustrations.† There are further comparisons of the two systems in Chapter VI.

2. The Electrical Double Layer

An important group of electrical phenomena has to do with the nature of the distribution of ions in solution in the presence of an electric field. To begin with, consider a plane surface bearing a uniform charge density, in

† The writer does not consider the SI system to be well suited to physical chemistry; see Adamson under General References.

contact with a solution phase containing positive and negative ions. To be specific, the surface is assumed to be positively charged. The electrical potential at the surface is taken to be ψ_0, and it decreases as one proceeds out into the solution in a manner to be determined. At any point the potential ψ determines the potential energy $ze\psi$ of the ion in the electric field, where z is the valence of the ion and e is the charge on the electron. The probability of finding an ion at some particular point will be proportional to the Boltzmann factor $e^{-ze\psi/kT}$, the situation being somewhat analogous to that of a gas in a gravitational field where the potential is mgh, and the variation in concentration with altitude is given by

$$n = n_0 e^{-mgh/kT} \tag{V-1}$$

where n_0 is the concentration at zero altitude. Equation V-1 is the familiar barometric formula.

For the case of an electrolyte solution consisting of two kinds of ions of equal and opposite charge, $+z$ and $-z$,

$$n^- = n_0 e^{ze\psi/kT} \qquad n^+ = n_0 e^{-ze\psi/kT} \tag{V-2}$$

There are some additional complications over the gravitational case. First, positive charges are repelled from the surface, whereas negative ones are attracted; second, the system as a whole should be electrically neutral so that far away from the surface $n^+ = n^-$. Close to the surface, however, there will be an excess of negative over positive ions so that a net charge exists; the total net charge in the solution is balanced by the equal and opposite net charge on the surface. Finally, a third complication over the gravitational case is that the local potential is affected by the local charge density, and the interrelation between the two must be considered.

The net charge density ρ at any point is given by

$$\rho = ze(n^+ - n^-) = -2n_0 ze \sinh \frac{ze\psi}{kT} \tag{V-3}$$

(see the footnote to Section IV-13).

The integral of ρ out to infinity gives the total excess charge in the solution, per unit area, and is equal in magnitude but opposite in sign to the surface charge density $\boldsymbol{\sigma}$:

$$\boldsymbol{\sigma} = -\int \rho \, dx \tag{V-4}$$

The situation is that of a double layer of charge, the one localized on the surface of the plane, and the other developed in a diffuse region extending into the solution.

The mathematics is completed by one additional theorem, given by Poisson's equation, which relates the divergence of the gradient of the electrical potential at a given point to the charge density at that point:

$$\nabla^2 \psi = -\frac{4\pi\rho}{D} \tag{V-5}$$

Here, ∇^2 is the Laplace operator $(\partial^2/\partial x^2 + \partial^2/\partial y^2 + \partial^2/\partial z^2)$ and D is the dielectric constant of the medium.

Various forms of the solutions to Eqs. V-3 and V-5 have been studied by Gouy (1), Chapman (2), and Debye and Hückel (3) and, for the present purposes, these have been well summarized by Verwey and Overbeek (4) and by Kruyt (5). Perhaps the best known treatment is that of the simple Debye–Hückel theory that concerns itself with interionic attraction effects in electrolyte solutions. Substitution of Eq. V-3 into Eq. V-5 gives

$$\nabla^2\psi = \frac{8\pi n_0 ze}{D} \sinh \frac{ze\psi}{kT} \tag{V-6}$$

It is then assumed that $ze\psi$ is small compared to kT so that the exponentials in Eq. V-6 may be expanded in series and only the first terms employed. On doing this, one obtains

$$\nabla^2\psi = \frac{8\pi n_0 z^2 e^2 \psi}{DkT} = \kappa^2\psi \tag{V-7}$$

Where ions of various charges are involved,

$$\kappa^2 = \frac{4\pi e^2}{DkT} \sum_i n_i z_i^2 \tag{V-8}$$

The solution to Eq. V-7 for the jth kind of ion is

$$\psi_j(r) = \frac{z_j e}{Dr} e^{-\kappa r}$$

and describes how the potential falls off with distance. The quantity κ is now associated with the size of the ion atmosphere around each ion, and $1/\kappa$ is commonly called the ion atmosphere radius. In addition, the work of charging an ion in its atmosphere leads to an electrical contribution to the free energy of the ion, usually expressed as an activity coefficient correction to its concentration. The detailed treatment of interionic attraction theory and of its various modifications and difficulties is, of course, outside the scope of interest here, and more can be found in the monograph by Harned and Owen (6).

The treatment in the case of a plane charged surface and the resulting diffuse double layer is due mainly to Gouy and Chapman. Here $\nabla^2\psi$ may be replaced by $d^2\psi/dx^2$ since ψ is now only a function of distance normal to the surface. It is convenient to define the quantities y and y_0 as

$$y = \frac{ze\psi}{kT} \quad \text{and} \quad y_0 = \frac{ze\psi_0}{kT} \tag{V-9}$$

Equations V-2 and V-3 combine to give the simple form

$$\frac{d^2y}{dx^2} = \kappa^2 \sinh y \tag{V-10}$$

Using the boundary conditions ($y = 0$ and $dy/dx = 0$ for $x = \infty$), the first integration gives

$$\frac{dy}{dx} = -2\kappa \sinh \frac{y}{2} \tag{V-11}$$

and with the added boundary condition ($y = y_0$ at $x = 0$), the final result is

$$e^{y/2} = \frac{e^{y_0/2} + 1 + (e^{y_0/2} - 1)e^{-\kappa x}}{e^{y_0/2} + 1 - (e^{y_0/2} - 1)e^{-\kappa x}} \tag{V-12}$$

(note Problem 3 of this chapter).

For the case of $y_0 \ll 1$ (or, for singly charged ions and room temperature, $\psi_0 \ll 25$), Eq. V-12 reduces to

$$\psi = \psi_0 e^{-\kappa x} \tag{V-13}$$

The quantity $1/\kappa$ is thus the distance at which the potential has reached the $1/e$ fraction of its value at the surface and coincides with the center of action of the space charge. The plane at $x = 1/\kappa$ is therefore taken as the effective thickness of the diffuse double layer.

For $y_0 \gg 1$ and $x \gg 1/\kappa$, Eq. V-12 reduces to

$$\psi = \frac{4kT}{ze} e^{-\kappa x} \tag{V-14}$$

This means that the potential some distance away appears to follow Eq. V-13, but with an apparent ψ_0 value of $4kT/ze$, which is independent of the actual value. For monovalent ions at room temperature this apparent ψ_0 would be 100 mV.

Once the solution for ψ has been obtained, Eq. V-2 may be used to give n^+ and n^- as a function of distance. Illustrative curves for the variation of ψ with distance and concentration are shown in Fig. V-1. Furthermore, use of Eq. V-4 gives a relationship between $\boldsymbol{\sigma}$ and y_0

$$\boldsymbol{\sigma} = -\int_0^\infty \rho \, dx = \frac{D}{4\pi} \int_0^\infty \frac{d^2\psi}{dx^2} dx = -\left(\frac{D}{4\pi}\right)\left(\frac{d\psi}{dx}\right)_{x=0} \tag{V-15}$$

Insertion of $(d\psi/dx)_{x=0}$ from Eq. V-11 gives

$$\boldsymbol{\sigma} = \left(\frac{2n_0 DkT}{\pi}\right)^{1/2} \sinh \frac{y_0}{2} \tag{V-16}$$

Equation IV-98 is the same as Eq. V-16, but solved for ψ_0 and assuming 20°C and $D = 80$. For *small* values of y_0, Eq. V-16 reduces to

$$\boldsymbol{\sigma} = \frac{D\kappa\psi_0}{4\pi} \tag{V-17}$$

By analogy with the Helmholtz condenser formula (see also Section IV-3B),

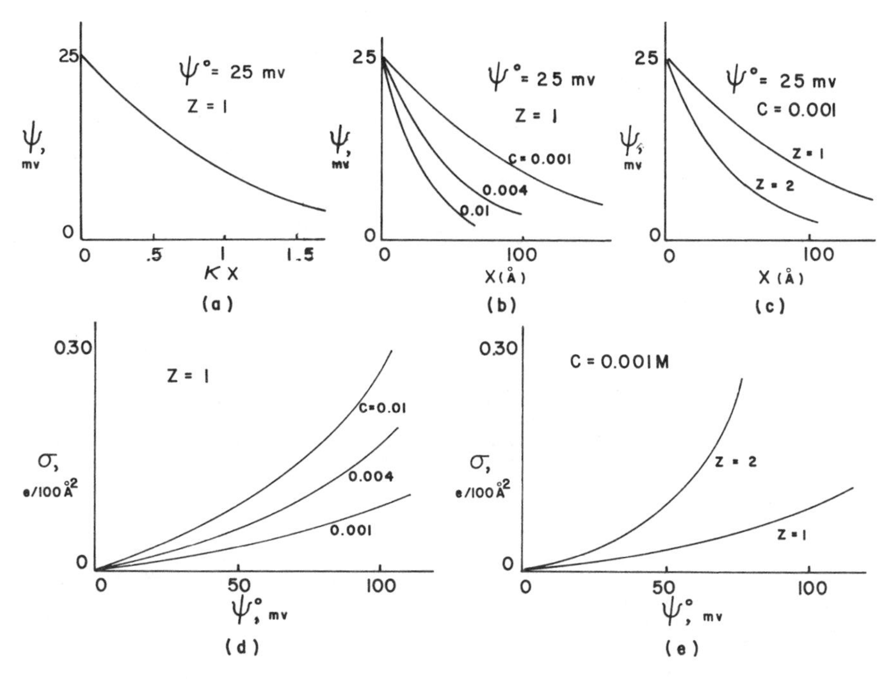

Fig. V-1. The diffuse double layer (from Ref. 7).

for small potentials the diffuse double layer can be likened to an electrical condenser of plate distance $1/\kappa$. For larger y_0 values, however, σ increases more than linearly with ψ_0, and the capacity of the double layer also begins to increase.

Several features of the behavior of the Gouy–Chapman equations are illustrated in Fig. V-1. Thus the higher the electrolyte concentration, the more sharply does the potential fall off with distance, Fig. V-1*b*, as follows from the larger κ value. Figure V-1*c* shows that for a given equivalent concentration, the double layer thickness decreases with increasing valence. Figure V-1*d* gives the relationship between surface charge density σ and surface potential ψ_0, assuming 0.001 *M* 1:1 electrolyte, and illustrates the point that these two quantities are proportional to each other at small ψ_0 values so that the double layer acts like a condenser of constant capacity. Finally, Figs. V-1*d* and *e* show the effects of electrolyte concentration and valence on the charge–potential curve.

3. Units—the SI System

The equations of Section V-2 are written in the cgs/esu system of units, and we outline here the forms that they take if cast in the SI system, along with some comparisons and illustrative calculations.

A. Potential

The term $e\psi/kT$, such as appears in Eq. V-3 is the same in both systems. In cgs/esu, e is in esu, $e = 4.8032 \times 10^{-10}$ esu, and if ψ is in esu, the product $e\psi$ gives energy in ergs, and k is therefor 1.3805×10^{-16} erg/molecule (°K). One esu of potential equals 300 volts practical, or 1 V_{esu} = 300 V. As a useful number, kT/e has the value 25.69 mV at 25°C.

In SI, e is given in coulombs (C), $e = 1.6021 \times 10^{-19}$ C, and the product $e\psi$ gives energy in joules (J) if ψ is in volts. The Boltzmann constant is now $k = 1.3805 \times 10^{-23}$ J/molecule (°K). Again, kT/e = 25.69 mV at 25°C.

B. Coulomb's Law and Equations of Electrostatics

Coulomb's law in cgs/esu is just

$$f = \frac{q_1 q_2}{x^2 D} \qquad \epsilon = \frac{q_1 q_2}{xD} \tag{V-18}$$

where f is the force in dynes between charges q_1 and q_2 in esu, separated by distance x in centimeters and in a medium of dielectric constant D; ϵ is the corresponding potential energy in ergs, attractive if the q's are of opposite sign. In SI, however, we must write

$$f = \frac{q_1 q_2}{4\pi\epsilon_0 x^2 D} \qquad \epsilon = \frac{q_1 q_2}{4\pi\epsilon_0 D x} \tag{V-19}$$

Charges are now in coulombs (C), x in meters (m), and f and ϵ come out in newtons (N) and J, respectively. Coulomb's law is a fundamental law of nature, however, and cannot be expressed in the new system without an adjusting constant ϵ_0, called the *permittivity* (of vacuum), $\epsilon_0 = 10^7/4\pi c^2 = 8.854 \times 10^{-12}$. The permittivity constant thus converts dynes to newtons (or ergs to joules), centimeters to meters, and electrostatic units (esu) to coulombs (C); the quantity 4π cancels the one in Eq. V-19 (placed there for essentially esthetic reasons).

Poisson's equation, Eq. V-5, is now

$$\nabla^2\psi = \frac{-\rho}{\epsilon_0 D} \tag{V-20}$$

with ψ in volts, and ρ in cubic centimeters. The change carries through in related equations. Thus Eq. V-6 becomes

$$\nabla^2\psi = \frac{2n_0 ze}{\epsilon_0 D} \sinh \frac{ze\psi}{kT} \tag{V-21}$$

where n_0 is now concentration in molecules per cubic meter. In general, an equation written in cgs/esu converts to one written in SI if ψ is replaced by $(4\pi\epsilon_0)^{1/2}\psi$ and e is replaced by $e/(4\pi\epsilon_0)^{1/2}$. For example, Eq. V-7 becomes with these substitutions,

$$\nabla^2\psi = \frac{2n_0 z^2 e^2 \psi}{\epsilon_0 DkT} \tag{V-22}$$

As noted in the footnote to Eq. IV-18, the equation for a parallel plate condensor is, in SI:

$$\Delta V = \frac{\boldsymbol{\sigma} d}{\epsilon_0 D} \tag{V-23}$$

As a test, application of the preceding substitution rules indeed converts Eq. IV-18 to Eq. V-23. Here, V has been replaced by $(4\pi\epsilon_0)^{1/2}V$ and charge density $\boldsymbol{\sigma}$ by $\boldsymbol{\sigma}/(4\pi\epsilon_0)^{1/2}$.

Alternatively, the capacity C of a parallel plate condensor is

$$C = \frac{D}{4\pi d}\ \text{(cgs/esu)} \qquad C = \frac{\epsilon_0 D}{d}\ \text{(SI)} \tag{V-24}$$

per unit area. If d is in centimeters, C will be in statfarads per square centimeter, 1 stratfarad $= 1.1126 \times 10^{-12}$ F; and if d is in meters, use of the SI form gives C in farads per square meter.

4. The Stern Treatment of the Electrical Double Layer

The Gouy–Chapman treatment of the double layer, just outlined, runs into difficulties at small κx values when ψ_0 is large. For example, if ψ_0 is 300 mV, y_0 is 12, and if C_0 is, say 10^{-3} mole/l, then the local concentration of negative ions near the surface, as given by Eq. V-2, would be $C^- = 10^{-3}e^{12} = 160$ mole/l! The trouble lies in the assumption of point charges and the consequent neglect of ionic diameters.

The treatment of the case of real ions is difficult, and Stern (8) suggested that it be handled by dividing the region near the surface into two parts, the first consisting of a layer of ions adsorbed at the surface and forming an inner, compact double layer, and the second consisting of a diffuse Gouy layer. Roughly, the potential is assumed to vary with distance as illustrated in Fig. V-2.

The crux of the Stern treatment is the estimation of the extent to which ions enter the compact layer and the degree to which ψ is therefore reduced, that is, the value of ψ_δ. Stern divided both the surface region and the bulk solution region into occupiable sites (much as is done in the treatment of polymer adsorption, Section IV-11A) and assumed that the fractions of sites in each region that were occupied by ions were related by a Boltzmann expression. If S_0 denotes the number of occupiable sites on the surface, then $\boldsymbol{\sigma}_0 = zeS_0$ and $\boldsymbol{\sigma}_S/(\boldsymbol{\sigma}_0 - \boldsymbol{\sigma}_S)$ is the ratio of occupied to unoccupied sites. For a dilute solution phase, the corresponding ratio is just the mole fraction N_S of the solute. Stern considered these two ratios to be related as

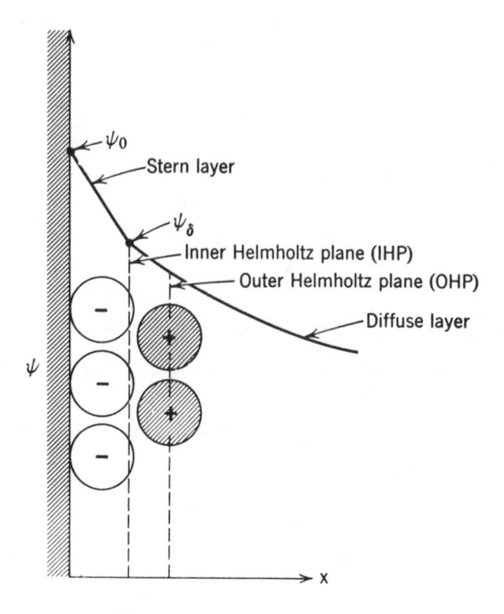

Fig. V-2. The Stern layer.

follows:

$$\frac{\boldsymbol{\sigma}_S}{\boldsymbol{\sigma}_0 - \boldsymbol{\sigma}_S} = N_s e^{(ze\psi_\delta + \phi)/kT} \tag{V-25}$$

where ψ_δ is the potential at the boundary between the compact and diffuse layers and ϕ allows for any additional chemical adsorption potential. The charge density for the compact layer is then

$$\frac{\boldsymbol{\sigma}_S}{\boldsymbol{\sigma}_0} = \frac{N_S e^{-(ze\psi_\delta + \phi)/kT}}{1 + N_S e^{-(ze\psi_\delta + \phi)/kT}} \tag{V-26}$$

The equation is often simplified by neglecting the second term in the denominator.

The picture is therefore one of a compact layer of thickness δ, in which $-d\psi/dx$ is approximated by $(\psi_0 - \psi_\delta)/\delta$ and, hence,

$$\boldsymbol{\sigma}_S = \frac{D'}{4\pi\delta}(\psi_0 - \psi_\delta) \tag{V-27}$$

The capacity of the compact layer $D'/4\pi\delta$ can be estimated from electrocapillarity and related studies (see Section V-8). For generality, D' is a local dielectric constant that may be different from that of bulk solvent. The compact layer is then followed by a diffuse Gouy layer given by Eqs. V-12 and V-15, with ψ_0 replaced by ψ_δ. The total surface charge density $\boldsymbol{\sigma}$ is the sum of $\boldsymbol{\sigma}_S$ and $\boldsymbol{\sigma}_d$ for the two layers, and the total electrical capacity may be considered as given by the formula

$$\frac{1}{C} = \frac{1}{C_S} + \frac{1}{C_d} \tag{V-28}$$

for capacities in series. In concentrated solutions, C_d, which by Eq. V-17 is given by $D\kappa/4\pi$, becomes sufficiently large so that, as an approximation, $C = C_S$.

5. Further Treatment of the Stern and Diffuse Layers

The Stern treatment breaks the "double" layer into two regions, or two double layers of charge, the first corresponding to the compact or adsorbed layer and the second to the diffuse or Gouy layer. It will be shown that the existence of some additional divisions can be argued.

First, consider a sol of silver iodide in equilibrium with saturated solution. There are essentially equal concentrations of silver and of iodide ions in solutions although the particles themselves are negatively charged due to a preferential adsorption of iodide ions. If, now, the Ag^+ concentration is increased tenfold (e.g., by the addition of silver nitrate), the thermodynamic potential of the silver ion increases by

$$\mu = kT \ln \frac{C}{C_0} \tag{V-29}$$

which, expressed as a potential amounts to 57 mV at 25°C. As a result, some Ag^+ ions are adsorbed on the surface of the silver iodide, but these are few (on a mole scale) and mingle with other surface Ag^+ ions so as to be indistinguishable from them. Thus the *chemical* potential of Ag^+ in the silver iodide is virtually unchanged. On the other hand, the total, or *electrochemical* potential of Ag^+, must be the same in both phases (see Section V-11), and this can be true only if the surface potential has increased by 57 mV. Thus changes in ψ_0 can be directly computed from Eq. V-29 and, in the preceding case, ψ_0 is 57 mV more positive than before. Since the Ag^+ concentration in solution can be easily changed by many powers of ten, ψ_0 can therefore correspondingly be changed by hundreds of millivolts. In this case Ag^+ is called the *potential-determining* ion. Furthermore, measurements of the emf of a cell having a silver–silver iodide electrode allows the determination of the Ag^+ concentration before and after addition of the silver nitrate, so the amount of adsorption and hence surface charge, as well as the change in surface potential, can be determined (9).

In the cases of oxides and of proteins and biological materials, H^+ is frequently a potential-determining ion because of the dependence of the degree of dissociation of acidic or basic groups on the pH of the solution. Hydrogen ion is not part of the Stern layer in the sense discussed in the preceeding section. Also, the potential-determining ion need not be one of a type already present in the colloidal particle. Thus Cl^- ion is potential-determining for gold sols, apparently through the formation of highly stable chloride complexes with the surface atoms. It seems reasonable to consider that such potential-determining ions completely leave the solution phase in the sense that they are desolvated and enter into tight chemical association with the solid.

Clearly, some further divisions of the surface region are in order. As illustrated in Fig. V-3, the surface of the solid has a potential ψ_0, which may be controlled by potential-determining ions that are indistinguishable from ions already in the solid, as in the case of Ag^+ or I^- and AgI, and for which, therefore, one cannot speak of a surface concentration or chemical (as distinct from electrochemical) potential. Next, there may be a layer of chemically bound, desolvated ions such as H^+ or OH^- on an oxide surface (or of Cl^- on Au). If such a layer is present, ψ_0 will be associated with it, as indicated in the figure, along with a surface charge density $\boldsymbol{\sigma}_0$. Next will be the Stern layer of compactly bound, but more or less normally solvated ions, at potential ψ_β and net surface charge density $\boldsymbol{\sigma}_\beta$. We thus have an inner, potential determining compact layer of capacity and dielectric constant C_1 and D_1, respectively, and an outer compact (Stern) layer of capacity and dielectric constant C_2 and D_2. Beyond this is the diffuse layer of Section V-2.

The Stern layer may be thought of as being rather immobile in the sense of mobility normal to the surface since, if the adsorption forces are strong, the lifetime of an ion in the layer will be rather long (see Section XVI-2). In addition, there is the question of lateral mobility, or resistance to shear. It

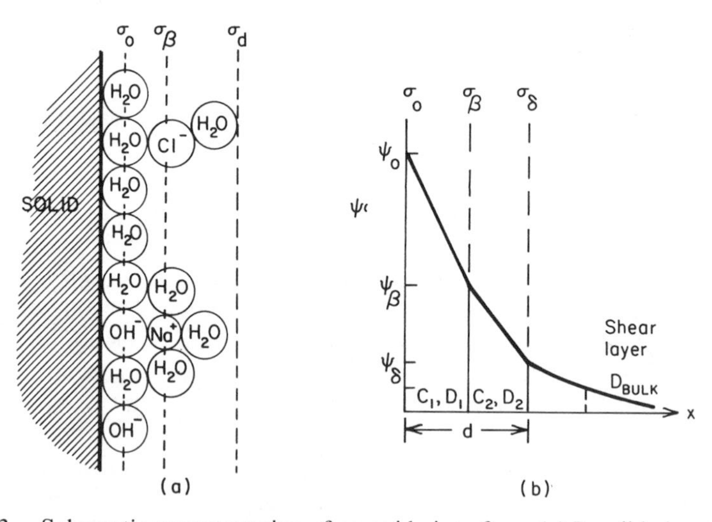

Fig. V-3. Schematic representation of an oxide interface. (*a*) Possible locations for molecules comprising the places of charge. (*b*) Potential decay away from the surface. (From Ref. 10.)

seems likely that the ions and surrounding medium in the Stern layer would be rather rigidly held and that the Stern layer itself would also be immobile in the sense of resisting shear. Since this type of immobility refers to the medium as a whole, and hence primarily to the solvent, there is no reason why the shear plane should coincide exactly with the Stern layer boundary and, as suggested in Fig. V-3*b*, it may well be located somewhere further out. The potential at this shear layer is known as the ξ *potential* and is the potential involved in the electrokinetic phenomena, discussed in Section V-7.

The picture of Fig. V-3 allows one to write acid–base equilibrium constants for the species in the inner compact layer, and ion-pair association constants for the outer compact layer. These are normal equilibrium constants except that the concentration or activity of an ion (H^+, OH^-, Na^+, etc.) in a given layer is related to that in the bulk solution by a term $\exp(-e\psi/kT)$, where ψ is the appropriate potential (10). The charge density in the inner and outer compact layers is given by the algebraic sum of the amounts of ions of either charge present, per unit area, which is related to amount of ions removed from solution as determined, for example, by a pH titration. If the capacity of the layers can be estimated, one has a relationship between charge density and potential and thence to the experimentally measurable zeta potential (see Ref. 11).

Other approaches are possible. An alternative one in which actual ionic volumes are recognized was proposed by Sparnaay (12), for example. Also, from an electrical point of view, the compact region can be structured into what is called an Inner Helmoltz Plane (the Helmholtz condenser formula being used in connection with it), located at the surface of the layer of Stern adsorbed ions, and an Outer Helmholtz Plane, located on the plane of centers of the next layer of ions, marking the beginning of the diffuse layer. These planes are marked *IHP* and *OHP*, respectively, in Fig. V-2, and are further discussed in Section V-9C. Levine and co-workers (13) in a series of papers introduce a *discrete charge* effect in which, essentially, the local Stern potential that determines adsorption is not the average potential but is that modified by an ion "self-potential." This can be visualized, alternatively, as a repulsion between adsorbed ions. One effect of the treatment is to allow a prediction that the potential at the *OHP* can go through a maximum as ψ_0 is increased, an effect deduced from electrocapillarity studies. (See also Problem 10 of this chapter.)

6. The Free Energy of a Diffuse Double Layer

The calculation involved here is conceptually a complex one, and for the necessarily detailed discussion needed to do it justice the reader is referred to Verwey and Overbeek (4) and Kruyt (5) or to Harned and Owen (6). Qualitatively, what must be done is to calculate the reversible electrostatic work for the process:

$$\left\{\begin{matrix}\text{charged surface plus diffuse} \\ \text{double layer of ions}\end{matrix}\right\} \rightarrow \left\{\begin{matrix}\text{uncharged surface plus normal} \\ \text{solution of uncharged particles}\end{matrix}\right\} \quad \text{(V-30)}$$

One way of doing this makes use of the Gibbs equation (Eq. III-78) in the form (see also Section V-9A)

$$dG^s \ (= d\gamma) = -\sum \Gamma_i \, d\mu_i = -\sigma \, d\psi_0 \tag{V-31}$$

Here, the only surface adsorption is taken to be that of the charge balancing the double layer charge, and the electrochemical potential change is equated to a change in ψ_0. Integration then gives

$$G^s_{\text{elect}} = G_d = -\int_0^{\psi_0} \sigma \, d\psi_0 \tag{V-32}$$

and with the use of Eq. V-16 one obtains for the electrostatic free energy per square centimeter of a diffuse double layer

$$G_d = -\frac{8n_0kT}{\kappa}\left(\cosh\frac{y_0}{2} - 1\right) \tag{V-33}$$

The ordinary Debye–Hückel interionic attraction effects have been neglected as being of second-order importance.

It should be noted that G_d is not the same as the mutual electrostatic energy of the double layer, but differs from it by an entropy term. This term can be visualized as the difference in entropy between the more random arrangement of ions in bulk solution and that in the double layer. A very similar mathematical process is involved in the calculation of the electrostatic part of the free energy of an electrolyte due to interionic attraction effects. That G_d is negative in spite of the adverse entropy term can be considered as due to the overriding effect of the energy of adsorption or of concentration of ions near a charged surface.

7. Repulsion between Two Plane Double Layers

If two parallel plane double layer systems approach each other, there develops a repulsion between them—a repulsion that plays an important role in determining the stability of colloidal particles against coagulation and in the force balance in a soap film (see Section VI-5B). The situation is as illustrated in Fig. V-4, which shows two ψ versus x curves, from planes a distance $2d$ apart, and how the actual potential variation should look.

As Langmuir pointed out (14), the total force acting on the planes can be regarded as the sum of an osmotic pressure force (since the ion-concentrations are different from those in the adjacent bulk medium) and that from the electrical field. The total force must be constant across the space between the planes, and since the field $d\psi/dx$ is zero at the midpoint, the total force is given by the net osmotic pressure at this point.

If the solution is dilute, then

$$(\pi_{\text{os}})_{\text{net}} = P = n_{\text{excess}}kT = [n_0(e^{ze\psi_M/kT} + e^{-ze\psi_M/kT}) - 2n_0]kT \tag{V-34}$$

where the last term corrects for the osmotic pressure of the bulk solution and ψ_M is the midpoint potential. The case of ψ large is fairly easy. Equation V-34 reduces

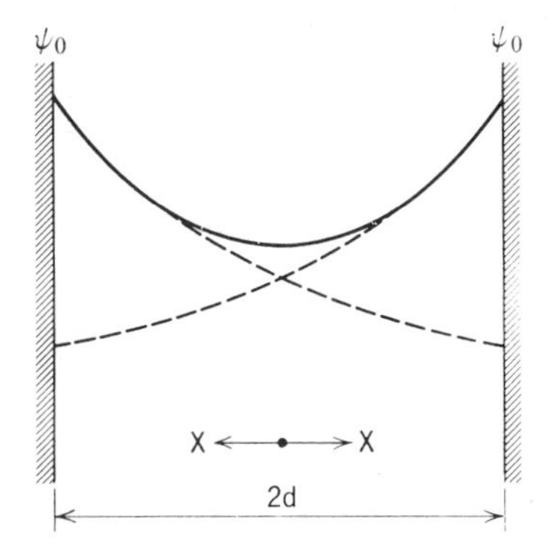

Fig. V-4. Two interacting double layers.

to

$$P = n_0 kT e^{e\psi_M/kT} \tag{V-35}$$

The detailed derivation is discussed in Kruyt (5). Integration of Eq. V-11 with the new boundary conditions and combination with Eq. V-35 gives

$$P = \left(\frac{2\pi}{\kappa d}\right)^2 n_0 kT = \frac{\pi}{2} D \left(\frac{kT}{ed}\right)^2 \tag{V-36}$$

or, for water at 20°C,

$$P = \frac{8.90 \times 10^{-7}}{d^2} \text{ dyne/cm}^2 \tag{V-37}$$

The potential energy of the two plates is given by

$$\epsilon_d = -2\int_\infty^d P\, dd = \frac{1.78 \times 10^{-6}}{d} \text{ erg/cm}^2 \tag{V-38}$$

The treatment above is restricted to the case of ψ_M large and $\kappa d >$ about 3. For the case of surfaces far apart, where ψ_M is small and the interaction weak, Kruyt (5) gives the equation

$$\epsilon_d = \frac{64 n_0 kT}{\kappa} \gamma^2 e^{-2\kappa d} \tag{V-39}$$

where

$$\gamma = \frac{e^{y_0/2} - 1}{e^{y_0/2} + 1}$$

The equations are transcendental for the general case, and their solution has been discussed recently in several contexts (15–17), including the case of surfaces having a layer of fixed dipoles (17). An important aspect of the general case is that the integration involved requires some assumption as to how ψ_0 or, alternatively, how $\boldsymbol{\sigma}_0$ varies with d. Traditionally, ψ_0 has been taken to be constant; it is equally plausible physically, however, that $\boldsymbol{\sigma}_0$ is constant, being fixed by the presence of charged sites on the surface. As illustrated in Fig. V-5, the results diverge considerably at small d. Interestingly, however, if the surface charge is due to weakly

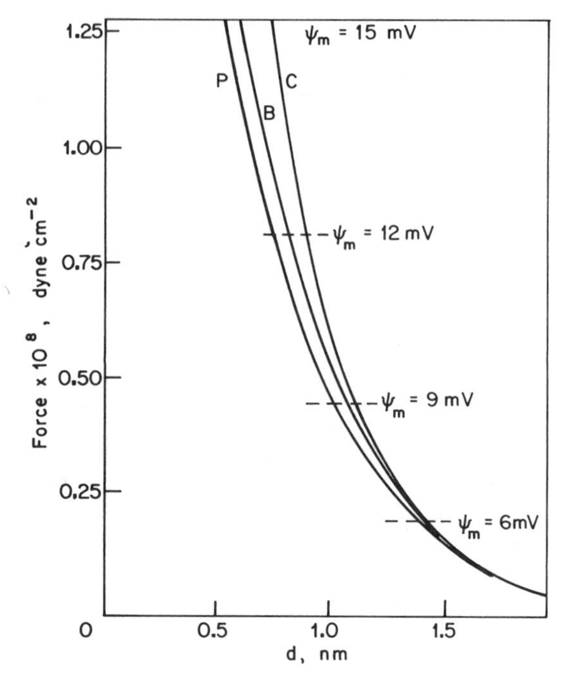

Fig. V-5. Force between charged surfaces as a function of d for constant ψ_0 (P), constant $\boldsymbol{\sigma}_0$ (C), and for an equilibrium α for Eq. V-40 of 0.5 at $d = \infty$. The horizontal dashed lines show the variation of d for various constant ψ_M. The electrolyte solution is taken to be similar to that for a biological system, with $1/\kappa = 0.8$ nm and ψ_0 for $d = \infty$ of 18 mV. (From Ref. 15.)

acidic sites, so that the equilibrium

$$S-\mathrm{H} = S + H^+ \tag{V-40}$$

prevails, where S denotes a surface group, then the degree of dissociation α will vary as ψ_0 varies, that is, $\boldsymbol{\sigma}_0$ varies with d in an *equilibrium* manner (15). As a consequence, as also shown in Fig. V-5, the result is now similar to that for fixed ψ_0.

8. The Zeta Potential

There are a number of electrokinetic phenomena that have in common the feature that relative motion between a charged surface and the bulk solution is involved. Essentially, a charged surface experiences a force in an electric field and, conversely, a field is induced by the relative motion of such a surface. In each case the plane of slip between the double layer and the medium is involved, and the results of the measurements may be interpreted in terms of its charge density. The ζ *potential* is not strictly a phase-boundary potential because it is developed wholly within the fluid region. It can be regarded as the potential difference in an otherwise practically uniform medium between a point some distance from the surface and a point on the plane of shear.

For convenience in seeing the relationships between the various electrokinetic effects, they are summarized in Table V-1.

A. Electrophoresis

The most familiar type of electrokinetic experiment consists of setting up a potential gradient in a solution containing charged particles and determining their rate of motion. If the particles are small molecular ions, the phenomenon is called ionic conductance; if they are larger units, such as protein molecules, or colloidal particles, it is called electrophoresis.

In the case of small ions, Hittorf transference cell measurements may be combined with conductivity data to give the mobility of the ion, that is, the velocity per unit potential gradient in solution, or its *equivalent conductance*. Alternatively, these may be measured more directly by means of the moving boundary method.

For ions, the charge is not in doubt, and the velocity is given by

$$v = ze\omega\mathbf{F} \tag{V-41}$$

where $\mathbf{F}$ is the field in volts per centimeter (here, esu/cm) and ω is the intrinsic mobility (the Stokes' law value of ω for a sphere is $1/6\pi\eta r$). If $\mathbf{F}$ is given in ordinary volts per centimeter, then

$$v = \frac{ze\omega\mathbf{F}}{300} = zu\mathbf{F} \tag{V-42}$$

where u is now an electrochemical mobility. From the definition of equivalent conductance, it follows that

$$\lambda = \mathcal{F}uz \tag{V-43}$$

where $\mathcal{F}$ is Faraday's number.

Thus in the case of ions, measurements of this type are generally used to obtain values of the mobility and, through Stokes' law or related equations, an estimate of the effective ionic size.

TABLE V-1
Electrokinetic Effects

	Nature of solid surface	
Potential	Stationary (e.g., a wall or apparatus surface)	Moving (e.g., a colloidal particle)
Applied	Electroosmosis	Electrophoresis
Induced	Streaming potential	Sedimentation potential

In the case of a charged particle, the total charge is not known, but if the diffuse double layer up to the plane of shear may be regarded as the equivalent of a parallel plate condenser, one may write

$$\boldsymbol{\sigma} = \frac{D\zeta}{4\pi\tau} \tag{V-44}$$

where τ is the effective thickness of the double layer from the shear plane out, usually taken to be $1/\kappa$. The force exerted on the surface, per square centimeter, is $\boldsymbol{\sigma}\mathbf{F}$, and this is balanced (when a steady state velocity is reached) by the viscous drag $\eta v/\tau$, where η is the viscosity of the solution. Thus

$$v = \frac{\boldsymbol{\sigma}\mathbf{F}\tau}{\eta} = \frac{\mathbf{F}\tau\boldsymbol{\sigma}}{300\eta} \tag{V-45}$$

where the factor $\frac{1}{300}$ converts $\mathbf{F}$ to practical volts, or

$$v = \frac{\zeta D\mathbf{F}}{4\pi\eta} \tag{V-46}$$

and the velocity per unit field is proportional to the ζ potential, or, alternatively, to the product $\boldsymbol{\sigma}\tau$, sometimes known as the *electric moment per square centimeter.*

The experimental measurement of the electrophoretic mobility of colloidal particles cannot be accomplished very easily through simple conductance measurements because of the swamping contribution of electrolytes present. It may be made by a moving boundary type of apparatus or by direct microscopic observation of the motion of a particle in an applied field.

There are, of course, a number of complications in the more detailed theory of electrophoretic motion (5, 18). The effective viscosity in the diffuse double layer is affected by the fact that the ions in it are also moving due to the field $\mathbf{F}$; this gives rise to what is known as the "*electrophoretic retardation.*" Briefly, since the net charge in the fluid region of the double layer is opposite in sign to that of the surface, these ions, on the whole, move relative to the solution in a direction opposite to that of the surface. They, in turn, entrain solvent with them, so that there is a local motion of the medium opposing the motion of the charged particle or surface. The observed velocity is then the velocity calculated by the simple treatment minus the local medium velocity.

There is also a relaxation effect in that, due to the motion of the particle, the double layer lags somewhat behind and, again, the effect is one of retardation. Another point is that the double layer region is a source of conductance, and so is the surface of the particle itself. This last is difficult to evaluate and, for reliable ζ potential measurements, nonconducting particles are necessary.

B. Electroosmosis

The effect known either as *electroosmosis* or *electroendosmosis* is a complement to that of electrophoresis. In the latter case, when a field $\mathbf{F}$ is applied, the surface or particle is mobile and moves relative to the solvent, which is fixed (in laboratory coordinates). If, however, the surface is fixed, it is the mobile diffuse layer that moves under an applied field, carrying solution with it. If one has a tube of radius r, whose walls possess a certain ζ potential and charge density, then Eqs. V-45 and V-46 again apply, with v now being the velocity of the diffuse layer.

Experimentally, ζ potentials may be measured by this method by means of an apparatus of the type sketched in Fig. V-6. The potential is supplied by electrodes, as shown in the figure, and the transport of liquid across the tube may be observed through the motion of an air bubble in the capillary providing the return path. For water at 25°C, a field of about 1500 V/cm is needed to produce a velocity of 1 cm/sec if ψ_0 is 100 mV.

The simple treatment of this and of other electrokinetic effects was greatly clarified by Smoluchowski (19); for electroosmosis it is as follows. The volume flow V (in cc/sec) for a tube of radius r is given by applying the linear velocity v to the body of liquid in the tube

$$V = \pi r^2 v \tag{V-47}$$

or

$$V = \frac{r^2 \zeta D \mathbf{F}}{4\eta} \tag{V-48}$$

Alternatively, as illustrated in Fig. V-7, one may have a porous diaphragm (the pores must have radii greater than the double layer thickness) separating the two fluid reservoirs. Here, the return path lies through the center of each tube or pore of the diaphragm and, as shown in the figure, the flow diagram has solution moving one way near the walls and the opposite way near the center. When the field is first established, electroosmotic flow occurs and, for a while, there is net transport of liquid through the diaphragm so that a difference in level develops in the standing tubes. As the hydrostatic head develops, a counterflow sets in and, finally, a steady state electroos-

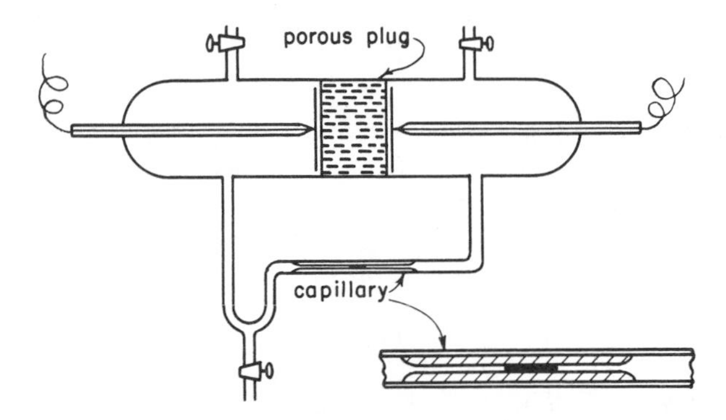

Fig. V-6. Capillary method for the measurement of electroosmosis.

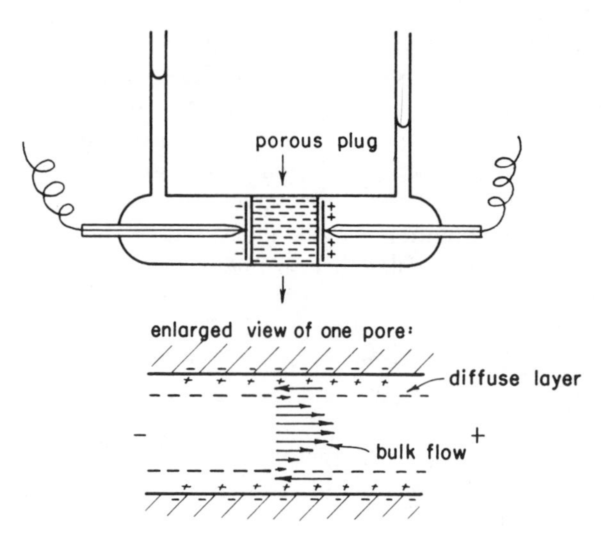

Fig. V-7. Alternative apparatus for the measurement of electroosomotic pressure.

motic pressure is reached such that the counterflow just balances the flow at the walls. The pressure P is related to the counterflow by Poiseuille's equation

$$V(\text{counterflow}) = \frac{\pi r^4 P}{8\eta l} \tag{V-49}$$

where l is the length of the tube. From Eqs. V-48 and V-49, the steady state pressure is

$$P = \frac{2\zeta D\mathbf{E}}{\pi r^2} \tag{V-50}$$

where $\mathbf{E} = \mathbf{F}l$ and is the total applied potential. For water at 25°C,

$$v = 7.8 \times 10^{-3}\mathbf{F}\zeta(\text{cm/sec}) \tag{V-51}$$

$$P = \frac{4.2 \times 10^{-8}\mathbf{E}\zeta}{r^2}(\text{cm Hg}) \tag{V-52}$$

where potentials are in practical volts and lengths are in centimeters. For example, if ζ is 100 mV, then 1 V applied potential can give a pressure of 1 cm of mercury only if the capillary radius is about 1 μ.

C. Streaming Potential

The situation in electroosmosis may be reversed in that the solution is caused to flow down the tube, and an induced potential, the *streaming potential*, is measured. The derivation, again due to Smoluchowski (19), is as follows. If streamline flow is assumed, then the velocity at a radius x

from the center of the tube is

$$v = \frac{P(r^2 - x^2)}{4\eta l} \quad \text{(V-53)}$$

The double layer is centered at $x = r - \tau$, and substitution into Eq. V-53 gives

$$v_d = \frac{\tau r P}{2\eta l} \quad \text{(V-54)}$$

if the term in τ^2 is neglected. The current due to the motion of the double layer is then

$$i = 2\pi r \boldsymbol{\sigma} v_d \quad \text{(V-55)}$$

or

$$i = \frac{\pi r^2 \boldsymbol{\sigma} \tau P}{\eta l} \quad \text{(V-56)}$$

If k is the specific conductance of the solution, then the actual conductance of the liquid in the capillary tube is $C = \pi r^2 k/l$, and by Ohm's law, the streaming potential **E** is given by $\mathbf{E} = i/C$. Combining these equations (including Eq. V-54),

$$\mathbf{E} = \frac{\tau \boldsymbol{\sigma} P}{\eta k} \quad \text{(V-57)}$$

or

$$\mathbf{E} = \frac{\zeta P D}{4\pi\eta k} \quad \text{(V-58)}$$

For a more detailed derivation, see Kruyt (5).

As with electrokinetic effects generally, streaming potentials are difficult to measure reproducibly. A type of apparatus that has been used involves simply forcing a liquid under pressure through a porous plug or capillary and measuring **E** by means of electrodes in the solution on either side (5, 8, 20, 21).

Alternatively, the flow may be between two parallel plates (22). Interestingly, the relation between **E** and ζ is the same as that given by Eq. V-58, that is, the relationship is independent of geometry.

Quite sizable streaming potentials can develop with liquids of very low conductivity. The effect, for example, poses a real problem in the case of jet aircraft where the rapid flow of jet fuel may produce sparks.

D. Sedimentation Potential

The final and less commonly dealt with member of the family of electrokinetic phenomena is that of sedimentation potential. If charged particles are caused to move relative to the medium as a result, say, of a centrifugal field, there again will be an induced potential **E**. The formula relating **E** to ζ and other parameters is (21, 23)

$$\mathbf{E} = \frac{C_m}{k}\frac{D\zeta}{6\pi n}\frac{\omega^2}{2}(R_2^2 - R_1^2) \tag{V-59}$$

where C_m is the apparent mass of the dispersed substance per cubic centimeter, ω is the angular velocity of the centrifuge, and R_2 and R_1 are the distances from the axis of rotation of the two points between which the sedimentation potential is measured.

E. Further Developments in the Theory of Electrokinetic Phenomena

It may be shown that in electroosmosis

$$\frac{V}{i} = \frac{\zeta D}{4\pi\eta k} \tag{V-60}$$

and an old relation, attributed to Saxén (24) is that V/i at zero pressure is equal to the streaming potential ratio ($\mathbf{E}/P$) at zero current. A further generalization of such cross effects and the general phenomenological development of electrokinetic and hydrodynamic effects has been given by Mazur and Overbeek (25) and by Lorenz (26); the subject constitutes an important example of the application of Onsager's reciprocity relationships.

Because these effects are frequently determined using porous plugs, it should be mentioned that a problem develops in connection with the correction for surface conductance. See Rutgers, de Smet, and Rigole for a discussion of this problem (27).

F. General Observation on ζ Potentials—Stability of Colloids

Zeta potentials have been measured for a rather wide variety of solid–solution interfaces, not always with concordant results. Clearly, Stern layer adsorption of impurities can have profound effects. Some illustrative data for the glass–water interface are shown in Fig. V-8. As might be ex-

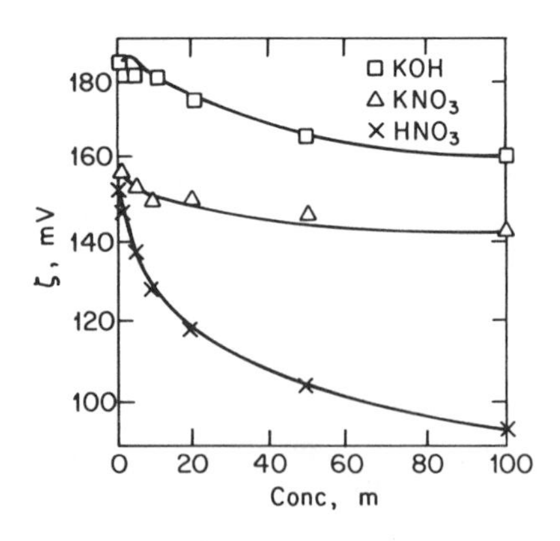

Fig. V-8. Zeta potential-concentration curves for aqueous solutions at the glass–solution interface. (From Ref. 27.)

TABLE V-2
Flocculation of Gold Sol

Concentration of Al^{3+} eq/l × 10^6	Electrophoretic velocity, $cm^2/VS \times 10^6$	Stability
0	3.30 (toward anode)	Indefinitely stable
21	1.71	Flocculated in 4 hr
—	0	Flocculates spontaneously
42	0.17 (toward cathode)	Flocculated in 4 hr
70	1.35	Incompletely flocculated in 4 days

pected, the ζ potential drops with increasing concentration of electrolyte since more and more reduction in ψ is occurring in the immobile layer.

The ζ potential of silver iodide may be either positive or negative, in the range of ± 75 mV, depending on the Ag^+ or I^- concentration, thus illustrating the point that the presence of a potential-determining ion can reverse the sign of the ζ potential. Metal sols generally have negative ζ potentials, showing no particular correlation with their work functions, or their standard electrode potentials, and thought perhaps to be due to a dipolar layer of oxide coating (5). Specific ions may have a strong potential-determining effect, however, as is the case mentioned earlier of Cl^- with gold sols.

The double layer system plays an important role in determining the stability of colloidal particles. Lyophobic sols are generally considered to be stabilized by the presence of a surface charge. To coalesce, two particles must approach closely and there is a deterring interaction of the two double layers (as discussed in Section V-7). Variations in ψ_0 will thus have a marked effect on the stability of such sols; for example, it is routinely observed in the titration of iodide ion with silver nitrate that coagulation occurs as soon as a slight excess of silver ion has been added. The data in Table V-2 due to Burton (28) illustrate clearly the general relation between stability and ζ potential, in the case of a gold sol in a solution containing added Al^{3+}.

The effectiveness of an electrolyte is generally measured in terms of its flocculation value expressed as the concentration (e.g., millimoles per liter) needed to coagulate the sol in some given time interval. There is roughly a 10- to 100-fold increase in effectiveness as one goes from mono- to di- to trivalent ions; this may be due partly to the decreasing thickness of the double layer, but it appears mainly to be due to the greater adsorption of highly charged ions into the Stern layer (14). This effect of valence is stated by what is known as the Schulze–Hardy rule. In addition, there is an order of effectiveness for ions of a given valence type, known as the Hofmeister series, in which flocculation values vary directly (or the effectiveness inversely) with the hydrated radius of the ion. See Section VI-5 for further discussion.

9. Electrocapillarity

It has long been known that the form of a curved surface of mercury in contact with an electrolyte solution depends on its state of electrification (29), and the earliest comprehensive investigation of the electrocapillary effect was made by Lippmann in 1875 (30). A sketch of his apparatus is shown in Fig. V-9.

Qualitatively, it is observed that the mercury surface initially is positively charged, and, on reducing this charge by means of an applied potential, it is found that the height of the mercury column and hence the interfacial tension of the mercury increases, goes through a maximum, and then decreases. This roughly parabolic curve of surface tension versus applied potential, illustrated in Fig. V-10, is known as the *electrocapillary curve*. For mercury, the maximum occurs at -0.48 V applied potential $\mathbf{E}_{max}$ (in the presence of an "inert" electrolyte, such as potassium carbonate, and referred to the mercury in a normal calomel electrode, NCE). Vos and Los (31) give the value of $\mathbf{E}_{max}$ as -0.5084 V versus NCE for 0.116 *M* KCl at 25°C, the difference from -0.48 V being presumably due to Cl^- adsorption at the Hg/solution interface.

The left-hand half of the electrocapillary curve is known as the ascending or anodic branch, and the right-hand half is known as the descending or cathodic branch. As is discussed in the following section, the left-hand branch is in general sensitive to the nature of the anion of the electrolyte, whereas the right-hand branch is sensitive to the nature of the cations present. A material change in solvent may affect the general shape and location of the entire curve and, with nonionic solutes, the curve is affected mainly in the central region.

It is necessary that the mercury or other metallic surface be polarized, that is, that there be essentially no current flow across the interface. In this way no chemical changes occur, and the electrocapillary effect is entirely associated with potential changes at the interface and corresponding changes in the adsorbed layer and diffuse layer.

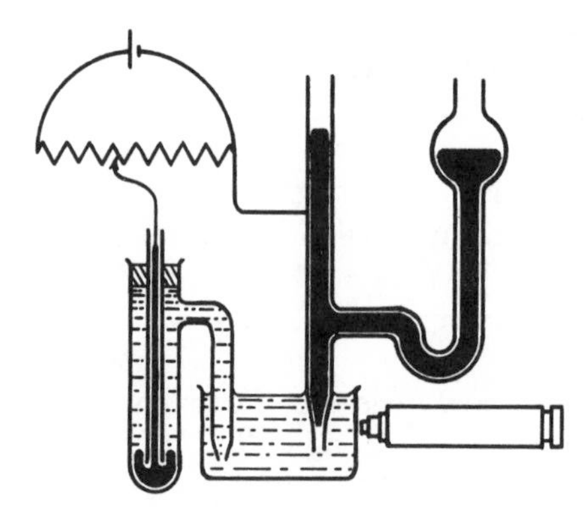

Fig. V-9. The Lippon apparatus for observing the electrocapillary effect.

Recalling the earlier portions of this chapter on Stern and diffuse layer theories, it can be appreciated that varying the potential at the mercury–electrolyte solution interface will vary the extent to which various ions adsorb. The great advantage in electrocapillarity studies is that the free energy of the interface is directly measurable; through the Gibbs equation, surface excesses can be calculated, and from the variation of surface tension with ap-

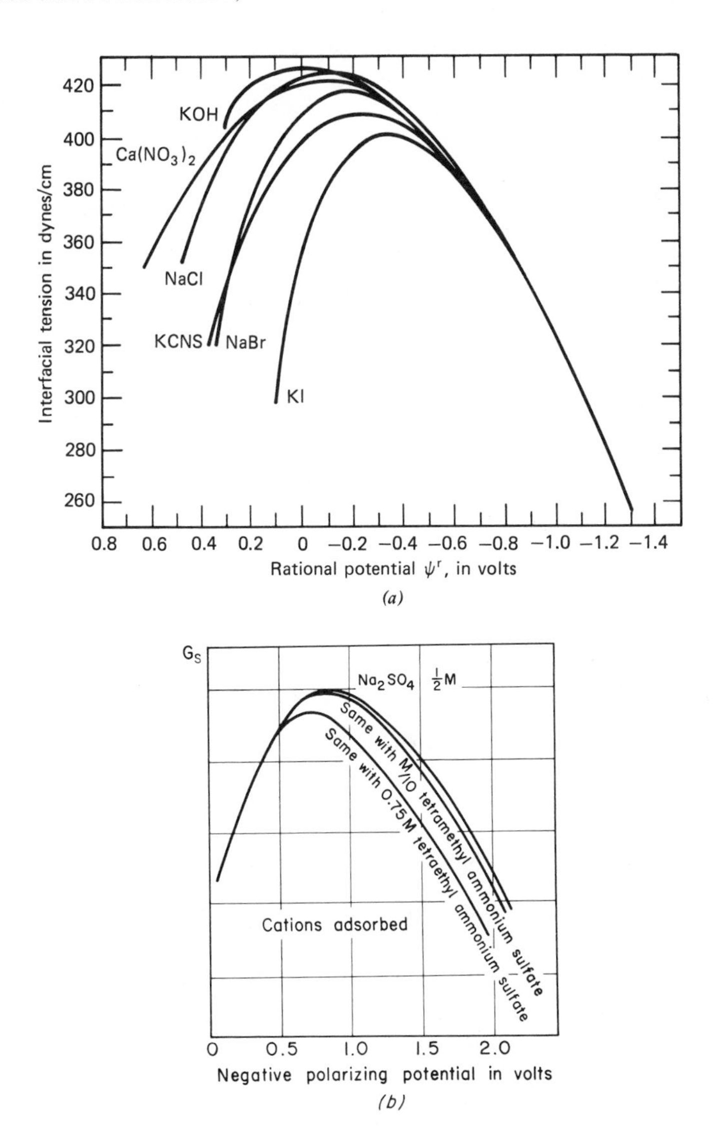

Fig. V-10. Electrocapillary curves: (*a*) adsorption of anions (from Ref. 32); (*b*) absorption of cations (from Ref. 5); (*c*) effect of varying solvent medium (from Ref. 33). (*d*) Electrocapillary curves for *n*-pentanoic acid in 0.1 *N* $HClO_4$. Solute activities from top to bottom are 0, 0.04761, 0.09096, 0.1666, and 0.500. (From Ref. 34.)

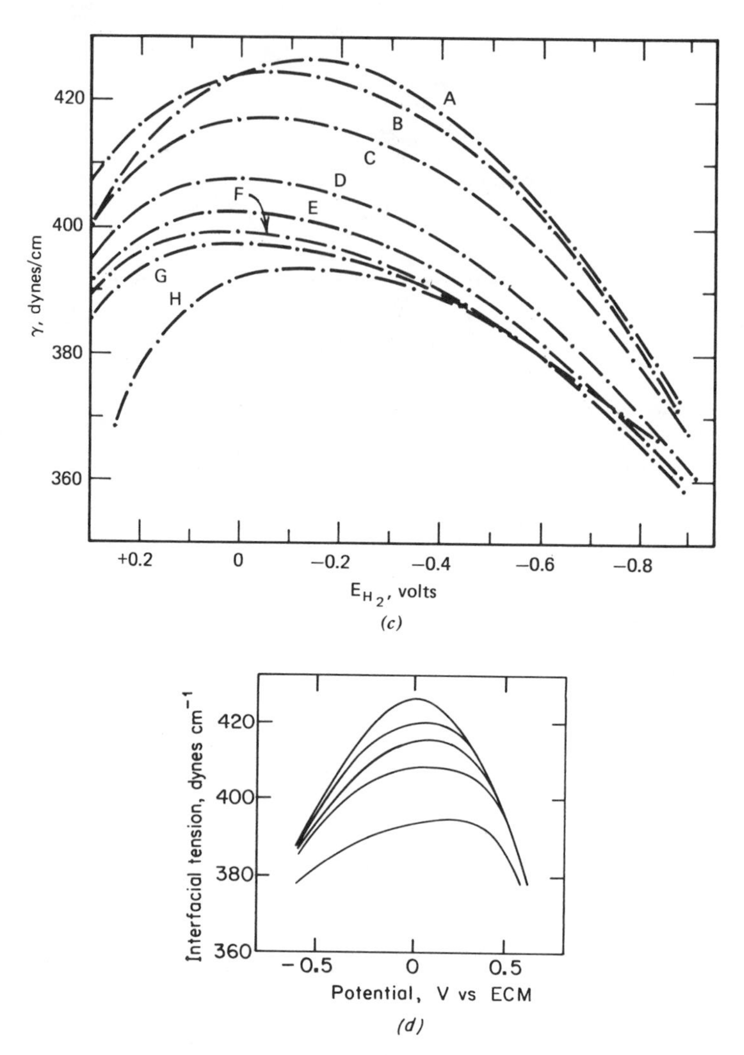

Fig. V-10. (*Continued*)

plied potential, the surface charge density can be calculated. Thus major parameters of double layer theory are directly determinable. This transforms electrocapillarity from what would otherwise be an obscure phenomenon to a powerful method for gaining insight into the structure of the interfacial region at all metal–solution interfaces.

A. *Thermodynamics of the Electrocapillary Effect*

The basic equations of electrocapillarity are the Lippmann equation (30)

$$\left(\frac{\partial \gamma}{\partial \mathbf{E}}\right)_{\mu} = -\boldsymbol{\sigma} \tag{V-61}$$

and the related equation

$$\left(\frac{\partial^2 \gamma}{\partial \mathbf{E}^2}\right)_\mu = -\frac{\partial \boldsymbol{\sigma}}{\partial \mathbf{E}} = C \tag{V-62}$$

where $\mathbf{E}$ is the applied potential, $\boldsymbol{\sigma}$ is the amount of charge on the solution side of the interface per square centimeter, C is the differential capacity of the double layer, and the subscript means that the solution (and the metal phase) composition is held constant.

A number of more or less equivalent derivations of the electrocapillary Eq. V-61 have been given, and these have been reviewed by Grahame (32). Lippmann based his derivation on the supposition that the interface was analogous to a parallel plate condenser, so that the reversible work dG, associated with changes in area and in charge, was given by

$$dG = \gamma\, d\mathcal{A} + \Delta\phi\, dq \tag{V-63}$$

where q is the total charge and $\Delta\phi$ is the difference in potential, the second term on the right thus giving the work to increase the charge on a condenser. Integration of Eq. V-63, keeping γ and $\Delta\phi$ constant, gives

$$G = \gamma\mathcal{A} + \Delta\phi q \tag{V-64}$$

Equation V-64 may now be redifferentiated, and on comparing the results with Eq. V-63, one obtains

$$0 = \mathcal{A}\, d\gamma + q\, d(\Delta\phi) \tag{V-65}$$

Since $q/\mathcal{A} = \boldsymbol{\sigma}$, Eq. V-65 rearranges to give Eq. V-61.

The treatments that are concerned in more detail with the nature of the adsorbed layer make use of the general thermodynamic framework of the derivation of the Gibbs equation (Section III-5B) but differ in the handling of the electrochemical potential and the surface excess of the ionic species (13, 35–37). The derivation given here is after that of Grahame and Whitney (37). Equation III-73 gives the combined first and second law statement for the surface excess quantities

$$dE^s = T\, dS^s + \sum_i \mu_i\, dm_i^s + \gamma\, d\mathcal{A} \tag{V-66}$$

Also,

$$\mu_i = \left(\frac{\partial E}{\partial m_i}\right)_{S,V,m_j,\mathcal{A}} \tag{V-67}$$

or, on a mole basis,

$$\bar{\mu}_1 = \frac{\partial E}{\partial n_i} \tag{V-68}$$

The chemical potential μ_i has been generalized to the *electrochemical* potential $\bar{\mu}_i$ since we will be dealing with phases whose charge may be varied.

The problem that now arises is that one desires to deal with individual ionic species and that these are not independently variable. In the present treatment, the difficulty is handled by regarding the electrons of the metallic phase as the dependent component whose amount varies with the addition or removal of charged components in such a way that electroneutrality is preserved. One then writes for the ith charged species,

$$\frac{\partial E}{\partial n_i} = \bar{\mu}_i + z_i\bar{\mu}_e \qquad \text{(V-69)}$$

where $\bar{\mu}_e$ is the electrochemical potential of the electrons in the system. The Gibbs equation (Eq. III-77) then becomes

$$d\gamma = -S^s\, dT - \sum_i \Gamma_i\, d\bar{\mu}_i - \sum_i \Gamma_i z_i\, d\bar{\mu}_e \qquad \text{(V-70)}$$

The electrochemical potentials $\bar{\mu}_i$ may now be expressed in terms of the chemical potentials μ_i and the electrical potentials (see Section V-11):

$$d\bar{\mu}_i = d\mu_i + z_i\mathcal{F}\, d\phi \qquad \text{(V-71)}$$

where $d\phi$ is the change in the electrical potential of the phase (for uncharged species $z_i = 0$, so the chemical and electrochemical potentials are equal). Since the metal phase α is at some uniform potential ϕ^α and the solution phase β is likewise at some potential ϕ^β, the components are grouped according to the phase they are in. The relationship Eq. V-71 is introduced into Eq. V-70 with this discrimination, giving

$$d\gamma = -S^s\, dT - \sum_i^c \Gamma_i\, d\mu_i - \sum_i^c \Gamma_i z_i\, d\mu_e - \mathcal{F}\sum_i^\alpha \Gamma_i z_i\, d\phi^\alpha$$
$$- \mathcal{F}\sum_i^\beta \Gamma_i z_i\, d\phi^\beta + \mathcal{F}\sum_i^\alpha \Gamma_i z_i\, d\phi^\alpha + \mathcal{F}\sum_i^\beta \Gamma_i z_i\, d\phi^\alpha \qquad \text{(V-72)}$$

In obtaining Eq. V-72, it must be remembered that $d\bar{\mu}_e = d\mu_e - \mathcal{F}\, d\phi^\alpha$, since electrons are confined to the metal phase. Canceling and combining terms,

$$d\gamma = -S^s\, dT - \mathcal{F}\sum_i \Gamma_i z_i\, d(\phi^\beta - \phi^\alpha) - \sum_i^c \Gamma_i\, d\mu_i - \sum_i^c \Gamma_i z_i\, d\mu_e \qquad \text{(V-73)}$$

or, because of the electroneutrality requirement,

$$\sum_i^c \Gamma_i z_i = \Gamma_e, \qquad d\gamma = -S^s\, dT - \boldsymbol{\sigma}\, d(\phi^\beta - \phi^\alpha) - \sum_i^{c+1} \Gamma_i\, d\mu_i \qquad \text{(V-74)}$$

In the preceding equations, the summation index c indicates summation over all components other than the electrons. The quantity $\mathcal{F}\sum_i\Gamma_i z_i$ in Eq. V-73 has been identified with $\boldsymbol{\sigma}$, the excess charge on the solution side of the interface. In general, however, this identification is not necessarily valid because the Γ's depend on the location of the dividing surface. For a *completely polarized* surface, where the term implies that no ionic species is to be found on *both* sides of the interface, the matter of locating the dividing surface becomes trivial and Eq. V-74 may be used. At constant temperature and composition, Eq. V-74 reduces to the Lippmann equation (Eq. V-61); it should be noted, however, that **E** in Eq. V-61 refers to the externally measured potential difference whereas $\Delta\phi$ is the difference in potential between the two phases in question. These two quantities are not actually the same but generally are thought to differ by some constant involving the nature of the reference electrode and of other junctions in the system; however, changes in them are taken to be equal so that **dE** may be substituted for $d(\Delta\phi)$.

One may also examine the variation of γ with solute concentration. The case of electrolyte solutes is discussed by Grahame (32); that of nonelectrolytes is examined in Section V-9D.

B. Experimental Methods

The various experimental methods associated with electrocapillarity may be divided into those designed to determine surface tension as a function of applied potential and those designed to measure directly either the charge or the capacity of the double layer.

The capillary electrometer, due to Lippmann (30) and illustrated in Fig. V-9, is the classical and still very important type of apparatus. It consists of a standing tube connected with a mercury reservoir and terminating in a fine capillary at the lower end; this capillary usually is slightly tapered and its diameter is of the order of 0.05 mm. There is an optical system, for example, a cathetometer for viewing the meniscus in the capillary. The rest of the system then consists of the solution to be studied, into which the capillary dips, a reference electrode, and a potentiometer circuit by means of which a variable voltage may be applied. The measurement is that of the height of the mercury column as a function of **E**, the applied potential. Since the column height is to be related to the surface tension it is essential that the solution wet the capillary so that the mercury–glass contact angle is 180°, and the meniscus can be assumed to be hemispherical, allowing Eq. II-11 to be used to obtain γ.

Usually one varies the head of mercury or applied gas pressure so as to bring the meniscus in the capillary to a fixed reference point (33). Koenig (38), Hansen and Williams (39), and Grahame and co-workers (40) have given more or less elaborate descriptions of the capillary electrometer apparatus. A more recent paper is that of Hills and Payne (41). Nowadays the capillary electrometer is customarily used in conjunction with capacitance measurements (see below).

Vos and Los (31) describe the use of sessile drop profiles (Section III-9B) for interfacial tension measurements, thus avoiding an assumption as to the solution–Hg–glass contact angle.

The surface charge density $\boldsymbol{\sigma}$ may be determined directly by means of an apparatus of the type shown schematically in Fig. V-11, in which a steady stream of mercury droplets is allowed to fall through the solution. Since for mercury the surface is positively charged, as each drop forms electrons flow back to the reservoir through the external circuit and to the bottom electrode. This apparatus was described by

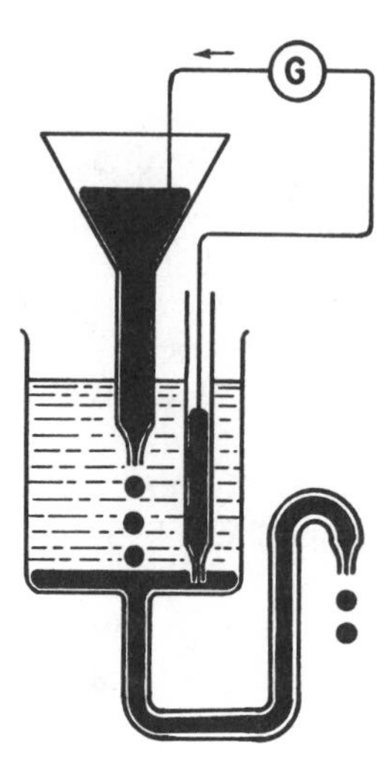

Fig. V-11. Lippmann apparatus for the direct determination of the surface charge.

Lippmann and by Grahame (42). If the total number of coulombs passing through the galvanometer is divided by the number of drops and then further by the surface area per drop, $\boldsymbol{\sigma}$ is obtained. One may, in addition, bias the dropping electrode with an applied voltage and observe the variation in current, and hence in $\boldsymbol{\sigma}$, with $\mathbf{E}$, allowing for the concomitant change in surface tension and hence in drop size.

Grahame (42), in addition, has described a capacitance bridge in which a direct determination of the double layer capacity could be made. In this case a slow and uniform rate of drop formation was established and, since the capacity of the system varied with drop size, the procedure adopted was to time the interval between the falling of one drop and the instant that the succeeding drop reached such a size that there was a moment of balance in the bridge. The capacity at that instant was therefore known and, by assuming that the drops went through a sequence of spherical shapes, the area of the drop at the point of balance could be calculated.

It should be noted that the capacity as given by $C_i = \boldsymbol{\sigma}/\mathbf{E}^r$, where $\boldsymbol{\sigma}$ is obtained from the current flow at the dropping electrode or from Eq. V-61, is an integral capacity ($\mathbf{E}^r$ is the potential relative to the electrocapillary maximum or ecm, and an assumption is involved here in identifying this with the potential difference across the interface). The differential capacity C given by Eq. V-62 is also then given by

$$C = C_i + \mathbf{E}^r\left(\frac{\partial C_i}{\partial \mathbf{E}^r}\right)_\mu \qquad \text{(V-75)}$$

C. *Results for the Mercury–Aqueous Solution Interface*

The shape of the electrocapillary curve is easily calculated if it is assumed that the double layer acts as a condenser of constant capacity C. In this case, double integration of Eq. V-62 gives

$$\gamma = \gamma_{\max} - \tfrac{1}{2}C(\mathbf{E}^r)^2 \qquad \text{(V-76)}$$

Incidentally, a quantity called the *rational potential* ψ^r is defined as $\mathbf{E}^r$ for the mercury–water interface (no added electrolyte) so, in general, $\psi^r = \mathbf{E} + 0.480$ V if a normal calomel reference electrode is used.

Equation V-76 is that of a parabola, and electrocapillary curves are indeed approximately parabolic in shape. Because $\mathbf{E}_{max}$ and γ_{max} are very nearly the same for certain electrolytes, such as sodium sulfate and sodium carbonate, it is generally assumed that specific adsorption effects are absent, and $\mathbf{E}_{max}$ is taken as a constant (−0.480 V) characteristic of the mercury–water interface. For most other electrolytes there is a shift in the maximum voltage, and ψ^r_{max} is then taken to be ($\mathbf{E}_{max}$ − 0.480). Some values for these quantities are given in Table V-3 (32). Much information of this type is due to Gouy (43), although additional results are to be found in most of the other references cited in this section.

The variation of the integral capacity with $\mathbf{E}$ is illustrated in Fig. V-12, as determined both by surface tension and by direct capacitance measurements; the agreement confirms the general correctness of the thermodynamic relationships. The differential capacity C shows a general decrease as $\mathbf{E}$ is made more negative, but may include maxima and minima; the case of nonelectrolytes is mentioned in the next subsection.

At the electrocapillary maximum, $d\gamma/d\mathbf{E}$ is zero and hence $\boldsymbol{\sigma}$ is zero. There may still be adsorption of ions, but in equal amounts, that is, $\Gamma_+^{max} = \Gamma_-^{max}$ (for a 1:1 electrolyte).

It follows from Eq. V-74 that

$$\frac{d\gamma^{max}}{d\mu} = -\Gamma_{salt}^{max} \tag{V-77}$$

Some of Grahame's values for Γ_{salt}^{max} and ψ'_{max} are included in Table V-3. For a common cation, the sequence of anions in order of increasing adsorption

TABLE V-3
Properties of the Electrical Double Layer at the Electrocapillary Maximum

Electrolyte	Concentration, M	$\mathbf{E}_{max}$	ψ'_{max}	Γ_{salt}^{max} $\mu C/cm^2$
NaF	1.0	−0.472	0.008	
	0.001	−0.482	−0.002	
NaCl	1.0	−0.556	−0.076	3.6
	0.1	−0.505	−0.025	1.1
KBr	1.0	−0.65	−0.17	10.6
	0.01	−0.54	−0.06	0.6
KI	1.0	−0.82	−0.34	15.2
	0.001	−0.59	−0.11	1.3
NaSCN	1.0	−0.72	−0.24	14.0
K_2CO_3	0.5	−0.48	0.00	−2.2
NaOH	1.0	−0.48	0.00	Small
Na_2SO_4	0.5	−0.48	0.00	Small

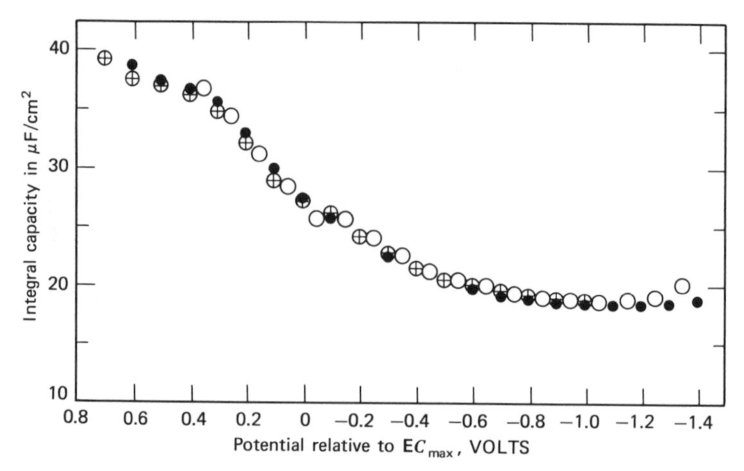

Fig. V-12. Variation of the integral capacity of the double layer with potential for 1 *N* sodium sulfate. ●, from differential capacity measurements; ⊕, from the electrocapillary curves; ○, from direct measurements. (From Ref. 32.)

is similar to that of the Hofmeister series in coagulation studies, and it is evident that specific adsorption properties are involved.

Some of the commonly accepted specific features are illustrated in Fig. V-13. There may be chemically adsorbed ions of the same charge as that of the interface, anions in this case; these are dehydrated and the Inner Helmholtz Plane (IHP) passes through their centers. A Stern type layer of mainly electrostatically adsorbed cations follows, the Outer Helmholtz Plane (OHP) passes through the centers of these ions. Note that most of the interface is occupied by solvent water; for even a highly charged interface the surface concentration of adsorbed ions is quite low—a typical value might be 100 $Å^2$ per ion. The surface layer of solvent molecules is shown as highly oriented; the result is that their effective dipole moment is quite low.

This writer has one caveat with respect to diagrams such as that of Fig. V-13. It is pointed out in a later chapter (Section XIV-4B) that a liquid surface can be heterogeneous in the sense that it is perturbed by the adsorbate. There seems no particular reason why the mercury surface should not better be visualized as dimpled rather than plane. Also, the IHP and OHP are merely planes of average electrical property; the actual local potentials, if they could be measured, must be varying wildly between locations near an adsorbed ion and locations where only solvent water is present in the surface layer.

D. Effect of Uncharged Solutes and Changes of Solvent

The location and shape of the entire electrocapillary curve are affected if the general nature of the medium is changed. This is illustrated by the data in Fig.

V-10*c* for the case of 0.01 *N* hydrochloric acid solution in mixed water–ethanol media of various compositions (33, 37). The surface adsorption of methanol, obtained by the use of Eq. V-78,

$$\Gamma = -\left(\frac{\partial \gamma}{\partial \mu}\right)_{\mathrm{E}} \tag{V-78}$$

was always positive and varied with the solution composition in a manner similar to the behavior at the solution–air interface.

Gouy (43) made a large number of determinations of the effect of added organic solutes, as have others since and, more recently, Hansen and co-workers. Fig. V-10*d* shows data for *n*-pentanoic acid solutions; similar results are obtained for other acids and for various alcohols. The data may be worked up as follows. One first

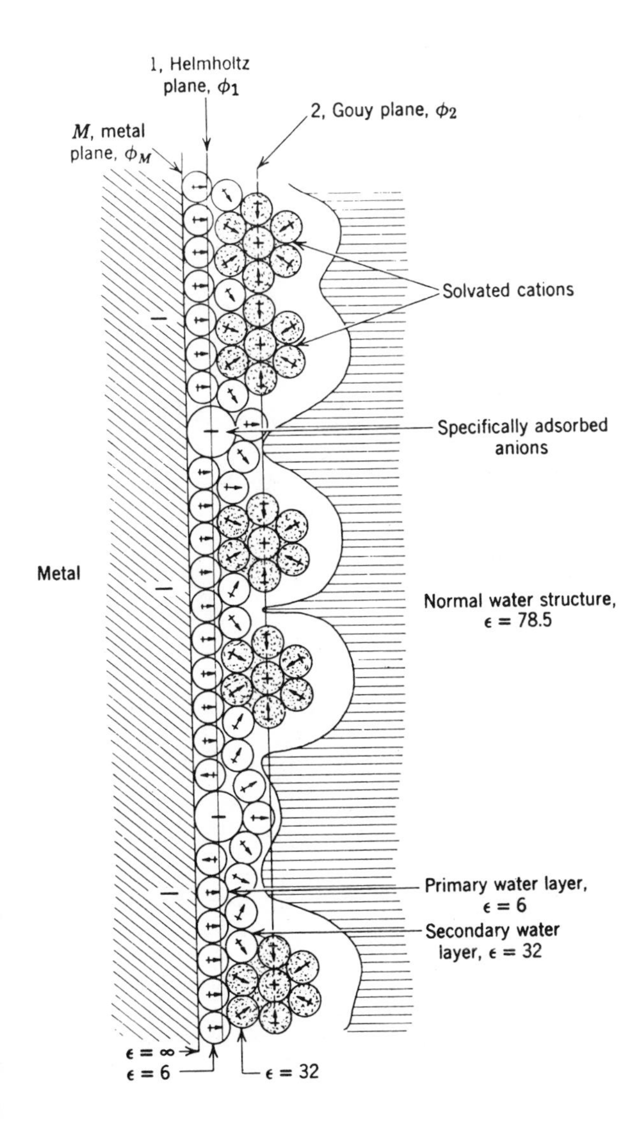

Fig. V-13. A detailed model of the double layer (from Ref. 48).

constructs plots of π versus ln a at constant potential, where a is the solute activity, defined as C/C_0 where C_0 is the solubility of the solute. For a given system, the plots were all of the same shape and superimposed if suitably shifted along the abscissa scale, that is, if a were multiplied by a constant (different for each voltage). The composite plot for 3-pentanol is shown in Fig. V-14 (44; see also Refs. 34, 45–47). Differentiation and use of Eq. V-78 then yields data from which plots of Γ versus a may be constructed, that is, adsorption isotherms.

The isotherms could be fit to the Frumkin equation,

$$\frac{\theta}{1 - \theta} = Bae^{2\alpha\theta} \tag{V-79}$$

where $\theta = \Gamma/\Gamma_m$, Γ_m being Γ at maximum coverage, and the constants B and α describe the adsorption equilibrium and lateral interaction, respectively. Later (34), a "Flory–Huggins" type of isotherm equation was tested,

$$\frac{\theta}{(1 - \theta)^x} = Bae^{2\alpha\theta} \tag{V-80}$$

The constant x allows for the possibility that the adsorbate displaces more than one solvent molecule at the interface. Empirical, that is best-fitting, x values may not have much real physical meaning, however; the added parameter simply allows better fitting of the data.

A typical differential capacity verus **E** plot is shown in Fig. V-15. If the Frumkin equation is used, and the interface modeled as consisting of capacities in parallel, one characteristic of the Hg–solvent interface and the other, of the Hg–adsorbate interface, then the differential capacity data could be fairly well represented. The maxima can be explained on the basis that, as the magnitude of the surface charge

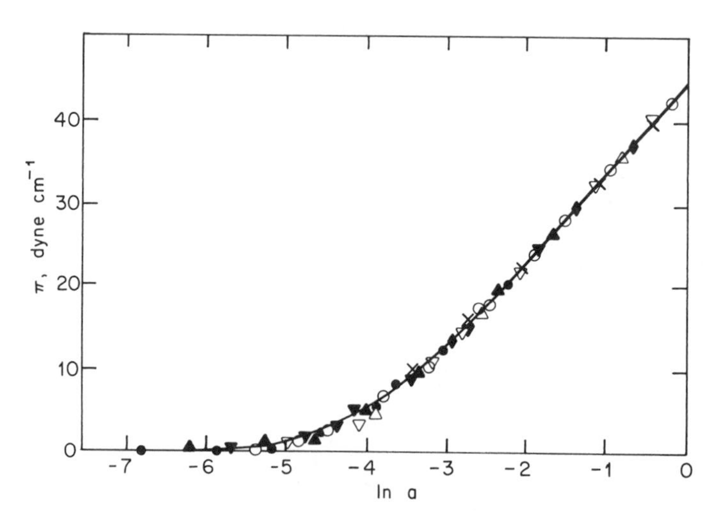

Fig. V-14. Composite π versus ln a curve for 3-pentanol. The various data points are for different **E**'s; each curve for a given **E** has been shifted horizontally to give the optimum match to a reference curve, for an **E** near the electrocapillary maximum. (From Ref. 44.)

and electric field increase, there is an increased preference for material of high dielectric constant in the compact or adsorption layer, leading to displacement of organic molecules by water. For additional and alternative discussion of such effects, see Frumkin and Damaskin (49).

E. Other Electrocapillary Systems

Scattered data are available on the electrocapillary effect when liquid metal phases other than pure mercury are involved. Frumkin and co-workers (50) have reported on the electrocapillary curves of amalgams with less noble metals than mercury, such as thallium or cadmium. The general effect is to shift the maximum to the right, and Koenig (36) has discussed the thermodynamic treatment for the adsorption of the metal solute at the interface. Liquid gallium gives curves similar to those for mercury, again shifted to the right; this and other systems such as those involving molten salts as the electrolyte are reviewed by Delahay (51). Narayan and Hackerman (52) studied adsorption at the In–Hg–electrolyte interface.

The equations of electrocapillarity become complicated in the case of the solid metal–electrolyte interface. The problem is that the work spent in a differential *stretching* of the interface is not equal to that in *forming* an infinitesimal amount of new surface, if the surface is under elastic strain. Couchman and co-workers (53, 54) and Mohliner and Beck (55) have, among others, discussed the thermodynamics of the situation, including some of the problems of terminology.

10. The Electrified Solid–Liquid Interface

A great deal of work has appeared on the behavior of electrified solid–liquid interfaces. The subject merges with the large fields of electrochemistry and solid state physics of semiconductors and is treated only briefly and in selected aspects in this and the following section.

There are, of course, problems relative to the case of the mercury–liquid interface. It is not possible to measure the interfacial tension directly, and electrocapillary curves are obtained by indirect means. Two principal approaches are the following. First, the solid, if conducting, may be studied as an electrode. Various techniques

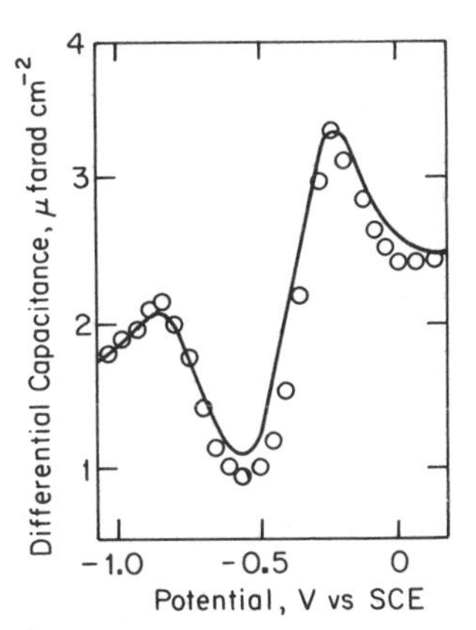

Fig. V-15. Differential capacitance curve for 0.1 N $HClO_4$ solution containing n-pentanoic acid, a = 0.04761. The points are from direct experimental measurements, and the solid line is calculated from the Frumkin equation (see text). (Reprinted with permission from Ref. 46. Copyright 1976 American Chemical Society.)

allow the measurement of the capacitance of the interfacial region as a function of potential, electrolyte, and so on. Second, the solid may be present as a finely divided suspension—essentially a colloidal phase. The absolute interfacial potential may not be measured directly, but changes in it can be made quantitatively by varying the concentration of a potential-determining ion. Also, the zeta potential can be determined by electrophoretic measurements and the results used to approximate the interfacial potential. Finally, being finely divided, there is enough surface area for the direct determination of the amount of electrolyte and other species adsorbed.

A. *Electrode–Solution Interface*

Of general interest is the applied potential such that there is zero charge at the interface (the *pzc*). Bockris and co-workers (56) [see also Bockris and Reddy (57)] compare three methods for determining the *pzc* (which is also the potential at the maximum of the electrocapillary curve, from Eq. V-61). Adsorption, even on the small area of an electrode, can be measured by using radioactive labeling of the adsorbate. A neutral adsorbate species should show a maximum (or minimum) in adsorption at the *pzc*. The interfacial capacitance should go through a minimum at the *pzc*, and, interestingly, *friction* at the metal–electrolyte solution interface appears to go through a maximum at the *pzc*. The three methods gave approximately the same value for the *pzc* as determined for four metals, Ag, Au, Ni, and Pt. It might be thought that since $\boldsymbol{\sigma}$ is zero at the *pzc*, the applied potential should correspond to the *absolute* potential difference between the metal and solution phases. This is not so since adsorption both of neutral solute species and of solvent produces potential effects at the metal–solution interface. The matter of absolute half-cell emf's has been discussed by Gomer and Tryson (58), see Section V-11C.

It is possible to measure the capacitance at a metal–solution interface from the potential versus time behavior, following a voltage pulse. In the case of iron with an oxide film, it was assumed that

$$\frac{1}{C} = \frac{1}{C_\mathrm{f}} + \frac{1}{C_\mathrm{d}} \tag{V-81}$$

where C_f is the capacitance associated with the oxide film and C_d is that assigned to the electrical double layer (estimated at 20 μ F/cm^2) (59). For a film of estimated thickness 100 Å, C_f as obtained from C and Eq. V-81 corresponded to a film dielectric constant of about 80.

B. *Silver Halide–Solution Interface*

Much of the work with the silver halide–solution interface has been with colloidal suspensions; considerable success has been achieved in preparing monodisperse silver halide sols (12) so that well defined interfaces may be studied. There are indications that the state of charge at the interface, and its variation with concentration of silver or halide ions in solution is complicated by the semiconductor nature of silver halides. The presence of defects (see Chapter VII) can allow the presence of a diffuse double layer in the solid phase, for example.

11. Types of Potentials and the Meaning of Potential Differences when Two Phases Are Involved

A. The Various Types of Potentials

Various kinds of potentials have been referred to in the course of this and the preceding chapter, and their interrelation is the subject of the present section. The chief problem is that certain types of potential differences are physically meaningful in the sense that they are operationally defined, whereas others that may be spoken of more vaguely are really conceptual in nature and may not be definable experimentally.

It is easy, for example, to define the potential difference between two points in a vacuum; this can be done by means of Coulomb's law in terms of the work required to transport unit charge from one point to the other. The potential of a point in space, similarly, is given by the work required to bring unit charge up from infinity. It is this type of definition that is involved in the term *Volta potential* ψ, which is the potential *just outside* or, practically speaking, about 10^{-3} cm from a phase. It is defined as $1/e$ times the work required to bring unit change from infinity up to a point close to the surface. In like manner, one may define the *difference* in potential between two points, *both within a given phase*, in terms of the work required to transport unit charge from one point to the other. An example would be that ζ potential or potential difference between bulk solution and the shear layer of a charged particle.

The electrostatic potential *within* a phase, that is, $1/e$ times the electrical work of bringing unit charge from vacuum at infinity into the phase, is called the *Galvani*, or *inner potential* ϕ. Similarly, the electrostatic potential difference between two phases is $\Delta\phi$. This quantity presents some subtleties, however. If one were to attempt to measure $\Delta\phi$ for a substance by bringing a charge from infinity in vacuum into the phase, the work would consist of two parts, the first being the work to bring the charge to a point just outside the surface, giving the Volta potential, and the second being the work to take the charge across the phase boundary into the interior, giving the *surface potential jump* χ. The situation is illustrated in Fig. V-16. The relationship between ϕ, ψ, and χ is simply

$$\phi = \chi + \psi \tag{V-82}$$

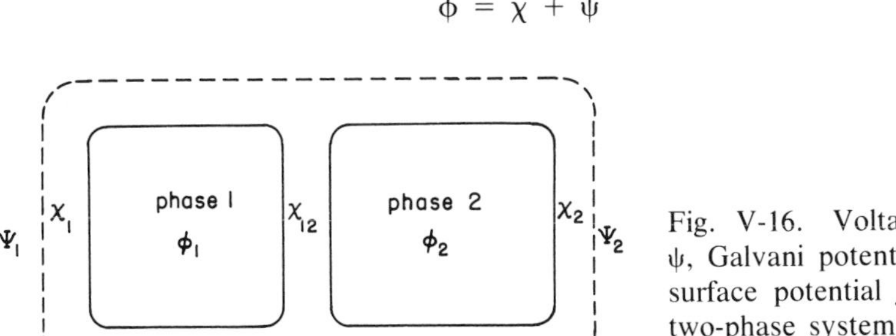

Fig. V-16. Volta potentials ψ, Galvani potentials ϕ, and surface potential jumps χ in two-phase system (from Ref. 60).

and the distinctions above between them are those introduced by Lange (60).

The problem is that ϕ, although apparently defined, is not susceptible to absolute experimental determination. The difficulty is that the unit charge involved must in practice be some physical entity such as an electron or an ion. When some actual charged species is transported across an interface, work will be involved, but there will be, in addition to the electrostatic work, a chemical work term that may be thought of as involving van der Waals forces, exchange forces, image forces, and so on. Some theoretical estimates of χ have been made, and Verwey (61) (see also Ref. 57) gives χ for the water–vacuum interface as -0.5 V, which has the physical implication that surface molecules are oriented with the hydrogens directed outward.

The *electrochemical potential* $\bar{\mu}_i^\alpha$ is defined as the total work of bringing a species i from vacuum into a phase α and is thus experimentally defined. It may be divided into a chemical work μ_i^α, the *chemical potential*, and the electrostatic work $z_i e\phi^\alpha$:

$$\bar{\mu}_i^\alpha = \mu_i^\alpha + z_i e\phi^\alpha$$

$$\bar{\mu}_i^\alpha = \mu_i^\alpha + z_i e(\chi^\alpha + \psi^\alpha) \quad \text{(V-83)}$$

Since ψ^α is experimentally measurable, it is convenient to define another potential, the *real potential* α_i^α:

$$\alpha_i^\alpha = \mu_i^\alpha + z_i e\psi^\alpha \quad \text{(V-84)}$$

or

$$\bar{\mu}_i^\alpha = \alpha_i^\alpha + z_i e\chi^\alpha \quad \text{(V-85)}$$

Thus $\bar{\mu}_i$, α_i, and ψ are experimentally definable, while the surface potential jump χ, the chemical potential μ, and the Galvani potential difference between two phases $\Delta\phi = \phi^\beta - \phi^\alpha$ are not. While $\bar{\mu}_i$ is defined, there is a practical difficulty in carrying out an experiment in which the *only* change is the transfer of a charged species from one phase to another; electroneutrality generally is maintained in a reversible process so that it is a *pair* of species that is transferred, for example, a positive and a negative ion, and it is the sum of their electrochemical potential changes that is then determined. As a special case of some importance, the difference in chemical potential for a species i between two phases of *identical* composition is given by

$$\bar{\mu}_i^\beta - \bar{\mu}_i^\alpha = z_i e(\phi^\beta - \phi^\alpha) = z_i e(\psi^\beta - \psi^\alpha) \quad \text{(V-86)}$$

since now $\mu_i^\beta = \mu_i^\alpha$ and $\chi^\beta = \chi^\alpha$.†

Finally, the difference in Volta potential between two phases is the *surface potential* ΔV, discussed in Chapter III:

$$\Delta V = \psi^\beta - \psi^\alpha \quad \text{(V-87)}$$

† Two phases at a different potential cannot have the identical chemical composition, strictly speaking. However, it would only take about 10^{-17} mole of electrons or ions to change the electrostatic potential of the 1 cc portion of matter by 1 V, so that little error is involved.

Another potential, mentioned in the next section, is the *thermionic work function* Φ, where $e\Phi$ gives the work necessary to remove an electron from the highest populated level in a metal to a point outside. We can write

$$\Phi = \mu + e\chi \tag{V-88}$$

Although certain potentials, such as the Galvani potential difference between two phases, are not experimentally well defined, *changes* in them may sometimes be related to a definite experimental quantity. Thus in the case of electrocapillarity, the imposition of a potential **E** on the phase boundary is taken to imply that a corresponding change in $\Delta\phi$ occurs, that is, $d\mathbf{E} = d(\Delta\phi)$. It might also be noted that the practice of dividing the electrochemical potential of a charged species into a chemical and an electrostatic part is actually an arbitrary one not strictly necessary to thermodynamic treatments. It has been used by Brønsted (62) and Guggenheim (63) as a useful way of separating out the feature common to charged species that their free energy does depend on the value of the electric field present. For further discussions of these various potentials, see Butler (64), Adam (65), de Boer (66). Case et al. (67) have reported on the determination of real potentials and on estimates of χ.

B. Volta Potentials, Surface Potential Differences, and the Thermionic Work Function

The thermionic work function for metals may be measured fairly accurately, and an extensive literature exists on the subject. A metal, for example, will spontaneously emit electrons, since their escaping tendency into space is enormously greater than that of the positive ions of the metal. The equilibrium state that is finally reached is such that the accumulated positive charge on the metal is sufficient to prevent further electrons from leaving. Alternatively, if a metal filament is negatively charged, electrons will flow from it to the anode; the rate of emission is highly temperature dependent, and Φ may be calculated from this temperature dependence (68). Φ may also be obtained from the temperature coefficient of the photoelectric emission of electrons or from the long wavelength limit of the emission. As the nature of these measurements indicates, Φ represents an energy rather than a free energy.

When two dissimilar metals are connected, as illustrated in Fig. V-17, there is a momentary flow of electrons from the metal with the smaller work function to the other so that the electrochemical potential of the electrons becomes the same. For the two metals α and β,

$$\bar{\mu}_e = -e(\Phi^\alpha + \psi^\alpha) = -e(\Phi^\beta + \psi^\beta) \tag{V-89}$$

The difference in Volta potential ΔV, which has been called the surface (or contact) potential in this book, is then given by

$$\Delta V = \psi^\alpha - \psi^\beta = \Phi^\beta - \Phi^\alpha \tag{V-90}$$

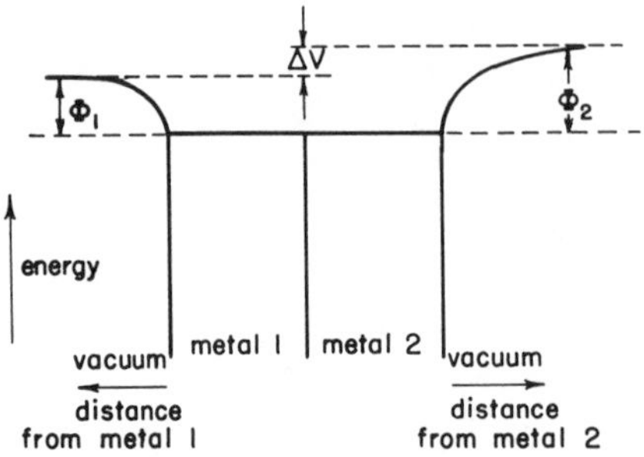

Fig. V-17. The surface potential difference ΔV.

It is this potential difference that is discussed in Chapter IV in connection with monomolecular films. Since it is developed in the space between the phases, none of the uncertainties of phase boundary potentials is involved.

In the case of films on liquids, ΔV was measured directly rather than through determinations of thermionic work functions. However, just as an adsorbed film greatly affects ΔV, so must it affect Φ. In the case of metals, the effect of adsorbed gases on ΔV is easily studied by determining changes in the work function, and this approach is widely used today (see Section VIII-2C). Much of Langmuir's early work on chemisorption on tungsten may have been stimulated by a practical appreciation of the important role of adsorbed gases in the behavior of vacuum tubes.

Semiconductors form a special class of substances, and one for which surface effects are very important. Their discussion, as a class, is outside the scope of this book, although results for semiconductor substances are mentioned from time to time in connection with other material. Holmes (69) has edited a monograph on the electrochemistry of semiconductors.

C. Electrode Potentials

We now consider briefly the matter of electrode potentials. The familiar Nernst equation was at one time treated in terms of the "solution pressure" of the metal in the electrode, but it is better to consider directly the net chemical change accompanying the flow of 1 faraday ($\mathcal{F}$), and to equate the electrical work to the free energy change. Thus, for the cell

$$\text{Zn}'/\text{solution containing Zn}^{2+} \text{ and Ag}^+/\text{Ag}/\text{Zn}'' \qquad \text{(V-91)}$$

the net cell reaction is

$$\tfrac{1}{2}\text{Zn} + \text{Ag}^+ = \tfrac{1}{2}\text{Zn}^{2+} + \text{Ag} \qquad \text{(V-92)}$$

and

$$-\mathcal{F}\mathcal{E} = \mu_{\text{Ag}}^{\text{Ag}} - \tfrac{1}{2}\mu_{\text{Zn}}^{\text{Zn}} - \mu_{\text{Ag}^+}^{\text{S}} + \tfrac{1}{2}\mu_{\text{Zn}^{2+}}^{\text{S}} \qquad \text{(V-93)}$$

where the superscripts denote the phase involved. Equation V-93 may be abbreviated

$$\mathcal{F}\mathcal{E} = \Delta\mu_0 - RT \ln \frac{a_{\text{Zn}^{2+}}^{1/2}}{a_{\text{Ag}^+}} \qquad \text{(V-94)}$$

Equation V-94 may be derived in a more detailed fashion by considering that the difference between the electrochemical potentials of the electrons at the Zn′ and

Zn″ terminals, the chemical potentials being equal, gives $\Delta\phi$ and hence $\mathscr{E}$. Thus

$$\bar{\mu}_e'' - \bar{\mu}_e' = -(\phi'' - \phi') = -\mathscr{F}\mathscr{E} \tag{V-95}$$

Since the electrons are in equilibrium,

$$\bar{\mu}_e' = \bar{\mu}_e^{\mathrm{Zn}} \qquad \bar{\mu}_e'' = \bar{\mu}_e^{\mathrm{Ag}} \tag{V-96}$$

Also, there is equilibrium between electrons, metallic ions, and metal atoms within each electrode:

$$\tfrac{1}{2}\bar{\mu}_{\mathrm{Zn}^{2+}}^{\mathrm{Zn}} + \bar{\mu}_e^{\mathrm{Zn}} = \tfrac{1}{2}\mu_{\mathrm{Zn}}^{\mathrm{Zn}} \tag{V-97}$$

$$\bar{\mu}_{\mathrm{Ag}^+}^{\mathrm{Ag}} + \bar{\mu}_e^{\mathrm{Ag}} = \bar{\mu}_{\mathrm{Ag}}^{\mathrm{Ag}} \tag{V-98}$$

Then

$$-\mathscr{F}\mathscr{E} = \mu_{\mathrm{Ag}}^{\mathrm{Ag}} - \tfrac{1}{2}\mu_{\mathrm{Zn}}^{\mathrm{Zn}} - \bar{\mu}_{\mathrm{Ag}^+}^{\mathrm{Ag}} + \tfrac{1}{2}\bar{\mu}_{\mathrm{Zn}^{2+}}^{\mathrm{Zn}} \tag{V-99}$$

In addition, since there is equilibrium between the metal ions in the two phases,

$$\bar{\mu}_{\mathrm{Zn}^{2+}}^{\mathrm{Zn}} = \bar{\mu}_{\mathrm{Zn}^{2+}}^{\mathrm{S}} \qquad \bar{\mu}_{\mathrm{Ag}^+}^{\mathrm{Ag}} = \bar{\mu}_{\mathrm{Ag}^+}^{\mathrm{S}} \tag{V-100}$$

Substitution of Eq. V-100 into Eq. V-99 gives Eq. V-93, since

$$\tfrac{1}{2}\bar{\mu}_{\mathrm{Zn}^{2+}}^{\mathrm{S}} - \bar{\mu}_{\mathrm{Ag}^+}^{\mathrm{S}} = \tfrac{1}{2}\mu_{\mathrm{Zn}^{2+}}^{\mathrm{S}} - \mu_{\mathrm{Ag}^+}^{\mathrm{S}}$$

This approach, much used by Guggenheim (63), seems elaborate, but in the case of more complex situations than the above, it can be a very powerful one.

A problem that has fascinated surface chemists is whether, through suitable measurements, one can determine *absolute* half-cell potentials. If some one standard half-cell potential can be determined on an absolute basis, then all others are known through the table of standard potentials. Thus, if we know $\mathscr{E}^0$ for

$$\mathrm{Ag} = \mathrm{Ag}^+\,(\mathrm{aq}) + e^- \tag{V-101}$$

we can immediately obtain $\mathscr{E}^0_{\mathrm{H_2/H^+}}$ since $\mathscr{E}^0_{\mathrm{A/Ag^+}} - \mathscr{E}^0_{\mathrm{H_2/H^+}}$ is -0.800 V at 25°C.

The standard states of Ag and of $\mathrm{Ag^+}$(aq) have the conventional definitions, but there is an ambiguity in the definition of the standard state of e^-. Suppose that a reference electrode R is positioned above a solution of $\mathrm{AgNO_3}$, which in turn is in contact with an Ag electrode. The Ag electrode and R are connected by a wire. Per Faraday, the processes are:

$$\begin{aligned} &e^-(\text{in } R) = e^-\ (\text{in air}) \\ &e^-(\text{in air}) = e^-\ (\text{in solution}) \\ &\underline{e^-[\text{in solution} + \mathrm{Ag}^+(\mathrm{aq})] = \mathrm{Ag}} \\ &e^-(\text{in } R) + \mathrm{Ag}^+(\mathrm{aq}) = \mathrm{Ag} \end{aligned} \tag{V-102}$$

The potential corresponding to the reversible overall process is the measurable quantity V_{obs}. If we know the work function for R, that is, the potential Φ_R for e^-(in R) = e^- (in air), then $V_{\mathrm{obs}} - \Phi_R$ is $\mathscr{E}$ for the process

$$e^-(\text{in air}) + \mathrm{Ag}^+(\mathrm{aq}) = \mathrm{Ag} \tag{V-103}$$

On correcting to unit activity $\mathrm{Ag^+}$(aq), we can obtain $\mathscr{E}^0_{\mathrm{Ag/Ag^+}}$. Electron solvation energy is neglected in this definition.

Fig. V-18 shows the apparatus used by Gomer and Tryson (58) for measurements of V_{obs}. While their analysis (and literature review) is much more detailed than the foregoing discussion, they concluded that $\mathscr{E}^0_{\mathrm{H_2/H^+}}$ is -4.73 ± 0.05 V at 25°C.

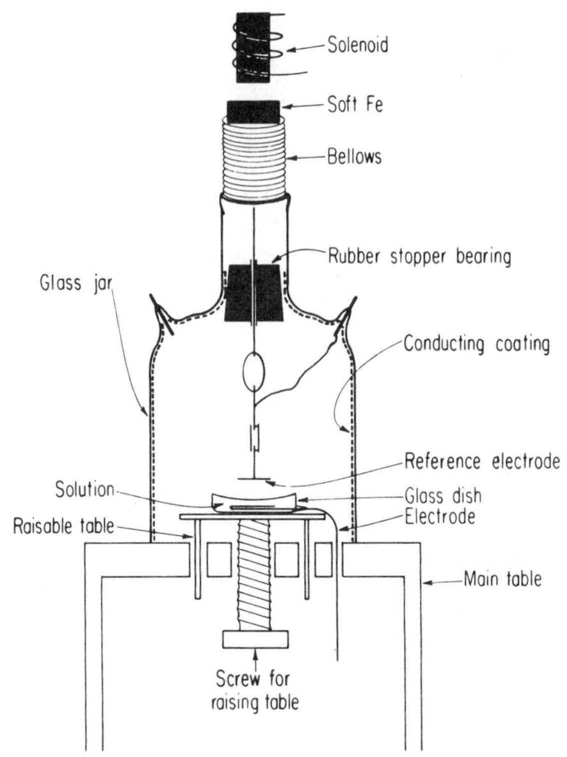

Fig. V-18. Schematic diagram for the apparatus for measurement of V_{obs} (see text). The vibrating reference electrode is positioned close to the surface of a $AgNO_3$ solution in which there is an Ag electrode, which in turn is in electrical contact with the reference electrode. (From Ref. 58.)

D. Irreversible Electrode Phenomena

An important and common phenomenon is that of overvoltage or the observation that in order to pass appreciable current through an electrochemical cell it is frequently necessary to apply a larger than reversible potential to the electrodes. A simple method for studying overvoltages is illustrated in Fig. V-19. The desired current is passed through the electrode to be studied E_x by means of the circuit $E_1 - E_x$, and the potential variation at E_x is measured by means of a potentiometer circuit $E_x - E_h$. One may either make the potential measurement while current is being passed through the circuit $E_1 - E_x$ or, by means of a rotating commutator or thyratron switching, make the measurement just after turning off the current.

Overvoltage effects may be divided roughly into three main classes with respect to causes. First, whenever current is flowing there will be chemical change at the electrode and a corresponding local accumulation or depletion of material in the adjacent solution. This effect, known as *concentration polarization*, can be serious. Usually, however, it is possible to eliminate it by suitable stirring. Second, overvoltage arises from the Ohm's law potential drop when appreciable current is flowing. This effect is not very important when only small currents are involved and may be further minimized by the use of the circuit $E_x - E_h$ with the electrode E_h in the form of a probe positioned close to E_x and on the opposite side to the one facing E_1. The commutator method also is designed to eliminate ohmic overvoltage. Finally, the most interesting class of overvoltage effect is that where some essential step in the electrode reaction itself is slow, presumably due to some activated chemical process being involved. In general, a further distinction between these three classes of overvoltage phenomena is that, on shutting off the polarizing current, ohmic currents cease instantly and concentration polarization decays slowly and in a complex way, whereas activation polarization often decays exponentially.

Considering now only the third type of overvoltage, most electrodes involving a metal and its ion in solution are fairly reversible and rather high current densities are required to produce appreciable overvoltages. Actually, the effect has been studied primarily in connection with gas electrodes, particularly with respect to hydrogen overvoltage. In this case rather small current flow ($\mu A/cm^2$) can produce a sizable effect in terms of the overvoltage to cause visible hydrogen evolution; values range from essentially zero for platinized platinum (and 0.09 for smooth platinum) to 0.78 V for mercury.

The first law of electrode kinetics, observed by Tafel in 1905 (70) is that overvoltage η varies with current density i, according to the equation

$$\eta = \alpha - b \ln i \tag{V-104}$$

where η is negative for a cathodic process and positive for an anodic process.

This law may be accounted for in a simple way as follows:

$$O(\text{oxidized state}) + e^- = R(\text{reduced state}) \tag{V-105}$$

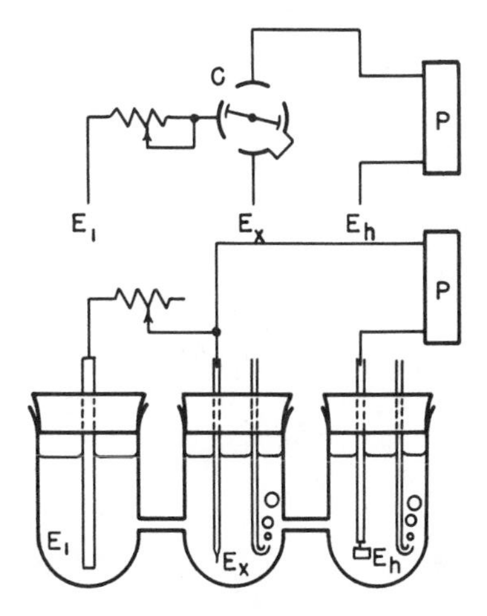

Fig. V-19. Apparatus for measuring overvoltages.

that is activated. Then the forward rate is given by

$$R_f = i_f = \frac{kT}{h} e^{-\Delta G_f^{0\ddagger}/RT}(O) \tag{V-106}$$

where $\Delta G_f^{0\ddagger}$ is the standard free energy of activation. Similarly, for the reverse rate.

$$R_b = i_b = \frac{kT}{h} e^{-\Delta G_f^{0\ddagger}/RT}(R) \tag{V-107}$$

It is now assumed that $\Delta G_f^{0\ddagger}$ consists of a chemical component and an electrical component and that it is only the latter that is affected by changing the electrode potential. The specific assumption is that

$$\Delta G_f^{0\ddagger} = (\Delta G_f^{0\ddagger})\,\text{chem} + \alpha\mathcal{F}(\phi_M - \phi_S) \tag{V-108}$$

that is, that a certain fraction α of the potential difference between the metal and solution phase contributes to the activation energy. This coefficient α is known as the *transfer coefficient* (see Ref. 57). Then

$$i_f = \frac{kT}{h} e^{-(\Delta G_{f\,\text{chem}}^{0\ddagger}/RT)} e^{-(\alpha\mathcal{F}/RT)(\phi_M - \phi_S)}(0) \tag{V-109}$$

Equations such as V-109 are known as Butler-Volmer equations (57). At equilibrium, there will be equal and opposite currents in both directions, $i_f^0 = i_b^0 = i^0$. By Eq. V-109 the apparent *exchange* current i^0 will be

$$i_0 = \frac{kT}{h} e^{-(\Delta G_{f\,\text{chem}}^{0\ddagger}/RT)} e^{-(\alpha\mathcal{F}/RT)(\phi_M{}^0 - \phi_S)}(0) \tag{110}$$

where $(\phi_M^0 - \phi_S)$ is now the equilibrium potential difference at the electrode. Combination of Eqs. V-109 and V-110 give

$$i_f = i^0 e^{-(\alpha\mathcal{F}/RT)(\phi_M - \phi_M{}^0)} = i^0 e^{-(\alpha\mathcal{F}/RT)\eta} \tag{V-111}$$

Equation V-111 may be written in the form

$$\ln i = \ln i^0 - \frac{\alpha\mathcal{F}}{RT}\eta \tag{V-112}$$

which in turn rearranges to the Tafel equation.

The treatment may be made more detailed by supposing that the rate determining step is actually from species O in the Outer Helmholtz Plane (at potential ϕ_2 relative to the solution) to species R similarly located. The effect is to make i^0 dependent on the value of ϕ_2 and hence on any changes in the electrical double layer. This type of analysis has permitted some detailed interpretations to be made of kinetic schemes for electrode reactions and also connects that subject to the general one of this chapter.

The measurement of α from the experimental slope of the Tafel equation may help to decide between rate determining steps in an electrode process. Thus in the reduction of water to evolve H_2 gas, if the slow step is the reaction of H_3O^+ with the metal M to form surface hydrogen atoms, M–H, α is expected to be about $\frac{1}{2}$. If, on the other hand, the slow step is the surface combination of two hydrogen atoms to form H_2, a second-order process, then α should be 2 (see Ref. 57).

A newly developing interest is in electrochemistry at "well defined" surfaces, that is, involving electrodes such as Pt, whose surface structure has been characterized by surface spectroscopic and diffraction methods (71). An old subject, but one that has received recent impetus because of the interest in solar energy conversion, is that of photo-assisted electrode processes.

12. Problems

1. ψ_0 is $-$ 50 mV for a certain silica surface in contact with 0.001 M aqueous NaCl at 25°C. Calculate, assuming simple Gouy-Chapman theory (*a*) ψ at 100 Å from the surface, (*b*) the concentrations of Na^+ and of Cl^- ions 5 Å from the surface, and (*c*) the surface charge density in electronic charges per unit area.

2. Using the Gouy-Chapman equations calculate and plot ψ at 10 Å from a surface as a function of ψ_0 (from 0 to 100 mV at 25°C) for 0.001 M and for 0.01 M 1:1 electrolyte.

3. Show that Eq. V-12 can be written in the equivalent form

$$y = 2 \ln \left[\frac{1 + e^{-\kappa x} \tanh (y_0/4)}{1 - e^{-\kappa x} \tanh (y_0/4)} \right]$$

4. Derive Eq. V-14.

5. Derive Eq. IV-99 (including verification of the numerical coefficient of 6.10).

6. Show what Eq. V-8 (for κ^2) becomes when written in the SI system. Calculate κ for 0.01 M 1:1 electrolyte at 25°C, using SI units; repeat the calculation in the cgs/esu system and show that the result is equivalent.

7. Show what Eq. V-16 (for $\boldsymbol{\sigma}$) becomes when written in the SI system. Calculate $\boldsymbol{\sigma}$ for 0.01 M 1:1 electrolyte at 25°C and $\psi_0 = -50$ mV.

8. Show whether or not Eq. V-34 applies to the P, B, and C plots of Fig. V-5. Assume a 1:1 electrolyte and 20°C.

9. Derive the general equation for the differential capacity of the diffuse double layer from the Gouy–Chapman equations. Make a plot of surface charge density $\boldsymbol{\sigma}$ versus this capacity. Show under what conditions your expressions reduce to the simple Helmholtz formula of Eq. V-17.

10. Using the Gouy–Chapman theory, evaluate the derivative

$$\left(\frac{\partial \psi_0}{\partial \ln n_0} \right)_{\boldsymbol{\sigma}}$$

Show under what conditions the derivative should be equal to $-kT/e$. The Esin and Markov effect consists of the observation that experimentally this coefficient can be more negative than $-$ kT/e, and this observation has been used to deduce the presence of a discrete charge effect (see Ref. 12). What is the qualitative or physical argument by which this conclusion is reached?

11. A circular metal plate of 10 cm^2 area is held parallel to and at a distance d above a solution, as in a surface potential measurement. The temperature is 25°C.

(*a*) First, assuming ψ_0 for the solution interface is 50 mV, calculate ψ as a function of x, the distance normal to the surface on the air side. Assume the air has 10^7 ions/cm^3. (Ignore the plate for this part.) Plot your result.

(*b*) Calculate the repulsive potential between the plate and the solution if d = 3 cm, assuming ψ_0 for the plate also to be 50 mV and using Eq. V-39. Are the assumptions of the equation good in this instance? Explain.

(*c*) Calculate the contribution of the solution–air interface to the surface potential if the plate is now used as a probe in a surface potential measurement and is 1 mm from the surface.

12. Using the conditions of the Langmuir approximation for the double layer repulsion, calculate for what size particles in water at 25°C the double layer repulsion energy should equal kT, if the particles are 80 Å apart.

13. For water at 25°C, Eq. V-46 can be written

$$\zeta = 12.9\, v/\mathbf{F}$$

Show what the units of ζ, v, and $\mathbf{F}$ must be.

14. Show what form Eq. V-46 takes if written in the SI system. Referring to the equation of Problem 13, show what number replaces 12.9 if ζ is in volts, v in meters per second, and $\mathbf{F}$ in volts per meter.

15. It was mentioned that streaming potentials could be a problem in jet aircraft. Suppose that a hydrocarbon fuel of dielectric constant 8 and viscosity 0.03 poise (P) is being pumped under a driving pressure of 30 atm. The potential between the pipe and the fuel is, say, 125 mV, and the fuel has a low ion concentration in it, equivalent to 10^{-8} *M* NaCl. Making and stating any necessary and reasonable assumption, calculate the streaming potential that should be developed. *Note*: Consider carefully the handling of units.

16. Make a calculation to confirm the numerical illustration following Eq. V-52. Show what form the equation takes in the SI system, and repeat the calculation in SI units. Show that the result corresponds to that obtained using cgs/esu units.

17. (*a*) Calculate the expected streaming potential **E** for pure water at 25°C flowing down a quartz tube under 10 atm pressure. Take ζ to be 125 mV. (*b*) Show what form Eqs. V-57 and V-58 take in the SI system and repeat the calculation of part (*a*) in SI units.

18. Streaming potential measurements are to be made, using a glass capillary tube and a particular electrolyte solution, for example, 0.01 *M* sodium acetate in water. Discuss whether the streaming potential should or should not vary appreciably with temperature.

19. While the result should not have very exact physical meaning, as an exercise, calculate the ζ potential of potassium ion, knowing that its equivalent conductivity is 50 cm^2/(equiv)(ohm) in water at 25°C.

20. Make semiquantitative conversion of Fig. V-12 from a plot of integral capacity versus $\mathbf{E}^s$ to a plot of charge density $\boldsymbol{\sigma}$ versus ψ^r. The points in the figure come from three types of experimental measurements. Explain clearly what the data are and what is done with the data, in each case, to get the $\boldsymbol{\sigma}$ versus ψ^r plot. What does the agreement between the three types of measurement confirm? Explain whether it confirms that ψ^r is indeed the correct absolute interfacial potential difference.

21. Assume that a salt, MX (1 : 1 type), adsorbs at the mercury–water interface according to the Langmuir equation:

$$\frac{x}{x_m} = \frac{bc}{1 + bc}$$

where x_m is the maximum possible amount, and x/x_m is 0.5 at $c = 0.1\ M$. Neglect activity coefficient effects and estimate the value of the mercury–water interfacial tension at 25°C at the electrocapillary maximum where 0.01 M salt solution is used.

22. Using the Gibbs equation, obtain from Eq. V-80 the corresponding expression for π as a function of θ. Calculate and plot π versus ln a for n-butanol for the mercury–0.1 N $HClO_4$ interface. Take Γ_m to be 5.186×10^{-10} mole/cm^2, and B, x, and α to be 7.770, 0.433, and 0.624, respectively. Assume 25 α.

23. Referring to Eq. V-81, calculate the value of C for the case mentioned, namely a 100-Å film of dielectric constant 80. Optional: repeat the calculation in the SI system.

24. Make a calculation to confirm the statement made in the footnote following Eq. V-86.

25. Derive the following equation, useful in electrocapillarity studies:

$$\Gamma_+ = -v_+ \boldsymbol{\sigma} \left(\frac{d\mathbf{E}^-}{d\mu} \right)_\gamma$$

(See Ref. 32.)

General References

A. W. Adamson, *J. Chem. Ed.*, **55,** 634 (1978); *J. Chem. Ed.*, **56,** 665 (1979) (regarding SI).

J. O'M. Bockris and A. K. N. Reddy, *Modern Electrochemistry*, Plenum, New York, 1970.

J. A. V. Butler, *Electrical Phenomena at Interfaces*, Methuen, London, 1951.

P. Delahay, *Double Layer and Electrode Kinetics*, Interscience, New York, 1965.

D. C. Grahame, *Chem. Rev.*, **41,** 441 (1947).

E. A. Guggenheim, *Thermodynamics*, Interscience, New York, 1949.

H. S. Harned and B. B. Owen, *The Physical Chemistry of Electrolyte Solutions*, Reinhold, New York, 1950.

G. Kortüm, *Treatise on Electrochemistry*, Elsevier, New York, 1965.

H. R. Kruyt, *Colloid Science*, Elsevier, New York, 1952.

H. van Olphen and K. J. Mysels, *Physical Chemistry: Enriching Topics from Colloid and Surface Chemistry*, Theorex (8327 La Jolla Scenic Drive), La Jolla, California, 1975.

J. Th. G. Overbeek, *Advances in Colloid Science*, Vol. 3, Interscience, New York, 1950.

M. J. Sparnaay, *The Electrical Double Layer*, Pergamon, New York, 1972.

E. J. W. Verwey and J. Th. G. Overbeek, *Theory of the Stability of Lyophobic Colloids*, Elsevier, New York, 1948.

Textual References

1. G. Gouy, *J. Phys.*, **9,** No. 4, 457 (1910); *Ann. Phys.*, **7,** No. 9, 129 (1917).
2. D. L. Chapman, *Phil. Mag.*, **25,** No. 6, 475 (1913).
3. P. Debye and E. Hückel, *Phys. Z.*, **24,** 185 (1923); P. Debye, *Phys. Z.*, **25,** 93 (1924).
4. E. J. W. Verwey and J. Th. G. Overbeek, *Theory of the Stability of Lyophobic Colloids*, Elsevier, New York, 1948.

5. H. R. Kruyt, *Colloid Science*, Elsevier, New York, 1952.
6. H. S. Harned and B. B. Owen, *The Physical Chemistry of Electric Solutions*, Reinhold, New York, 1950.
7. K. J. Mysels, *An Introduction to Colloid Chemistry*, Interscience, New York, 1959.
8. O. Stern, *Z. Elektrochem.*, **30,** 508 (1924).
9. J. Lyklema and J. Th. G. Overbeek, *J. Colloid Sci.*, **16,** 595 (1961).
10. J. A. Davis, R. O. James, and J. A. Leckie, *J. Colloid Interface Sci.*, **63,** 480 (1978).
11. R. O. James, J. A. Davis, and J. O. Leckie, *J. Colloid Interface Sci.*, **65,** 331 (1978).
12. M. J. Sparnaay, *The Electrical Double Layer*, Pergamon, New York, 1972.
13. See S. Levine and G. M. Bell, *J. Colloid Sci.*, **17,** 838 (1962); *J. Phys. Chem.*, **67,** 1408 (1963).
14. I. Langmuir, *J. Chem. Phys.*, **6,** 873 (1938); R. Defay and A. Sanfeld, *J. Chim. Phys.*, **60,** 634 (1963).
15. A. Dunning, J. Mingins, B. A. Pethica, and P. Richmond, *J. Chem. Soc., Faraday Trans. I,* **74,** 2617 (1978).
16. W. Olivares and D. A. McQuarrie, *J. Phys. Chem.*, **84,** 863 (1980).
17. P. L. Levine and G. M. Bell, *J. Colloid Interface Sci.*, **74,** 530 (1980).
18. See J. Th. Overbeek, *Advances in Colloid Science*, Vol. 3, Interscience, New York, 1950, p. 97.
19. N. von Smoluchowski, *Bull. Int. Acad. Polon. Sci., Classe Sci. Math. Nat.*, **1903,** 184.
20. P. B. Lorenz, *J. Phys. Chem.*, **57,** 430 (1953).
21. J. T. Davies and E. K. Rideal, *Interfacial Phenomena*, Academic, New York, 1963.
22. R. A. Van Wagenen and J. D. Andrade, *J. Colloid Interface Sci.*, **76,** 305 (1980).
23. A. J. Rutgers and P. Nagels, *Nature,* **171,** 568 (1953).
24. U. Saxén, *Wied Ann.*, **47,** 46 (1892).
25. P. Mazur and J. Th. G. Overbeek, *Rec. Trav. Chim.*, **70,** 83 (1951).
26. P. B. Lorenz, *J. Phys. Chem.*, **57,** 430 (1953).
27. A. J. Rutgers, M. de Smet, and W. Rigole, *Physical Chemistry: Enriching Topics from Colloid and Surface Chemistry*, H. Van Olphen and K. J. Mysels, Eds., Theorex (8327 La Jolla Scenic Drive), La Jolla, California, 1975.
28. E. F. Burton, *Phil. Mag.*, **11,** No. 6, 425 (1906); **17,** 583 (1909).
29. W. Henry, *Nicolson's J.*, **4,** 224 (1801).
30. G. Lippmann, *Ann. Chim. Phys.*, **5,** 494 (1875).
31. H. Vos and J. M. Los, *J. Colloid Interface Sci.*, **74,** 360 (1980).
32. D. C. Grahame, *Chem. Rev.*, **41,** 441 (1947).
33. R. Parsons and M. A. V. Devanathan, *Trans. Faraday Soc.*, **49,** 673 (1953).
34. K. G. Baikerikar and R. S. Hansen, *Surf. Sci.*, **50,** 527 (1975).
35. A. Frumkin, *Z. Phys. Chem.*, **103,** 55 (1923).
36. F. O. Koenig, *J. Phys. Chem.*, **38,** 111 and 339 (1934).
37. D. C. Grahame and R. B. Whitney, *J. Am. Chem. Soc.*, **64,** 1548 (1942).
38. F. O. Koenig, *Z. Phys. Chem.*, **154,** 454 (1931).
39. L. A. Hansen and J. W. Williams, *J. Phys. Chem.*, **39,** 439 (1935).
40. D. C. Grahame, R. P. Larsen, and M. A. Poth, *J. Am. Chem. Soc.*, **71,** 2978 (1949).

41. G. J. Hills and R. Payne, *Trans. Faraday Soc.*, **61,** 317 (1963).
42. D. C. Grahame, *J. Am. Chem. Soc.*, **63,** 1207 (1941).
43. G. Gouy, *Ann. Phys.*, **6,** 5 (1916); **7,** 129 (1917).
44. D. E. Broadhead, R. S. Hansen, and G. W. Potter, Jr., *J. Colloid Interface Sci.*, **31,** 61 (1969).
45. R. S. Hansen and K. G. Baikerikar, *J. Electroanal. Chem.*, **82,** 403 (1977).
46. D. E. Broadhead, K. G. Baikerikar, *J. Phys. Chem.*, **80,** 370 (1976).
47. K. G. Baikerikar and R. S. Hansen, *J. Colloid Interface Sci.*, **52,** 277 (1975).
48. J. O'M. Bockris, M. A. V. Devanathan, and K. Müller, *Proc. Roy. Soc.* (*London*), **274,** 55 (1963).
49. A. Frumkin and B. B. Damaskin, in *Modern Aspects of Electrochemistry*, J. O'M. Bockris and B. E. Conway, Eds., Butterworths, London, 1954.
50. A. Frumkin and A. Gorodetzkaja, *Z. Phys. Chem.*, **136,** 451 (1928); A. Frumkin and F. J. Cirves, *J. Phys. Chem.*, **34,** 74 (1930).
51. P. Delahay, *Double Layer and Electrode Kinetics*, Interscience, New York, 1965.
52. R. Narayan and N. Hackerman, *J. Electrochem. Soc.*, **118,** 1426 (1971).
53. P. R. Couchman, D. H. Everett, and W. A. Jesser, *J. Colloid Interface Sci.*, **52,** 410 (1975).
54. P. R. Couchman and C. R. Davidson, *J. Electroanal. Chem.*, **85,** 407 (1977).
55. D. M. Mohliner and T. R. Beck, *J. Phys. Chem.*, **83,** 1169 (1979).
56. J. O'M. Bockris, S. D. Argade, and E. Gilead, *Electrochim. Acta,* **14,** 1259 (1969).
57. J. O'M. Bockris and A. K. N. Reddy, *Modern Electrochemistry*, Plenum, New York, 1970.
58. R. Gomer and G. Tryson, *J. Chem. Phys.*, **66,** 4413 (1977).
59. K. S. Yun, S. M. Wilhelm, S. Kapusta, and N. Hackerman, *J. Electroanal. Soc.*, **127,** 85 (1980).
60. See E. Lange and F. O. Koenig, *Handbuch der Experimentalphysik*, Vol. 12, Part 2, Leipzig, 1933, p. 263.
61. E. J. W. Verwey, *Rec. Trav. Chim.*, **61,** 564 (1942).
62. J. N. Brønsted, *Z. Phys. Chem.*, **A143,** 301 (1929).
63. E. A. Guggenheim, *Thermodynamics*, Interscience, New York, 1949; *J. Phys. Chem.*, **33,** 842 (1929); **34,** 1540 and 1758 (1930).
64. J. A. V. Butler, *Electrical Phenomena at Interfaces*, Methuen, London, 1951.
65. N. K. Adam, *The Physics and Chemistry of Surfaces*, 3rd ed., Oxford University Press, London, 1941.
66. J. H. de Boer, *Electron Emission and Adsorption Phenomena*, Macmillan, New York, 1935.
67. B. Case, N. S. Hush, R. Parsons, and M. E. Peover, *J. Electroanal. Chem.*, **10,** 360 (1965).
68. N. K. Adam, *Physical Chemistry*, Oxford University Press, London, 1958; see also Ref. 66.
69. P. J. Holmes, Ed., *The Electrochemistry of Semiconductors*, Academic, New York, 1962.
70. J. Tafel, *Z. Phys. Chem.*, **50,** 641 (1905).
71. A. T. Hubbard, *Acc. Chem. Res.*, **13,** 177 (1980).

CHAPTER VI

Long-Range Forces

1. Introduction

It has certainly been evident that surface phenomena can be related to forces between molecules, and, in particular, to an asymmetry or unbalance of forces at an interface. The calculation of Section III-2B, for example, is based on the Lennard-Jones potential function illustrated in Fig. III-5. Functions of this type find much use in the next chapter, in which surface energies of crystalline solids are calculated. On a more qualitative level, the subject of orientation at liquid interfaces (Section III-3) involves an appraisal of intermolecular forces. Forces between ions, mostly electrostatic in nature, form a major portion of the material of Chapter V.

In this chapter, the various fundamental types of forces are discussed briefly as a unified subject. Generally speaking, interatomic forces are short range; were this not the case, the energy of a portion of matter would depend on its size in macroscopic systems; it would not be possible, for example, to tabulate molar enthalpies of formation without specifying the scale of the experiment. On the other hand, there are some ways in which forces across or between interfaces can be rather long range in their action, and the discussion of such situations constitutes the chief subject of the present chapter.

Although we know of gravitational, magnetic, and electrical forces, it is only the last that usually are of any importance in chemistry. Electrical forces may be divided, for the moment, into those of repulsion and those of attraction. Two types of repulsive forces have been considered in the preceding chapters: the coulomb repulsion between like-charged ions and the quite general repulsion that arises if any two atoms are brought too close together. As illustrated in Fig. III-5, this last repulsion is very short range, rising rapidly as atoms or molecules come closer than a certain distance. In fact, this critical distance essentially defines atomic or molecular "diameters." Such repulsion may be expressed mathematically by means of an inverse twelfth power of intermolecular distance, as in Eq. III-46. Basically, the effect is due to the reluctance of the electron clouds of two atoms to overlap each other, and since wave functions have a radial portion that is exponential, the overlap, and hence also the repulsion effect, is closer to being exponential than to an inverse power dependence on distance, and is often so represented (note Eq. VII-10, for example).

Attractive forces are again electrostatic in nature. Much of chemistry, of course, is concerned with the wave-mechanical or exchange force responsible for the chemical bond, and this again is short range. The emphasis here, however, is on less specific, essentially electrostatic, attractive forces that give rise to condensation of a vapor to a liquid and, for this reason, are often called *van der Waals* forces.

An immediate complication is the presence of two systems of units, cgs/esu and SI. As in Section V-1, the various equations are given in the conventional cgs/esu system. This is followed by a rephrasing in terms of SI, as a sequel to Section V-3.

2. Forces between Atoms and Molecules

It is best to start with Coulomb's law, according to which the force between two point charges is given by

$$f = \frac{q_1 q_2}{x^2} \qquad \text{(VI-1)}$$

We assume the dielectric constant for vacuum, that is, unity in the cgs system; x is the distance of separation. The potential energy of interaction $\epsilon = \int f dx$ is given by

$$\epsilon = \frac{q_1 q_2}{x} \qquad \text{(VI-2)}$$

where ϵ is in ergs if q is in electrostatic units and x in centimeters. The potential V of a charge q is defined as

$$V = \frac{q}{x} \qquad \text{(VI-3)}$$

with the meaning that unit opposite charge q_0 will experience a potential energy $-qq_0/x$ at a separation x. The sign of V is negative if that of q is negative. Here V is in volts esu, 1 V_{esu} = 300 V (ordinary volts). The potential energy of a charge q is then

$$\epsilon(q, V) = qV \qquad \text{(VI-4)}$$

We next consider a molecule having a dipole moment μ, that is, one in which charges q^+ and q^- are separated by a distance d, giving a dipole moment $\mu = qd$. A dipole experiences no net interaction with a uniform potential, since the two charges are affected equally and oppositely. If there is a gradient of the potential, or a field, defined as $F = dV/dx$, then there *is* a net effect. The potential energy of a dipole aligned with a field is

$$\epsilon(\mu, F) = -Vq + \left(V - \frac{dV}{dx} d\right) q = -\mu F \qquad \text{(VI-5)}$$

Again, ϵ is in ergs if μ is in esu and V in esu. The conventional unit of dipole moment is the *debye*, 1×10^{-18} esu-cm, corresponding to unit electronic charges 0.21 Å apart. Conversely, a dipole produces a potential. Thus for distances large compared to d, a test charge experiences the potential energy

$$\epsilon(q_0, \mu) = \frac{\mu}{x^2} q_0 \tag{VI-6}$$

so that the potential of a dipole is

$$V(\mu) = \frac{\mu}{x^2} \tag{VI-7}$$

Note Problem 1 of this chapter. The field of a dipole (along the line of the dipole) is just

$$F(\mu) = \frac{2\mu}{x^3} \tag{VI-8}$$

(Potentials and fields are reported with a positive sign; the interaction energy of a charge or a dipole is minus if attractive and positive if repulsive).

Two dipoles interact each with the field of the other to give

$$\epsilon(\mu, \mu) = -\mu \frac{2\mu}{x^3} = -\frac{2\mu^2}{x^3} \tag{VI-9}$$

This is the interaction for dipoles end-on. In a liquid, thermal agitation tends to make the relative orientations random, while the interaction energy acts to favor alignment. The analysis is similar to that for molar polarization and leads to the result (due to Keesom in 1912)

$$\epsilon(\mu, \mu)_{av} = -\frac{2\mu^4}{3kTx^6} \tag{VI-10}$$

This *orientation* attraction thus varies inversely with the sixth power of the distance between the dipoles. Remember, however, that the derivation has assumed separations large compared to d.

A further type of interaction is that in which a field induces a dipole moment in a polarizable molecule or atom. We have

$$\mu_{ind} = \alpha F \tag{VI-11}$$

where α is the polarizability and has units of volume in the cgs system. It follows from Eq. VI-5 that

$$\epsilon(\alpha, F) = -(\mu_{ind})(F) = -\tfrac{1}{2}\alpha F^2 \tag{VI-12}$$

(The factor $\frac{1}{2}$ enters because, strictly speaking, we must integrate $\int_0^F \mu_{ind} dF$). The induced dipole is instantaneous (as compared to molecular motions), and the potential energy between a dipole and a polarizable species is therefore independent of temperature:

$$\epsilon(\mu, \alpha) = -\frac{1}{2}\alpha\left(\frac{2\mu}{x^3}\right)^2 = -\frac{2\alpha\mu^2}{x^6} \tag{VI-13}$$

On averaging the interaction over all orientations of the dipole, one obtains the final result, worked out by Debye in 1920:

$$\epsilon(\mu, \alpha) = -\frac{\alpha\mu^2}{x^6} \tag{VI-14}$$

As an exercise, it is not hard to show that the interaction of a polarizable molecule with a charge q is just

$$\epsilon(\alpha, q) = -\frac{(q)^2\alpha}{2x^4} \tag{VI-15}$$

None of the above account for the existence of the quite general attraction between atoms and molecules that are not charged and do not have a dipole moment. After all, CO and N_2, as similar-sized molecules, have roughly comparable heats of vaporization and hence intermolecular attraction, although only the former has a dipole moment.

London (1, 2), in 1930, showed the existence of an additional type of electrical force between atoms, which has the required characteristics. This is known as the *dispersion* force, or the London–van der Waals force. It is always attractive and arises through the fact that even neutral atoms constitute systems of oscillating charges because of the presence of a positive nucleus and negative electrons. The derivation may be sketched as follows. The energy of an atom 1 in a field F is given by

$$\epsilon(x) = -\tfrac{1}{2}\alpha_1 F^2$$

and in this case F can be written

$$F = \frac{2\bar{\mu}_2}{x^3} \tag{VI-16}$$

where $\bar{\mu}_2$ is the average dipole moment (or, really, the root mean square average) for the oscillating electron–nucleus system of the second atom. Now the polarizability of an atom can be expressed as a sum over all excited states of the transition moment squared divided by the energy, and if approximated by the largest term, then for atom 2

$$\alpha_2 \cong \frac{(\overline{ed})^2}{h\nu_0} \tag{VI-17}$$

where $(ed)^2$ is electron charge times displacement and averages to $\bar{\mu}_2^2$; $h\nu_0$ is approximately the ionization energy. Thus, the energy of atom 1 in the average dipole field of atom 2 becomes

$$\epsilon(x) = -\frac{\frac{3}{4}h\nu_0\alpha_1\alpha_2}{x^6} \quad \text{(VI-18)}$$

(the $\frac{3}{4}$ factor comes from the detailed derivation).

At the same level of approximation, the corresponding form for two different atoms is (3):

$$\epsilon(x) = -\frac{3}{2}\frac{\alpha_1\alpha_2}{x^6[(1/h\nu_1) + (1/h\nu_2)]} \quad \text{(VI-19)}$$

where $h\nu_1$ and $h\nu_2$ denote the characteristic (roughly, ionization) energies for atoms 1 and 2. A useful version (4) is

$$\epsilon(x) = -\frac{363\alpha_1\alpha_2}{x^6[(\alpha_1/n_1)^{1/2} + (\alpha_2/n_2)^{1/2}]} \quad \text{(VI-20)}$$

n_1 and n_2 are the numbers of electrons in the outer shells, x is in angstrom units, $\epsilon(x)$ is in kilocalories per mole, and α is in cubic angstrom units. A widely used alternative, known as the Kirkwood–Müller form is

$$\epsilon(x) = -6mc^2\frac{\alpha_1\alpha_2}{(\alpha_1/\chi_1) + (\alpha_2/\chi_2)}\frac{1}{x^6} \quad \text{(VI-21)}$$

where m is the electronic mass, c the velocity of light, and χ_1 and χ_2 are the diamagnetic susceptibilities (5; see also Ref. 6).

At small distances or for molecules with more complicated charge distributions, the more general form is

$$\epsilon(x) = -C_1x^{-6} - C_2x^{-8} - C_3x^{-10} - \cdots \quad \text{(VI-22)}$$

Equations VI-19–21 give only the C_1 or dipole–dipole interaction term. The C_2 term is called dipole–quadrupole [a quadrupole may be represented as $(-q_1) - (q_2) - (-q_1)$ where charge q_2 is twice the magnitude of charge q_1], and the term with C_3 can arise from dipole–octupole and quadrupole–quadrupole interactions.

An evaluation by Fontana (7) gives C_1, C_2, and C_3 as 1400, 3000, and 7900 for Ar, respectively, and 1750, 5800, and 24,000 for Na; x is to be expressed in angstrom units, and the resulting $\epsilon(x)$ will be in units of kT at 25°C.

A recent and very important approach to the study of van der Walls forces is through the spectroscopy of van der Waals molecules. That is, molecules such as Ar_2, O_2–O_2, and Ar–N_2, (8).

Table VI-1 (Ref. 1) shows the approximate values for the Keesom ($\mu - \mu$), the dipole–polarizable molecule $\mu - \alpha$, and the dispersion $\alpha - \alpha$ interactions for several molecules. Even for highly polar molecules the last is very important.

We will call "van der Waals" forces those intermolecular interactions that give rise to an attractive potential varying with the inverse sixth power

TABLE VI-1
Contributions to the Interaction Energies between Neutral Molecules[a]

Molecule	$10^{18}\mu$ (esu-cm)	$10^{24}\alpha$ (cm^3)	$h\nu_0$ (eV)[b]	$10^{60}\phi x^6$ ($erg \cdot cm^6$)		
				$\mu-\mu$[c]	$\mu-\alpha$	$\alpha-\alpha$
He	0	0.2	24.7	0	0	1.2
Ar	0	1.6	15.8	0	0	48
CO	0.12	1.99	14.3	0.0034	0.057	67.5
HCl	1.03	2.63	13.7	18.6	5.4	105
NH_3	1.5	2.21	16	84	10	93
H_2O	1.84	1.48	18	190	10	47

[a] Adapted from J. O. Hirschfelder, C. F. Curtiss, and R. B. Bird, *Molecular Theory of Gases and Liquids*, corrected ed., Wiley, New York, 1964, p. 988.
[b] One electron-volt (eV) corresponds to 1.602×10^{-12} erg (or 23 kcal/mole).
[c] Calculated for 20°C.

of the intermolecular distance. This is the dependence indicated by the a/V^2 term in the van der Waals equation for a nonideal gas,

$$\left(P + \frac{a}{V^2}\right)(V - b) = RT \tag{VI-23}$$

where V is the volume per mole, and a and b are constants, the former giving the measure of the attractive potential and the latter, the actual volume of a mole of molecules. The three types of interactions given in Table VI-1 are of this van der Waals type. The first two, $\mu - \mu$ and $\mu - \alpha$, are difficult to handle in the case of condensed systems since they are sensitive to molecular orientation, that is, to structure.

The dispersion or $\alpha - \alpha$ interaction, however, is independent of structure (in first order) and, moreover, should be approximately additive in the case of a collection of molecules. The qualitative reason is that the dispersion attraction arises from a rather small perturbation of electronic motions so that many such perturbations can add without serious mutual interaction. It is both because of this simplifying aspect and because of the undoubted general importance of the dispersion attraction that this type of force seems to have largely dominated the thinking of surface and colloid chemists. The way in which the dispersion effect can lead to long-range forces is taken up in Section V-4.

3. The SI System

We continue here the discussion of Section V-3. Coulomb's law is restated as

$$\epsilon = \frac{q_1 q_2}{4\pi\epsilon_0 D x} \tag{V-19}$$

TABLE VI-2
Correspondences for Converting from Cgs/Esu to SI Equations[a]

Parameter in Cgs/Esu	Parameter in SI	Parameter in Cgs/Esu	Parameter in SI
V	$(4\pi\epsilon_0)^{1/2}V$	μ	$\mu/(4\pi\epsilon_0)^{1/2}$
q	$q/(4\pi\epsilon_0)^{1/2}$	α	$\alpha/(4\pi\epsilon_0)$
F	$(4\pi\epsilon_0)^{1/2}F$		

[a] See Appendix in J. D. Jackson, *Classical Thermodynamics*, Wiley, New York, 1962. Also, Bryce Crawford, private communication.

where ϵ_0 is the "permittivity" of vacuum; ϵ will be in joules if q is in coulombs and x, in meters. More than a change in units is involved; the quantity $(4\pi\epsilon_0)$ may appear in various equations. As noted in Section V-3, equations in cgs/esu convert to equations in SI if V is replaced by $(4\pi\epsilon_0)^{1/2}V$ and q, by $q/(4\pi\epsilon_0)^{1/2}$. This approach is extended to the quantities of the preceding section in Table VI-2. As examples of the use of this table, Eq. (VI-4) is unchanged, the factors $(4\pi\epsilon_0)^{1/2}$ canceling, Eq. (VI-9) becomes

$$\epsilon(\mu, \mu) = -\frac{2\mu^2}{(4\pi\epsilon_0)x^3} \tag{VI-24}$$

Eq. (VI-12) is unchanged, and Eq. (VI-18) becomes

$$\epsilon(\alpha, \alpha) = -\frac{\frac{3}{4}h\nu_1\alpha_1\alpha_2}{(4\pi\epsilon_0)^2x^6} \tag{VI-24a}$$

In using the SI equations, remember that energy is in joules, charge in coulombs, and dipole moment in coulomb-meters. Polarizability has the dimensions As^4kg^{-1} in SI, that is, $\alpha_{SI} = (4\pi\epsilon_0)(10^{-6})\alpha_{cgs}$. See Problems 3 and 5 of this chapter.

4. Long-Range Forces

No fundamentally new forces are involved in this discussion, but when one considers the interaction between an extended portion of matter such as a colloidal particle and a molecule or a second particle, the summation of elementary interactions can lead to forces of importance in affecting the surface chemical and colloidal behavior of a system. The two situations under most active development currently are that involving the interaction of electrical double layer systems and that involving dispersion forces. Both are long range, in a sense, since both are treated in terms of fields, stationary or oscillating, mutually propagated between the particles.

A. Attraction Due to the Dispersion Effect

The total interaction between an atom and a slab of infinite extent and depth can be obtained by a summation over all atom–atom interactions, if simple additivity of forces can be assumed. While not strictly correct, this is the usual assumption made at this point and, further, if the distance from an atom to the surface of the slab is large compared to the atomic diameter, the summation may be replaced by a triple integration. This has been done by de Boer (9), using the simple dispersion formula, Eq. VI-18. Following Eq. VI-22, we write

$$\epsilon(x) = -\frac{C_1}{x^6} \qquad \text{(VI-25)}$$

where C_1 is taken to be $(\frac{3}{4})h\nu_0\alpha^2$. Since ionization potentials are in the range of 10 to 20 eV or 20 to 40 $\times$ 10^{-12} erg, and polarizabilities, about 1 to 2 $\times$ 10^{-24} cm^3 (note Table VI-1), C_1 will have a value in the range of 10 to 100 $\times$ 10^{-60} erg-cm^6 per atom.

The triple integration has the effect of changing the dependence on x and of introducing the quantity n, the number of atoms per cubic centimeter:

$$\epsilon(x)_{\text{atom-slab}} = -\frac{(\pi/6)nC_1}{x^3} \qquad \text{(VI-26)}$$

The attractive *force* experienced by the atom is given by the derivative of $\epsilon(x)$ with respect to x. Now, if one desires the attractive energy between two slabs, a further integration over the depth of the second slab is needed, as illustrated in Fig. VI-1. This energy will now vary as x^{-2},

$$\epsilon(x)_{\text{slab-slab}} = -\left(\frac{\pi}{12}\right)\left(\frac{n^2C_1}{x^2}\right) \qquad \text{(VI-27)}$$

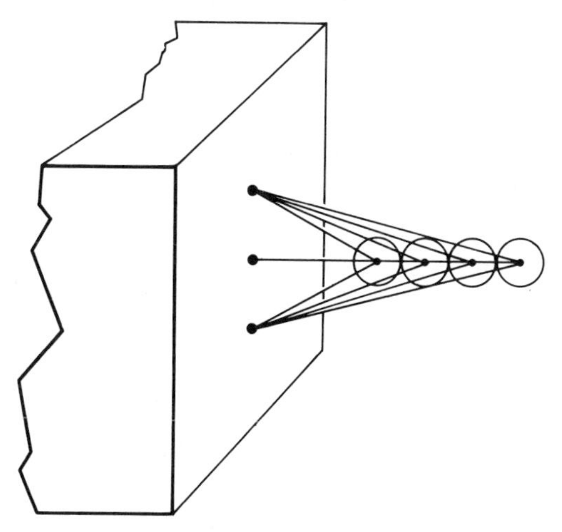

Fig. VI-1. Van der Waals forces between a surface and a column of molecules.

or,

$$\epsilon(x)_{\text{slab-slab}} = -\left(\frac{1}{12\pi}\right)\left(\frac{H}{x^2}\right) \tag{VI-27a}$$

where ϵ is in ergs per square centimeter and H is known as the *Hamaker constant* after its user (10),† and is equal to $\pi^2 n^2 C_1$. It might be noted that H contains the quantity $(n\alpha)^2$ or $(\alpha/V)^2$ where V is the molecular volume. For most elements, this ratio is about 0.1, so that Hamaker constant is of the order of 10^{-12} erg. For two square slabs of side d,

$$\frac{\epsilon}{kT} = -\frac{200}{T}\left(\frac{d}{x}\right)^2 \tag{VI-28}$$

The interaction energy is thus comparable to that of thermal motion for slabs separated by distances of the order of their linear dimension.

Equation VI-24 may be integrated for various macroscopic shapes. A case important in colloid chemistry is that of two spheres of radius r, for which

$$\epsilon(x) = -\frac{rH}{12x} \tag{VI-29}$$

if $x < r$, where x is the surface-to-surface distance. In the direct experimental measurement of H it is easier to use a sphere (or, for practical purposes, a lens or segment of a sphere) and a flat plate, for which we have

$$\epsilon(x) = -\frac{Hr}{6x} \tag{VI-30}$$

The integrals are such that the same equation is obtained in the case of crossed cylinders; if these are of different radii, then $r = (r_1 r_2)^{1/2}$. Note that $\epsilon(x)$ has a simple inverse dependence on x, so that the attraction is indeed long range. In all of these cases, the *force* between the two objects will be just $d\epsilon(x)/dx$.

B. The Retarded Dispersion Attraction

The dispersion force depends on the propagation of oscillating electric fields or, alternatively, on the interchange of virtual photons between atoms; presumably, it is therefore propagated at the velocity of light in the medium. The frequency, about that given by ν_0 of Eq. VI-18 or about 10^{15}/sec, would allow light to travel about 1000 Å during one oscillation period, so that atoms that far apart might be expected to get out of phase. The effect is known as one of *retardation*, and an early treatment by Casimir and Polder (11) gives the retarded potential between two atoms as

$$\epsilon(x) = -\frac{23hc\alpha^2}{8\pi^2 x^7} \tag{VI-31}$$

† A more commonly used symbol for the Hamaker constant is A.

where h is Planck's constant and c is the velocity of light. The energy of interaction between two slabs becomes

$$\epsilon(x) = -\left(\frac{23hc}{240\pi x^3}\right) \sum \alpha_i n_i \tag{VI-32}$$

where the summation is over all the kinds of atoms present. This summation can be related to dielectric constant by the Clausius–Mosotti equation, and with this substitution and on differentiation, Eq. VI-32 becomes

$$f(x) = \frac{207hc}{1280\pi^3 x^4}\left(\frac{D-1}{D+2}\right)^2 \tag{VI-33}$$

The reader is also referred to a paper by McLachlan (12).

Dielectric constant, polarizability, and the reflection of light are theoretically related, and it is not surprising (post facto) that rather different appearing alternative treatments exist for dispersion type attractions. An early theory by Lifshitz (13) has gained considerable contemporary usage (see, e.g., Refs. 14–16). The fluctuating electromagnetic fields that give rise to the dispersion attraction also determine the reflectivity of the material. For two macroscopic slabs the Lifshitz formulation gives for the attractive force

$$f(x) = \frac{\pi hc}{480x^4}\left(\frac{D-1}{D+1}\right)^2 \phi = \frac{B}{x^4} \tag{VI-34}$$

where D is a dielectric constant average approximately equal to the square of the index of refraction of the material, and ϕ is a function of dielectric constant whose value is around unity. In more general form, an integral of the optical properties of the material over all frequencies is involved (15). Note, however, the similarity between Eqs. VI-34 and VI-33.

For the cases of a lens and a plate and of two crossed cylinders, the retarded force becomes

$$f(x) = \frac{2\pi Br}{3x^3} \tag{VI-35}$$

with a typical theoretical value of B of around 10^{-19} erg-cm. The corresponding equation for the unretarded force is, from Eq. VI-30,

$$f(x) = \frac{Hr}{6x^2} \tag{VI-36}$$

Notice the difference in dependence on x of the retarded and unretarded regimes.

C. Experimental Verification

Beginning in the 1950s, a number of attempts were made to verify directly the long-range dispersion attraction predicted for two macroscopic objects. Several investigators (17–19) reported measurements of the force either between parallel glass or quartz plates or between a flat plate and a spherical lens (18). Forces of the order of 0.01 dyne/cm² are involved if the separation is about 1 μ, and it can be appreciated that the experimental difficulties are considerable. The surfaces must be quite smooth, absolutely dust-free, and free of any electrostatic charge. Moreover, since the force increases rapidly as the surfaces approach each other, the balancing system used must have a considerable restoring moment to avoid seizure of the surfaces. Overbeek and Sparnaay and co-workers used a stiff spring and measured distances by means of the change in capacity of a parallel plate condenser, whereas Derjaguin and co-workers have used a delicate balance having an associated feedback circuit that gave the necessary restoring force without impairing sensitivity.

The use of two flat plates is very difficult experimentally since the slightest surface roughness or adventitious dust causes serious error, and, while the first results confirmed the theoretically predicted order of magnitude of the

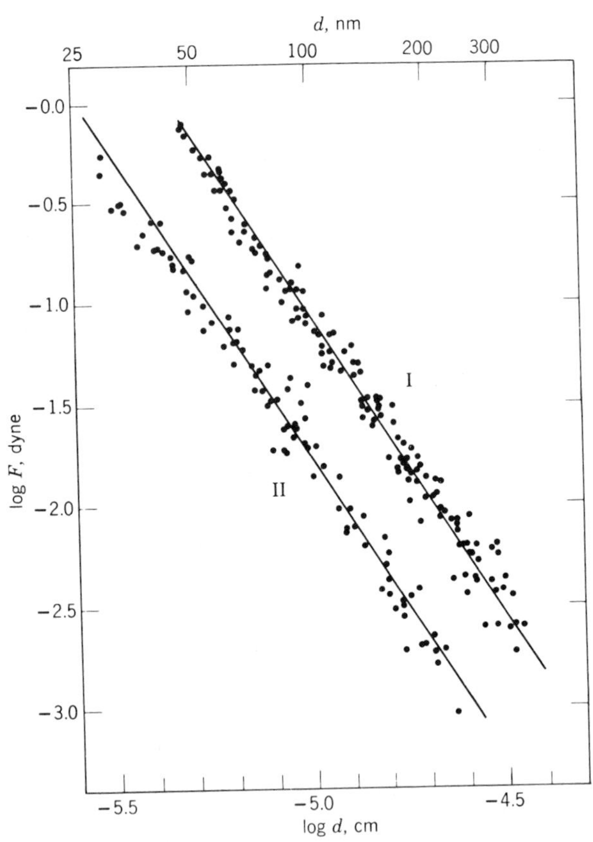

Fig. VI-2. Attraction between a flat plate and a sphere of radius 413.5 cm (case I) or 83.75 cm (case II), all of fused silica. The lines are drawn with a slope of −3.00. (From Ref. 20.)

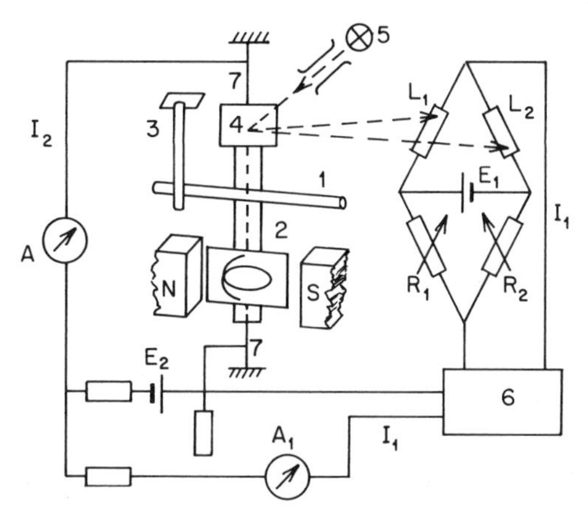

Fig. VI-3. Schematic diagram of apparatus for measuring the dispersion attraction between crossed filaments. One of the filaments (1) is attached to the movable frame (2) of a galvanometer, which is placed in a magnetic field. The other filament (3) is fixed. The frame (2) of the galvanometer is suspended by means of thin bands or strips (7). A light beam from a fixed source (5) is reflected from the mirror (4) attached to the galvanometer frame, and the reflected beams impinge on the photoresistances L_1 and L_2, which form a bridge circuit with the variable resistances R_1 and R_2. The separation x between the filaments can be determined from the R_1 and R_2 settings and the current from the variable source E_2 required to make I_1 zero. The force of attraction is then determined from the current I_2 measured by means of a nanoammeter A_2. The microammeter A_1 and amplifier (6) monitor the current I_1 in the diagonal of the bridge. (See Ref. 16 for details.) (Reprinted by permission from Nature.)

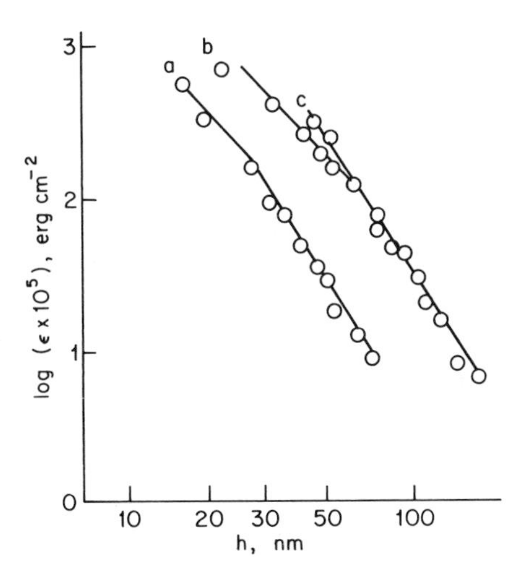

Fig. VI-4. Dependence of ϵ for hypothetical flat plates as calculated from measurements of the attractive force between crossed filaments in air. (*a*) Quartz, (*b*) platinum, and (*c*) gold. (From Ref. 16.) (Reprinted by permission from Nature.)

attraction, it was not easy to distinguish between Eqs. VI-35 and VI-36. Later work has made use of a flat plate and a lens, or of two crossed cylinders or wires. Fig. VI-2 shows results obtained by Rouweler and Overbeek (20) which confirmed the slope of 3 predicted by Eq. VI-35 for a plot of log f versus log x, in the case of x large.

By using crossed cylinders of mica, of about 1-cm radius, whose surfaces were essentially atomically smooth, Tabor and Winterton (21) were able to measure f well into the unretarded region and thus obtain values both of H and of B. More recently, it has been possible to extend such measurements to other substances by using crossed wires or filaments, now of radius in the 0.1–0.5 mm range. Fig. VI-3 shows the apparatus used by Derjaguin, Rabinovich, and Churaev (see Ref. 16), and some of their results are shown in Fig. VI-4. Note the transition from the unretarded to the retarded behavior as indicated by the change in slope at about 30 nm separation distance.

5. Long-Range Forces in Solution

The confirmation of the predicted long-range dispersion attraction between objects in air has been a major experimental triumph. Of more general interest, however, is the matter of long-range forces between particles in solution, a subject of great importance in colloid chemistry. There are now two kinds of theoretical complications. The first is that of treating van der Waals forces between objects in a condensed medium, and the second is that of the electrostatic repulsion due to the electrical double layer. The combined treatment has come to be known as the DLVO (Derjaguin, Landau, Vervey, Overbeek) theory. Experimentally, of course, one can only determine the net interaction potential as a function of the separation distance x. These various aspects are discussed briefly in the following sections.

A. Dispersion Attraction in a Condensed Medium

The potential energy of two slabs immersed in a medium is again given by Eq. VI-27. The new aspect is that the dispersion interaction between bodies 1 and 2 immersed in medium 3 is determined by a net Hamaker constant H_{132}. It can be shown (22, 23) that as an approximation

$$H_{132} = H_{12} - (H_{13} + H_{23} - H_{33}) \qquad \text{(VI-37)}$$

Thus H_{132} should exceed H_{12} if H_{33} is larger than $(H_{13} + H_{23})$. The term *hydrophobic bonding* has been used to describe the enhanced attraction between two particles in a solvent if the solvent–particle interaction is weaker than the solvent–solvent interaction. Conversely, H_{132} should be negative for sufficiently large H_{13} and H_{23}, so that the two slabs or, in general, two particles, should appear to repel each other.

Hamaker constants can be related to works of adhesion. If two slabs in contact are separated to infinite distance, then the work required, $\epsilon(x = 0)$, is just the work of adhesion, provided that the process is done reversibly and, of course, only dispersion forces are involved (note Ref. 24). As a simple illustration, consider two slabs of oily or hydrophobic material immersed in water. Separated, the surface free energy per square centimeter is 2 γ_{HW}, where γ_{HW} is the surface free energy of the hydrocarbon–water interface. If brought together so that their interfaces coalesce, this interfacial free energy is lost. There is thus a gain of 2 γ_{HW} driving the slabs together. There are various modelistic treatments, generally centering on the difference in energy of water near a hydrophobic species and water surrounded by water (see Ref. 25 and references therein). There is also at this point a considerable similarity of the Hamaker constant approach to that of the Good–Fowkes model; see Sections IV-2 and VIII-4. Note, however, Problem 18 of this chapter.

B. Electrical Double Layer Repulsion

There are many situations where van der Waals attraction is balanced by electrical double layer repulsion. An outstanding example is that of the flocculation of lyophobic colloids. A sol consisting of charged particles experiences both the double layer repulsion and the van der Waals attraction effects, and the balance of these determines the ease and hence the rate with which the particles can approach sufficiently close to stick together. Verwey and Overbeek (26, 27) considered the case of two spheres of colloidal size, and by combining the appropriate equations they found sets of net potential energy versus distance of separation curves of the type illustrated in Fig. VI-5 for the case of ψ_0 = 25.6 mV (i.e., $\psi_0 = kT/e$ at 25°C). At low ionic strength, as measured by κ, double layer repulsion is overwhelming except at very small separations, but as κ is increased, another limiting condition of net attraction at all distances is reached. There is a critical region of κ value such that a small potential minimum, of about $\frac{1}{2}kT$, occurs at a distance of separation x about equal to a particle diameter.

Qualitatively, then, it can be seen why increasing the ionic strength of a solution promotes flocculation. The net potential is given approximately as the combination of Eqs. V-29 and VI-27 (for two slabs):

$$\epsilon(x)_{\text{net}} = \frac{64 n_0 kT}{\kappa}\gamma^2 e^{-2\kappa x} - \frac{(1/12\pi)H}{x^2} \qquad \text{(VI-38)}$$

If the condition for rapid flocculation is taken to be that no barrier exists, the requirement is essentially that $\epsilon(x) = 0$ and $d\epsilon(x)/dx = 0$ for some value of x. This leads to the condition (after substituting for κ by means of Eq. V-8).

$$n_0 = \left[\frac{2^7 3^2}{\exp(4)}\right]\left[\frac{D^3 k^5 T^5 \gamma^4}{e^6 H^2}\right]\frac{1}{z^6} \qquad \text{(VI-39)}$$

Thus for a z–z electrolyte, equivalent conditions of concentration would be

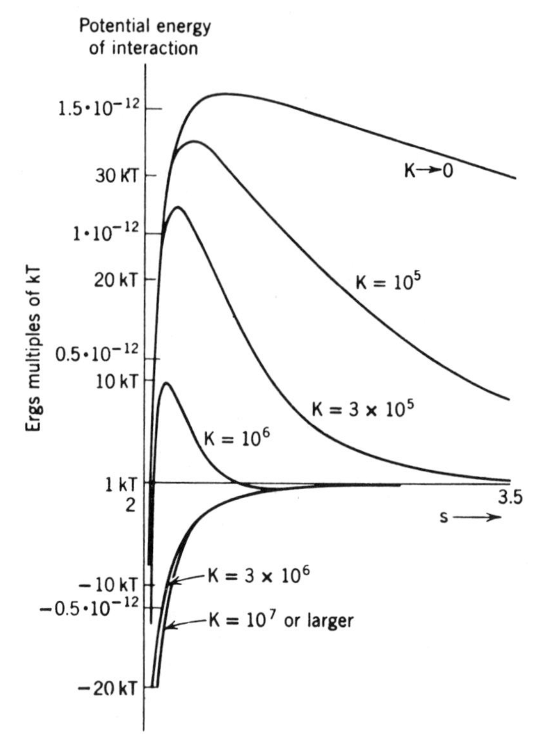

Fig. VI-5. The effect of electrolyte concentration on the interaction potential energy between two spheres. (From Ref. 26.)

in the order $1:(\frac{1}{2})^6:(\frac{1}{3})^6$ or 100:1.6:0.13 for a 1–1, 2–2, and 3–3 electrolyte, respectively, which is about the experimental observation as embodied in the Schulze–Hardy rule for the effect of valence type on the flocculating ability of ions. Actually, with the higher charged ions, there is an increasing tendency for specific adsorption, so that flocculation ability becomes a matter of reduction in ψ_0 as well as of reduction in the double layer thickness.

Quantitative studies on flocculation rates have provided estimates of Hamaker constants in approximate agreement with theory. One assumes that the particles flocculate by diffusion, but in doing so must have the activation energy to pass a potential energy barrier of the type shown in Fig. VI-5. The barrier height is estimated from the measured flocculation rate; other measurements give the zeta potential (assumed to be constant during the approach of particles), and the two data allow an estimate of the Hamaker constant (see Refs. 28–30).

It is possible now to measure directly the force between objects in an electrolyte solution and thus obtain experimental curves of the type shown in Fig. VI-5. Israelachvili and co-workers (see Ref. 31) have determined the force between crossed mica cylinders in aqueous solutions, and with 1:1

electrolytes found good agreement with DLVO theory. Similar measurements have been made by Derjaguin and co-workers using crossed filaments of various materials, as illustrated by Fig. VI-6 for the case of glass in 0.1 *M* KCl. It was possible to estimate the value of *H* for this system (1.2×10^{-13} erg) from the attractive force at the minimum. Note Problem 14 in this chapter.

C. *Other Experimental Approaches*

Hamaker constants have been estimated in a number of ways less direct than those just described. There is, for example, considerable evidence that platelike colloidal particles recognize each other's presence at rather large distances. Thus in iron oxide (Fe_2O_3) and tungsten trioxide (hydrated WO_3 sols), the platelike particles on settling arrange themselves in horizontal layers, separated by layers of wafer as much as 8000 Å thick (32, 33). The formation of these "Schiller," or iridescent, layers is suggestive of the presence of some kind of long-range interaction between the colloidal platelets. The nature of the Brownian motion suggests long-range attractive forces (34).

It appears, moreover, that many similar phenomena frequently occur when one has a solution of large and highly asymmetric particles. Thus solutions of tobacco mosaic virus are truly remarkable in their behavior. The virus itself is a protein unit roughly cyclindrical in shape and about 3420 Å long and 150 Å in diameter (35). Solutions containing 1 or 2 wt % of this virus separate into two liquid phases, and the more concentrated of the two layers appears to have a crystalline regularity (36). The virus particles are lined up within a few minutes of arc of being parallel with

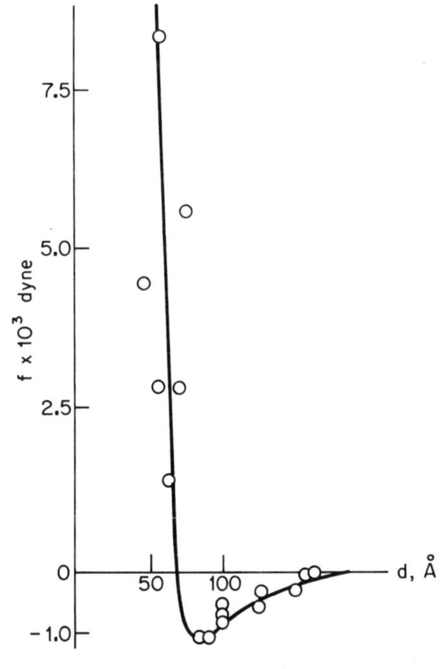

Fig. VI-6. Variation of interaction force with separation distance for crossed 1-mm glass filaments in 0.1 *M* KCl. (From Ref. 16.) (Reprinted by permission from Nature.)

each other, although there may be as much as 500 Å of solvent between them. On the other hand, the x-ray and other evidence show that the upper or more dilute of the two solutions (in which the average distance apart is 1500 Å) does not exhibit any ordering of the virus molecules.

The ability of a dilute and a concentrated solution to coexist indicates that more than long-range forces are involved. Onsager (37) has suggested an entropy-related explanation. Briefly, the transition from dilute to concentrated solution is viewed as one in which very asymmetric particles lose one or two degrees of rotational freedom.

Apart from the foregoing complication, the tendency for platelike particles to seek an equilibrium separation is very suggestive of the presence of a secondary minimum in the $\epsilon(x)$ function (illustrated in Figs. VI-5 and VI-6. As a more quantitative example, certain clays have a sheet crystal structure, and these sheets appear to find an equilibrium separation when the crystal is equilibrated with an aqueous electrolyte solution. This separation, which may be as large as 100 Å, varies with the electrolyte concentration in the manner expected if the separation represented a balance between van der Waals attraction and electrostatic repulsion (38). Another approach has been to measure the osmotic pressure of water in a suspension of platelets. The method has been applied to montmorillionite suspensions (39).

The equilibrium thickness of a soap film will depend on the net of at least three forces. First, the film–air interfaces are charged so that an electrical double layer repulsion provides an outward pressure. Second, the long-range van der Waals attraction effect provides a compressive pressure; this is just the attraction between two slabs separated by a distance equal to the film thickness (see Problem 9), and so is given by Eq. VI-27. Third, if the film is connected with bulk liquid it is experimentally possible to apply suction on the bulk liquid and hence to change the hydrostatic pressure in the film. In addition, of course, any propagated structure could seriously alter the equilibrium thickness that the foregoing pressure balance would otherwise determine.

Studies in which equilibrium film thickness is measured as a function of ionic strength (which varies the double layer repulsion) and of applied hydrostatic suction may provide a means of experimental testing of both double layer repulsion theory and of long-range attractive forces. For a review of the subject, see Ref. 40.

As noted earlier in this section, Hamaker constants can be related to interfacial tensions—see Section VIII-5B. An interesting experimental approach to the estimation of solid–solid and solid–liquid interfacial tensions is, incidentally, that of ''engulfing'' studies (24). An advancing freezing or solidification front is established for some solvent 1. Small spheres of suspended solid 2 may either be swept along by the S_1–L_1 interface or may be engulfed in the solid solvent S_1. The result yields information about the various interfacial tensions involved (note Section VIII-4).

6. The Disjoining Pressure

The discussion of long-range forces has so far been couched in terms of interaction energies or forces. Deriaguin and co-workers (see Ref. 16) have proposed a related quantity called the *disjoining pressure,* Π. In the case of two plates immersed in a medium, illustrated in Fig. VI-7, Π is the pressure, in excess of the external pressure, that must be applied to the medium

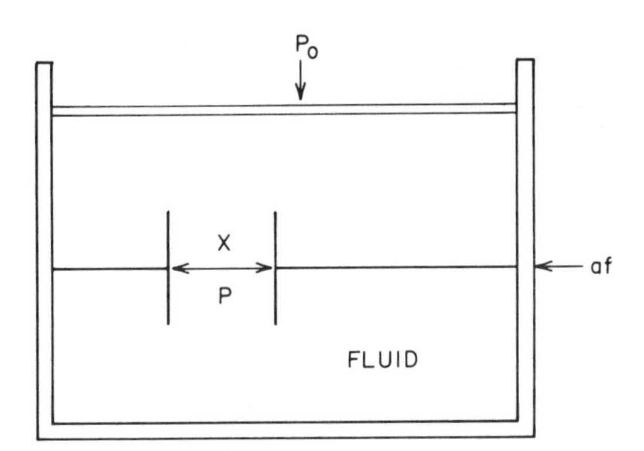

Fig. VI-7. Illustration of disjoining pressure. In this case, $\Pi = P - P_0$. (From Ref. 41.)

between the plates to maintain a given separation. In this case, Π is numerically just the force f of attraction (or repulsion) between the plates, per unit area. A more general definition is that

$$\Pi = -\frac{1}{\mathcal{A}}\left(\frac{\partial G}{\partial x}\right)_{\mathcal{A}, T, V} \tag{VI-39}$$

where x is the thickness of the film or intersurface region (see Ref. 41).

The concept may also be applied to soap films (note Problem 15), and to an adsorbed layer at the solid–vapor interface (see Section VIII-4). Especially in this last case, a useful alternative definition of Π is that it is the mechanical pressure that would have to be applied to a bulk substance to bring it into equilibrium with a given film of the same substance. As another example, Ter-Minassian-Saraga and co-workers have studied the disjoining pressure of thin films of liquid crystals (41).

Derjaguin (16) has recognized that just as with interaction energies, Π may be regarded as a net of several components. These include Π_m, due to dispersion interaction, Π_e due to overlapping of diffuse double layers in the case of charged surfaces, Π_a due to overlapping of adsorbed layers of neutral molecules, and Π_s due to alteration of solvent structure in the boundary layer region. The distinction between these last two components may be difficult to make experimentally; collectively, they may be regarded as a "structural" contribution to Π. The matter of structure in thin interfacial regions is considered in Section VI-8.

7. Dipole-Induced Dipole Propagation

The discussion so far has emphasized the relatively nonspecific long-range effects of the dispersion potential and electrical double layer repulsion. A rather different basis for a long-range attraction is that of propagation of polarization, that is, of the dipole-induced dipole effect. This may be described in terms of the situation in which there is an adsorbed layer of atoms on a surface. If the surface is considered to contain surface charges of individual value ze (where z might be the formal charge on a polar surface atom), then by Eq. VI-7 the interaction energy with a polarizable atom absorbed on the surface is

$$\epsilon_{01} = -\frac{(ze)^2\alpha}{2d^4} = -\frac{(ze)^2}{2d}\frac{\alpha}{d^3} \tag{VI-40}$$

where d is a distance of separation; the induced dipole moment for the atom is $ze\alpha/d^2$, which corresponds to equal and opposite charge $ze\alpha/d^3$ having been induced in the atom along an axis normal to the surface. If a second layer of atoms is present, the field of this charge $ze\alpha/d^3$ induces a dipole moment in an adjacent second-layer atom, with corresponding interaction energy

$$\epsilon_{12} = -\frac{(ze)^2}{2d}\left(\frac{\alpha}{d^3}\right)^3 \tag{VI-41}$$

The situation is illustrated in Fig. VI-8. Continued step-by-step propagation leads to the general term

$$\epsilon_{i,(i+1)} = \epsilon_{(i-1),i}\left(\frac{\alpha}{d^3}\right)^2 \tag{VI-42}$$

The locus of successive ϵ values can be represented by an exponential relationship

$$\epsilon(x) = \epsilon_0 e^{-ax} \tag{VI-43}$$

where x denotes distance from the surface, and a is given by

$$a = -\frac{1}{d_0}\ln\left(\frac{\alpha}{d^3}\right)^2 \tag{VI-44}$$

where d_0 is the atomic diameter.

The distance d may be regarded as an effective value, smaller than d_0.

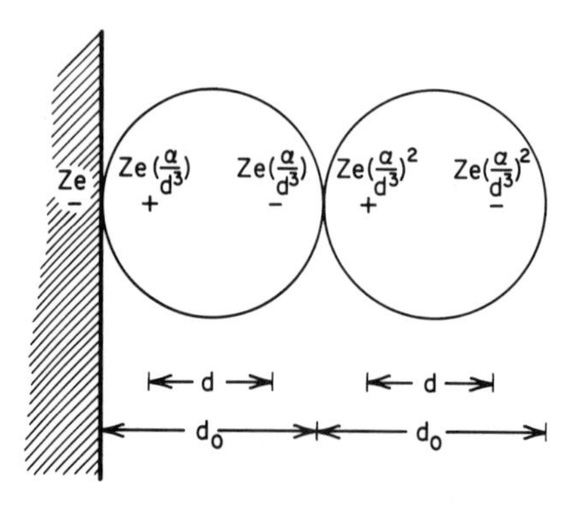

Fig. VI-8. Long-range interaction through propagated polarization.

It is thus seen that the dipole-induced dipole propagation gives an exponential rather than an inverse x cube dependence of $\epsilon(x)$ with x. As with the dispersion potential, the interaction depends on the polarizability, but unlike the dispersion case, it is only the polarizability of the adsorbed species that is involved. The application of Eq. VI-43 to physical adsorption is considered in Section XVI-7D. For the moment, the treatment illustrates how a "long-range" interaction can arise as a propagation of short-range interactions.

8. Evidence for Deep Surface Orientation

There are many indications that a thin film of material has properties different from those of the bulk condensed phase. Adsorbed films, for example, must be multimolecular before their surface approaches bulk phase properties (see Chapter XVI). Especially striking are the data on films adsorbed from the vapor phase at pressures approaching P^0, the condensation pressure. In a now classic paper, Derjaguin and Zorin (41a) found that with certain polar liquids the adsorbed film on glass reached a *limiting* thickness of about 100 Å at P^0, that is, that the presence of the glass substrate made a thick film an alternative state capable of coexisting in equilibrium with bulk liquid. More recent studies (42–43) report measurements of similarly thick films *and* of the film–bulk liquid–solid substrate contact angle. As discussed further in Section X-6C, such films *must* be structurally perturbed relative to the bulk phase.

It has recently been possible to extend measurements of the force between surfaces down to the small nanometer range of separation, using the crossed cylinder technique (Section VI-5). With a large molecule solvent $[(CH_3)_2SiO]_4$, Horn and Israelachvili (44) observed that the force between crossed mica cylinders showed a damped oscillation of about 10 Å period and extending out to some 50 Å separation, attributed to structuring of the solvent near the mica surfaces. Again, for aqueous electrolyte media, relatively short-range repulsions were observed between the mica cylinders, attributed to a layer of hydrated counterions (44a, 44b).

It is, of course, well established that certain liquids have the property of being oriented in the form of "liquid crystals." The compound *p*-azoanisole is outstanding is this respect; the liquid is birefringent, and in other ways is definitely anisotropic. Interestingly, the direction of its anisotropy is very sensitive to the nature of the walls of the vessel containing it. Merely stroking a glass plate will insure that liquid, heated on the plate to above its anisotropy temperature, will show a prefrential orientation on cooling. It is thus not surprising that thin films of liquid crystalline material are highly oriented (or have a structural component to their disjoining pressure)—note Ref. 41. Thick films of long chain fatty acids are anisotropic in the liquid state, to a few degrees above the melting point, and to a depth of several hundred molecules.

The interesting results of Shereshefsky and co-workers (see Section III-1C) on the vapor pressure of liquids in capillaries suggest that the capillary wall is capable of inducing changes in the liquid structure over distances of the order of a micron. The anomalous water history (Section VI-9) suggests, unfortunately, that the abnormally low vapor pressures may have been primarily due to impurities.

Drost-Hansen (45), however, does make a case that water at an interface may be structurally perturbed to a depth of several hundred angstroms. He terms such water "vicinal" water. In some cases, there appear to be sharp, almost periodic variations of properties with temperature. Wiggins (46), for example, reports quite significant up-and-down variations with temperature of the K^+/Na^+ distribution ratio between silica gel and bulk water; there are biological implications (45, 46). In this general connection, Benson (47) has shown that the temperature dependence of many of the bulk properties of water can be accounted for in terms of an equilibrium between two isomeric structures, one of these being icelike. The proportion of the two structures might then be significantly altered in an interfacial region.

There is evidence that the viscosity of thin liquid films is anomalous. The thinning of films by a jet of air indicated that lubricating oils in films several thousand angstroms thick have 10 times the bulk viscosity (48). The viscosity of water in small capillaries is reported to be 40% above the bulk value (49). By contrast, thin films of polydimethylsiloxane have, by the blow-off method, below bulk viscosity values (50). There are some caveats concerning the method, however. See Mysels (40), the discussion following Ref. 51, and some recent results of Bascom and Singleterry (52) which suggest that anomalous thinning profiles may be due to surface roughness or to a nonzero contact angle. Derjaguin and co-workers, however, report a higher than normal viscosity for the liquid water film at the ice–quartz interface from studies on the rate of movement of an ice column in a capillary (53).

Electron and x-ray diffraction studies of films of long chain compounds on metal surfaces have indicated the presence of structure well above the bulk melting point, and in layers 100 molecules thick, and there have been many observations (54) that the melting points and other physical properties of relatively thick films may differ considerably from the corresponding bulk characteristics (see Section IV-15). Returning to the water–quartz or glass interface, infrared studies have indicated the presence of a deep (weakly) structurally perturbed layer (55). Dielectric constant and nuclear magnetic resonance (nmr) measurements have indicated that adsorbed water films on alumina and other solids are structurally perturbed to a considerable depth (see Ref. 51 and also Section XVI-14).

For additional summaries of evidences for deep structurally perturbed layers at an interface see Henniker (56), Kitchener (57), Derjaguin (58), and Drost-Hansen (51). Much of such evidence is merely suggestive, being subject either to experimental questioning or to alternative explanation. There remains, however, a sufficiently solid body of observations that there seems no doubt that significant structural perturbations are present and that these are especially important in the case of hydrogen bonded liquids such as water. The theoretical prediction of such structure is not easy. Certainly, more specific forces than those of dispersion and electrical double layer repulsion are involved. In addition to hydrogen bonding, these may be of the dipole-induced dipole type discussed in Section VI-7.

9. Anomalous Water

Some mention must be made of perhaps the major topic of conversation among surface and colloid chemists during the period 1966–1973. The story beings with reports of unusual properties of liquids, especially water, in small capillaries. One of the early reports is that of Shereshefsky and co-workers in 1955 of unexpectedly low vapor pressures for liquids in capillaries (Section III-1C), but the real start of the "anomalous" water chain of events is with a paper by Fedyakin in 1962, followed closely by a series of papers with Deryaguin. (For a detailed bibliography up to 1970–1971 see Ref. 59.)

The reports were amazing. Water condensed from the vapor phase into 10 to 100 μ diameter quartz or pyrex capillaries had a density of 1.4 g/cm^3, about 10 times of viscosity of ordinary water, a low and not sharp freezing point, a very different thermal expansion behavior through the -40 to 20°C temperature region, a high surface tension (around 75 dyne/cm), an abnormal nmr spectrum, and, very important, a lower vapor pressure than normal water. Various molecular weight measurements indicated a molecular weight of around 180, corresponding to $(H_2O)_{10}$. This last, plus the vapor pressure results, meant that a new molecular form of water existed, which was *more stable* than ordinary water. Furthermore, the vapor could be distilled through a 700 to 800°C tube and be recondensed (in a capillary) to return to the same unusual properties. The heat of vaporization was estimated at 6 kcal/mole (as compared to about 10 for ordinary water).

The Russian scientists called this new form of water "anomalous water" or "water II." While most of the foregoing work appeared in 1962–1965 publications in the USSR, it was not until visits by Deryaguin to Great Britain and to the United States in 1966 and 1967 that Western scientists really became interested. There was much controversy—how *could* nature have hidden for so long the existence of a more stable form of water than the one we know? Perhaps the data were wrong. (This writer held from the beginning that the data were probably correct, but that they were being somehow misinterpreted and could not possibly mean that a new, extrastable form of water existed.)

By 1968 various laboratories in England and in the United States had repeated the experiments of Deryaguin and co-workers, and largely *confirmed* his results. Strong government research funding was directed towards anomalous water research. Excited stories began to appear in the popular press. There was talk of preparing gallons and then thousands of gallons of anomalous water for all kinds of great technical benefits.

Scientific reinforcement mounted further. The first infrared spectra were reported in 1969 (60), unlike those of any known substance, and a hexagonal polymeric structure was proposed. The new name of "polywater" was suggested, a name that became very popular in the United States. Theory entered. CNDO-type calculations were stated to "establish the existence and characterize" polywater (the newer name "cyclimetric water" was proposed) (61). A new bonding scheme was presented (62). A major portion of the 1970 National Colloid Symposium was organized around anomalous water. (The reader is referred to the August, 1971 issue of the *Journal of Colloid and Interface Science* for the papers of this symposium.)

By late 1969 the gathering of the storm of refuting evidence began; this evidence was to demolish the idea that a new, stable form of water existed. The microamounts of anomalous water available made analysis difficult, but new techniques (see Chapter VIII) made it possible to get elemental analyses on samples extruded onto a flat substrate. A landmark paper in early 1970 reported large impurity contents in samples that essentially duplicated the earlier infrared spectra (63). The point was crucial; previously, reports of impurities could be dismissed on the grounds that preparations of impure materials did not disprove the ability of others to prepare pure ones. All attempts to make large scale preparations of anomalous water were failing. The unusual circumstance that only *freshly drawn* capillaries could be used began to receive more attention. It had been thought that the fresh surface had the special catalytic properties needed to form polywater, but now it was becoming appreciated that it also provided a host of leachable impurities.

A mass of scientific (and anecdotal) detail has necessarily been omitted. The story ends here with the honest and brave paper by Deryaguin and co-workers in 1974 (64) confirming and acknowledging that anomalous water is a mixture of colloidal and molecularly dissolved impurities.

The history provides an outstanding example of how science reacts to an announcement of great discovery and eventually confirms or denies it. There are lessons. The problem was not one of data, which were real, but was rather one of interpretation. There was a tendency to ignore small but disturbing clues as well as the limits to things possible set by the laws of thermodynamics. We should have learned that, in chemistry, spectroscopy and wave mechanics are at best confirmatory disciplines, useful in understanding what has been established by other means.

10. Problems

1. Derive Eq. VI-7. Show explicitly what approximation is made. For what x/d ratio would an error of 10% be involved?

2. Calculate $\epsilon(\alpha, \alpha)$ for two Ar atoms 10 Å apart, and compare to kT at 77 K (around the b.p. of Ar). Do the problem in both cgs/esu and SI units.

3. The atomic unit (a.u.) of dipole moment is that of a plus and a minus electronic charge separated by a distance equal to the radius of the first Bohr orbit a_0. Similarly, the a.u. of polarizability is a_0^3 (64). Express μ and α for HCl in atomic units, using both the cgs/esu and the SI approach.

4. Derive Eq. VI-15.

5. Give the last three columns of Table VI-1 in SI units.

6. Derive Eq. VI-25 from Eq. VI-24 by carrying out the required triple integration.

7. Explain qualitatively, that is, without doing the actual detailed integrations, why Eq. VI-30 applies both to the case of crossed cylinders and to the case of a lens or sphere and a flat plate.

8. The Hamaker constant for the case of a substance having two kinds of atoms is obtained by replacing the quantity $n^2\alpha^2$ by $(n_1^2\alpha_1^2 + n_2^2\alpha_2^2 + 2n_1n_2\alpha_1\alpha_2)$. Explain why this formulation should be correct and calculate the H appropriate for ice,

taking the polarizabilities of H and of O to be 0.67×10^{-24} and 3.0×10^{-24} cm^3, respectively. Calculate the expected force between two thick slabs of ice that are 100 Å apart. Do the same for two spheres of ice of 1 μm radius and again 100 Å apart. Assume for $h\nu_0$ ice to be $\sim 10^{-11}$ erg.

9. The long-range van der Waals interaction provides a cohesive pressure for a thin film that is equal to the mutual attractive force per square centimeter of two slabs of the same material as the film and separated by a thickness equal to that of the film. Consider a long column of the material, of unit cross section. Let it be cut in the middle and the two halves separated by d, the film thickness. Then, from one outside end of one of each piece, slice off a layer of thickness d; insert one of these into the gap. The system now differs from the starting point by the presence of an isolated thin layer, as illustrated in Fig. VI-9. Show by suitable analysis of this sequence that the opening statement is correct. *Note*: About the only assumptions needed are that interactions are superimposable and that they are finite in range.

10. Calculate the B of Eq. VI-35 from the data of Fig. VI-2. Compare with the theoretical value of B, taking D to be 2.1 and ϕ to be 0.35.

11. There are theoretical indications (53) that Hamaker constants can be interpolated by means of a geometric mean law,

$$H_{12} = (H_{11}H_{22})^{1/2} \qquad \text{(VI-45)}$$

Derive an expression for H_{132} of Eq. VI-37 in which all H_{ij} type terms have been eliminated.

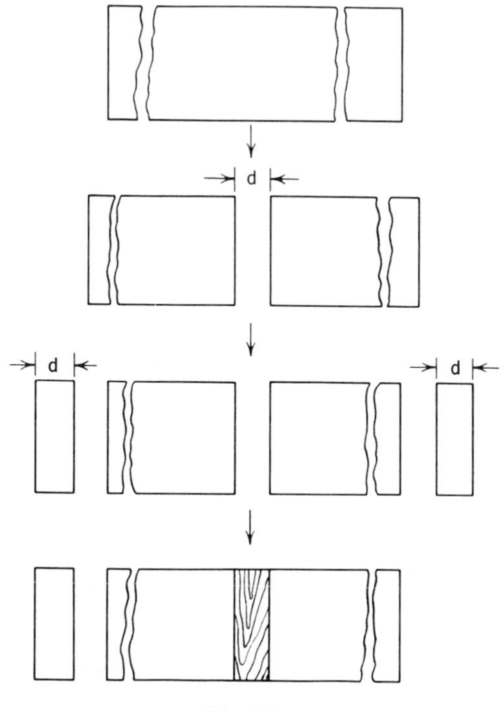

Fig. VI-9.

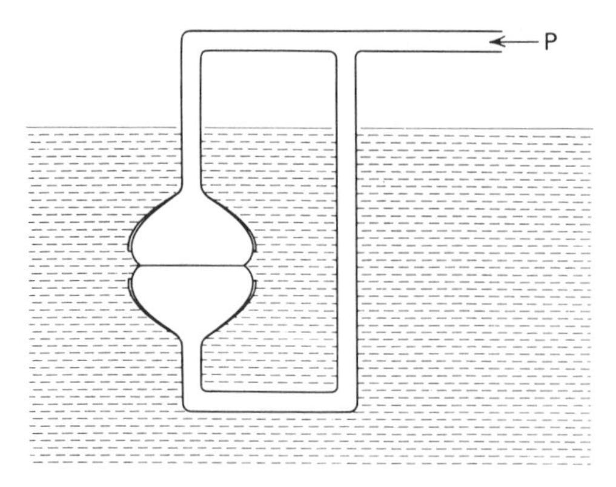

Fig. VI-10. A measurement of disjoining pressure.

12. Two spherical colloidal particles have a radius of 200 Å, a potential ψ^0, of 25.7 mV, and their Hamaker constant is 3×10^{-14} erg. They are in a 0.002*M* solution of 1:1 electrolyte at 25°C. Calculate $\epsilon(x)$ as a function of x for the double layer repulsion and for the van der Waals attraction. Plot your results, along with that of the net potential of interaction.

13. Write Eq. VI-37 for the case of two particles (e.g. slabs) of material 1 in solvent s. An analysis resembling that in Problem 9 provides a proof of your equation. Formulate this proof; it is due to Hamaker (note Ref. 22).

14. Calculate H and B from the data of Fig. VI-4.

15. Make a semiquantitative conversion of Fig. VI-6 to a plot of $\epsilon(x)$ versus x.

16. Consider the situation illustrated in Fig. VI-10 in which two air bubbles, formed in a liquid, are pressed against each other so that a liquid film is present between them. Relate the disjoining pressure of the film to the Laplace pressure P in the air bubbles.

17. Derjaguin and Zorin report that at 25°C, water at 0.98 of the saturation vapor pressure adsorbs on quartz to give a film 40 Å thick. Calculate the value of the disjoining pressure of this film and give its sign.

18. Show that Eq. VI-44 indeed follows from Eq. VI-42.

19. There are some contradictory aspects to statements in Section VI-5A. On one hand, it seems reasonable from Eq. VI-37 that H_{131} can be negative, so that two slabs of the same material might repel each other when in a liquid medium 3. The work of adhesion, however, is just $W = 2\gamma_{13}$, and should always be positive, so that it should always take work to separate the two slabs. Can you resolve the contradiction? (Refs. 10 and 23 may be helpful.)

General References

A. W. Adamson, *A Textbook of Physical Chemistry*, 2nd ed., Academic, New York, 1979.

J. N. Israelachvili and B. W. Ninham, *J. Colloid Interface Sci.*, **58,** 14 (1977) (a discursive review of intermolecular forces).

J. Mahanty and B. W. Ninham, *Dispersion Forces*, Academic, New York, 1976.

H. van Olphen and K. J. Mysels, Eds., *Physical Chemistry: Enriching Topics from Colloid and Surface Science*, Theorex (8327 La Jolla Scenic Dr.), 1975.

J. Th. G. Overbeek, *Colloid Science*, H. R. Kruyt, Ed., Elsevier, Amsterdam, 1952.

M. J. Sparnaay, *The Electrical Double Layer*, in the *International Encyclopedia of Physical Chemistry and Chemical Physics*, D. H. Everett, Ed., Vol. 4, Pergamon, New York, 1972.

E. J. W. Verwey and J. Th. G. Overbeek, *Theory of the Stability of Lyophobic Colloids*, Elsevier, Amsterdam, 1948.

Textual References

1. See A. W. Adamson, *A Textbook of Physical Chemistry*, 2nd ed., Academic, New York, 1979.
2. F. London, *Z. Phys. Chem.*, **B11,** 222 (1930).
3. See R. J. Munn, *Trans. Faraday Soc.*, **57,** 187 (1961).
4. J. C. Slater and J. G. Kirkwood, *Phys. Rev.*, **37,** 682 (1931).
5. A. Müller, *Proc. Roy. Soc.* (*London*), **A154,** 624 (1936).
6. A. D. Crowell, in *The Solid–Gas Interface*, E. A. Flood, Ed., Marcel Dekker, New York, 1967.
7. P. R. Fontana, *Phys. Rev.*, **123,** 1865 (1961); see also *ibid.*, **125,** 1597 (1962).
8. G. E. Ewing, *Accts. Chem. Res.*, **8,** 185 (1975).
9. J. H. de Boer, *Trans. Faraday Soc.*, **32,** 10 (1936); M. Polanyi and F. London, *Naturwiss.* **18,** 1099 (1930).
10. H. C. Hamaker, *Physica,* **4,** 1058 (1937).
11. H. B. G. Casimir and D. Polder, *Phys. Rev.*, **73,** 360 (1948).
12. A. D. McLachlan, *Mol. Phys.*, **7,** 381 (1963–64).
13. E. M. Lifshitz, *Sov. Phys. JETP* (*Engl. Transl.*), **2,** 73 (1956).
14. M. van den Tempel, *Adv. Colloid Interface Sci.*, **3,** 137 (1972).
15. J. Mahanty and B. W. Ninham, *Dispersion Forces*, Academic, New York, 1976.
16. B. V. Derjaguin, Y. I. Rabinovich, and N. V. Churnaev, *Nature,* **272,** 313 (1978).
17. See W. Black, J. G. V. de Jongh, J. Th. G. Overbeek, and M. J. Sparnaay, *Trans. Faraday Soc.*, **56,** 1597 (1960) and preceding papers.
18. See I. I. Abrikossova and B. V. Derjaguin, *Proc. 2nd Int. Congr. Surface Activity,* Butterworths, London, Vol. 3, 1957, p. 398; see also B. V. Derjaguin and I. I. Abrikossova, *Phys. Chem. Solids,* **5,** 1 (1958); B. V. Derjaguin, I. I. Abrikossova, and E. M. Lifshitz, *Q. Rev.* (*London*), **10,** 292 (1956).
19. J. A. Kitchener and A. P. Prosser, *Proc. Roy. Soc.* (*London*), **A242,** 403 (1957).
20. G. C. J. Rouweler and J. Th. G. Overbeek, *Trans. Faraday Soc.*, **67,** 2117 (1971).
21. D. Tabor, *J. Colloid Interface Sci.*, **31,** 364 (1969); D. Tabor and R. H. S. Winterton, *Proc. R. Soc.*, **A312,** 435 (1969).
22. D. Bargeman and F. van Voorst Vader, *J. Electroanal. Chem.*, **37,** 45 (1972).
23. J. Visser, *Adv. Colloid Interface Sci.*, **3,** 331 (1972).
24. See A. W. Neumann, S. N. Omenyi, and C. J. van Oss, *Colloid Polym. Sci.*, **257,** 413 (1979); C. J. van Oss, S. N. Omenyi, and A. W. Neumann, *ibid,* **257,** 737 (1979).
25. R. B. Hermann, *J. Phys. Chem.*, **79,** 163 (1975).
26. E. J. W. Verwey and J. Th. G. Overbeek, *Theory of the Stability of Lyophobic Colloids,* Elsevier, Amsterdam, 1948.

27. E. J. W. Verwey and J. Th. G. Overbeek, *Trans. Faraday Soc.,* **42B,** 117 (1946).
28. R. H. Ottewill and A. Watanabe, *Kolloid-Z.,* **173,** 7 (1960).
29. G. D. Parfitt and N. H. Picton, *Trans. Faraday Soc.,* **64,** 1955 (1968); G. D. Parfitt and D. G. Wharton, *Proc. Brit. Ceram. Soc.,* p. 13, 1969.
30. C. G. Force and E. Matijević, *Kolloid-Z,* **224,** 51 (1968).
31. J. N. Israelachvili and B. W. Ninham, *J. Colloid Interface Sci.,* **58,** 14 (1977); J. N. Israelachvili and G. E. Adams, *Nature,* **262,** 774 (1976).
32. P. Bergmann, P. Low-Beer, and H. Zocher, *Z. Phys. Chem.,* **A181,** 301 (1938).
33. S. Levine, *Trans. Faraday Soc.,* **42B,** 102 (1946).
34. S. Hachisu and K. Furusawa, *Sci. Light (Tokyo),* **12,** No. 1, 157 (1963); see *Chem. Abstr.,* **61,** 65g (1964).
35. C. T. O'Konski and A. J. Haltner, *J. Am. Chem. Soc.,* **78,** 3604 (1956).
36. J. D. Bernal and I. Fankuchen, *J. Gen. Physiol.,* **25,** 111 (1941).
37. L. Onsager, *Ann. N.Y. Acad. Sci.,* **51,** 627 (1949).
38. A. V. Blackmore and R. D. Miller, *Soil Sci. Soc. Am. Proc.,* **25,** 169 (1961); A. V. Blackmore and B. P. Warkentin, *Nature,* **186,** 823 (1960).
39. L. Barclay, A. Herrington, and R. H. Ottewill, *Kolloid-Z.,* **250,** 655 (1972).
40. K. J. Mysels, *J. Phys. Chem.,* **68,** 3441 (1964).
41. E. Perez, J. E. Proust, and L. Ter-Minassian-Saraga, *Colloid and Polym. Sci.,* **256,** 784 (1978); E. Manev, J. E. Proust, and L. Ter-Minassian-Saraga, *Colloid and Polym. Sci.,* **255,** 1133 (1977); J. E. Proust and L. Ter-Minassian-Saraga, *J. Phys.,* **40,** C3-490 (1980).
41a. B. V. Derjaguin and Z. M. Zorin, *Proc. 2nd Int. Congr. Surface Activity (London),* **2,** 145 (1957).
42. M. E. Tadros, P. Hu, and A. W. Adamson, *J. Colloid Interface Sci.,* **49,** 184 (1974).
42a. J. Tse and A. W. Adamson, *J. Colloid Interface Sci.,* **72,** 515 (1979).
43. M. Hirata, M. Ichikawa, and S. Iwai, *Abs., ACS/CSJ Chem. Congr.,* Honolulu, April, 1979.
44. R. G. Horn and J. N. Israelachvili, *Chem. Phys. Lett.,* **71,** 192 (1980).
44a. R. M. Pashley, *Colloids and Surfaces,* in preparation.
44b. R. M. Pashley and J. N. Israelachvili, *Colloids and Surfaces,* in preparation.
45. W. Drost-Hansen, *J. Colloid Interface Sci.,* **58,** 251 (1977); *Phys. Chem. Liq.,* **7,** 243 (1978).
46. P. M. Wiggins, *Clin. Exp. Pharm., Phys.,* **2,** 171 (1975); P. M. Wiggins, *Cell-Associated Water,* W. Drost-Hansen and J. S. Clegg, Eds., Academic, New York, 1979.
47. S. W. Benson, *J. Am. Chem. Soc.,* **100,** 5640 (1978).
48. B. V. Deryaguin and V. V. Karassev, *2nd Int. Conf. Surface Act.,* **3,** 531 (1957).
49. N. V. Churayev, V. D. Sobolev, and Z. M. Zorin, *Discuss Far. Soc.,* 213 (1971).
50. B. V. Deryaguin, V. V. Karasev, I. A. Lavygin, I. I. Skorokhodov, and E. N. Khromova, *Spec. Discuss. Faraday Soc.* **98,** No. 1 (1970).
51. W. Drost-Hansen, *Ind. Eng. Chem.,* **61,** 10 (1969).
52. W. D. Bascom and C. R. Singleterry, *J. Colloid Interface Sci.,* **66,** 559 (1978).
53. S. S. Barer, N. V. Churaev, B. V. Derjaguin, O. A. Kiseleva, and V. D. Sobolev, *J. Colloid Interface Sci.,* **74,** 173 (1980).
54. G. Karagounis, *Helv. Chim. Acta,* **37,** 805 (1954).

55. N. K. Roberts and G. Zundel, *J. Phys. Chem.,* **85,** 2706 (1981).
56. J. C. Henniker, *Rev. Mod. Phys.,* **21,** 322 (1949).
57. J. A. Kitchener, *Endeavor,* **22,** 118 (1963).
58. B. V. Deryaguin, *Pure Appl. Chem.,* **10,** 375 (1965).
59. L. C. Allen, *J. Colloid Interface Sci.,* **36,** 554 (1971).
60. E. R. Lippincott, R. R. Stromberg, W. H. Grant, and G. L. Cessac, *Science,* **164,** 1482 (1969).
61. L. C. Allen and P. A. Kollman, *Science,* **167,** 1443 (1970); note, however, *Nature,* **233,** 550 (1971).
62. J. W. Linnett, *Science,* **167,** 1719 (1970).
63. D. L. Rousseau and S. P. S. Porto, *Science,* **167,** 1715 (1970).
64. B. V. Deryaguin, Z. M. Zorin, Ya. I. Rabinovich, and N. V. Churaev, *J. Colloid Interface Sci.,* **46,** 437 (1974).
65. J. Gready, G. B. Bacskay, and N. S. Hush, *Chem. Phys.,* **31,** 467 (1978).

CHAPTER VII

Surfaces of Solids

1. Introduction

The interface between a solid and its vapor (or an inert gas) is discussed in this chapter from an essentially phenomenological point of view. We are interested in surface energies and free energies and in how they may be measured or estimated theoretically. The study of solid surfaces at the molecular level, through the methods of spectroscopy and diffraction, has become a large field of its own and is taken up in Chapter VIII.

A. *The Surface Mobility of Solids*

A solid, by definition, is a portion of matter that is rigid and resists stress. Although the surface of a solid must in principle be characterized by surface free energy and total energy quantities, it is evident that the usual methods of capillarity are not very useful. These, after all, depend on measurements of properties connected with equilibrium or equipotential surfaces, as given by Laplace's equation (Eq. II-7). A solid, generally speaking, deforms in response to applied forces in an elastic manner, and its shape will be determined more by its past history than by surface tensional forces.

Practically speaking, however, substances commonly considered as solid may possess sufficient plasticity to flow at least slowly, and in such cases variations of the methods of capillarity may be applicable. As an example, a thin copper wire near its melting point has been observed to shorten (even under small loads) and, from the stress such that the strain rate was zero, a surface tension of 1370 dyne/cm was calculated (1). Moreover, the process of sintering is possible because of the ability of metals and other solids to exhibit some bulk and surface mobility. For example, if a powdered metal, usually under some pressure, is heated to a temperature above about two-thirds of the melting point, a fusion of the particles is found to occur at regions of contact. The reduction of total surface area and hence of surface free energy is the principal driving force. Thus both bulk and surface diffusion generally become appreciable at the temperatures involved in sintering. For example, scratches on a silver surface fill up when it is heated to near the melting point (2), and if copper or silver spheres are placed on a flat surface of the same metal, then the crack between the sphere and the surface fills up on heating to a temperature again somewhat below the melting point (3). The term "heating" is a relative one, of course. Small ice spheres at $-10°C$ develop a connecting neck, if touching, and detailed study of the kinetics of the process provides an indication of whether, depending on conditions, the process is occurring by bulk or surface diffusion (4, 5). Sintering in metals has been reviewed by Herring (6) and Kuczynski (7).

It is instructive to consider just how mobile the surface atoms of a solid might be expected to be. Following the approach in Section III-2, one may first consider the matter of the evaporation–condensation equilibrium. The gas kinetic theory equation for the number of moles hitting 1 cm^2 of surface per second is

$$Z = P(\tfrac{1}{2}\pi MRT)^{1/2} \qquad \text{(VII-1)}$$

For saturated water vapor at room temperature, Z is about 1×10^{22} molecules/(cm^2)(sec) and at equilibrium is balanced by an equal evaporation rate. This leads to the conclusion that the average lifetime of a molecule at the water–gas interface is about 1 μsec. Now if the same considerations are applied to a very nonvolatile metal such as tungsten, whose room temperature vapor pressure is estimated to be 10^{-40} atm, then Z becomes about 10^{-17} atoms/(cm^2)(sec), and the average lifetime of a surface atom becomes about 10^{32} sec! Even for much more volatile but still refractory solids, such as most other metals and most salts, the lifetimes are very long at room temperature. As with sintering, the matter is relative. Copper at 725°C has an estimated vapor pressure of 10^{-8} mm Hg (8), which leads to an estimated surface lifetime of 1 hr. Low melting solids, such as ice, iodine, and various organic solids, can have surface lifetimes comparable to those for liquids, at temperatures that would be considered low.

Not all molecules striking a surface necessarily condense, and Z in Eq. VII-1 gives an upper limit to the rate of condensation and hence to the rate of evaporation. Alternatively, actual measurement of the evaporation rate gives, through Eq. VII-1, an effective vapor pressure P_e, which may be less than the actual vapor pressure, P^0. The ratio P_e/P^0 is called the vaporization coefficient α. As a perhaps extreme example, α is only 8.3×10^{-5} for (111) surfaces of arsenic (9).

The general picture is similar if one looks at diffusion rates. For copper at 725°C, the bulk self-diffusion coefficient is about 10^{-11} cm^2/sec (10), and the use of the Einstein equation

$$\mathcal{D} = \frac{x^2}{2t} \qquad \text{(VII-2)}$$

where x is the average Brownian displacement in time t, leads to a time of about 0.1 sec for a displacement of 100 Å. At room temperature, however, the time would be about 10^{27} sec, since the apparent activation energy for diffusion is about 54 kcal/mole in this case. Surface diffusion constitutes a second type of diffusion process, and one that often is of more importance in surface chemical effects than is bulk diffusion. Surface diffusion on copper becomes noticeable at around 700°C (11) and, because of its lower activation energy, tends to be the dominant transport process over an important temperature region. By means of the field emission microscope (see Section VIII-2), surface migration is found to become noticeable at as low as half the temperature at which evaporation from the solid is appreciable. Finally, for solids near their melting point, the surface region may actually be liquidlike (12).

It thus appears that there is a very great range in the nature of solid surfaces. Those most often studied at room temperature fall into the refractory class, that is, are far below their melting points. For these, surface atoms are relatively immobile, although vibrating about equilibrium or quasi-equilibrium positions; the surface is highly conditioned by its past history and cannot be studied by the usual methods of capillarity. Solids at temperatures near their melting points show surface mobility in the form of traffic with the vapor phase and with the interior, but perhaps especially in the form of lateral mobility in the surface region.

B. Effect of Past History on the Condition of Solid Surfaces

The immobility, at least the large scale immobility, of the surface atoms of a refractory solid has the consequence that the surface energy quantities and other physical properties of the surface depend greatly on the immediate history of the material. A clean cleavage surface of a crystal will have a different (and probably lower) surface energy than a ground or abraded surface of the same material, and different also from that after heat treatment. In particular, polishing of surfaces drastically affects their nature. The mechanical procedure involved in a polishing operation differs considerably from that used in grinding; in the latter case, a material as hard as or harder than the surface to be abraded must be employed, whereas, in the former case, the polishing material is relatively soft (e.g., rouge or iron oxide) and is best held by a backing of soft material such as leather or fabric.

A valuable tool for the study of the surface region of a solid is that of electron diffraction, using low energy electrons whose range in the solid may be only a few tens of angstrom units, so that the resulting diffraction pattern is primarily that of the surface region (13, 14). For example, the results of investigations using this technique suggest that, whereas grinding leads primarily to a mechanical attrition of the surface without greatly changing its molecular crystallinity, polishing leaves a fairly deep and nearly amorphous surface layer. This surface layer resulting from polishing is generally known as the Beilby layer, since Beilby showed that such layers appear amorphous under the microscope and have the general appearance of a film of viscous liquid that not only covers the surface smoothly but also flows into surface irregularities such as cracks and scratch marks (15, 16). It appears that the polish layer is formed through a softening if not an actual melting of the metal surface (17). Studies indicate that the depth of a Beilby layer is in the range of 20 to 100 Å for metals such as gold and nickel (18).

The cold-working of metals also affects the nature of the surface region. For example, with iron, copper, and aluminum surfaces, first mechanically polished and then electropolished in phosphoric or perchloric acid baths, the diffuse electron diffraction halos characteristic of the Beilby layer gave way to a more normal crystal pattern either of the metal or the metal oxide (19). Apparently, electropolished surfaces are nearly normal in nature.

Surface defects constitute another history-dependent aspect of the condition of a surface, discussed in Section VII-4C. Such imperfections may to some extent be reversibly affected by such processes as adsorption so that it is not safe to regard even a refractory solid as having surface atoms fixed in position. Finally, it must be remembered that solid surfaces are very easily contaminated; in the case of liquid mercury, where direct surface tension measurements are possible, the importance of surface contamination and the ease with which it can occur are quite apparent. The detection of contamination in the case of solid surfaces is less easy (note Ref. 20). and whole areas of older surface chemistry have had to be revised in the light of modern ultrahigh vacuum and surface cleaning techniques.

Before considering some of the methods for determining the surface properties of solids and the nature of the results obtained, a brief discussion of the theoretical aspects of surface energy quantities for solids is desirable. This follows in the next two sections.

2. Thermodynamics of Crystals

A. *Surface Tension and Surface Free Energy*

Unlike the situation with liquids, in the case of a solid, the surface tension is not necessarily equal to the surface stress. As Gibbs (21) pointed out, the former is the work spent in *forming* unit area of surface (and may alternatively be called the surface free energy, see Sections II-1 and III-2), while the latter involves the work spent in *stretching* the surface. It is helpful to imagine that the process of forming a fresh surface of a monatomic substance is divided into two steps: first, the solid or liquid is cleaved so as to expose a new surface, keeping the atoms fixed in the same positions that they occupied when in the bulk phase; second, the atoms in the surface region are allowed to rearrange to their final equilibrium positions. In the case of the liquid, these two steps occur as one, but with solids the second step may occur only slowly because of the immobility of the surface region. Thus, with a solid it may be possible to stretch or to compress the surface region without changing the number of atoms in it, only their distances apart.

In Chapter III, surface free energy and surface stress were treated as equivalent, and both were discussed in terms of the energy to form unit additional surface. It is now desirable to consider an independent, more mechanical definition of surface stress. If a surface is cut by a plane normal to it, then, in order that the atoms on either side of the cut remain in equilibrium, it will be necessary to apply some external force to them. The total such force per unit length is the surface stress, and half the sum of the two surface stresses along mutually perpendicular cuts is equal to the surface tension. (Similarly, one-third of the sum of the three principal stresses in the body of a liquid is equal to its hydrostatic pressure.) In the case of a liquid or isotropic solid the two surface stresses are equal, but for a nonisotropic solid or crystal this will not be true. In such a case the partial surface stresses or stretching tensions may be denoted as τ_1 and τ_2.

Shuttleworth (22) gives a relation between surface free energy and stretching tension as follows. For an anisotropic solid, if the area is increased in two directions by $d\mathcal{A}_1$ and $d\mathcal{A}_2$, as illustrated in Fig. VII-1, then the total increase in free energy is given by the reversible work against the surface stresses, that is,

$$\tau_1 = G^s + \mathcal{A}_1 \left(\frac{dG^s}{d\mathcal{A}_1}\right) \quad \text{and} \quad \tau_2 = G^s + \mathcal{A}_2 \left(\frac{dG^s}{d\mathcal{A}_2}\right) \tag{VII-3}$$

where G^s is the free energy per unit area. If the solid is isotropic, Eq. VII-3 reduces to

$$\tau = \frac{d(\mathcal{A}G^s)}{d\mathcal{A}} = G^s + \frac{\mathcal{A}dG^s}{d\mathcal{A}} \tag{VII-4}$$

For liquids, the last term in Eq. VII-4 is zero, so that $\tau = G^s$ (or $\tau = \gamma$, since we will use G^s and γ interchangeably); the same would be true of a solid if the change in area $d\mathcal{A}$ were to occur in such a way that an equilibrium surface configuration was always maintained. Thus the stretching of a wire under reversible conditions would imply that interior atoms would move into the surface as needed so that the increased surface area was not accompanied by any change in specific surface properties. If, however, the stretching were done under conditions such that full equilibrium did not prevail, a surface stress would be present, whose value would differ from γ by an amount that could be time dependent and would depend on the term $\mathcal{A}\ dG^s/d\mathcal{A}$.

B. The Equilibrium Shape of a Crystal

The problem of the equilibrium shape for a crystal is an interesting one. Since the surface free energy for different faces is usually different, the question becomes one of how to construct a shape of specified volume such that the total surface free energy is a minimum. A general, quasi-geometrical solution has been given by Wulff (23) and may be described in terms of the following geometric construction.

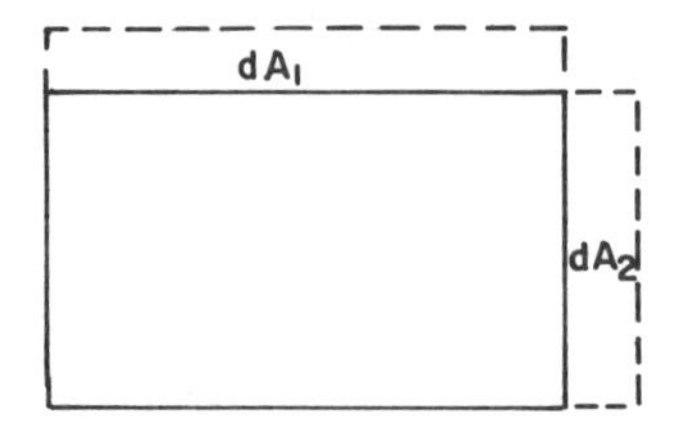

Fig. VII-1

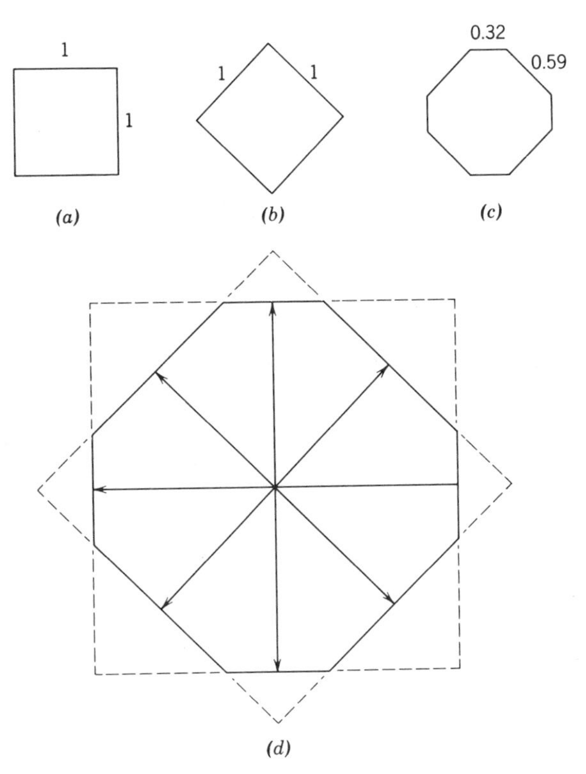

Fig. VII-2. Conformation for a hypothetical two-dimensional crystal. (*a*) (10) type planes only. For a crystal of 1-cm^2 area, the total surface free energy is $4 \times 1 \times 250 = 1000$ erg. (*b*) (11) type planes only. For a crystal of 1-cm^2 area, the total surface free energy is $4 \times 1 \times 225 = 900$ erg. (*c*) For the shape given by the Wulff construction, the total surface free energy of a 1-cm^2 crystal is $[(4 \times 0.32 \times 250) + (4 \times 0.59 \times 225)] = 851$ erg. (*d*) Wulff construction considering only (10) and (11) type planes.

Given the set of surface free energies for the various crystal planes, draw a set of vectors from a common point, of length proportional to the surface free energy and of direction normal to that of the crystal plane. Construct the set of planes normal to each vector and positioned at its end. It will be possible to find a geometric figure whose sides are made up entirely from a particular set of such planes that do not intersect any of the other planes. The procedure is illustrated in Fig. VII-2 for a two-dimensional crystal for which γ_{10} is 250 erg/cm and γ_{11} is 225 erg/cm. Note that the optimum shape is one making use of both types of planes. For the particular surface tensions given, there is a free energy gain in rounding off the corners of a square all (11) sides crystal, to reduce its perimeter. Note Problem VII-1 of this chapter.

A statement of Wulff's theorem is that for an equilibrium crystal there exists a point in the interior such that its perpendicular distance h_i from the

ith face is proportional to γ_i. This is, of course, the basis of the construction of Fig. VII-2.

Herring (24), while agreeing with the Wulff construction, pointed out that one would not expect infinitely sharp edges of actual crystals, but he concluded that in practice the equilibrium radii of curvature could be quite small. More recently, Benson and Patterson (25) have given an analytical proof of Wulff's theorem, and Drechsler and Nicholas have made calculations on the equilibrium shapes of face-centered and body-centered cubic crystals (26). The seeking of such a minimum free energy polyhedron seems actually to occur. Thus initially irregular cavities in rock salt (27) and metals (28) assume a regular shape on heating. If small enough, irregular crystals take on an equilibrium shape on annealing (29).

Since crystal habit can be determined by kinetic and other nonequilibrium effects, it can happen that an actual crystal has faces that are not the equilibrium ones of the Wulff construction. For example, if a (100) plane is a stable or singular plane, but by, say, grinding, the actual facet deviates from this by a small angle so as to nominally be describable as a (x00) plane where x is a large number, then a local reduction in free energy will occur if that facet decomposes into a set of (100) steps and (010) risers. The general criteria that determine whether a given plane should spontaneously undergo such a local decomposition have been worked out (see Ref. 30).

C. The Kelvin Equation

The Kelvin equation (Eq. III-20), which gives the increase in vapor pressure for a curved surface and hence of small liquid drops, should also apply to crystals. Thus

$$\frac{RT \ln P}{P^0} = \frac{2\gamma \bar{V}}{r} \tag{VII-5}$$

Since an actual crystal will be polyhedral in shape, and may well expose faces of different surface tension, the question is what value of γ and of r should be used. As noted in connection with Fig. VII-2, the Wulff theorem states that γ_i/r_i is invariant for all faces of an equilibrium crystal. In Fig. VII-2, r_{10} is the radius of the circle inscribed to the set of (10) faces and r_{11} is the radius of the circle inscribed to the set of (11) faces. (See also Ref. 31.) Equation VII-5 may also be applied to the solubility of small crystals (see Section VIII-2). In the case of relatively symmetrical crystals, Eq. VII-5 can be used as an approximate equation with r and γ regarded as mean values.

3. Theoretical Estimates of Surface Energies and Free Energies

A. Covalently Bonded Crystals

The nature of the theoretical approach to the calculation of surface energy quantities necessarily varies with the type of solid considered. Perhaps the simplest case is that of a covalently bonded crystal whose sites are occupied by atoms; in this instance no long-range interactions need be considered. The example of this type of calculation, *par excellence*, is that for the surface energy of diamond. Harkins (32) considered the surface energy at 0°K to be simply one-half of the energy to rupture that number of bonds passing through 1 cm^2, that is,

$$E^s = \frac{1}{2}E_{\text{cohesion}}$$

The unit cell for diamond is shown in Fig. VII-3, and it is seen that three bonds would be broken by a cleavage plane parallel to (111) planes. From the (111) interplanar distance of 2.32 Å, and the density of diamond 3.51 g/cm^3, one computes that 1.83×10^{15} bonds/cm^2 are involved, and, using 90 kcal/mole as the bond energy, the resulting value for the surface energy is 5650 erg/cm^2. For (100) planes, the value is 9820 erg/cm^2. Since these figures are for 0°K, they are also equal to the surface free energy at that temperature.

Harkins then estimated T_c for diamond to be about 6700°K and, using Eq. III-10, found the entropy correction at 25°C to be negligible so that the preceding values also approximate the room temperature surface free energies. These values cannot be strictly correct, however, since no allowance has been made for surface distortion (see Section VII-3B,C).

B. Rare Gas Crystals

A common type of solid is that held together primarily by van der Waals forces or by these in combination with coulomb forces. If the lattice sites are occupied by atoms or monatomic ions, then orientation effects are absent, and calculations of the surface energy may be made with a minimum of difficulty.

We consider in moderate detail the case of rare gas crystals because the procedure is simple enough to be described briefly. The lattice is face-centered cubic and can be considered a simple cubic lattice with alternate sites

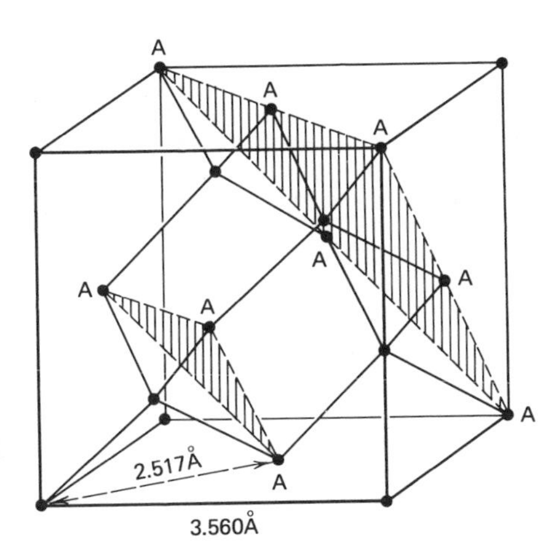

Fig. VII-3. Diamond structure. (From Ref. 32.)

unoccupied. The principal problem is now to calculate the net energy of interaction across a plane, such as the one indicated by the dotted line in Fig. VII-4. In other words, as was the case with diamond, the surface energy at 0°K is essentially the excess potential energy of molecules near the surface.

The calculation is actually made in two steps. First, as indicated, one separates the two parts, holding the atoms fixed in their positions, thus obtaining the principal contribution $E^{s\prime}$, to the surface energy. Next, one calculates the decrease from this value, due to the subsequent rearrangement of the surface layer to its equilibrium position. Thus

$$E^s = E^{s\prime} - E^{s\prime\prime} \tag{VII-6}$$

It is seen from an examination of Fig. VII-4 that the mutual interaction for a pair of planes occurs once if the separation of the plane is a, twice if it is $2a$, and so on, so if the planes are labeled by the index l, as shown in the figure, the mutual potential energy is

$$2u' = \sum_{l \geq 1} l\epsilon(r) \tag{VII-7}$$

Here, u' is the surface energy per atom in the surface plane, and $\epsilon(r)$ is the potential energy function in terms of the distance r between two atoms.

Various functions for $\epsilon(r)$ have been proposed; a classic form is that used by Shuttleworth (33):

$$\epsilon(r) = \lambda r^{-s} - \mu r^{-t} \tag{VII-8}$$

The first term gives the repulsion between atoms and, since s is about 12, this is important only at small distances; the second term corresponds to the van der Waals attraction, and the best value for t is about 6. (Experimental values for s and t are obtainable from the virial coefficients in the rare gases.) The summation indicated in Eq. VII-7 must be carried out over all interatomic distances. The distance from the origin to a point in the lattice

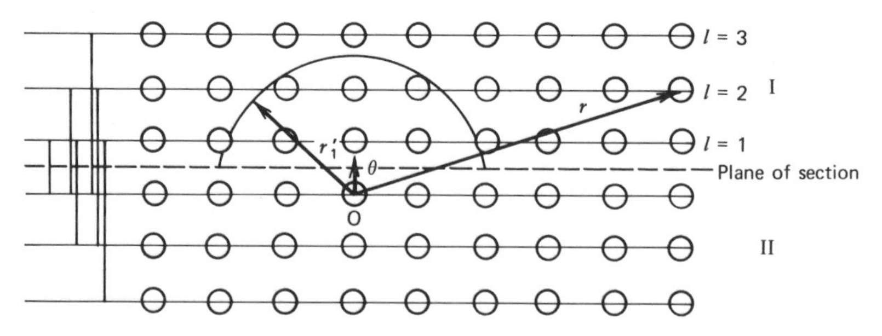

Fig. VII-4. Interactions across a dividing surface for a rare gas crystal. (From Ref. 33.)

is

$$d = (x^2 + y^2 + z^2)^{1/2}$$

where x, y, and z are given by m_1a, m_2a, and m_3a; a is the side of the simple cubic unit cell; and the m's are integers. Thus

$$d = a(m_1^2 + m_2^2 + m_3^2)^{1/2} \quad \text{(VII-9)}$$

Equation VII-9 then becomes

$$-2u' = \lambda a^{-s} \sum_{(m_1+m_2+m_3)\text{even}, l \geq 1} \frac{l}{(m_1^2 + m_2^2 + m_3^2)^{s/2}} - \mu a^{-t} \sum_{(m_1+m_2+m_3)\text{even}, l \geq 1} \frac{l}{(m_1^2 + m_2^2 + m_3^2)^{t/2}} \quad \text{(VII-10)}$$

Only even values of $(m_1 + m_2 + m_3)$ are used since every other site is vacant. The numerical values of these lattice sums are dependent on the exponents used for $\epsilon(r)$, and Eq. VII-10 may be written

$$-2u' = B_s\lambda a^{-s} - B_t\mu a^{-t} \quad \text{(VII-11)}$$

Similarly, the energy of evaporation ϵ_0 is given by

$$-2\epsilon_0 = \lambda a^{-s} \sum_{(m_1+m_2+m_3)\text{even}} \frac{1}{(m_1^2 + m_2^2 + m_3^2)^{s/2}} - \mu a^{-t} \sum_{(m_1+m_2+m_3)\text{even}} \frac{1}{(m_1^2 + m_2^2 + m_3^2)^{t/2}} \quad \text{(VII-12)}$$

or

$$-2\epsilon_0 = A_s\lambda a^{-s} - A_t\mu a^{-t} \quad \text{(VII-13)}$$

In these equations, the sums give twice the desired quantity because each atom is counted twice. In addition, the condition that the atoms be at their equilibrium distances gives, from differentiation of Eq. (VII-13),

$$sA_s\lambda a^{-s} = tA_t\mu a^{-t} \quad \text{(VII-14)}$$

Eliminating λa^{-s} and μa^{-t} from Eqs. VII-11, VII-13, and VII-14,

$$u' = \frac{\epsilon_0(sB_t/A_t - tB_s/A_s)}{s - t} \quad \text{(VII-15)}$$

If the interaction between atoms that are not nearest neighbors is neglected, then the ratios B/A are each equal to the ratio of the number of nearest neighbors to a surface atom (across the dividing plane) to the number of nearest neighbors for an interior atom. The calculation then reduces to that given by Eq. III-17.

Returning to the complete calculation, $E^{s\prime}$ is then given by u' multiplied by the number of atoms per unit area in the particular crystal plane.

Shuttleworth calculated the lattice distortion term $E^{s''}$ by allowing the surface plane to move relative to the adjacent interior plane until a position of minimum energy was reached, and he found a several percent increse in the first interplanar distance. Benson and Claxton (34), using computer methods, carried out the procedure for the first five layers and found that the distortion dropped off rapidly; it was 3.5% for the first layer and only 0.04% for the fifth. Moreover, the more elaborate calculation made very little difference in the approximately 1% correction to $E^{s'}$ that $E^{s''}$ amounts to in the case of rare gas crystals. They also made the calculation using the potential function

$$\epsilon(r) = be^{-r/\rho} - cr^{-6} \tag{VII-16}$$

in which the form of the repulsive term was changed; again, the effect on the calculated γ's was small. Their results (using the Lennard-Jones 6–12 potential, Eq. VII-8) are given in Table VII-1.

Shuttleworth (33) also calculated surface stresses using Eq. VII-4, obtaining negative values of about one-tenth of those for the surface energies.

C. *Ionic Crystals*

1. Surface Energies at 0°K. The complete history of the successive attempts to calculate the surface energy of simple ionic crystals is too long and complex to give here, and the following constitutes only a brief summary. The classical procedure entirely resembles that described for the rare gas crystals, except that charged atoms occupy the sites, and a more complicated potential energy function must be used. For the face-centered cubic alkali halide crystals, if some particular ion is selected as the origin, all ions with the same charge have coordinates for which $(m_1 + m_2 + m_3)$ is even, while ions with opposite charges have coordinates for which the sum is odd. As with the rare gas crystals, the first step is to calculate the mutual potential energy across two halves of a cleavage plane, and very similar lattice sums are involved, using, however, the appropriate potential energy function for like or unlike ions, as the case may be. Early calculations of this nature were made by Born and co-workers (35).

TABLE VII-1
Surface Energies at 0°K for Rare Gas Crystals

Rare gas	$2a$, Å	E_{vap}, erg/atom	E^s, erg/cm² (100)	(110)	(111)	E^s_{liq}, erg/cm²
Ne	4.52	4.08×10^{-14}	21.3	20.3	19.7	15.1
Ar	5.43	13.89	46.8	44.6	43.2	36.3
Kr	5.59	19.23	57.2	54.5	52.8	
Xe	6.18	26.87	67.3	64.1	62.1	

Refinements were made by Lennard-Jones, Taylor, and Dent (36–38), including an allowance for surface distortion. Their value of E^s for (100) planes of sodium chloride at 0°K was 77 erg/cm^2. Subsequently, Shuttleworth obtained a value of 155 erg/cm^2 (33).

There have been important developments in several directions. The type of potential function used has been refined and, currently, the treatment of Huggins and Mayer (39) is in use; this takes the form

$$\epsilon_{ij}(r) = \frac{z_i z_j e^2}{r} - \frac{C_{ij}}{r^6} - \frac{d_{ij}}{r^8} + bb_i b_j e^{-r/\rho} \qquad \text{(VII-17)}$$

where the first term on the right gives the coulomb energy between ions i and j, the next two terms give the van der Waals attraction in the form of dipole–dipole and dipole–quadrupole interactions, and the last term gives the electronic repulsion. It turns out that, although the van der Waals terms are unimportant contributors to the *total lattice energy*, the dipole–dipole (inverse r^6) term makes a 20–30% contribution to the *surface energy* (33, 40). In the case of multivalent ion crystals, such as CaF_2, CaO, etc. the results appear to be quite sensitive to the form of the potential function used. Benson and McIntosh (41) obtained values for the surface energy of (100) planes of MgO ranging from −298 to 1362 erg/cm^2 depending on this variable!

Dynamic models for ionic lattices recognize explicitly the force constants between ions, and their polarization. In "shell" models, the ions are represented as a shell and a core, coupled by a spring (see Refs. 42–44), and parameters are evaluated by matching bulk elastic and dielectric properties. Application of these models to the surface region has allowed calculation of surface vibrational modes (45) and LEED patterns (46–48) (see Section VIII-3).

A very important matter is the complete evaluation of the surface distortion associated with the unsymmetrical field at the surface. The problem is a difficult one to treat, in general, and can be simplified by assuming that distortion is limited to motions normal to the plane. This approach was taken by Verwey (49) and, more recently, by Benson and co-workers (50); their calculated displacements for the first five planes in the (100) face of sodium chloride are shown in Fig. VII-5. The distortion correction to $E^{s\prime}$ amounted to about 100 erg/cm^2 or about half of $E^{s\prime}$ itself. The displacements shown in Fig. VII-5 suggest a tendency toward ion-pair formation, and Molière and Stranski (51) suggested that lateral displacements to give ion-doublet structures should also be considered. Calculations by Tasker (52), however, yielded much smaller displacements than those shown in Fig. VII-5.

The uncertainties in choice of potential function and in how to approximate the surface distortion contribution combine to make the calculated surface energies of ionic crystals rather uncertain. Some results are given in Table VII-2, but comparison between the various references cited will yield major discrepancies. Experimental verification is difficult (see Section

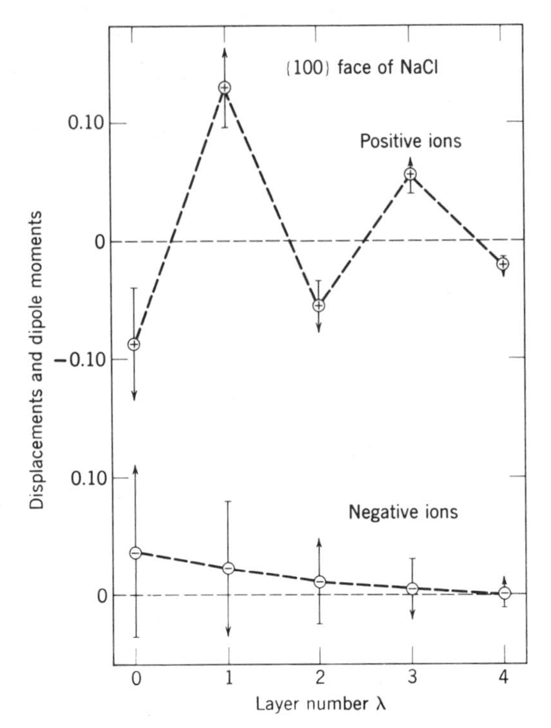

Fig. VII-5. Equilibrium configuration of the first five layers of the (100) face of NaCl. Displacements in units of a, positive values indicate moments in direction of outward normal. Direction and magnitude of dipole movements (debye units) indicated by arrows. For negative ions, the lengths of the arrows correspond to one-tenth of the dipole moment. (From Ref. 50.)

VII-5). Qualitatively, one expects the surface energy of a solid to be distinctly higher than the surface tension of the liquid and, for example, the value of 212 erg/cm^2 for (100) planes of NaCl given in the table is indeed higher than the surface tension of the molten salt, 190 erg/cm^2 (33). Practically speaking, an extrapolation to room temperature of the value for the molten salt using semiempirical theory such as that of Ref. 56 is probably still the most reliable procedure.

2. *Surface Stresses and Edge Energies.* Some surface tension values, that is, values of the surface stress τ are included in Table VII-2. These are obtained by applying Eq. VII-4 to the appropriate lattice sums. The calculation is very sensitive to the form of the lattice potential. Earlier calculations have given widely different results, including negative τ's (33, 36, 37).

It is possible to set up the lattice sums appropriate for obtaining the edge energy, that is, the mutual potential energy between two cubes having a common edge. The dimensions of the edge energy k are ergs per centimeter, and calculations by Lennard-Jones and Taylor (36) gave a value of k of about 10^{-5} for the various alkali halides. A more recent estimate is 3×10^{-6} for the edge energy between (100) planes of sodium chloride (57). As with surface stress values, these figures can only be considered as tentative estimates.

D. Molecular Crystals

A molecular crystal is one whose lattice sites are occupied by molecules, as opposed to atoms or monatomic ions, and the great majority of solids belong in this category. Included would be numerous salts, such as $BaSO_4$, involving a molecular ion; most nonionic inorganic compounds as, for example, CO_2 and H_2O; and all organic compounds. Theoretical treatment is difficult. Relatively long-range van der Waals forces (see Chapter VI) are involved; surface distortion now includes surface reorientation effects.

One molecular solid to which a great deal of attention has been given is ice. A review by Fletcher (58) cites calculated surface tension values of 100–120 erg/cm^2 (see Ref. 59) as compared to an experimental measurement of 109 erg/cm^2 (60). There is much evidence that a liquidlike layer develops at the ice–vapor interface, beginning around -35°C, and thickening with increasing temperature (58, 61, 62).

E. Metals

The calculation of the surface energy of metals has been along two rather different lines. The first has been that of Skapski, outlined in Section III-1B. In its simplest form, the procedure involves simply prorating the surface energy to the energy of vaporization on the basis of the ratio of the number

TABLE VII-2
Calculated Surface Energies and Surface Stresses (in erg/cm^2)

	(100) Planes		(110) Planes	
Crystal[a]	E^s	τ	E^s	τ
LiF	480	1530	1047	407
LiCl	294	647	542	252
NaF	338	918	741	442
NaCl	212	415	425	256
NaBr	187	326	362	221
NaI	165	231	294	182
KCl	170	295	350	401
CsCl[b]			219	
CaO[b]	509–879			
MgO[b]	360–924			
CaF_2[b]			1082	

[a] From Ref. 52 unless otherwise indicated.
[b] From 53; see also Refs. 54 and 55.

of nearest neighbors for a surface atom to that for an interior atom. The effect is to bypass the theoretical question of the exact calculation of the cohesional forces of a metal and, of course, to ignore the matter of surface distortion.

Empirically, however, the results are reasonably accurate, and the approach is a very useful one. An application of it to various Miller index planes is given by MacKenzie and co-workers (63). Related is a statistical mechanical treatment by Reiss and co-workers (64) [see also Schonhorn (65)].

A related approach carries out lattice sums using a suitable interatomic potential, much as has been done for rare gas crystals (66). One may also obtain the dispersion component to E^s by estimating the Hamaker constant H by means of the Lifshitz theory (Eq. VI-34), but again using lattice sums (67). Thus for a *fcc* crystal the dispersion contributions are

$$E^s(100) = \frac{0.09184\ H}{a^2} \quad \text{and} \quad E^s(110) = \frac{0.09632\ H}{a^2} \qquad \text{(VII-18)}$$

where a is the nearest neighbor distance. The value so calculated for mercury is about half of the surface tension of the liquid and close to the γ^d of Fowkes (68; see Section IV-2).

The broken bond approach has been extended by Nason and co-workers (see Ref. 69) to calculate E^s as a function of surface composition for alloys. The surface free energy follows on adding an entropy of mixing term, and the free energy is then minimized.

The second type of model is that of free electrons in a box whose sides correspond to the surfaces of the metal; the treatment is thus quantum mechanical and is essentially independent of the type of lattice involved. In the simplest version, Brager and Schuchowitsky (70) assumed the walls to be impenetrable, which meant that the standing electron waves were required to have nodes at the walls. This requirement eliminated a certain number of otherwise permissible states, and the kinetic energy corresponding to these rejected points gave the surface energy. The approach was refined by Huang and Wyllie (71), and further by Huntington (72) and by Ewald and Juretschke (73). Agreement with experimental estimates has been considerably improved in more recent (and more complex) electron gas models (see Refs. 74 and 75).

As with rare gas and ionic crystals, there should be surface distortion in the case of metals. Burton and Jura (76) have estimated theoretically the increased interplanar spacing expected at the surface of various metals.

4. Factors Affecting the Surface Energies and Surface Tensions of Actual Crystals

A. State of Subdivision

Surface chemists are very often interested in finely divided solids having high specific surface area, and it is worthwhile to consider briefly an illustrative numerical example of how surface properties should vary with particle size. We will refer the calculation to a 1-g sample of sodium chloride, of density 2.2g/cm^3 and with assumed surface energy of 200 erg/cm^2 and edge energy 3×10^{-6} erg/cm. The original 1-g cube is now considered to be successively divided into smaller cubes, and the number of such cubes, their area and surface energies, and edge lengths and edge energies are summarized in Table VII-3.

It might be noted that only for particles smaller than about 1 μ or of surface area greater than a few square meters per gram does the surface energy become significant. Only for very small particles does the edge energy become important, at least with the assumption of perfect cubes.

B. Deviations from Ideal Considerations

The numerical illustration given is so highly idealized that any experimental agreement with the numbers quoted could hardly be more than coincidental. It seems worthwhile to collect together at this point the several layers of complications that are involved.

1. The figures in the table are theoretical estimates for 100 planes at 0°K, and even on this basis are highly uncertain for reasons discussed in the preceding section.
2. To obtain the surface energy at room temperature, a correction is needed, based on the integral of C_p^s, the surface contribution to the heat capacity (see Eq. VII-22); the correction is difficult to make theoretically but should be small for refractory substances. Any attempt to estimate surface free energies by third-law-type heat capacity integrations would be extremely difficult to defend as accurate.
3. Continuing, for the moment, the restriction to an equilibrium surface, the most stable shape for a crystal will be some polyhedron, as predicted by the Wulff theorem (Section VII-2B), so that the specific surface energy would be a weighted average of that of the various equilibrium faces.

TABLE VII-3
Variation of Specific Surface Energy with Particle Size[a]

Side, cm	Total area, cm^2	Total edge, cm	Surface energy, erg/g	Edge energy, erg/g
0.77	3.6	9.3	720	2.8×10^{-5}
0.1	28	550	5.6×10^3	1.7×10^{-3}
0.01	280	5.5×10^4	5.6×10^4	0.17
0.001	2.8×10^3	5.5×10^6	5.6×10^5	17
10^{-4} (1 μ)	2.8×10^4	5.5×10^8	5.6×10^6 (0.1 cal)	1.7×10^3
10^{-6} (100 Å)	2.8×10^6	5.5×10^{12}	5.6×10^8 (13 cal)	1.7×10^7 (0.4 cal)

[a] 1 g NaCl, $E^s = 200$ erg/cm^2, $k = 3 \times 10^{-6}$ erg/cm.

4. It appears that the *equilibrium* surface for any given plane is *not* smooth. Temperley (77) points out that, although the total surface energy is indeed a minimum for a given apparent area if the surface is plane, this requires an improbably ordered arrangement, and the minimum free energy is for a saw-toothed surface (with teeth or waves as much as several hundred angstroms in height) representing an optimum energy-entropy balance. Burton and Cabrera (78) predict a similar effect, from a different analysis, which leads them to expect there to be a surface melting point only above which will an equilibrium molecular roughness appear. By either approach, the surface area and surface energy should be larger than that given by the simple procedure summarized in Table VII-3.

5. Several additional factors come in if nonequilibrium crystals are involved. Even though perfect cleavage planes are supposed to be present, those that predominate may be determined by experimental conditions. Selective adsorption of some constituent of the mother liquor may change the crystal habit by retarding the outward growth of certain planes; thus sodium chloride crystallizes from urea solutions in the form of octahedra instead of cubes. The shape of small silver crystals annealed in air is different from those annealed in nitrogen apparently because of the effect of adsorbed oxygen (30). Moreover, as noted in Section VII-2B, given sufficient surface mobility, planes vicinal to a stable one in their orientation may spontaneously decompose into stepped or grooved surfaces.

6. Actual crystal planes tend to be incomplete and imperfect in many ways. A nonequilibrium surface stress may be relieved by surface imperfections. There may be overgrowths, incomplete planes, steps, dislocations, and so on, as illustrated in Fig. VII-6. Correspondingly, in addition to the ideal energy for removing a surface atom or ion, there will be additional and different values for the positional energy of one in an interior corner, an interior side, an edge, and so forth. Stranski (80, 81) considers several *dozen* different types of such positions and estimates the energies for the removal of a sodium chloride unit; these vary enormously. Dunning (82) has pointed out that the presence of adsorbed molecules changes the energy of the various types of site and, again depending on surface mobility, should alter their distribution.

7. In addition to all of the preceding points, the surface condition of a crystal can be markedly affected by abrasion, sintering, or polishing, so that the mechanical history is very important.

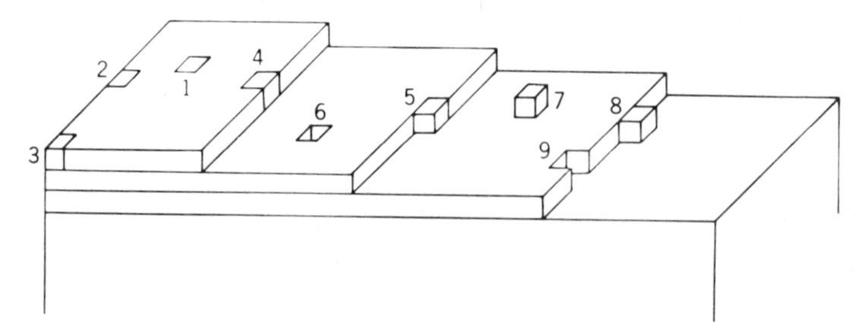

Fig. VII-6. Types of surface locations. (From Ref. 79.)

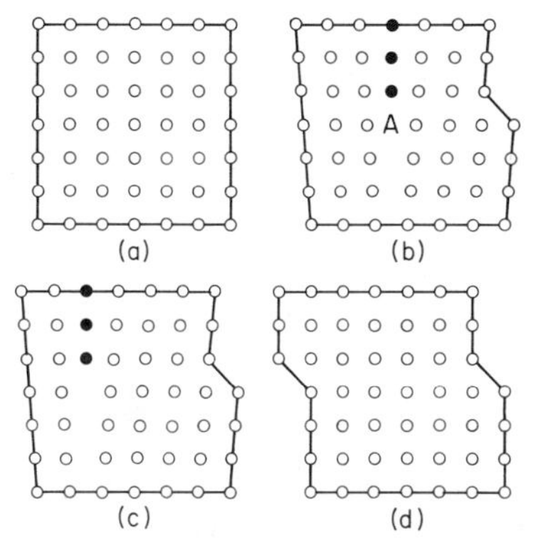

Fig. VII-7. Motion of an edge dislocation in a crystal undergoing slip deformation. (*a*) The undeformed crystal. (*b, c*) Successive stages in the motion of the dislocation from right to left. (*d*) The undeformed crystal. (From Ref. 86, with permission.)

It is because of these complications, both theoretical and practical, that it is doubtful that calculated surface energies for solids will ever serve as more than a guide as to what to expect experimentally. Corollaries are that different preparations of the same substance may give different E^s values and that widely different experimental methods may yield different apparent E^s values for a given preparation. In this last connection, see Section VII-5 especially.

C. Dislocations

Dislocation theory as a portion of the subject of solid-state physics is somewhat beyond the scope of this book, but it is desirable to examine the subject briefly in terms of its implications in surface chemistry. Perhaps the most elementary type of defect is that of an extra or interstitial atom—Frenkel defect (83)—or a missing atom or vacancy—Schottky defect (84). Such point defects play an important role in the treatment of diffusion and electrical conductivities in solids and the solubility of a salt in the host lattice of another or different valence type (85). Point defects have a thermodynamic basis for their existence, in terms of the energy and entropy of their formation, the situation is similar to the formation of isolated holes and erratic atoms on a surface. Dislocations, on the other hand, may be viewed as an organized concentration of point defects; they are lattice defects and play an important role in the mechanism of the plastic deformation of solids. Lattice defects or dislocations are not thermodynamic in the sense of the point defects; their formation is intimately connected with the mechanism of nucleation and crystal growth (see Section IX-4), and they constitute an important source of surface imperfection.

One type of dislocation is the *edge dislocation*, illustrated in Fig. VII-7. We imagine that the upper half of the crystal is pushed relative to the lower half, and the sequence shown is that of successive positions of the dislocation. An extra plane, marked as full circles, moves through the crystal until it emerges at the left. The process is much like moving a rug by pushing a crease in it.

The dislocation may be characterized by tracing a counterclockwise circuit around the point A in Fig. VII-7*b*, counting the same number of lattice points in the plus and minus directions along each axis or row. Such a circuit would close if the crystal were perfect, but if a dislocation is present, it will not, as illustrated in the figure. This circuit is known as a *Burgers* circuit (87); its failure to close distinguishes a dislocation from a point imperfection. The ends of the circuit define a vector, the *Burgers vector b*, and the magnitude and angle of the Burgers vector are used to define the magnitude and type of a dislocation.

The second type of dislocation is the *screw dislocation*, illustrated in Fig. VII-8*a*—from Frank (88)—and in *b*; each cube represents an atom or lattice site. The geometry of this may be imagined by supposing that a block of rubber has been sliced part way through and one section bent up relative to the other one. The crystal has a single plane, in the form of a spiral ramp; the screw dislocation can be produced by slip on *any* plane containing the dislocation line *AB*—Fig. VII-8*b*. The distortion around a screw dislocation is mostly shear in nature, as suggested in Fig. VII-8*a* by showing unit cells as cubes displaced relative to one another in the direction of the slip vector. Combinations of screw and edge dislocations may also occur, of course. A photomicrograph of a Carborundum crystal is shown in Fig. VII-9, illustrating spiral growth patterns (89).

The density of dislocations is usually stated in terms of the number of dislocation lines intersecting unit area in the crystal; it ranges from $10^8/cm^2$ for "good" crystals to $10^{12}/cm^2$ in cold worked metals. Thus, dislocations are separated by 10^2 to 10^4 Å, or every crystal grain larger than about 100 Å will have dislocations on its surface; one surface atom in a thousand is

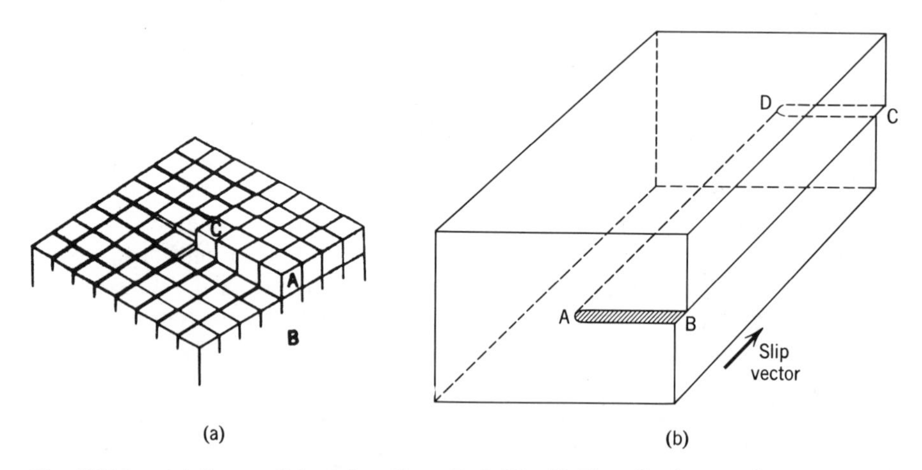

Fig. VII-8. (*a*) Screw dislocation (from Ref. 88). (*b*) The slip that produces a screw-type dislocation. Unit slip has occurred over *ABCD*. The screw dislocation *AD* is parallel to the slip vector. (From W. T. Read, Jr., *Dislocations in Crystals*, McGraw-Hill, New York, 1953, p. 15.)

Fig. VII-9. Screw dislocation in a carborundum crystal. (From Ref. 89.)

apt to be near a dislocation. By elastic theory, the increased potential energy of the lattice near a dislocation is proportional to $|b|^2$. The core or dislocation line is severely strained, and the chemical potential of the material in it may be sufficiently high to leave it hollow. Frank (90) related the rigidity modulus μ, the surface tension, and the Burgers vector b by

$$r = \frac{\mu b^2}{8\pi^2 \gamma} \tag{VII-19}$$

where r is the radius of the hollow cylinder. A crystal grown in a solvent medium may be especially prone to have hollow dislocations because of the probably relatively low solid–liquid interfacial tension. Adsorption of a gas may sufficiently lower γ to change a dislocation that emerges flush with the surface into a pit (82). Etching of crystals tends to remove material preferentially from dislocation sites, producing etch pits, and the production of etch pits is in fact a way of seeing and counting dislocations by microscopy. In general, then, the surface sites of dislocations can be numerous enough to mar seriously the uniformity of a surface and in a way that is sensitive to past history and that may interact with the very surface phenomenon being studied, such as adsorption; in addition, surface dislocation appears to play a major role in surface kinetic processes such as crystal growth and catalyzed reactions.

D. Surface Heterogeneity

It has been made evident in this section that the total surface energy depends on the state of subdivision of a substance and that the average specific surface energy of a particular substance can be variable and different from its ideal value. The remaining point to give specific emphasis to is that the average E^s does not adequately describe a real surface.

If the surface of an actual solid is examined portion by portion on a molecular scale, the local E^s will vary greatly, and this variation could be characterized statistically by a normalized distribution function, $f(E^s)$, such that

$$\bar{E}^s = \int_0^\infty f(E^s)\, dE^s \qquad \text{(VII-20)}$$

where $\bar{E}^s$ denotes the average value. The use of an equation of this type will be necessary in the discussion of adsorption phenomena (Section XVI-15). The next more detailed statistical picture would provide a correlation function giving the distribution of local E^s values as a function of distance from a particular site i of surface energy E_i^s.

The statistical approach constitutes a coarse and macroscopic way of describing a real surface, as opposed to the molecular picture of imperfections, dislocations, and so on. The approach does provide at least an approximate way of applying thermodynamics to surface phenomena involving real systems.

5. Experimental Estimates of Surface Energies and Free Energies

There is a rather limited number of methods for obtaining experimental surface energy and free energy values, and many of them are peculiar to special solids or situations. The only general procedure is the rather empirical one of estimating a solid surface tension from that of the liquid. Evidence from a few direct measurements (see Section VII-1A) and from nucleation studies (Section IX-3) suggests that a solid near its melting point generally has a surface tension 10 to 20% higher than the liquid—about in the proportion of the heat of sublimation to that of liquid vaporization, and a value estimated at the melting point can then be extrapolated to another temperature by means of an equation such as III-10.

A. Methods Depending on the Direct Manifestation of Surface Tensional Forces

It was mentioned during the discussion on sintering (Section VII-1A) that some bulk flow can occur with solids near their melting point; solids may act like viscous liquids, in that their strain rate is proportional to the applied stress. As shown in Fig. VII-10, the strain rate (fractional elongation per unit time) versus stress (applied load per unit area) for 1-mil gold wires near

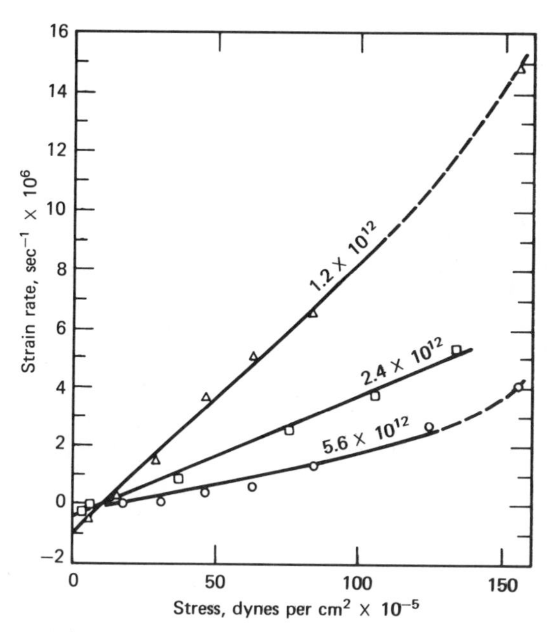

Fig. VII-10. Strain rate versus stress for 1-mil gold wire at various temperatures: Δ, 1020°C; □, 970°C; ○, 920°C. (From Ref. 91.)

1000°C is nearly linear at small loads (91). The negative strain rate at zero load is considered to be due to the surface tension of the metal and, therefore, the stress such as to give zero strain rate must just balance the surface tensional force along the circumference of the wire. Udin and co-workers (1) obtained a value of 1370 dyne/cm for copper near its melting point by this method, and Alexander and co-workers (91) found values of 1300 to 1700 dyne/cm for gold at 1000°C. Interestingly, wire that had been heated for long periods of time at low stresses showed alternating waists and bulges; the similar phenomenon in the case of a column of liquid was discussed in Section II-3. Greenhill and McDonald (92) similarly found the surface tension of solid paraffin to be about 50 dyne/cm at about 50°C, as opposed to that of about 25 dyne/cm for the liquid state.

An indirect estimate of surface tension may be obtained from the change in lattice parameters of small crystals owing to surface tensional compression. Nicholson (93), for example, reported positive values for magnesium oxide and sodium chloride. The result may have represented a nonequilibrium surface stress rather than surface tension (see Ref. 53). With less soft materials, however, surface stress may lead to wrinkling (93a).

A direct measurement of surface tension is sometimes possible from the work of cleaving a crystal. Mica, in particular, has such a well-defined cleavage plane that it can be split into large sheets of fractional millimeter thickness. Orowan (94), in reviewing the properties of mica, gives the equation

$$2\gamma = \frac{T^2 x}{2E} \tag{VII-21}$$

where T is the tension in the sheet of thickness x that is being split and E is the modulus of elasticity. Essentially, one balances the surface tension energy for the two surfaces being created against the elastic energy per square centimeter. The process is not entirely reversible, so the values obtained, again, are only approximate. Interestingly, they were 375 erg/cm^2 in air and 4500 erg/cm^2 in a vacuum, the former apparently representing the work of cohesion of two surfaces with adsorbed water on them.

Gilman (95) and Westwood and Hitch (96) have applied the cleavage technique to a variety of crystals. The salts studied (with cleavage plane and best surface tension value in parentheses) were LiF (100, 340), MgO (100, 1200), CaF_2 (111, 450), BaF_2 (111, 280), $CaCO_3$ (001, 230), Si (111, 1240), Zn (0001, 105), Fe (3% Si) (100, about 1360), and NaCl (100, 110). Both authors note that their values are in much better agreement with a very simple estimate of surface energy by Born and Stern in 1919, which used only coulomb terms and a hard sphere repulsion. In more recent work, however, Becher and Freiman (96a) have reported distinctly higher values of γ, the "critical fracture energy." Westwood and Hitch suggest, incidentally, that the cleavage experiment, not being fully reversible, may give only a bond breaking or nearest neighbor type of surface energy with little contribution from surface distortion.

B. Surface Energies and Free Energies from Heats of Solution

The illustrative data presented in Table VII-3 indicate that the total surface energy may amount to a few tenths of a calorie per gram for particles of the order of 1 μ in size. When the solid interface is destroyed, as by dissolving, the surface energy appears as an extra heat of solution, and with accurate calorimetry it is possible to measure the small difference between the heat of solution of coarse and of finely crystalline material.

An excellent example of work of this type is given by the investigations of Benson and co-workers (97, 98). They found, for example, a value of $E^s = 276$ erg/cm^2 for sodium chloride. Accurate calorimetry is required since there is only a few calories per mole difference between the heats of solution of coarse and finely divided material. The surface area of the latter may be determined by means of the BET gas adsorption method (see Section XVI-5).

Brunauer and co-workers (99, 100) found values of E^s of 1310, 1180, and 386 erg/cm^2 for CaO, $Ca(OH)_2$, and tobermorite (a calcium silicate hydrate). Jura and Garland (101) reported a value of 1040 erg/cm^2 for magnesium oxide. They also measured the heat capacity of powdered versus coarse material down to low temperatures, thus allowing surface entropy estimates; thus

$$S^s = \frac{1}{\mathscr{A} \int_0^T \Delta C_p \, d \ln T} \tag{VII-22}$$

A question in the use of Eq. VII-22 is that reversibility is implied and yet the surfaces of the small crystals especially could hardly have been equilibrium ones.

It would seem, then, that what was obtained was the entropy of a particular, and undoubtedly a nonequilibrium, surface configuration. The question was put in one sense by Bauer (102), who expressed concern as to whether the experimental surface effects on heats of solution and heat capacities were really extensive properties and therefore independent of crystal size and shape. Patterson and co-workers (103), using various particle size fractions of finely divided sodium chloride prepared by a volatilization method, did find the surface contribution to the low temperature heat capacity to vary approximately in proportion to the area, as determined by gas adsorption. It would have been of interest to make similar measurements on samples of sodium chloride prepared by other methods.

A similar set of questions comes up in the thermodynamic treatment of adsorption equilibrium on a finely divided solid whose surface may not be an equilibrium surface. This aspect is discussed in more detail in Section X-3.

Finally, some theoretical considerations of the heat capacity of powders at low temperature by Jura and Pitzer (104) suggest that around a few degrees absolute, the principal contribution to the apparent surface heat capacity of very small particles (around 100 Å) may come from the translation and rotation of the particles themselves.

C. *Relative Surface Tensions from Equilibrium Shapes of Crystals*

It was noted in Section VII-2B that, given the set of surface tension values for various crystal planes, the Wulff theorem allowed the construction of the equilibrium or minimum free energy shape. The procedure may be applied in reverse. Small crystals will gradually take on their equilibrium shape on annealing near their melting point and, likewise, small air pockets in a mass of the substance will form equilibrium-shaped "negative" crystals. The latter procedure offers the possible advantage that adventitious contamination of the solid–air interface is less likely.

Some instances of this type of equilibration were noted in Section VII-2B. Specifically, Nelson et al. (28), using the equilibrated hole approach, determined that the ratio $\gamma_{100}/\gamma_{110}$ was 1.2, 0.98, and 1.14 for copper at 600°C, aluminum at 550°C, and molybdenum at 2000°C, respectively, and 1.03 for $\gamma_{100}/\gamma_{111}$ for aluminum at 450°C. Metal tips in field emission studies (see Section VIII-2C) tend to take on an equilibrium faceting, and these shapes have been found to agree fairly well with calculations (105).

Sundquist (29), studying small crystals of metals, noted a great tendency for rather rounded shapes and concluded that for such metals as silver, gold, copper, and iron there was not more than about 15% variation in surface tension between different types of crystal planes. This observation is supported on theoretical grounds by Herring (24), who concluded that under some conditions the polar γ-plot (or Wulff plot) should consist of smooth regions (which are portions of spheres) meeting in cusps. The question of whether the equilibrium shape is rounded or polyhedral is very likely a matter of temperature, however; the results cited here were obtained under temperature conditions such that surface mobility was probably high. In fact, as noted in Section VII-1A, there is even the possibility that the surface region of a solid near its melting point is liquidlike.

Some further discussion relative to crystal facets is given in Section X-4.

D. Dependence of Other Physical Properties on Surface Energy Changes at a Solid Interface

A few additional phenomena are worth mentioning. First, there are several that are not included here because they are discussed in more detail elsewhere. Thus solid–liquid interfacial tensions can be estimated from solubility measurements (Section X-2) and also from nucleation studies (Section IX-3). In addition, changes or differences in the interfacial free energy at a solid–vapor interface can be obtained from gas adsorption studies (Section X-3), and the corresponding differences in surface energies, from heat of immersion data (Section X-3). Contact angle measurements give a difference between a solid–liquid and a solid–vapor interfacial free energy (Section X-4), and Chapter V shows how to estimate the effect of potential on the free energy of a solid–electrolyte solution interface.

1. Expansion of a Solid as a Result of Adsorption. There are many solids that show marked swelling as a result of the uptake of a gas or a liquid. In certain cases involving the adsorption of a vapor by a porous solid, a linear relationship has been found between the percent linear expansion of the solid and the film pressure of the adsorbed material (106). This last is calculated from the adsorption isotherm using the Gibbs equation; as discussed in Section XVI-6,

$$\pi = (\text{constant}) \int v \, d \ln P \qquad \text{(VII-23)}$$

where π is the film pressure, v is the volume of vapor adsorbed, and P is the pressure of the vapor. A study by Yates (107) confirmed the linear relationship for a series of inert gases adsorbed on porous glass. A form derived by him is

$$\left(\frac{\partial \pi}{\partial V}\right)_T = \tfrac{3}{2}K \qquad \text{(VII-24)}$$

where V is the volume of the porous material and K is the bulk modulus of the adsorbent substance.

2. Tensile Strengths of Solids. Most solid surfaces are marred by small cracks, and it appears clear that it is often because the presence of such surface imperfections that observed tensile strengths fall below the theoretical ones. Thus for sodium chloride, the theoretical tensile strength is about 200 kg/mm^2, as given by Stranski (108). The tensile strength, or breaking stress τ, can be related to the work of cohesion 2γ by assuming some distance d for the range of action of the forces between the separating planes

$$2\gamma = d\tau \qquad \text{(VII-25)}$$

Polanyi (109) took d to be about 10 Å, and on this basis τ for sodium chloride would be roughly $2 \times 10^7 \times 190 = 4 \times 10^9$ dyne/cm^2 or about 40 kg/mm.^{2} The actual breaking stress may be a hundredth or a thousandth of this, depending on the surface condition of the rock salt. Stranski found that the measured tensile strength of rock salt crystals varied markedly with their size, increasing as the dimensions decreased, and maximum values close to the theoretical were obtained with crystals of the order of 0.02 mm in edge. Coating the crystals with saturated solution so that surface deposition of small crystals occurred resulted in a much lower tensile strength, but not if the solution contained some urea.

Of course, it is common knowledge that glass cracks easily at a scratch line but, in addition, a superficially flawless glass surface is marred by many fine cracks (110, 111), and very freshly drawn glass shows a high initial tensile strength that decreases rapidly with time.

In view of this marked effect of surface cracks on tensile strength, plus the relationship Eq. VII-25 that indicates how tensile strength and *actual* surface tension may be related, it is not surprising that the experimental tensile strengths of solids may be affected by their environment in a way directly related to changes in their surface tensions. In general, any reduction in the solid interfacial tension should also reduce the observed tensile strength. Thus the tensile strength of glass in various liquids decreases steadily with decreasing polarity of the liquid (112).

Frumkin (113) discusses work by Rehbinder and co-workers showing that the measured hardness of a metal immersed in an electrolyte solution varies with applied potential in the manner of an electrocapillary curve. A dramatic demonstration of surface tension effects is the easy deformation of single crystals of tin and of zinc if the surface is coated with an oleic acid monolayer (114).

6. Reactions of Solid Surfaces

Perhaps the simplest case of reaction of a solid surface is that where the reaction product is continuously removed, as in the dissolving of a soluble salt in water or that of a metal or metal oxide in an acidic solution. This situation is discussed in Section XV-2C, in connection with surface area determination.

More complex in its kinetics is the reaction of a solid with a liquid or gas to give a second solid as, for example,

$$CuSO_4 \cdot 5H_2O = CuSO_4 \cdot H_2O + 4\,H_2O \qquad \text{(VII-26)}$$

$$CaCO_3 = CaO + CO_2 \qquad \text{(VII-27)}$$

$$2\,Cu + COS\,(g) = Cu_2S + CO \qquad \text{(VII-28)}$$

$$6\,Fe_2O_3 = 4\,Fe_3O_4 + O_2 \qquad \text{(VII-29)}$$

The usual situation, true for the first three cases, is that in which the reactant and product solids are mutually insoluble. Langmuir (115) pointed out that such reactions undoubtedly occur at the linear interface between the two solid phases. The rate of reaction will thus be small when either solid phase is practically absent. Moreover, since both forward and reverse rates will depend on the amount of this common solid–solid interface, its extent cancels out at equilibrium, in harmony with the thermodynamic conclusion that for the reactions such as Eqs. (26–28) the equilibrium constant is given simply by the gas pressure and does not involve the amounts of the two solid phases.

Qualitative examples abound. Perfect crystals of sodium carbonate, sulfate, or phosphate may be kept for years without efflorescing, although if scratched, they begin to do so immediately. Too strongly heated or "burned" lime or plaster of Paris takes up the first traces of water only with difficulty. Reactions of this type tend to be autocatalytic. The initial rate is slow, due to the absence of the necessary linear interface, but the rate accelerates as more and more product is formed. See Refs. 116–122 for other examples. A *topochemical* reaction is one in which the boundary between the two phases moves in a regular way. A *topotactic* reaction is one where the product or products retain the external crystalline shape of the reactant crystal (123). More often, however, there is a complicated morphology with pitting, cracking, and pore formation, as with calcium carbonate (124).

In the case of reaction VII-29, the reactant and product are mutually soluble. Langmuir argued that in this case, escape of oxygen is easer from bulk Fe_2O_3 than from such units as an Fe_2O_3–Fe_3O_4 interface. The reaction therefor proceeds by a gradual escape of oxygen randomly from all portions of the Fe_2O_3, thus producing a solid solution of the two oxides.

The kinetics of reactions in which a new phase is formed may be complicated by the interference of that phase with the ease of access of the reactants to each other. This is the situation in corrosion and tarnishing reactions. Thus in the corrosion of a metal by oxygen the increasingly thick coating of oxide that builds up may offer more and more impedance to the reaction. Typical rate expressions are the logarithmic law

$$y = k_1 \log (k_2 t + k_3) \qquad \text{(VII-30)}$$

where y denotes the thickness of the film, the parabolic law

$$y^2 = k_1 t + k_2 \qquad \text{(VII-31)}$$

and, of course, the simple law for an unimpeded reaction

$$y = kt \qquad \text{or} \qquad y = k_1 + k_2 t \qquad \text{(VII-32)}$$

Thus the oxidation of light metals such as sodium, calcium, or magnesium follows Eq. VII-32, the low temperature oxidation of iron follows Eq. VII-30, and the high temperature oxidation follows Eq. VII-31. The controlling factor seems to be the degree of protection offered by the coating of oxide (125). If, as in the case of the light metals, the volume of the oxide produced is less than that of the metal consumed, then the oxide tends to be porous and nonprotective, and the rate, consequently, is constant. Evans (125) suggests that the logarithmic equation results when there is discrete mechanical breakdown of the film of product. In the case of the heavier metals, the volume of the oxide produced is greater than that of the metal consumed and, although this tends to give a dense protective coating, if the volume difference is too great, flaking or other forms of mechanical breakdown may occur as a result of the compressional stress produced.

Studies have been made on the rate of growth of oxide films on different crystal faces of a metal, using ellipsometric methods. The rate was indeed different for (100), (101), (110), and (311) faces of copper (126); moreover, the film on a (311) surface was anisotropic in that its apparent thickness varied with the angle of rotation about the film normal.

The rate law may change with temperature. Thus for reaction VII-28 the rate was paralinear (i.e. linear after an initial curvature) below about 470°C and parabolic above this temperature (127), presumably because the CuS_2 product was now adherent. Nonsimple rate laws are reported for the oxidation of iron coated with an iron–silicon solid solution (128) and that of zirconium coated with carbonitride (129); several layers of material are present during these reactions.

The parabolic law may be derived on the basis of a uniform coating of product being formed through which a rate-controlling diffusion of either one or both of the reactants must occur (130). For example, in the case of the reaction between liquid sulfur and silver, it was determined that silver diffused through the Ag_2S formed and that the reaction occurred at the Ag_2S–S rather than at the Ag–Ag_2S boundary (131). The rate constant k_1 in the parabolic rate law generally shows an exponential dependence on temperature, related to that of the rate-determining diffusion process. As emphasized by Gomes (132), however, activation energies computed from the temperature dependence of rate constants are meaningless except in terms of a specific mechanism.

The foregoing discussion indicates how corrosion products may form a protective coating over a metal and thus protect it from extensive reaction. Perhaps the best illustration of this is the fact that aluminum, which should react vigorously with water, may be used to make cooking utensils. If, however, the coating is impaired then the expected reactivity of the metal is observed. Thus an amalgamated aluminum surface does react with water, and aluminum foil wet with mercurous nitrate or chloride solution and rolled up tightly may actually reach incandescence as a result of the heat of the reaction. For some representative studies in the field of passivation and corrosion inhibition see Refs. 133 and 134 by Hackerman and co-workers. Photo-induced corrosion, especially of semiconductors, is of current interest in connection with solar energy conversion systems; see Ref. 135.

7. Problems

1. The surface tensions for a certain cubic crystalline substance are $\gamma_{100} = 150$ erg/cm^2, $\gamma_{110} = 120$ erg/cm^2, and $\gamma_{210} = \gamma_{120} = 120$ erg/cm^2. Make a Wulff construction and determine the equilibrium shape of the crystal in the xy plane. (If the

plane of the paper is the *xy* plane, then all the ones given are perpendicular to the paper, and the Wulff plot reduces to a two-dimensional one. Also, $\gamma_{100} = \gamma_{010}$, etc.)

2. Consider a hypothetical two-dimensional crystal having a simple square unit cell. Given that γ_{10} is 400 erg/cm and γ_{11} is 200 erg/cm, make a Wulff construction to show whether the equilibrium crystal should consist of (10) type of (11) type edges. Calculate directly the edge energy for both cases to verify your conclusion.

3. Referring to Fig. VII-2, assume the surface tension of 10 type planes to be 350 erg/cm. (*a*) For what surface tension value of 11 type planes should the stable crystal habit just be that of Fig. VII-2*a*, and (*b*) for what surface tension value of 11 type planes should the stable crystal habit be just that of Fig. VII-2*b*? Explain your work.

4. An enlarged view of a crystal is shown in Fig. VII-11; assume for simplicity that the crystal is two-dimensional. Assuming equilibrium shape, calculate γ_{11} if γ_{10} is 275 dyne/cm. Crystal habit may be changed by selective adsorption. What percent reduction in the value of γ_{10} must be effected (by, say, dye adsorption selective to that face) in order that the equilibrium crystal exhibit only 10 faces? Show your calculation.

5. A crystal that is in the cubic system has as equilibrium cross-sectional shape a regular octahedron whose sides are of equal length and consist alternatively of (100) and (110) type planes. Calculate the ratio $\gamma_{100}/\gamma_{110}$.

6. Bikerman (136) has argued that the Kelvin equation should not apply to crystals, that is, in terms of increased vapor pressure or solubility of small crystals. The reasoning is that perfect crystals of whatever size will consist of plane facets whose radius of curvature is therefore infinite. On a molecular scale, it is argued that local condensation–evaporation equilibrium on a crystal plane should not be affected by the extent of the plane, that is, the crystal size, since molecular forces are short range. This conclusion is contrary to that in Section VII-2C. Discuss the situation. There is a problem here requiring some serious thought.

7. According to Beamer and Maxwell (137) the element Po has a simple cubic structure with $a = 3.34$ Å. Estimate by Harkins' method the surface energy for 100 and 111 planes. Take the energy of vaporization to be 50 kcal/mole.

8. Make the following approximate calculations for the surface energy per square centimeter of solid krypton (nearest neighbor distance 3.97 Å), and compare your results with those of Table VII-1. (*a*) Make the calculations for 100, 110, and 111 planes, considering only nearest neighbor interactions. (*b*) Make the calculation for 100 planes, considering all interactions within a radius defined by the sum ($m_1^2 + m_2^2 + m_3^2$) being 8 or less.

9. Calculate the surface energy at 0 K 100 planes of radon, given that its energy of vaporization is 35×10^{-14} erg/atom and that the crystal radius of the radon atom is 2.5 Å. The crystal structure many be taken to be the same as for other rare gases. You may draw on the results of calculations for other rare gases.

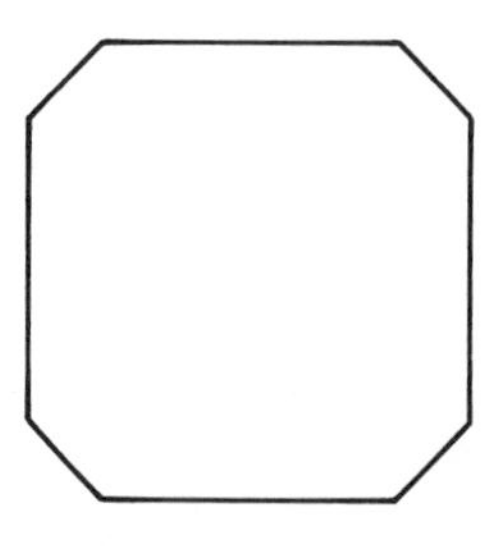

Fig. VII-11. Wulff construction of a two-dimensional crystal.

10. Using the nearest neighbor approach, and assuming a face-centered cubic structure for the element, calculate the surface energy per atom for (*a*) an atom sitting on top of a 111 plane and (*b*) against a step of a partially formed 111 plane (e.g., position 8 in Fig. VII-6). Assume the energy of vaporization to be 60 kcal/mole.

11. Taking into account only nearest neighbor interactions, calculate the value for the line or edge tension k for solid argon at 0°K. The units of k should be in ergs per centimeter.

12. Calculate the Hamaker constant for Ar crystal, using Eq. VII-18. Compare your value with the one that you can estimate from the data and equations of Chapter VI.

13. Metals A and B form an alloy or solid solution. To take a hypothetical case, suppose that the structure is simply cubic, so that each interior atom has six nearest neighbors and each surface atom has five. A particular alloy has a bulk mole fraction $x_A = 0.50$, the side of the unit cell is 4.0 Å, and the energies of vaporization E_A and E_B are 30 and 35 kcal/mole for the respective pure metals. The A–A bond energy is E_{AA} and the B–B bond energy is E_{BB}; assume that $E_{AB} = \frac{1}{2}(E_{AA} + E_{BB})$. Calculate the surface energy E^s as a function of surface composition. What should the surface composition be at 0 K? In what direction should it change on heating, and why?

14. Calculate the surface tension of gold at 1020°C. Take your data from Fig. VII-10.

15. The excess heat of solution of sample A of finely divided sodium chloride is 18 cal/g, and that of sample B is 12 cal/g. The area is estimated by making a microscopic count of the number of particles in a known weight of sample, and it is found that sample A contains 15 times more particles per gram than does sample B. Are the specific surface energies the same for the two samples? If not, calculate their ratio.

16. In a study of tarnishing the parabolic law, Eq. VII-31, is obeyed, with $k_2 = 0$. The film thickness y, measured after a given constant elapsed time, is determined in a series of experiments, each at a different temperature. It is found that y so measured varies exponentially with temperature, and from $d \ln y/d(1/T)$ an apparent activation energy of 10 kcal/mole is found. If k_1 in Eq. VII-31 is in fact proportional to a diffusion coefficient, show what is the activation energy for the diffusion process.

General References

H. E. Buckley, *Crystal Growth*, Wiley, New York, 1951.

D. D. Eley, Ed., *Adhesion*, The Clarendon Press, Oxford, 1961. E. Passaglia, R. R. Stromberg, and J. Kruger, Eds., *Ellipsometry in the Measurement of Surfaces and Thin Films*, National Bureau of Standards Miscellaneous Publication 256, Washington, D.C., 1964.

R. C. Evans, *An Introduction to Crystal Chemistry*, Cambridge University Press, 1966.

P. P. Ewald and H. Juretschke, *Structure and Properties of Solid Surfaces*, University of Chicago Press, Chicago, 1953.

R. S. Gould and S. J. Gregg, *The Surface Chemistry of Solids*, Reinhold, New York, 1951.

W. T. Read, Jr., *Dislocations in Crystals*, McGraw-Hill, New York, 1953.

"Solid Surfaces," (*Advan. Chem. Ser.*, No. 33), American Chemical Society, Washington, D.C., 1961.

G. A. Somorjai, *Principles of Surface Chemistry*, Prentice-Hall, Englewood Cliffs, New Jersey, 1972.

Textual References

1. H. Udin, A. J. Shaler, and J. Wulff, *J. Met.*, **1**, No. 2; *Trans. AIME*, 186 (1949).
2. B. Chalmers, R. King, and R. Shuttleworth, *Proc. Roy. Soc.* (*London*), **A193,** 465 (1948).
3. G. C. Kuczynski, *J. Met.*, **1,** 96 (1949).
4. P. V. Hobbs and B. J. Mason, *Phil. Mag.*, **9,** 181 (1964).
5. H. H. G. Jellinek and S. H. Ibrahim, *J. Colloid Interface Sci.*, **25,** 245 (1967).
6. C. Herring, in *Structure and Properties of Solid Surfaces*, R. Gomer and C. S. Smith, Eds., University of Chicago Press, Chicago, 1953, p. 1.
7. G. C. Kuczynski, *Acta Met.*, **4,** 58 (1956).
8. H. N. Hersh, *J. Am. Chem. Soc.*, **75,** 1529 (1953).
9. G. M. Rosenblatt, *Acc. Chem. Res.*, **9,** 169 (1976).
10. G. Cohen and G. C. Kuczynski, *J. Appl. Phys.*, **21,** 1339 (1950).
11. E. Menzel, *Z. Phys.* **132,** 508 (1952).
12. N. H. Fletcher, *Phil. Mag.*, **7,** No. 8, 255 (1962).
13. L. H. Germer, in *Frontiers in Chemistry*, R. E. Burk and O. Grummitt, Eds., Vol. 4, Interscience, New York, 1945.
14. See L. H. Germer and A. U. MacRae, *J. Appl. Phys.*, **33,** 2923 (1962).
15. G. Beilby, *Aggregation and Flow of Solids*, Macmillan, New York, 1921.
16. R. C. French, *Proc. Roy. Soc.* (*London*), **A140,** 637 (1933).
17. F. P. Bowden and D. Tabor, *The Friction and Lubrication of Solids*, The Clarendon Press, Oxford, 1950.
18. S. J. Gregg, *The Surface Chemistry of Solids*, Reinhold, New York, 1951.
19. J. J. Trillat, *CR*, **224,** 1102 (1947).
20. M. L. White, *Clean Surfaces, Their Preparation and Characterization for Interfacial Studies*, G. Goldfinger, Ed., Marcel Dekker, New York, 1970.
21. J. W. Gibbs, *The Collected Works of J. W. Gibbs*, Longmans, Green, New York, 1931, p. 315.
22. R. Shuttleworth, *Proc. Phys. Soc.* (*London*), **63A,** 444 (1950).
23. G. Wulff, *Z. Krist.*, **34,** 449 (1901).
24. C. Herring, *Phys. Rev.*, **82,** 87 (1951).
25. G. C. Benson and D. Patterson, *J. Chem. Phys.*, **23,** 670 (1955).
26. M. Drechsler and J. F. Nicholas, *J. Phys. Chem. Solids*, **28,** 2609 (1967).
27. H. G. Müller, *Z. Phys.*, **96,** 307 (1935).
28. R. S. Nelson, D. J. Mazey, and R. S. Barnes, *Phil. Mag.*, **11,** 91 (1965).
29. B. E. Sundquist, *Acta Met.*, **12,** 67, 585 (1964).
30. W. M. Mullins, *Phil. Mag.*, **6,** 1313 (1961).
31. W. J. Dunning, in *Structure of Surfaces*, D. Fox, M. M. Labes, and A. Weissberg, Eds., Wiley-Interscience, New York, 1963.
32. W. D. Harkins, *J. Chem. Phys.*, **10,** 268 (1942).
33. R. Shuttleworth, *Proc. Phys. Soc.* (*London*), **A62,** 167 (1949); **A63,** 444 (1950).
34. G. C. Benson and T. A. Claxton, *Phys. Chem. Solids,* **25,** 367 (1964).

35. See M. Born and W. Heisenberg, *Z. Phys.*, **23,** 388 (1924); M. Born and J. E. Mayer, *Z. Phys.*, **75,** 1 (1932).
36. J. E. Lennard-Jones and P. A. Taylor, *Proc. Roy. Soc.* (*London*), **A109,** 476 (1925).
37. J. E. Lennard-Jones and B. M. Dent, *Proc. Roy. Soc.* (*London*), **A121,** 247 (1928).
38. B. M. Dent, *Phil. Mag.*, **8,** No. 7, 530 (1929).
39. M. L. Huggins and J. E. Mayer, *J. Chem. Phys.*, **1,** 643 (1933).
40. F. van Zeggeren and G. C. Benson, *J. Chem. Phys.*, **26,** 1077 (1957).
41. G. C. Benson and R. McIntosh, *Can. J. Chem.*, **33,** 1677 (1955).
42. B. G. Dick and A. W. Overhauser, *Phys. Rev.*, **112,** 90 (1958).
43. M. J. L. Sangster, G. Peckham, and D. H. Saunderson, *J. Phys. C,* **3,** 1026 (1970).
44. C. R. A. Catlow, K. M. Diller, and M. J. Norgett, *J. Phys. C,* **10,** 1395 (1977).
45. T. S. Chen and F. W. de Wette, *Surf. Sci.*, **74,** 373 (1978).
46. M. R. Welton-Cook and M. Prutton, *Surf. Sci.*, **74,** 276 (1978).
47. A. J. Martin and H. Bilz, *Phys. Rev. B,* **19,** 6593 (1979).
48. C. B. Duke, R. J. Meyer, A. Paton, and P. Mark, *Phys. Rev. B,* **8,** 4225 (1978).
49. E. J. W. Verwey, *Rec. Trav. Chim.*, **65,** 521 (1946).
50. G. C. Benson, P. I. Freeman, and E. Dempsey, *Adv. Chem. Ser.*, No. 33, American Chemical Society, 1961, p. 26.
51. K. Molière, W. Rathje, and I. N. Stranski, *Discuss. Faraday Soc.*, **5,** 21 (1949); K. Molière and I. N. Stranski, *Z. Phys.*, **124,** 429 (1048).
52. P. W. Tasker, *Phil. Mag. A,* **39,** 119 (1979).
53. G. C. Benson and K. S. Yun, in *The Solid-Gas Interface*, E. A. Flood, Ed., Marcel Dekker, New York, 1967. See also G. C. Benson and T. A. Claxton, *J. Chem. Phys.*, **48,** 1356 (1968).
54. G. C. Benson, *J. Chem. Phys.*, **35,** 2113 (1961).
55. G. C. Benson and T. A. Claxton, *Can. J. Phys.*, **41,** 1287 (1963).
56. H. Reiss and S. W. Mayer, *J. Chem. Phys.*, **34,** 2001 (1961).
57. H. P. Schreiber and G. C. Benson, *Can. J. Phys.*, **33,** 534 (1955).
58. N. H. Fletcher, *Rep. Progr. Phys.*, **34,** 913 (1971).
59. A. U. S. de Reuck, *Nature,* **179,** 1119 (1957).
60. W. M. Ketcham and P. V. Hobbs, *Phil. Mag.*, **19,** No. 162, 1161 (1969).
61. H. H. G. Jellinek, *J. Colloid Interface Sci.*, **25,** 192 (1967).
62. M. W. Orem and A. W. Adamson, *J. Colloid Interface Sci.*, **31,** 278 (1969).
63. J. K. MacKenzie, A. J. W. Moore, and J. F. Nicholas, *J. Phys. Chem. Solids,* **23,** 185 (1962).
64. H. Reiss, H. L. Frisch, E. Hefland, and J. L. Lebowitz, *J. Chem. Phys.* **32,** 119 (1960).
65. H. Schonhorn, *J. Phys. Chem.*, **71,** 4878 (1967).
66. J. F. Nicholas, *Australian J. Phys.*, **21,** 21 (1968).
67. T. Matsunaga and Y. Tamai, *Surf. Sci.*, **57,** 431 (1976).
68. F. M. Fowkes, *Ind. Eng. Chem.*, **56,** 40 (1964).
69. F. W. Williams and D. Nason, *Surf. Sci.*, **45,** 377 (1974).
70. A. Brager and A. Schuchowitsky, *Acta Physiochim.* (*USSR*), **21,** 13, 1001 (1946).
71. K. Huang and G. Wyllie, *Proc. Phys. Soc.* (*London*), **A62,** 180 (1949).
72. H. B. Huntington, *Phys. Rev.*, **81,** 1035 (1951).

73. P. P. Ewald and H. Juretschke, in *Structure and Properties of Solid Surfaces*, R. Gomer and C. S. Smith, Eds., University of Chicago Press, Chicago, 1952, p. 82.
74. N. D. Land and W. Kohn, *Phys. Rev. B,* **1,** 4555 (1970).
75. J. H. Rose and J. F. Dobson, *Solid State Communications*, in preparation.
76. J. J. Burton and G. Jura, *J. Phys. Chem.,* **71,** 1937 (1967).
77. H. N. V. Temperley, *Proc. Cambridge Phil. Soc.,* **48,** 683 (1952).
78. W. K. Burton and N. Cabrera, *Discuss. Faraday Soc.,* **5,** 33 (1949).
79. S. Brunauer, *Pure Appl. Chem.,* **10,** 293 (1965).
80. I. N. Stranski, *Z. Phys. Chem.,* **136,** 259 (1928).
81. See also W. D. Harkins, *The Physical Chemistry of Surface Films*, Reinhold, New York, 1952.
82. W. J. Dunning, *J. Phys. Chem.,* **67,** 2023 (1963).
83. J. Frenkel, *Z. Phys.,* **35,** 652 (1926).
84. C. Wagner and W. Schottky, *Z. Phys. Chem.,* **11B,** 163 (1930).
85. See F. A. Kruger and H. J. Vink, *Phys. Chem. Solids,* **5,** 208 (1958).
86. R. C. Evans, *An Introduction to Crystal Chemistry*, Cambridge University Press, 1964.
87. J. M. Burgers, *Proc. Phys. Soc.* (*London*), **52,** 23 (1940).
88. F. C. Frank, *Discuss. Faraday Soc.,* **5,** 48 (1949).
89. A. R. Verma, *Phil. Mag.,* **42,** 1005 (1951).
90. F. C. Frank, *Acta Cryst.,* **4,** 497 (1951).
91. B. H. Alexander, M. H. Dawson, and H. P. Kling, *J. Appl. Phys.,* **22,** 439 (1951).
92. E. B. Greenhill and S. R. McDonald, *Nature,* **171,** 37 (1953).
93. M. M. Nicolson, *Proc. Roy. Soc.* (*London*), **A228,** 507 (1955).
93a. A. I. Murdoch, *Int. J. Eng. Sci.,* **16,** 131 (1978).
94. E. Orowan, *Z. Phys.,* **82,** 235 (1933); see also A. I. Bailey, *Proc. Int. Congr. Surf. Act., 2nd,* Vol. 3, p. 406 (1957).
95. J. J. Gilman, *J. Appl. Phys.,* **31,** 2208 (1960).
96. A. R. C. Westwood and T. T. Hitch, *J. Appl. Phys.,* **34,** 3085 (1963).
96a. P. F. Becher and S. W. Freiman, *J. Appl. Phys.,* **49,** 3779 (1978).
97. For a description of the apparatus see G. C. Benson and G. W. Benson, *Rev. Sci. Instr.,* **26,** 477 (1955).
98. G. C. Benson, H. P. Schreiber, and F. van Zeggeren, *Can. J. Chem.,* **34,** 1553 (1956).
99. S. Brunauer, D. L. Kantro, and C. H. Weise, *Can. J. Chem.,* **34,** 729 (1956).
100. S. Brunauer, D. L. Kantro, and C. H. Weise, *Can. J. Chem.,* **37,** 714 (1959).
101. G. Jura and C. W. Garland, *J. Am. Chem. Soc.,* **74,** 6033 (1952); **75,** 1006 (1953).
102. S. H. Bauer, *J. Am. Chem. Soc.,* **75,** 1004 (1953).
103. D. Patterson, J. A. Morrison, and F. W. Thompson, *Can. J. Chem.,* **33,** 240 (1955).
104. G. Jura and K. S. Pitzer, *J. Am. Chem. Soc.,* **74,** 6030 (1952).
105. M. Drechsler and J. F. Nicholas, *J. Phys. Chem. Solids,* **28,** 2609 (1967).
106. D. H. Bangham and N. Fakhoury, *J. Chem. Soc.,* **1931,** 1324; see also *Proc. Roy. Soc.* (*London*), **A147,** 152, 175 (1934).
107. D. J. C. Yates, *Proc. Roy. Soc.* (*London*), **A224,** 526 (1954).
108. I. N. Stranski, *Bericht* **75B,** 1667 (1942).

109. M. Polanyi, *Z. Phys.*, **7,** 323 (1921).
110. A. A. Griffith, *Phil. Trans. Roy. Soc.* (*London*), **A221,** 163 (1920).
111. W. C. Hynd, *Sci. J. Roy. Coll. Sci.*, **17,** 80 (1947).
112. D. McCammond, A. W. Newmann, and N. Natarajan, *J. Am. Ceram. Soc.*, **58,** 15 (1975).
113. A. Frunkin, *J. Colloid Sci.*, **1,** 277 (1946).
114. P. A. Rehbinder, V. I. Likhtman, and V. M. Maslinnikov, *CR* Acad. Sci. U.R.S.S., **32,** 125 (1941).
115. See I. Langmuir, *J. Am. Chem. Soc.*, **38,** 2221 (1916).
116. See W. D. Harkins, *The Physical Chemistry of Surface Films*, Reinhold, New York, 1952.
117. T. S. Renzema, *J. Appl. Phys.*, **23,** 1412 (1952).
118. B. Reitzner, *J. Phys. Chem.*, **65,** 948 (1961).
119. B. Reitzner, J. V. R. Kaufman, and F. E. Bartell, *J. Phys. Chem.*, **66,** 421 (1962).
120. F. P. Bowden and H. M. Montague-Pollock, *Nature*, **191,** 556 (1961).
121. P. W. M. Jacobs, F. C. Tompkins, and V. R. P. Verneker, *J. Phys. Chem.*, **66,** 1113 (1962).
122. F. L. Hirshfield and G. M. J. Schmidt, *J. Polym. Sci.*, **2(A),** 2181 (1964).
123. L. S. D. Glasser, F. P. Glasser, and H. F. W. Taylor, *Q. Rev.*, **16,** 343 (1962).
124. R. Sh. Mikhail, S. Hanafi, S. A. Abo-el-enein, R. J. Good, and J. Irani, *J. Colloid Interface Sci.*, **75,** 74 (1980).
125. U. R. Evans, *J. Chem. Soc.*, **1946,** 207.
126. J. V. Cathcart, J. E. Epperson, and G. F. Peterson, *Acta Met.*, **10,** 699 (1962); see also *Ellipsometry in the Measurement of Surfaces and Thin Films*, E. Passaglia, R. R. Stromberg, and J. Kruger, Eds., *Natl. Bur. Stand.* (*U.S.*), Misc. Publ., **256** (1964).
127. P. Hadjisavas, M. Caillet, A. Galerie, and J. Besson, *Rev. Chim. Miner.*, **14,** 572 (1977).
128. A. Abba, A. Galerie, and M. Caillet, *Mater. Chem.*, **5,** 147 (1980).
129. M. Caillet, H. F. Ayedi, and J. Besson, *J. Less-Common Met.*, **51,** 323 (1977).
130. G. Cohn, *Chem. Rev.*, **42,** 527 (1948).
131. N. F. Mott and R. W. Gurney, *Electronic Processes in Ionic Crystals*, The Clarendon Press, Oxford, 1940.
132. W. Gomes, *Nature*, **192,** 865 (1961).
133. N. Hackerman, D. D. Justice, and E. McCafferty, *Corrosion*, **31,** 240 (1975).
134. F. M. Delnick and N. Hackerman, *J. Electrochem. Soc.*, **126,** 732 (1979).
135. K. W. Frese, Jr., M. J. Madou, and S. R. Morrison, *J. Phys. Chem.*, **84,** 3172 (1980).
136. J. J. Bikerman, *Phys. Stat. Sol.*, **10,** 3 (1965).
137. W. H. Beamer and C. R. Maxwell, *J. Chem. Phys.*, **17,** 1293 (1949).

CHAPTER VIII

Surfaces of Solids: Microscopy and Spectroscopy

1. Introduction

In recent years a number of methods have come into use that provide information about the structure of a solid surface, its composition, and the oxidation states present. In essentially all cases the solid-high vacuum interface is probed with a beam of ions, electrons, or electromagnetic radiation, and various diffraction, scattering, and other atomic processes observed.

The function of this chapter is to review these various methods with emphasis on the types of phenomenology involved and information obtained. Many of the effects are complicated and not yet fully unraveled theoretically. Perhaps for this reason there has been a tendency to give each type of experiment a separate name and there is now a veritable alphabet soup of designations, many of which are contrived acronyms. A short catalog is given in Table VIII-1. Not included are spectroscopic methods not requiring high vacuum, such as nuclear magnetic resonance, electron paramagnetic resonance, and infrared absorption; some of these are discussed in Chapter XV.

Following Table VIII-1, a few of the more commonly used techniques are discussed briefly; representative references to those not covered are given in the table. Also, many of the various measurements have found advantageous use in the study of the adsorbed state, and further examples of their use are to be found in Chapters XV, XVI, and XVII.

2. Microscopy of Surfaces

A. Optical and Electron Microscopy

Conventional optical microscopy can resolve features down to about the wavelength of visible light, or about 5000 Å. Surface faceting and dislocations may be seen (as in Fig. VII-9).

Ellipsometry (see Section IV-3D) does not resolve features but does allow film thicknesses to be measured. Thickness resolution can be as good as 0.5 Å.

Electron microscopy can resolve features down to about 10 Å. Since

TABLE VIII-1
Techniques for Studying Surface Structure and Composition

	Technique	Atomic process	Type of information	Illustrative reference
Microscopy				
FEM	Field emission microscopy	Electrons are emitted from a metal tip in a high field	Surface structure	1
FIM	Field ion microscopy	He ions formed in a high field at a metal tip	Surface structure	2
SEM	Scanning electron microscopy	A beam of electrons scattered from a surface is focused	Surface morphology	3, 4
Diffraction				
HEED	High energy electron diffraction	Diffraction of elastically backscattered electrons (around 20 keV, grazing incidence)	Surface structure	5
RHEED	Reflection high energy electron diffraction	Similar to HEED	Surface structure, composition	6, 7
SHEED	Scanning high energy electron diffraction	Similar to RHEED; intensity of diffraction pattern in scanned	Surface heterogeneity	7
LEED	Low energy electron diffraction	Elastic backscattering of electrons (10–200 eV)	Surface structure	8
ELEED	Elastic low energy electron diffraction	Same as LEED		
PhD	Photoelectron diffraction	X-rays (40–1500 eV) eject photoelectrons whose intensity is measured as a function of energy and angle	Surface structure	9, 9a, 9b
NPD	Normal PhD			
APD	Azimuthal PhD			

TABLE VIII-1 (*Continued*)

	Technique	Atomic process	Type of information	Illustrative reference
Spectroscopy—of emitted x-rays or light photons				
IRE	Infra-red emission	Infra-red emission from a metal surface is affected in angular distribution by adsorbed species	Orientation of adsorbed molecules	10
OSEE	Optically stimulated exoelectron emission	Light falling on a surface in a potential field leads to electron emission	Qualitative presence and nature of adsorbed species	11
SXES	Soft x-ray emission spectroscopy	An x-ray or electron beam ejects K-electrons and the spectrum of the consequent x-rays is determined	Energy levels and chemical state of adsorbed molecules; surface composition	12
XES	X-ray emission spectroscopy	Same as SXES		
Spectroscopy—of emitted electrons				
AES	Auger electron spectroscopy	An incident high energy electron ejects an inner electron from an atom. An electron further out (e.g. *L*) falls into the vacancy, and the released energy is given to an outer electron, which is the ejected Auger electron	Surface composition	7, 13, 14
CELS	Characteristic energy loss spectroscopy	Incident electrons are scattered inelastically	Surface energy states; composition and energy states of adsorbed species	
EELS	ELS Electron loss spectroscopy	Same as CELS		15, 16
EIS	Electron impact spectroscopy	Same as CELS		16
HRELS	High resolution electron energy loss spectroscopy	Same as CELS	Identification of adsorbed species through their	17

ESCA	Electron spectroscopy for chemical analysis	Monoenergetic x-rays eject electrons from various atomic levels; The electron energy spectrum is measured	Surface composition, oxidation state	7, 18
PES	Photoelectron spectroscopy	Similar to ESCA using uv light		
UPS	Ultraviolet photoemission spectroscopy	Same as PES		
XPS	X-ray photoelectron spectroscopy			
INS	Ion neutralization spectroscopy	An inert gas ion hitting the surface is neutralized with the ejection of an Auger electron from a surface atom	Kinetics of surface reactions; changes as chemisorption occurs	19
Spectroscopy—of emitted ions or molecules				
EID	Electron impact desorption	An electron beam (100–200 eV) strikes a surface, ejecting ions whose energy is measured	Characterization of surface sites and adsorbed species	20
ISS	Ion scattering spectroscopy	Inelastic backscattering of ions (about 1 keV ion beam)	Surface composition	7, 18, 21
LEIS	Low energy ion scattering	A mono-energetic beam of rare gas ions is scattered elastically by surface atoms	Surface composition	22
MBRS	Molecular beam spectroscopy	A modulated molecular beam hits the surface and the time lag for reaction products to appear is measured	Kinetics of surface reactions; changes as chemisorption occurs	23
SIMS	Secondary ion mass spectroscopy	Ionized surface atoms are ejected by impact of around 1 keV ions, and subjected to mass analysis	Surface composition	7, 7a

TABLE VIII-1 (*Continued*)

	Technique	Atomic process	Type of information	Illustrative reference
Spectroscopy—of emitted ions or molecules (*Continued*)				
TPRS	Temperature programmed reaction spectroscopy	As an adsorbant is progressively heated, chemisorbed species leave the surface at characteristic temperatures	Characterization of sursites and adsorbed species	24
FDS	Flash desorption spectroscopy	Same as TPRS		25
TDS	Thermal desorption spectroscopy	Same as TPRS		
Spectroscopy—of emitted neutrons				
	Neutron scattering	High energy neutrons are scattered inelastically	Surface vibrational states	26
Incident beam spectroscopy				
APS	Appearance potential spectroscopy	Intensity of emitted x-rays or Auger electrons is measured as a function of the incident electron energy	Surface composition	7
AEAPS	Auger electron appearance potential spectroscopy	Same as APS		
SXAPS	Soft x-ray appearance potential spectroscopy	Same as APS		
EXAFS	Extended x-ray adsorption fine structure	Variation of x-ray adsorption as a function of x-ray energy beyond an adsorption edge; the probability is affected by backscattering of the emitted electron from an adjacent atom	Number and separation distance of surface atoms	27
SEXAFS	Surface EXAFS			

electrons are transmitted through the sample, however, the technique is not well suited for studying surfaces.

B. Scanning Electron Microscope

Scanning electron microscopy is widely used to examine surfaces; resolution down to a few thousand Angstroms is possible, depending on the nature of the sample. As indicated in Fig. VIII-1, the surface is scanned by a focused electron beam, and the intensity of *secondary* electrons is monitored. The output from the secondary electron detector modulates the raster of a cathode-ray tube, which is scanned in synchronization with the focused electron beam. Each point on the cathode-ray tube (essentially a television type tube) raster (or image forming area) corresponds to a point on the surface of the sample, and the strength of the image at each point varies according to the intensity of secondary electron production from the corresponding point on the surface. As in television, image quality depends on having a high intensity of signal so that a wide variation in signal is possible and hence good contrast in the image, and on having a large number of lines scanned so as to give good resolution.

Scanning electron microscope images characteristically have a wide range of contrast; that is, detail can be seen both in very dark and in very light areas. The images also have great depth of focus; they are sharp at both very low and very high points of the surface. The result is that even quite rough surfaces show in startling clarity and feeling of depth (note Ref. 4). Figure VIII-2 shows how clearly small NaCl crystals stand out. See Ref. 28 for additional examples.

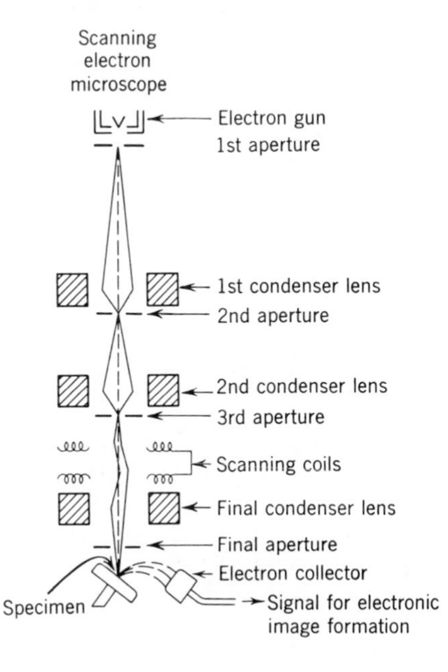

Fig. VIII-1. Schematic of image formation in scanning electron microscope. (From Ref. 4.)

Fig. VIII-2. Scanning electron microscope picture of small NaCl crystals. (Courtesy Dr. R. F. Baker.)

C. Field Emission and Field Ion Microscopy

The field emission microscope was invented by Müller in 1936 (31), and the technique and related developments have become major contributors to the structural study of surfaces. The subject has been reviewed by Ehrlich (29); see also Somorjai's monograph (8). The subject is highly developed, and only a selective, introductory presentation is given here. Some further aspects that relate directly to chemisorption are discussed in Chapter XVII.

The basic device can be very simple, as illustrated in Fig. VIII-3 from Ref. 30. A tip of refractory metal, such as tungsten, is carefully (and often unsuccessfully!) electrically heat polished to yield a nearly hemipsherical

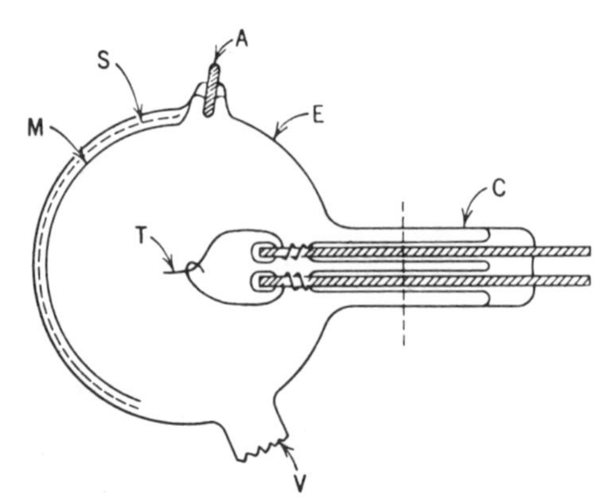

Fig. VIII-3. Schematic drawing of one form of the field emission microscope: E, glass envelope; S, phosphorescent screen; M, metal backing; A, anode lead-in; T, emitter tip; C, tip support structure; V, vacuum lead. (From Ref. 30.)

end of about 10^{-5} cm radius. A potential of about 10,000 V is applied between this tip and the hemispherical fluorescent screen, and if the two radii of curvature are a and b, respectively, the field at the tip can be calculated as follows. The field F must be equal to kr^2, as an inverse square falling off with distance, and the total potential difference V is then

$$V = \int_a^b F\,dr = -k\left(\frac{1}{b} - \frac{1}{a}\right) \tag{VIII-1}$$

If $1/b$ is neglected in comparison to $1/a$, then $k = aV$ and $F_{\text{at } a} = V/a$. Thus in the present example, F would be 10^9 V/cm. This very high value, equivalent to 10 V/ A, is sufficient to pull electrons from the metal, and these now accelerate along radial lines to hit the fluorescent screen. Individual atoms serve as emitting centers, and different crystal faces emit with different intensities, depending on packing density and work function. Since the magnification factor (b/a) is enormous (about 10^6), it might be imagined that individual atoms could be seen but, actually, resolution is limited by the kinetic energy of the motion of electrons in the metal at right angles to the emission line, and obtainable resolutions are about 30 to 50 Å. Even so, the technique produces miraculous pictures showing the various crystal planes that formed the tip in patterns of light and dark, as illustrated in Fig. VIII-4, which also shows how deposited nitrogen increases the emission so that the migration of nitrogen atoms over the surface can be followed.

A very ingenious modification of the field emission microscope, also due to Müller (31), is known as the field ion microscope and makes use of the fact that if the tip is positively charged, a gas molecule that approaches it is stripped of an electron, and the resulting positive ion accelerates out radially to hit the fluorescent screen. Helium is now the most commonly used gas, although other gases have been used and, in fact, early observations of the ion emission effect were of the field-induced emission of adsorbed atoms on the surface, such as hydrogen. The equipment used is similar to that for field emission, but now provides for the admission of very low but controlled pressures of the helium and for cooling of the tip.

The subject is again a highly developed one; for details see the monograph by Müller and Tsong (1). One very great advantage of field ion microscopy is that low temperatures, even cryoscopic, may be used, and the resolution is now a few angstroms so that *individual atoms* are seen. A good example of a field ion emission photograph for a tungsten tip is shown in Fig. VIII-5 (from Ref. 2). The spots represent individual tungsten atoms, and the patterns are due to the geometries for different crystallographic planes. Atoms on certain planes, such as those on (110) and (211) planes, have local field strengths such as not to ionize helium at the voltage used and hence do not appear in the pattern on the fluorescent screen. A ball model of the arrangement of tungsten atoms on a roughly hemispherical tip is shown in Fig. VIII-6 (2). This illustrates that while a very rounded contour can be made up, it is through the use of some rather high Miller index planes that

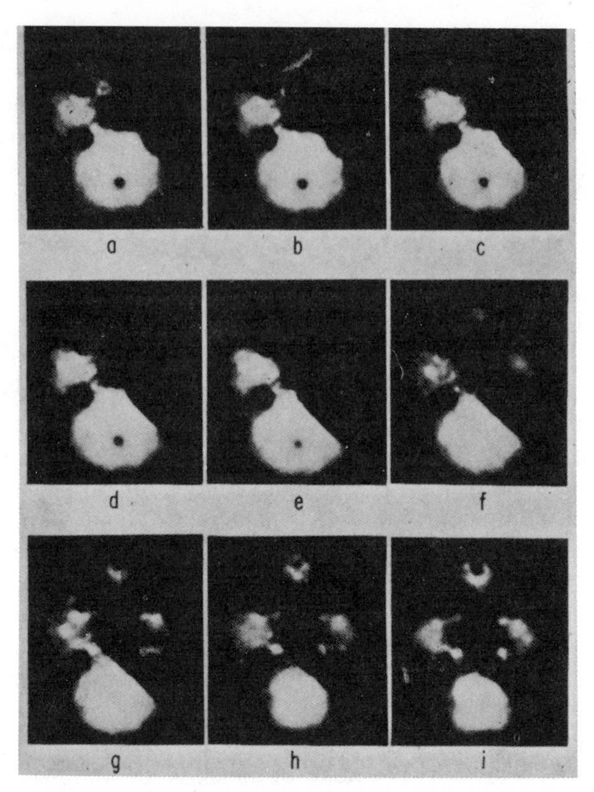

Fig. VIII-4. Surface migration of nitrogen deposited at 20 K on tungsten. The deposit, initially in the upper left region of the tip, spreads selectively over various facets as the tip is allowed to warm up in stages. (From Ref. 29.)

may have a low density of surface atoms, often arranged in such a way that wide troughs or pockets are present. These different types of planes not only ionize helium with different efficiencies (protruding atoms are most efficient) but also the evaporation probabilities at a given applied voltage are different for different planes (see Ref. 34). Also, as might be expected, planes vary in adsorption ability. This aspect of field microscope work is discussed in more detail later (Section XVII-2D).

While field ion microscopy has provided a very effective tool for seeing the pattern of surface atoms and for individual adsorbed atoms on a surface and their motions, it does not provide physical measurements of the energy properties of the surface. In this respect, field emission is superior. The central equation concerning this aspect is due to Fowler and Nordheim (35) and gives the effect of an applied field on the rate of electron emission. The situation, in simplified form, is shown in Fig. VIII-7. In the absence of a field, a barrier, corresponding to the thermionic work function Φ, prevents the escape of electrons from the Fermi sea. An applied field causes this barrier to be accordingly reduced in proportion to distance out, instead of being flat, and the net barrier is $(\Phi - V)$ where the potential V decreases linearly with distance according to $V = xF$ where F is the field in volts per centimeter. The net potential barrier is now finite. A quantum-mechanical tunneling process is now possible, and the solution for the case of an electron in a finite potential box gives

$$P = \text{const} \exp\left[-\frac{2^{2/3}m^{1/2}}{\hbar}\int^{l}(\Phi - V)\,dx\right] \qquad \text{(VIII-2)}$$

where P is the probability of escape, m is the mass of the electron, and l is the width of the barrier Φ/F. On performing the indicated integration, the approximate equation results

$$P = \text{const}\left[\exp\frac{2^{1/2}m^{1/2}}{\hbar}\Phi^{3/2}F\right] \qquad \text{(VIII-3)}$$

which may be put in the convenient experimental form

$$\frac{i}{V^2} = A\exp -\frac{B\Phi^{3/2}}{V} \qquad \text{(VIII-4)}$$

where i is the total emission current, V is the applied voltage, and A and B are constants that depend on Φ and geometry factors. The V^2 dependence of i arises from more detailed considerations, including the rate of arrival of electrons at the

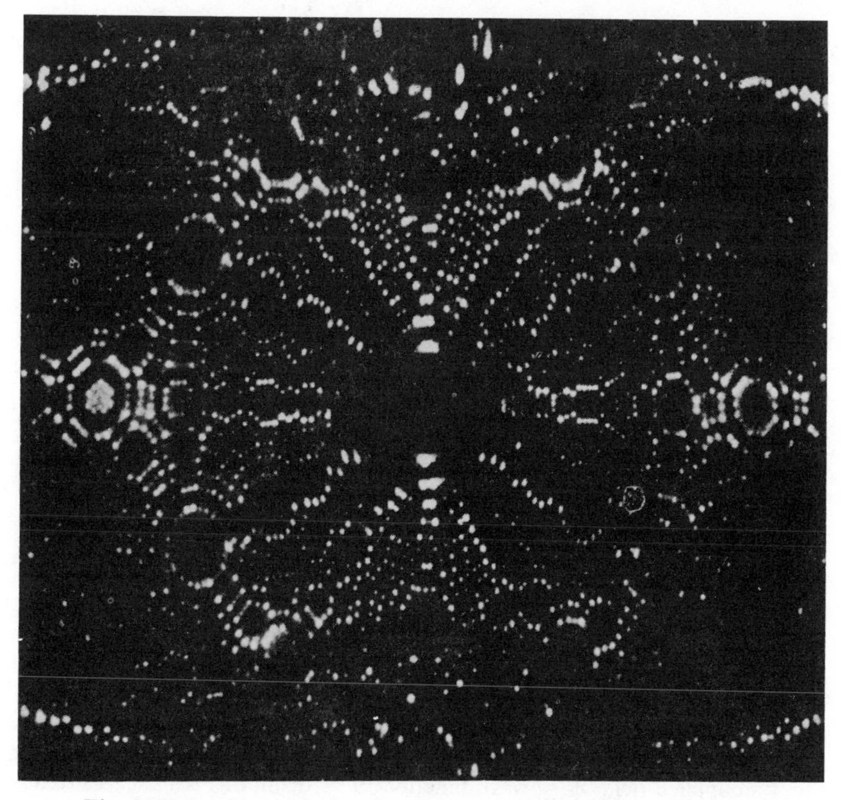

Fig. VIII-5. Ion emission from clean tungsten. (From Ref. 33.)

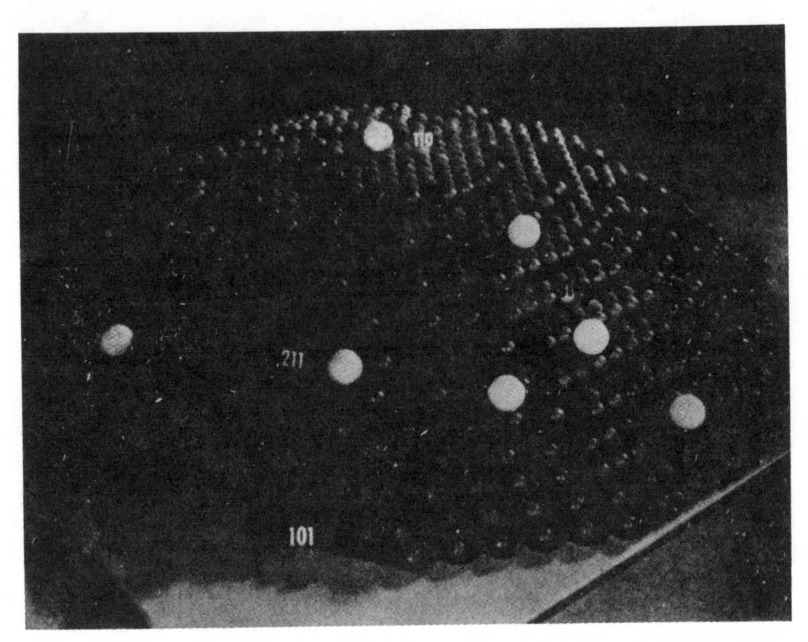

Fig. VIII-6. Xe atoms adsorbed on a hard-sphere model of a tungsten tip. (From Ref. 2.)

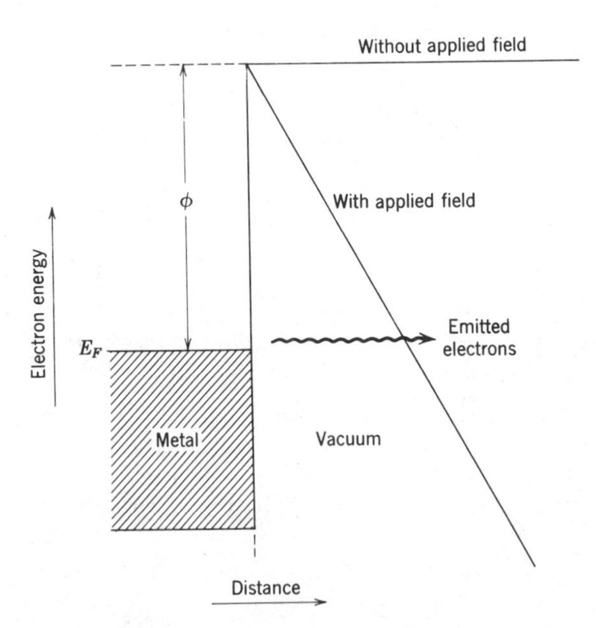

Fig. VIII-7. Schematic potential energy diagram for electrons in a metal with and without an applied field. ϕ = work function; μ = depth of Fermi sea. (From Ref. 8.)

surface. However, the important aspect of Eq. VIII-4 is that the emission current depends exponentially on $1/V$ and with a coefficient from which the work function can be evaluated.

Experimentally, it is possible to observe the electron emission from a particular atom (i.e., the intensity of a particular spot on the fluorescent screen) as V is varied, and thus to obtain the work function for that atom. Also, the change in work function when the site is covered by an adsorbed atom can be determined. Thus on adsorption of nitrogen on tungsten, the work function for the (100) plane decreased from 4.71 to 4.21 V (112). In effect, one determines on a nearly atomic basis the surface potential change ΔV, of Section IV-3B (36), and chemical conclusions about the magnitude and direction of the dipole contribution of the adsorbed atom can be drawn. Thus nitrogen adsorbed on 100 planes of tungsten appears to be at the negative end of a surface dipole, whereas if it is adsorbed on 111 planes, the reverse appears to be the case. For more details see the monograph by Gomer (37); also Ref. 104. Certain aspects are discussed further in connection with chemisorption, Chapter XVII.

Some information about the relative surface tensions of different crystal planes can be obtained by observing the relative development of various facets in field ion microscopy. Brenner (38), in studying iridium surfaces, observed also that adsorbed oxygen preferentially lowered the surface free energy of the 110 and 113 facets, since they increased in prominence.

3. Low Energy Electron Diffraction (LEED)

It was recognized very early in the history of diffraction studies that just as x-rays and relatively high energy electrons (say 50 keV) gave information about the bulk periodicity of a crystal, low energy electrons (around 100 eV) whose penetrating power is only a few atomic diameters, should give information about the surface structure of the solid. The first reported experiment is that of Davisson and Germer in 1927 (39), with marginal results. Work was handicapped by considerable experimental difficulties in generating a beam of monoenergetic electrons and detecting its scattering and even more by the fact that at that time ultrahigh vacuum techniques had not been developed. Even at a pressure of 10^{-6} torr (mm Hg) a surface becomes covered with a monolayer of adsorbed gas in about 1 sec, and to be sure of dealing with clean surfaces, pressures down to 10^{-10} torr are needed.

Much of the experimental development was carried out by MacRae and co-workers (3). A contemporary experimental arrangement is shown in Fig. VIII-8 (8, 40, 41). Electrons from a hot filament are given a uniform acceleration, striking the crystal normal to its surface. The scattered electrons may have been either elastically or inelastically scattered; only the former are used in the diffraction experiment. Of the several grids shown, the first is at the potential of the crystal, and the second is a repelling grid which allows only electrons of original energy to pass—those inelastically scattered, and hence of lower energy are stopped. The final grid is positively charged to accelerate the accepted electrons onto the fluorescent screen. The diffraction pattern may then be photographed.

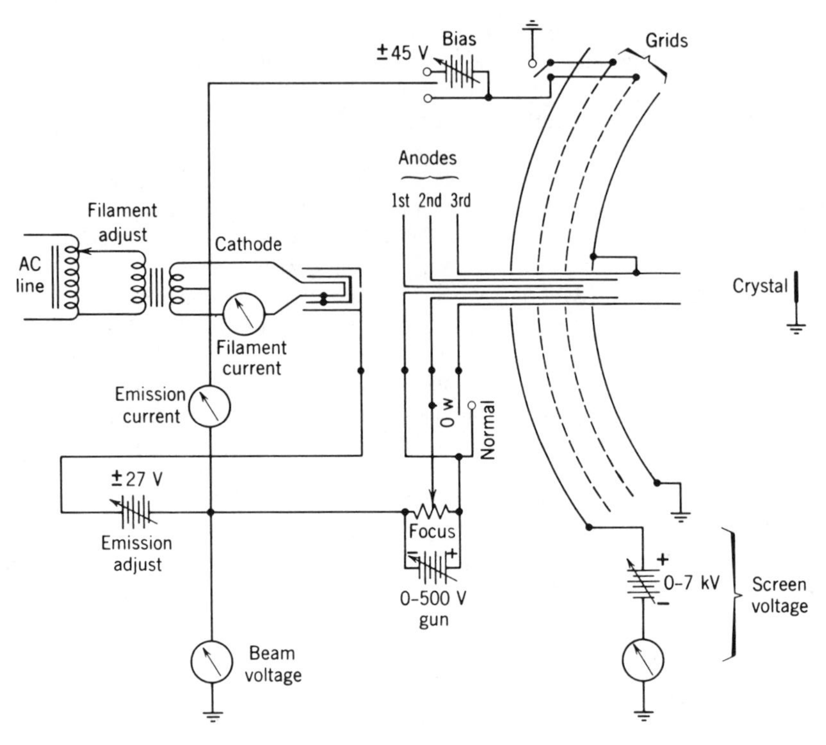

Fig. VIII-8. Low energy electron diffraction (LEED) apparatus. (From Ref. 8.)

The diffraction is essentially that of a grating. As illustrated in Fig. VIII-9, the Laue condition for incidence normal to the surface is

$$a \cos \alpha = n_1\lambda \tag{VIII-5}$$

where a is the repeat distance in one direction, n_1 is an integer, and λ is the wavelength of the electrons. For a second direction, in the usual case of a two-dimensional grid of atoms,

$$b \cos \beta = n_2\lambda \tag{VIII-6}$$

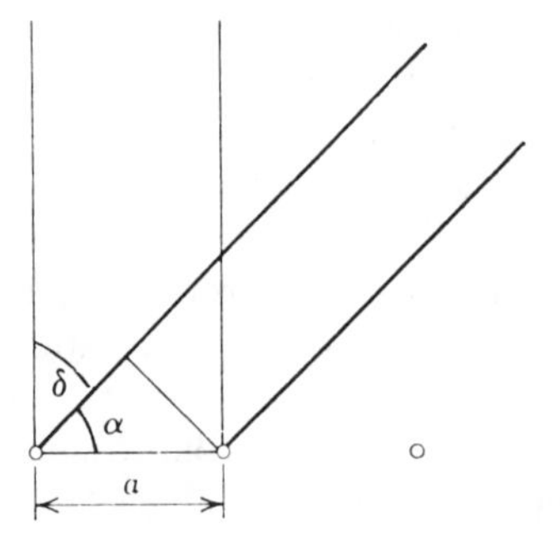

Fig. VIII-9

The diffraction pattern consists of a small number of spots whose symmetry of arrangement is that of the surface grid of atoms. Figure VIII-10 shows one of MacRae's patterns, for a surface formed by (110) planes of Ni. A pattern such as that shown in Fig. VIII-10 will expand or contract as the energy of the electron beam is decreased or increased. The pattern is primarily due to the first layer of atoms because of the small penetrating power of the low energy electrons (or, in HEED, because of the grazing angle of incidence used); there may, however, be weak indications of scattering from a second or third layer.

The intensity of a diffraction spot is temperature dependent because of the vibration of the surface atoms. As an approximation,

$$d \ln \frac{I}{dT} = \frac{12\, h^2(\cos^2 \phi)}{mk\lambda^2\, \theta_D^2} \tag{VIII-7}$$

where I is the intensity of a given spot, θ is the angle between the direction of the incident beam and the diffracted beam (usually small), λ is the wavelength of the electron, and θ_D is the debye temperature of the lattice (43).

The intensity of a LEED spot will vary with the energy of the electron beam. This is to be expected, of course, since the wavelength of the electrons is being varied, but the detailed shape of the intensity–voltage plot is sensitive to the interatomic spacings and to the type of atom. Alternatively, to calculate such plots one

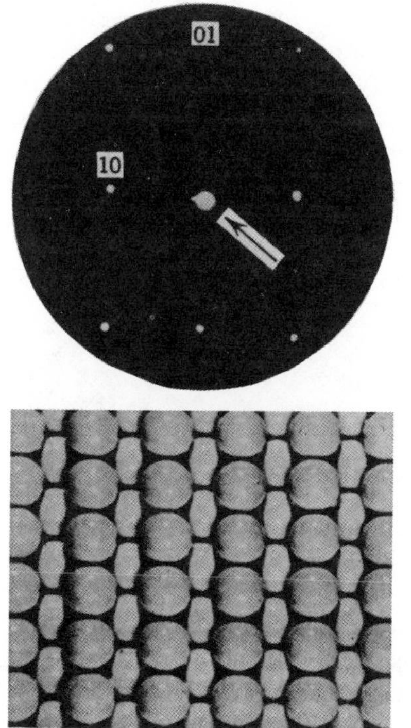

Fig. VIII-10. Top: diffraction pattern from a clean Ni (110) surface, with 76-V electrons. The arrow indicates the position of the oo spot. Bottom: model of the (110) surface of a full-centered crystal. (From Ref. 3.)

represents the lattice by a periodic potential function. As illustrated in Fig. VIII-11 experiment and theory can be brought into reasonable agreement; in other cases this has not been easy (note Ref. 44).

It might be imagined that the structure of a clean surface of, say, a metal single crystal, would simply be that expected from the bulk structure. This is not necessarily so; recall that in the case of the alkali metal halides the ions of the surface layer are displaced differently (Fig. VII-5). The surface structure, however, should "fit" on the bulk structure, and it has become customary to describe the former in terms of the latter. A (1×1) surface structure has the same periodicity or "mesh" as the corresponding crystallographic plane of the bulk structure, as illustrated in Fig. VIII-12(*a*). If alternate rows of atoms are shifted, the surface periodicity is reduced to a $C(2 \times 1)$ structure, the C denoting the presence of a center atom, as shown in Fig. VIII-12*b*. Figure VIII-12*c* shows a set of structures derived from a (111) plane.

The usual object of a LEED study is the determination of the surface structure. This is not always easy even though one may know what crystallographic plane forms the crystal facet on which the electron beam impinges. The problem is that the LEED pattern is not simply a picture of the surface structure; rather, the pattern is one of the reciprocal lattice. That is, a LEED pattern displays the repeat distances and the various angles between them. Usually, there will be more than one possible surface structure giving the same LEED pattern and in order to decide between possibilities, it is necessary to make calculations for not only each of the positions expected for the spots but also of the intensity–voltage plots (as in Fig. VIII-11) (see Refs. 8, 41, 44, 48, 49). There can be problems in allowing for multiple scattering. Also, if a surface is nominally that of a high Miller index plane, what may actually be present is a series of steps and risers made up of low Miller index planes (41), and this may show up as a splitting of the diffraction spots.

Most LEED studies have been made on elemental crystals—metals, Si,

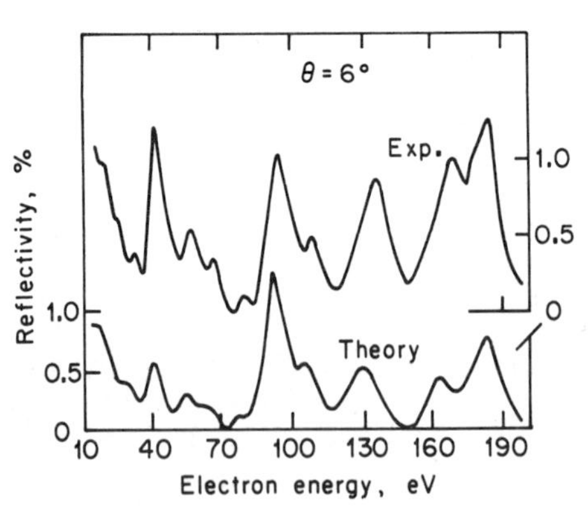

Fig. VIII-11. An example of experimental LEED intensity–voltage profiles and their comparison to theoretical behavior. (From Ref. 41, with citations to Refs. 45 and 46, with permission.)

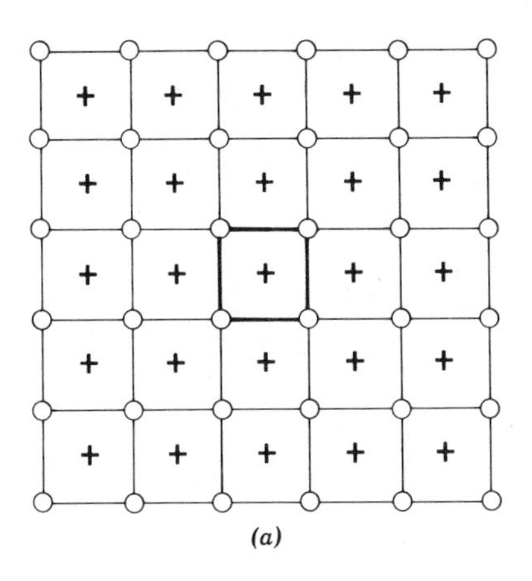

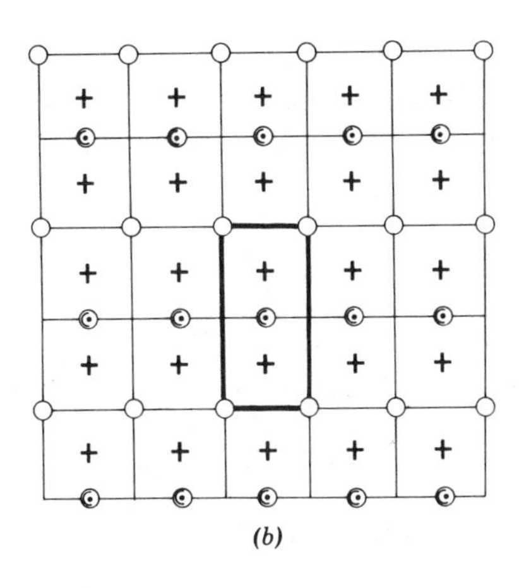

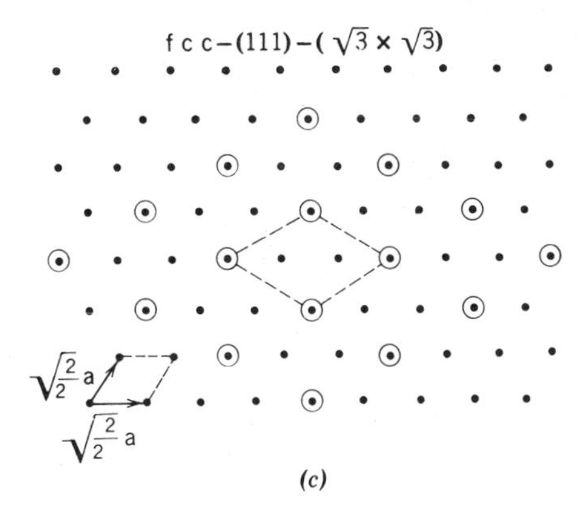

Fig. VIII-12. Surface structures. (*a*) (1 × 1) structure on the (100) surface of a fcc crystal (from Ref. 47). (*b*) *C* (2 × 1) surface structure on the (100) surface of fcc crystal (from Ref. 47). In both cases the unit cell is indicated with heavy lines, and the atoms in the second layer with pluses. In (*b*) the shaded circles mark shifted atoms. (*c*) Common surface structures on substrates with sixfold rotational symmetry. (From Ref. 14.)

Ge, and for these some theoretical treatment has been attempted. Thus beginning with the (experimentally) hypothetical case of (100) Ar surfaces, Burton and Jura (47) estimated theoretically the free energy for a surface transition from a (1 × 1) to a $C(2 \times 1)$ structure as given by

$$\Delta G = 4.83 - 0.0592T(\mathrm{erg/cm^2}) \qquad \text{(VIII-8)}$$

Above 81.5 K the $C(2 \times 1)$ structure becomes the more stable. Two important points are first that a change from one surface structure to another can occur without any bulk phase change being required, and second, that the energy difference between alternative surface structures may not be very large, and the free energy difference can be quite temperature dependent.

A probable consequence of the preceding points is that adsorption of some foreign species onto a surface may alter the surface structure. This aspect is discussed further in Chapters XVI and XVII, but, as an example, Orent and Hansen (43) report a series of LEED structures as O_2 or NO are adsorbed on Ru at various temperatures.

Finally, it has been possible to obtain LEED patterns from films of molecular solids deposited on a metal backing. Examples include ice and naphthalene (50) and various phthalocyanines (51). (The metal backing helps to prevent surface charging.)

4. Spectroscopic Methods

If a surface, typically a metal surface, is irradiated with a probe beam of photons, electrons, or ions (usually positive ions), one generally finds that photons, electrons, and ions are produced in various combinations. A particular method consists of using a particular type of probe beam and detecting a particular type of produced species. The method becomes a spectroscopic one if the intensity or efficiency of the phenomenon is studied as a function of the energy of the produced species, at constant probe beam energy, or vice versa. Quite a few combinations are possible, as is evident from the listing in Table VIII-1, and only a few are considered here.

The various spectroscopic methods do have in common that they typically allow analysis of the surface composition. Some also allow an estimation of the chemical state of the atom in question. A common problem, however, is that it may be difficult to estimate the surface contribution versus that from second, third, and so on layers.

A. Auger Electron Spectroscopy (AES)

The physics of the method is as follows. A probe electron (2 to 3 keV usually) ionizes an inner electron of a surface atom, creating a vacancy. Suppose that a K electron has been ejected by the incident electron (or x-ray) beam. A more outward-lying electron may now drop into this vacancy.

If, say, it is an L_{I} electron that does so, the energy $E_K - E_{L_{\text{I}}}$ is now available and may appear as a characteristic x-ray. Alternatively, however, this energy may be given to an outer electron, expelling it from the atom. This *Auger electron* might, for example, come from the L_{III} shell, in which case its kinetic energy would be $(E_K - E_{L_{\text{I}}}) - E_{L_{\text{III}}}$. The $L_{\text{I}} - K$ transition is the probable one, and an Auger spectrum is mainly that of the series $(E_K - E_{L_{\text{I}}}) - E_i$, where the ith type of outer electron is ejected.

The principal use of Auger spectroscopy is in the determination of surface composition, although peak positions are secondarily sensitive to the valence state of the atom. See Refs. 52 and 53 for reviews.

Experimentally, it is common for LEED and Auger capabilities to be combined; the basic equipment is the same. For Auger measurements, a grazing angle of incident electrons is needed, to maximize the contribution of surface atoms. The voltage on the retarding grid (Fig. VIII-8) may be modulated to make the detector signal responsive to electrons of just that energy. As shown in Fig. VIII-13, one then obtains a *derivative* plot. Note in the figure how the cleaning of the tungsten surface removed the carbon and oxygen impurities.

A final aspect of Auger spectroscopy is that the intensity of the Auger electrons varies with the angle of observation. There is generally a falling off in intensity with cos θ, where θ is the angle from the normal to the surface, but with strong ripples due to diffraction effects (49). The effect can give information about surface geometry and composition.

B. Photoelectron Spectroscopy (ESCA)

In photoelectron spectroscopy monoenergetic x-radiation ejects inner (1*s*, 2*s*, 2*p*, etc.) electrons. The electron energy is then $E_0 - E_i$ where E_0 is the x-ray quantum energy and E_i is that of the ith type of electron. The energy of the ejected electrons is determined by means of an electron spectrometer, thus obtaining a spectrum both of the primary photoelectrons and of Auger electrons. The method is more accurate than is Auger spectroscopy and, because of this, one can observe that a given type of electron has an energy that is dependent on the valence state of the atom. Thus for 1*s* sulfur electrons, there is a *chemical shift* of over 5 V, the ionization energy increasing as the valence state of sulfur varies from -2 to $+6$. The effect is illustrated in Fig. VIII-14 for the case of aluminum, showing how it is possible to analyze for oxidized aluminum on the surface. An example involving fluorocarbon polymer surfaces is given in Ref. 54. Because the method is often used for chemical analysis, it is sometimes termed ESCA, for Electron Spectroscopy for Chemical Analysis.

Because of the use of x-rays, which are penetrating, special techniques are used to emphasize the contribution from surface atoms. One, for example, is set the x-ray beam at a grazing angle to the surface.

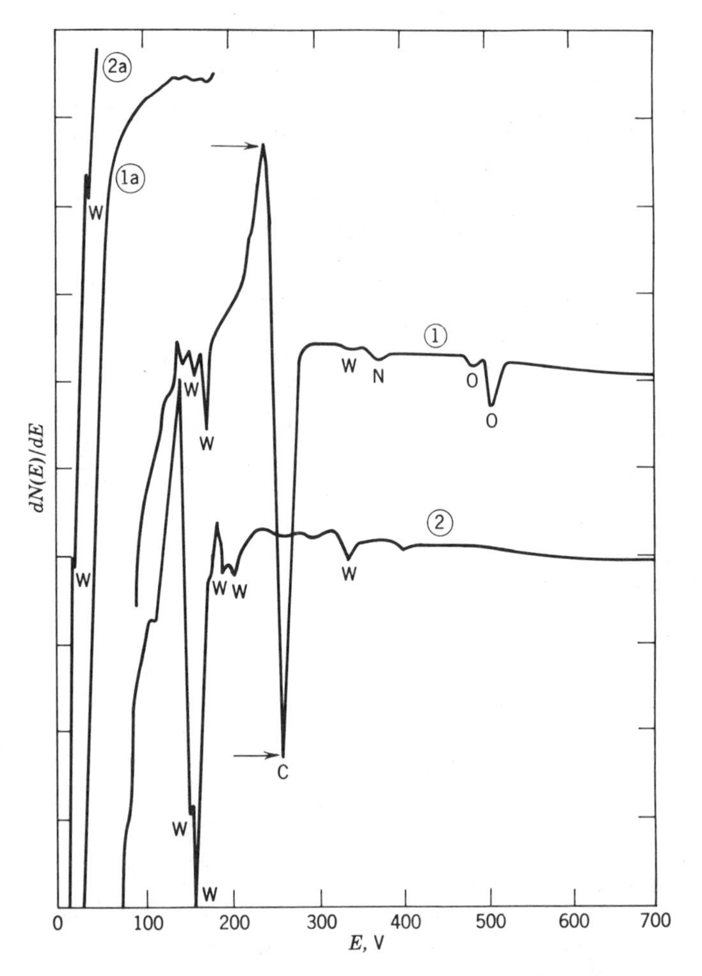

Fig. VIII-13. Auger spectrographs of the W (110) crystal face before (curve 1) and after (curve 2) ion bombardment and heating to 1300°C. (From Ref. 53.)

C. *Ion Scattering (ISS, LEIS)*

If, as illustrated in Fig. VIII-15, a beam of monoenergetic ions of mass M_i is elastically scattered by atoms in the surface, of mass M_a, conservation of momentum and energy requires that

$$E_s = \left[\frac{\cos\theta + (r^2 - \sin^2\theta)^{1/2}}{1 + r}\right]^2 E_i \tag{VIII-9}$$

E_s is the energy of the scattered ion, E_i, its initial energy, and $r = M_a/M_i$. For the case of $\theta = 90°$, Eq. VIII-9 reduces to

$$E_s = \frac{(M_a - M_i)}{(M_a + M_i)} E_i \tag{VIII-10}$$

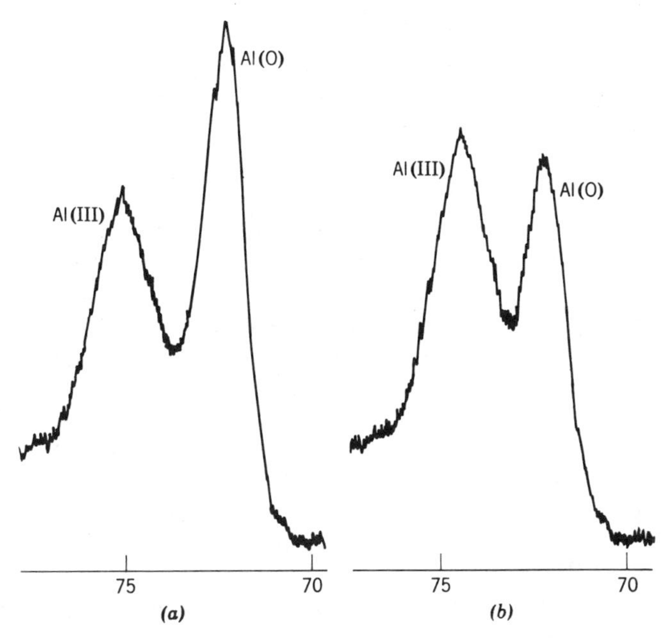

Fig. VIII-14. ESCA spectrum of Al surface showing peaks for the metal. Al(0) and for surface oxidized aluminum, Al(III). (*a*) Freshly abraided sample; (*b*) sample after five days of ambient temperature air exposure showing increased Al(III)/Al(0) ratio due to surface oxidation. (From Ref. 55.)

These equations indicate that the energy of the scattered ions is sensitive to the mass of the scattering atom s in the surface. By scanning the energy of the scattered ions, one obtains a kind of mass spectrometric analysis of the surface composition. Figure VIII-16 shows an example of such a spectrum. Neutral, that is, molecular, as well as ion beams may be used, although for the former a velocity selector is now needed to define E_i.

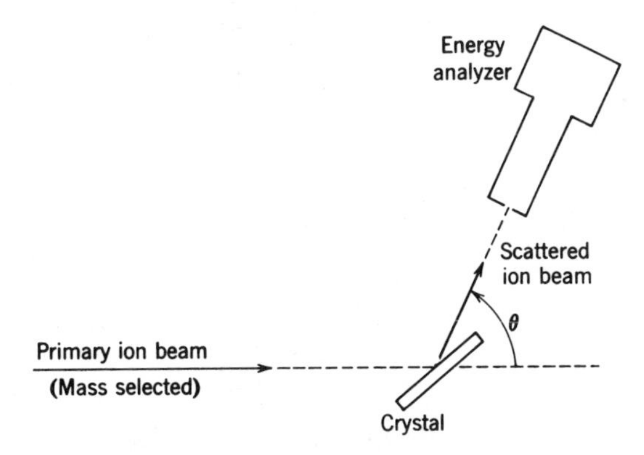

Fig. VIII-15. An ion-scattering experiment. (From Ref. 56.)

Equations VIII-9 and VIII-10 are useful in relating scattering to M_a. It is also of interest to study the variation of scattering intensity with scattering angle. It is now better to recognize that surface atoms cannot easily move parallel to the surface, although they can move normal to it. This restriction implies that the velocity component of the incident ion or molecule beam which is normal to the surface remains unchanged on scattering. Analysis (see Ref. 8) shows the scattering to be essentially specular, the intensity peaking sharply for an angle of reflection equal to the angle of incidence; this is illustrated by the curve for H_2 in Fig. VIII-17.

A useful complication is that if kinetic energy is not conserved, that is, if the collision is inelastic, there should be a quite different angular scattering distribution. As an extreme, if the impinging molecule sticks to the surface for a while before evaporating from it, memory of the incident direction is lost, and the most probable "scattering" is now normal to the surface, the probability decreasing with the cosine of the angle to the normal. In intermediate cases, some but not complete exchange of rotational and vibrational energy may occur between the molecule and the surface. The experiment shown in Fig. VIII-17 indicates that D_2 and HD, but not H_2, were able to exchange rotational energy with the surface; the sharp maximum shown for H_2 and the specular angle has largely disappeared for the other two molecules. The inability of H_2 to exchange rotational energy efficiently is probably due to the relatively large energy separation of rotational excited states in this case.

Studies of inelastic scattering are of considerable interest in heterogeneous catalysis. The degree to which molecules are scattered specularly gives information about their residence time on the surface. Often new chemical species appear, whose trajectory from the surface correlates to

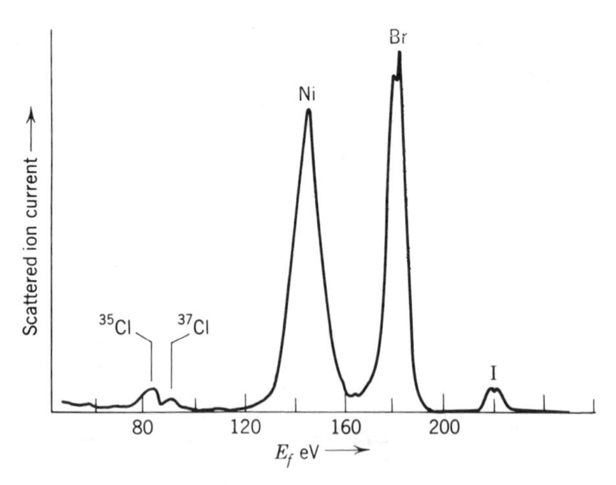

Fig. VIII-16. Energy spectrum of Ne^+ ions that are scattered over 90° by a halogenated nickel surface. The incident energy of the ions is 300 V. (From Ref. 57.)

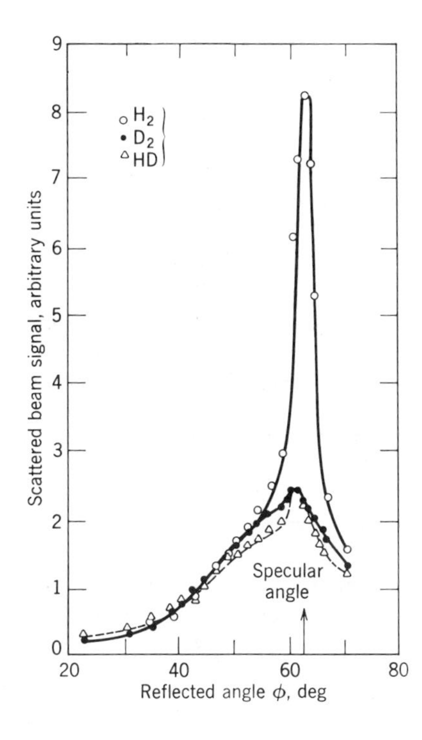

Fig. VIII-17. Angular distribution of H_2, D_2, and HD molecular beams scattered from an oriented silver (111) film. (From Ref. 8.)

some degree with that of the incident beam of molecules. The study of such *reactive* scattering gives mechanistic information about surface reactions.

5. Problems

1. A LEED pattern is obtained for the 111 surface of an element that crystallizes in the face-centered close-packed system. Show what the pattern should look like in symmetry appearance. Consider only first-order nearest neighbor diffractions.

2. Derive Eqs. VIII-5 and VIII-6.

3. Derive Eq. VIII-10.

4. Ehrlich and co-workers found that in a study of a tungsten surface, plots of $\log(i/V^2)$ versus $10^4/V$ gave straight lines, as expected from the Fowler–Nordheim equation. Their slope for a clean tungsten surface was -2.50, from which they calculate Φ to be 4.50 V. Six minutes after admitting a low pressure of nitrogen gas, the slope of the same type of plot was found to be -2.35. Calculate Φ for the partially nitrogen covered surface.

5. The relative intensity of a certain LEED diffraction spot is 0.25 at 300°K and 0.050 at 570°K, using 390 eV electrons. Calculate the debye temperature of the crystalline surface (in this case of Ru metal).

6. Calculate the magnification of Fig. VIII-2 if the sample consists of uniform cubes of the size shown in the foreground, and the specific surface area is 2.8×10^4 cm²/g.

7. Calculate the energies of x-rays, electrons, $^4He^+$ ions, and H_2 molecules that correspond to a wavelength of 2.0 Å.

8. Coin an acronym for the experiment in which neutrons are scattered inelastically from microcrystals.

General References

R. Gomer, *Field Emission and Field Ionization,* Harvard University Press, Cambridge, 1961.

C. S. Fadley, *Electron Spectroscopy, Theory, Techniques, and Applications,* C. R. Brundle and A. D. Baker, Eds., Vol. 2, Pergamon, New York, 1978.

N. B. Hannay, *Treatise on Solid State Chemistry,* Vol. 6A, Surfaces I, Plenum, New York, 1976.

E. W. Müller and T. T. Tsong, *Field Ion Microscopy,* American Elsevier, New York, 1969.

G. A. Somorjai, *Principles of Surface Chemistry,* Prentice-Hall, Englewood Cliffs, New Jersey, 1972.

G. A. Somorjai, *Science,* **201,** 489 (1978).

J. P. Thomas and A. Cachard, Eds., *Material Characterization Using Ion Beams,* Plenum, New York, 1976.

P. R. Thornton, *Scanning Electron Microscopy,* Chapman and Hall Ltd., 1968. See also *Scanning Electron Microscopy: Systems and Applications 1973,* The Institute of Physics, London, 1973.

Textual References

1. E. W. Müller and T. T. Tsong, *Field Ion Microscopy,* American Elsevier, New York, 1969.
2. G. Ehrlich, *Ad. Catal.,* **14,** 255 (1963).
3. A. U. MacRae, *Science,* **139,** 379 (1963); see also L. H. Gerner and A. U. MacRae, *J. Appl. Phys.,* **33,** 2923 (1962).
4. F. L. Baker and L. H. Princen, *Encyclopedia of Polymer Science and Technology,* Vol. 15, Wiley, New York, 1971, p. 498.
5. M. R. Leggett and R. A. Armstrong, *Surf. Sci.,* **24,** 404 (1971).
6. P. B. Sewell, D. F. Mitchell, and M. Cohen, *Surf. Sci.,* **33,** 535 (1972).
7. E. G. MacRae and H. D. Hagstrom, *Treatise on Solid State Chemistry,* N. B. Hannay, Ed., Vol. 6A, Part I, Plenum, New York, 1976.

7a. R. J. Colton, *J. Vac. Sci. Technol.,* **18,** 737 (1981).

8. G. A. Somorjai, *Principles of Surface Chemistry,* Prentice-Hall, Englewood Cliffs, New Jersey, 1972.
9. S. D. Kevan, D. H. Rosenblatt, D. R. Denley, B. C. Lu, and D. A. Shirley, *Phys. Rev. B,* **20,** 4133 (1979).

9a. L. G. Petersson, S. Kono, N. F. T. Hall, S. Goldberg, J. T. Lloyd, and C. S. Fadley, *Mater. Sci. Eng.,* **42,** 111 (1980).

9b. S. Kono and C. S. Fadley, *Nucl. Instrum. Methods,* **177,** 207 (1980).

10. R. G. Greenler, *Surf. Sci.,* **69,** 647 (1977).
11. Y. Momose, T. Ishll, and T. Namekawa, *J. Phys. Chem.,* **84,** 2908 (1980).
12. S. Evans, J. Pielaszek, and J. M. Thomas, *Surf. Sci.,* **55,** 644 (1976).
13. J. L. Gland and V. N. Korchak, *Surf. Sci.,* **75,** 733 (1978).

14. G. A. Somerjai and F. J. Szalkowski, *Adv. High Temp. Chem.*, **4,** 137 (1971).
15. Y. Sakisaka, K. Akimoto, M. Nishijima, and M. Onchi, *Solid State Commun.*, **29,** 121 (1979).
16. S. Trajmar, *Acc. Chem. Res.*, **13,** 14 (1980).
17. G. B. Fisher and B. A. Sexton, *Phys. Rev. Lett.*, **44,** 683 (1980).
18. D. S. Zingg, L. E. Makovsky, R. E. Tischer, F. R. Brown, and D. M. Hercules, *J. Phys. Chem.*, **84,** 2898 (1980).
19. G. A. Somorjai, *Treatise on Solid State Chemistry,* N. B. Hannay, Ed., Vol. 6A, Part I, Plenum, New York, 1976.
20. M. Nishijima, K. Fujiwara, and T. Murotani, *J. Appl. Phys.*, **46,** 3089 (1975).
21. H. H. Brongersma, M. J. Sparnaay, and T. M. Buck, *Surf. Sci.*, **71,** 657 (1978).
22. E. T. Taglauer and W. Heiland, *Surf. Sci.*, **47,** 234 (1975).
23. R. J. Madix and J. A. Schwarz, *Surf. Sci.*, **24,** 264 (1971).
24. Y. Viswanath and L. D. Schmidt, *J. Chem. Phys.*, **59,** 4184 (1973).
25. L. W. Anders and R. S. Hansen, *J. Chem. Phys.*, **62,** 4652 (1975).
26. K. H. Rieder, *Surf. Sci.*, **26,** 637 (1971).
27. Boon-Teng Teo, *Acc. Chem. Res.*, **13,** 412 (1980).
28. L. H. Princen, *Treatise on Coatings,* R. R. Meyers, Ed., Vol. 2, Part II, Chapter 7, Marcel Dekker, New York, 1976.
29. G. Ehrlich and F. G. Hudda, *J. Chem. Phys.*, **35,** 1421 (1961).
30. R. Gomer, *Adv. Catal.*, **7,** 93 (1955).
31. E. W. Müller, *Z. Phys.*, **120,** 261 (1942).
32. E. W. Müller and T. T. Tsong, *Field Ion Microscopy,* American Elsevier, New York, 1969.
33. G. Ehrlich, *J. Appl. Phys.*, **15,** 349 (1964); G. Ehrlich and F. G. Hudda, *J. Chem. Phys.*, **36,** 3233 (1962).
34. A. J. W. Moore and J. A. Spink, *Surf. Sci.*, **44,** 198 (1974).
35. R. H. Fowler and L. W. Nordheim, *Proc. Roy. Soc.* (*London*), **A112,** 173 (1928).
36. T. A. Delchar and G. Ehrlich, *J. Chem. Phys.*, **42,** 2686 (1965).
37. R. Gomer, *Field Emission and Field Ionization,* Harvard University Press, Cambridge, Mass., 1961.
38. S. S. Brenner, *Surf. Sci.*, **2,** 496 (1964).
39. C. J. Davisson and L. H. Germer, *Phys. Rev.*, **30,** 705 (1927).
40. G. A. Somorjai, *Angew. Chem., Int. Ed. Engl.*, **16,** 92 (1977).
41. L. K. Kesmodel and G. A. Somorjai, *Acc. Chem. Res.*, **9,** 392 (1976).
42. E. W. Müller, *Z. Phys.*, **37,** 838 (1936).
43. T. W. Orent and R. S. Hansen, *Surf. Sci.*, **67,** 325 (1977).
44. C. B. Duke, R. J. Meyer, A. Paton, and P. Mark, *Phys. Rev.*, **18,** 4225 (1978).
45. S. Y. Tong and L. L. Kesmodel, *Phys. Rev. B,* **8,** 3753 (1973).
46. J. E. Demuth and T. N. Rhodin, *Surf. Sci.*, **42,** 261 (1974).
47. J. J. Burton and G. Jura, *Structure and Chemistry of Solid Surfaces,* G. Somorjai, Ed., Wiley, New York, 1969.
48. M. J. Sparnaay, *J. Radioanal. Chem.*, **12,** 101 (1972).
49. A. J. Martin and H. Bilz, *Phys. Rev. B,* **19,** 6593 (1979).
50. L. E. Firment and G. A. Somorjai, *J. Chem. Phys.*, **63,** 1037 (1975).
51. J. C. Buchholz and G. A. Somorjai, *J. Chem. Phys.*, **66,** 573 (1977).
52. G. A. Somorjai and F. J. Szalkowski, *Adv. High Temp. Chem.*, **4,** 137 (1971).

53. T. Smith, *J. Appl. Phys.*, **43,** 2964 (1972).
54. D. W. Dwight and W. M. Riggs, *J. Colloid Interface Sci.*, **47,** 650 (1974).
55. From Instrument Products Division, E. I. du Pont de Nemours Co., Inc.
56. H. H. Brongersma and P. M. Mul, *Chem. Phys. Lett.*, **14,** 380 (1972).
57. H. H. Brongersma and P. M. Mul, *Surf. Sci.*, **35,** 393 (1973).

CHAPTER IX

The Formation of a New Phase—Nucleation and Crystal Growth

1. Introduction

A rather difficult and yet very interesting and important matter is that of the kinetics and mechanism of the formation of a new phase, as, for example, the condensation of a vapor, the freezing of a liquid, or the precipitation of a solute from solution. The topic is included at this point because while the *verification* of nucleation theory has been primarily in terms of situations involving the liquid–vapor interface, an important *application* of it has been to the estimation of solid–liquid interfacial free energies.

It is recognized that in the formation of a new phase, and in the absence of participating foreign surfaces, the general sequence of events is that small clusters of molecules form and that these grow by accretion to the point of becoming recognizable droplets or crystallites that may finally coalesce or grow to yield massive amounts of the new phase. The normal observation, furthermore, is that this sequence does not take place if the vapor pressure is only just slightly over the saturation value or if the liquid is only slightly undercooled. Instead, the vapor pressure usually can be increased considerably over the equilibrium value without anything happening, until, at some fairly sharp limit, general condensation in the form of a fog of droplets takes place. Similarly, solutions or liquids may be considerably supersaturated or supercooled. Very pure liquid water, for example, can be cooled to $-40°C$ before spontaneous freezing occurs. Some early observations of this type were made by Fahrenheit in 1714 (see Ref. 1).

The impedance to the form of a new phase clearly is associated with the extra surface energy of small clusters that make their formation difficult. Such clusters, if small, have been called *germs* (2) and, if somewhat larger and recognizable as precursors to the new phase, they are called *nuclei*. Considerable attention has been devoted to the mechanism and kinetics of nucleation, and in the following sections a brief description of the results is presented.

Crystal growth, either from the vapor or the melt, may also involve nucleation at each successive lattice plane. In addition, the shape or *habit* of a growing crystal often is determined by surface chemical factors.

2. Classical Nucleation Theory

Some important thermodynamic relationships may be written concerning the free energy of formation of a cluster. Consider the process

$$nA\ (\text{gas}, P) = A_n\ (\text{small liquid drop}) \tag{IX-1}$$

In the absence of surface tensional effects, ΔG would be given by the free energy to transfer n moles from the vapor phase at activity or pressure P, to the liquid phase, at activity or pressure P^0, that is,

$$\Delta G = -nkT \ln \frac{P}{P^0} \tag{IX-2}$$

The drop, however, possesses a surface energy $4\pi r^2\gamma$, so that ΔG actually is

$$\Delta G = -nkT \ln x + 4\pi r^2\gamma \tag{IX-3}$$

where x denotes P/P^0. If the density of the liquid is ρ, and the molar volume is therefore M/ρ, Eq. IX-3 may be written

$$\Delta G = -\frac{4}{3}\pi r^3 \frac{\rho}{M} RT \ln x + 4\pi r^2\gamma \tag{IX-4}$$

Since the terms are of opposite sign if x is greater than unity and depend differently on r, a plot of ΔG versus r goes through a maximum. This is illustrated in Fig. IX-1 for the case of water at 0°C with $x = 4$(3). At the maximum, $r = r_c$, and by setting $d(\Delta G)/dr = 0$, one obtains from Eq. IX-4

$$RT \ln x = \frac{2\gamma M}{r_c\rho} = \frac{2\gamma \bar{V}}{r_c} \tag{IX-5}$$

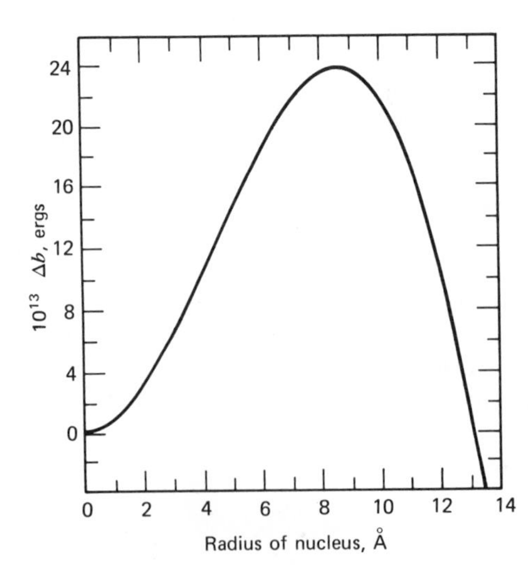

Fig. IX-1. Variation of ΔG with drop size. Ordinate represents 10^{13} ΔG, in ergs. (From Ref. 3.)

Since the free energy to form a drop of size r_c has the maximum positive value, this radius is known as the critical radius; in the preceding example it is about 8 Å, and the critical drop contains about 90 molecules of water. Equation IX-5 is also the Kelvin equation (Eq. III-20) and gives the equilibrium vapor pressure for a drop of radius r.

By combining Eqs. IX-4 and IX-5, one obtains for $\Delta G_{\max}$:

$$\Delta G_{\max} = \frac{4\pi r_c^2 \gamma}{3} \tag{IX-6}$$

The conclusion that $\Delta G_{\max}$ is equal to one-third of the surface free energy for the whole nucleus was given by Gibbs (4). An alternative form is obtained through the elimination of r_c:

$$\Delta G_{\max} = \frac{16\pi\gamma^3 M^2}{3\rho^2(RT \ln x)^2} \tag{IX-7}$$

Equation IX-7 may also be applied to crystals, although in such a case small changes in the numerical factor may result because of the nonspherical shape that may be involved.

The dynamic picture of a vapor at a pressure near P^0 is then somewhat as follows. If P is less than P^0, then ΔG for a cluster increases steadily with size, and although in principle all sizes would exist, all but the smallest would be very rare, and their numbers would be subject to random fluctuations. Similarly, there will be fluctuations in the number of embryonic nuclei of size less than r_c, in the case of P greater than P^0. Once a nucleus reaches the critical dimension, however, a favorable fluctuation will cause it to grow indefinitely. The experimental maximum supersaturation pressure is such that a large traffic of nuclei moving past the critical size develops with the result that a fog of liquid droplets is produced.

It follows from the foregoing discussion that the essence of the problem is that of estimating the rate of formation of nuclei of critical size, and a semirigorous treatment has been given by Becker and Doring (5, 6). The simplifying device that was employed, which made the treatment possible, was to consider the case of a steady state situation such that the average number of nuclei consisting of 2, 3, 4, . . . , N molecules, although different in each case, did not change with time. A detailed balancing of evaporation and condensation rates was then set up for each size nucleus, and by a clever integration procedure, the flux I, or the rate of formation of nuclei containing n molecules from ones containing $(n - 1)$ molecules, was estimated. The detailed treatment is rather lengthy, although not difficult, and the reader is referred to Refs. 1, 3, and 5 through 7 for details.

The final equation obtained by Becker and Doring may be written down immediately by means of the following qualitative argument. Since the flux I is taken to be the same for any size nucleus, it follows that it is related to the rate of formation of a cluster of two molecules, that is, to Z, the gas kinetic collision frequency (collision per cubic centimeter-second).

For the steady state case, Z should also give the forward rate of formation or flux of critical nuclei, except that the positive free energy of their formation amounts to a free energy of activation. If one correspondingly modifies the rate Z by the term $e^{-\Delta G_{\max}/kT}$, an approximate value for I results:

$$I = Z \exp\left(\frac{-\Delta G_{\max}}{kT}\right) \quad \text{(IX-8)}$$

While Becker and Doring obtained a more complex function in place of Z, its numerical value is about equal to Z, and it turns out that the exponential term, which is the same, is the most important one. Thus the complete expression is

$$I = \frac{Z}{n_c}\left(\frac{\Delta G_{\max}}{3\pi kT}\right)^{1/2} \exp\left(\frac{-\Delta G_{\max}}{kT}\right) \quad \text{(IX-9)}$$

where n_c is the number of molecules in the critical nucleus.

The full equation for I is obtained by substituting into Eq. IX-9 the expression for $\Delta G_{\max}$ and the gas kinetic expression for Z:

$$I = 2n^2\sigma^2\left(\frac{RT}{M}\right)^{1/2} \exp\left[\frac{-16\pi\gamma^3 M^2}{3kT\rho^2(RT\ln x)^2}\right] \quad \text{(IX-10)}$$

where σ is the collision cross section and n is the number of molecules of vapor per cubic centimeter. Since Z is roughly $10^{23}P^2$, where P is given in millimeters of mercury, Eq. IX-10 may be simplified as

$$I = 10^{23}P^2 \exp\left[\frac{-17.5\bar{V}^2\gamma^3}{T^3(\ln x)^2}\right] \quad \text{[in nuclei/(cm}^3\text{)(sec)]} \quad \text{(IX-11)}$$

By way of illustration, the various terms in Eq. IX-11 are evaluated for water at 0°C in Table IX-1. Taking $\bar{V}$ as 20 cm^3/mole, γ as 72 erg/cm^2, and P^0 as 4.6 mm, Eq. IX-11 becomes

$$I = 2 \times 10^{24}x^2 \exp\left[-\frac{118}{(\ln x)^2}\right] \quad \text{(IX-12)}$$

TABLE IX-1
Evaluation of Equation IX-12 for Various Values of x

x	$\ln x$	$118/(\ln x)^2 = A$	e^{-A}	I, nuclei/(cm^3)(sec)
1.0	0		0	0
1.1	0.095	1.31×10^4	10^{-5700}	10^{-5680}
1.5	0.405	720	10^{-310}	10^{-286}
2.0	0.69	246	10^{-107}	10^{-82}
3.0	1.1	95.5	10^{-42}	10^{-17}
3.5	1.25	75.5	10^{-33}	2×10^{-8}
4.0	1.39	61.5	$10^{-26.7}$	0.15
4.5	1.51	51.8	$10^{-22.5}$	10^3

The figures in the table show clearly how rapidly I increases with x, and it is generally sufficient to define the critical supersaturation pressure such that ln I is some arbitrary value such as unity.

Frequently, vapor phase supersaturation is studied not by varying the vapor pressure P directly but rather by cooling the vapor and thus changing P^0. If T_0 is the temperature at which the saturation pressure, is equal to the actual pressure P, then at any temperature T, $P/P^0 = x$ is given by

$$\ln x = \frac{\Delta H_v}{R}\left(\frac{1}{T} - \frac{1}{T_0}\right) \tag{IX-13}$$

where ΔH_v is the latent heat of vaporization. Then, by Eq. IX-5, it follows that

$$\Delta T = T_0 - T = \frac{2\gamma M T_0}{r_c \Delta H_v \rho} = \frac{2\gamma \bar{V} T_0}{r_c \Delta H_v} \tag{IX-14}$$

and, by means of Eq. IX-6, ΔG_{max} may be expressed in terms of ΔT. An analogous equation may also be written for the supercooling of a melt, where the heat of fusion ΔH_f replaces ΔH_v, and the quantity x is now the ratio of the liquid to the solid vapor pressure. As was the case when P was varied, the rate of nucleation increases so strongly with the degree of supercooling that a fairly sharp critical value exists for ΔT.

At a sufficiently low temperature the phase nucleated will be crystalline rather than liquid. The theory is reviewed in Ref. 1; it is similar to that for the nucleation of liquid drops but is complicated by the need to consider that the rate of growth of nuclei may be different in different directions (note Section IX-4).

The case of nucleation from a condensed phase, usually that of a melt, may be treated similarly. The chief modification to Eq. IX-8 that ensues is in the frequency factor; instead of free collisions between vapor molecules, one now has a closely packed liquid phase. The rate of accretion of clusters is therefore related to the diffusion process, and the situation was treated by Turnbull and Fisher (8). Again, the reader is referred to the original literature for the detailed derivation, and the final equation is justified here only in terms of a qualitative argument. If one considers a crystalline nucleus that has formed in a supercooled melt, then the rate at which an additional molecule may add can be regarded as determined by the frequency with which a molecule may jump from one position in the liquid to another just at the surface of the solid. Such a jump is akin to those involved in diffusion, and the frequency may be approximated by means of absolute rate theory as being equal to the frequency factor kT/h times an exponential factor containing the free energy of activation for diffusion. The total rate of such occurrences per cubic centimeter of liquid is

$$Z = n\frac{kT}{h}\exp\left(-\frac{\Delta G_D}{kT}\right) \tag{IX-15}$$

where n is the number of molecules of liquid per cubic centimeter. The steady state treatment is again used, and the final result is analogous to Eq. IX-8:

$$I = n\frac{kT}{h}\exp\left(-\frac{\Delta G_D}{kT}\right)\exp\left(-\frac{\Delta G_{\max}}{kT}\right) \tag{IX-16}$$

For liquids that are reasonably fluid around their melting points, the kinetic factors in Eq. IX-16 come out about $10^{33}/(\text{cm}^3)(\text{sec})$, so that Eq. IX-16 becomes

$$I = 10^{33}\exp\left(-\frac{\Delta G_{\max}}{kT}\right) \tag{IX-17}$$

where $\Delta G_{\max}$ is given by Eq. IX-7, or by a minor modification of it, which allows for a nonspherical shape for the crystals.

Another important point in connection with the rate of nuclei formation in the case of melts or of solutions is that the rate reaches a maximum with degree of supercooling. To see how this comes about, r_c is eliminated between Eq. IX-16 and the one for liquids analogous to Eq. IX-14, giving

$$I = \frac{nkT}{h}e^{-A/kT} \tag{IX-18}$$

where

$$A = \Delta G_D + \frac{4\pi\gamma}{3}\left(\frac{2\bar{V}\gamma T_0}{\Delta H_f \Delta T}\right)^2$$

On setting dI/dT equal to zero, one obtains for y at the maximum rate the expression

$$\frac{(y-1)^3}{y^2(3-y)} = \frac{4\pi\gamma}{3\Delta H_D}\left(\frac{2\gamma\bar{V}}{\Delta H_f}\right)^2 \tag{IX-19}$$

where ΔH_D is the activation energy for diffusion, and $y = T_0/T$. Qualitatively, while the concentration of nuclei increases with increasing supercooling, their rate of formation decreases due to the activation energy for diffusion or, essentially, due to the increasing viscosity of the medium. Glasses, then, result from a cooling so rapid that the temperature region of appreciably rapid nucleation is passed before much actual nucleation occurs.

Still another situation is that of a supersaturated or supercooled solution, and straightforward modifications can be made in the preceding equations. Thus in Eq. IX-5, x now denotes the ratio of the actual solute activity to that of the saturated solution. In the case of a nonelectrolyte, $x = S/S_0$, where S denotes the concentration. Equation IX-14 now contains ΔH_s, the molar heat of solution

The nucleation theory outlined at the beginning of this section is a relatively simple macroscopic one, and it contains potentially dangerous assumptions and simplifying mathematics. The assumption that small clusters can be treated as portions of a bulk phase has been discussed by Cahn and Hilliard (9) who concluded that the interface of the critical nucleus becomes more and more diffuse as it decreases in size with increasing supersaturation. Related concerns are that of the expected change in interfacial tension at a highly curved surface (10, 11) and the matter of the definition of the location of the surface of tension (see Section III-2A).

In principle, nucleation should occur for any supersaturation, given time, and the critical supersaturation ratio is arbitrarily defined in terms of the condition needed to make nucleation rapid on a humanly subjective basis. As was illustrated in Table IX-1, the nucleation rate I changes so rapidly with degree of supersaturation that, fortunately, even a few powers of 10 error in the preexponential term (e.g., of Eq. IX-12) made little difference. However, in 1962 Lothe and Pound (12) added two factors, hitherto largely ignored, that amounted to an increase of 10^{20} (!) in expected nucleation rate. Basically, it was proposed that ΔG_{max} for the critical nucleus was incomplete and should include translational and rotational entropy contributions. The problem is discussed by Oriani and Sundquist (11) and Dunning (1). Some subtle statistical thermodynamics are involved, and after a period of uncertainty it appears that the original treatment (which agrees with data) is correct. An analysis by Deryaguin (13) confirms this conclusion; see also Reiss (14).

Classical nucleation theory must be modified in the case of nucleation near a critical temperature. Observed supercooling and superheating is far in excess of that predicted by conventional theory, and McGraw and Reiss (15a) pointed out that if a usually neglected excluded volume term is retained, the effect is to considerably increase the free energy of the critical nucleus. Another type of complication is that in binary systems, the composition of nuclei will differ from the bulk composition; this case has also been treated, see Ref. 15b.

An interesting and important case is that of nucleation by ions. This is the action in a cloud chamber; the phenomenon can be important in atmospheric physics. The theory has been reexamined (16a, 16b). Interestingly, it appears that the full heat of solvation of an ion is approached after only five to ten water molecules have associated with it (17a).

3. Results of Nucleation Studies

One-Component Systems. As might be expected, the most easily interpretable studies are those for the nucleation of a vapor and, in general, the experimental findings are quite consistent with the Becker and Doring theory. Thus Volmer and Flood (17b) found the critical value of x to be 5.03 for water vapor at 261°K, using the cloud chamber technique of adiabatically expanding vapor-laden air. The theoretical value of x for $\ln I$ equal to about unity is 5.14. Sander and Damköhler (18) found the critical value of x for water vapor as well as its temperature dependence to be in reasonable agreement with theory.

Some aspects of nucleation theory, the use of a turbulent jet for measurements, and a review of the then-current data are given by Higuchi and O'Konski (19). Some results given by them and by Volmer and Flood (17b) are summarized in Table IX-2, and it can be seen that, in general, agreement

TABLE IX-2
Critical Supersaturation Pressures for Vapors

Substance	Temperature K	γ, erg/cm^2	Critical value of x	
			Observed	Calculated
Water	275		4.2	4.2
	264	77.0	4.85	4.85
	261		5.0	5.1
Methanol	270	24.8	3.0, 3.2	1.8
Ethanol	273	24.0	2.3	2.3
n-Propanol	270	25.4	3.0	3.2
Isopropyl alcohol	265	23.1	2.8	2.9
n-Butyl alcohol	270	26.1	4.6	4.5
Nitromethane	252	40.6	6.0	6.2
Ethyl acetate	242	30.6	8.6–12	10.0
Dibutyl phthalate	332	29.4	26–29[a]	
Triethylene glycol	324	42.8	27–37[a]	

[a] Values of interfacial tension of nucleus from turbulent jet measurements, by various equations (19).

between theory and experiment is quite good. As noted by LaMer and Pound (20), this implies that use of the macroscopic surface tension is valid even for nuclei, that is, even for clusters only about 10 Å in radius. This contradicts Tolman's equation, which gives the effect of drop size on surface tension. As pointed out in Section III-1D, Tolman's treatment assumes a continuous medium and therefore is not exact for very small drops; this is also true of the Becker and Doring treatment insofar as it uses Eq. IX-5. Nontheless, it may well be true that the specific surface free energy changes but little with size down to rather small clusters. Benson and Shuttleworth (21) found that even for a crystallite containing *13* molecules, E^s was only 15% less than the value for a plane surface. However, the reliability of such calculations is difficult to assess; Walton (22), with more allowance for distortion, found a 35% increase in the surface energy of a KCl crystal four ions on the edge. It is this assumption of the essential identity of nuclear and macroscopic interfacial tensions that provides the basis for calculating solid interfacial tensions from nucleation data. Condensation from a pure supersaturated vapor (i.e., with no inert gas present) does not agree well with theory (23), perhaps because of lack of thermal equilibrium.

Turning to the situations involving solids, the supercooling of liquids has generally been carried out on small drops because it is very difficult to eliminate impurities and effects due to foreign surfaces if ordinary volumes are used. A conventional rule is that $\Delta T_{\max}$, the maximum degree of supercooling possible, is about 0.18 of the melting point (3). Some recent studies, however, report that much higher degrees of supercooling can be achieved with the use of metal in oil emulsions. As an example, a ΔT of

84 K or 0.36 of the melting point was achieved with Hg (23a). For nonmetallic liquids, T/T_m is more variable, but generally lies within the range 0.1 to 0.2; for water it is 0.14, corresponding to $\Delta T = 40°$. Staveley and co-workers (24, 25) used Eq. IX-16 to calculate the interfacial tensions for various inorganic and organic substances and concluded that the ratio of the molar surface tension (Eq. III-16) to ΔH_f was nearly constant at about 0.3 but exhibited a tendency toward smaller values for lower T_m's. Mason (26) estimated from the supercooling of water drops that the interfacial tension between ice and water was about 22 erg/cm^2. This compares well with the value of 26 erg/cm^2 from the melting point change in a capillary (Section X-2) and with Gurney's theoretical estimate of 10 erg/cm^2 (27). Turnbull and Cormia (28) give values of 7.2, 9.6, and 8.3 erg/cm^2 for the solid–liquid interfacial tensions of *n*-heptadecane, *n*-octadecane, and *n*-tetracosane, respectively.

An interesting technique is that of the "heat development" of nuclei. The liquid is held at the desired temperature for a prescribed time, and nuclei gradually accumulate; they are then made visible as crystallites by quickly warming the solution to a temperature just below T_0, under which circumstances no new nuclei form but existing ones grow rapidly.

Solutions. Supersaturation phenomena in solutions are, of course, very important, but, unfortunately, data in this field tend to be not too reliable. Not only is it difficult to avoid accidental nucleation by impurities, but also solutions, as well as pure liquids, can exhibit memory effects whereby the attainable supersaturation depends on the past history, especially the thermal history, of the solution. In the case of slightly soluble salts, where precipitation is brought about by the mixing of reagents, it is difficult to know the effective degree of supersaturation.

If x is replaced by S/S_0, then Eq. IX-16 takes on the form

$$\ln t = k_1 - \frac{k_2\gamma^3}{T^3(\ln S/S_0)^2} \tag{IX-20}$$

and Stauff (29) found fair agreement with this equation in the case of potassium chlorate solutions. Gindt and Kern (30) took (N/t), the average rate of nucleation, as proportional to I, where N is the number of crystals produced, and from the slope of plots of log (N/t) versus $1/(\ln S/S_0)^2$ they deduced the rather low solid–solution interfacial tension value of 2 to 3 erg/cm^2 for KCl and other alkali halides. Alternatively, one may make estimates from the critical supersaturation ratio, that is, the value of S/S_0 extrapolated to infinite rate of precipitation or zero induction time. Enüstün and Turkevich (31) found a value of 7.5 for $SrSO_4$ and from their solubility estimated interfacial tension (see Section X-2) deduced a critical nucleus of size of about 18 Å radius.

4. Crystal Growth

The visible crystals that develop during a crystallization procedure are built up as a result of growth either on nuclei of the material itself or surfaces of foreign material serving the same purpose. Neglecting for the moment the matter of impurities, nucleation theory provides an explanation for certain qualitative observations in the case of solutions.

Once nuclei form in a supersaturated solution, they begin to grow by accretion and, as a result, the concentration of the remaining material drops. There is thus a competition for material between the processes of nucleation and of crystal growth. The more rapid the nucleation, the larger the number of nuclei formed before relief of the supersaturation occurs and the smaller the final crystal size. This, qualitatively, is the basis of what is known as von Weimarn's law (32):

$$\frac{1}{d} = \frac{kS}{S_0} \tag{IX-21}$$

where d is some measure of the particle size. Although essentially empirical, the law appears to hold approximately.

A beautiful illustration of a regulated balance between rates of nucleation and of crystal growth is provided by the investigations of LaMer and co-workers on monodisperse sols (33). It is possible to produce accurately monodisperse sulfur sols through the acid decomposition of $S_2O_3^{-2}$. The decomposition is slow, and dissolved sulfur builds up slowly to the critical nucleation point; nucleation then develops and proceeds over a short time interval until the sulfur concentration drops below the critical value. From this stage on, the new sulfur produced by the decomposition of the thiosulfate is taken up by the growth of the sulfur nuclei, and the rate constant for crystal growth is large enough so that the sulfur concentration never again reaches the nucleation level. By this means a single crop of nuclei is produced, which then grows uniformly to give a monodisperse sol.

The mechanism of crystal growth has been a topic of considerable interest. In the case of a perfect crystal, the starting of a new layer involves a kind of nucleation since the first few atoms added must occupy energy-rich positions. Becker and Doring (5), in fact, have treated crystal growth in terms of such surface nucleation processes. A more recent treatment is due to Sholl and Fletcher (34). Dislocations may also serve as surface nucleation sites, and, in particular, Frank (35) suggested that crystal growth might occur at the step of a screw dislocation (see Section VII-4C) so that the surface would advance in spiral form. However, while crystallization phenomena now occupy a rather large section of the literature, it is still by no means clear what mechanisms of crystal growth predominate. Buckley (36) comments that spiral patterns are somewhat uncommon and, moreover, occur on well-developed and hence *slowly* growing faces. Some interferometric studies of the concentration gradients around a growing crystal (37, 38) showed that, depending on the crystal, the maximum gradient may occur either near the center of a face or near the edges, and the pattern of fringes around a given crystal may change considerably from time to time and without necessarily any direct correlation with local growth rates. Clearly, the possibility of surface deposition at one point, followed by surface migration to a final site, must be considered. On the other hand, the Frank mechanism is widely accepted, and in individual cases it has been possible to observe the slow turning of a spiral pattern as crystal growth occurred (39).

The nucleation of a system by means of foreign bodies is, of course, a well-known phenomenon. Most chemists are acquainted with the practice of scratching the side of a beaker to induce crystallization. Of special interest, in connection with artificial rainmaking, is the nucleation of ice. Silver iodide, whose crystal structure is the same as ice and whose cell dimensions are very close to it, will nucleate water below −4°C (40). Other agents have been found. A fluorophlogopite (a mica) does somewhat better than AgI (41) (see also Ref. 42).

An interesting application of the nucleation phenomenon has been to the detection of traces of atmospheric SO_2. Photooxidation in the nucleation chamber produces SO_3, a powerful nucleating agent for water (43; see also Ref. 44). Nucleation may occur on a liquid or liquid solution surface, such as a CCl_4 solution of symmetrical trichlorobenzene (45). Fletcher (46) comments that the free energy barrier to the growth of an ice cluster on the surface of a foreign particle should be minimized if the particle–ice–water contact angle is small, which implies that the surface should be hydrophobic to water. The whole subject is, of course, of great interest in artificial rainmaking.

As a final comment, there are situations where it is desirable not to have nucleation, that is, to form glasses. A mixed solvent system will often give glasses on cooling, and the general criterion seems to be that the components strongly mutually solvate in the liquid state (e.g., water and glycerine) but do not form mixed crystals.

5. Problems

1. Calculate ΔG of Eq. IX-4 for the case of *n*-octane and plot against r over the range r = 5Å to 300 Å and for the two cases of x = 3 and x = 300. Assume 20°C.

2. Assuming that for water ΔG_D is 7 kcal/mole, calculate the rate of nucleation for ice nuclei for several temperatures and locate the temperature of maximum rate. Discuss in terms of this result why glassy water might be difficult to obtain.

3. Calculate what the critical supersaturation ratio should be for water if the frequency factor in Eq. IX-11 were indeed too low by a factor of 10^{20}. Alternatively, taking the observed value of the critical supersaturation ratio as 4.2, what value for the surface tension of water would the "corrected" theory give?

4. Verify Fig. IX-1.

General References

R. S. Bradley, "Nucleation in Phase Changes," *Q. Rev. (London)*, **5**, 315 (1951).
H. E. Buckley, *Crystal Growth*, H. Publisher, New York, 1951.
P. P. Ewald and H. Juretschke, *Structure and Properties of Solid Surfaces*, University of Chicago Press, Chicago, 1953.
W. E. Garner, *Chemistry of the Solid State*, Academic, New York, 1955.
K. Nishioka and G. M. Pound, Surface and Colloid Science, E. Matijevic, Ed., Wiley, New York, 1976.
Nucleation, A. C. Zettlemoyer Ed., Marcel Dekker, New York, 1969.
The Physics of Rainclouds, Cambridge University Press, Cambridge, England, 1962.

Textual References

1. W. J. Dunning, *Nucleation*, A. C. Zettlemoyer Ed., Marcel Dekker, New York, 1969.

2. J. A. Christiansen and A. E. Nielsen, *Acta Chem. Scand.*, **5,** 673 (1951).
3. R. S. Bradley, *Q. Rev. (London)*, **5,** 315 (1951).
4. J. W. Gibbs, *Collected Works of J. W. Gibbs*, Longmans, Green, New York, 1931, p. 322.
5. R. Becker and W. Doring, *Ann. Phys.*, **24,** 719 (1935).
6. See M. Volmer, *Kinetik der Phasenbildung*, Edwards Brothers, Ann Arbor, Michigan, 1945.
7. J. Frenkel, *Kinetic Theory of Liquids*, The Clarendon Press, Oxford, 1946.
8. D. Turnbull and J. C. Fisher, *J. Chem. Phys.*, **17,** 71 (1949).
8a. R. C. Reid, *Am. Sci.*, **64,** 146, (1976).
9. J. W. Cahn and J. E. Hilliard, *J. Chem. Phys.*, **31,** 688 (1959).
10. See I. W. Plesner, *J. Chem. Phys.*, **40,** 1510 (1964).
11. R. A. Oriani and B. E. Sundquist, *J. Chem. Phys.*, **38,** 2082 (1963).
12. J. Lothe and G. M. Pound, *J. Chem. Phys.*, **36,** 2080 (1962).
13. B. V. Deryaguin, *J. Colloid Interface Sci.*, **38,** 517 (1972).
14. H. Reiss, Nucleation II, A. C. Zettlemoyer, Ed., Marcel Dekker, New York, 1976.
15a. R. McGraw and H. Reiss, *J. Stat. Phys.*, **20,** 385 (1979).
15b. H. Reiss and M. Shugard, *J. Chem. Phys.*, **65,** 5280 (1976).
16a. A. W. Castleman, Jr., P. M. Holland, and R. G. Keese, *J. Chem. Phys.*, **68,** 1760 (1978).
16b. A. I. Rusanov, *J. Colloid Interface Sci.*, **68,** 32 (1979).
17a. N. Lee, R. G. Keese, and A. W. Castleman, Jr., *J. Colloid Interface Sci.*, **75,** 555 (1980).
17b. M. Volmer and H. Flood, *Z. Phys. Chem.*, **A170,** 273 (1934).
18. A. Sander and G. Damköhler, *Naturwiss.*, **31,** 460 (1943).
19. W. L. Higuchi and C. T. O'Konski, *J. Colloid Sci.*, **15,** 14 (1960).
20. V. K. LaMer and G. M. Pound, *J. Chem. Phys.*, **17,** 1337 (1949).
21. G. C. Benson and R. Shuttleworth, *J. Chem. Phys.*, **19,** 130 (1951).
22. A. G. Walton, *J. Chem. Phys.*, **36,** 3162 (1963).
23. B. Barschdorff, W. J. Dunning, P. P. Wegener, and B. J. C. Wu, *Nat. Phys. Sci.*, **240,** 166 (1972).
23a. J. H. Perepezko and D. H. Rasmussen, *Metall. Trans. A*, **9A,** 1490 (1978) (and later papers).
24. H. J. de Nordwall and L. A. K. Staveley, *J. Chem. Soc.*, **1954,** 224.
25. D. G. Thomas and L. A. K. Staveley, *J. Chem. Soc.*, **1952,** 4569.
26. B. J. Mason, *Proc. Roy. Soc. (London)*, **A215,** 65 (1952).
27. R. Shuttleworth, *Proc. Phys. Soc. (London)*, **63A,** 444 (1950).
28. D. Turnbull and R. L. Cormia, *J. Chem. Phys.*, **34,** 820 (1961).
29. J. Stauff, *Z. Phys. Chem.*, **A187,** 107, 119 (1940).
30. R. Gindt and R. Kern, *CR*, **256,** 4186 (1963).
31. B. V. Enüstün and J. Turkevich, *J. Am. Chem. Soc.*, **82,** 4502 (1960).
32. P. P. von Weimarn, *Chem. Rev.*, **2,** 217 (1925).
33. V. K. LaMer and R. H. Dinegar, *J. Am. Chem. Soc.*, **72,** 4847 (1950).
34. C. A. Sholl and N. H. Fletcher, *Acta Metall.*, **18,** 1083 (1970).
35. F. C. Frank, *Discuss. Faraday Soc.*, **5,** 48 (1949).
36. H. E. Buckley, in *Structure and Properties of Solid Surfaces*, R. Gomer and C. S. Smith, Eds., University of Chicago Press, Chicago, 1952, p. 271.
37. G. C. Krueger and C. W. Miller, *J. Chem. Phys.*, **21,** 2018 (1953).

38. S. P. Goldsztaub and R. Kern, *Acta Cryst.*, **6,** 842 (1953).
39. W. J. Dunning, private communication.
40. B. Vonnegut, *J. Appl. Phys.*, **18,** 593 (1947).
41. J. H. Shen, K. Klier, and A. C. Zettlemoyer, *J. Atmos. Sci.*, **34,** 957 (1977).
42. V. A. Garten and R. B. Head, *Nature*, **205,** 160 (1965).
43. D. C. Marvin and H. Reiss, *J. Chem. Phys.*, **69,** 1897 (1978).
44. A. W. Gertler, B. Almeida, M. A. El-Sayed, and H. Reiss, *Chem. Phys.*, **42,** 429 (1979).
45. J. Rosinski, *J. Phys. Chem.*, **84,** 1829 (1980).
46. N. H. Fletcher, private communication.

CHAPTER X

The Solid–Liquid Interface—Contact Angle

1. Introduction

The study of the solid–liquid interface is certainly no easier than that of the solid–gas interface. All of the complexities noted in Section VII-4 are present. Neither, generally, can the surface structural and spectroscopic methods of Chapter VIII be applied since the liquid phase now present prevents the use of electron or molecular beams; note, however, Ref. 1. Moreover, the solid–liquid interface is of central importance in a host of applied situations, and its study has inevitably led surface chemists to a far greater *variety* of systems than involved in Chapters VII and VIII. "Solids" now include polymers, for example, and all manner of liquids must be considered.

There is, perforce, some retreat to phenomenology and much emphasis on empirical rules and semiempirical models. We tend to treat the solid as a uniform medium; at best, heterogeneity and the variable surface restructuring that different liquids might induce are handled by means of some modifying parameter that averages the complexities. Also, differences or changes in quantities take practical dominance over the now usually unattainable absolute values.

Sections X-2 and X-3 take up methods for estimating solid–liquid interfacial quantities, which logically belong in this chapter. The main emphasis, however, is on contact angle measurements and their interpretation. The measurements are relatively easy, the phenomenon is of great practical importance, and a number of instructive and pleasing (although modelistic) analyses are possible.

2. Surface Free Energies from Solubility Changes

This section represents a continuation of Section VII-5, which dealt primarily with the direct estimation of surface quantities at a solid–gas interface. Although in principle some of the methods described there could be applied at a solid–liquid interface, very little has been done apart from the study of the following Kelvin effect, and nucleation studies, discussed in Chapter IX.

The Kelvin equation may be written

$$\frac{RT \ln (a/a_0)}{a_0} = \frac{2\gamma \bar{V}}{r} \tag{X-1}$$

in the case of a crystal, where γ is the interfacial tension of a given face and r is the radius of the inscribed sphere (see Section VII-2C); a denotes the activity, as measured, for example, by the solubility of the solid. Since equilibrium of the whole crystal is implied, we assume interfacial tension and free energy to be equal.

In principle, then, small crystals should show a higher solubility in a given solvent than should large ones. A corollary is that a mass of small crystals should eventually recrystallize to a single crystal. This is the basis for the coarsening of fine precipitates by digestion. Strictly speaking, Eq. X-1 should be applied only to the case of a single crystal in its saturated solution, but Enüstün and Turkevich (2) give evidence that in the case of a mixed mass of crystals it is the finest particle size group that determines the solubility.

In the case of a sparingly soluble salt that dissociates into ν^+ positive ions M and ν^- negative ions A, the solubility S is given by

$$S = \frac{(M)}{\nu^+} = \frac{(A)}{\nu^-} \tag{X-2}$$

and the activity of the solute is given by

$$a = (M)^{\nu+}(A)^{\nu-} = S^{(\nu^+ + \nu^-)}(\nu^+)^{\nu+}(\nu^-)^{\nu-} \tag{X-3}$$

if activity coefficients are neglected. On substituting into Eq. X-1,

$$\frac{RT(\nu^+ + \nu^-)\ln(S/S_0)}{S_0} = \frac{2\gamma\bar{V}}{r} \tag{X-4}$$

Most studies of the Kelvin effect have been made with salts, and with rather variable and contradictory results; the reader is referred to the paper by Enüstün and Turkevich (2) for more detail. They report, for example, a value of 85 erg/cm^2 for γ at the $SrSO_4$–water interface, confirmed by a more recent study (3). A value of 171 erg/cm^2 has been reported for NaCl in alcohol (4). Unresolved questions include that of the thermodynamic meaning of a solubility measurement for a nonuniform and nonequilibrium collection of crystals, as well as the role of other factors such as that of the electrical double layer presumably present. Knapp (5; see also 6) gives for this last effect the equation

$$\frac{RT\ln(S/S_0)}{S_0} = \frac{2\gamma\bar{V}}{r} - \frac{q^2\bar{V}}{8\pi D r^4} \tag{X-5}$$

supposing the particle to possess a fixed double layer of charge q. Other potential difficulties are in the estimation of solution activity coefficients and the presence of defects affecting the chemical potential of the small crystals. See Refs. 7 and 8.

An interesting and somewhat related method for estimating the interfacial tension between a solid and its melt is described by Skapski et al. (9). In a capillary, the freezing zone shows a meniscus, hemispherical in the cases described, with the consequence that a freezing point depression occurs, as given by

$$T = T_m - \frac{2T_m\gamma_{SL}}{rq_f\rho_s} \tag{X-6}$$

where T_m is the normal melting point, q_f is the heat of fusion per gram, and ρ_s is the density of the solid. The value reported for the ice–water interface was, for example, 26 erg/cm^2. For comparison, a value of 33 erg/cm^2 was obtained by the grain boundary contact angle method (10), and one of 22 erg/cm^2 from a study of the morphological stability of a cylindrical interface (11).

3. Surface Energy and Free Energy Differences from Immersion, Adsorption, and Engulfment Studies

A. Heat of Immersion

If a clean solid surface is immersed in a liquid, there is generally a liberation of heat, and this heat of immersion may be written

$$q_{\mathrm{imm}} = E_{\mathrm{S}} - E_{\mathrm{SL}} \tag{X-7}$$

where E denotes the total surface energy (see Section III-1A) of the designated interface. Also, one can define an energy of adhesion, analogous to the work of adhesion (Eq. III-47),

$$E_{\mathrm{A(SL)}} = w_{\mathrm{SL}} - T\frac{\partial w_{\mathrm{SL}}}{\partial T} = E_{\mathrm{S}} + E_{\mathrm{L}} - E_{\mathrm{SL}} \tag{X-8}$$

Condensed phases are involved, and no distinction is being made here between enthalpy and energy; the difference is small. See Ref. 12 for a more rigorous treatment.

The heat of immersion may be determined calorimetrically by measuring the heat evolved on immersion of a clean solid or solid powder in the liquid in question, and the experimental technique is described by several authors (13–17) and also in Section XVI-4. Since the values are of the order of a few hundred ergs per square centimeter, it has been necessary to use finely divided solids whose absolute surface area must then be estimated by some independent means, usually gas adsorption.

Some q_{imm} data are given in Table X-1. A polar solid will show a large heat of immersion in a polar liquid and a smaller one in a nonpolar liquid; nonpolar solids, such as Graphon or Teflon, have low heats of immersion with little dependence on the nature of the solid. Chessick and co-workers (22) found that heats of immersion relative to that for water were practically independent of the nature of the solid. Conversely, Zettlemoyer (18) notes that for a given solid, q_{imm} is essentially a linear function of the dipole moment of the wetting liquid.

There are complexities. The wetting of carbon blacks is very dependent on the degree of surface oxidation; Healey et al. (23) found that q_{imm} in water varied with the fraction of hydrophilic sites as determined by water adsorption isotherms. In the case of oxides such as TiO_2 and SiO_2, q_{imm} can vary considerably with pretreatment and with the specific surface area (20). Quartz finely divided by crushing showed about half the heat of immersion per square centimeter than did the coarser material (24). It was suggested that crushing and grinding tend to produce an amorphous surface.

One may obtain the difference between the heat of immersion of a clean surface and one with a preadsorbed film of the same liquid into which

TABLE X-1
Heats of Immersion at 25°C

	q_{imm} erg/cm^2				
Solid	H_2O	C_2H_5OH	n-Butylamine	CCl_4	n-C_6H_{14}
TiO_2 (rutile)	550[a]	400[a]	330[a]	240[b]	135[a]
Al_2O_3	400–600[c]				100[c]
SiO_2	400–600[c]			270[b]	100[c]
$BaSO_4$	490[b]			220[b]	
Graphon	32[a]	110[a]	106[a]		103[a]
Teflon 6	6[a]				47[d]

[a] Ref. 18.
[b] Ref. 19.
[c] Ref. 20 (there was considerable variation in q_{imm}, depending on particle size and outgassing conditions).
[d] Ref. 21.

immersion is carried out. This difference can now be related to the heat of adsorption of the film, and this aspect of immersion data is discussed further in Section XVI-4.

B. *Surface Energy and Free Energy Changes from Adsorption Studies*

It turns out to be considerably easier to obtain fairly precise measurements of a change in the surface free energy of a solid than it is to get an absolute experimental value. The procedures and methods may now be clear-cut, and the calculation has a thermodynamic basis, but there remain some questions about the physical meaning of the change. This point is discussed further in the following material and in Section X-6.

There is always some degree of adsorption of a gas or vapor at the solid–gas interface; for vapors at pressures approaching the saturation pressure, the amount of adsorption can be quite large and may approach or exceed the point of monolayer formation. This type of adsorption, that of vapors near their saturation pressure, is called *physical adsorption*; the forces responsible for it are similar in nature to those acting in condensation processes in general and may be somewhat loosely termed ''van der Waals'' forces, discussed in Chapter VI. The very large volume of literature associated with this subject is covered in some detail in Chapter XVI.

The present discussion is restricted to an introductory demonstration of how, in principle, adsorption data may be employed to determine changes in the solid–gas interfacial free energy. A typical adsorption isotherm (of the physical adsorption type) is shown in Fig. X-1. In this figure, the amount adsorbed v, expressed as cubic centimeters at standard pressure and temperature of gas adsorbed per gram of solid, is plotted against P/P^0, where P is the actual pressure of the gas and P^0 is its saturation pressure (i.e., the

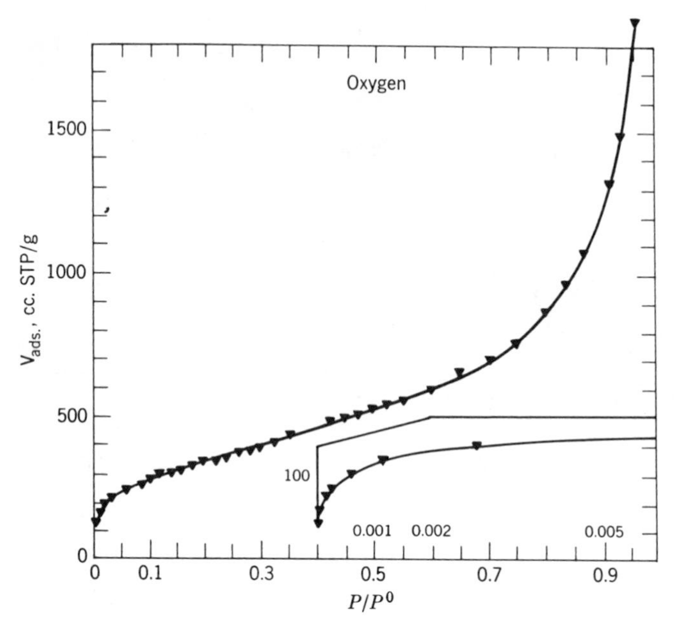

Fig. X-1. Adsorption of oxygen on titanium dioxide at 78.2°K. (From Ref. 25.)

vapor pressure of pure liquid adsorbate). Since the relationship is for a particular temperature, the plot is referred to as an *adsorption isotherm*.

The surface excess per square centimeter Γ is just n/Σ where n is the moles adsorbed per gram and Σ is the specific surface area. By means of the Gibbs equation (III-80), one can write the relationship

$$d\gamma = -\Gamma RT\, d \ln a \tag{X-9}$$

For an ideal gas, the activity can be replaced by the pressure, and

$$d\gamma = -\Gamma RT\, d \ln P \tag{X-10}$$

or

$$\pi = -\int d\gamma = RT \int \Gamma d \ln P \tag{X-11}$$

where π is the film pressure. It then follows that

$$\pi = \gamma_S - \gamma_{SV} = \frac{RT}{\Sigma} \int_0^P n\, d \ln P \tag{X-12}$$

$$\pi^0 = \gamma_S - \gamma_{SV^0} = \frac{RT}{\Sigma} \int_0^{P^0} n\, d \ln P \tag{X-13}$$

where n is the moles adsorbed.

Equations X-12 and X-13 thus provide a thermodynamic evaluation of the change in interfacial free energy accompanying adsorption. As discussed further in Section X-5C, typical values of π for adsorbed films on solids range up to 100 erg/cm^2.

A somewhat subtle point of difficulty is the following. Adsorption isotherms are quite often entirely reversible in that adsorption and desorption curves are identical. On the other hand, the solid will not generally be an equilibrium crystal and, in fact, will often have quite a heterogeneous surface. The quantities γ_S and γ_{SV} are therefore not very well defined as separate quantities. It seems preferable to regard π, which *is* well defined in the case of reversible adsorption, as simply the change in interfacial free energy and to leave its further identification to treatments accepted as modelistic.

The film pressure can be subjected to further thermodynamic manipulation, as discussed in Section XVI-13. Thus

$$\pi - T\frac{d\pi}{dT} = H_S - H_{SV} \qquad \text{(X-14)}$$

where the subscript SV denotes solid having an adsorbed film (in equilibrium with some vapor pressure P of the gaseous adsorbate). Equation X-14 is analogous to Eq. III-8 for a one-component system.

A heat of immersion may refer to the immersion of a clean solid surface, $q_{S,imm}$, or to the immersion of a solid having an adsorbed film on the surface. If the immersion of this last is into liquid adsorbate, we then report $q_{SV,imm}$; if the adsorbed film is in equilibrium with the saturated vapor pressure of the adsorbate (that is, the vapor pressure of the liquid adsorbate P^0), we will write $q_{SV^0,imm}$. It follows from these definitions that

$$q_{S,imm} - q_{SV,imm} = \pi - T\frac{d\pi}{dT} + \Gamma\Delta H_v \qquad \text{(X-15)}$$

where the last term is the number of moles adsorbed (per square centimeter of surface) times the molar enthalpy of vaporization of the liquid adsorbate (see Ref. 12).

This discussion of gas adsorption applies in similar manner to adsorption from solution, and this topic is taken up in more detail in Chapter XI.

C. Engulfment

Engulfment is a term used to describe what can happen when a foreign particle is overtaken by an advancing interface such as that between freezing solid and its melt. The experimental observation is that the particle may "ride" along with the advancing interface, or it may engulfed by the new phase. The relevent algebra for a spherical particle is given in Section XIII-4A. The criterion for engulfment is just that γ_{PS} be less than γ_{PL} where P denotes particle and S and L denote the solid and its melt, respectively. Conversely, if $\gamma_{PS} > \gamma_{PL}$ the particle should be rejected, remaining in the liquid phase. Experiments of this type have been useful in testing semiempirical theories relating solid–liquid, solid–vapor, and liquid–vapor interfacial tensions (26). See also Section X-6B.

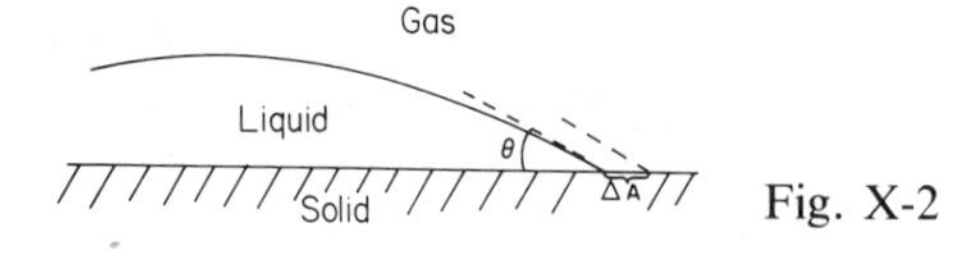

Fig. X-2

4. Contact Angle

A. Young's Equation

It is observed that in most instances a liquid placed on a solid will not wet it but remains as a drop having a definite angle of contact between the liquid and solid phases. The situation, illustrated in Fig. X-2, is similar to that for a lens on a liquid (Section IV-2C), and a simple derivation leads to a very useful relationship.

The change in surface free energy ΔG^s, accompanying a small displacement of the liquid such that the change in area of solid covered $\Delta\mathcal{A}$, is

$$\Delta G^s = \Delta\mathcal{A}(\gamma_{SL} - \gamma_{SV^0}) + \Delta\mathcal{A}\gamma_{LV}\cos(\theta - \Delta\theta) \tag{X-16}$$

At equilibrium

$$\lim_{\Delta\mathcal{A}\to 0} \frac{\Delta G^s}{\Delta\mathcal{A}} = 0$$

and

$$\gamma_{SL} - \gamma_{SV^0} + \gamma_{LV}\cos\theta = 0\dagger \tag{X-17}$$

or

$$\gamma_{LV}\cos\theta = \gamma_{SV^0} - \gamma_{SL} \tag{X-18}$$

Alternatively, in combination with the definition of work of adhesion (Eq. III-47),

$$w_{SLV} = \gamma_{LV}(1 + \cos\theta) \tag{X-19}$$

Equation X-17 was stated in qualitative form by Young in 1805 (27), and current usage is to call it Young's equation; this usage will be accepted here. The equivalent equation, Eq. X-19, was stated in algebraic form by Dupré in 1869 (28), along with the definition of work of adhesion. An alternative designation for both equations, which are really the same, is that of the Young and Dupré equation (see Ref. 29 for an emphatic dissent).

It is important to keep in mind that the phases are mutually in equilibrium. In particular, the designation γ_{SV^0} is a reminder that the solid surface must be in equilibrium with the saturated vapor pressure P^0 and therefore be

† It can be shown that, regardless of the macroscopic geometry of the system, $\Delta\theta/\Delta\mathcal{A}$ behaves as a second order differential and drops out in taking the limit of $\Delta\mathcal{A} \to 0$.

covered with an adsorbed film of film pressure π^0 (see Section X-3B). Thus

$$\gamma_{LV} \cos \theta = \gamma_S - \gamma_{SL} - \pi^0 \quad \text{(X-20)}$$

This distinction between γ_S and γ_{SV^0} seems first to have been made by Bangham and Razouk (30); it was also stressed by Harkins and Livingstone (31). Another quantity, introduced by Bartell and co-workers (32) is the *adhesion tension* **A** which will be defined here as

$$\mathbf{A}_{SLV} = \gamma_{SV^0} - \gamma_{SL} = \gamma_{LV} \cos \theta \quad \text{(X-21)}$$

Both here and in Eq. X-19 the subscript SLV serves as a reminder that the work of adhesion and the adhesion tension involve γ_{SV^0} rather than γ_S.

For practical purposes, if the contact angle is greater than 90°, the liquid is said not to wet the solid—in such a case drops of liquid tend to move about easily on the surface and not to enter capillary pores. On the other hand, a liquid is considered to wet a solid only if the contact angle is zero. It must be understood that this last is a limiting extreme only in a geometric sense. If θ is zero, Eq. X-17 ceases to hold, and the imbalance of surface free energies is now given by a spreading coefficient (see Section IV-2A)

$$S_{L/S(V)} = \gamma_{SV^0} - \gamma_{LV} - \gamma_{SL} \quad \text{(X-22)}$$

Alternatively, the adhesion tension now exceeds γ_{LV} and may be written as K_{SLV}.

The preceding definitions have been directed toward the treatment of the solid–liquid–gas contact angle. It is also quite possible to have a solid–liquid–liquid contact angle where two mutually immiscible liquids are involved. The same relationships apply, only now more care must be taken to specify the extent of mutual saturations. Thus for a solid and liquids A and B, Young's equation becomes

$$\gamma_{AB} \cos \theta_{SAB} = \gamma_{SB(A)} - \gamma_{SA(B)} = \mathbf{A}_{SAB} \quad \text{(X-23)}$$

Here, θ_{SAB} denotes the angle as measured in liquid A, and the phases in parentheses have saturated the immediately preceding phase. A strictly rigorous nomenclature would be yet more complicated, we simply assume that A and B are saturated by the solid and further take it for granted that the two phases at a particular interface are mutually saturated. *If* mutual saturation effects are neglected, then the combination of Eqs. X-23 and X-21 gives

$$\mathbf{A}_{SA} - \mathbf{A}_{SB} = \gamma_{SB} - \gamma_{SA} = \mathbf{A}_{SAB} \quad \text{(X-24)}$$

If in applying Eq. X-24 an adhesion tension corresponding to $\theta = 0$ is obtained, then spreading occurs, and the result should be expressed by K or by a spreading coefficient. Thus adhesion tension is a unifying parameter covering both contact angle and spreading behavior. A formal, rigorous form of Eq. X-24 would be

$$\mathbf{A}_{SAV} - \mathbf{A}_{SBV} = \mathbf{A}_{SAB} + (\pi_{SV(B)} - \pi_{SV(A)}) + (\pi_{SB(A)} - \pi_{SA(B)}) \quad \text{(X-25)}$$

where, for example, $\pi_{SB(A)}$ is the film pressure or interfacial free energy lowering at the solid–liquid B interface due to saturation with respect to A.

There are some subtleties with respect to the physical chemical meaning of the contact angle equation, and these are taken up in Section X-6. The preceding, however, serves to introduce the conventional definitions to permit discussion of the experimental observations.

B. Nonuniform Surfaces

The derivation given for the contact angle equation can be adapted in an empirical manner to the case of a nonuniform solid surface. First, the surface may be *rough* with a coefficient r giving the ratio of actual to apparent or projected area. We now have $\Delta\mathcal{A}_{SL(true)} = r\Delta\mathcal{A}_{SL(apparent)}$ and similarly for $\Delta\mathcal{A}_{SV}$, so that Eq. X-18 becomes

$$\cos\theta_r = r\cos\theta_{true} \tag{X-26}$$

Equation X-26 was given in 1948 by several authors (33–35); a more recent derivation is by Good (36). Note that by this equation, if θ is less than 90°, it is decreased by roughness, while if θ is greater than 90°, it is *increased.*

Alternatively, the surface may be *composite,* that is, consist of small patches of various kinds. *If* it may be assumed that contact angle equilibrium is in practice determined by sufficiently macroscopic fluctuations such that $\Delta\mathcal{A}_{SL}$ and $\Delta\mathcal{A}_{SV}$ effectively average the heterogeneties, then for the case of two kinds of patches occupying fractions f_1 and f_2 of the surface, it follows that

$$\gamma_{LV}\cos\theta_c = f_1(\gamma_{S_1V} - \gamma_{S_1L}) + f_2(\gamma_{S_2V} - \gamma_{S_2L}) \tag{X-27}$$

or

$$\cos\theta_c = f_1\cos\theta_1 + f_2\cos\theta_2 \tag{X-28}$$

An interesting application of Eq. X-28 is to the case of a mesh or screen of material; f_2 is now the fraction of open area, γ_{S_2V} is zero, and γ_{S_2L} is simply γ_{LV}. The relationship then becomes

$$\cos\theta_c = f_1\cos\theta_1 - f_2 \tag{X-29}$$

Wenzel (33), Baxter and Cassie (37), and Dettre and Johnson (38) found that the apparent contact angle for water drops on paraffin metal screens, on textile fabrics, and on embossed polymer surfaces did vary with f_2 in about the manner predicted by Eq. X-29.

The subject is, of course, of direct interest in the field of waterproofing fabrics, and there is a natural example in the structure of feathers. As pointed out by Cassie and Baxter (37), the main stem of a feather carries barbs on either side, and the barbs in turn support fine fibers known as barbules. The barbules on one side of a barb are notched, and those on the other side are hooked so that the two sets engage to form a resilient framework of high porosity. For typical feathers, f_2 is about 0.5, and the apparent contact angle is about 150° (receding) as opposed to a true one of about 100°.

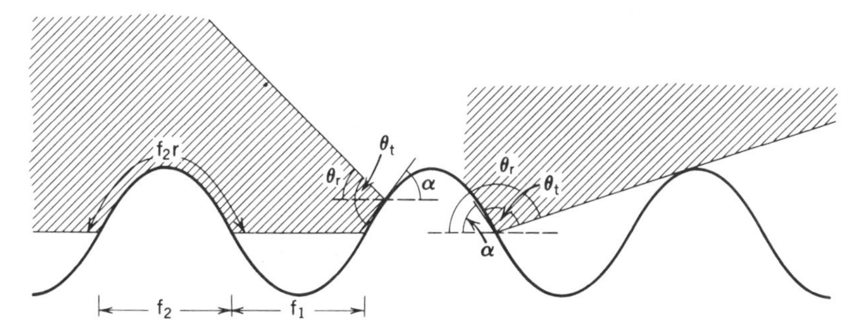

Fig. X-3. Drop edge on a rough surface.

If the contact angle is large and the surface sufficiently rough, the liquid may trap air so as to give a composite surface effect, as illustrated in Fig. X-3. Equation X-29 becomes

$$\cos \theta_{\text{apparent}} = rf_1 \cos \theta_1 - f_2 \tag{X-30}$$

A possible mechanism for such air trapping is suggested in the figure. If it is assumed that the local or true angle θ_t remains invariant as the liquid advances over a roughness asperity, then if θ_t is large, the liquid surface can be so reentrant that it intercepts the next asperity and traps air between the two.

It should be emphasized that the equations of this subsection are quite empirical and modelistic. It is by no means clear, for example, whether for a composite surface it is $\cos \theta$ that is averaged as in Eq. X-28, or θ, or some other function of θ. Roughness is not adequately defined by r, but is also a matter of topology; the same roughness in the form of parallel grooves gives an entirely different behavior than one in the form of pits. These and some other aspects have been discussed by Neumann (39).

5. Experimental Methods and Results of Contact Angle Measurements

A. Measurement of Contact Angle

The various techniques for measuring contact angle have been reviewed in detail by Neumann and Good (40). The most commonly used method is that measuring θ directly for a drop of liquid resting on a flat surface of the solid, as illustrated in Fig. X-4. Zisman and co-workers (41) simply viewed a sessile drop through a comparator microscope fitted with a goniometer scale, thus measuring the angle directly. Leja and Poling (42) photographed sessile or clinging bubbles at a slight angle so that a portion of the bubble was reflected from the surface, the angle of meeting of the direct and reflected images then being twice the contact angle. Ottewill (43) made use of a captive bubble method wherein a bubble formed by manipulation of a micrometer syringe is made to contact the solid surface, as illustrated in Fig. X-5. The contact angle may be measured from photographs of the

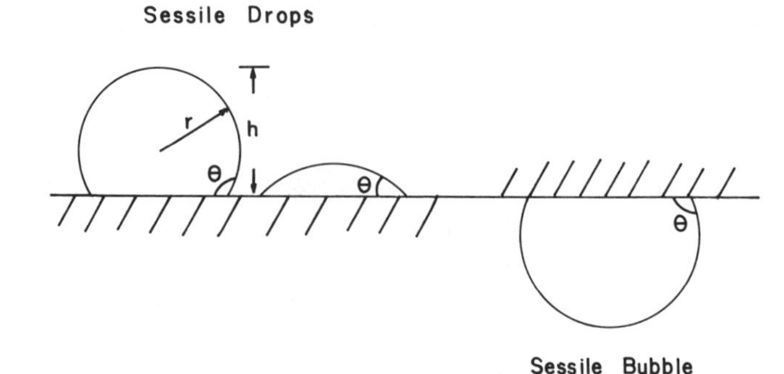

Fig. X-4. Use of sessile drops or bubbles for contact angle determination.

bubble profile, or directly, by means of a goniometer telemicroscope (44). The method has the advantages that it is easy to swell or to shrink the bubble to obtain receding or advancing angles, adventitious contamination is minimized, and there can be no question that the solid–vapor interface is in equilibrium with the saturated vapor pressure of the liquid. There may be difficulty in getting the bubble to adhere if θ is small, however.

Neumann (45) has developed the Wilhelmy slide technique (Section II-8) into a method capable of giving contact angles to 0.1° precision (as compared to the usual 1°). As shown in Fig. X-6 the meniscus at a partially immersed

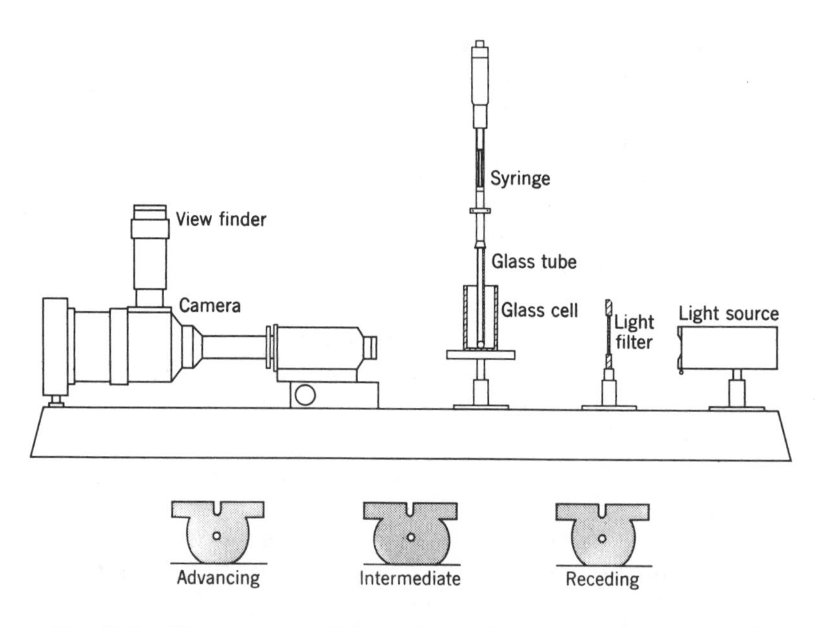

Fig. X-5. The captive bubble method. (Courtesy of R. H. Ottewill.)

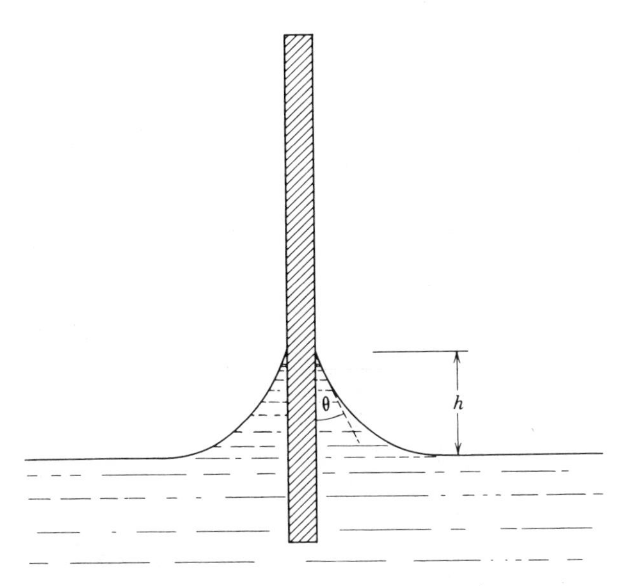

Fig. X-6. Neumann's method for contact angle measurement.

plate rises to a definite height h if θ is finite. The equation is

$$\sin\theta = 1 - \frac{\rho g h^2}{2\gamma} = 1 - \left(\frac{h}{a}\right)^2 \tag{X-31}$$

where a is the capillarity constant (Eq. II-10). The termination of the meniscus is quite sharp under proper illumination (unless θ is small), and h can be measured by means of a traveling microscope. The method is well suited to obtaining the temperature dependence of contact angle; some illustrative data are shown in Fig. X-7.

The Langmuir–Schaeffer method (46) is useful (and underused). The method consists of finding that angle of incident light beam such that the reflected beam from the edge of the meniscus on a vertical plate or a capillary tube exactly returns along the line of the incident beam; see Ref. 47.

A classic method for obtaining accurate results is that of Adam and Jessop (47a), known as the tilting plate method (see also Ref. 48). A several centimeter-wide plate of the solid dips into the liquid, and its tilt is altered until the angle is such that the liquid surface appears to remain perfectly flat right up to the surface of the solid. The method has limited use, however, because both a large, smooth sample of solid and a large volume of liquid are required.

Contact angle may also be obtained indirectly, from the measurements of a sessile drop. Thus, referring to Fig. X-4, if a spherical shape can be assumed (50, 51),

$$\tan\frac{\theta}{2} = \frac{h}{r} \tag{X-32}$$

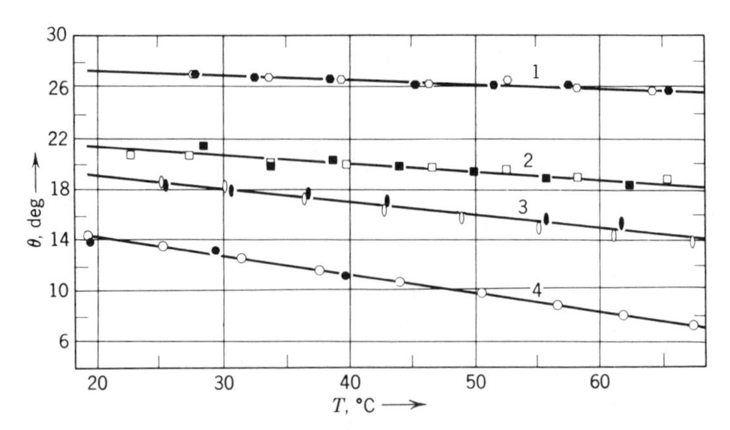

Fig. X-7. Temperature dependence for θ for various *n*-alkanes on a siliconed glass slide. (1) *n*-hexadecane; (2) *n*-tetradecane; (4) *n*-decane. (After Ref. 49.)

An alternative procedure requires the measurement of the diameter of a drop of known volume; see Ref. 52, 53.

The solid for which a contact angle measurement is desired may be available only in finely divided form, and it may not be possible to compact it to a smooth enough surface for one of the preceding methods to be used. An alternative procedure, due to Bartell and co-workers (54), is to compress the material to a porous plug and to measure the capillary pressure toward the liquid in question.

If the porous plug is regarded as equivalent to a bundle of capillaries of average radius r, then from the Laplace equation (Eq. II-7) it follows that

$$\Delta \mathbf{P} = \frac{2\gamma_{LV} \cos \theta}{r} \tag{X-33}$$

where, depending on the value of θ, **ΔP** is the pressure required to force entry of the liquid or to restrain its entry. For a wetting liquid,

$$\Delta \mathbf{P}_0 = \frac{2\gamma_{LV,0}}{r} \tag{X-34}$$

The principle of the method is to obtain the effective capillary radius r by measuring the pressure required to prevent a wetting liquid from entering. The measurement is then repeated with the nonwetting liquid, and by elimination of r from Eqs. X-33 and X-34,

$$\cos \theta = \frac{\Delta \mathbf{P}}{\Delta \mathbf{P}_0} \frac{\gamma_{LV,0}}{\gamma_{LV}} \tag{X-35}$$

B. Hysteresis in Contact Angle Measurements

It is a frequent observation that advancing and receding contact angles may be very different; an everyday example is given by the appearance of a raindrop on a dirty windowpane (note Refs. 54a and 54b, which deal with the case of a drop on an inclined plane). The effect can be quite large; for water on surfaces of minerals the advancing angle may be as much as 50°

larger than the receding one, and for mercury on steel, a difference of 154° has been reported. In situations where it is of critical importance that the contact angle be as small as possible, it is therefore wise to establish a receding angle.

There appear to be three types of causes for hysteresis. First, it seems clear that contamination of either the liquid or the solid surface is apt to give rise to hysteresis. Suppose, for example, that the solid surface initially has an oily contamination. On contact with water, much of the oil would be spread on the water surface, and the solid emerging from the water during a receding angle measurement would have a lower π value or higher γ_{SV} than would fresh solid onto which liquid was advanced. Inspection of Eq. X-18 shows that the effect would be to give a smaller receding than advancing angle. Fowkes and Harkins (48), in their experiments with graphite and talc, found that rigorous cleaning of the liquid and solid surfaces practically eliminated hysteresis. They felt that there was no contact angle hysteresis for a pure liquid on a pure, insoluble, and smooth solid surface.

Second, hysteresis effects are definitely associated with rough surfaces. Mason and co-workers (55) review past observations and report measurements on a variety of well-characterized rough surfaces. Earlier data by Dettre and Johnson (55a), shown in Fig. X-8, is fairly typical. Note that it is only the advancing angle that behaves in the general way predicted by Eq. X-26. However, for systems with $\theta < 90°$, the advancing angle also increased with r, *contrary* to Eq. X-26. The sharp upturn in the receding angle occurred at a point where the surface became composite due to trapped air.

Dettre and Johnson (55a) [see also Good (37)] made calculations on a mathematical model consisting of sinusoidal grooves such as those in Fig. X-3 in cross section, concentric with a drop of spherical shape (i.e., gravity was neglected). A surface free energy minimization, assuming the Young equation to give the local contact angle, led to a drop configuration such that the apparent angle, θ, in Fig. X-3, was that given by Eq. X-26. Moreover, the free energy of the system went through maxima as the constant volume drop was given successively flatter shapes so that the liquid front moved over successive ridges. The actual barrier heights were quite small, but their presence allowed the suggestion that hysteresis was due to there being insufficient macroscopic vibrational energy for the drop to surmount them. More qualitative arguments, but of a similar nature, have been given by Bikerman (52), Shuttleworth and Bailey (36), and Schwartz and Minor (56).

Johnson and Dettre (57) made calculations on a mathematical model of a heterogeneous surface, with conclusions similar to those preceding; the model is further discussed by Neumann and Good (58). Earlier, a more qualitative version of the same analysis was given by Pease (59). As shown in Fig. X-9, Dettre and Johnson (60) found that slides partly coated with TiO_2 and partly with trimethyloctadecylammonium chloride showed considerable hysteresis, and about as expected. The hysteresis with no TiO_2 treatment was attributed to incomplete adsorption of the hydrophobic agent and its presence in patches. Qualitatively, one supposes that the advancing angle is largely determined by the less polar portions of a heterogeneous

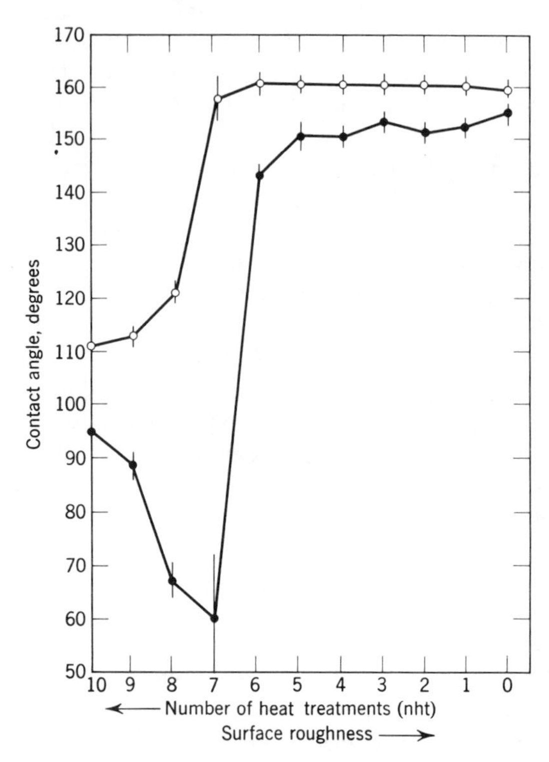

Fig. X-8. Water contact angles on TFE-methanol telomer wax surface as a function of roughness; ○, advancing angles; ●, receding angle. (From Ref. 55a.)

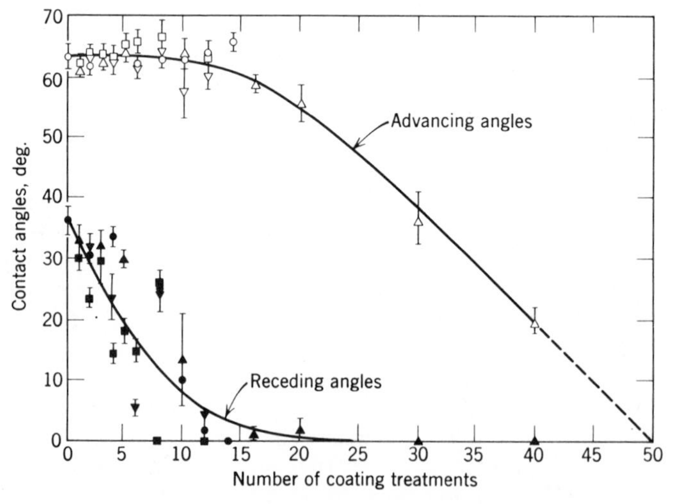

Fig. X-9. Water wettability of titania-coated glass after treatment with trimethyloctadecylammonium chloride solution plotted as a function of the number of coating treatments with 1.1.% polydibutyl titanate. (From Ref. 60.)

surface, and the receding angle, by the more polar portions. As might be expected, hysteresis tends to be small for nearly wetting liquids (note Ref. 61).

It appears to this writer that the concept of a macroscopic free energy barrier probably does provide an adequate explanation for hysteresis if roughness or heterogeneity is on a scale not greatly smaller than that of the drop itself. Surface nonuniformity on a microscopic scale should provide barriers easily surmountable through the ambient vibrations in any laboratory so that hysteresis present in such cases should either be due to impurities, as discussed, or to the third type of cause, discussed in the next paragraph. See Section X-6 for some further discussion.

The third cause of hysteresis that appears to have emerged is surface immobility on a macromolecular scale. As an indication of this point, methylene iodide shows a large hysteresis on agar aquagels (66° versus 30°) (62), unexplainable in terms of surface roughness or heterogeneity in a chemical sense (the agar strands are sheathed with water). However, there would be patches of immobilized surface, and if motion of the three-phase line involves substrate drag (as viscosity measurements on films suggest, see Section IV-3C), there would be a barrier to motion that the increased force component of the nonequilibrium angle would overcome. In the case of a liquid on a solid surface, the corresponding requirement is that the adsorbed film of vapor be mobile. Where the liquid contains a surfactant, low mobility—both in the form of slow desorption from the solid–liquid interface into bulk liquid and hindered spreading past the three-phase line to form an equilibrium composition solid–vapor film—again can cause hysteresis. An example is that of paraffin oil on silica gel, with hexadecanol surfactant, illustrated in Fig. X-10 (63). A somewhat similar thought has been advanced in connection

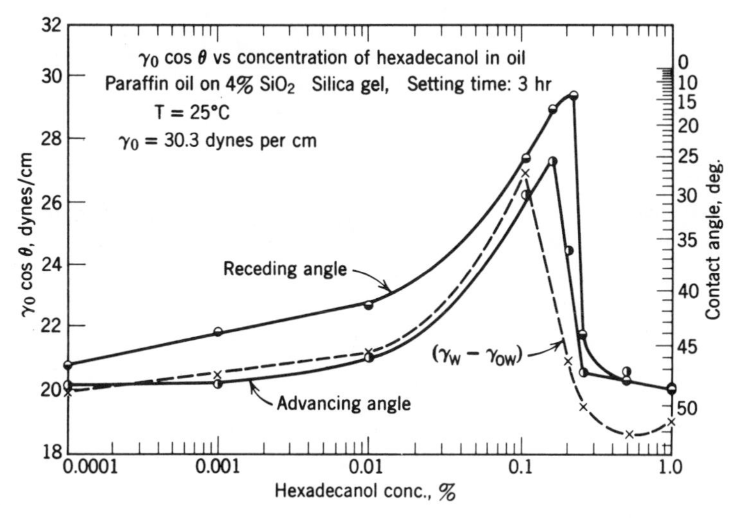

Fig. X-10. A test of the Young equation. γ_{oil} cos θ for paraffin oil on silica gel versus concentration of surfactant—values calculated from the aqueous solution and aqueous solution—oil interfacial tensions. (From Ref. 63.)

with contact angle studies using stretched (and hence anisotropic) polymer films (64). See Refs. 65 and 66 for some additional general comments on hysteresis.

Contact angle may also depend on the *speed* with which the three-phase line is advanced or receded. See Refs. 67 and 68 for some discussion of this and other dynamic effects.

Finally, there have been experimental observations that contact angle varies with the *size* of the liquid drop (69, 70) or, in the captive bubble method, with the size of the bubble (43). One does not expect size *per se,* that is, gravity effects, to influence contact angle and an alternative possible explanation is that linear tension f (Section IV-2C) becomes increasingly important with decreasing size. The relevant modification of Eq. X-18 is (69):

$$\gamma_{LV} \cos\theta = \gamma_{SV^0} - \gamma_{SL} - \frac{f}{r} \qquad \text{(X-36)}$$

where r is the radius of curvature of the three-phase line. Linear tension can in principle be either positive or negative (71), and the reported effects would call for negative f. On the other hand, experimental and theoretical indications are that f should generally be small and positive (71, 71a). Further study is needed.

C. Results of Contact Angle Measurements

It is clear from the foregoing discussion that there is room for a good deal of variability in reported contact angle values. The data collected in Table X-2, therefore, are intended mainly as a guide to the type of behavior to be expected. The older data consist mainly of results for refractory and relatively polar solids, while much of the newer data are for low energy polymer type surfaces.

Bartell and co-workers (72), employing mainly solids such as graphite and stibnite, observed a number of regularities in contact angle behavior but were hampered by the fact that most organic liquids appear to wet such solids. Zisman and co-workers (73) were able to take advantage of the appearance of low surface energy polymers to make extensive contact angle studies with homologous series of organic liquids. Table X-2 includes some data on the temperature dependence of contact angle and some values for π^0, the film pressure of adsorbed vapor at its saturation pressure P^0.

There is appreciable contact angle hysteresis for many of the systems reported in Table X-2; the customary practice of reporting advancing angles has been followed.

As a somewhat anecdotal aside, there has been an interesting question as to whether gold is or is not wet by water, with many publications on either side. This history has been reviewed by Smith (74). The present consensus seems to be that absolutely pure gold is water-wet, and that the reports of nonwetting are a documentation of the ease with which gold surface becomes contaminated (see Ref. 75, but also 76). The detection and control of surface contaminants has been discussed by White (76); see also Gaines (76a).

TABLE X-2

Liquid γ, erg/cm^2	Solid	θ, deg	$d\theta/dT$, deg/°K	π^0, erg/cm^2	Reference
Advancing contact angle, 20–25°C					
Mercury (484)	PTFE[a]	150			73
	Glass	128–148			47, 79, 80
Water (72)	*n*-H[b]	111			77
	Paraffin	110			81
	PTFE[a]	112			81
		108			73
		98		8.8	82
	FEP[c]	108	−0.05		83
	Polypropylene	108	−0.02		84
	Polyethylene	103	−0.01		85
		96	−0.11		83
		94			86
		93			87
		88		14	88
	Human skin	90			89
		75[d]			90
	Naphthalene	88[e]	−0.13		91
	Stibnite (Sb_2S_3)	84			48
	Graphite	86		19	92
				59	93
	Graphon	82			86
	Pyrolytic carbon	72		228	88
	Stearic acid[f]	80		98	88
	Gold	0			74
	Platinum	40			86
	Silver iodide	17			94
	Glass	Small		ca. 20[g]	95
CH_2I_2 (67, 50.8[h])	PTFE	85, 88			90, 109
	Paraffin	61			81
		60			90
	Talc	53			50
	Polyethylene	46, 51.9			90, 109
		40[e]			87
Formamide (58)	FEP[c]	92	−0.06		83
	Polyethylene	75	−0.01		83
CS_2 (ca. 35[h])	Ice[i]	35	0.35		45
Benzene (28)	PTFE[a]	46			73
	n-H[b]	42			77
	Paraffin	0			96
	Graphite	0			96
n-Propanol (23)	PTFE[a]	43		8.8	91
	Paraffin	22			97
	Polyethylene	7		5	98

TABLE X-2 (*Continued*)

Advancing contact angle, 20–25°C					
Liquid γ, erg/cm^2	Solid	θ, deg	$d\theta/dT$, deg/°K	π^0, erg/cm^2	Reference
n-Decane (23)	PTFE[a]	40	−0.11	ca. 1.0	99, 100
		35			101
		32	−0.12		102
n-Octane (21.6)	PTFE[a]	30	−0.12		102
		26		1.8, 3.0	82, 103
		26			101

Solid-liquid-liquid contact angles				
Solid	Liquid 1	Liquid 2	Contact angle, θ_{s12}	Reference
Stibnite	Water	Benzene	130	72
Aluminum oxide	Water	Benzene	22	72
PTFE[a]	Water	*n*-Decane	ca. 180	100
	Benzyl alcohol	Water	30	104
PE[b]	Water	*n*-Decane	ca. 180	100
	Paraffin oil	Water	30	104
Mercury	Water	Benzene	ca. 100[d]	105
Glass	Mercury	Gallium	ca. 0	106

[a] Polytetrafluoroethylene (teflon).
[b] *n*-Hexatriacontane.
[c] Polytetrafluoroethylene-co-hexafluoropropylene.
[d] Not cleaned of natural oils.
[e] Single crystal.
[f] Langmuir–Blodgett film deposited on copper.
[g] From graphical integration of the data of Ref. 95.
[h] See Ref. 109.
[i] At approximately −10°C.

A major contribution to the rational organization of contact angle data was made by Zisman and co-workers. They observed that cos θ (advancing angle) is usually a monotonic function of γ_L for a homologous series of liquids. The proposed function was

$$\cos \theta_{SLV} = a - b\gamma_L = 1 - \beta(\gamma_L - \gamma_c) \quad \text{(X-37)}$$

Figure X-11 shows plots of cos θ versus γ_L for various series of liquids on Teflon (polytetrafluoroethylene) (73). Each line extrapolates to zero θ at a certain γ_L value, which Zisman has called the critical surface tension γ_c; since various series extrapolated to about the same value, he proposed that

γ_c was a quantity characteristic of a given solid. For Teflon, the representative γ_c was taken to be about 18 and was regarded as characteristic of a surface consisting of —CF_2— groups.

The critical surface tension concept has provided a useful means of summarizing wetting behavior and allowing predictions of an interpolative nature. A schematic summary of γ_c values is given in Fig. X-12 (77). In addition, actual contact angles for various systems can be estimated since β in Eq. X-37 usually has a value of about 0.03 to 0.04.

The effect of temperature on contact angle is not usually very great, as a practical observation. Some values of $d\theta/dT$ are included in Table X-2; a common figure is about -0.1 deg/K (but note the case of CS_2 on ice; also rather large temperature changes may occur in L_1–L_2–S systems (see Ref. 78).

Knowledge of the temperature coefficient of θ provides a means of calculating the heat of immersion. Differentiation of Eq. X-18 yields

$$q_{\text{imm}} = E_{\text{SV}} - E_{\text{SL}} = E_{\text{L}} \cos\theta - \frac{T\gamma_{\text{L}}\, d\cos\theta}{dT} \tag{X-38}$$

Since both sides of Eq. X-38 can be determined experimentally, from heat of immersion measurements on the one hand and contact angle data on the other hand, a test of the thermodynamic status of Young's equation is possible. A comparison

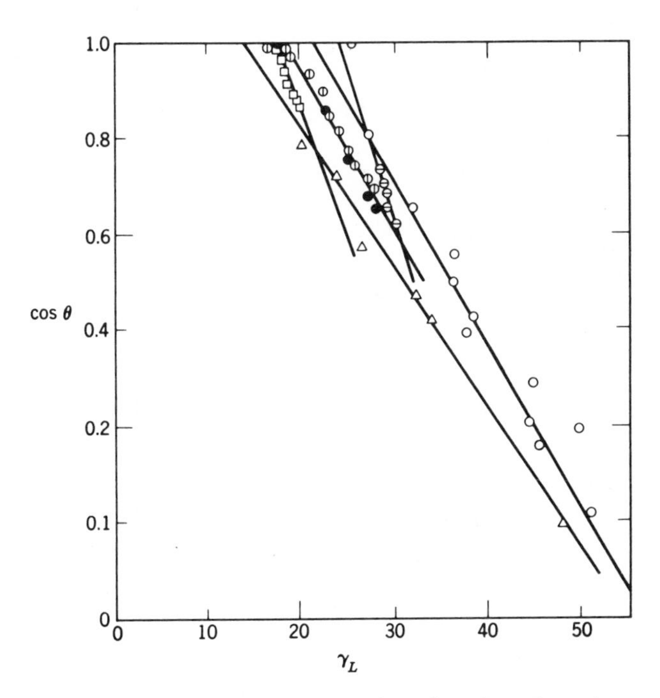

Fig. X-11. Zisman plots of the contact angles of various homologous series on Teflon: ○, RX; ⊖, alkylbenzens; ⦶, *n*-alkanes; ●, dialkyl ethers; □, siloxanes; △, miscellaneous polar liquids. (Data from Ref. 73.)

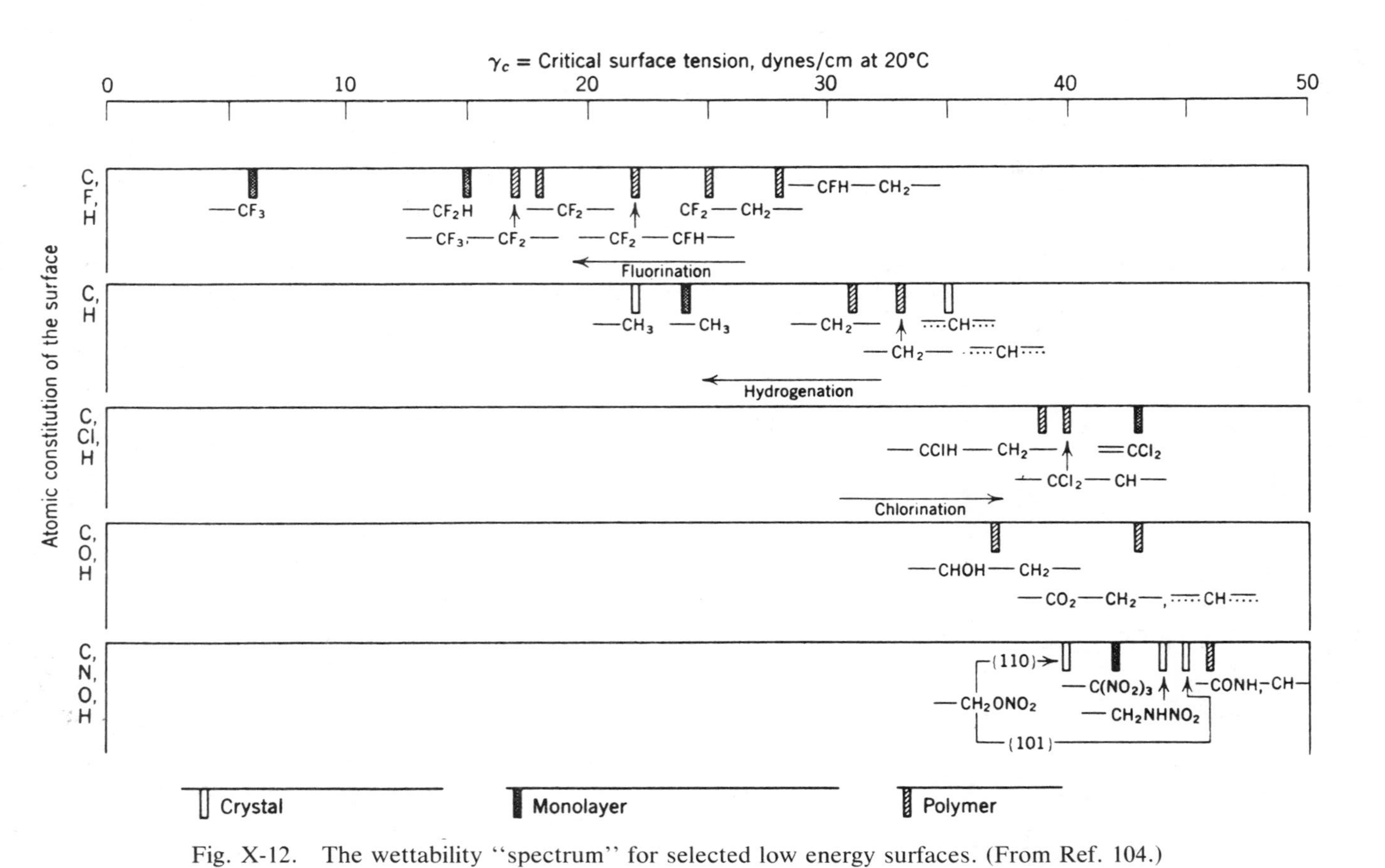

Fig. X-12. The wettability "spectrum" for selected low energy surfaces. (From Ref. 104.)

of calorimetric data for *n*-alkanes (21) with contact angle data (40) is shown in Fig. X-13. The agreement is certainly encouraging.

Values for π^0, the film pressure of the adsorbed film of the vapor (of the liquid whose contact angle is measured) are scarce. Vapor phase adsorption data, required by Eq. X-13, cannot be obtained in this case by the usual volumetric method (see Chapter XVI). This method requires the use of a powdered sample (to have sufficient adsorption to measure), and interparticle capillary condensation grossly distorts the isotherms at pressures above about 0.9 P^0 (107). Also, of course, contact angle data are generally for a smooth, macroscopic surface of a solid, and there is no assurance that the surface properties of the solid remain the same when it is in powdered form. More useful and more reliable results may be obtained by the gravimetric method (see Chapter XVI) in which the amount adsorbed is determined by a direct weighing procedure. In this case it has been possible to use stacks of 200 to 300 thin sheets of material, thus obtaining sufficient adsorbent surface (see Ref. 108). Probably the most satisfactory method, however, is that using ellipsometry to measure adsorbed film thickness (Section IV-3D); there is no possibility of capillary condensation effects, and the same smooth and macroscopic surface can be used for contact angle measurements. Most of the π^0 data in Table X-2 were obtained by this last procedure; the corresponding contact angles are for the identical surface. In addition to the values in Table X-2, Whalen and Hu quote ones of 5 to 7 erg/cm^2 for *n*-octane, carbon tetrachloride, and benzene on Teflon (108a); Tamai and co-workers (108b) find values of 5 to 11 erg/cm^2 for various hydrocarbons on Teflon.

A recent approach has been to measure a contact angle with and without the presence of vapor of a third component, one insoluble in the contact angle forming liquid (109). The π^0 value is then inferred from any change in contact angle. By this means Fowkes and co-workers obtained a value of 7 erg/cm^2 for cyclohexane on polyethylene, but zero for water and methylene iodide. The discrepancy with the value for water in Table X-2 could be due to differences in the polyethylene samples. It was suggested (109), moreover, that the vapor adsorption method reflects polar sites while contact angle is determined primarily by the nonpolar regions. A possibility, however, is that the interpretation of the effect of the third component vapor on contact angle is in error. The method balances the absorbing ability of the third component against that of the contact angle forming liquid and so does not actually give a simple π^0 for a single species.

Contact angle will vary with liquid composition, often in a regular way, as illustrated in Fig. X-14 (note also Ref. 89!). The presence of surfactant solutes may, of course, alter the contact angle drastically. The case of hexadecanol in paraffin oil, using 4% silica gel as the "solid," has already been mentioned (see Fig. X-10). The sharp rise in $\gamma_{oil} \cos \theta$ paralleled very well the dotted line behavior predicted from

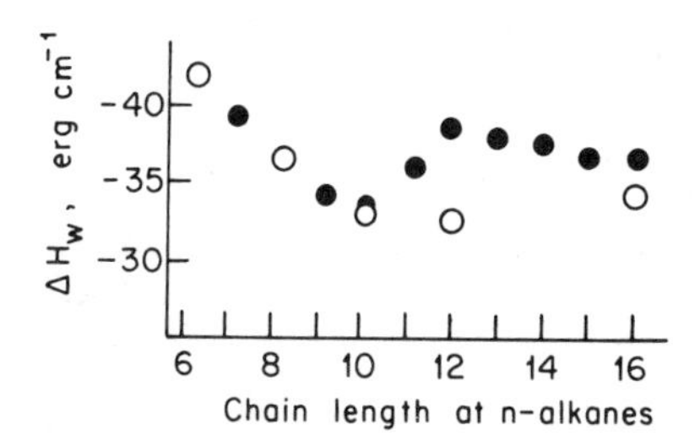

Fig. X-13. Heats of wetting from θ, ●, and calorimetric heats of immersion. ○, of PTFE in *n*-alkanes. (From Ref. 39.)

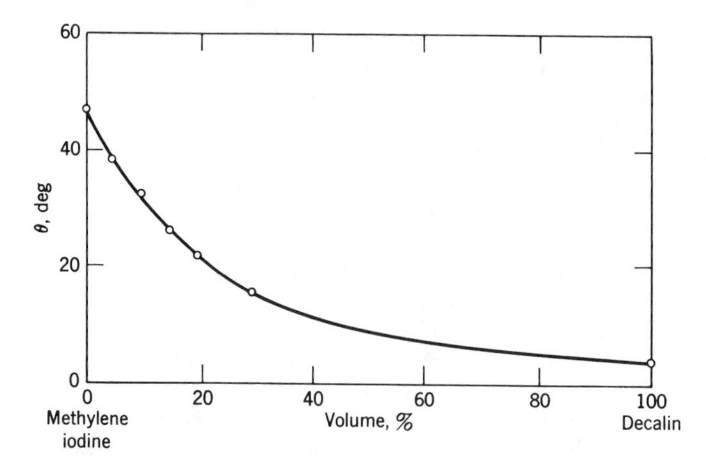

Fig. X-14. Contact angles for methylene iodide–decalin mixtures on polyethylene (advancing angles). (Data from Ref. 111.)

the separate measurements of $\mathbf{A}_{wo} = (\gamma_{water} - \gamma_{oil-water})$. If it is accepted that the silica gel surface was essentially waterlike, apart from being macroscopically rigid, so that $\mathbf{A}_{wo} = \mathbf{A}_{gel-oil}$, the data show very clearly how contact angle is affected by the separate influences of surfactant adsorption at all of the various interfaces. Zisman and co-workers (110) have treated the effect of surfactants on water contact angles largely in terms of Eq. X-37, that is, the reduction in γ_L, but recognized that adsorption at the other two interfaces played an important role; thus the values of $\gamma_L \cos \theta$ were not at all constant, as would be expected if the only action of the surfactant were through the reduction in γ_L. Fowkes and Harkins (48), in fact, used the variation in $\gamma_L \cos \theta$ to calculate film pressures for *n*-butyl alcohol adsorption at the aqueous solution–solid interface of such solids as paraffin, graphite, and talc. The implicit and undoubtedly incorrect assumption, however, was that no change occurred at the solid–vapor interface. Ruch and Bartell (112), studying the aqueous decylamine–platinum system, made direct estimates of the adsorption at the platinum–solution interface, and from this and the contact angle data, application of the Young equation indicated that up to 40 erg/cm^2 change in the solid–vapor interfacial free energy occurred due to decylamine adsorption (see Ref. 112). The autophobic systems of Zisman and co-workers (73) (Section IV-15B) also illustrate the potentially dominant role of the nature of the film adsorbed at the solid–gas interface. A complete study of systems of this type would require separate adsorption measurements for the surfactant at all three interfaces; this has not so far been done.

6. Some Theoretical Aspects of Contact Angle Phenomena

A. Thermodynamics of the Young Equation

The extensive use of the Young equation (Eq. X-18) in the preceding section reflects its general acceptance. Curiously, however, the equation has never been verified experimentally (if the work of Michaels and Dean (63), Section X-5B, is discounted on the grounds that a dilute silica gel surface is not that of a rigid solid). The problem, of course, is that surface tensions of solids are not easy to measure. Fowkes and Sawyer (92) claimed a verification in the case of some liquids on a glassy fluorocarbon polymer, but based on the assumption that the surface tension of the glass was the same as that of fluid, less polymerized fractions. However, nucleation studies indicate that the interfacial tension between a solid and its liquid is appreciable, and it is difficult to affirm that this is not the case here. On the other hand, the test illustrated in Fig. X-13 is reasonably successful. Another possible experimental test is that of comparing the observed effect of a solute on contact angle with the effect calculated from separate adsorption studies. See Ref. 113 for a successful partial test of this type.

Bikerman (114) has criticized the derivation of Eq. X-18 out of concern for the ignored vertical component of γ_L given by $\gamma_L \sin \theta$. This component is real; on soft surfaces a circular ridge is raised at the periphery of a drop (e.g., see Ref. 39). On harder solids, there is no visible effect, but the stress is there. The thickness of the three-phase line at which this force is located is hard to estimate, but if it is of molecular dimension, this vertical component could produce local stresses approaching the yield pressures of even very rigid solids. It has been suggested that contact angle is determined by the balance of surface stresses rather than by one of surface free energies, the two not necessarily being the same for a nonequilibrium solid (see Section VII-2).

It is in fact possible to consider at least *three* different contact angles for a given system! Let us call them θ_m, θ_{th}, and θ_{app}. The first, θ_m, is the microscopic angle between the liquid and the ridge of solid; it is the angle determined by the balance of surface stresses taking into account local deformations. The second θ_{th}, is the thermodynamic angle; this is the angle obtained in the derivation of Eq. X-18 and in general is the angle obtained in a free energy minimization for the overall system. Finally, if the surface is rough or heterogeneous the line of three-phase junction will be scalloped, and one may see experimentally an apparent angle θ_{app}, which is some kind of average. Equation X-28 averages $\cos \theta$, for example.

Returning to θ_{th}, there is no dearth of thermodynamic proofs that the Young equation represents a condition of thermodynamic equilibrium at a three-phase boundary involving a smooth, homogeneous, and incompressible solid; see, for example, Johnson (115). Overall or system free energy minimizations are more difficult because of the problem of analytical representation of the surface area of a deformed drop; however, the case of an infinite drop is not hard, and the Young equation again results (116). See Ref. 70 for a brief discussion of the thermodynamic status of Young's equation.

A somewhat different point of view is the following. Since γ_{SV} and γ_{SL} always occur as a difference, it is possible that it is this *difference* (the adhesion tension) that is the fundamental parameter. The adsorption isotherm for a vapor on a solid may be of the form shown in Fig. X-1, and the asymptotic approach to infinite adsorption as saturation pressure is approached means that at P^0 the solid is in equilibrium with bulk liquid. As Derjaguin and Zorin (95) note (see also Ref. 116), in a contact angle system, the adsorption isotherm must cross the P^0 line and have an unstable region, as illustrated in Fig. X-15. See Section XVI-12. A Gibbs integration to the first crossing gives

$$\pi^0 = \gamma_S - \gamma_{SV^0} = RT \int_{\Gamma=0}^{\Gamma^0} \Gamma \, d \ln P \tag{X-39}$$

while that to the limiting condition of an infinitely thick and hence duplex film gives

$$I = \gamma_S - \gamma_{SL} - \gamma_L = RT \int_{\Gamma=0}^{\infty} \Gamma \, d \ln P \tag{X-40}$$

From the difference in Eqs. X-39 and X-40, it follows that

$$\gamma_{SV^0} - (\gamma_{SL} + \gamma_L) = RT \int_{\Gamma=\Gamma^0}^{\infty} \Gamma \, d \ln P = \Delta I \tag{X-41}$$

where ΔI is given by the net shaded area in the figure. Also, ΔI is just the spreading coefficient S_{LSV} or,

$$\Delta I = \gamma_L(\cos\theta - 1) \tag{X-42}$$

The integral ΔI, while expressible in terms of surface free energy differences, is defined independently of such individual quantities. A contact angle situation may thus be viewed as a consequence of the ability to coexist of two states: bulk liquid and thin film.

Equation X-42 may, alternatively, be given in terms of a disjoining pressure (Section VI-6) integral (117):

$$\gamma_L \cos\theta = \gamma_L + \int_{x^0}^{\infty} \Pi \, dx \tag{X-43}$$

Here, x denotes film thickness and x^0 is that corresponding to Γ^0.

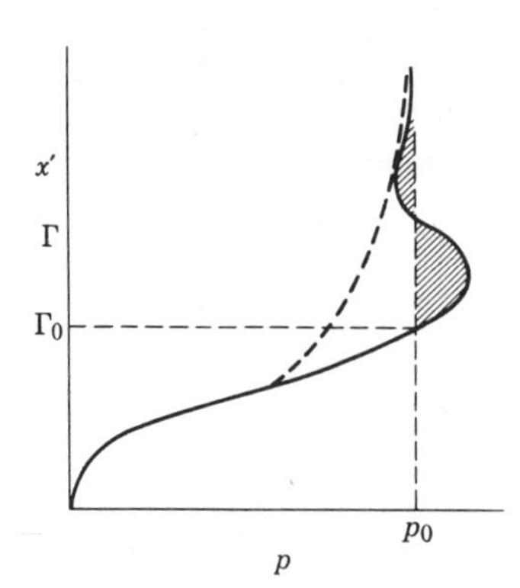

Fig. X-15. Variation of adsorbed film thickness with pressure. (From Ref. 116.)

B. Semiempirical Models—The Girifalco–Good–Fowkes–Young Equation

A model that has proved to be very stimulating to research on contact angle phenomena begins with a proposal by Girifalco and Good (118) (see Section IV-2A). If it is assumed that the two phases are mutually entirely immiscible and interact only through additive dispersion forces whose constants obey a geometric mean law, that is, $C_{1,AB} = (C_{1,AA}C_{1,BB})^{1/2}$ in Eq. VI-22, then the interfacial free energies should obey the equation

$$\gamma_{AB} = \gamma_A + \gamma_B - 2\Phi(\gamma_A\gamma_B)^{1/2} \tag{X-44}$$

The parameter Φ should be unity if molecular diameters also obey a geometric mean law (119) and is often omitted. Equation X-44, if applied to the Young equation with omission of Φ, leads to the relationship (118)

$$\cos\theta = -1 + 2\left(\frac{\gamma_S}{\gamma_L}\right)^{1/2} - \frac{\pi^0}{\gamma_L} \tag{X-45}$$

The term $-\pi^0/\gamma_L$ corrects for the adsorption of vapor on the solid. (To the extent that contact angle is an equilibrium property, the solid *must* be in equilibrium with the saturated vapor pressure P^0 of the bulk liquid.) The term has generally been neglected, mainly for lack of data; also, however, there has been some tendency to affirm that the term is always negligible if θ is large (e.g. Ref. 40; see the earlier discussion of π^0, however, and further discussion following).

Neglecting the π^0 term in Eq. X-45, $\cos\theta$ should be a linear function of $1/\sqrt{\gamma_L}$, and data for various liquids on PTFE do cluster reasonably well along such a line (120). Later, Good (see Ref. 121) improved his agreement by adding dipole interaction terms to his potential function.

Neumann (122) observed a systematic variation of $\gamma_L \cos\theta$ with γ_L for a series of liquids on various polymers and from this was able to infer (with neglect of π^0) a linear variation of Φ of Eq. X-44 with γ_{SL}. In combination with Eq. X-44, this led to

$$\gamma_{SL} = \frac{[\gamma_S^{1/2} - \gamma_L^{1/2}]^2}{1 - 0.015\,(\gamma_S\gamma_L)^{1/2}} \tag{X-46}$$

Equation X-46 may then be used in combination with Young's equation for predictive purposes.

Some additional comments are the following. The manner of summing dispersion interactions may be questioned, including the neglect of structural or entropy aspects, see Ref. 123. It has been recognized that other than dispersion forces may be important in determining interfacial tensions, such as dipole–dipole and dipole–induced dipole forces (see Section VI-2) and hydrogen bonding. One may add terms to Eq. X-44 (see Ref. 123) or elaborate on the definition of Φ (see Ref. 124). A different combining law for both dispersion and polar interactions may be used (125, 126) or an empirical term added to Eq. X-44 (see Ref. 127). The basic approach may be varied, as in using the van der Waals model (128). As may be gathered, the problem is not an easy one, and many semiempirical approaches are being tried.

Fowkes (129) added a very fruitful suggestion. He noted that for polar liquids (e.g., water) much of the intermolecular potential generating γ_L is due to hydrogen bonding and various dipole interactions and that these should not be important at, say, a water–hydrocarbon interface. He proposed that only dispersion interactions would be important across such an interface, and modified Eq. X-44 to read

$$\gamma_{AB} = \gamma_A + \gamma_B - 2(\gamma_A^d \gamma_B^d)^{1/2} \qquad \text{(X-47)}$$

where γ^d denotes an effective surface tension due to the dispersion or general van der Waals component of the interfacial tension. For water (W) and a saturated hydrocarbon (H), it was assumed that $\gamma_H^d = \gamma_H$, from which γ_W^d was found to be 22 erg/cm^2. Similarly, for mercury, $\gamma_{Hg}^d = 200$ erg/cm^2. It was then assumed that γ_A^d was a property of substance A which held at any AB interface; this made possible the calculation of other γ^d values and the interfacial tension between two polar liquids. Equation X-45 now becomes (neglecting the term in π_{SV^0})

$$\cos\theta = -1 + \frac{2(\gamma_S^d \gamma_L^d)^{1/2}}{\gamma_L} \qquad \text{(X-48)}$$

Equation X-48 may appropriately be called the *Girifalco–Good–Fowkes–Young* equation. For liquids whose $\gamma_L^d = \gamma_L$, Eq. X-48 suggests that Zisman's critical surface tension corresponds to γ_S^d. This treatment provides, for example, an interpretation of why water, a polar liquid, behaves toward hydrocarbons as though the surface were nonpolar (see however Section XVI-13 for an indication that polar interactions *are* important between water and hydrocarbons).

In a fairly detailed review, Fowkes relates γ_d's to the theoretical treatment of dispersion interactions (see Chapter VI) and thus to Hamaker constants (130).

We return to the matter of whether π^0 may safely be neglected in Eqs. X-45 and X-48. Values of the type given in Table X-2 can, in fact, cause serious error in estimations of γ_S^d. Consider the case of *n*-octane on PTFE, for which $\theta = 26°$ and $\pi^0 = 1.8$ erg/cm^2. We take $\gamma_L^d = \gamma_L$ and, omitting the π^0 term, find $\gamma_S = 19.5$ erg/cm^2. If the π^0 term is included, the result is 21.2 erg/cm^{-2}, or close to 2 erg/cm^{-2} higher. Further, the 1.8 erg/cm^{-2} value for π^0 is probably a minimum one since the low pressure portion of the adsorption isotherm was not measured and its contribution to π^0 therefore neglected.

If we assume the "conventional" value of 19.5 erg/cm^2 for γ^d for Teflon, and proceed to calculate γ_L^d for water from contact angle data, we find the result to be very sensitive to the value for θ chosen. For $\theta = 98°$, 112°, and 115°, one gets γ_L^d for water to be 49.3, 26.0, and 22.1 erg/cm^2, respectively, with neglect of π^0, and 64.2, 37.1, and 32.5 erg/cm^2, again respectively, if the π^0 term is included. The effect of this term is thus quite large in all three cases.

The preceding examples illustrate that the matter of π^0 is of some importance to the estimation of solid interfacial tensions; it is a matter that is not yet resolved. As noted in Section X-5C there is dispute on the experimental value in the case of water on polyethylene. Nor is there theoretical agreement. On the one hand, it has been shown that the Girifalco and Good model carries the implication of π^0's being generally significant (116) and, on the other hand, both Good (119) and Fowkes (see Ref. 109) propose equations that predict π^0 to be zero for water on low energy polymers. It is certainly desirable that accurate and uncontested π^0 values be established.

Needed also is a better understanding of the thermodynamic and structural properties of adsorbed films as well as of the boundary layer at the liquid–solid interface.

C. Potential–Distortion Model

A rather different approach is to investigate possible adsorption isotherm forms for use with Eq. X-42. As is discussed more fully in Section XVI-7, Polanyi in about 1914 proposed that adsorption be treated as a compression of a vapor in the potential field $\epsilon(x)$ of the solid; with sufficient compression, condensation to liquid adsorbate would occur. If $\epsilon_0(x)$ denotes the field necessary for this, then

$$\epsilon_0 = kT \ln \frac{P^{0\prime}}{P}$$

where $P^{0\prime}$ is the vapor pressure of the liquid film. If the bulk of the observed adsorption is attributed to the condensed liquidlike layer, then $\Gamma = x/d^3$ where d is a molecular dimension such that d^3 equals the molecular volume. Halsey and coworkers (131) took $\epsilon(x)$ to be given by Eq. VI-25, but for the present purpose it is convenient to use the exponential form given by Eq. VI-43, $\epsilon_0 = \epsilon^0 e^{-ax}$. Equation X-48 then becomes

$$kT \ln \frac{P^{0\prime}}{P} = \epsilon^0 e^{-ax}; \qquad \Gamma = \frac{x}{d^3} \tag{X-49}$$

and insertion of these relationships into Eq. X-11 gives an analytical expression for π (see Ref. 116).

The adsorption isotherm corresponding to Eq. X-49 is of the shape shown in Fig. X-1, that is, it cannot explain contact angle phenomena. The ability of a liquid film to coexist with bulk liquid in a contact angle situation suggests that the film structure has been modified by the solid and is different from that of the liquid, and, in an empirical way, this modified structure corresponds to an effective vapor pressure $P^{0\prime}$. $P^{0\prime}$ represents the vapor pressure that bulk liquid would have were its structure that of the film. Such a structural perturbation should relax with increasing film thickness, and this can be represented in exponential form

$$kT \ln \frac{P^{0\prime}}{P^0} = \beta e^{-\alpha x} \tag{X-50}$$

where P^0 is now the vapor pressure of normal liquid adsorbate.

Combination of Eqs. X-49 and X-50 gives

$$kT \ln \frac{P^0}{P} = \epsilon^0 e^{-ax} - \beta e^{-\alpha x} \tag{X-51}$$

This isotherm is now of the shape shown in Fig. X-15, with x_0 determined by the condition $\epsilon^0 e^{-ax_0} = \beta e^{-\alpha x_0}$. The integral of Eq. X-41 then becomes

$$\Delta I = \left(\frac{\epsilon^0}{d^3}\right) e^{-ax_0} \left(\frac{1}{a} - \frac{1}{\alpha}\right) \tag{X-52}$$

The condition for a finite contact angle is then that $\Delta I/\gamma_L < 1$, or that the adsorption potential field decay more rapidly than the structural perturbation.

Equation X-51 has been found to fit data on the adsorption of various vapors on low-energy solids, the parameters a and α being such as to predict the observed θ (88, 98).

There is no reason why the distortion parameter β should not contain an entropy as well as an energy component, and one may therefore write $\beta = \beta_0 - sT$. The entropy of adsorption, relative to bulk liquid, becomes $\Delta S^0 = s \exp(-\alpha x)$. A critical temperature is now implied, $T_c = \beta_0/s$, at which contact angle goes to zero (100). T_c was calculated to be 174°C, for example, by fitting adsorption and contact angle data for the n-octane–PTFE system.

An interesting question that arises is what happens when a thick adsorbed film [such as reported at P^0 for various liquids on glass (95) and for water on pyrolytic carbon (88)] is layered over with bulk liquid. That is, if the solid is immersed in the liquid adsorbate, is the same distinct and relatively thick interfacial film still present, forming some kind of discontinuity or interface with bulk liquid, or is there now a smooth gradation in properties from the surface to the bulk region? This type of question seems not to have been studied, although the answer should be of importance in fluid flow problems and in formulating better models for adsorption phenomena from solution (see Section XI-1).

D. The Microscopic Meniscus Profile

The microscopic contour of a meniscus or a drop is a matter that presents some mathematical problems even with the simplifying assumption of a uniform, rigid solid. Since bulk liquid is present, the system must be in equilibrium with the local vapor pressure so that an equilibrium adsorbed film must also be present. The likely picture for the case of a nonwetting drop on a flat surface is shown in Fig. X-16a. There is a region of negative

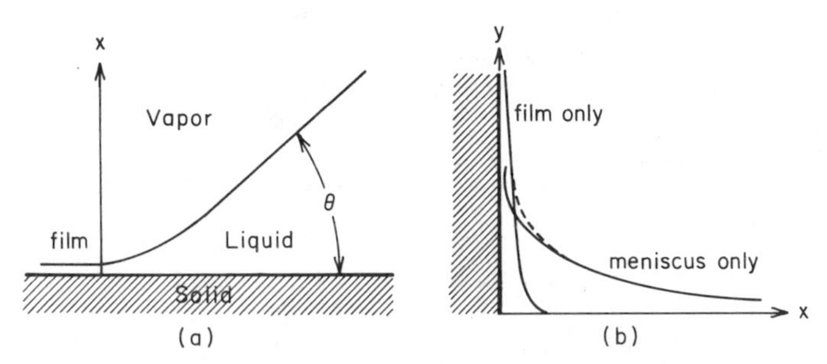

Fig. X-16. (a) Microscopic appearance of the three-phase contact region. (b) Wetting meniscus against a vertical plate showing the meniscus only, adsorbed film only, and joined profile. (From Ref. 133 with permission. Copyright 1980 American Chemical Society.)

curvature as the drop profile joins the plane of the solid. Analyses of the situation have been reviewed by Good (124). In a recent mathematical analysis by Wayner (132) a characteristic distance x_0 appears, which is identified with the adsorbed film thickness and found to be 1 to 2 Å. Adsorption was taken as due to dispersion forces only, however, and some approximations were made in the analysis.

The preceding x_0 estimate contrasts with the values of 5 to 60 Å found by direct ellipsometric measurements of adsorbed film thicknesses for various organic vapors on PTFE or polyethylene (82, 98). The matter needs more attention. On the theoretical side, the adsorption potential function used is of the type of Eq. VI-25, which cannot be correct for distances of the order of an atomic diameter, and a rigorous analysis is needed. On experimental side, the adsorbed film thicknesses observed could reflect "lakes" filling in irregularities in the surface or on polar sites, in which case the appropriate contact angle to use might be closer to the receding than to the advancing one.

A detailed mathematical analysis has been possible for a second situation, that of a wetting meniscus against a flat plate, illustrated in Fig. X-16*b*. The relevent equation is (133)

$$y''[1 + y'^2]^{-1.5} = \frac{2y}{a^2} - \frac{\epsilon}{\gamma V} \qquad \text{(X-53)}$$

where a^2 is the capillary constant, V the molar volume of the liquid, and ϵ, the adsorption potential function. Equation X-53 has been solved by a numerical method for the cases of $\epsilon = \epsilon_0/x^n$, $n = 2, 3$. Again, a characteristic distance x_0 is involved, but of the order of 1000 Å if the ϵ function is to match observed adsorption isotherms. Less complete analyses have been made for related situations, such as a wetting meniscus between horizontal flat plates, the upper plate extending beyond the lower (134, 135).

7. Problems

1. The increased solubility of microcrystals of $SrSO_4$ formed in porous glass has been measured. For crystals of 25 Å diameter the solubility product at 25°C was 6.58 times that for large crystals. Calculate the surface tension of the $SrSO_4$–H_2O interface.

2. If a preparation of $SrSO_4$ shows a 3% increase in solubility, what is the apparent spherical radius of the particles? Use 85mJ/m^2 as the interfacial tension at 25°C. Equating surface tensions and surface energies, estimate the increase in heat of solution of $SrSO_4$ in joules per kilogram due to the surface energy effect.

3. Reference 3 gives the equation $\log(a/a_0) = 16/x$ where a is the solubility activity of a crystal and a_0 is the normal value, and x is the crystal size measured in angstroms. Derive this equation.

4. Given the following data for the water–graphite system, calculate for 25°C (*a*) the energy of immersion of graphite in water, (*b*) the adhesion tension of water on graphite, (*c*) the work of adhesion of water to graphite, and (*d*) the spreading coefficient of water on graphite. Energy of adhesion: 280 mJ/m^2 at 25°C; surface

tension of water at 20, 25, and 30°C is 72.8, 72.0, and 71.2 mJ/m^2, respectively; contact angle at 25°C is 90°.

5. Calculate the heat of immersion of Teflon in *n*-decane at 25°C, given that the contact angle is 32° at 25°C and that $d\theta/dT$ is -0.12 deg/°C.

6. Calculate from data available in the text the value of q_{imm} for Teflon in *n*-octane at around 20°C.

7. Show by means of a suitable cycle that the $\Gamma\,\Delta H_v$ term indeed belongs in Eq. X-15.

8. A particular drop of a certain liquid rests on a flat surface with a contact angle of 40°. The drop has the shape corresponding to $\beta = 80$ of the Bashforth and Adams' tables, and its basal diameter is 1.5 cm. Calculate (*a*) the height of the drop at its center and (*b*) the value of a^2 for the liquid.

9. It has been estimated that for the *n*-decane-PTFE system, π^0 is 0.82 mJ/m^2 at 15°C and 1.54 mJ/m^2 at 70°C (101). Make a calculation for the difference in heat of immersion in *n*-decane of 1 m^2 of clean PTFE surface and of 1 m^2 of surface having an adsorbed film in equilibrium with P^0. Assume 25°C.

10. Derive Eq. X-31. *Hint*: The algebraic procedure used in obtaining Eq. II-15 is suggested as a guide.

11. As an extension of Problem 10, integrate a second time to obtain the equation for the meniscus profile in the Newmann method. Plot this profile as (y/a) versus (x/a) where y is the vertical elevation of a point on the meniscus (above the flat liquid surface), x is the distance of the point from the slide, and a is the capillary constant. (All meniscus profiles, regardless of contact angle, can be located on this plot).

12. Bartell and co-workers report the following capillary pressure data in porous plug experiments, using powdered carbon. Benzene, which wets carbon, showed a capillary pressure of 6200 g/cm^2. For water, the pressure was 12,000 g/cm^2, and for benzene displacing water in the plug, the entry pressure was 5770 g/cm^2. Calculate the water–carbon and carbon–benzene–water contact angles and the adhesion tension at the benzene–carbon interface.

13. Using the data of Table X-2, estimate the contact angle for benzene on stibnite and the corresponding adhesion tension.

14. Using the data of Table X-2, estimate the contact angle for gallium on glass and the corresponding adhesion tension. The gallium–mercury interfacial tension is 37 mJ/m^2 at 25°C, and the surface tension of gallium is about 700 mJ/m^2.

15. Water at 20°C rests on solid naphthalene with a contact angle of 90°, while a water–ethanol solution of surface tension 35 dyne/cm shows an angle of 30°. Calculate (*a*) the work of adhesion of water to naphthalene, (*b*) the critical surface tension of naphthalene, and (*c*) γ^d for naphthalene.

16. (*a*) Estimate the contact angle for ethanol on Teflon. Make an educated guess, that is, with explanation, as to whether the presence of 10% hexane in the ethanol should appreciably affect θ and, if so, in what direction. (*b*) Answer the same questions for the case where nylon rather than Teflon is the substrate [nylon is essentially $(CONH\text{-}CH_2)_x$]. (*c*) Discuss in which of these cases the hexane should be the more strongly adsorbed at the solution–solid interface.

17. Given the following data for the water–solid naphthalene system: contact angle at 25°C, 90°; surface tension and surface energy of water at 25°C, 72 and 200 mJ/m^2, respectively; energy of adhesion of water to naphthalene, 290 mJ/m^2 at 25°C. Calculate (*a*) the energy of immersion of naphthalene crystals in water, per square

centimeter, (*b*) the work of adhesion of water to naphthalene, (*c*) the adhesion tension of water on naphthalene, and (*d*) the spreading coefficient of water on naphthalene.

Neglect any tendency of water vapor to adsorb on solid naphthalene in answering the preceding questions. As a final question, however, (*e*) sketch the likely appearance of actual adsorption isotherm data for water adsorption on naphthalene, with emphasis on the range of P/P^0 from about 0.3 to close to 1.0. Certain qualitative features are all that are expected.

18. Fowkes and Harkins reported that the contact angle of water on paraffin is 111° at 25°C. For a 0.1*M* solution of butylamine, of surface tension 56.3 mJ/m^2, the contact angle was 92°. Calculate the film pressure of the butylamine absorbed at the paraffin–water interface. State any assumptions that are made.

19. The surface tension of liquid sodium at 100°C is 220 erg/cm^{-2}, and its contact angle on glass is 66°. (*a*) Estimate, with explanation of your procedure, the surface tension of glass at 100°C. (*b*) If given further that the surface tension of mercury at 100°C is 460 erg/cm^2, and that its contact angle on glass is 143°, estimate by a different means than in (*a*) the surface tension of the glass. By "different" is meant a different conceptual procedure.

20. Using appropriate data from Table II-8, calculate the water–mercury interfacial tension using the simple Girifalco and Good equation and then using Fowkes' modification of it.

21. A 0.1-mm diameter fiber (essentially a rod) is lowered until it just touches the surface of methylene iodide, at which point a meniscus forms. Derive the appropriate simple equation and calculate the gain in weight of the fiber if methylene iodide has a contact angle of 45° with the fiber.

22. Referring to Fig. X-5, explain which bubble shows an advancing angle, which a receding angle.

23. Suppose that the linear tension for a given three-phase line is 1×10^{-2} dyne. Calculate θ for drops of radius 0.1, 0.01, and 0.001 cm, if the value for a large drop is 60°. Assume water at 20°C.

24. The contact angle of ethylene glycol on paraffin is 83° at 25°C, and γ_L^d for ethylene glycol is 28.6 mJ/m^2. γ_L^d for water is 22.1 mJ/m^2, and the surface tensions of ethylene glycol and of water are 48.3 and 72.8 mJ/m^2, respectively. Neglecting π^0's, calculate from the above data what the contact angle should be for water on paraffin.

25. An astronaut team has, as one of its assigned experiments, the measurement of contact angles for several systems (to test the possibility that these may be different in gravity free space). Discuss some methods that would be appropriate and some that would not be appropriate to use.

26. What is the critical surface tension for human skin? Look up any necessary data and make a Zisman plot of contact angle on skin versus surface tension of water–alcohol mixtures. (Note Ref. 89).

27. Calculate γ_L^d for *n*-propanol from its contact angle on PTFE, using γ = 19.4 mJ/m^2 for PTFE and (*a*) including and (*b*) neglecting π_{SL}^0.

28. Calculate γ^d for napthalene; assume that it interacts with water only with dispersion forces.

29. Derive from Eq. X-46 an equation relating θ, γ_{SV}, and γ_L.

30. Derive Eq. X-42.

31. Show that according to the Girifalco and Good treatment of interfacial ten-

sions (with $\Phi = 1$), it follows that γ_s, the surface tension of the solid, is equal to γ_c, the Zisman critical surface tension. Note any major assumptions made in this demonstration.

32. (*a*) Estimate the contact angle for butyl acetate on Teflon (essentially a CF_2 surface) at 20°C using data and semiempirical relationships in the text. (*b*) Use the Zisman relationship to obtain an expression for the spreading coefficient of a liquid on a solid, $S_{L/S}$, involving only γ_L as the variable. Show what $S_{L/S}$ should be when $\gamma_L = \gamma_c$ (neglect complications involving π^0).

General References

Advances in Chemistry, No. 43, American Chemical Society, Washington, D.C., 1964.

J. T. Davies and E. K. Rideal, *Interfacial Phenomena*, Academic, New York, 1963.

R. Defay and I. Prigogine, *Surface Tension and Adsorption*, Wiley, New York, 1966.

R. E. Johnson, Jr., and R. H. Dettre, *Surface and Colloid Science*, Vol. 2, E. Matijević, Ed., Wiley-Interscience, New York, 1969.

J. Mahanty and B. W. Ninham, *Dispersion Forces*, Academic, New York, 1976.

K. L. Sutherland and I. W. Wark, *Principles of Flotation*, Australian Institute of Mining and Technology, Inc., Melbourne, 1955.

Textual References

1. A. T. Hubbard, *Acc. Chem. Res.*, **13,** 177 (1980).
2. B. V. Enüstün and J. Turkevich, *J. Am. Chem. Soc.*, **82,** 4502 (1960).
3. B. V. Enüstün, M. Enuysal, and M. Dösemeci, *J. Colloid Interface Sci.*, **57,** 143 (1976).
4. F. van Zeggeren and G. C. Benson, *Can. J. Chem.*, **35,** 1150 (1957).
5. L. F. Knapp, *Trans. Faraday Soc.*, **18,** 457 (1922).
6. H. E. Buckley, *Crystal Growth*, Wiley, New York, 1951.
7. R. I. Stearns and A. F. Berndt, *J. Phys. Chem.*, **80,** 1060 (1976).
8. See *J. Phys. Chem.* **80,** 2707–2709 for comments on Ref. 5.
9. A. Skapski, R. Billups, and A. Rooney, *J. Chem. Phys.*, **26,** 1350 (1957).
10. W. M. Ketcham and P. V. Hobbs, *Phil. Mag.*, **19,** 1161 (1969).
11. S. C. Hardy and S. R. Coriell, *J. Cryst. Growth*, **5,** 329 (1969).
12. J. C. Melrose, *J. Colloid Interface Sci.*, **24,** 416 (1967).
13. A. C. Makrides and N. Hackerman, *J. Phys. Chem.*, **63,** 594 (1959).
14. J. W. Whalen, *Adv. Chem. Ser.*, No. 33, 281 (1961); *J. Phys. Chem.*, **65,** 1676 (1961).
15. S. Partya, F. Rouquerol, and J. Rouquerol, *J. Colloid Interface Sci.*, **68,** 21 (1979).
16. A. C. Zettlemoyer and J. J. Chessick, *Adv. Chem. Ser.*, No. 43, 88 (1964).
17. M. Topic, F. J. Micale, H. Leidheiser, Jr., and A. C. Zettlemoyer, *Rev. Sci. Instr.*, **45,** 487 (1974).
18. A. C. Zettlemoyer, *Ind. Eng. Chem.*, **57,** 27 (1965).
19. W. D. Harkins, *The Physical Chemistry of Surfaces*, Reinhold, New York, 1952.
20. W. H. Wade and N. Hackerman, *Adv. Chem. Ser.*, No. 43, 222 (1964).

21. J. W. Whalen and W. H. Wade, *J. Colloid Interface Sci.*, **24,** 372 (1967).
22. J. J. Chessick, A. C. Zettlemoyer, F. H. Healey, and G. J. Young, *Can. J. Chem.*, **33,** 251 (1955).
23. F. H. Healey, Yung-Fang Yu, and J. J. Chessick, *J. Phys. Chem.*, **59,** 399 (1955).
24. See R. L. Venable, W. H. Wade, and N. Hackerman, *J. Phys. Chem.*, **69,** 317 (1965).
25. J. R. Arnold, *J. Am. Chem. Soc.*, **71,** 104 (1949).
26. S. N. Omenyi and A. W. Neuman, *J. Appl. Phys.*, **47,** 3956 (1976).
27. T. Young, *Miscellaneous Works*, Vol. 1. G. Peacock, Ed., Murray, London, 1855, p. 418.
28. A. Dupré, *Theorie Mecanique de la Chaleur*, Paris, 1869, p. 368.
29. J. C. Melrose, *Adv. Chem. Ser.*, No. 43, 158 (1964).
30. D. H. Bangham and R. I. Razouk, *Trans. Faraday Soc.*, **33,** 1459 (1937).
31. W. D. Harkins and H. K. Livingston, *J. Chem. Phys.*, **10,** 342 (1942).
32. See F. E. Bartell and L. S. Bartell, *J. Am. Chem. Soc.*, **56,** 2205 (1934).
33. R. N. Wenzel, *Ind. Eng. Chem.*, **28,** 988 (1936); *J. Phys. Colloid Chem.*, **53,** 1466 (1949).
34. A. B. D. Cassie, *Discuss. Faraday Soc.*, **3,** 11 (1948).
35. R. Shuttleworth and G. L. J. Bailey, *Discuss. Faraday Soc.*, **3,** 16 (1948).
36. R. J. Good, *J. Am. Chem. Soc.*, **74,** 5041 (1952); see also J. D. Eick, R. J. Good, and A. W. Neumann, *J. Colloid Interface Sci.*, **53,** 235 (1975).
37. S. Baxter and A. B. D. Cassie, *J. Text. Inst.*, **36,** T67 (1945); A. B. D. Cassie and S. Baxter, *Trans. Faraday Soc.*, **40,** 546 (1944).
38. R. H. Dettre and R. E. Johnson, Jr., *Symp. Contact Angle, Bristol,* **1966.**
39. A. W. Neumann, *Adv. Colloid Interface Sci.*, **4,** 105 (1974).
40. A. W. Neumann and R. J. Good, *Surface and Colloid Science*, Vol. 2, R. J. Good and R. R. Stromberg, Eds., Plenum, New York, 1979.
41. W. C. Bigelow, D. L. Pickett, and W. A. Zisman, *J. Colloid Sci.*, **1,** 513 (1946); see also R. E. Johnson and R. H. Dettre, *J. Colloid Sci.*, **20,** 173 (1965).
42. J. Leja and G. W. Poling, *Preprint*, International Mineral Processing Congress, London, April, 1960.
43. R. H. Ottewill, private communication; see also A. M. Gaudin, *Flotation*, McGraw-Hill, New York, 1957, p. 163.
44. A. W. Adamson, F. P. Shirley, and K. T. Kunichika, *J. Colloid Interface Sci.*, **34,** 461 (1970).
45. A. W. Neumann and D. Renzow, *Z. Phys. Chemie Neue Folge*, **68,** 11 (1969); W. Funke, G. E. H. Hellweg, and A. W. Neumann, *Angew. Makromol. Chemie*, **8,** 185 (1969).
46. I. Langmuir and V. J. Schaeffer, *J. Am. Chem. Soc.*, **59,** 2405 (1937).
47. R. J. Good and J. K. Paschek, *Wetting, Spreading, and Adhesion*, J. F. Padday, Ed., Academic, 1978.
47a. N. K. Adam and G. Jessop, *J. Chem. Soc.*, **1925,** 1863.
48. F. M. Fowkes and W. D. Harkins, *J. Am. Chem. Soc.*, **62,** 3377 (1940).
49. A. W. Neumann, *Z. Phys. Chem. Neue Folge*, **41,** 339 (1964).
50. F. E. Bartell and H. H. Zuidema, *J. Am. Chem. Soc.*, **58,** 1449 (1936).
51. P. Rehbinder, M. Lipetz, M. Rimskaja, and A. Taubmann, *Kollid-Z.*, **65,** 268 (1933).
52. J. J. Bikerman, *Ind. Eng. Chem., Anal. Ed.*, **13,** 443 (1941).

53. L. R. Fisher, *J. Colloid Interface Sci.,* **72,** 200 (1979).
54. F. E. Bartell and C. W. Walton, Jr., *J. Phys. Chem.,* **38,** 503 (1934); F. E. Bartell and C. E. Whitney, *J. Phys. Chem.,* **36,** 3115 (1932); F. E. Bartell and H. J. Osterhof, *Colloid Symposium Monograph*, The Chemical Catalog Company, New York, 1928, p. 113.
54a. H. Lomas, *J. Colloid Interface Sci.,* **33,** 548 (1970). See also H. M. Princen, *J. Colloid Interface Sci.,* **36,** 157 (1971) and H. Lomas, *J. Colloid Interface Sci.,* **37,** 247 (1971).
54b. R. A. Brown, F. M. Orr, Jr., and L. E. Scriven, *J. Colloid Interface Sci.,* **73,** 76 (1980).
55. J. F. Oliver, C. Huh, and S. G. Mason, *Colloid Surf.,* **1,** 79 (1980).
55a. R. H. Dettre and R. E. Johnson, Jr., *Adv. Chem. Ser.,* No. 43, 112, 136 (1964). See also Ref. 50.
56. A. M. Schwartz and F. W. Minor, *J. Colloid Sci.,* **14,** 584 (1959).
57. R. E. Johnson, Jr. and R. H. Dettre, *J. Phys. Chem.,* **68,** 1744 (1964).
58. A. W. Neumann and R. J. Good, *J. Colloid Interface Sci.,* **38,** 341 (1972).
59. D. C. Pease, *J. Phys. Chem.,* **49,** 107 (1945).
60. R. H. Dettre and R. E. Johnson, Jr., *J. Phys. Chem.,* **69,** 1507 (1965).
61. L. S. Penn and B. Miller, *J. Colloid Interface Sci.,* **78,** 238 (1980).
62. A. S. Michaels and R. C. Lummis, private communication.
63. A. S. Michaels and S. W. Dean, Jr., *J. Phys. Chem.,* **66,** 1790 (1962).
64. R. J. Good, J. A. Kvikstad, and W. O. Bailey, *J. Colloid Interface Sci.,* **35,** 314 (1971).
65. T. D. Blake and J. M. Haynes, Prog. Surf. Membrane Sci., **6,** 125 (1973).
66. I. W. Work, *Australian J. Chem.,* **30,** 205 (1977).
67. R. E. Johnson, Jr., R. H. Dettre, and D. A. Brandreth, *J. Colloid Interface Sci.,* **62,** 205 (1977).
68. W. Radigan, H. Ghiradella, H. L. Frisch, H. Schonhorn, and T. K. Kwei, *J. Colloid Interface Sci.,* **49,** 241 (1974).
69. R. J. Good and M. N. Koo, *J. Colloid Interface Sci.,* **71,** 283 (1979).
70. B. A. Pethica, *J. Colloid Interface Sci.,* **62,** 567 (1977).
71. S. Torza and S. G. Mason, *Kolloid-Z Z. Polym.* **246,** 593 (1971).
71a. F. P. Buff and H. J. Saltzburg, *J. Chem. Phys.,* **26,** 23 (1957).
72. F. E. Bartell and H. J. Osterhof, *Colloid Symposium Monograph*, The Chemical Catalog Company, New York, 1928, p. 113.
73. See W. A. Zisman, *Adv. Chem. Ser.,* No. 43, (1964).
74. T. Smith, *J. Colloid Interface Sci.,* **75,** 51 (1980).
75. M. E. Schrader, *J. Phys. Chem.,* **74,** 2313 (1970).
76. M. L. White, *Clean Surfaces, Their Preparation and Characterization for Interfacial Studies*, G. Goldfinger, Ed., Marcel Decker, New York, 1970.
76a. G. L. Gaines, Jr., *J. Colloid Interface Sci.,* **79,** 295 (1981).
77. H. W. Fox and W. A. Zisman, *J. Colloid Sci.,* **7,** 428 (1952).
78. M. C. Phillips and A. C. Riddiford, *Nature,* **205,** 1005 (1965).
79. *International Critical Tables*, Vol. 4, McGraw-Hill, New York, 1928, p. 434.
80. H. K. Livingston, *J. Phys. Chem.,* **48,** 120 (1944).
81. J. R. Dann, *J. Colloid Interface Sci.,* **32,** 302 (1970).
82. P. Hu and A. W. Adamson, *J. Colloid Interface Sci.,* **59,** 605 (1977).
83. F. D. Petke and B. R. Ray, *J. Colloid Interface Sci.,* **31,** 216 (1969).
84. H. Schonhorn, *J. Phys. Chem.,* **70,** 4086 (1966).

85. H. Schonhorn, *Nature,* **210,** 896 (1966).
86. A. C. Zettlemoyer, *J. Colloid Interface Sci.,* **28,** 343 (1968).
87. H. Schonhorn and F. W. Ryan, *J. Phys. Chem.,* **70,** 3811 (1966).
88. M. E. Tadros, P. Hu, and A. W. Adamson, *J. Colloid Interface Sci.,* **49,** 184 (1974).
89. A. W. Adamson, K. Kunichika, F. Shirley, and M. Orem. *J. Chem. Ed.,* **45,** 702 (1968).
90. A. El-Shimi and E. D. Goddard, *J. Colloid Interface Sci.,* **48,** 242 (1974).
91. J. B. Jones and A. W. Adamson, *J. Phys. Chem.,* **72,** 646 (1968).
92. F. M. Fowkes and W. M. Sawyer, *J. Chem. Phys.,* **20,** 1650 (1952).
93. G. E. Boyd and H. K. Livingston, *J. Am. Chem. Soc.,* **64,** 2383 (1942).
94. J. A. Koutsky, A. G. Walton, and E. Baer, *Surf. Sci.,* **3,** 165 (1965).
95. B. V. Derjaguin and Z. M. Zorin, *Proc. 2nd Int. Congr. Surf. Act., London, 1957,* Vol. 2, p. 145.
96. H. W. Fox, E. F. Hare, and W. A. Zisman, *J. Phys. Chem.,* **59,** 1097 (1955); O. Levine and W. A. Zisman, *J. Phys. Chem.,* **61,** 1068, 1188 (1957).
97. W. R. Good, *J. Colloid Interface Sci.,* **44,** 63 (1973).
98. J. Tse and A. W. Adamson, *J. Colloid Interface Sci.,* **72,** 515 (1979).
99. A. W. Newmann, G. Haage, and D. Renzow, *J. Colloid Interface Sci.,* **35,** 379 (1971).
100. A. W. Adamson, *J. Colloid Interface Sci.,* **44,** 273 (1973).
101. H. W. Fox and W. A. Zisman, *J. Colloid Interface Sci.,* **5,** 514 (1950).
102. C. L. Sutula, R. Hautala, R. A. Dalla Betta, and L. A. Michel, Abstracts, 153rd Meeting, American Chemical Society, April, 1967.
103. J. W. Whalen, *Vacuum Microbalance Techniques*, A. W. Czanderna, Ed., Vol. 8, Plenum Press, 1971.
104. E. G. Shafrin and W. A. Zisman, *J. Phys. Chem.,* **64,** 519 (1960).
105. W. D. Bascom and C. R. Singleterry, *J. Phys. Chem.,* **66,** 236 (1962).
106. H. Peper and J. Berch, *J. Phys. Chem.,* **68,** 1586 (1964).
107. W. D. Wade and J. W. Whalen, *J. Phys. Chem.,* **72,** 2898 (1968).
108. T. D. Blake and W. H. Wade, *J. Phys. Chem.,* **75,** 1887 (1971).
108a. J. W. Whalen and P. C. Hu, *J. Colloid Interface Sci.,* **65,** 460 (1978).
108b. Y. Tamai, T. Matsunaga, and K. Horiuchi, *J. Colloid Interface Sci.,* **60,** 112 (1977).
109. F. M. Fowkes, D. C. McCarthy, and M. A. Mostafa, *J. Colloid Interface Sci.,* **78,** 200 (1980).
110. See M. K. Bernett and W. A. Zisman, *J. Phys. Chem.,* **62,** 1241 (1959).
111. A. Baszkin and L. Ter-Minassian-Saraga, *J. Colloid Interface Sci.,* **43,** 190 (1973).
112. R. J. Ruch and L. S. Bartell, *J. Phys. Chem.,* **64,** 513 (1960).
113. R. Williams, *J. Phys. Chem.,* **79,** 1274 (1975).
114. J. J. Bikerman, *Proc. 2nd Int. Congr. Surf. Act., London, 1957,* Vol. 3, p. 125.
115. R. E. Johnson, Jr., *J. Phys. Chem.,* **63,** 1655 (1959).
116. A. W. Adamson and I. Ling, *Adv. Chem. Ser.,* No. 43, (1964).
117. Z. M. Zorin, V. P. Romanov, and N. V. Churaev, *Colloid Poly. Sci.,* **257,** 968 (1979).
118. L. A. Girifalco and R. J. Good, *J. Phys. Chem.,* **61,** 904 (1957). See also R. J. Good, *Adv. Chem. Ser.,* No. 43, 74 (1964).

119. R. J. Good and E. Elbing, *J. Colloid Interface Sci.,* **59,** 398 (1977).
120. R. J. Good and L. A. Girifalco, *J. Phys. Chem.,* **64,** 561 (1960).
121. R. J. Good, *Ind. Eng. Chem.,* **62,** 54 (1970).
122. A. W. Newmann, R. J. Good, C. J. Hope, and M. Sejpal, *J. Colloid Interface Sci.,* **49,** 291 (1974).
123. F. M. Fowkes, *J. Phys. Chem.,* **72,** 3700 (1968); J. F. Padday and N. D. Uffindell, *ibid.*, 3700 (1968).
124. R. J. Good, *Surface and Colloid Science*, Vol. 11, R. J. Good and R. R. Stromberg, Eds., Plenum, 1979.
125. S. Wu, *J. Polym. Sci., Part C,* **34,** 19 (1971); *J. Adhes.,* **5,** 39 (1973).
126. A. El-Shimi and E. D. Goddard, *J. Colloid Interface Sci.,* **48,** 242 (1974).
127. Y. Tamai, T. Matsunaga, and K. Horiuchi, *J. Colloid Interface Sci.,* **60,** 112 (1977). See also Y. Tamai, *J. Phys. Chem.,* **79,** 965 (1975).
128. D. E. Sullivan, *J. Chem. Phys.,* **74,** 2604 (1981).
129. F. M. Fowkes, *J. Phys. Chem.,* **67,** 2538 (1963); *Adv. Chem. Ser.,* No. 43, 99 (1964).
130. F. M. Fowkes, *Chemistry and Physics of Interfaces*, S. Ross, Ed., American Chemical Society, 1971.
131. G. D. Halsey, *J. Chem. Phys.,* **16,** 931 (1948).
132. P. C. Wayner, Jr., *J. Colloid Interface Sci.,* **77,** 495 (1980).
133. A. W. Adamson and A. Zebib, *J. Phys. Chem.,* **84,** 2619 (1980).
134. F. Renk, P. C. Wayner, Jr., and B. M. Homsy, *J. Colloid Interface Sci.,* **67,** 408 (1978).
135. See B. V. Derjaguin, V. M. Starov, and N. V. Churaev, *Colloid J.,* **38,** 875 (1976).

CHAPTER XI

The Solid–Liquid Interface—Adsorption from Solution

This chapter on adsorption from solution is intended only to develop the more straightforward and important aspects of adsorption phenomena that prevail when a solvent is present. The general subject has a vast literature, and it is both necessary and reasonable in a textbook to limit the presentation to the more important characteristic features and theory.

With nonelectrolytes, a logical division is made according to whether the adsorbate solution is dilute or concentrated. In the first case treatment is very similar to that for gas adsorption, whereas in the second case the role of solvent becomes more explicit. The adsorption of electrolytes is treated briefly, mainly in terms of the exchange of components in an electrical double layer either at the surface of a nonporous particle or in an ion exchanger or zeolite.

A very important application of adsorption phenomena is that of chromatography, where the adsorbed material is held in a fixed bed or conformation, and solution passes down its length. This subject is not discussed here, partly because chromatographic conditions involve aspects not strictly related to surface chemistry and partly because the field is an applied one too large to treat adequately in a limited space.

1. Adsorption of Nonelectrolytes from Dilute Solution

The adsorption of nonelectrolytes at the solid–solution interface may be viewed in terms of two somewhat different physical pictures. The first is that the adsorption is essentially confined to a monolayer next to the surface, with the implication that succeeding layers are virtually normal bulk solution. The picture is similar to that for the chemisorption of gases (see Chapter XVII) and similarly carries with it the assumption that solute–solid interactions decay very rapidly with distance. Unlike the chemisorption of gases, however, the heat of adsorption from solution is usually fairly small and is more comparable with heats of solution than with chemical bond energies.

The second picture is that of an interfacial layer or region, multimolecular in depth (perhaps even 100 Å deep), over which a more slowly decaying interaction potential with the solid is present (note Section X-6C). The situation would then be more like that in the physical adsorption of vapors (see Chapter XVI), which become multilayer near the saturation vapor pressure (see Fig. X-15, for example). Adsorption from solution, from this point of view, corresponds to a partition between a bulk and an interfacial phase.

While both models find some experimental support, the monolayer one has been much more amenable to simple analysis. As a consequence, most of the discussion in this chapter is in terms of it, although occasional *caveats* are entered. We consider first the case of adsorption from dilute solution, as this corresponds to the usual experimental situation and also because the various adsorption models take on a simpler algebraic form and are thus easier to develop than those for concentrated solutions.

A. *Adsorption Isotherms*

The moles of solute species adsorbed per gram of adsorbent is given experimentally by $\Delta C_2 V_{\text{sol}}/m$ where ΔC_2 is the change in concentration of the solute following adsorption, V_{sol} is the total volume of solution, and m is the grams of adsorbent. It will be convenient in the following development to suppose that mole numbers and other extensive quantities are on a *per gram of adsorbent basis*, so that n_2^s, the moles of solute adsorbed per gram, is given by

$$n_2^s = V\,\Delta C_2 = n_0\,\Delta N_2 \qquad \text{(XI-1)}$$

where n_0 is the total moles of solution per gram of adsorbent and ΔN_2 is the change in mole fraction of solute following adsorption. In dilute solution, both forms are equivalent, although, as seen in Section XI-4, this is not the case otherwise. See the discussion of definitions and terminology by Everett (1) and Schay (2); n_2^s has been called the *specific reduced surface excess*, for example.

The quantity n_2^s is in general a function of C_2, the equilibrium solute concentration, and temperature for a given system, that is $n_2^s = f(C_2, T)$. At constant temperature, $n_2^s = f_T(C_2)$, and this is called the *adsorption isotherm function*. The usual experimental approach is to determine this function, that is, to measure adsorption as a function of concentration at a given temperature.

Various functional forms for f have been proposed either as a result of empirical observation, or in terms of specific models, and a particularly important example of the last is that known as the *Langmuir* adsorption equation (3). By analogy with the derivation for the case of gas adsorption (see Section XVI-3), the Langmuir model assumes the surface to consist of adsorption sites, the area per site being σ^0; all adsorbed species interact only with a site and not with each other, and adsorption is thus limited to a monolayer. In the case of adsorption from solution, however, it seems more plausible to consider an alternative phrasing of the model. Adsorption is still limited to a monolayer, but this layer is now regarded as an ideal two-dimensional solution of equal size solute and solvent molecules of area σ^0.

Thus lateral interactions, absent in the site picture, cancel out in the ideal solution layer picture because of being independent of composition. However, in the first version σ^0 is a property of the solid lattice, while in the second it is a property of the adsorbed species; both versions attribute differences in adsorption behavior entirely to differences in absorbate–solid interactions. Both present adsorption as a competition between solute and solvent.

It is perhaps fortunate that both versions lead to the same algebraic formulations, but we will imply a preference for the two-dimensional solution picture by expressing surface concentrations in terms of mole fractions. The adsorption process can now be written as

$$\text{A(solute in solution, } N_2) + \text{B(adsorbed solvent, } N_1^s) = \text{A(adsorbed solute, } N_2^s) + \text{B(solvent in solution, } N_1) \quad \text{(XI-2)}$$

The equilibrium constant for this process is

$$K = \frac{N_2^s a_1}{N_1^s a_2} \quad \text{(XI-3)}$$

where a_1 and a_2 are the solvent and solute activities in solution and, by virtue of the model, the activities in the adsorbed layer are given by the respective mole fractions N_1^s and N_2^s. Since the treatment is restricted to dilute solutions, a_1 is constant, and we can write $b = K/a_1$; also, $N_1^s + N_2^s = 1$ so that Eq. XI-2 becomes

$$N_2^s = \frac{ba_2}{(1 + ba_2)} \quad \text{(XI-4)}$$

Since $n_2^s = N_2^s n^s$ where n^s is the number of moles of adsorption sites per gram, Eq. XI-3 can also be written

$$n_2^s = \frac{n^s ba_2}{(1 + ba_2)} \quad \text{(XI-5)}$$

or

$$\theta = \frac{ba_2}{(1 + ba_2)}$$

where $\theta = n_2^s/n^s$ is the fraction of surface occupied. Also,

$$n^s = \frac{\Sigma}{N\sigma^0} \quad \text{(XI-6)}$$

where Σ denotes the surface area per gram. In sufficiently dilute solution, activity coefficient effects will be unimportant, so that in Eq. IX-4, a_2 may be replaced by C_2.

The equilibrium constant K can be written

$$K = e^{\Delta S^0/R} e^{-\Delta H^0/RT} \quad \text{(XI-7)}$$

where ΔH^0 is the net enthalpy of adsorption, often denoted by $-Q$, where Q is the heat of adsorption. Thus the constant b can be written

$$b = b' \, e^{Q/RT} \tag{XI-8}$$

The entropies and enthalpies of adsorption may be divided, in a formal way, into separate quantities for each component:

$$K = \frac{K_2}{K_1}, \qquad \Delta S^0 = \Delta S_2^0 - \Delta S_1^0, \qquad \Delta H^0 = \Delta H_2^0 - \Delta H_1^0 \tag{XI-9}$$

It is not necessary to limit the model to that of idealized sites; Everett (4) has extended the treatment by incorporating surface activity coefficients as corrections to N_1^s and N_2^s.

Returning to Eq. XI-4, with C_2 replacing a_2, at low concentrations n_2^s will be proportional to C_2, with a slope $n^s b$. At sufficiently high concentrations, n_2^s approaches the limiting value n^s. Thus n^s is a measure of the capacity of the adsorbent and b, of the intensity of the adsorption. In terms of the ideal model, n^s should not depend on temperature, while b should show an exponential dependence, as given by Eq. XI-8. The two constants are conveniently evaluated by putting Eq. XI-5 in the form

$$\frac{C_2}{n_2^s} = \frac{1}{n^s b} + \frac{C_2}{n^s} \tag{XI-10}$$

That is, a plot of C_2/n_2^s versus C_2 should give a straight line of slope $1/n^s$ and intercept $1/n^s b$.

An equation algebraically equivalent to Eq. XI-4 results if instead of site adsorption the surface region is regarded as an interfacial solution phase, much as in the treatment in Section III-7C. The condition is now that $v^s = n_1^s \bar{V}_1 + n_2^s \bar{V}_2$. If a_1^s and a_2^s, the activities of the two components in the interfacial phase, are represented by the volume fractions V_1^s and V_2^s, the result is

$$v_2^s = \frac{v^s b a_2}{(1 + b a_2)} \tag{XI-11}$$

Here v_2^s is the volume of adsorbed solute, and v^s is the (constant) volume of the interfacial solution (5).

Most surfaces are heterogeneous so that b in Eq. XI-8 will vary with θ. The adsorption isotherm may now be written

$$\Theta(C_2, T) = \int_0^\infty f(b)\theta(C_2, b, T)\, db \tag{XI-12}$$

where $f(b)$ is the distribution function for b, $\theta(C_2, b, T)$ is the adsorption isotherm function (e.g., Eq. XI-5), and $\Theta(C_2, T)$ is the experimentally observed adsorption isotherm. In a sense, this approach is an alternative to the use of surface activity coefficients.

The solution to this integral equation is discussed in Section XVI-15, but one particular case is of interest here. If $\theta(C_2, b, T)$ is given by Eq. XI-5 and the variation in b and θ is attributed entirely to a variation in the heat of adsorption Q, and $f(Q)$ is taken to be

$$f(Q) = \alpha e^{-Q/nRT} \qquad \text{(XI-13)}$$

then the solution to Eq. XI-12 is of the form (6, 7)

$$\Theta = \frac{n_2^s}{n^s} = aC_2^{1/n} \qquad \text{(XI-14)}$$

where $a = \alpha RTnb'$ and b' is as defined in Eq. XI-8. Equation XI-14 is known as the *Freundlich adsorption isotherm* after its user (8).

The Freundlich equation, unlike the Langmuir, does not become linear at low concentrations but remains convex to the concentration axis; nor does it show a saturation or limiting value. The constants (an^s) and n may be obtained from a plot of log n_2^s versus log C_2 and, roughly speaking, the intercept an^s gives a measure of the adsorbent capacity, and the slope $1/n$, of the intensity of adsorption. As just mentioned, the shape of the isotherm is such that n is a number greater than unity.

There is no assurance that the derivation of the Freundlich equation is unique; consequently, if data fit the equation it is only likely, but not proven, that the surface is heterogeneous. Basically, the equation is an empirical one, limited in its usefulness to its ability to fit data.

B. Qualitative Results of Adsorption Studies—Traube's Rule

The nonelectrolytes that have been studied are for the most part organic compounds; these include fatty acids, aromatic acids, esters, and other single functional group compounds, plus a great variety of more complex species such as porphyrins, bile pigments, carotenoids, lipids, and dyestuffs. Frequently these more complex substances have been studied only in terms of their chromatographic behavior, so that qualitative information concerning relative adsorbabilities may be known but not actual isotherms.

Typical adsorbents, expecially of the older literature, include alumina, silica gel, various forms of carbon (blood charcoal, sugar charcoal, etc., and carbon blacks), and various organic compounds such as sugars and starches. In the case of hydrous oxides and carbons, not only are the composition and state of subdivision important, but also the adsorptive properties are strongly dependent on the moisture content and degree of heating or activation used. As to solvents, a great deal of work has been done with aqueous systems, but since organic absorbates are common, one also finds data for solutions in a variety of common organic solvents.

The behavior of a given system may be predicted very qualitatively in terms of the separate adsorption constants of Eqs. XI-9. The rule is that a polar (nonpolar) adsorbent will preferentially adsorb the more polar (nonpolar) component of a nonpolar (polar) solution. Polarity is used here in the general sense of ability to engage in hydrogen bonding or dipole–dipole type interactions as opposed to nonspecific dispersion interactions. A semiquantitative extension of the foregoing is known as Traube's rule (9) which, as given by Freundlich (8), states: "The adsorption of organic substances from aqueous solutions increases strongly and regularly as we ascend the homologous series."

Data illustrating Traube's rule are shown in Fig. XI-1*a*, in which it is seen that the initial slopes, and hence *b* values in Eq. XI-4, increase in the order: formic acid, acetic acid, propionic acid, and butyric acid. The adsorption in this case was on carbon and from aqueous solution. Holmes and McKelvey (10) made the logical extension of Freundlich's statement by noting that the situation really is a relative one and that a reversal of order should occur if a polar adsorbent and a nonpolar solvent were used. Thus as illustrated in Fig. XI-1*b*, the reverse sequence was indeed observed for fatty acids adsorbed on silica gel from toluene solution. Bartell and Fu (11) further noted that this reversal occurred with silica gel even if aqueous solutions were used.

As discussed in Chapter III, the uniform progression in adsorbabilities in proceeding along a homologous series can be understood in terms of a constant increment in the work of adsorption with each additional CH_2 group. The film pressure π may be calculated from the adsorption isotherm

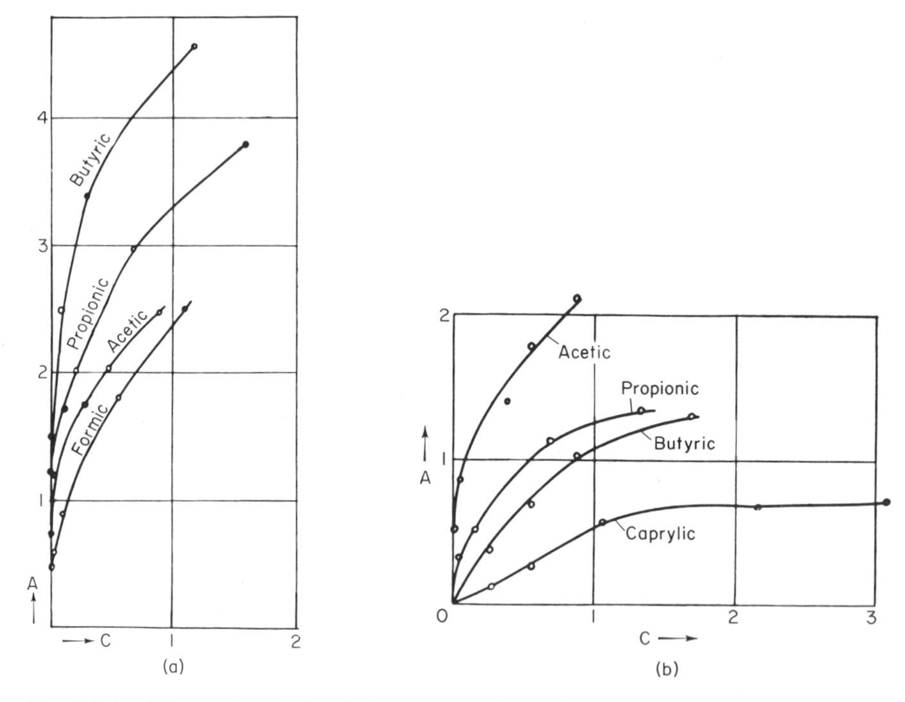

Fig. XI-1. Illustration of Traube's rule: (*a*) adsorption of fatty acids on carbon from aqueous solution; (*b*) adsorption of fatty acids on silica gel from toluene. (From Ref. 9.)

by means of Eq. X-11 as modified for the case of adsorption from a dilute solution:

$$\pi = \frac{RT}{\Sigma} \int_0^{C_2} n_2^s \, d \ln C_2 \tag{XI-15}$$

If the Langmuir equation is obeyed, then combination of Eqs. XI-5, XI-6, and X-15 gives

$$\pi = \frac{RT}{N\sigma^0} \ln (1 + bC_2) \tag{XI-16}$$

so that equal values of bC_2 correspond to equal values of π. Eq. XI-16 is known as the *Szyszkowski* equation (12). The sequence shown in Fig. XI-1*a* may thus be interpreted as meaning that, as the homologous series is ascended, successively lower concentrations suffice to give the same film pressure. This last statement is now entirely parallel to the usual form of Traube's rule as applied to the surface tension of solutions (Section III-7E).

Another observation is that there is generally an inverse relationship between the extent of adsorption of a species and its solubility in the solvent used, that is, the less soluble the material, the more strongly will it tend to be adsorbed. For example, Hansen and Craig (13) found that the adsorption isotherms of members of a homologous series of fatty acids or alcohols were superimposable on each other if plotted as grams adsorbed per gram of adsorbent versus the *reduced concentration* C_2/C_2^0. Here, C_2^0 denotes the solubility of the adsorbate in the solvent. The adsorbents used were Graphon and Spheron, the former being a rather uniform surface carbon obtained by partial graphitization of carbon black, and the solvent was water. A similar superimposability is observed in the adsorption of vapors (see Section XVI-9). One of Hansen and Craig's plots is shown in Fig. XI-2; the deviations of the lower members at high C_2/C_2^0 values stem from the fact that their solubilities were rather high so that a secondary effect, discussed in Section XI-4, was present.

The correlating role of C_2^0 provides emphasis to the view of adsorption as a partition between the solution and interfacial phases; K_1 and K_2 in Eq. XI-9 can be regarded as the separate partition coefficients. Thus, a good solvent affects ΔS_2 and ΔH_2 so as to reduce K_2, and, qualitatively, $K_2C_2^0$ tends to be a constant. The effect of increasing temperature, which is usually to decrease the adsorption, that is, to decrease b or K in Eq. XI-9, can be understood either in terms of adsorption normally being exoergic, or in terms of the preceding, as reflecting an increase in C_2^0. For example, Bartell (14) found that, although the adsorption of *n*-butyl alcohol on charcoal from dilute solutions increased with temperature, the reverse dependence developed with more concentrated solutions. This reversal was attributed to the decreasing solubility of butyl alcohol in water with increasing temperature.

The Polanyi adsorption theory may be extended to adsorption from solution (see Section XVI-7 for the detailed presentation of this theory). Briefly, the adsorbed material is regarded as, in effect, precipitated from solution as a consequence of an attractive adsorption potential. One writes

$$RT \ln \frac{C^0}{C} = \epsilon_{SL} \tag{XI-17}$$

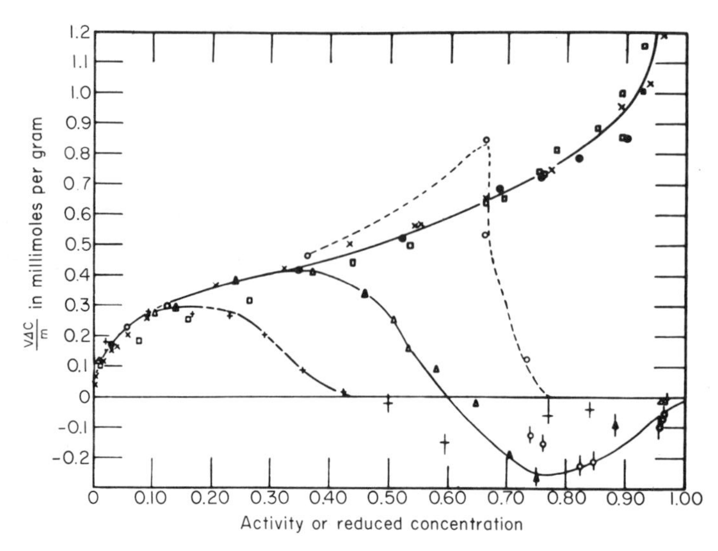

Fig. XI-2. Adsorption of fatty acids on Spheron 6: +, acetic; △, propionic; ○, *n*-butyric; ×, *n*-valeric; ●, *n*-caproic; □, *n*-heptylic. (From Ref. 13.)

where C^0 is the solubility of the adsorbate and ϵ_{SL} is the adsorption potential. This last is a function of the amount adsorbed, decreasing with increasing thickness of the adsorbed layer. The model is thus one for *multilayer* adsorption. See Ref. 16 for a discussion of this approach.

The preceding discussion helps to underline the point that adsorption from solution is a relatively complex phenomenon; it depends on the nature of solute–solvent interactions in the solution phase and in the interfacial region, as well as on their interactions with the absorbent. It therefore can be appreciated that it is difficult to be very much more specific on constitutive effects than in the discussion of Traube's rule. However, generally speaking, the adsorption constant K can be expected to be large if there is a specific opportunity for adsorbent–adsorbate hydrogen bonding. Kipling (16) cites as examples the relative affinities of silica gel for a series of nitro and nitroso derivatives of diphenylamine and *N*-ethylaniline (17) and the much stronger adsorption of phenol on charcoal than of the diorthotertiarybutyl derivative (18). It should be noted that many charcoals have partially oxidized surfaces. Thus Spheron 6, which has surface oxygens (19), adsorbs alcohol preferentially over benzene, but after heating at 2700°C (to give Graphon), it prefers benzene (20). Aromatic ring compounds tend to adsorb preferentially to aliphatic ones, for example, on carbon, presumably because of π electron interactions or, alternatively, because of the higher polarizability of such rings. Bulky substituents reduce this preference, perhaps because of their preventing a close approach of the ring to the adsorbent surface (18). High molecular weight materials such as sugars, dyes, and polymers tend to be more strongly adsorbed than low molecular weight species. In chromatography the order of elution is normally inverse as the K value for adsorption so that even the qualitative literature provides a host of comparisons. The reader is referred to reviews such as that by Neher (21).

Purely geometric effects apparently can be important in that Linde molecular sieve 5Å adsorbs hexane preferentially to benzene presumably because only the former can pass into the pores; the large pore size 10 and 13Å sieves show much stronger adsorption of benzene (22). Quite apart from such specific effects, it has been speculated that the fundamental mechanism of adsorption in a porous solid may be more akin to capillary condensation (see Section XVI-16B) than to surface adsorption. Hansen and Hansen (23) have supported this point of view; the picture would be one of relatively thick pockets of adsorbed phase held by virtue of a low solution–adsorbed phase–solid contact angle and a finite adsorbed phase–solution interfacial tension. This is a definite possibility in the case of systems near a solubility limit (see Section XI-1C), but as a general explanation this mechanism has been argued against fairly effectively (5, 24).

Finally, the adsorbent–adsorbate interaction may be so specific that the adsorption may properly be called chemisorption; the isotherm will tend to be of the simple Langmuir type, with a very large K value, and may be slow. The adsorption of fatty acids on metals is often of this type, probably due to salt formation with the oxide coating of the metal. Thus Hackerman and co-workers (25) found that fatty acids, nitriles, and so forth, showed a partially irreversible adsorption on iron and steel powder, obeying the Langmuir isotherm, while Smith and Allen (26) noted that the adsorption of *n*-nonadecanoic acid on copper, nickel, iron, and aluminum was quite dependent on whether the surface used had been exposed by machining in air or under solvent, away from oxygen. The subject, incidentally, is of considerable interest in the study of rust-preventive coatings (27, 28).

C. Multilayer Adsorption

Equation XI-5, the Langmuir equation, applies to a large number of adsorption systems where dilute solutions are involved, but some interesting cases of sigmoid isotherms have been reported. Hansen et al. (29) found that, for a number of higher acids and alcohols (four or more carbon atoms) adsorbed on various carbons from aqueous solution, the isotherms showed no saturation effect but rather the general shape characteristic of multilayer adsorption (see Section XVI-5). The final, marked increase in adsorption took place, significantly, as the saturation concentration was approached.

It is appealing to accept the parallel to low temperature gas adsorption where, after a semiplateau region in the isotherm plot, one finds rapidly increasing adsorption occurring as the saturation pressure P^0 is approached. With such gas adsorption isotherms, the rational variable is the reduced pressure P/P^0, in the case of solution adsorption, it was noted in the preceding section that C/C^0 is the correlating variable. Some of the data reported by Hansen et al. are shown in Fig. XI-2. There seems to be no doubt that some form of multilayer adsorption was occurring; otherwise, impossibly small areas per molecule would be implied.

Once the general possibility of multilayer adsorption from solution is accepted, the whole array of potential isotherms discussed in Chapter XVI becomes available for consideration. Little has been done with their application to solution adsorption, however, and mostly it is the more common BET (Brunauer, Emmett, and Teller) equation (Section XVI-5) that has been used. In the foregoing example a modified form of the BET equation, with adsorption limited to about three layers, would fit the data. The Polanyi model, mentioned in the preceeding subsection, appears to treat multilayer adsorption very well (18).

In solution adsorption, two potentially adsorbing components must be present, unlike the case with gas adsorption, and there is really no good reason to suppose that multilayer adsorption of a solute occurs with complete exclusion of solvent. In other words, the situation might more profitably be regarded as one of a phase separation induced by the interactions with the solid surface or as a capillary effect. As a potential example of the first case, Kiselev and co-workers (30) found a sigmoid isotherm for the adsorption of methanol in heptane by silica gel that rose to very high values as C_2/C_2^0 approached unity. On the other hand, Bartell and Donahue (31) reported isotherms for the adsorption of water by silica gel from solutions in hexyl alcohol in which the rapid terminal increase in adsorption took place at C_2/C_2^0 values around 0.8. Such behavior is characteristic of capillary condensation, in the case of gas adsorption, and it is possible here that a capillary induced phase separation occurred. Since water is preferentially adsorbed, there would be a tendency, with increasing concentration, for a water-rich layer to deepen and to go over to meniscus formation in a pore if the capillary pressure from the resulting curvature provided a positive driving force. This would be the case if the effect of the pressure were to make the local value of C_2^0 smaller.

A perhaps simpler type of multilayer formation is that in which physical adsorption of a species occurs on top of its own chemisorbed layer. The observed isotherm may then be a sum of two Langmuir isotherms, but if the chemisorption is complete at a low concentration, the effect is that of an isotherm that is ordinary in appearance except that it originates from a point a way up on the adsorption axis of the isotherm plot. Behavior of this type was observed in the adsorption of caproic and stearic acids on steel (32). While this interpretation is plausible in the case cited, the same behavior could in principle result from the presence of two kinds of surface, one much more strongly adsorbing than the other.

2. Adsorption of Polymers

The study of adsorption of polymers is so interwoven with the general field of polymer chemistry, and hence so relatively specialized, that only a brief summary presentation is attempted here. First, the requirement that the polymer be soluble limits solution adsorption studies mainly to linear macromolecules. As summarized in two short review articles by Eirich and coworkers (33, 34) and by Kipling (16), these include synthetic rubber polymers, cellulose-type polymers, and methacrylate, vinyl, styrene, and so on, polymers, mostly in fairly polar organic solvents and mostly with carbon as the adsorbent (perhaps because of the bias of the rubber industry). A second point is that polymers as prepared are generally polydisperse, and their adsorption is more that of a multicomponent system in which fractionation effects can be important. The more recent work has stressed the use of at least relatively narrow molecular weight fractions. Third, as was true at the water–air interface (Section IV-11), a large number of configurations at the solid–solution interface are possible and probably for this reason adsorption equilibrium can be exceedingly slow in attainment; adsorption that appears to have leveled off after an hour or two may actually be subject to continued drift upward for days or months (note Ref. 35). Heller (36) give the equation

$$\frac{1}{(x/m)} = k + k't \tag{XI-18}$$

where x/m is grams adsorbed per gram of adsorbent and t is time. Fourth, it takes several parameters to describe the state of a polymer at an interface. These include the number of points of attachment, the horizontal spread as given by the average radius $(\bar{r}^2)^{1/2}$, and the thickness Δr, as illustrated in Fig. XI-3. There are, in fact, more parameters than realistically can be extracted from adsorption data, and this has made it difficult to confirm or deny the various proposed models.

A very simple model is that derived from the mass action approach in which Eq. XI-2 is modified by writing that v molecules of solvent are displaced per polymer molecule. This introduces $(N_1^s)^v$ so that we have (37)

$$\frac{\theta}{v(1-\theta)^v} = bC_2 \tag{XI-19}$$

Next, it is possible to introduce various assumptions concerning the adsorption statistics, for example, that there is a Gaussian distribution of end-to-end distances. An approach of this type led to the equation (38)

$$\frac{\theta}{(1-\theta)e^{2K_2\theta}} = (KC)^{1/v} \tag{XI-20}$$

An even more detailed approach is that of Silberberg (39) (note Fig. IV-41) and DiMarzio and co-workers (40). Everett and co-workers have made a statistical mechanical calculation of the adsorption of monomers and r-mers (41).

In view of the sophistication of the various treatments it is almost embarrassing

Fig. XI-3. Hypothetical conformation of an adsorbed chain molecule. (From Ref. 34.)

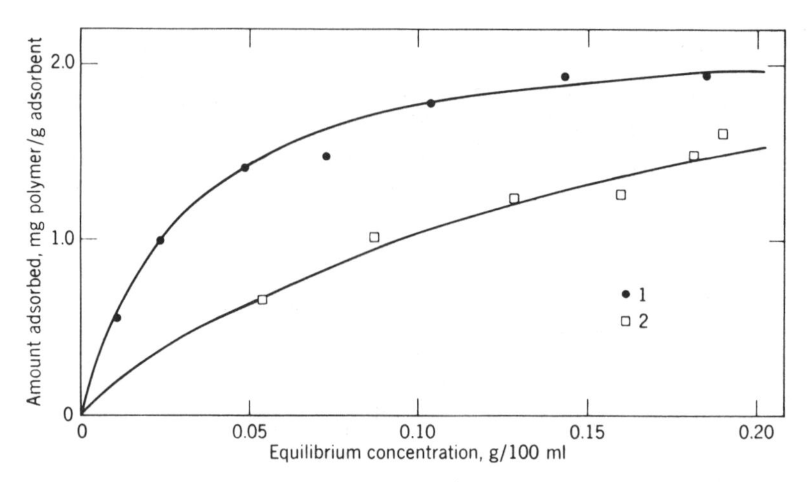

Fig. XI-4. Adsorption of polystyrene from benzene onto pyrex glass at 30°C. Curve 1: polymer molecular weight 950,000; curve 2: polymer molecular weight 110,000. (From Ref. 43.)

that most polymer adsorption data fit the simple Langmuir equation, Eq. XI-5, as well as any other, within experimental error (see Refs. 36, 42).

Representative adsorption isotherms are shown in Fig. XI-4. The situation is sufficiently complex in the case of polymers that an analysis of why the mass action expectation, Eq. XI-19, should apparently fail has not been easy. One expectation that is met, however, is that illustrated in Fig. XI-5. The root-mean-square limiting thickness of adsorbed polymer films does tend to increase with the square root of the polymer length, indicating that long segments of polymer do extend into the solution and are coiled much like dissolved polymer is.

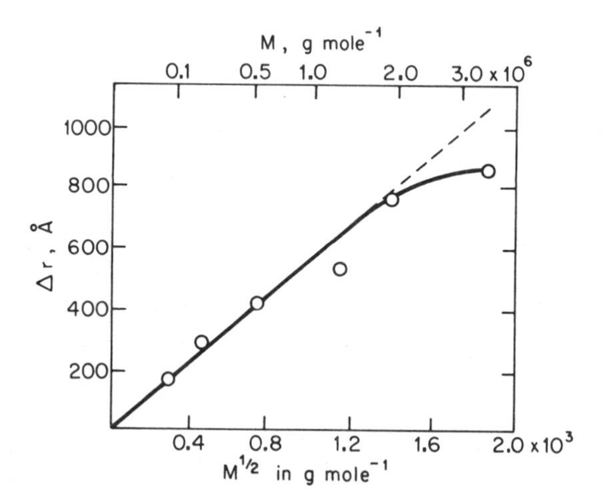

Fig. XI-5. Root-mean-square thickness in the plateau region of polystyrene adsorbed on chrome ferrotype plate plotted against the square root of the molecular weight. The solvent is cyclohexane. (From Ref. 33.)

Another deviant aspect of polymer adsorption is that adsorption may *increase* with increasing temperature; the adsorption must then be entropy rather than energy favored. Since the polymer must surely lose entropy on adsorption, the effect must come from a gain in solvent entropy. Yet this gain apparently is not so simply explained as in terms of release of adsorbed solvent to the solution, in view of the failure of Eq. XI-19. The interfacial phase point of view is probably more realistic; adsorbed polymer films can be quite thick, as illustrated in Fig. XI-5. Film thicknesses, Δr in Fig. XI-3, are often measured hydrodynamically, in terms of the increase in the apparent adsorbent particle radius as given by viscosity–concentration measurements. Alternatively, the change in effective pore diameter in capillary flow, following adsorption, may be measured. Ellipsometry has also been used (44).

As a reminder that polymer adsorption data may rarely represent final equilibrium, hysteresis is usually observed on desorption (see Refs. 35 and 43, for example). The form taken seems to be more that of residual difficultly desorbable polymer, that is, an open loop, rather than that of the closed loops found in capillary condensation systems (see Section XVI-17). The situation is an interesting one. Desorption is slow if, after reaching an adsorption equilibrium, the solution is diluted or replaced by pure (but the same) solvent. Yet the adsorbed polymer is *not* irreversibly bound to the surface! If labeled polymer is added to an equilibrium mixture, exhange with the adsorbed polymer occurs. This means that adsorbed polymer can be displaced by another polymer molecule, but not by solvent. Also, desorption can be rapid if a good solvent replaces an original poor solvent ("good" in this context means a solvent in which solvent–polymer interaction is strong and the polymer molecule has a large radius of gyration in solution; the reverse applies in the case of a "poor" solvent). (See Ref. 45.)

Polymer adsorption on suspended particles can have an important stabilizing effect. When two particles approach, the extended loops of adsorbed polymers become compressed, leading to an adverse entropy effect. Also, solution polymer becomes excluded, with much the same consequence (46).

3. Surface Area Determination

The estimation of surface area from solution adsorption studies is subject to many of the same considerations as in the case of gas adsorption, but with the added complication that larger molecules are involved, whose surface orientation and pore penetrability may be uncertain. A first condition is that a definite adsorption model be obeyed, which in practice means that area determinations are limited to cases in which the simple Langmuir equation, Eq. XI-5, holds. The constant n^s is found, for example, from a plot of the data according to Eq. XI-10, and the specific surface area Σ then follows from Eq. XI-6. The problem is to pick the correct value of σ^0.

In the case of gas adsorption where the BET method is used (Section XVI-5) it is reasonable to use the van der Waals area of the adsorbate molecule; moreover, being small or even monatomic, surface orientation is not a major problem. In the case of adsorption from solution, however, the adsorption may be chemisorption, with σ^0 determined by the spacing of adsorbent sites, or physical adsorption, with σ^0 more determined by the area of the adsorbate molecule in its particular orientation.

Fatty acid adsorption has been used for surface area estimation, because of evidence that in many cases the orientation is perpendicular to the surface and with about the close-packed area per molecule of 20.5 Å. This seems to be true for adsorption on such diverse solids as carbon black and not too electropositive metals, and for TiO_2. In all of these cases, the adsorption is probably chemisorption in type, involving hydrogen bonding or actual salt formation with surface oxygen. Fairly polar solvents are used to avoid multilayer formation on top of the first layer, but even so, the apparent area obtained may vary with the solvent used. In the case of stearic acid on a graphitized carbon surface, Graphon, the adsorption, while still obeying the Langmuir equation, appears to be physical, with the molecules lying flat on the surface. In brief, the method must be applied with caution, and with confirmatory evidence. As another example, Bisio and co-workers (46a), in studying the adsorption of surfactants on polycarbonate, concluded that depending on surfactant and concentration the adsorbed molecules might be lying flat on the surface, perpendicular to it, or might form a bilayer.

A second class of adsorbates of which much use has been made is that of dyestuffs; the method is appealing because of the ease with which analysis may be made colorimetrically. The adsorption generally follows the Langmuir equation, but can be multilayer. Graham found an apparent molecular area of 197 $Å^2$ for methylene blue on Graphon (47) or larger than the actual molecular area of 175 $Å^2$, but the apparent value for the more oxidized surface of Spheron was about 105 $Å^2$ per molecule (48). As discussed by Padday (49), problems may arise because of dye association in solution and on the surface. As with fatty acid adsorption, the dye method must be used with caution.

Rahman and Ghosh (49a) have used pyridine adsorption on various oxides to obtain surface areas. Adsorption followed the Langmuir equation; the effective molecular area of pyridine is about 24 $Å^2$ per molecule.

Two quite different approaches are the following. Everett (50) has proposed a method using binary liquid systems (see the next section); the approach has been discussed more recently by Schay and Nagy (51). Surface areas may be estimated from the exclusion of like charged ions from a charged interface (52). The method, discussed further in Section XI-5, is intriguing in that no estimation of either site or molecular area is called for.

In general, however, surface area determination by means of solution adsorption studies, while convenient experimentally, cannot be considered free of systematic error. See also Section XV-2 for a discussion of how to define surface area phenomenologically. Nonetheless, if a solution adsorption procedure has been standardized for a given system, by means of independent checks, it can be very useful for determining relative areas of a series of similar materials.

4. Adsorption in Binary Liquid Systems

A. Adsorption at the Solid–Solution Interface

The discussion so far has been confined to systems in which the solute species are dilute, so that adsorption was not accompanied by any significant change in the activity of the solvent. In the case of adsorption from binary liquid mixtures, where the complete range of concentration, from pure liquid A to pure liquid B, is available, a more elaborate analysis is needed. The terms solute and solvent are no longer meaningful, but it is nonetheless convenient to cast the equations around one of the components, arbitrarily designated here as component 2.

Adsorption is still as defined by Eq. XI-1, but only in the form

$$n_2^s \text{ (apparent)} = n_0 \, \Delta N_2^l \tag{XI-21}$$

since in concentrated solutions, concentration units become awkward to use because density is now also a function of composition; the superscript l will be used where helpful to make it clear that the quantity is for the solution phase. Furthermore, the adsorption defined by Eq. XI-21 is now an apparent adsorption, that is, is no longer the actual moles of the component adsorbed. However, it does turn out that the apparent adsorption is simply related to a surface excess quantity, as shown in the following demonstration.

We suppose that the Gibbs dividing surface (see Section III-5) is located at the surface of the solid (with the implication that the solid itself is not soluble). It follows that the surface excess Γ_2^s, according to this definition, is given by (see Problem XI-7)

$$\Gamma_2^s = \frac{n^s}{\Sigma} (N_2^s - N_2^l) \tag{XI-22}$$

Here, n^s denotes the total number of moles associated with the adsorbed layer, and N_1^s and N_1^l are the respective mole fractions in that layer and in solution at equilibrium. As before, it is assumed, for convenience, that mole numbers refer to that amount of system associated with one *gram of adsorbent*. Equation XI-22 may be written

$$\Gamma_2^s = \frac{n^s}{\Sigma} \left(\frac{n_2^s}{n^s} - \frac{n_2^l}{n^l} \right) \tag{XI-23}$$

where n_2^s and n_2^l are the moles of component 2 in the adsorbed layer and in solution. Since $n_2^s + n_2^l = n_2^0$, the total number of moles of component 2 present, and $n^s + n^l = n_0$, the total number of moles in the system, substitution into Eq. XI-23 yields

$$\Gamma_2^s = \frac{n_0}{\Sigma} (N_2^0 - N_2^l) = \frac{n_0 \, \Delta N_2^l}{\Sigma} \tag{XI-24}$$

where N_2^0 is the mole fraction of component 2 before adsorption.

Another form of Eq. XI-24 may be obtained (from Eq. IX-23 and remembering that $N_1^l + N_2^l = 1$ and $n^s = n_1^s + n_2^s$):

$$\Gamma_2^s = \frac{n_0 \, \Delta N_2^l}{\Sigma} = \frac{n_2^s N_1^l - n_1^s N_2^l}{\Sigma} \tag{XI-25}$$

It is important to note that the experimentally defined or *apparent adsorption* $n_0 \, \Delta N_2^l / \Sigma$, while it gives Γ_2^s, does *not* give the amount of component 2 in the adsorbed layer n_2^s. Only in dilute solution where $N_2^l \to 0$ and $N_1^l \simeq 1$ is this true. The adsorption isotherm, Γ_2^s plotted against N_2, is thus a *composite isotherm* or, as it is sometimes called, the *isotherm of composition change*.

Equation XI-25 shows that Γ_2^s can be viewed as related to the difference between the individual adsorption isotherms of components 1 and 2. Figure

XI-6 (53) shows the composite isotherms resulting from various combinations of individual ones. Note in particular Fig. XI-6*a*, which shows that even in the absence of adsorption of component 1, that of component 2 must go through a maximum (due to the N_1^l factor in Eq. XI-25), and that in all other cases the apparent adsorption of component 2 will be negative in concentrated solution.

Everett and co-workers (54) describe an improved experimental procedure for obtaining Γ_i^s quantities. Some of their data are shown in Fig. XI-7. Note the negative region for n_1^s at the lower temperatures.

A elegant and very interesting approach was that of Kipling and Tester (55), who determined the separate adsorption isotherms for the vapors of benzene and of ethanol on charcoal, that is, the adsorbent was equilibrated with the vapor in equilibrium with a given solution, and from the gain in weight of the adsorbent and the change in solution composition, following adsorption, the amounts of each component in the adsorbed film could be calculated. These individual component isotherms could then be inserted in Eq. XI-25 to give a calculated apparent solution adsorption isotherm, which in fact agreed well with the one determined directly. Their data are illustrated in Fig. XI-8. They also determined the separate adsorption isotherms for benzene and charcoal and ethanol and charcoal; these obeyed

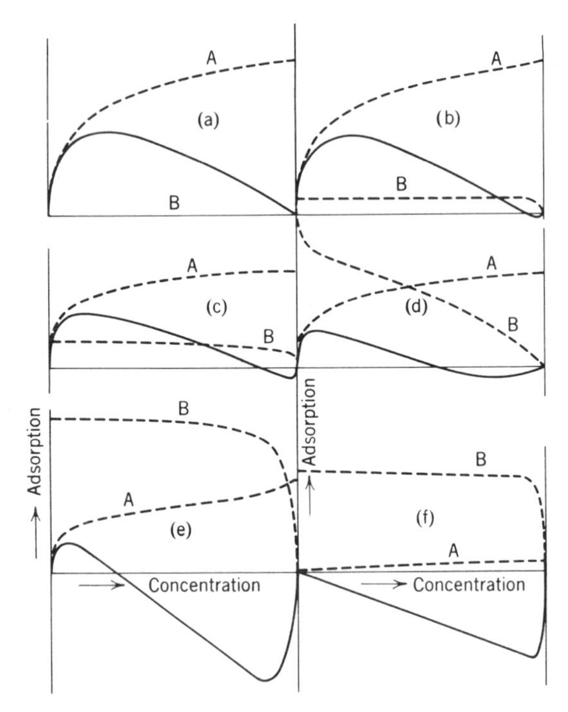

Fig. XI-6. Composite adsorption isotherms; – – –, individual isotherms; ———, isotherms of composition change. (From Ref. 53.)

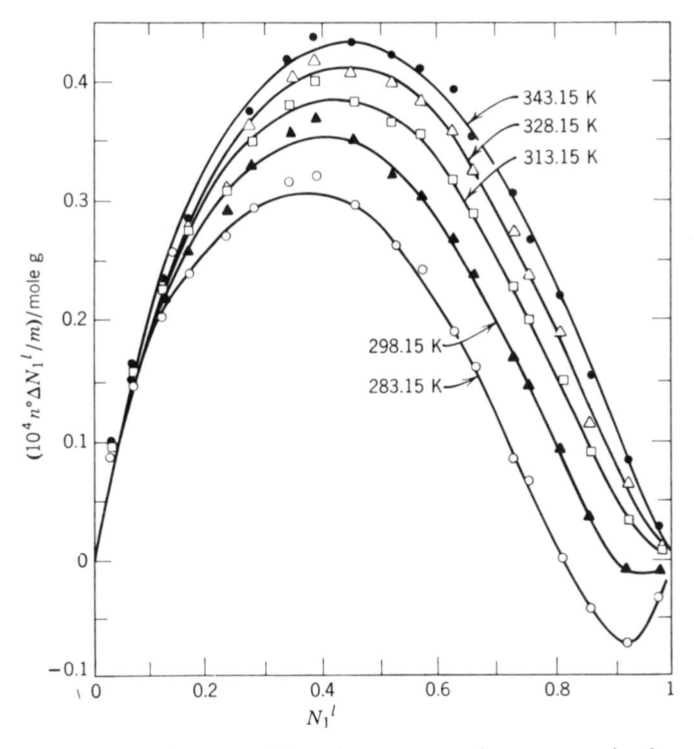

Fig. XI-7. Isotherm of composition change or surface excess isotherm for the adsorption of (1) benzene and (2) *n*-heptane on graphon. (From Ref. 54.)

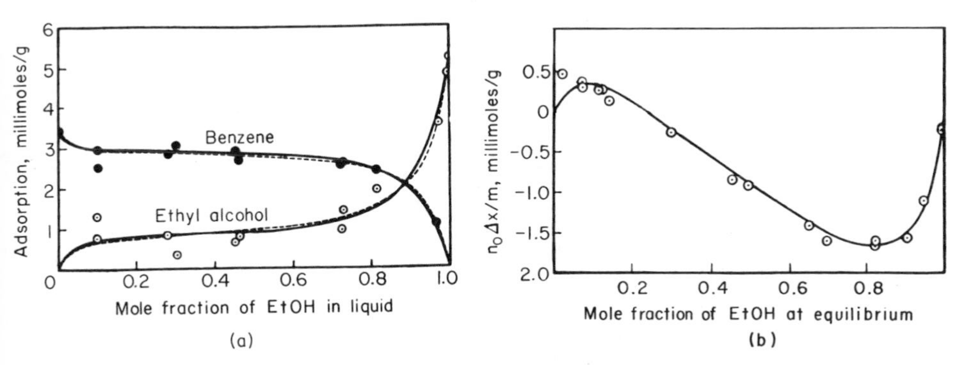

Fig. XI-8. Relation of adsorption from binary liquid mixtures to the separate vapor pressure adsorption isotherms system: ethanol–benzene–charcoal; (*a*) separate mixed vapor isotherms; (*b*) calculated and observed adsorption from liquid mixtures. (From Ref. 55.)

Langmuir equations for gas adsorption (Eq. XI-4)

$$\theta_1 = \frac{b_1 P_1}{1 + b_1 P_1}, \qquad \theta_2 = \frac{b_2 P_2}{1 + b_2 P_2} \tag{XI-26}$$

from which the constants b_1 and b_2 were thus separately evaluated. The composite vapor adsorption isotherms of Fig. XI-8*a* were then calculated, using these constants and the added assumption that in this case no bare surface was present. The Langmuir equation for the competitive adsorption of two gas-phase components is (see Section XVI-3)

$$\theta_2 = \frac{b_2 P_2}{1 + b_1 P_1 + b_2 P_2} \tag{XI-27}$$

and the effect of their assumptions was to put Eq. XI-27 in the form

$$\theta_2 = \frac{b_2 P_2}{b_1 P_1 + b_2 P_2} \simeq \frac{b_2 P_2^0 N_2^l}{b_1 P_1^0 N_1^l + b_2 P_2^0 N_2^l} \tag{XI-28}$$

and to identify b_2 and b_1 with the separately determined values from Eqs. XI-26. Again, good agreement was found, as shown by the dotted lines in Fig. XI-8*a*.

The Langmuir model as developed in Section XI-1 may be applied directly to Eq. XI-22 (4). We replace a_1 and a_2 in Eq. XI-3 by N_1 and N_2 (omitting the superscript l as no longer necessary for clarity), and solve for N_2^s (with N_1 replaced by $1 - N_2$):

$$N_2^s = N_2 \frac{K}{1 + (K - 1)N_2} \tag{XI-29}$$

This is now substituted into Eq. XI-22, giving

$$\Gamma_2^s = \frac{n^s}{\Sigma} \frac{(K - 1)N_1 N_2}{1 + (K - 1)N_2} \tag{XI-30}$$

Or, using the apparent adsorption, $n^0 \, \Delta N_2$, Eq. XI-30 may be put in the linear form

$$\frac{N_1 N_2}{n_0 \, \Delta N_2} = \frac{1}{n^s(K - 1)} + \frac{1}{n^s} N_2 \tag{XI-31}$$

which bears a close resemblance to that for the simple Langmuir equation, Eq. XI-10. Note that for $K = 1$, $\Gamma_2^s = 0$, that is, no fractionation occurs at the interface. Everett (4) found Eq. XI-31 to be obeyed by several systems, for example, that of benzene and cyclohexane on Spheron 6.

An interesting development that avoids the assumption that activities in the adsorbed phase can be represented by mole fractions is the following. First, if the two adsorbing species are of different molecular sizes, it seems reasonable to write Eq. XI-2 as

$$\frac{\text{(species 1 in solution)}}{\sigma_1} + \frac{\text{(species 2 in adsorbed phase)}}{\sigma_2} = \frac{\text{(species 1 in adsorbed phase)}}{\sigma_1} + \frac{\text{(species 2 in solution)}}{\sigma_2} \qquad \text{(XI-32)}$$

where σ denotes molecular area (or volume, if an adsorbed phase of constant thickness is assumed). Equation XI-3 may then be written

$$K = \frac{a_1^l}{a_1^s}\left(\frac{a_2^s}{a_2^l}\right)^{\sigma_1/\sigma_2} \qquad \text{(XI-33)}$$

Through the use of equations like Eq. III-53 the activity ratios can be written in terms of surface tension differences which, in turn, can be eliminated by means of the Gibbs–Duhem and Gibbs adsorption equations (Eqs. III-81 and III-78) and integration:

$$\ln K = -\sigma_1 \int_{a_1=0}^{a_1=1} \frac{\Gamma_2^s}{N_2}\, d\ln a_1 \qquad \text{(XI-34)}$$

(see Refs. 4, 52, and 56). Shay has made similar developments (57).

Isotherms of type a in Fig. XI-6 are relatively linear for large N_2, that is,

$$n_0\,\Delta N_2^l = a - bN_2^l \qquad \text{(XI-35)}$$

Now, Eq. XI-25 can be written in the form

$$n_0\,\Delta N_2^l = n_2^s - n^s N_2^l \qquad \text{(XI-36)}$$

(since $n^s = n_1^s + n_2^s$ and $N_1 + N_2 = 1$). Comparing Eqs. XI-35 and XI-36, the slope b gives the monolayer capacity n^s. The surface area Σ follows if the molecular area can be estimated. The treatment assumes that in the linear region the surface is mostly occupied by species 2 so that n_2^s is nearly constant. See Refs. 58 and 59.

There is a number of very pleasing and instructive relationships between adsorption from a binary solution at the solid–solution interface and that at the solution–vapor and the solid–vapor interfaces. The subject is sufficiently specialized, however, that the reader is referred to the general references and, in particular, to Ref. 17.

The preceding material has contained the tacit assumption that the surfaces are homogeneous, that is, uniform. This will not in general be true, and the separate adsorption isotherm equations will be correspondingly more complex. Also, where surface heterogeneity can be varied, major effects can result on the isotherms of composition change. This was true, for example in the methanol–benzene and n-butanol benzene adsorption on carbon blacks on varying the amount of edge atoms or of surface oxide groups (60).

B. Heat of Adsorption at the Solid–Solution Interface

Rather little has been done on heats of wetting of a solid by a solution, but two examples suggest a fairly ideal type of behavior. Young et al. (59) studied the Graphon–aqueous butanol system, for which monolayer adsorption of butanol was completed at a fairly low concentration. From direct adsorption studies, they determined θ_b, the fraction of surface covered by butanol, as a function of concentration. They then prorated the heat of immersion of Graphon in butanol, 113 erg/cm^2, and of Graphon in water, 32 erg/cm^2, according to θ_b. That is, each component in the adsorbed film was considered to interact with its portion of the surface independently of the other,

$$q_{\mathrm{imm}} = N_1^s q_1 + N_2^s q_2 \tag{XI-37}$$

where q_{imm} is the heat of immersion in the solution and q_1 and q_2 are the heats of immersion in the respective pure liquid components. To this prorated q_{imm} was added the heat effect due to concentrating butanol from its aqueous solution to the composition of the interfacial solution, using bulk heat of solution data. They found that their heats of immersion calculated in this way agreed very well with the experimental values.

As a numerical illustration, from the solution adsorption data, $\theta = 0.5$ at a butanol concentration of 0.3 g/100 ml. The heat of solution was about 25 cal/g of dissolved butanol. The molecular area of butanol was taken to be 40 Å^2, corresponding to 3×10^{-8} g/cm^2 or 1.5×10^{-8} g/cm^2 at $\theta = 0.5$. To remove this much butanol from solution then required about 15 erg, and the heat of interaction with the Graphon was $(113 + 32)/2$ or 72 erg, giving a net calculated heat of immersion of about 57 erg/cm^2, which was close to what was observed experimentally. A similar approach worked fairly well in the case of the benzene–cyclohexane–charcoal system (61). The success of this type of calculation suggests that no very important adsorbate–adsorbate interactions are present in the mixed surface phase.

As a quite different and more fundamental approach, the isotherms of Fig. XI-7 allowed a calculation of K as a function of temperature, through Eq. XI-34. The plot of ln K versus $1/T$ gave an enthalpy quantity which should be just the difference between the heats of immersion of the Graphon in benzene and in *n*-heptane, or 2.6×10^{-3} cal/m^2 (54). The experimental heat of immersion difference is 2.4×10^{-3} cal/m^2, or probably indistinguishable.

5. Adsorption of Electrolytes

The interaction of an electrolyte with an adsorbent may take one of several forms. Several of these are discussed, albeit briefly, in what follows. The electrolyte may be adsorbed *in toto*, in which case the situation is similar to that for molecular adsorption. It is more often true, however, that ions of one sign are held more strongly, with those of the opposite sign forming a diffuse or secondary layer. The surface may be polar, with a potential ψ, so that primary adsorption can be treated in terms of the Stern model (Section V-4), or the adsorption of interest may involve exchange of ions in the diffuse layer.

In the case of ion exchangers, the primary ions are chemically bonded into the framework of the polymer, and the exchange is between ions in the secondary layer. A few illustrations of these various types of processes follows.

A. Stern Layer Adsorption

Adsorption at a charged surface where both electrostatic and specific chemical forces are involved has been discussed to some extent in connection with various other topics. These examples are drawn together here for a brief review along with some more specific additional material. The Stern equation, Eq. V-25, may be put in a form more analogous to the Langmuir equation, Eq. XI-5:

$$\frac{\theta}{1 - \theta} = C_2 \exp\left(\frac{ze\psi + \phi}{kT}\right) \qquad \text{(XI-38)}$$

The effect is to write the adsorption free energy or, approximately, the energy of adsorption Q as a sum of electrostatic and chemical contributions.

Stern layer adsorption was involved in the discussion of the effect of ions on ζ potentials (Section V-8), electrocapillary behavior (Section V-9), and electrode potentials (Section V-10) and enters into the effect of electrolytes on charged monolayers (Section IV-13). More specifically, this type of behavior occurs in the adsorption of electrolytes by ionic crystals. A large amount of work of this type has been done, partly because of the importance of such effects on the purity of precipitates of analytical interest, and partly because of the role of such adsorption in coagulation and other colloid chemical processes.

As an example, Weiser (62) studied the adsorption of various ions by barium sulfate, concluding that it is the solubility of the barium salt of the ion rather than ion charge that more nearly correlates with the extent of adsorption. Paneth, Hahn, and Fajans, in their studies of this type of adsorption, concluded that an ion tends to be strongly adsorbed on a crystalline solid if it forms a difficultly soluble or weakly dissociated compound with the oppositely charged ion of the crystal (63). This type of adsorption thus seems more controlled by the ϕ than by ψ term of Eq. XI-38.

Adsorption on ionic crystals may easily be complicated by *aging* effects. The surface area of freshly formed precipitates generally decreases steadily with time, along with a decrease in *specific* surface energy due to reduction in the number of edges, corners, and other high energy features. One example of a detailed study is that of Kolthoff and co-workers (64).

It is possible, next, for neutral species to be adsorbed even where specific chemical interaction is not important, owing to a consequence of an electrical double layer. As discussed in connection with Fig. V-4, the osmotic pressure of solvent is reduced in the region between two charged plates. There is, therefore, an equilibrium with bulk solution that can be shifted by changing the external ionic strength. As a typical example of such an effect, the interlayer spacing of montmorillonite (a sheet or layer type alumino-silicate see Fig. XI-9) is very dependent on the external ionic strength. A spacing of about 19 Å in dilute electrolyte reduces to about 15 Å in 1 to 2 *M* 1:1 electrolyte (65). Phenomenologically, a strong water adsorption is repressed by added electrolyte.

If specific chemical interactions are not dominant, the adsorption of an ionic species is largely determined by its charge. This is the basis for an early conclusion (66) that the order of increasing adsorption of ions by sols is that of increasing charge. The extent of adsorption in turn determines the power of such ions to coagulate sols and accounts for the related statement that the coagulating ability of an ion will be greater the higher its charge. This rule, the Schulze–Hardy rule, is discussed in Section VI-5B.

Very often both chemical and electrical interactions are important. For example, Connor and Ottewill concluded that the adsorption of long chain quaternary ammonium ions on latex particles was at first largely electrostatic (67). The surface initially is negatively charged, owing to surface carboxyl groups. At the knee of the isotherm shown in Fig. XI-10, this surface charge has been neutralized (the direction of electrophoretic motion of the particles reverses), and the further adsorption is due to attraction of the alkyl chains to the surface. At the highest concentrations some association may be occurring. Fuerstenau and co-workers have in fact proposed that in systems of this kind (a specific one for them being that of sodium dodecylbenzene

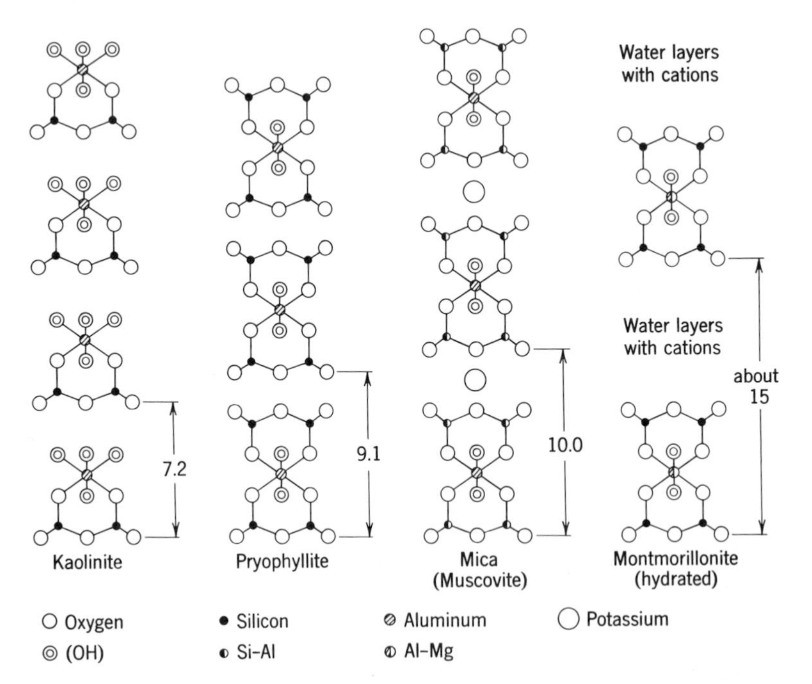

Fig. XI-9. End-on view of the layer structures of clays, pyrophillite, and mica. (From Ref. 68.)

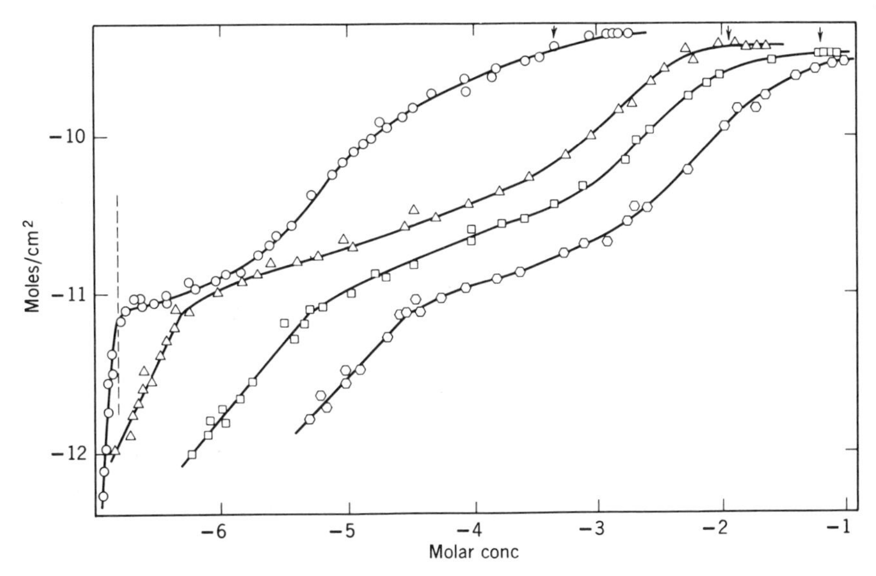

Fig. XI-10. Adsorption isotherms on particles of Latex-G at pH 8 in 10^{-3} *M* KBr solution. ○, hexadecyltrimethylammonium ion, △, dodecyltrimethylammonium ion; □, decyltrimethylammonium ion; ⬡, octyltrimethylammonium ion. Arrows mark the cmc values; the vertical dashed line marks the reversal of charge for the hexadecyltrimethylammonium ion case. (From Ref. 67.)

sulfonate on alumina) a surface aggregation occurs to give *hemimicelles* (69). The allusion is to the micelles that colloidal electrolytes form in solution above a certain concentration—the critical micelle concentration (cmc) (see Section XIII-5). While it is reasonable to suppose that surface association can occur, it is questionable whether the structure could be like that of solution micelles.

The effect of adsorption on the charge of the adsorbent particles may be determined from electrophoretic measurements and then expressed as changes in ζ potential. An example is provided by Fig. XI-11, showing that adsorption of cations by quartz eventually reduces the ζ to zero. If, much in the manner of Traube's rule studies (Section XI-1B), the concentration required to give ζ potential is regarded as determined primarily by ϕ in Eq. XI-38, then the observation that log C_2 showed a linear dependence of chain length can be accounted for. Moreover, the slope $(\partial \ln C_2/\partial n)_{\zeta=0}$, where n is the chain length, gave an energy increment of about 600 cal/CH_2 group, or about the same as at the water–air interface (see Section III-7D). Fuerstenau interprets this as evidence for surface association or "hemimicelle" formation.

Surface charge may be controlled or fixed by a potential determining ion. Table XI-1 (from Ref. 71) lists the potential determining ion and its concentration giving zero charge on the mineral. There is a large family of

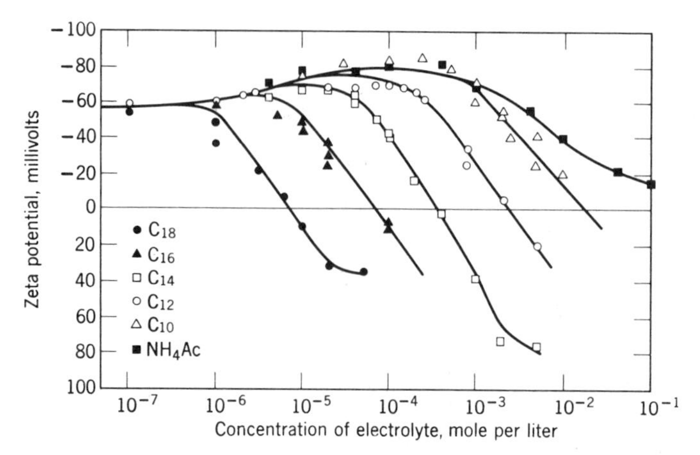

Fig. XI-11. Effect of hydrocarbon chain length on the ζ potential of quartz in solutions of alkylammonium acetates and in solutions of ammonium acetate. (From Ref. 70.)

minerals for which hydrogen (or hydroxide) ion is potential determining—oxides, silicates, phosphates, carbonates, and so on. For these, adsorption of surfactant ions is highly pH dependent. An example is shown in Fig. XI-12. This type of behavior has important applications in flotation and is discussed further in Section XIII-4C.

Electrolyte adsorption on metals is important in electrochemistry. One study reports the adsorption of various anions on Ag, Au, Rh, and Ni electrodes, using ellipsometry. The interpretations were complicated by the effect of charge on the optical properties of the metals (72).

TABLE XI-1
Potential Determining Ion and Point of Zero Charge[a]

Material	Potential determining ion	Point of zero charge
Fluorapatite, $Ca_5(PO_4)_3(F, OH)$	H^+	pH 6
Hydroxyapatite, $Ca_5(PO_4)_3(OH)$	H^+	pH 7
Alumina, Al_2O_3	H^+	pH 9
Calcite, $CaCO_3$	H^+	pH 9.5
Fluorite, CaF_2	Ca^{2+}	*p*Ca 3
Barite (synthetic), $BaSO_4$	Ba^{2+}	*p*Ba 6.7
Silver iodide	Ag^+	*p*Ag 5.6
Silver chloride	Ag^+	*p*Ag 4
Silver sulfide	Ag^+	*p*Ag 10.2

[a] From Ref. 71.

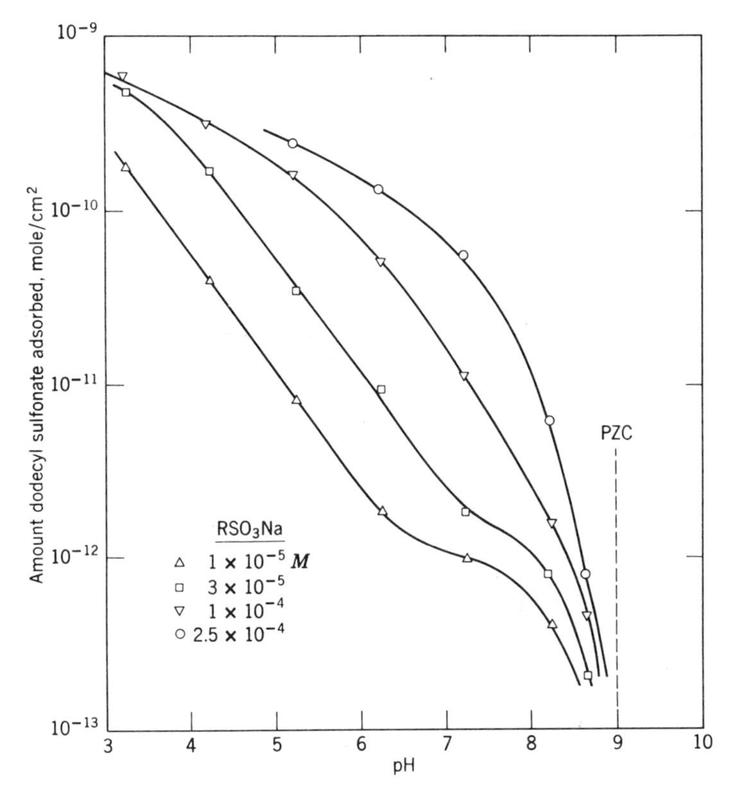

Fig. XI-12. Adsorption of sodium dodecyl sulfonate on alumina as a function of pH, in 2×10^{-3} *M* NaCl solution. (From Ref. 71.)

B. Surface Areas from Negative Adsorption

An interesting application of electrical double layer theory has been to the estimation of surface area from the degree of *exclusion* from the solid–solution interface of ions having the same charge as that at the interface. According to Eq. V-2 and as illustrated in Fig. XI-13, the concentration of negative ions should diminish near a negatively charged interface. As a consequence of this exclusion, the solution concentration should increase from n_0 to n_0' on equilibration with the solid. By material balance,

$$\mathcal{A}\Gamma^- = V\,\Delta n_0 \tag{XI-39}$$

where $\mathcal{A}$ is the surface area of solid added to volume V of solution, Γ^- is the surface (negative) adsorption of the negative ions, and $\Delta n_0 = n_0' - n_0$ is the increase in concentration.

The negative adsorption, given by the shaded area shown in the figure, is

$$\Gamma^- = \int_0^\infty (n_0' - n^-)\,dx \tag{XI-40}$$

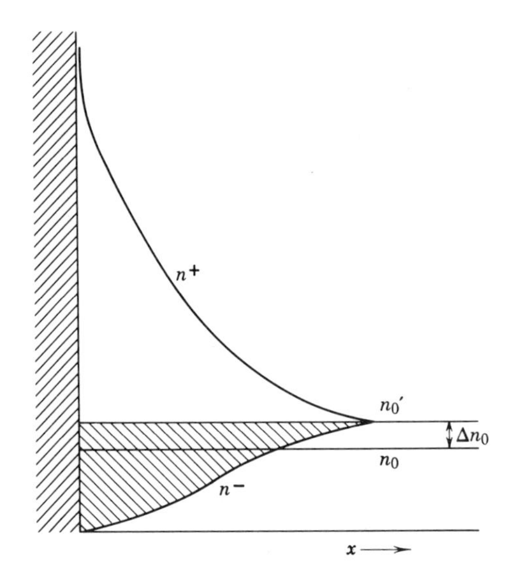

Fig. XI-13. Distribution of ions at an uncharged surface n_0 and at a negatively charged surface assuming ψ_0 is very large. (From Ref. 52.)

A final expression results if n^- is replaced by the appropriate Eq. V-2, and dx expressed in terms of $d\psi$ by means of Eq. V-11. In simplifying the result, Van den Hul and Lyklema (52) obtain the approximate equation for aqueous 1:1 electrolyte solutions at 20°C:

$$\mathscr{A} = \frac{0.52 \times 10^9 V \, \Delta n_0}{\sqrt{n_0'}} \tag{XI-41}$$

where V is in cubic centimeters and concentrations are in moles per cubic centimeter. The method was applied successfully to the measurement of the surface area of AgI suspensions; the required assumption that ψ_0 was constant (and high) was met by addition of suitable concentration of I^- ions.

The negative adsorption method was introduced by Schofield (73, 74) who developed equations similar to the foregoing, but with the assumption of constant surface charge for the solid. With this assumption, the method is particularly suitable to zeolites and clays whose charge is likely due to discreet charged sites. A determination of $\mathscr{A}$ for a montmorillonite by chloride ion exclusion was given by Edwards and Quirk (75).

C. Counterion Adsorption—Ion Exchange

A very important class of adsorbents consists of those having charged sites due to ions or ionic groups bound into the lattice. The montmorillonite clays, for example, consist of layers of tetrahedral SiO_4 units sharing corners with octahedral Al^{3+}, having coordinated oxygen and hydroxyl groups, as illustrated in Fig. XI-9 (68). In not too acid solution, cation exchange with protons is possible, and because of the layer structure swelling effects occur, which are understandable in terms of ionic strength effects on double layer repulsion (see Section VI-5B and Ref. 76).

The aluminum silicate can be regarded as networks of silicate tetrahedra with some replacement by aluminum, so that electroneutrality requires the inclusion of hydroxyl groups (i.e., protons) or of other cations. A tremendous variety of structures is known, and some of the three-dimensional network ones are porous enough to show the same type of swelling phenomena as the layer structures—and also ion exchange behavior. The zeolites fall in this last category and have been studied extensively, both as ion exchangers and as gas adsorbents (e.g., Ref. 77 and 78). As a recent example, Goulding and Talibudeen have reported on isotherms and calorimetric heats of Ca^{2+}–K^+ exchange for several aluminosilicates (79).

Organic ion exchangers were introduced in 1935, and a great variety is now available. The first ones consisted of phenol–formaldehyde polymers into which natural phenols had been incorporated, but now various polystyrene polymers are much more common. Here RSO_3^- groups, inserted by sulfonation of the polymer, are sufficiently acidic that ion exchange can occur even in quite acid solution. The properties can be controlled by varying the degree of sulfonation and of cross-linking. Other anionic groups, such as $RCCO^-$, may be introduced to vary the selectivity. Also, anion exchangers having RNH_2 groups are in wide use. Here addition of acid gives the $RNH_3^+X^-$ function, and anions may now exchange with X^-.

Both the kinetics and the equilibrium aspects of ion exchange involve more than purely surface chemical considerations. Thus, the formal expression for the exchange,

$$AR + B = BR + A \tag{XI-42}$$

where R denotes the matrix and A and B the exchanging ions, suggests a simple mass action treatment. The AR and BR centers are distributed throughout the interior of the exchanger phase and can be viewed as forming a nonideal solution. One may represent their concentrations in terms of mole fractions, in which case sizable activity coefficient corrections are generally needed or, alternatively, the exchanger phase may be treated as essentially a concentrated electrolyte solution so that volume concentrations are used, but again with activity coefficient corrections. The nonideality may be approached by considering the exchanger phase to act as a medium permeable to cations but not to the R or lattice ions, so that ion exchange appears as a Donnan equilibrium (see Section IV-10B), and specific recognition can then be given to the swelling effects that occur.

The rates of ion exchange are generally determined by diffusion processes; the rate-determining step may either be that of diffusion across a boundary film of solution or that of diffusion in and through the exchanger base (80). The whole matter is complicated by electroneutrality restrictions governing the flows of the various ions (81, 82).

As may be gathered, the field of ion exchange adsorption and chromatography is far too large to be treated here in more than this summary fashion. The reader is referred to Refs. 83 through 86, which include representative texts and review articles.

6. Problems

1. One hundred milliters of an aqueous solution of methylene blue contains 3.0 mg dye per liter and has an optical density (or molar absorbency) of 0.80 at a certain wavelength. The solution is then equilibrated with 25 mg of a charcoal, and the supernatant solution is now found to have an optical density of 0.20. Estimate the specific surface area of the charcoal.

2. The adsorption of stearic acid from *n*-hexane solution on a sample of steel powder is measured with the following results:

Concentration, mM/liter	Adsorption, mg/g	Concentration, mM/liter	Adsorption, mg/g
0.01	0.786	0.15	1.47
0.02	0.864	0.20	1.60
0.04	1.00	0.25	1.70
0.07	1.17	0.30	1.78
0.10	1.30	0.50	1.99

Explain the behavior of this system, and calculate the specific surface area of the steel.

3. Dye adsorption from solution may be used to estimate the surface area of a powdered solid. Suppose that if 2.0 g of a bone charcoal is equilibrated with 100 cm^3 of initially 10^{-4} M methylene blue, the final dye concentration is 0.4×10^{-4} M, while if 4.0 g of bone charcoal had been used, the final concentration would have been 0.2×10^{-4} M.

Assuming that the dye adsorption obeys the Langmuir equation, calculate the specific surface area of the bone charcoal in square meters per gram. The molecular area of methylene blue in a monolayer may be taken to be 65 $Å^2$.

4. The adsorption of Aerosol OT on Vulcan R was found to obey the Langmuir equation (87). The plot of C/x versus C was linear, where C is in millimoles per liter and x is in micromoles per gram. For $C = 0.5$, C/x was 100; the plot went essentially through the origin. Calculate the saturation adsorption in micromoles per gram.

5. The adsorption of stearic acid on Spheron 6 was measured, using various solvents, with the results shown.

	Millimoles adsorbed per gram at:	
Solvent	$N_2 = 0.001$	$N_2 = 0.004$
Cyclohexane	0.030	0.050
Ethyl alcohol	0.015	0.025
Benzene	0.008	0.010

Calculate for each case the apparent specific surface area of the Spheron, assuming the Langmuir equation to hold. State any other assumptions made; discuss the significance of your results.

6. Referring to Eq. XI-1, show that $\Sigma n_i^s = 0$ if the sum is over all of the solution components.

7. Derive Eq. XI-22.

8. Vapors A and B both adsorb on a certain solid according to the Langmuir equation, Eq. XI-26. The pressure for half coverage is 10 mm Hg for gas A and 4 mm Hg for gas B at 25°C, and the respective vapor pressures of the pure liquids are 40 mm Hg and 20 mm Hg. Calculate and plot (*a*) the two separate vapor adsorption isotherms, (*b*) each component adsorption in the composite vapor adsorption isotherm obtained by equilibrating the solid with the vapor from solutions of various compositions from pure A to pure B, and (*c*) the composite adsorption isotherm for the apparent adsorption from solutions of A and B. Assume the bulk solution, the interfacial layer, and the vapors to be ideal.

9. For the adsorption on Spheron 6 from benzene–cyclohexane solutions, the plot of $N_1 N_2 / n_0 \, \Delta N_2$ versus N_2 (cyclohexane being component 2) has a slope of 2.3 and an intercept of 0.4. (*a*) Calculate K. (*b*) Taking the area per molecule to be 40 Å^2, calculate the specific surface area of the Spheron 6. (*c*) Make a plot of the isotherm of composition change. Note: assume n^s is in mmoleg^{-1}.

10. Equation XI-30,

$$n_2^s\,(\text{apparent}) = n^s \frac{(K - 1)N_1 N_2}{1 + (K - 1)N_2}$$

will, under certain conditions, predict negative apparent adsorption. When such conditions prevail, explain to which of the isotherms of composition change shown in Figure XI-6 the calculated isotherm will most closely correspond.

11. Calculate the heat of immersion of Graphon in an aqueous solution containing 0.1 g/100 ml of butanol (see Section XI-4B).

12. The heat of adsorption measurements described in Section XI-4B really referred to heats of immersion. A closer analogy to the heat of adsorption in gas adsorption would be given by the process:

$$\begin{pmatrix}\text{solid in contact with solution,} \\ \text{but with no adsorption so that} \\ \Gamma_2^s = 0\end{pmatrix} \rightarrow \begin{pmatrix}\text{solid in contact with solution and} \\ \text{having the equilibrium} \\ \text{adsorbed layer}\end{pmatrix}$$

Referring to the numerical example in the section cited, calculate the value of this heat of adsorption, in ergs per square centimeter for 0.3 g Graphon in 100 ml butanol solution.

13. The adsorption of polystyrene polymer, of molecular weight 300,000, on carbon from toluene solution was studied. The carbon used had a specific surface area of 120 m^2/g, and a saturation adsorption of 33 mg of polymer per gram of carbon was found. The adsorption was 28 mg/g at a concentration of 0.1 mg of polymer per milliliter. (*a*) Assuming the Langmuir equation to be obeyed, calculate the Langmuir b constant. (*b*) Do the same, using Eq. XI-19, and assuming that $\upsilon = 50$. (*c*) Plot the complete isotherms, as calculated according to (*a*) and to (*b*) and comment on the degree of experimental precision needed to distinguish between them. (*d*) Calculate the number of polymer molecules adsorbed per particle of carbon at saturation.

14. Discuss how it might be that desorption of a polymer is very slow if an equilibrium solution is diluted, yet that same equilibrium system shows rapid exchange with labeled polymer.

15. Referring to Section XI-5B and to Fig. XI-13, the effect of the exclusion of co-ions (ions of like charge to that of the interface) results in an increase in solution concentration from n_0 to n_0'. Since the solution must remain electrically neutral, this means that the counter ions (ions of charge opposite to that of the interface) must also increase in concentration from n_0 to n_0'. Yet Fig. XI-13 shows the counter ions to be positively adsorbed. Shouldn't their concentration therefore *decrease* on adding the adsorbent to the solution? Explain.

General References

Advances in Chemistry, Vol. 79, American Chemical Society, Washington, D.C., 1968.

R. Defay and I. Prigogine, *Surface Tension and Adsorption*, transl. by D. H. Everett, Wiley, New York, 1966.

W. Eitel, *Silicate Science*, Vol. 1, Academic, New York, 1964.

F. Helfferich, *Ion Exchange*, McGraw-Hill, 1962.

J. X. Khym, *Analytical Ion-Exchange Procedures in Chemistry and Biology*, Prentice-Hall, Englewood Cliffs, New Jersey, 1974.

J. J. Kipling, *Adsorption from Solutions of Non-Electrolytes*, Academic, New York, 1965.

E. Lederer and M. Lederer, *Chromatography*, Elsevier, New York, 1955.

J. A. Marinsky and Y. Marcus, Eds., *Ion Exchange and Solvent Extraction*, Marcel Dekker, New York, 1973.

K. L. Mittal, *Adsorption at Interfaces, ACS Symposium Series* No. 8, American Chemical Society, Washington, D.C., 1975.

G. H. Osborn, *Synthetic Ion Exchangers*, 2nd ed., Chapman and Hall, London 1961.

G. Schay, *Surface and Colloid Science*, E. Matijevic, Ed., Wiley-Interscience, New York, 1969.

H. van Olphen and K. J. Mysels, Eds., *Physical Chemistry: Enriching Topics from Colloid and Surface Science*, Theorex, La Jolla, California, 1975.

Textual References

1. D. H. Everett, *Pure Appl. Chem.*, **31,** 579 (1972).
2. G. Schay, *Pure Appl. Chem.*, **48,** 393 (1976).
3. I. Langmuir, *J. Am. Chem. Soc.*, **40,** 1361 (1918).
4. D. H. Everett, *Trans. Faraday Soc.*, **60,** 1803 (1964); **61,** 2478 (1965). Also S. G. Ash, D. H. Everett, and G. H. Findenegg, *Trans. Faraday Soc.*, **64,** 2645 (1968).
5. A. Klinkenberg, *Rec. Trav. Chim.*, **78,** 83 (1959).
6. J. Zeldowitsh, *Acta Physicochim.* (*USSR*) **1,** 961 (1934).
7. G. Halsey and H. S. Taylor, *J. Chem. Phys.*, **15,** 624 (1947).
8. H. Freundlich, *Colloid and Capillary Chemistry*, Methuen, London, 1926.
9. I. Traube, *Annals,* **265,** 27 (1891), and preceding articles.
10. H. N. Holmes and J. B. McKelvey, *J. Phys. Chem.*, **32,** 1522 (1928).

11. F. E. Bartell and Y. Fu, *J. Phys. Chem.*, **33**, 676 (1929).
12. B. von Szyszkowski, *Z. Phys. Chem.*, **64**, 385 (1908); H. P. Meissner and A. S. Michaels, *Ind. Eng. Chem.*, **41**, 2782 (1949).
13. R. S. Hansen and R. P. Craig, *J. Phys. Chem.*, **58**, 211 (1954).
14. F. E. Bartell, T. L. Thomas, and Y. Fu, *J. Phys. Colloid Chem.*, **55**, 1456 (1951).
15. M. Manes and L. J. E. Hofer, *J. Phys. Chem.*, **73**, 584 (1969).
16. J. J. Kipling, *Adsorption from Solutions of Non-Electrolytes*, Academic, New York, 1965.
17. W. A. Schroeder, *J. Am. Chem. Soc.*, **73**, 1122 (1951).
18. O. H. Wheeler and E. M. Levy, *Can. J. Chem.*, **37**, 1235 (1959).
19. W. R. Smith and W. D. Schaeffer, *Proc. Rubber Technol. Conf., 2nd, London,* **1948**.
20. C. G. Gasser and J. J. Kipling, *Proc. Conf. Carbon, 4th Buffalo,* **1959**, p. 55.
21. R. Neher, *Chromatog. Rev.*, **1**, 99 (1959).
22. S. P. Zhdanov, A. V. Kiselev, and L. F. Pavolova, *Kinet. Catal.* (*USSR*), **3**, 391 (1962).
23. R. D. Hansen and R. S. Hansen, *J. Colloid Sci.*, **9**, 1 (1954).
24. J. L. Morrison and D. M. Miller, *Can. J. Chem.*, **33**, 350 (1955).
25. N. Hackerman and A. H. Roebuck, *Ind. Eng. Chem.*, **46**, 1481 (1954). See also F. A. Matsen, A. C. Makrides, and N. Hackerman, *J. Chem. Phys.*, **22**, 1800 (1954).
26. H. A. Smith and K. A. Allen, *J. Phys. Chem.*, **58**, 449 (1954).
27. H. F. Finley and N. Hackerman, *J. Electrochem. Soc.*, **107**, 259 (1960).
28. G. Kar, T. W. Healy, and D. W. Fuerstenau, *Corros. Sci.*, **13**, 375 (1973).
29. R. S. Hansen, Y. Fu, and F. E. Bartell, *J. Phys. Colloid Chem.*, **53**, 769 (1949).
30. O. M. Dzhigit, A. V. Kiselev, and K. G. Krasilnikov, *Dokl. Akad. Nauk SSSR,* **58**, 413 (1947).
31. F. E. Bartell and D. J. Donahue, *J. Phys. Chem.*, **56**, 665 (1952).
32. E. L. Cook and N. Hackerman, *J. Phys. Colloid Chem.*, **55**, 549 (1951).
33. F. R. Eirich, *J. Colloid Interface Sci.*, **58**, 423 (1977).
34. F. Rowland, R. Bulas, E. Rothstein, and F. R. Eirich, *Ind. Eng. Chem.*, September 1965, p. 46.
35. C. Peterson and T. K. Kwei, *J. Phys. Chem.*, **65**, 1330 (1961).
36. W. Heller, *Pure Appl. Chem.*, **12**, 249 (1966).
37. H. L. Frisch, M. Y. Hellman, and J. L. Lundberg, *J. Polym. Sci.*, **38**, 441 (1959).
38. R. Simha, H. Frisch, and F. Eirich, *J. Phys. Chem.*, **57**, 584 (1953); *J. Chem. Phys.*, **25**, 365 (1953); *J. Polym. Sci.*, **29**, (1958).
39. See A. Silverberg, *Faraday Discuss. Chem. Soc.*, **59**, 203 (1975).
40. E. Di Marcio and R. Rubin, *Am. Chem. Soc. Polym. Prepr.*, **11**, 1239 (1970); E. Di Marcio, *J. Chem. Phys.*, **42**, 2101 (1965).
41. S. G. Ash, D. Everett, and G. H. Findenegg, *Trans. Faraday Soc.*, **66**, 708 (1970).
42. B. J. Fontana and J. R. Thomas, *J. Phys. Chem.*, **65**, 480 (1961); also, F. McCrackin, *J. Chem. Phys.*, **47**, 1980 (1967).
43. F. W. Rowland and F. R. Eirich, *J. Polym. Sci.*, **4**, 2421 (1966).
44. R. R. Stromberg, D. J. Tutas, and E. Passaglia, *J. Phys. Chem.*, **69**, 3955 (1965).

45. F. R. Eirich, *Interface Conversion for Polymer Coatings*, P. Weiss and G. D. Cheever, Eds., American Elsevier, 1969.
46. R. I. Feigin and D. H. Napper, *J. Colloid Interface Sci.*, **75**, 525, 567 (1980).
46a. P. D. Bisio, J. G. Cartledge, W. H. Keesom, and C. J. Radke, *J. Colloid Interface Sci.*, **78**, 225 (1980).
47. D. Graham, *J. Phys. Chem.*, **59**, 896 (1955).
48. J. J. Kipling and R. B. Wilson, *J. Appl. Chem.*, **10**, 109 (1960).
49. J. F. Padday, *Pure and Applied Chemistry, Surface Area Determination*, Butterworths, London, 1969.
49a. M. A. Rahman and A. K. Ghosh, *J. Colloid Interface Sci.*, **77**, 50 (1980).
50. D. H. Everett, *Trans. Faraday Soc.*, **61**, 2478 (1965).
51. G. Schay and L. G. Nagy, *J. Colloid Interface Sci.*, **38**, 302 (1972).
52. H. J. Van den Hul and J. Lyklema, *J. Colloid Interface Sci.*, **23**, 500 (1967).
53. J. J. Kipling, *Q. Rev.* (*London*), **5**, 60 (1951).
54. S. G. Ash, R. Bown, and D. H. Everett, *J. Chem. Thermodyn.*, **5**, 239 (1973).
55. J. J. Kipling and D. A. Tester, *J. Chem. Soc.*, **1952**, 4123.
56. G. D. Parfitt and P. C. Thompson, *Trans. Faraday Soc.*, **67**, 3372 (1971).
57. G. Schay, *Surf. Colloid Sci.*, **2**, 155 (1969); *J. Colloid Interface Sci.*, **42**, 478 (1973).
58. J. J. Kipling, *Proc. Int. Congr. Surf. Act., 3rd, Mainz,* **1960**, Vol. 2, p. 77.
59. G. J. Young, J. J. Chessick, and F. H. Healey, *J. Phys. Chem.*, **60**, 394 (1956).
60. M. I. Coltharp and N. Hackerman, *J. Colloid Interface Sci.*, **43**, 177, 185 (1973).
61. D. F. Billett, D. H. Everett, and E. E. H. Wright, *Proc. Chem. Soc.* (*London*), **1964**, July, p. 216.
62. H. B. Weiser, *Colloid Chemistry*, Wiley, New York, 1950.
63. K. Fajans, *Radio Elements and Isotopes; Chemical Forces and Optical Properties of Substance*, McGraw-Hill, New York, 1931.
64. I. M. Kolthoff and R. C. Bowers, *J. Am. Chem. Soc.*, **76**, 1503 (1954).
65. A. M. Posner and J. P. Quirk, *J. Colloid Sci.*, **19**, 798 (1964).
66. W. D. Bancroft, *J. Phys. Chem.*, **19**, 363 (1915).
67. P. Connor and R. H. Ottewill, *J. Colloid Interface Sci.*, **37**, 642 (1971).
68. F. F. Aplan and D. W. Fuerstenau, in *Froth Flotation*, D. W. Fuerstenau, Ed., American Institute of Mining and Metallurgical Engineering, New York, 1962.
69. S. G. Dick, D. W. Fuerstenau, and T. W. Healy, *J. Colloid Interface Sci.*, **37**, 595 (1971).
70. P. Somasundaran, T. W. Healy, and D. W. Fuerstenau, *J. Phys. Chem.*, **68**, 3562 (1964).
71. D. W. Fuerstenau, *Chem. Biosurfaces,* **1**, 143 (1971).
72. W. Paik, M. A. Genshaw, and J. O'M. Bockris, *J. Phys. Chem.*, **74**, 4266 (1970).
73. R. K. Schofield, *Nature,* **160**, 408 (1947).
74. R. K. Schofield and O. Talibuddin, *Discuss. Faraday Soc.*, **3**, 51 (1948).
75. D. G. Edwards and J. P. Quirk, *J. Colloid Sci.*, **17**, 872 (1962).
76. G. H. Bolt and R. D. Miller, *Soil Sci. Am. Proc.*, **19**, 285 (1955); A. V. Blackmore and R. D. Miller, *ibid.*, **25**, 169 (1961). H. van Olphen, *J. Phys. Chem.*, **61**, 1276 (1957).
77. R. M. Barrer and R. M. Gibbons, *Trans. Faraday Soc.*, **59**, 2569 (1963), and preceding references.
78. G. L. Gaines, Jr., and H. C. Thomas, *J. Chem. Phys.*, **23**, 2322 (1955).

79. K. W. T. Goulding and O. Talibudeen, *J. Colloid Interface Sci.*, **78,** 15 (1980).
80. G. E. Boyd, A. W. Adamson, and L. S. Myers, Jr., *J. Am. Chem. Soc.*, **69,** 2836 (1947).
81. J. J. Grossman and A. W. Adamson, *J. Phys. Chem.*, **56,** 97 (1952).
82. R. Schlögl and F. Helfferich, *J. Chem. Phys.*, **26,** 5 (1957).
83. J. A. Marinsky and Y. Marcus, Eds., *Ion Exchange and Solvent Extraction*, Marcel Dekker, New York, 1973.
84. G. H. Osborn, *Synthetic Ion Exchangers*, 2nd ed., Chapman and Hall, London, 1961.
85. F. C. Nachod and J. Schubert, *Ion Exchange Technology*, Academic, New York, 1956.
86. R. Kunin and R. J. Myers, *Ion Exchange Resins*, Wiley, New York, 1950.
87. J. C. Abram and G. D. Parfitt, *Proc. 5th Conf. Caron*, Pergamon, New York, 1962, p. 97.

CHAPTER XII

Friction and Lubrication—Adhesion

1. Introduction

This chapter and the two that follow are introduced at this time to illustrate some of the many extensive areas in which there are important applications of surface chemistry. Friction and lubrication as topics properly deserve mention in a textbook on surface chemistry, partly because these subjects do involve surfaces directly and partly because many aspects of lubrication depend on the properties of surface films. The subject of adhesion is treated briefly in this chapter mainly because it, too, depends greatly on the behavior of surface films at a solid interface and also because friction and adhesion have some interrelations.

The subject of friction and its related aspects has, incidentally, come to be known as *tribology*. Tribology is the study of interacting surfaces in relative motion, the Greek root *tribos* meaning rubbing.

2. Friction Between Unlubricated Surfaces

A. Amontons' Law

The coefficient of friction μ between two solids is defined as F/W, where F denotes the frictional force and W is the load or force normal to the surfaces, as illustrated in Fig. XII-1. There is a very simple law concerning the coefficient of friction μ, which is amazingly well obeyed. This law, known as *Amontons' law,* states that μ is independent of the apparent area of contact; it means that, as shown in the figure, with the same load W the frictional forces will be the same for a small sliding block as for a large one. A corollary is that μ is independent of load. Thus if $W_1 = W_2$, then $F_1 = F_2$.

Although friction between objects is a matter of everyday experience, it is curious that Amontons' law, although of fairly good general validity, seems quite contrary to intuitive thinking. Indeed when Amontons, a French army engineer, presented his findings to the Royal Academy in 1699 (*Mem. Acad. Roy. Sci,* **1699,** 206), they were met with some skepticism. Amontons anticipated some of this in his paper and attempted to explain his findings by qualitative reasoning. As is often the case, however, it is the law as given

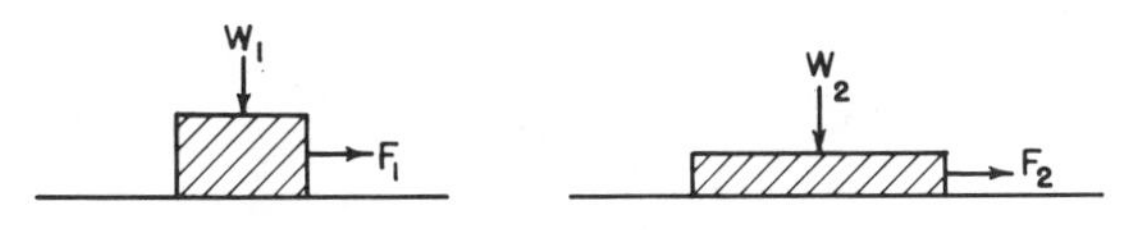

Fig. XII-1. Amontons' law.

by the data that has stood up over the course of time rather than the associated explanations. Amontons was thinking partly of friction as due to work associated with vertical motions as a result of surface roughness and, as is shown, the actual explanation (or at least the one now accepted) is quite different. Amontons went on to assert that the coefficient of friction was always $\frac{1}{3}$; this is indeed about the usual value but, as discussed in Section XII-4, the frictional coefficient between really clean surfaces can be much larger; in the case of some plastics, it may be much smaller.

The basic law of friction has been known for some time. Amontons was, in fact, preceded by Leonardo da Vinci, whose notebook illustrates with sketches that the coefficient of friction is independent of the apparent area of contact (see Refs. 1 and 2). It is only relatively recently, however, that the probably correct explanation has become generally accepted.

B. Nature of the Contact Between Two Solid Surfaces

There is abundant evidence that even very smooth-appearing surfaces are irregular on a molecular scale of distances. The nature of such surface irregularities may be studied by electron microscopy either by low angle reflection or of carbon replicas. Resolution down to perhaps 10 Å can be achieved (see Ref. 3), which allows a very direct appreciation of the presence of imperfections, steps, and so forth on a surface (see Section VII-4), as well as of grooves left by one surface sliding over another. Low energy electron diffraction and field emission techniques have also provided much information about surface structures (see Section VIII-3). Optical interference methods can detect changes in surface elevation of as little as 10 Å, and on a coarser scale, surface roughness can be observed from the magnified motion of a diamond-tipped stylus passing slowly over the surface.

The results of such studies have made it clear that the surfaces of crystalline materials may have quite irregular steps of hundreds or thousands of angstrom units in depth and that lapped or ground surfaces may be quite jagged on this scale of distances. Even polished metal surfaces are not really smooth. As discussed in Section VII-1B, polishing appears to bring about local melting, and the effect is that ragged asperities are smoothed out; but even so, the resulting surface has a waved appearance. In the case of metals, it also is apt to have an oxide coating, as illustrated in Fig. XII-2(3).

As a result of the irregular nature of even the smoothest surfaces available, two surfaces brought into contact will touch only in isolated regions. In fact, on initial contact, one would expect to have only three points of touching, but even for minute total loads, the pressure at these points would be suf-

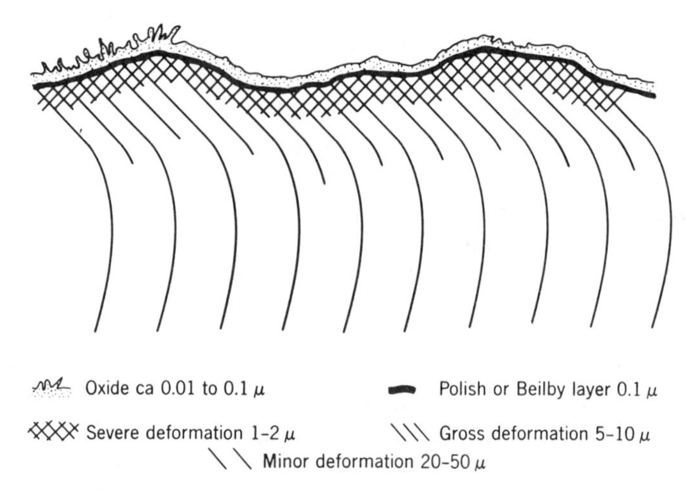

Fig. XII-2. Schematic diagram showing topography and structure of a typical polished metal specimen. (From Samuels, as given in Ref. 3.)

ficient to cause deformation leading to multiple contacts. Actual contact regions might appear as in Fig. XII-3. Case *a* in the figure applies to metal–metal contacts where the local pressure is high enough to cause plastic flow of both metals, with local welding. Case *b* illustrates the situation if a soft material, such as a plastic, is pressed against a hard one, such as a metal.

The true area of contact is clearly much less than the apparent area. The former can be estimated directly from the resistance of two metals in contact. It may also be calculated if the statistical surface profiles are known from roughness measurements. As an example, the true area of contact, A, is about 0.01% of the apparent area in the case of two steel surfaces under a 10-kg load (4).

Another indication that friction is confined to a real area that is much smaller than the apparent area is in the very high local temperatures that can prevail during a sliding motion. These can be estimated by rubbing a

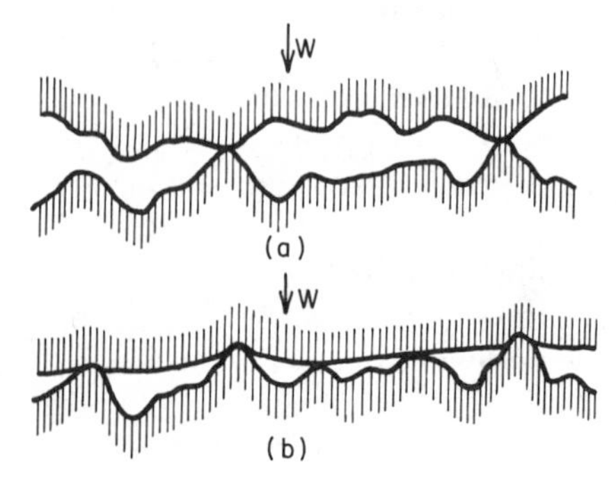

Fig. XII-3. Comparison of actual contact areas for (*a*) metal-on-metal, and (*b*) plastic-on-metal. (From Ref. 1.)

slider of one metal over the surface of another one and noting the potential difference that develops (4). An alternative procedure that has been used when one of the surfaces is transparent, such as glass, is to view the local hot spots with an infrared sensitive cell and estimate the local temperature from the spectrum of the emitted infrared light (5). Local temperature increases of hundreds of degrees have been recorded by these means; the final limitation may be the melting point of the lower melting material. In the case of plastics, microscopic examination of a cross section along the track of the slider reveals that melting has occurred to some depth, as shown in Fig. XII-4 (6). Note that if the plastic is rubbing against glass, the melt zone is much deeper than if it is rubbing against the more heat conducting steel.

In summary, it has become quite clear that contact between two surfaces is limited to a small fraction of the apparent area, and, as one consequence of this, rather high local temperatures can develop during rubbing. Another consequence discussed in more detail later, is that there are also rather high local pressures. Finally, there is direct evidence (7, 8) that the two surfaces do not remain intact when sliding past each other. Microscopic examination of the track left by the slider shows gouges and irregular pits left by the bodily plucking out of portions of the softer metal by the harder ones. Similarly, a radioautograph obtained by pressing a photographic emulsion against the harder metal shows spots of darkening corresponding to specks of the radioactively labeled softer metal. This type of evidence is interpreted to mean that fairly strong seizure or actual welding occurs at the sites of contact.

C. *Role of Shearing and Plowing—Explanation of Amontons' Law*

As two surfaces are brought together, the pressure is extremely large at the initial few points of contact, and deformation immediately occurs to allow more and more to develop. This plastic flow continues until there is a total area of contact such that the local pressure has fallen to a characteristic yield pressure $\mathbf{P}_m$ of the softer material. Thus, as illustrated in Fig.

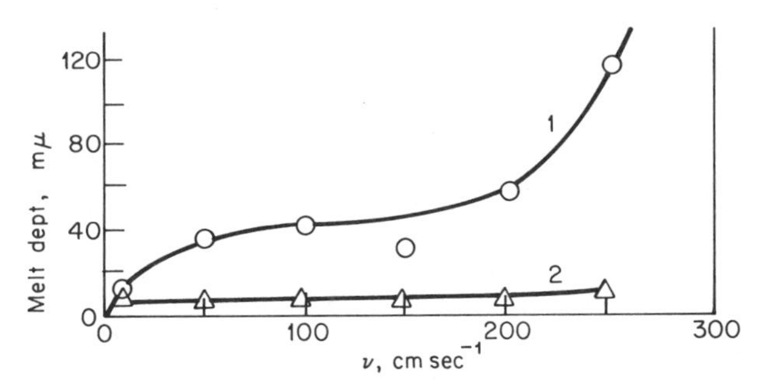

Fig. XII-4. Effect of sliding speed on the depth of melting of nylon rubbing against (1) glass and (2) steel. (From Ref. 6.)

XII-5, around each region of contact there is a plastic zone, with further elastic deformation outside.

Normally, then, the actual contact area is determined by the yield pressure, so that

$$A = \frac{W}{\mathbf{P}_m} \tag{XII-1}$$

For most metals $\mathbf{P}_m$ values are in the range of 10 to 100 kg/mm^2, so that in a friction experiment with a load of 10 kg, the true contact area would indeed be about the 10^{-3} cm^2 estimated from conductivity measurements. There are considerable uncertainties in the absolute values of A calculated in this way, however, since work hardening may occur at the contact points, with a consequent increase in $\mathbf{P}_m$ over that for the original metal. Nevertheless, although the correct value on $\mathbf{P}_m$ may be in doubt, the relationship given by Eq. XII-1 should still apply. For very light loads, such that the elastic limit is *not* exceeded, analysis indicates that for a hemispherical rider on a flat surface

$$A = kW^{2/3} \tag{XII-2}$$

so that in this case the area of contact is no longer directly proportional to the load. This is for a single contact region, however. If there is a Gaussian distribution of asperities, it turns out that A is again proportional to W (1).

In a typical measurement of friction, a slider is pressed against a stationary block, and the force F required to move the slider is measured. This force in general will consist of two terms. First, there is the force F required to shear the junctions at the points of actual contact. This is given by

$$F = As_m \tag{XII-3}$$

where s_m is the shear strength per unit area. The second term F' is the force required to displace the softer material from the front of the harder one. With metals of different hardness, the harder one, if used as a slider, will plow a track in the softer, and there is therefore a work term associated with this plowing action. In a general way, one expects F' to be proportional to the cross section of the slider (see Fig. XII-6),

$$F' = kA' \tag{XII-4}$$

where A' denotes the cross section of the plowed track.

The plowing contribution can be estimated by employing a rider that is very thin, Fig. XII-6*b*, so that the shear contribution is minimized. Usually

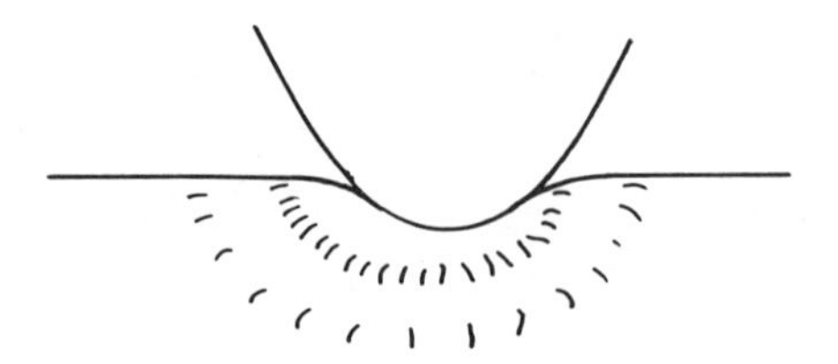

Fig. XII-5. Plastic deformation around a region of contact.

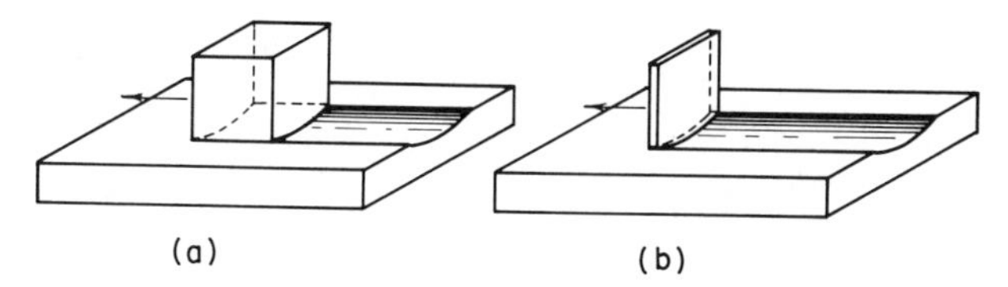

Fig. XII-6. Illustration of shearing (*a*) and plowing (*b*) actions.

the plowing term is important only for the case of a hard material rubbing against a soft one; if both are hard, friction is due mostly to the shear term. As an approximation, then, A may be eliminated from Eqs. XII-1 and XII-3 to give

$$F = W\frac{s_m}{\mathbf{P}_m} \quad \text{(XII-5)}$$

or

$$\mu = \frac{s_m}{\mathbf{P}_m} = \text{constant} \quad \text{(XII-6)}$$

This is Amontons' law as stated in the opening paragraphs of this chapter.

Another point in connection with Eq. XII-6 is that both the yielding and the shear will involve mainly the softer material, so that μ is given by a ratio of properties of the same substance. This ratio should be nearly independent of the nature of the metal itself since s_m and $\mathbf{P}_m$ tend to vary together in agreement with the observation that for most frictional situations, the coefficient of friction lies between about 0.5 and 1.0. Also, temperature should not have much effect on μ, as is observed.

As is suggested by the preceding analysis, Amontons' law does indeed hold well if neither of the two surfaces is too soft. Thus Bowden and Tabor (4) cite data for the coefficient of friction of aluminum on aluminum giving a nearly constant value of μ for loads ranging from 0.037 to 4000 g. The case of polymers is not so simple, however, in that plastic deformation tends to be fairly important in determining A, as discussed in Section XII-5C.

The coefficient of friction may also depend on the relative velocity of the two surfaces. This will, for example, affect the local temperature, the extent of work hardening of metals, and the relative importance of the plowing and shearing terms. These facts work out such that the coefficient of friction tends to decrease with increasing sliding speeds (2, 9), contrary to what is sometimes known as Coulomb's law, which holds that μ should be independent of sliding velocity. At the very low speeds commonly used, about 0.01 cm/sec, the effect is small and so is not usually involved in discussions of friction (see the next section, however).

D. Static and "Stick-Slip" Friction

In general it is necessary to distinguish static and kinetic coefficients of friction, μ_s and μ_k. The former is given by the force needed to initiate motion, and the latter, by the force to maintain a given sliding speed. There is a general argument to show that $\mu_s > \mu_k$. Suppose that μ_s were less than μ_k, then, for a load W, $F_s = W\mu_s$, and $F_k = W\mu_k$, that is, $F_s < F_k$. Now if a force were applied that lay between these two values, a paradoxical situation would exist. Since F would be greater than F_s, the rider *should* move, but since F is less than F_k, the rider *should not* move: the contradiction is inescapable and leads to the conclusion that the static coefficient must be equal to or greater than the sliding or kinetic coefficient. The static coefficient of friction tends to increase with time of contact, and it has been argued that μ_s for zero time should be equal to μ_k. Actually, μ_k is a function of sliding speed and appears to go through a maximum with increasing speed. This maximum occurs below 10^{-8} cm/sec for titanium, at about 10^{-4} cm/sec for indium, and at about 1 cm/sec for the very soft plastic, Teflon (11). Thus most materials show a negative coefficient of μ_k with sliding speed at speeds in the usual range of 0.01 cm/sec. It might be noted that most junctions caused by the seizure of asperities seem to be about 0.001 cm in size so that it is necessary to traverse about this distance to obtain a meaningful value of μ_k. A practical limit is thus set on experiments at very small sliding speeds.

A phenomenon that may develop, especially at small sliding speeds, is that known as stick-slip friction, in which the slider moves in jumps that may be of quite high frequency. This effect is partly a consequence of having some play and lack of rigidity in the mechanism holding the slider. Once momentary sticking occurs, the slider is pushed back against the elastic restoring force of the holding mechanism as a result of the continued motion of the latter. When the restoring force exceeds that corresponding to μ_s, the slider moves forward rapidly, overshoots if there is sufficient play, and sticks again. The greater the difference between static and sliding frictional coefficients, the more prone is the system to stick-slip friction; with lubricated surfaces, there may be a fairly well-defined temperature above which the phenomenon is observed. Stick-slip friction is encouraged if μ_k *decreases* with sliding speed (why?) and with increasing W (10). A very common, classroom illustration of stick–slip friction is, incidentally, the squeaking of chalk on a chalkboard.

E. Rolling Friction

The slowing down of a wheel or sphere rolling on a surface is not, properly speaking, a frictional effect since there is no sliding of one surface against another. The effect, however, is as though there were a frictional force, and should therefore be mentioned. As illustrated in Fig. XII-7, a hard sphere rolling on a soft material travels in a moving depression. The material is compressed in front and rebounds

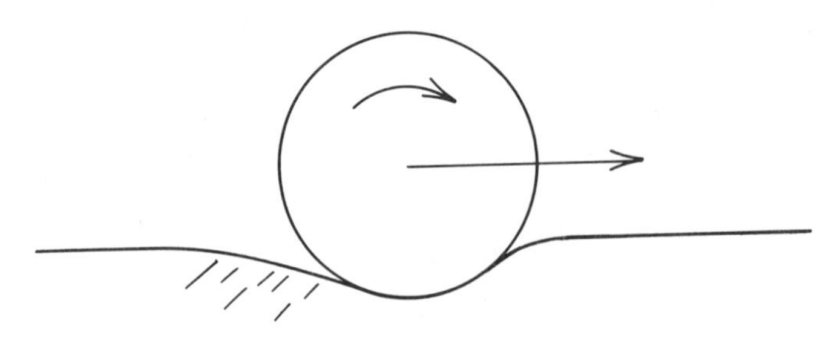

Fig. XII-7

at the rear, and were it perfectly elastic, the energy stored in compression would be returned to the sphere at its rear. Actual materials are not perfectly elastic, however, so energy dissipation occurs, the result being that kinetic energy of the rolling is converted into heat.

3. Two Special Cases of Friction

A. Use of Skid Marks to Estimate Vehicle Speeds

An interesting and very practical application of Amontons' law occurs in the calculation of the minimum speed of a vehicle from the length of its skid marks.

In the absence of skidding, the coefficient of *static* friction applies; at each instant, the portion of the tire that is in contact with the pavement has zero velocity. Rolling tire "friction" is more of the type discussed in Section XII-2E. If, however, skidding occurs, then since rubber is the softer material, the coefficient of friction as given by Eq. XII-6 is determined mainly by the properties of the rubber used and will be nearly the same for various types of pavement. Actual values of μ turn out to be about unity.

If μ is taken to be a constant during a skid, application of Amontons' law leads to a very simple relationship between the initial velocity of the vehicle and the length of the skid mark. The initial kinetic energy is $mv^2/2$, and this is to be entirely dissipated by the braking action, which amounts to a force F applied over the skid distance d. By Amontons' law,

$$F = \mu W = mg\mu \qquad \text{(XII-7)}$$

Then

$$\frac{mv^2}{2} = Fd = mg\,\mu d \qquad \text{(XII-8)}$$

or

$$v = (2d\mu g)^{1/2} \qquad \text{(XII-9)}$$

Thus if Amontons' law is obeyed, the initial velocity is entirely determined by the coefficient of friction and the length of the skid marks. The mass of the vehicle is not involved, neither is the size or width of the tire treads, nor how hard the brakes were applied, so long as the application is sufficient to maintain skidding.

The situation is complicated, however, because some of the drag on a skidding tire is due to the elastic hysteresis effect discussed in Section XII-2E. That is, asperities in the road surface produce a traveling depression in the tire with energy loss due to imperfect elasticity of the tire material. In fact, tires made of high elastic hysteresis material will tend to show superior skid resistance and coefficient of friction.

As might be expected, this simple picture does not hold perfectly. The coefficient of friction tends to increase with increasing velocity and also is smaller if the pavement is wet (12). On a wet road, μ may be as small as 0.2, and, in fact, one of the principal reasons for patterning the tread and sides of a tire is to prevent the confinement of a water layer between the tire and the road surface. Similarly, the texture of the road surface is important to the wet friction behavior. Properly applied, however, measurements of skid length provide a conservative estimate of the speed of the vehicle when the brakes are first applied, and it has become a routine matter for data of this kind to be obtained at the scene of a serious accident.

B. Ice and Snow

The coefficient of friction between two unlubricated solids is generally in the range of 0.5 to 1.0, and it has therefore been a matter of considerable interest that very low values, around 0.03, pertain to objects sliding on ice or snow. The first explanation, proposed by Reynolds in 1901, was that the local pressure caused melting, so that a thin film of water was present. Qualitatively, this explanation is supported by the observation that the coefficient of friction rises rapidly as the temperature falls, especially below about $-10°C$, if the sliding speed is small. Moreover, there is little doubt that formation of a water film is actually involved (2, 3).

Although the theory of pressure melting is attractive, certain difficulties develop when it is given more quantitative consideration. Thus if the freezing point is to be lowered only to $-10°C$, a local pressure of about 1000 kg/cm^2 must prevail, corresponding to an actual contact area of a few hundredths of a square centimeter for a person of average weight. In the case of skis, this represents an unreasonably small fraction of the total area. In fact, at $-10°C$, μ is about 0.4 for small speeds. At about 5 m/sec sliding speed, however, μ was found to drop rapidly to about 0.04 (3).

An alternative explanation that has been advanced is that at the lower temperatures the water film that leads to low friction is produced as a result of local heating. Here, a trial calculation shows that the frictional energy is sufficient to melt an appreciable layer of the ice or snow without having to assume inordinately small areas of actual contact. Moreover, experiments with skis constructed of different materials or waxed with different preparations have indicated that smaller coefficients of friction are obtained if the surface of the ski has a low heat conductivity than if it has a high value (13). Since the area of actual contact is still expected to be only a fraction of the apparent area, the frictional melting theory would require that the heat due to friction not be conducted away too rapidly by the body of the ski.

Another indication of the probable correctness of the pressure melting explanation is that the variation of the coefficient of friction with temperature for ice is much the same for other solids, such as solid krypton and carbon dioxide (14) and benzophenone and nitrobenzene (3). In these cases the density of the solid is greater than that of the liquid, so the drop in μ as the melting point is approached cannot be due to pressure melting.

While pressure melting may be important for snow and ice near 0°C, it is possible that even here an alternative explanation will prove important. Ice is a substance of unusual structural complexity, and it has been speculated that a liquidlike surface layer is present near the melting point (15, 16); if this is correct, the low μ values observed at low sliding speeds near 0°C may be due to a peculiarity of the surface nature of ice rather than to pressure melting.

4. Metallic Friction—Effect of Oxide Films

The study of the friction between metals turns out to be almost two subjects—that relating to truly clean metal surfaces and that involving metals with adsorbed gases or oxide coatings. If metal surfaces are freed of all surface contamination by electron bombardment of the heated surface in a vacuum, then, in the cases of tungsten, copper, nickel, and gold, the coefficients of friction can be quite large. Similarly, nickel surfaces so treated seized when placed in contact and rubbed slightly against each other, that is, a firm metal–metal weld occurred such that it was necessary to pry the pieces apart (17); iron surfaces gave a μ value of 3.5 at room temperature, but seized at 300°C. Machlin and Yankee (18) using fresh surfaces formed by machining in an inert atmosphere, again found that seizure usually occurred. With two dissimilar metals, however, this may not always happen if the two metals are mutually insoluble. Thus cadmium and iron, both mutually insoluble, did weld together on rubbing, but not silver and iron. The welds that form when seizure occurs appear to be of full strength, that is, for two like metals the strength of the weld is essentially that of the metal itself. An interesting general correlation has been reported by Buckley (19). As shown in Fig. XII-8, the coefficient of friction decreases smoothly with increasing *d*-bond character.

The behavior in the presence of air is quite different. For example, Tingle (20) found that the friction between copper surfaces decreased from a μ value of 6.8 to one of 0.80 as progressive exposure of the clean surfaces led to increasingly thick oxide layers. As noted by Whitehead (21), several behavior patterns are possible. At very light loads, the oxide layer may be effective in preventing metal–metal contact (and electrical conductivity); the coefficient of friction will then tend to be in the range of 0.6 to 1.0. At heavier loads, depending on the metal, the film may break down, and μ increases (and the conductivity) as metal–metal contacts form. This was true for copper but not for aluminum or silver. In the case of aluminum, the

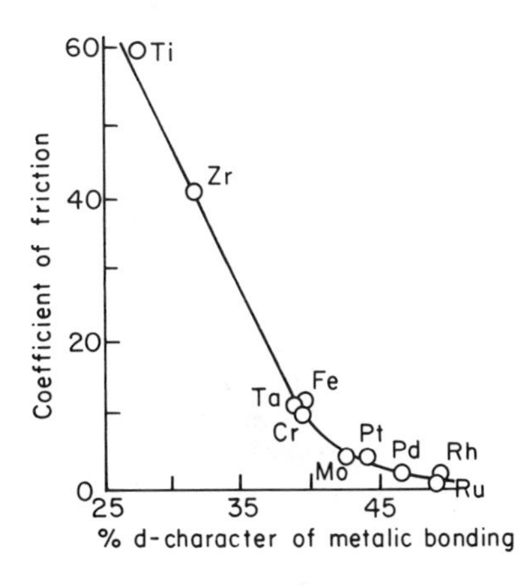

Fig. XII-8. Coefficient of friction of the indicated metals against a gold (111) surface at 25°C. (From Ref. 19.)

oxide film is broken even with very small loads, perhaps because the substrate metal is softer than its oxide so that the latter cracks easily. In the case of silver, oxide formation is negligible, so again friction is relatively independent of load. In all cases, even though metal–metal contacts may be present, the coefficient of friction is less than that between really pure surfaces, either because of adsorbed gases or possibly patches and fragments of oxide material.

Bowden and Tabor (22) proposed, for such heterogeneous surfaces, that

$$F = A(\alpha s_m + (1 - \alpha)s_0) \tag{XII-10}$$

where α is the fraction of surface contact that is of the metal-to-metal type, and s_m and s_0 are the shear strengths of the metal and the oxide, respectively. As may be gathered, the effect of oxide coatings on frictional properties is, in a way, a special case of the effect of films in general. Boundary lubrication, discussed in Section XII-7, extends the subject further.

5. Friction Between Nonmetals

A. *Relatively Isotropic Crystals*

Substances in this category include krypton, sodium chloride, and diamond, as examples, and it is not surprising that differences in detail as to frictional behavior do occur. The softer solids tend to obey Amontons' law with μ values in the "normal" range of 0.5 to 1.0, provided they are not too near their melting points. Ionic crystals, such as sodium chloride, tend to show irreversible surface damage, in the form of cracks, owing to their brittleness, but still tend to obey Amontons' law. This suggests that the area of contact is mainly determined by plastic flow rather than by elastic deformation.

Diamond behaves somewhat differently in that μ is low *in air,* about 0.1. It is dependent, however, on which crystal face is involved, and rises severalfold in vacuum (after heating) (1). The behavior of saphire is similar (23). Diamond surfaces, incidentally, can have an oxide layer. Naturally occurring ones may be hydrophilic or hydrophobic, depending on whether they are found in formations exposed to air and water. The relation between surface wettability and friction seems not to have been studied.

B. *Layer Crystals*

A number of substances such as graphite, talc, and molybdenum disulfide have sheetlike crystal structures, and it might be supposed that the shear strength along such layers would be small and hence the coefficient of friction. It is true that μ for graphite is about 0.1 (23, 24). However, adsorbed gases play a role since on thorough outgassing, μ rises to about 0.6 (1, 23, 25). Also, microscopic examination of slider tracks shows that strips of graphite layers have rolled up into miniature rollers about 0.05 μ in diameter (26), and it may well be that it is these that give the low μ values. The role of adsorbed gases may be to facilitate the detachment of layer fragments.

The structurally similar molybdenum disulfide also has a low coefficient of friction, but now not increased in vacuum (1, 27). The interlayer forces are, however,

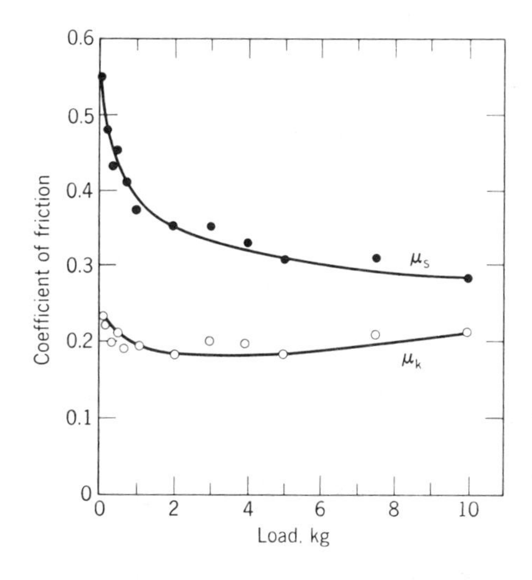

Fig. XII-9. Coefficient of friction of steel sliding on hexafluoropropylene as a function of load (first traverse). Velocity 0.01 cm/sec; 25°C. (From Ref. 28.)

much weaker than for graphite, and the mechanism of friction may be different. With molecularly smooth mica surfaces, the coefficient of friction is very dependent on load and may rise to extremely high values at small loads (3); at normal loads and in the presence of air, μ drops to a near normal level.

C. Plastics

A number of friction studies have been carried out on organic polymers in recent years. Coefficients of friction are for the most part in the normal range, with values about as expected from Eq. XII-6. The detailed results show some serious complications, however. First, μ is very dependent on load, as illustrated in Fig. XII-9, for a copolymer of hexafluoroethylene and hexafluoropropylene (28), and evidently the area of contact is determined more by elastic than by plastic deformation. The difference between static and kinetic coefficients of friction was attributed to transfer of an oriented film of polymer to the steel rider during sliding and to low adhesion between this film and the polymer surface. Tetrafluoroethylene (Teflon) has a low coefficient of friction, around 0.1, and in a detailed study, this lower coefficient and other differences were attributed to the rather smooth molecular profile of the Teflon molecule (29).

More recently emphasis has been given to adhesion between the polymer and substrate surface as a major explanation for friction (30), the other contribution being from elastic hysteresis (see Section XII-2E). The adhesion may be mostly due to van der Waals forces, but in some cases there is a contribution from electrostatic charging.

6. Some Further Aspects of Friction

An interesting aspect of friction is the manner in which the area of contact changes as sliding occurs. This change may be measured either by conductivity, proportional to $A^{1/2}$ if as in the case of metals it is limited primarily by a number of small metal-to-metal junctions, or by the normal adhesion, that is, the force to separate the two substances. As an illustration of this last, a steel ball pressed briefly against indium with a load of 15 g required about the same 15 g for its subsequent detachment (31). If relative motion was set in, a μ value of 5 was observed and, on stopping, the normal force for separation had risen to 100 g. The ratio of 100 to 15 g may thus be taken as the ratio of junction areas in the two cases.

Even if no perceptible motion occurs (see later, however), application of a force leads to microdisplacements of one surface relative to the other and, again, often a large increase in area of contact. The ratio F/W in such an experiment will be called ϕ, since it does not correspond to either the usual μ_s or μ_k; ϕ can be related semiempirically to the area change, as follows (32). We assume that for two solids pressed against each other at rest the area of contact A_0 is given by Eq. XII-1, $A = W/\mathbf{P}_m$. However, if shear as well as normal stress is present, then a more general relation for threshold plastic flow is

$$\mathbf{P}^2 + \alpha s^2 = \mathbf{P}_m^2 = \alpha s_0^2 \qquad \text{(XII-11)}$$

where s is the shear strength of the interfacial region, and s_0 is that of the bulk metal; the constant α is partly geometric, depending on the nature of the asperities. If we let $s = ms_0$, and write $\mu = s/\mathbf{P}$, we obtain (33)

$$\mu = \left(\frac{m^2}{\alpha(1 - m^2)}\right)^{1/2} \qquad \text{(XII-12)}$$

The constant α can be estimated by applying a load so as to establish an A_0 and then removing the load and measuring the force F_0 to cause motion. Under these conditions, $\mathbf{P} = 0$, $A_0 s = F_0$, and $A_0\mathbf{P}_m = W$, so that from Eq. XII-11,

$$\alpha = \left(\frac{W}{F_0}\right)^2 \qquad \text{(XII-13)}$$

For steel on indium, α was about 3, while for platinum against platinum, it was about 12 (3).

For very clean metal surfaces, m should approach unity, and μ becomes very large, as observed; with even a small decrease in m, μ falls to about unity, or to the type of value found for practically "clean" surfaces. And if a boundary film is present, making $m < 0.2$, Eq. XII-12 reduces to

$$\mu = \left(\frac{m^2}{\alpha}\right)^{1/2} = \left(\frac{m^2 s_0^2}{\alpha s_0^2}\right)^{1/2} = \frac{s}{\mathbf{P}_m} \qquad \text{(XII-14)}$$

which is a more general form of Eq. XII-6.

Deryrguin and co-workers (34) have proposed the equation

$$F = \mu W + \mu A\mathbf{p}_0 \qquad \text{(XII-15)}$$

where, in effect, $A\mathbf{p}_0$ adds to the external force W, owing to an internal or adhesion pressure $\mathbf{p}_0$. In fact, a number of the many systems showing deviations from Amontons' law would fit Eq. XII-15 approximately, in addition to the confirming behavior cited (paraffin on glass).

An early law known as Coulomb's law states that μ is independent of sliding speed. This may be approximately true over a small range, but the general observation is that μ decreases with increasing sliding speed, sometimes almost exponentially (1). At very high speeds there may be local melting. Finally, if the sliding surfaces are in contact with an electrolyte solution, an analysis indicates that the coefficient of friction should depend on the applied potential (35).

7. Friction between Lubricated Surfaces

A. Boundary Lubrication

Two limiting conditions exist where lubrication is used. In the first case, the oil film is thick enough so that the surface regions are essentially independent of each other, and the coefficient of friction depends on the hydrodynamic properties, especially the viscosity, of the oil. Amontons' law is not involved in this situation, nor is the specific nature of the solid surfaces.

As load is increased and relative speed is decreased, the film between the two surfaces becomes thinner, and increasing contact occurs between the surface regions. The coefficient of friction rises from the very low values possible for fluid friction to some value that usually is less than that for unlubricated surfaces. This type of lubrication, that is, where the nature of the surface region is important, is known as *boundary lubrication*. The general picture is illustrated in Fig. XII-10 by what is known as the Stribeck

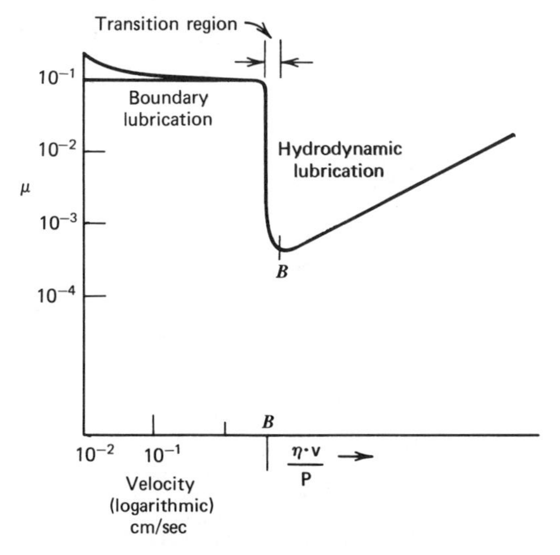

Fig. XII-10. Regions of hydrodynamic and boundary lubrications. (From Ref. 36.)

curve (33); the abscissa quantity, (viscosity)(speed)/(load), is known as the generalized Sommerfeld number (1).

Much of the classic work with boundary lubrication was carried out by Sir William Hardy (37, 38). He showed that boundary lubrication could be explained in terms of adsorbed films of lubricants and proposed that the hydrocarbon surfaces of such films reduced the fields of force between the two parts.

Hardy studied a number of lubricants of the hydrocarbon type, such as long chain paraffins, alcohols, and acids. An excess of lubricant was generally used, either as the pure liquid or as a solution of the solid in petroleum ether. In some cases the metal surface was then polished. The general observation was that μ values of 0.05 to 0.15 were obtained, that is, much lower than for unlubricated surfaces. Also, for a given homologous series, μ decreased nearly linearly with increasing molecular weight, although with the fatty acids, μ leveled off at about 0.05 at a molecular weight of 200. These relationships are shown in Fig. XII-11.

Levine and Zisman (39) confirmed and extended Hardy's results, using films on glass and on metal surfaces that were deposited by adsorption either from solution or from the molten compound. With alcohols the leveling off occurred at a molecular weight of about 280, again at $\mu = 0.05$, so that the behavior is much the same as for the acids, but shifted toward larger molecular weights. With hydrocarbons on copper, some data by Russell and co-workers (40) indicate that the leveling off occurs at about a molecular weight of 400 at $\mu = 0.10$. These points of no further decrease in μ were considered by Levine and Zisman to correspond to the formation of a condensed monolayer. This conclusion was supported by the observation that

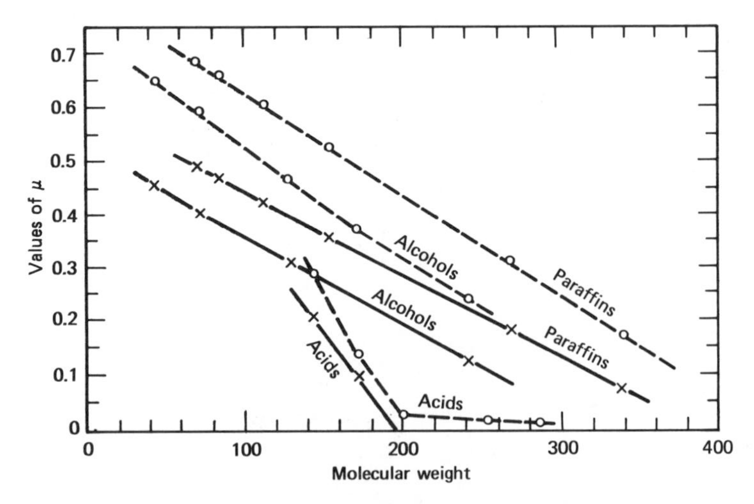

Fig. XII-11. Variations of μ with molecular weight of the boundary lubricant: ———, curve for spherical slider; – – –, curve for plane slider. (From Ref. 37.)

the air–methylene iodide–film contact angles ceased to increase with molecular weight at about the same stage.

There is no doubt that the reduction of μ by such as the foregoing lubricants is due to a thin and perhaps monomolecular film of adsorbed material. Thus Frewing (41) found that, using a hemispherical steel rider, the coefficients of friction for various alcohols, esters, and acids were about the same whether an excess of lubricant was used or whether one or more monolayers were built up by means of the Langmuir–Blodgett technique (Section IV-15). Later, films of low molecular weight alcohols, aldehydes, paraffins, and water formed by vapor phase adsorption on a steel surface were found to exhibit μ values that decreased to a minimum as the monolayer point was reached (42). Similarly, monolayers of halogenated compounds adsorbed on metal or glass slides from solution in aqueous or organic solvents or from the vapor phase led to μ values typical for boundary lubrication (40).

It is evident, moreover, that the coefficient of friction under boundary lubrication conditions is considerably dependent on the state of the monolayer. Frewing found that, on heating, the value of μ rose rather sharply at a characteristic temperature from values around 0.1 to values around 0.4. In other cases, the transition temperature was marked more by a change from smooth to stick-slip sliding than by any very great change in μ itself. For hydrocarbons, alcohols, and ketones, the transition temperature was close to the bulk melting point. Similarly, the leveling off in μ as one goes up a homologous series occurs with that member of the series whose melting point corresponds to the temperature of the measurements (see Ref. 39). Very likely these points of change correspond to the transition between an expanded and a condensed film; a similar correlation with bulk melting point exists for films on water.

A further indication that condensed films are required for good boundary lubrication is in the behavior of fatty acid monolayers. With films on glass (39), on platinum (42), and on various polymers (9), the change in μ correlates with the bulk melting point in about the same manner as for the other films, but for lauric acid on copper, the change in μ occurs at about 110°C (4) and for stearic acid on zinc, at about 130°C (43; see also Ref. 33). These temperatures are much higher than the respective bulk melting points of 43 and 69°C, but they do correspond to the softening points of the corresponding *metal salts* of the fatty acids. In fact, it is known from adsorption studies that fatty acids chemisorb on the more electropositive metals by salt formation with the oxide coating (see Section XI-1B).

B. The Mechanism of Boundary Lubrication

Hardy's explanation that the small coefficients of friction observed under boundary lubrication conditions were due to the reduction in the force fields between the surfaces as a result of adsorbed films is undoubtedly correct

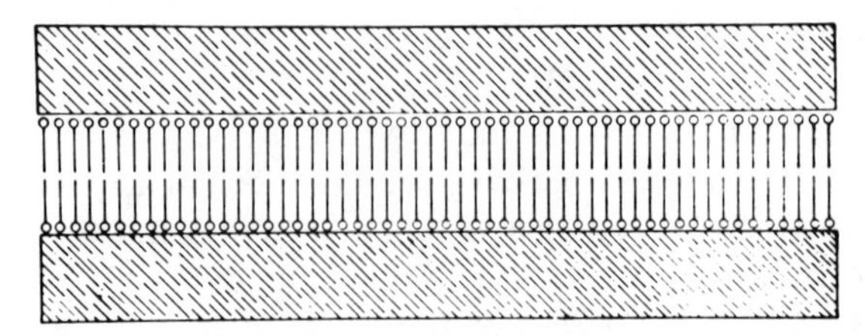

Fig. XII-12. Contact region in boundary lubrication according to Hardy.(From Ref. 38.)

in a general way. The explanation leaves much to be desired, however, and it is of interest to consider more detailed proposals as to the mechanism of boundary lubrication.

It has been pointed out that the value of μ in boundary lubrication depends greatly on the state of the adsorbed film and that, generally speaking, the film must be in a condensed state to give a low coefficient of friction. In combination with Hardy's explanation, the picture of a contact region might then be supposed to look as shown schematically in Fig. XII-12. However, this cannot be correct for at least two reasons. First, it is generally found that with very light loads, the coefficient of friction under boundary lubrication conditions *rises* to near normal values (3, 21); such an effect cannot be accounted for in terms of Hardy's model. The effect is illustrated in Fig. XII-13. Parenthetically, and more understandably, boundary lubrication fails under very high loads (44), and with repeated traverses of the same track (45).

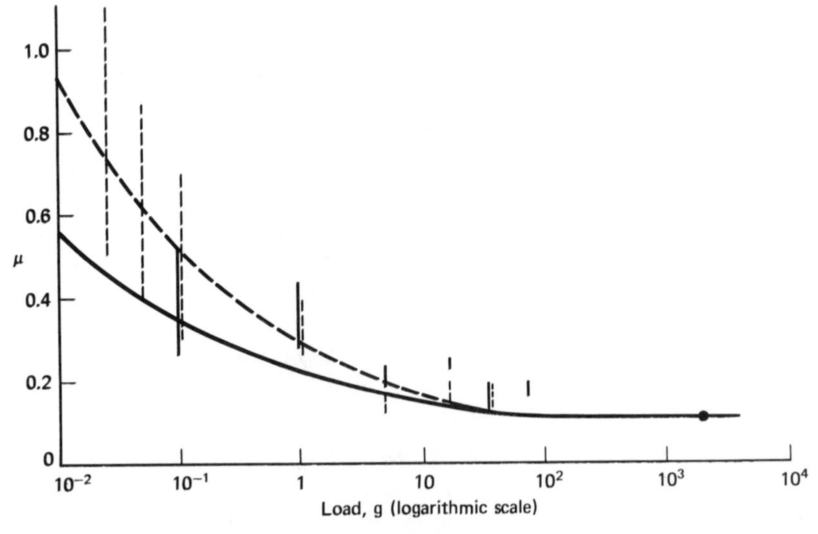

Fig. XII-13. Variation of μ with load. The data are the friction of copper lubicated with lauric acid (———) and with octacosanoic acid (– – –). (From Ref. 21.)

Second, it is found that metal–metal contacts are still present even under normal boundary lubrication conditions where μ is small. Very clear evidence for this has been provided by radioautographic studies (43). The transfer of metal from a radioactive slider is mapped by placing the surface against a photographic emulsion and noting where darkening occurs. The significant aspect of the results is that, while the total amount of transfer is greatly reduced when boundary lubrication is present over that in the absence of lubrication, the reduction is more in the *size* of each fragment than in the number of fragments. In other words, radioautographs for lubricated and unlubricated surfaces differ more in the intensity of the spots than in their number. Bowden and Tabor (3) note that the number of spots may be reduced by a factor of 3 or 4 with a fatty acid boundary lubricant while the total amount of metal transfer drops by a factor of perhaps 10^5. This observation bears on a matter of practical importance, namely, that various boundary lubricants can give about the same coefficient of friction yet still differ enormously between themselves in the amount of wear allowed.

The preceding observation suggests a second model, shown in Fig. XII-14 (4). In conformity with the radioautographic work, it is supposed that the load is supported over area A and that some portion of this area αA consists of metal–metal contacts of shear strength s_m, whereas the rest consists of film–film contacts of shear strength s_f. Thus in a manner analogous to Eq. XII-10, one writes for the total frictional force F,

$$F = A[\alpha s_m + (1 - \alpha)s_f] \qquad \text{(XII-16)}$$

or

$$F = A_1 s_m + A_2 s_f \qquad \text{(XII-17)}$$

For boundary lubrication, α must be of the order of 10^{-4} to account for the great reduction in metal pickup. A corollary is that most of the friction under boundary lubrication conditions must be due to film–film interactions.

This second picture, while an advance over Hardy's, again encounters difficulties. It does not suggest how A_2 could be so much greater than A_1

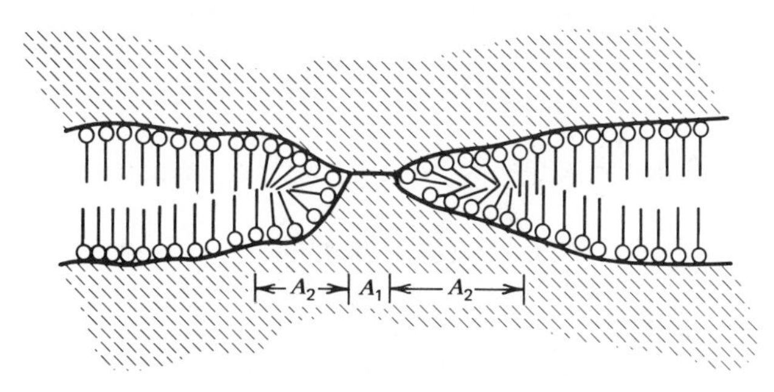

Fig. XII-14. Nature of the contact region in boundary lubrication according to Bowden and Tabor. (From Ref. 4.)

or why s_f should be higher with small loads. Also, it does not explain the behavior of the electrical conductivity. At small loads, the conductivity is very small, as would be expected in terms of the model, because of the small values of A_1 and the probably high electrical resistance of the film. However, at loads such that μ falls to its normal low value, the electrical conductivity rises to about the *same* as for unlubricated surfaces. There is still very little metal–metal contact, so the film must now have become conducting. Also, to explain why Amontons' law holds for boundary lubrication over the usual range of loads, it is necessary to suppose that A_2 is proportional to load. The model does not suggest how this could be true.

A third model, proposed by the author some years ago (46), seems to meet these objections. It is supposed that as two surfaces are pressed together under increasing load, it is *only* for very light loads that the situation is that of Fig. XII-14. That is, with small loads, there will be regions of slight metal–metal contact with a surrounding fringe of essentially normal monolayer. The load is then being supported both by the metal–metal contact region and that of film–film contact. Since A_2 corresponds to the area of the surrounding fringes of film–film contact, the geometry of the situation requires that A_2/A_1 increase as A_1 decreases. Amontons' law should therefore not be obeyed at small loads. In addition, μ should be at near normal values for unlubricated surfaces to the extent that it arises from A_1. In addition, s_f might be fairly large itself as a steric consequence of the entangling of the hydrocarbon chains.

Turning now to the situation for normal loads, there can be no doubt that most of the load is being supported by the boundary film and that therefore the film itself is under mechanical pressure. This must come about physically through deformation of the uneven metal surface—not enough to form metal–metal contacts by actually displacing the film, but enough to put some constriction and hence pressure on it. The effect of applying mechanical pressure on the film will be to contribute an additional term to its free energy, essentially the integral of $\bar{V}\,d\mathbf{P}$, where $\bar{V}$ is the molar volume and $\mathbf{P}$ is the mechanical pressure. In much the same manner, the escaping tendency and the vapor pressure of a liquid increases under mechanical pressure. Thus for an incompressible liquid,

$$RT \ln \frac{P}{P^0} = \bar{V}\mathbf{P} \qquad \text{(XII-18)}$$

where P and P^0 denote the new vapor pressure and the normal one, respectively.

In the case of boundary lubrication, it must be remembered that the area of actual contact is undoubtedly only a small fraction of the total area of apparent contact, so that only occasional small patches of film are put under mechanical pressure. As a result, it seems likely that the increased escaping tendency of the pressurized film material will manifest itself as a transfer of film molecules from the pressurized region to adjacent, normal regions, as illustrated in Fig. XII-15. There seems little doubt that transport of fatty

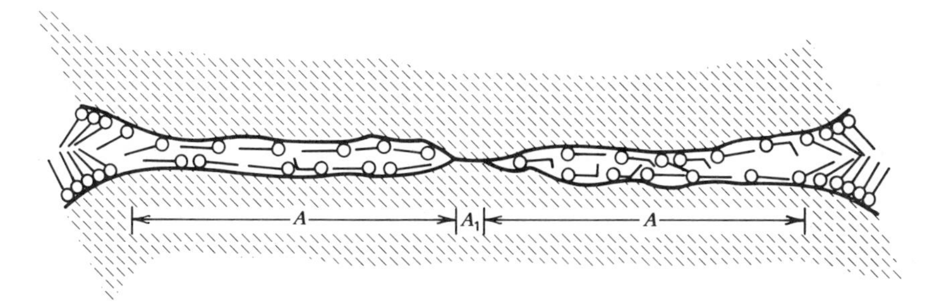

Fig. XII-15. Pressurized film model for boundary lubrication.

acid films can occur rapidly over short distances, either through surface mobility or by way of vapor phase hopping (47). Thus, as a result of the mechanical pressure on part of the film, the new equilibrium condition for a molecule of film material becomes

$$\int_0^{\mathbf{P}} V\, d\mathbf{P} + \int_0^{\pi'} \sigma'\, d\pi' = \int_0^{\pi} \sigma\, d\pi \qquad \text{(XII-19)}$$

where π and π' are the film pressures of the normal and pressurized films, respectively, and σ and σ' are the corresponding molecular areas. The second and third integrals give, according to the Gibbs equation (see Eq. X-11), the vapor pressures associated with the two film states, so that Eq. XII-19 amounts to the statement that the lower intrinsic vapor pressure of the film at π' is sufficiently enhanced by the mechanical pressure to be equal to that of the film at π.

As has been discussed, it appears that good boundary lubrication films are in a condensed or solid state, while, as illustrated in the figure, the pressurized film might consist of molecules lying more or less flat, and would therefore resemble the L_1 state of monolayers. Equation XII-19 may be written

$$\mathbf{P}\bar{V} \cong \int_{\pi'}^{\pi} \sigma\, d\pi; \qquad \pi \cong \pi_s,\ \pi' \cong \pi_{L_1} \qquad \text{(XII-20)}$$

where, as approximations the two film states are taken to lie on the same π–σ plot, and $\bar{V}$ is assumed constant. Inspection of Fig. IV-19 for a fatty acid monolayer on water suggests the further approximation

$$\mathbf{P}\sigma\tau = (\pi - \pi')\sigma; \qquad \mathbf{P}\tau = \Delta\pi \qquad \text{(XII-21)}$$

where τ is the thickness of the condensed film and σ is its molecular area, so that $\bar{V} = \sigma\tau$.†

† An alternative approximate derivation is the following. Consider a small displacement from equilibrium in which the thickness of the pressurized film decreases by $d\tau'$. As a result, dn molecules are transferred from pressurized to normal film, where $dn = d\tau'/\bar{V} = d\tau'/\sigma\tau$, per unit area. The mechanical work is $\mathbf{P}\, d\tau'$ and is balanced by the work of transferring the dn molecules against the surface pressure difference. Thus $\mathbf{P}\, d\tau' = (\pi - \pi')\sigma\, dn = \Delta\pi\, d\tau'/\tau$, which rearranges to Eq. XII-21.

If π' is neglected in comparison to π, then Eq. XII-21 becomes

$$\mathbf{P}\tau = \pi \tag{XII-22}$$

or, since $\mathbf{P} = W/A$,

$$A = \frac{\tau}{\pi} W \tag{XII-23}$$

and

$$\mu = \frac{\tau s'_f}{\pi} \tag{XII-24}$$

If π is taken to be 60 dyne/cm and τ to be 10 Å, then π/τ corresponds to about 6 kg/mm^2. For comparison, the elastic limits for lead, copper, and mild steel are about 2, 31, and 65, respectively, in the same units.

The mechanism of boundary lubrication may then be pictured as follows. At the unusually prominent asperities, the local pressure exceeds the yield pressure of the metal, film material is completely displaced, and an area of metal–metal contact develops. A much smaller, and probably elastic, deformation of the metal is sufficient, however, to place a relatively large area of film under varying mechanical pressure, so that most of the load is "floating" on pressurized film. To the approximation of Eqs. XII-21 or XII-22, the area of the pressurized film is proportional to the load, so that Amontons' law should be obeyed. In addition, the coefficient of shear for the film–film fraction s'_f now applies to film molecules that are lying flat, and it is taken to be much smaller than s_m or s_f, corresponding to the observed low μ values under boundary lubrication conditions with normal loads.

Bowden and Tabor (3) report an estimate for s'_f of 250 g/mm^2, using boundary lubricated mica surfaces in the form of crossed cylinders and assuming the apparent and real contact areas to be the same in this case. The value may therefore be in error, but used in Eq. XII-23 with the other numbers of the numerical example, the estimated value of μ becomes 0.04, or very close to what is observed with boundary lubricants generally. These authors also cite further support for the general idea that film molecules are forced to a horizontal configuration in load-bearing regions, and the general idea was proposed by Wilson in 1955 (48). It might be noted that the yield pressure of the solid does not enter in this treatment; it is required only that it be low enough that deformation will occur in preference to exceeding the pressure to produce the L_1 film (6 kg/mm^2 in the numerical example above). Boundary lubrication values of μ should then be relatively independent of the nature of the solid, as is observed.

It must not be forgotten that any mechanism for boundary lubrication must be a dynamic one, since it is a kinetic rather than a static coefficient of friction that is generally involved. In the case of nonlubricated surfaces, it is supposed that contact regions form and are sheared as the surfaces move past each other, with new regions constantly forming as new asperities pass each other. There is thus a steady state situation. Presumably, a similar situation is involved in the case of boundary lubrication, with regions such as shown in Fig. XII-15 forming and disappearing. It is doubtful if the equilibrium Eqs. XII-19 and XII-23 can apply very accurately at high sliding speeds since a finite time would be required for solid deformation and for migration of film molecules from a pressurized to an unpressurized region. The following numerical estimate is helpful with respect to this last requirement.

It is known that even condensed films must have surface diffusional mobility; Rideal and Tadayon (49) found that stearic acid films transferred from one surface to another by a process that seemed to involve surface diffusion to the occasional points of contact between the solids. Such transfer, of course, is observed in actual friction experiments in that an uncoated rider quickly acquires a layer of boundary lubricant from the surface over which it is passed (39). However, there is little quantitative information available about actual surface diffusion coefficients. One value that may be relevant is that of Ross and Good (50) for butane on Spheron 6, which, for a monolayer, was about 5×10^{-3} cm^2/sec. If the average junction is about 10^{-3} cm in size, this would also be about the average distance that a film molecule would have to migrate, and the time required would be about 10^{-4} sec. This rate of junctions passing each other corresponds to a sliding speed of 10 cm/sec so that the usual speeds of 0.01 cm/sec should not be too fast for pressurized film formation. See Ref. 47 for a recent study of another mechanism for surface mobility, that of evaporative hopping.

The foregoing analysis makes the assumption that the general film pressure π is quite high, corresponding to a condensed film. However, if π is small to start with, then $\Delta\pi$ is small and the area of L_1-type pressurized film to support the load W could become inordinately large. Alternatively, the pressurized film would have to be so dilute that it would offer little barrier to metal–metal contact. This is apparently what happens in boundary lubrication as the temperature is raised (or the molecular weight of the lubricant lowered). The film goes over to an expanded state, and the pressurized film becomes so dilute that it gives way to considerably more metal–metal contact with a consequent increase in μ.

The behavior of sebacic acid ($HOOC—C_8H_{16}—COOH$) as a boundary lubricant is interesting. The orientation in the film is presumably flat at all film pressures, and the film is probably highly condensed so that π is large and the film pressurization model should still apply, although there is now a possibility that surface diffusion would be so slow that the equilibrium relationships would not hold. However, since the molecules lie flat, one would expect the same value of s'_f to apply as for the pressurized films of straight chain compounds. The coefficient of friction is indeed at about the usual value for a boundary lubricant, but the significant observation is that, unlike the other cases, μ does not increase at small loads (4). This suggests that even at small loads the load is largely supported by the film rather than by metal–metal contacts. In turn, the implication is that the same may be true for straight chain lubricants, which means that s_f is indeed large, perhaps, as suggested, in consequence of a steric or chain tangling effect.

There is also a breakdown of boundary lubrication under extreme pressure conditions. The effect is considered to be related to that of increasing temperature (44); this is not unreasonable since the amount of heat to be dissipated will increase with load and a parallel increase in the local temperature would be expected. Also, of course, the film pressurization mechanism would predict a rise in μ for the same reason it does if the film is expanded.

As a final comment, the pressurized film model might account for the lack of increase in electrical resistance between boundary lubricated surfaces as compared to unlubricated metal ones. The point here is that the electrical resistivity (or, perhaps, the potential of dielectric breakdown) of the pressurized film should be much lower than for bulk film material owing to the proximity of the polar heads attached to the opposing faces; Wilson (48) suggested that conduction by the tunnel effect might be possible.

8. Adhesion

The term adhesion has been used so far in this book in a restricted and ideal sense, as in the work of adhesion concept (Section III-3). It also has the broader and more practical meaning of being simply the failing load, in some specified test, of a particular joint. This last definition now includes the very extensive subject of the practical performance of cements and glues, which are considered briefly under the term practical adhesion.

A. Ideal Adhesion

Ideal adhesion simply means the adhesion expected under one or another model situation. For example, the work and energy of adhesion as defined by Eqs. III-47 and X-8 refer to the reversible work and the energy, per square centimeter, to separate two phases that initially have a common interface. End effects are not considered. Although the energy rather than the reversible work may be more relevant to real situations, we will confine ourselves to the latter.

Equation III-47 may be written in the more complete form

$$w_{A(B)B(A)} = \gamma_{A(B)} + \gamma_{B(A)} - \gamma_{AB} = w_{AB} - \pi_{A(B)} - \pi_{B(A)} \qquad \text{(XII-25)}$$

where a phase in parenthesis saturates the one preceding it. Recalling the numerical example of Section IV-2A, for water A and benzene B, $w_{A(B)B(A)} = 56\ \text{erg/cm}^2$ and $w_{AB} = 76\ \text{erg/cm}^2$. These are fairly representative values for works of adhesion, and it is instructive to note that they correspond to fairly large separation forces. The reason is that these energies should be developed largely over the first few molecular diameters of separation, since intermolecular forces are rather short range. If the effective distance is taken to be 10 Å, for example, then the figure of 56 erg/cm^2 becomes an average force of 5.6×10^8 dyne/cm^2 or 560 kg/cm^2, which is much larger than most failing loads of adhesive joints.

Equation XII-25 may be combined with various semiempirical equations. Thus if Antonow's rule applied (Eq. IV-13), one obtains

$$w_{A(B)B(A)} = 2\gamma_{B(A)} \qquad \text{(XII-26)}$$

(which agrees well with the benzene–water example). The Girifalco and Good equation, Eq. IV-7, gives

$$w_{AB} = 2(\gamma_A\gamma_B)^{1/2} \qquad \text{(XII-27)}$$

(which does not agree as well). If substance B has a finite contact angle against substance A, treated as a solid, then

$$w_{A(B)B(A)} = \gamma_{B(A)}(1 + \cos \theta_B) \qquad \text{(XII-28)}$$

and in combination with Zisman's relationship, Eq. X-37,

$$w_{A(B)B(A)} = \gamma_{B(A)}(2 + \beta\gamma_c) - \beta\gamma^2_{B(A)} \qquad \text{(XII-29)}$$

The interesting implication of Eq. XII-29 is that for a given solid, the work of adhesion goes through a maximum as $\gamma_{B(A)}$ is varied (51). For the low energy surfaces Zisman and co-workers studied, β is about 0.04, and $w_{\max}$ is approximately equal to the critical surface tension γ_c itself; the liquid for this optimum adhesion has a fairly high contact angle.

There has been considerable elaboration of the simple Girifalco and Good relationship, Eq. XII-27. As noted in Sections IV-2A and X-6B, the surface free energies that appear under the square root sign may be supposed to be expressible as a sum of dispersion, polar, and so on, components. This type of approach as been developed by Dann (52) and Kaelble (53) as well as by Schonhorn and co-workers (see Ref. 54). Good (see Ref. 55) has preferred to introduce polar interactions into a detailed analysis of the meaning of Φ in Eq. IV-7. While there is no doubt that polar interactions are important, these are orientation dependent and hence structure-sensitive. This writer doubts that such interactions can be estimated other than empirically without a fairly accurate knowledge of the structure in the interfacial region.

A second ideal model for adhesion is that of a liquid wetting two plates, forming a circular meniscus, as illustrated in Fig. XII-16. Here a Laplace pressure $\mathbf{P} = 2\gamma_L/x$ draws the plates together and, for a given volume of liquid,

$$f = \frac{2\gamma_L V}{x^2} \qquad \text{(XII-30)}$$

so that the force f to separate the plates can be very large if the film is thin (note Problem XII-12).

B. Practical Adhesion

The usual practical situation is that in which two solids are bonded by means of some kind of glue or cement. A relatively complex joint is illustrated in Fig. XII-17. The strength of a joint may be measured in various ways. A common standard method is the *peel test* in which the normal force to separate the joint is measured, as illustrated in Fig. XII-18. (See Ref. 58, in which a more elaborate cantilever beam device is also used, and Ref. 59). Alternatively, one may determine the shear adhesive strength as the force

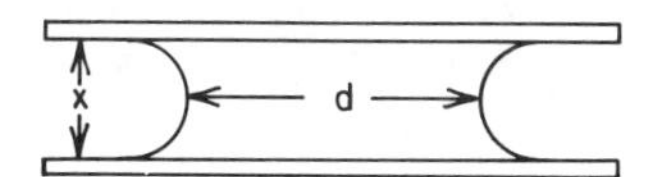

Fig. XII-16. Illustration of adhesion between two plates due to a meniscus of a wetting liquid.

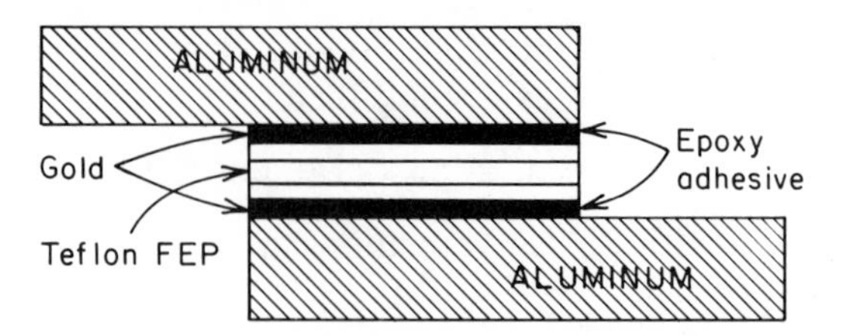

Fig. XII-17. A composite adhesive joint. (From Ref. 56.)

parallel to the surface that is needed to break the joint, as illustrated in Fig. XII-19.

Two models for adhesive joint failure are the following. One is essentially a classical, mechanical picture, relating stored elastic energy and the work of crack propagation, known as the Griffith–Irwin criterion (see Ref. 61). Bikerman (62), however, has argued that the actual situation is usually one of a *weak boundary layer*. This was thought to be a thin layer (but of greater than molecular dimensions) of altered material whose mechanical strength was less than that of either bulk phase. While oxides, contamination, and so on, could constitute a weak boundary layer, it was also possible that this layer could be structurally altered but pure material. Interestingly, Schonhorn and Ryan (63) found that proper cleaning of a plastic greatly improved the ability to bond it to metal substrates. Good (64, but see Ref. 61 also) has argued that the *molecular* interface may often be the weakest layer and that adhesive failure is at the true interface. A rather different idea, especially applicable to the peel test procedure, is due to Deryaguin (see Ref. 65). The peeling of a joint can produce static charging (sometimes visible as a glow or flashing of light), and much of the work may be due to the electrical work of, in effect, separating the plates of a charged condensor.

Returning to more surface chemical considerations, most literature discussions that relate adhesion to work of adhesion or to contact angle deal with surface free energy quantities. It has been pointed out that structural distortions are generally present in adsorbed layers and must be present if bulk liquid adsorbate forms a finite contact angle with the substrate (see Ref. 66). Thus both the entropy and the energy of adsorption are important (relative to bulk liquid). The same is likely true at a liquid–solid or a solid–solid interface, and it has been proposed that gradations of two extreme classes exist (67). In class A systems there is little structural perturbation at the interface and therefore little entropy of adhesion. In class B

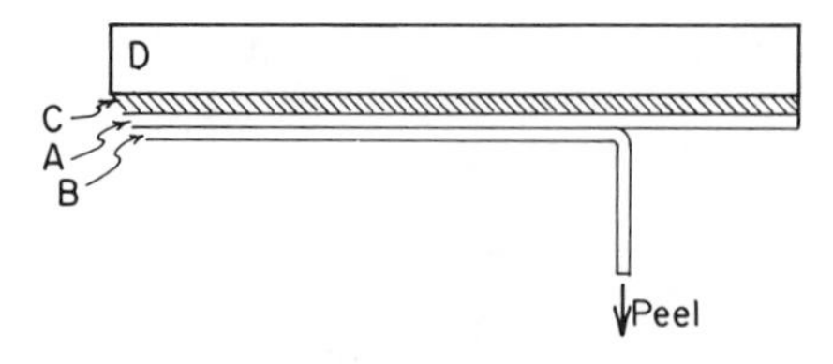

Fig. XII-18. Diagram of peel test. A and B, adhesive joint; C, double scotch adhesive tape, and D, rigid support. (From Ref. 57. By permission of IBC Business Press, Ltd.)

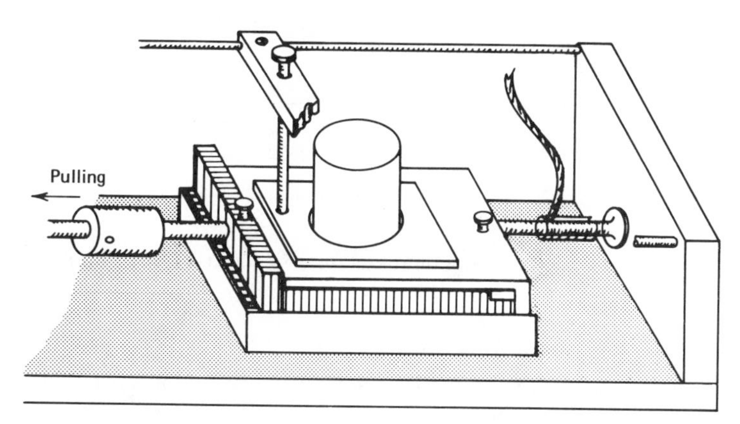

Fig. XII-19. Measurement of shear adhesion. (From Ref. 60.)

systems structural changes in one or both phases are important in the interfacial region, and there is an adverse entropy of adhesion, compensated by a greater than expected energy of adhesion. The distinction between energy and entropy may be important in practical adhesion situations. It is possible that the practical adhesive strength of a joint is determined more by the *energy* than by the *free energy* of adhesion; this would be true if the crucial point of failure occurred as an irreversible rather than a reversible process. Other aspects being the same, including the free energy of adhesion, the practical adhesion should be less for a class A than for a class B system. See Section X-6 for related discussion.

Quite other aspects are often dominant in practical adhesion situations. In contrast to the situation in friction, quite large areas of actual interfacial contact are desired for good adhesion. A liquid advancing over a rough surface can trap air, however, so that only a fraction of the apparent area may be involved in good interfacial contact. A low contact angle assists in preventing such composite surface formation and, with this in mind, it has been advocated that one criterion for good adhesion is that the adhesive spread on the surface it is to bond (68)—even though, as shown, this may not give the optimum ideal work of adhesion.

However, the actual strengths of "good" adhesive joints are only about a tenth of the ideal values (69), so that failure to wet completely does not seem to be an adequate explanation. Since adhesive joints are subject to shear as well as to normal force, there can be a "peeling" action in which the applied force concentrates at an edge to produce a crack that then propagates. Such stress concentrations can occur without deliberate intent simply because of imperfections and trapped bubbles in the interfacial region. Zisman (51) remarks that adhesives may give stronger joints with rough surfaces simply because such imperfections, including trapped air bubbles, will not be coplanar, and the propagation of cracks will then be less easy. This is illustrated in Fig. XII-20.

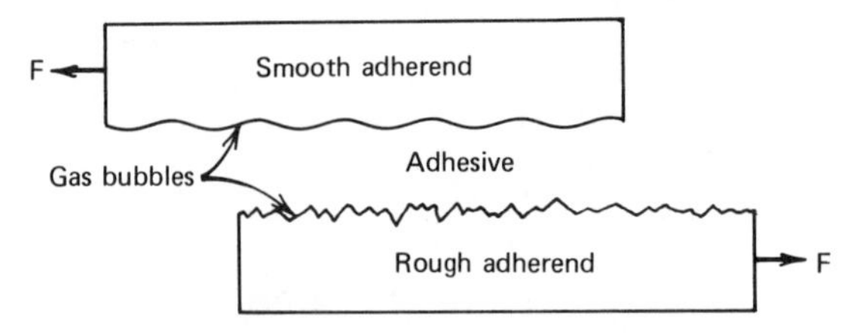

Fig. XII-20. Effects of surface roughness on coplanarity of gas bubbles. (From Ref. 51.)

The adhesion of polymers may be improved by using either ultraviolet irradiation or electron bombardment to increase the degree of surface cross-linking (70, 71). An adhesive joint may be weakened by penetration of water (54); the effect is of great importance in the "stripping" of blacktop road surfaces where water penetrates the interface between the rock aggregate and the asphalt.

Adhesion in biological systems constitutes a very important and challenging, but as yet rather diffuse, topic. Good adhesion is desired for bandages, certain tissue implants, in dentistry, and so on. Poor adhesion between blood constituents is, however, important for artificial arteries, heart valves, and other prosthetics. Baier (72, 73) and Newmann (74) have reviewed aspects of this subject.

As mentioned, there are cases where lack of adhesion is desirable. The coined word for this attribute is *abhesion*. As a nonbiological example, good abhesion is often desired between ice and either a metal or a polymer surface (as on airplane wings, or in ice-trays). *Ab*hesion of ice to metals is poor, and failure tends to occur in the ice itself. It is good with polytetrafluoroethylene (Teflon), apparently in part because in freezing against a surface that it wets poorly, air bubbles are produced at the ice–plastic interface. These allow stress concentration to lead to interfacial crack propagation. However, if water is cycled through several freezing and melting operations, dissolved air is removed and abhesion becomes poor (See Refs. 75–79). Fairly good abhesion has been reported for ice with polysiloxane–polycarbonate co-polymers, adhesive strengths of as low as 200 g/cm^2 being found (60).

9. Problems

1. The coefficient of friction for copper on copper is about 0.9. Assuming that asperities or junctions can be represented by cones of base and height each about 10^{-3} cm, and taking the yield pressure of copper to be 30 kg/mm^2, calculate the local temperature that should be produced. Suppose the frictional heat to be confined to the asperity, and take the sliding speed to be 10 cm/sec and the load to be 15 kg.

2. The resistance due to a circular junction is given by $R = 1/2ak$, where a is the radius of the junction and k is specific conductivity of the metal. For the case of

two steel plates, the measured resistance is $5 \times 10^{-5}\ \Omega$ for a load of 35 kg; the yield pressure of steel is 60 kg/mm^2, and the specific resistance is $5 \times 10^{-5}\ \Omega$/cm. Calculate the number of junctions, assuming that it is their combined resistance that is giving the measured value.

3. Deduce from Fig. XII-9, using the data for μ_s, how the contact area appears to be varying with load, and plot A versus W.

4. Calculate the angle of repose for a solid block on an inclined plane if the coefficient of friction is 0.40.

5. In a series of tests a car is brought to a certain speed, then braked by applying a certain force F on the brake pedal, and the deceleration a is measured. The pavement is dry concrete, and force F_0 is just sufficient to cause skidding. Sketch roughly how you think the plot of a versus F should look, up to F values well beyond F_0.

6. Derive the equation corresponding to Eq. XII-9 for the case where the road is inclined to the horizontal by an angle θ.

7. Discuss why stick-slip friction is favored if μ decreases with sliding speed.

8. The friction of steel on steel is being studied, using as boundary lubricants (*a*) $CH_3(CH_2)_8COOH$ and (*b*) $HOOC(CH_2)_8COOH$. Explain what your estimated value is for the coefficient of friction in case (*a*) and whether the value for case (*b*) should be about the same, or higher, or lower. The measurements are under conventional conditions, at 25°C. Discuss more generally what differences in frictional behavior might be found between the two cases.

9. Using the pressurized film model (Fig. XII-15) estimate the required film pressure if for a certain boundary lubricant $\mu = 0.10$ and $s_f' = 250$ g/mm^2.

10. The statement was made that the work of adhesion between two dissimilar substances should be larger than the work of cohesion of the weaker one. Demonstrate a basis on which this statement is correct and a basis on which it could be argued that the statement is incorrect.

11. Derive from Eq. XII-29 an expression for the maximum work of adhesion involving only β and γ_c. Calculate this maximum work for $\gamma_c = 18$ dyne/cm^1 and $\beta = 0.030$, as well as γ_L for this case, and its contact angle.

12. As illustrated in Fig. XII-16, a drop of water is placed between two large parallel plates; it wets both surfaces. Both the capillary constant a and d in the figure are much greater than the plate separation x. Derive an equation for the force between the two plates and calculate the value for a 2-cm^3 drop of water at 20°C, with $x = 1$ mm.

General References

J. J. Bikerman, *The Science of Adhesive Joints*, Academic, New York, 1961.

F. P. Bowden and D. Tabor, *The Friction and Lubrication of Solids*, Part I (1950) and Part II (1964), The Clarendon Press, Oxford.

F. P. Bowden and D. Tabor, *Friction, An Introduction to Tribology*, Anchor Books, New York, 1973.

P. J. Bryant, M. Lavik, and G. Salomon, Eds., *Mechanisms of Solid Friction*, New York, 1964.

Contact Angle, Wettability and Adhesion, Advances in Chemistry Series No. 43, American Chemical Society, Washington, D.C., 1964.

D. D. Eley, Ed., *Adhesion*, The Clarendon Press, Oxford, 1961.

R. Houwink and G. Salomon, Eds., *Adhesion and Adhesives,* Elsevier, New York, 1965.

D. H. Kaelble, *Physical Chemistry of Adhesion,* Wiley-Interscience, New York, 1971.

D. F. Moore, *Principles and Applications of Tribology,* Pergamon, New York, 1975.

J. A. Schey, *Metal Deformation Processes,* Marcel Dekker, New York, 1970.

F. P. Tabor, *Surface and Colloid Science,* E. Matijevic, Ed., Wiley-Interscience, New York, 1972.

Textual References

1. D. F. Moore, *Principles and Applications of Tribology,* Pergamon, New York, 1975.
2. F. P. Bowden and D. Tabor, *Friction,* Anchor Books, New York, 1973.
3. F. P. Bowden and D. Tabor, *The Friction and Lubrication of Solids,* Part II, The Clarendon Press, Oxford, 1964.
4. F. P. Bowden and D. Tabor, *The Friction and Lubrication of Solids,* Oxford University Press, New York, 1950.
5. F. P. Bowden and P. H. Thomas, *Proc. Roy. Soc.* (*London*), **A223,** 29 (1954).
6. K. Tanaka and Y. Uchiyrma, *Advances in Polymer Friction and Wear,* Lieng-Huang Lee, Ed., Vol. 5B of *Polymer Science and Technology,* Plenum, New York, 1974.
7. See "A Discussion on Friction," *Proc. Roy. Soc.* (*London*), **A212,** 439 (1952).
8. B. W. Sakmann, J. T. Burwell, and J. W. Irvine, *J. Appl. Phys.,* **15,** 459 (1944).
9. T. Fort, Jr., *J. Phys. Chem.,* **66,** 1136 (1962).
10. B. J. Briscoe, D. C. B. Evans, and D. Tabor, *J. Colloid Interface Sci.,* **61,** 9 (1977).
11. E. Rabinowicz, *Friction and Wear,* R. Davies, Ed., Elsevier, New York, 1959; see also J. T. Burwell and E. Rabinowicz, *J. Appl. Phys.,* **24,** 136 (1953).
12. C. E. O'Hara and J. W. Osterburg, *An Introduction to Criminalistics,* Macmillan, New York, 1959, p. 310.
13. F. P. Bowden, *Proc. Roy. Soc.* (*London*), **A217,** 462 (1953).
14. F. P. Bowden and G. W. Rowe, *Proc. Roy. Soc.* (*London*), **A228,** 1 (1955).
15. N. H. Fletcher, *Phil. Mag.,* **7,** No. 8, 255 (1962).
16. A. W. Adamson, L. M. Dormant, and M. Orem, *J. Colloid Interface Sci.,* **25,** 206 (1967).
17. F. P. Bowden and J. E. Young, *Nature,* **164,** 1089 (1949).
18. E. S. Machlin and W. R. Yankee, *J. Appl. Phys.,* **25,** 576 (1954).
19. D. H. Buckley, *J. Colloid Interface Sci.,* **58,** 36 (1977).
20. E. D. Tingle, *Trans. Faraday Soc.,* **46,** 93 (1950).
21. J. R. Whitehead, *Proc. Roy. Soc.* (*London*), **A201,** 109 (1950).
22. F. P. Bowden and D. Tabor, *Ann. Rep.,* **42,** 20 (1945).
23. F. P. Bowden and J. E. Young, *Proc. Roy. Soc.* (*London*), **A208,** 444 (1951).
24. I. M. Feng, *Lub. Eng.,* **8,** 285 (1952).
25. F. P. Bowden, J. E. Young, and G. Rowe, *Proc. Roy. Soc.* (*London*), **A212,** 485 (1952).
26. J. Spreadborough, *Wear,* **5,** 18 (1962).
27. See G. W. Rowe, *Wear,* **3,** 274 (1960).
28. R. C. Bowers and W. A. Zisman, *Mod. Plast.,* **41** (December 1963).

29. C. M. Pooley and D. Tabor, *Proc. Roy. Soc. London,* **A329,** 251 (1972).
30. D. Tabor, *Advances in Polymer Friction and Wear,* Lieng-Huang Lee, Ed., Vol. 5A, of *Polymer Science and Technology,* Plenum, New York, 1974.
31. J. S. McFarlane and D. Tabor, *Proc. Roy. Soc. (London),* **A202,** 244 (1950).
32. J. T. Burwell and E. Rabinowicz, *J. Appl. Phys.,* **24,** 136 (1953).
33. C. H. Riesz, *Metal Deformation Processes,* J. A. Schey, Ed., Marcel Dekker, New York, 1970.
34. B. V. Deryaguin, V. V. Karassev, N. N. Zakhavaeva, and V. P. Lazarev, *Wear,* **1,** 277 (1957–58).
35. J. O'M. Bockris and S. D. Argade, *J. Chem. Phys.,* **50,** 1622 (1969).
36. A. Bondi, *Physical Chemistry of Lubricating Oils,* Reinhold, New York, 1951.
37. W. B. Hardy and I. Bircumshaw, *Proc. Roy. Soc. (London),* **A108,** 1 (1925).
38. W. B. Hardy, *Collected Works,* Cambridge University Press, Cambridge, England, 1936.
39. O. Levine and W. A. Zisman, *J. Phys. Chem.,* **61,** 1068, 1188 (1957).
40. J. A. Russell, W. E. Campbell, R. A. Burton, and P. M. Ku, *ASLE Trans.,* **8,** 48 (1958).
41. J. J. Frewing, *Proc. Roy. Soc. (London),* **A181,** 23 (1942).
42. A. V. Fraioli, F. H. Healey, A. C. Zettlemoyer, and J. J. Chessick, *Abstracts of Papers, 130th Meeting of the American Chemical Society,* Atlantic City, New Jersey, September 1956.
43. D. Tabor, *Proc. Roy. Soc. (London),* **A212,** 498 (1952).
44. C. G. Williams, *Proc. Roy. Soc. (London),* **A212,** 512 (1952).
45. R. L. Cottington, E. G. Shafrin, and W. A. Zisman, *J. Phys. Chem.,* **62,** 513 (1958).
46. A. W. Adamson, *The Physical Chemistry of Surfaces,* Interscience, New York, 1960; also, unpublished work, 1959.
47. A. W. Adamson and V. Slawson, *J. Phys. Chem.,* **85,** 116 (1981).
48. R. W. Wilson, *Proc. Phys. Soc.,* **68B,** 625 (1955).
49. E. Rideal and J. Tadayon, *Proc. Roy. Soc. (London),* **A225,** 346, 357 (1954).
50. J. W. Ross and R. J. Good, *J. Phys. Chem.,* **60,** 1167 (1956).
51. W. A. Zisman, *Advances in Chemistry,* No. 43, American Chemical Society, Washington, D.C., 1964, p. 1.
52. J. R. Dann, *J. Colloid Interface Sci.,* **32,** 302 (1970).
53. D. H. Kaelble, *Physical Chemistry of Adhesion,* Wiley-Interscience, New York, 1971.
54. H. Schonhorn and H. L. Frisch, *J. Polym. Sci.,* **11,** 1005 (1973).
55. R. J. Good, *Adhesion Science and Technology,* Vol. 9A, L. H. Lee, Ed., Plenum, New York, 1975.
56. R. F. Roberts, F. W. Ryan, H. Schonhorn, G. M. Sessler, and J. E. West, *J. Appl. Polym. Sci.,* **20,** 255 (1976).
57. A. Baszkin and L. Ter-Minassian-Saraga, *Polymer,* **19,** 1083 (1978).
58. W. D. Bascom, P. F. Becher, J. I. Bitner, and J. S. Murday, Special Technical Publication 640, American Society for Testing and Materials, 1978.
59. D. Maugis and M. Barquins, *J. Phys. D; Appl. Phys.,* **11,** 1989 (1978).
60. H. H. G. Jellinek, H. Kachi, S. Kittaka, M. Lee, and R. Yokota, *Colloid Polym. Sci.,* **256,** 544 (1978).
61. R. J. Good, Special Technical Publication 640, American Society for Testing and Materials, 1978.

62. J. J. Bikerman, *The Science of Adhesive Joints,* 2nd ed., Academic, New York, 1968.
63. H. Schonhorn and F. W. Ryan, *J. Polym. Sci.,* **7,** 105 (1969).
64. R. J. Good, *J. Adhes.,* **4,** 133 (1972).
65. H. Schonhorn, *Polymer Surfaces,* D. T. Clark and W. J. Frost, Eds., Wiley-Interscience, New York, 1978.
66. A. W. Adamson, *J. Colloid Interface Sci.,* **44,** 273 (1973).
67. A. W. Adamson, F. P. Shirley, and K. T. Kunichika, *J. Colloid Interface Sci.,* **34,** 461 (1970).
68. J. R. Huntsberger, *Advances in Chemistry,* No. 43, American Chemical Society, Washington, D.C., 1964, p. 180; also, L. H. Sharpe and H. Schonhorn, *ibid.,* p. 189.
69. See D. D. Eley, Ed., *Adhesion,* The Clarendon Press, Oxford, 1961.
70. G. M. Sessler, J. E. West, F. W. Ryan, and H. Schonhorn, *J. Appl. Polym. Sci.,* **17,** 3199 (1973).
71. H. Schonhorn and F. W. Ryan, *J. Appl. Polym. Sci.,* **18,** 235 (1974).
72. R. E. Baier, *Adhesion in Biological Systems,* Academic, New York, 1970; R. E. Baier and L. Weiss, *Adv. Chem.,* **145,** 300 (1975).
73. R. E. Baier, *Adsorption of Microorganisms to Surfaces,* G. Bitton and K. C. Marshall, Eds., Wiley, New York, 1980.
74. A. W. Newmann, *Wetting, Spreading and Adhesion,* J. F. Padday, Ed., Academic, New York, 1978.
75. W. A. Zisman, *Ind. Eng. Chem.,* **55,** No. 10, 18 (1963).
76. H. H. G. Jellinek, *J. Colloid Interface Sci.,* **25,** 192 (1967).
77. W. D. Bascom, R. L. Cottington, and C. R. Singleterry, *J. Adhes.,* **1,** 246 (1969).
78. M. Landy and A. Freiberger, *J. Colloid Interface Sci.,* **25,** 231 (1967).
79. P. Barnes, D. Tabor, and J. C. F. Walker, *Proc. Roy. Soc. London,* **A324,** 127 (1971).

CHAPTER XIII

Wetting, Flotation, and Detergency

1. Introduction

We continue, in this chapter, the discussion of various areas of applied surface chemistry that was initiated in Chapter XII. The topics to be taken up here are so grouped because they have certain aspects in common. In each case contact angles are important—usually the contact angle between a liquid and a solid—and in each of these subjects much of the development that has occurred has involved the use of surface active materials designed to adsorb at one or more of the interfaces present and thus alter the corresponding interfacial tension.

As was true in the case of friction and lubrication, a great deal of work has been done in the fields of wetting and flotation, to a point that a certain body of special theory has developed, as well as a very large literature of applied surface chemistry. It is not the intent here, however, to cover these fields in any definitive way, but rather to use them to illustrate the role of surface chemistry in areas of practical importance.

The topic of detergency is of great importance, of course. It is more properly a topic of colloid chemistry, however, and is taken up in this chapter in only a cursory manner.

2. Wetting

A. Wetting as a Contact Angle Phenomenon

The terms *wetting* and *nonwetting* as employed in various practical situations tend to be defined in terms of the effect desired. Usually, however, wetting means that the contact angle between a liquid and a solid is zero or so close to zero that the liquid spreads over the solid easily, and nonwetting means that the angle is greater than 90° so that liquid tends to ball up and run off the surface easily.

To review briefly, a contact angle situation is illustrated in Fig. XIII-1, and the central relationship is the Young equation (see Section X-4A):

$$\cos\theta = \frac{\gamma_{SV} - \gamma_{SL}}{\gamma_{LV}} \qquad \text{(XIII-1)}$$

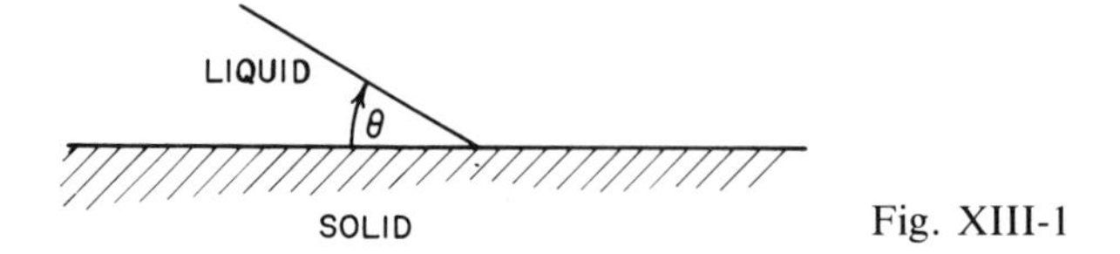

Fig. XIII-1

for the case of a finite contact angle, and the spreading coefficient $S_{L/S}$,

$$S_{L/S} = \gamma_{SV} - \gamma_{LV} - \gamma_{SL} \quad \text{(XIII-2)}$$

should wetting occur.

Qualitatively speaking, γ_{SL} and γ_{LV} should be made as small as possible if spreading is to occur. From a practical viewpoint, this is best done by adding to the liquid phase a surfactant that is adsorbed at both the solid–liquid and the liquid–air interfaces and therefore lowers these interfacial tensions. If the surfactant is nonvolatile, it is presumed not to affect γ_{SV} (see, however, Section X-5C).

There are various situations where good contact is desired between a liquid, usually an aqueous one, and an oily, greasy, or waxy surface. Examples would include sprays of various kinds, such as insecticidal sprays, which should wet the waxy surface of leaves or the epidermis of insects; animal dips, where wetting of greasy hair is desired; inks, which should wet the paper properly; scouring of textile fibers, including the removal of unwanted natural oils and the subsequent wetting of the fibers by desirable lubricants; the laying of dust, where a fluid must penetrate between dust particles as on roads or in coal mines.

As stated, wetting action is generally accomplished by the use of surfactant additives. It might be thought that it would be sufficient merely to insure a good lowering of γ_{LV} and that a rather small variety of additives would suffice to meet all needs. Actually, it is equally if not more important that the surfactant lower γ_{SL}, and each type of solid will make its own demands.

It is true that wetting agents or, in general, surfactant materials, typically consist of polar–nonpolar type molecules. The nonpolar portion is usually hydrocarbon in nature but may also be a fluorocarbon; it may be aliphatic or aromatic. The polar portion may involve almost any of the functional groups of organic chemistry. The group may be oxygen-containing, as in carboxylic acids, esters, ethers, or alcohols; sulfur-containing, as in sulfonic acids, their esters, or sulfates or their esters; it may contain phosphorus, nitrogen, or a halogen; it may or may not be ionic.

Another problem in wetting is that of hysteresis in contact angle (see Section X-5B). Usually it is the advancing angle that is important, but in the case of sheep dips it is the receding angle as the animal is immersed, and it is the degree of retention of the dip that is important. As noted in the section cited it is probably helpful in minimizing hysteresis that the adsorbed film of surfactant be mobile in nature. This means that the films should be of a liquid rather than a solid type.

Zettlemoyer and co-workers have made an interesting study of the various configurations that can be assumed in four-phase system (1). Typically, a solid surface is covered with a thick water film and a drop of oil added. Depending on the system, and on the thickness of the water film and the volume of the oil drop, the various configurations illustrated in Fig. XIII-2 can be obtained. Behavior of this type is of importance in such diverse fields as detergency, lithographic printing, and flotation.

The *speed* of wetting has been measured by running a tape of material that is wetted either downward through the liquid–air interface, or upward through the interface. For a polyester tape and a glycerol–water mixture, a wetting speed of about 20 cm/sec and a de-wetting speed of about 0.6 cm/sec[1] is reported (2).

B. Wetting as a Capillary Action Phenomenon

For some types of wetting more than just the contact angle is involved in the basic mechanism of the action. This is true in the laying of dust and the wetting of a fabric since in these situations the liquid is required to penetrate between dust particles or between the fibers of the fabric. The phenomenon is related to that of capillary rise, where the driving force is that of the pressure difference across the curved surface of the meniscus. The relevant equation is then Eq. X-33,

$$\Delta \mathbf{P} = \frac{2\gamma_{LV} \cos \theta}{r} \tag{XIII-3}$$

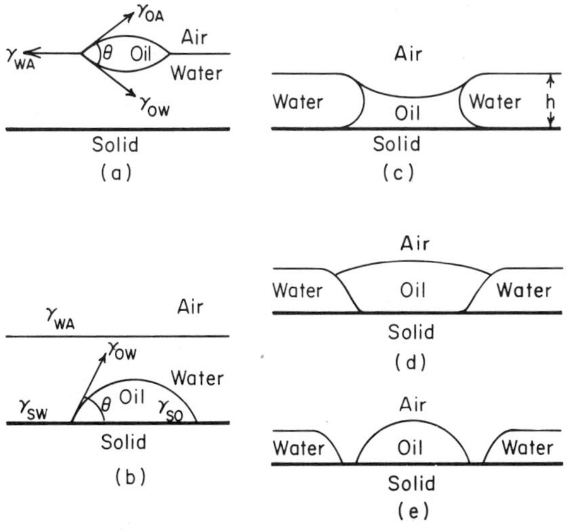

Fig. XIII-2. Configuration that four-phase systems can adopt (*a*) oil lens on water, (*b*) submerged oil drop, (*c*) partially submerged oil drop and hydrophobic solid, (*d*) same as (*c*) but hydrophilic solid, (*e*) oil and water not in contact. (From Ref. 1.)

where r denotes the radius (or equivalent radius) of the capillary. It is helpful to write Eq. XIII-3 in two separate forms. If θ is not zero, then

$$\mathbf{\Delta P} = \frac{2(\gamma_{SV} - \gamma_{SL})}{r} \tag{XIII-4}$$

so that the principal requirement for a large $\mathbf{\Delta P}$ is that γ_{SL} be made as small as possible, since for practical reasons it is not usually possible to choose γ_{SV}. On the other hand, if θ is zero, Eq. XI-3 takes the form

$$\mathbf{\Delta P} = \frac{2\gamma_{LV}}{r} \tag{XIII-5}$$

and the requirement for a large $\mathbf{\Delta P}$ is that γ_{LV} be large. The net goal is then to find a surfactant that reduces γ_{SL} without the same time reducing γ_{LV}. Since any given surfactant affects both interfacial tensions, the best agent for producing such opposing effects can be expected to vary from one system to another even more than with ordinary wetting agents. Capillary phenomena in assemblies of parallel cylinders have been studied by Princen (3).

In addition to $\mathbf{\Delta P}$ being large, it is also desirable in promoting capillary penetration that the *rate* of entry be large. For the case of horizontal capillaries or, in general, where gravity can be neglected, Washburn (4) gives the following equation for the rate of entry of a liquid into a capillary:

$$v = \frac{r\gamma_{L_1L_2} \cos \theta_{SL_1L_2}}{4(\eta_1 l_1 + \eta_2 l_2)} \tag{XIII-6}$$

The equation is for the general case of wetting liquid L_1 displacing liquid L_2 where the respective lengths of the liquid columns are l_1 and l_2 and the viscosities are η_1 and η_2. For a single liquid displacing air, the quantity $\gamma_L(\cos\theta)/\eta$ has the dimensions of velocity and thus gives a measure of the penetrating power of the liquid in a given situation.

The Washburn equation has most recently been confirmed for water and cyclohexane in glass capillaries ranging from 0.3 μm to 400 μm in radii (5). The contact angle formed by a moving meniscus may differ, however, from the static one (5, 6). Good and Lin (7) found a difference in penetration rate between an outgassed capillary and one with a vapor adsorbed film, and they propose that the driving force be modified by a film pressure term.

C. Tertiary Oil Recovery

Currently very important is *tertiary oil recovery*. Typically, a new oil well first produces spontaneously and then is pumped until oil flow becomes uneconomical. Additional oil may be recovered by pumping water into the formation (*water flooding*) and pumping out (at other points) a mixture of oil and water. Eventually this, too, becomes uneconomical, yet 30 to 50% of the original oil may remain in the formation. Tertiary oil recovery refers to processes designed to extract this residual oil—oil that typically is distributed in an unconnected way through the capillary system of the porous oil-bearing stratum. Processes using surfactants are becoming important

(10), but the physical chemistry of their action is not fully understood. Reducing the pressure drop across each oil–water meniscus appears to be important, and this is best done by reducing the oil–water interfacial tension. Aspects of the problem have been discussed by Melrose and Brandner (8) and by Wade and co-workers (9).

3. Water Repellency

Complementary to the matter of wetting is that of water repellency. Here, the desired goal is to make θ as large as possible. For example, in steam condensers, heat conductivity is improved if the condensed water does not wet the surfaces, but runs down in drops.

Fabrics may be made water-repellent by reversing the conditions previously discussed in the promotion of the wetting of fabrics. In other words, it is again a matter of capillary action, but now a large negative value of $\Delta \mathbf{P}$ is desired. As illustrated in Fig. XIII-3*a*, if $\Delta \mathbf{P}$ is negative (and hence if the contact angle is greater than 90°), the liquid will tend not to penetrate between the fibers, whereas if $\Delta \mathbf{P}$ is positive, liquid will pass through easily. It should be noted that a fabric so treated that it acts as in Fig. XIII-3*a* is only water-repellent, not waterproof. The fabric remains porous, and water will run through it if sufficient hydrostatic pressure is applied.

Since a finite contact angle is present, Eq. XIII-4 is the one involved. That is, the surface tension of the liquid is not directly involved, but rather the quantity ($\gamma_{SV} - \gamma_{SL}$), which must be made negative. This is done by coating the solids to reduce γ_{SV} as much as possible; in terms of the critical surface tension concept (Section X-5C), this means that the γ_c for the solid should be reduced to less than about 40, and lower if possible.

A factor that helps in fabric waterproofing is that the mesh or screen structure leads to larger than otherwise contact angles (Eq. X-29). Also, application of the waterproofing agent to give a microscopic roughness to the surface will lead to an increase in contact angle (Eq. X-26).

A different type of water repellency is that required to prevent the deterioration of blacktop roads, which consist of crushed rock coated with bituminous materials. Here the problem is that water tends to spread into the stone–oil interface, detaching

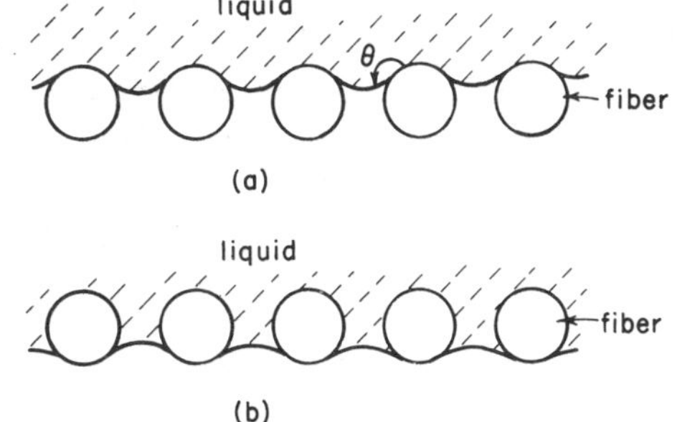

Fig. XIII-3. Effect of contact angle in determining water repellency of fabrics.

the aggregate from its binder (10, 11). No entirely satisfactory solution has been found, although various detergent-type additives have been found to help. Much more study of the problem is needed.

Contemporary concern about pollution has made it important to dispose of oil slicks from spills. The suitable use of surfactants may reverse the spreading of the slick, thereby concentrating the slick for easier removal.

4. Flotation

A very important but rather complex application of surface chemistry is to the separation of various types of solid particles from each other by what is known as flotation. The general method is of enormous importance to the mining industry; it permits large scale and economic processing of crushed ores whereby the desired mineral is separated from the *gangue* or nonmineral-containing material Originally applied only to certain sulfide and oxide ores, flotation methods now are used not only for these, but also in many other cases as well. A partial list of ores so treated commercially would include those of nickel and gold, as well as calcite, fluorite, barite (barium sulfate), scheelite (calcium tungstate), manganese carbonate and oxides, iron oxides, garnet, iron titanium oxides, silica and silicates, coal, graphite, sulfur, and soluble salts such as sylvite (potassium chloride). It has been estimated that 10^9 tons annually of ore are processed by flotation methods (12, 13)!

Prior to about 1920, flotation procedures were rather crude and rested primarily on the observation that copper and lead–zinc ore *pulps* (crushed ore mixed with water) could be *benefacted* (improved in mineral content) by treatment with large amounts of fatty and oily materials. The mineral particles collected in the oily layer and could be separated from the gangue and the water. Since then this oil flotation procedure has been largely displaced by what is known as froth flotation. Here, only minor amounts of oil are used, and a froth is formed by agitating or bubbling air through the material. The oily froth contains a concentration of mineral particles and may then be skimmed off.

It was observed very early that rather minor variations in the composition of the oils used could make great differences in performance, and there are today many secret recipes in use. The field is unusual in that empirical practice has been in the lead, with theory struggling to explain. Some basic aspects are fairly well understood, however, and a large variety of special additives are available. These include *collectors*, which adsorb on the mineral particles to make the basic modification in contact angle that is desired; *activators*, which enhance the selective action of the collector; *depressants*, which selectively reduce its action; and *frothing agents*, to promote foam formation. Later it is shown that frothing agents can play a direct role in the flotation itself. A brief review by Wark (14) summarizes much of the current status.

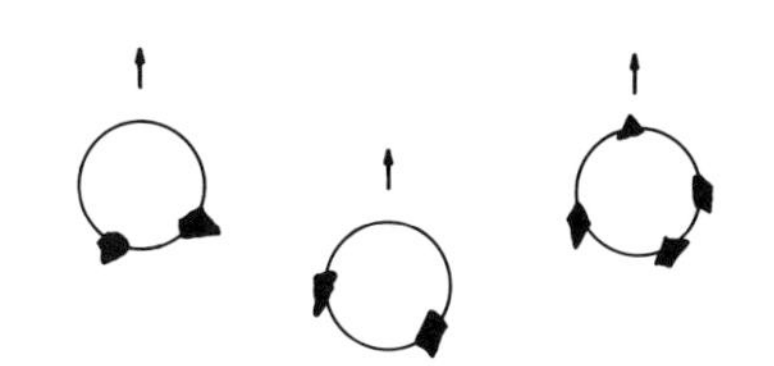

Fig. XIII-4. Flotation by mineral-laden air bubbles.

A. The Role of Contact Angle in Flotation

The basic phenomenon involved is that particles of ore are carried upward and held in the froth by virtue of their being attached to an air bubble, as illustrated in Fig. XIII-4. To see how a particle may occupy a stable position at an interface, let us consider first the gravity-free situation illustrated in Fig. XIII-5. The particle of solid is taken to be spherical, and both it and the two liquid phases A and B are assumed to have the same density. The particle may be entirely in phase A, or it may lie entirely in phase B. Alternatively, the particle may be located in the interface, in which case both γ_{SA} and γ_{SB} contribute to the total free energy. Also, however, some liquid–liquid interface has been eliminated.

Considering the detailed drawing in Fig. XIII-6, the condition for equilibrium is that a small displacement of the particle be accompanied by a zero net free energy change, or that the total surface free energy for the position be at a minimum. Let h denote the distance of penetration of the particle into phase A. The area of the solid–liquid A interface is $2\pi rh$, and that of the solid–liquid B interface is $(4\pi r^2 - 2\pi rh)$. The area of liquid–liquid interface occupied by the solid is πl^2, where $l = [r^2 - (r - h)^2]^{1/2}$, or $\mathcal{A}_l = \pi(2rh - h^2)$. The condition for equilibrium is that

$$dG = 0 = \gamma_{SA}(2\pi r\, dh) + \gamma_{SB}(-2\pi r\, dh) - \gamma_{AB}\pi(2r - 2h)\, dh \qquad \text{(XIII-7)}$$

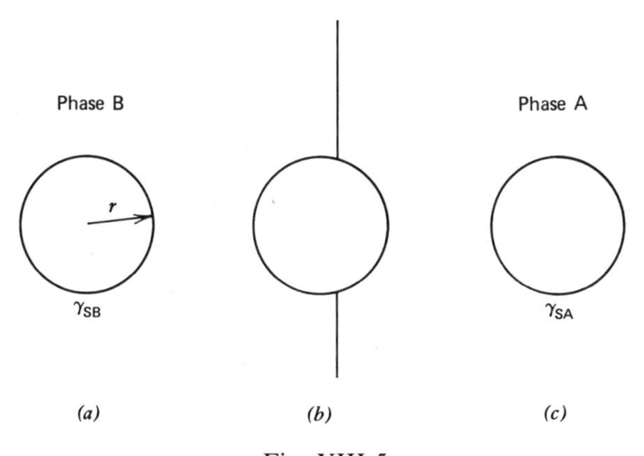

Fig. XIII-5

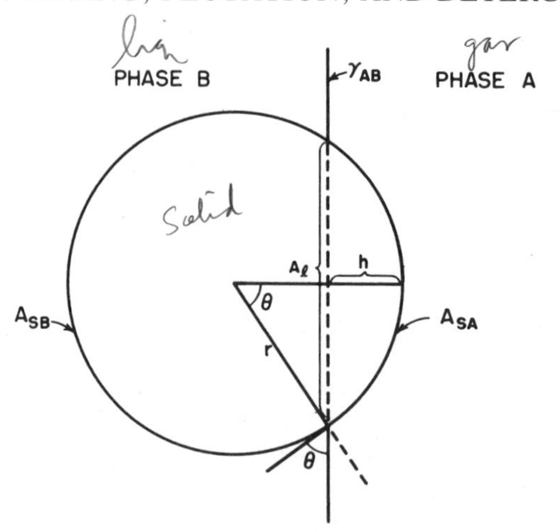

Fig. XIII-6

Equation XIII-7 simplifies to

$$\gamma_{SA} - \gamma_{SB} = \left(1 - \frac{h}{r}\right)\gamma_{BB} = \gamma_{AB}\cos\theta \qquad \text{(XIII-8)}$$

Equation XIII-8 is the Young equation for contact angle equilibrium, and the foregoing analysis leads to the conclusion that a particle will seek a position at a liquid–liquid interface such that the angle θ becomes the equilibrium contact angle; alternatively stated, if θ is finite, the particle is positively stable at the interface.

It is helpful to consider qualitatively the numerical magnitude of the surface tensional stabilization of a particle at a liquid–liquid interface. For simplicity, we will assume $\theta = 90°$, or that $\gamma_{SA} = \gamma_{SB}$. Also, with respect to the interfacial areas, $\mathcal{A}_{SA} = \mathcal{A}_{SB}$, since the particle will lie so as to be bisected by the plane of the liquid–liquid interface, and $\mathcal{A}_{AB} = \pi r^2$. The free energy to displace the particle from its stable position will then be just $\pi r^2 \gamma_{AB}$. For a particle of 1-mm radius, this would amount to about 1 erg, for $\gamma_{AB} = 40$ erg/cm^2. Also, this corresponds roughly to a restoring force of 10 dyne, since this work must be expended in moving the particle out of the interface, and this amounts to a displacement equal to the radius of the particle.

The usual situation is that illustrated in Fig. XIII-7, in which the particle is supported at a liquid–air interface against gravitational attraction. As was seen, the restoring force stabilizing the particle at the interface varies approximately with the particle radius; for the particle to remain floating, this restoring force must be equal

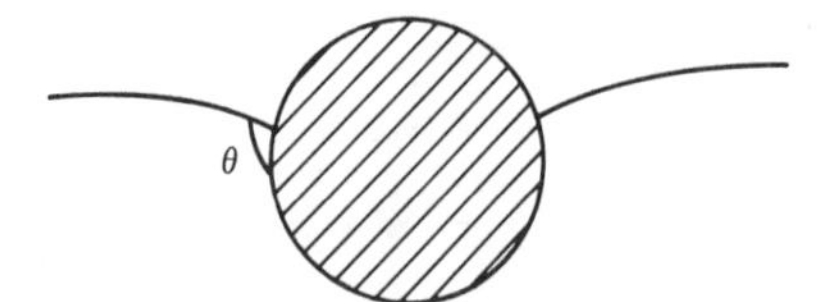

Fig. XIII-7

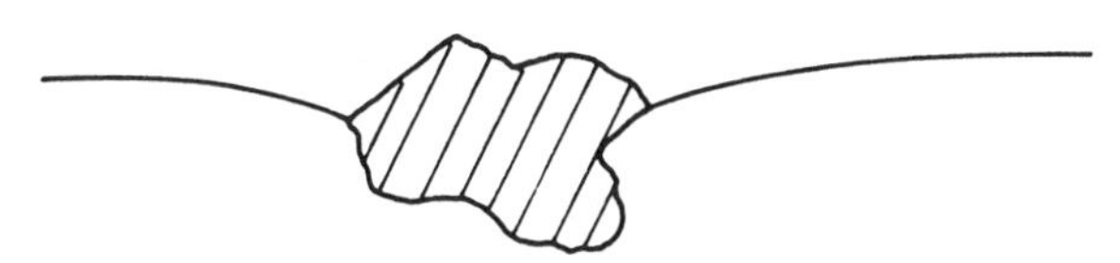

Fig. XIII-8

to or exceed that of gravity. Since the latter varies with the cube of the radius, it is clear that there will be a maximum size of particle that can remain at the interface. Thus referring to the preceding example, if the particle density is 3 g/cc, then the maximum radius is about 0.1 cm. The preceding analysis has been on an elementary basis; Rapacchietta and Neumann (15) have investigated the difficult problem of determining the actual surface profile of Fig. XIII-7.

In practice, it may be possible with care to float somewhat larger particles than those corresponding to the theoretical maximum. As illustrated in Fig. XIII-8, if the particle has an irregular shape, it will tend to float such that the three-phase contact occurs at an asperity since the particle would have to be depressed considerably for the line of contact to advance further. The resistance to rounding a sharp edge has been investigated by Mason and co-workers (16).

The preceding upper limit to particle size can be exceeded if more than one bubble is attached to the particle.† A matter relating to this and to the barrier that exists for a bubble to attach itself to a particle is discussed by Leja and Poling (17; see also Refs. 18 and 19). The attachment of a bubble to a surface may be divided into steps, as illustrated in Fig. XIII-9*a*, *b*, *c*, in which the bubble is first distorted, then allowed to adhere to the surface. Step 1, the distortion step, is not actually unrealistic, as a bubble impacting a surface does distort, and only after the liquid film between it and the surface has sufficiently thinned does adhesion suddenly occur (see Refs. 18 and 20–22). For step 1, the surface free energy change is

$$\Delta G_1 = \Delta \mathcal{A}_{LV} \gamma_{LV} \tag{XIII-9}$$

where $\Delta \mathcal{A}_{LV} = (\mathcal{A}'_{LV} + \mathcal{A}_{SV} - \mathcal{A}_{LV})$, and, for the second step,

$$\Delta G_2 = (\gamma_{SV} - \gamma_{SL} - \gamma_{LV}) \mathcal{A}_{SV} = -w_a \mathcal{A}_{SV} \tag{XIII-10}$$

so that for the overall process

$$\Delta G = -w_a \mathcal{A}_{SV} + \gamma_{LV} \Delta \mathcal{A}_{LV} = -w_{\text{pract}} \tag{XIII-11}$$

where w_a is the work of adhesion (*not* the same as w_{SL}). The quantity $-\Delta G$

† An instructive and decorative illustration of this is found in the following parlor experiment. Some water is poured into a glass bowl and about 1% by weight of sodium bicarbonate is added and then some moth balls. About one-third of the volume of vinegar is then poured in carefully. The carbon dioxide that is slowly generated clings as bubbles to the moth balls, and each ball rises to the surface when a net buoyancy is reached. On reaching the surface, some of the bubbles break away, the ball sinks, and the process is repeated.

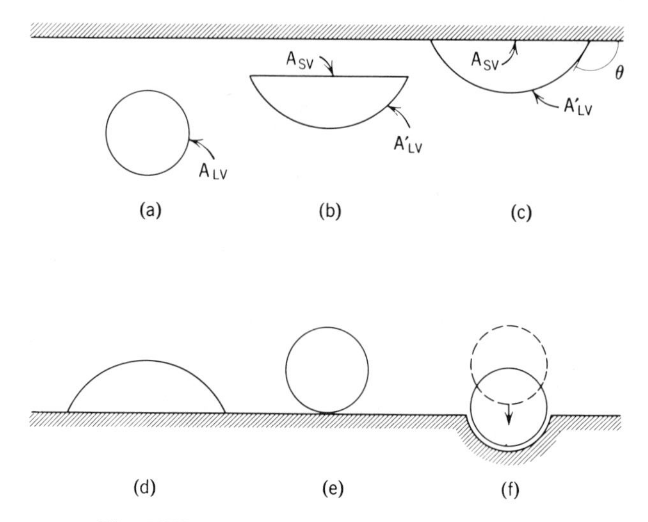

Fig. XIII-9. Bubble distortion and adhesion.

thus represents the practical work of adhesion of a bubble to a flat surface, while ΔG_1 is approximately the activation energy barrier to the adhesion.

As a numerical illustration, consider a bubble 0.1 cm in diameter in a system for which $w_a = 30$ erg/cm^2, $\gamma_{LV} = 30$ erg/cm^2, and $\theta = 90°$. The distorted bubble is just hemispherical, and we find $w_{\text{pract}} = 0.19$ erg, and $\Delta G_1 = 0.18$ erg. The gravitational energy available from the upward motion is only 0.066 erg or less than the ΔG_1 barrier, so that adhesion might be difficult. Conversely, if the bubble were clinging to the upper side of a surface, as in Fig. XIII-9*d*, ΔG_2 gives a measure of the energy barrier for detachment and is 0.38 erg; again, the barrier is more than the gravitational energy available, this time from the upward motion of the center of gravity of the bubble in the reverse step 1 plus step 2. One can thus account qualitatively for the reluctance of bubbles to attach to a surface and, once attached, to leave it.

If the contact angle is zero, as in Fig. XIII-9*e*, there should be no tendency to adhere to a flat surface. Leja and Poling point out, however, that, as shown in Fig. XIII-9*f*, if the surface is formed in a hemispherical cup of the same radius as the bubble, then for step 1a, the free energy change of attachment is

$$\Delta G_{1a} = -w_a \mathscr{A}_{SV} \qquad \text{(XIII-12)}$$

that is, there is now no distortion energy. In terms of the foregoing numerical example, the work of adhesion would be 0.47 erg, or greater than that for a contact angle of 90° and a flat surface.

These illustrations bear on the two principal mechanisms that have been advanced for the mineralization of bubbles. Taggart (23) thought that particles were nucleated and then grew at the particle surface, for example, a result of supersaturation in low pressure regions created by the vortices of the impeller blades of the mechanical agitator. Bubbles thus grown in a cavity would be especially stable toward detachment. Gaudin (24), on the other hand, proposed that the rising bubbles collided with the particles. Klassen (see Ref. 10) noted that a combination process was possible in that collisional encounters were more efficient if the particle already had a microbubble growing on its surface. Generally speaking, the encounter mechanism seems to be the more important of those proposed (14).

B. Flotation of Metallic Minerals

Clearly, it is important that there be a large contact angle as the solid particle–solution–air interface. Some minerals, such as graphite, are naturally hydrophobic, but even with these it has been advantageous to add materials to the system that will adsorb to give a hydrophobic film on the solid surface. The use of such collectors is, of course, essential in the case naturally hydrophilic minerals such as silica.

In the case of lead and copper ores, the use of xanthates,

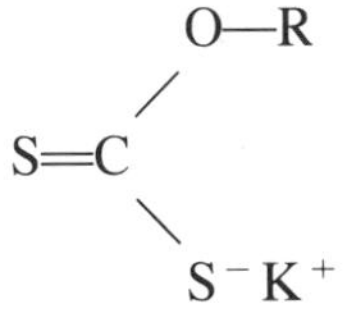

$$S{=}C\begin{matrix} \diagup O{-}R \\ \diagdown S^-K^+ \end{matrix}$$

has been widespread, and a reasonable explanation was that a reaction of the type

$$Pb(OH)_2 + 2ROCS_2^- = Pb\left[-S{-}C\begin{matrix} \diagup\!\!\diagup S \\ \diagdown OR \end{matrix}\right]_2 + 2OH^- \qquad \text{(XIII-13)}$$

occurred. However, early empirical observations that found that dissolved oxygen played a role have now been recognized as correct. Thus ethyl xanthate does not adsorb on copper in deaerated systems (25, 26); not only does it appear that oxidation of the surface is important, but also the actual surfactant may be an oxidation product of the xanthate, dixanthogen,

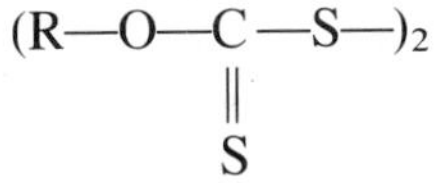

$$(R{-}O{-}\underset{\underset{S}{\|}}{C}{-}S{-})_2$$

Perxanthate ion may also be implicated (14). Even today, the exact nature of the surface reaction is clouded (14, 27, 18), although Gaudin (29) notes that the role of oxygen is very determinative of the chemistry of the mineral-collector interaction.

Various chemical tricks are possible. Thus zinc ores are not well floated with xanthates, but a pretreatment with dilute copper sulfate rectifies the situation by electrodepositing a thin layer of copper on the mineral particles. As another example, treatment of an ore containing a mixture of iron, zinc, and lead minerals with dilute cyanide will inhibit adsorption of the collector on the first two, but not on the last. In this case, cyanide would be called a *depressant*.

In addition to a large contact angle, it is also desirable that the mineral-laden bubbles not collapse when they reach the surface of the slurry, that is, that a stable (although not *too* stable) froth of such bubbles be possible. With this in mind, it is common practice to add various frothing agents to the system such as long chain alcohols and pine oils. However, it turns out that the actions of frothing agent and the collector are not independent of each other.

For example, since a complete monolayer of collector should give the greatest contact angle (and does), one would expect that this condition would also give the best flotation. Yet Gaudin and Sun (30) found that flotation was optimum when only 5 to 15% of a complete monolayer was present and that beyond this point there was a suppression of bubble adhesion. Data summarized by Schulman and Leja (31) for the xanthate–lauryl alcohol–copper system provided a satisfying explanation in showing that lauryl alcohol penetrated or formed mixed films with the adsorbed ethyl xanthate (see Section IV-8). When the bubble and the particle are separate, the frothing agent is concentrated at the liquid–air interface of the bubble, and the collector, at the liquid–solid interface of the particle. When the bubble and the particle interfaces merge, penetration of the collector film by that of the frother occurs, with consequent great stabilization of the solid–liquid–air contact. This "locking-in" effect does not occur well if the concentration of the collector is so high that a complete monolayer is formed at the solid–liquid interface, since penetration by the frothing agent is then inhibited. After all, the prime requirement for a large contact angle is that $(\gamma_{SV} - \gamma_{SL})$ be negative, and collector adsorption at the particle–liquid interface helps only in the indirect sense that the adsorbed film–liquid interfacial tension may be higher than the film–air value. Penetration by the frothing agent of the film at this last interface has the immediate effect of lowering γ_{SV} without necessarily affecting γ_{SL}.

The prime function of the frothing agent may therefore be to stabilize the attachment of the particle to the bubble, and its frothing ability, per se, may well be a matter of secondary importance. Also, as mentioned in the next chapter, well-mineralized bubbles themselves constitute a stable froth system.

C. Flotation of Nonmetallic Minerals

The examples in the preceding section, of the flotation of lead and copper ores by xanthates, was one in which chemical forces predominated in the adsorption of the collector. Flotation processes have been applied to a number of other minerals that are either ionic in type, such as potassium chloride, or are insoluble oxides, such as quartz or iron oxide. In these cases additional factors appear to be of importance.

As one example, sylvite (KCl) may be separated from halite (NaCl) by selective flotation in the saturated solutions, using long chain amines such as a dodecyl ammonium salt. There has been some mystery as to why such similar salts should be separable by so straightforward a reagent. One suggestion has been that since the $R{-}NH_3^+$ ion is small enough to fit into a K^+ ion vacancy, but too large to replace the Na^+ ion, the strong surface adsorption in the former case may be due to a kind of isomorphous surface substitution (19, 32, and especially 33). Barytes ($BaSO_4$) may be separated from unwanted oxides by means of oleic acid as a collector; the same is true of calcite (CaF_2). Flotation is widely used to separate calcium phosphates from siliceous and carbonate minerals (34).

The flotation of the insoluble oxide minerals turns out to be best under-

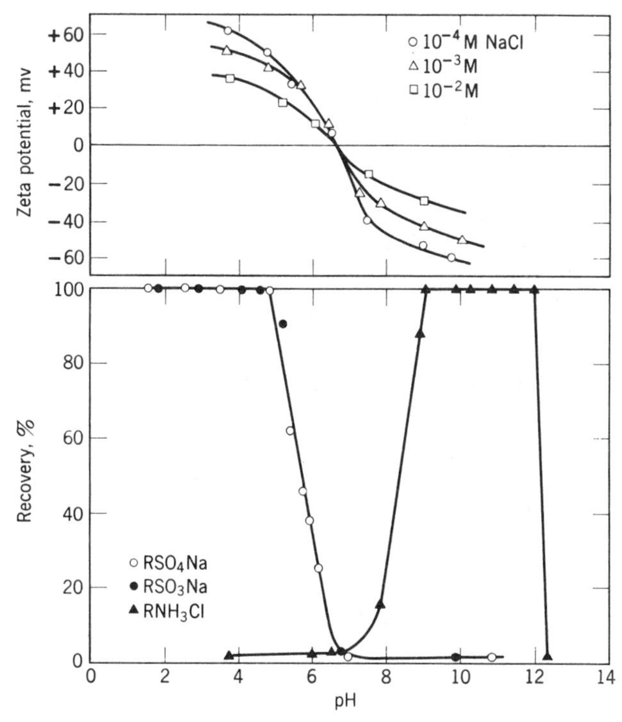

Fig. XIII-10. The dependence of the flotation properties of goethite on surface charge. Upper curves are ζ potential as a function of pH at different concentrations of sodium chloride: lower curves are the flotation recovery in 10^{-3} M solutions of dodecylammonium chloride, sodium dodecyl sulfate, or sodium dodecyl sulfonate. (From Ref. 35.)

stood in terms of electrical double layer theory. That is, the potential ψ of the mineral is as important as are specific chemical interactions. Hydrogen ion is potential determining (see Section V-5); for quartz, the pH of zero charge is about 3 while for, say, goethite, FeO(OH), it is 6.7 (35). As illustrated in Fig. XVI-10, anionic surfactants are effective for goethite below pH 6.7 since the mineral is then positively charged, and cationic surfactants work at higher pH's since the charge is now negative.

The adsorption appears to be into the Stern layer; as was illustrated in Fig. V-2. That is, the adsorption itself reduces the ζ potential of such minerals; in fact, at higher surface coverages of surfactant, the ζ potential can be reversed, indicating that chemical forces are at least comparable to electrostatic ones. The rather sudden falling off of ζ potential beyond a certain concentration suggested to Gaudin and Fuerstenau that some type of near phase transition can occur in the adsorbed film of surfactant (37). They proposed, in fact, that surface micelle formation set in, reminiscent of Langmuir's explanation of intermediate type film on liquid substrates (Section IV-6C).

In addition to the collector, polyvalent ions may show sufficiently strong adsorption on oxide minerals to act as potential-determining ions, either reinforcing or inhibiting the adsorption of the collector. Judicious addition of various electrolytes, then, as well as pH control, can permit a considerable amount of selectivity to be achieved.

5. Detergency

Detergency may be defined as the theory and practice of the removal of foreign material from solid surfaces by surface chemical means. This definition covers the extensive and important subjects of soil removal from fabrics, metal surfaces, and so on. It excludes, however, *purely* mechanical cleansing (e.g., by abrading down the surface) or *purely* chemical processes (e.g., by the chemical dissolving of impurities). Common soap is undoubtedly the oldest and best-known detergent, and its use in the washing of clothes is the best example of detergent action.

A. General Aspects of Soil Removal

The soil that accumulates on fabrics is generally of an oily nature and contains as well particles of dust, soot, and the likes. The oily material consists of animal fats and fatty acids, petroleum hydrocarbons, and a residuum of quite miscellaneous substances (38, 39).

Because of the complexity and irreproducibility of natural soil, it has been necessary to develop more or less standard soils and soiling procedures, as well as standard washing operations. Standard soils usually consist of a mixture of lampblack and some grease such as vaseline, and a standard washing apparatus is that known as a "launderometer." By means of the launderometer, standard swatches of cloth

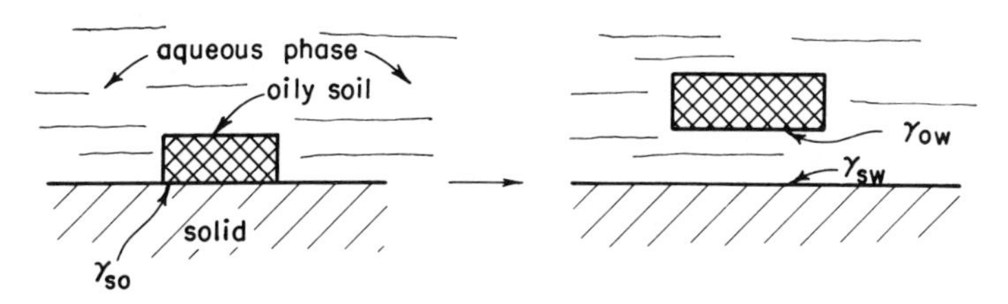

Fig. XIII-11. Surface tensional relationships in soil removal.

may be agitated with detergent solutions under fairly reproducible conditions of degree of agitation, temperature, and so on. By controlling such variables, one can then obtain performance data for various detergents. Usually, the amount of soil on the fabric is measured by determining its reflectance, using an empirical relationship between the amount of solid and the whiteness of the cloth. In some cases radioactive soil has been used, and the amount of soil present in the cloth has been determined by radioactivity assay.

If the stereotype of soil removal is taken to be as shown in Fig. XIII-11, namely, an oily particle adhering by surface tensional forces, then simple requirements can be stated. This change in surface free energy for the detachment of the soil is just

$$\Delta G = \gamma_{\mathrm{WO}} + \gamma_{\mathrm{SW}} - \gamma_{\mathrm{SO}} \tag{XIII-14}$$

The condition for the process to be spontaneous is that $\Delta G \leq 0$, or

$$\gamma_{\mathrm{SO}} \geq \gamma_{\mathrm{WO}} + \gamma_{\mathrm{SW}} \tag{XIII-15}$$

Alternatively, the soil may be considered to be a fluid, and the situation becomes then a contact angle problem in which θ is to be made as small as possible, as defined in Fig. XIII-12. For the case of $\theta = 0$, that is, a zero or positive spreading coefficient for the aqueous phase, Eq. XIII-15 again results.

Examination of Eq. XIII-15 shows that, in order to make the adhesion of the soil to the solid zero or negative, it is desirable to decrease γ_{WO} and γ_{SW} as much as possible, with a minimum of concomitant change in γ_{SO}. By this reasoning, a surfactant that adsorbs both at the oil–water and at the solid–water interface should be effective. On the other hand, a mere decrease in the surface tension of the water–air interface, as evidenced, say, by foam formation, is *not* a direct indication that the surfactant will function well as a detergent.

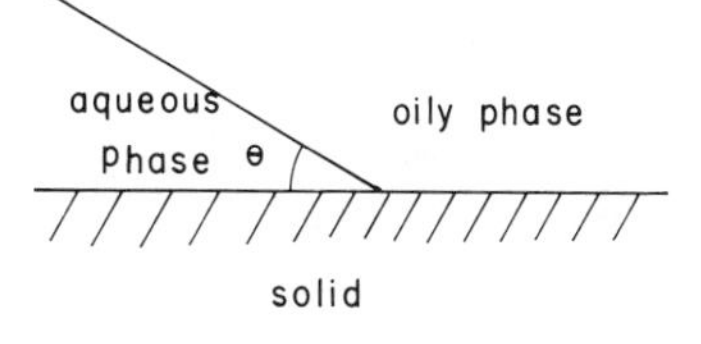

Fig. XIII-12. Surface tensional relationships in liquid soil removal.

As might be expected, however, a number of interrelated effects seem to be involved. These are reviewed briefly in the following material.

B. Properties of Colloidal Electrolyte Solutions

Solutions of detergents generally exhibit a rather special and interesting set of properties characteristic of what are called "colloidal electrolytes." These properties play an important role in determining indirectly, if not directly, the detergent ability of a given material, and they are therefore described here briefly before returning to a more detailed consideration of the mechanism of detergent action.

A general display of the various physical properties of a solution of a typical colloidal electrolyte such as sodium dodecyl sulfate is shown in Fig. XIII-13 (see Refs. 40 and 41). It is seen that striking alterations in the various physical properties occur in the region of what is called the *critical micelle concentration* (cmc). The near constancy of the osmotic pressure beyond the cmc suggests that something akin to a phase separation is occurring and,

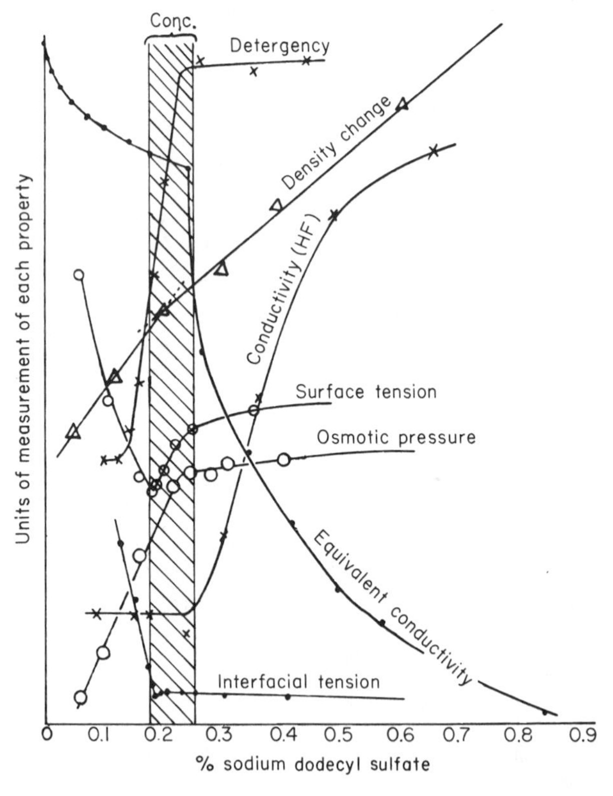

Fig. XIII-13. Properties of colloidal electrolyte solutions—sodium dodecyl sulfate. (From Ref. 41.)

although no gross phase separation can be observed, the sudden increase in light scattering indicates that the system is becoming colloidal in nature. Actually, the well-documented explanation is that in the region of cmc aggregation of the long chain electrolyte into fairly large, charged units begins to occur. These units are commonly called micelles. It is somewhat apart from the assumed scope of this book to consider the physical chemistry of micelle formation in any detail, but the phenomenon is so characteristic of detergent solutions that a brief discussion of it is necessary.

The existence of aggregates in detergent solutions above the cmc has been recognized since about 1913 (42), and since then a voluminous literature has developed on solutions of colloidal electrolytes; see Ref. 43 for a current review. Micelles are fairly narrowly dispersed in size and contain 50 to 100 monomer units although for nonionic detergents aggregation numbers up to 370 have been reported (see Ref. 44). Important evidence has come from light scattering studies (see Refs. 43, 45–46), including now more advanced techniques using laser light (see Ref. 47) and fluorescence quenching (47a). Micellar molecular weights are in the 12,000 to 40,000 g/mol range for surfactants such as sodium lauryl sulfate and sodium dodecyl sulfate, increasing with increasing ionic strength.

Another line of evidence comes from diffusion studies. In this case, a very interesting property of colloidal electrolyte solutions is made use of, namely, that of solubilization. Solutions of colloidal electrolytes above the cmc are able to bring into solution otherwise insoluble organic molecules such as benzene and various dyes. Orange OT, for example, barely colors pure water, but it gives brilliantly deep red solutions with sodium dodecyl sulfate. It appears that the solubilized material is incorporated into the micelle itself, since the presence of micelles is necessary for the effect, and because of other evidence. It has been possible to measure the diffusion rate of solubilized dye in an otherwise uniform solution of detergent and thus to obtain a close approximation to the self-diffusion coefficient for the micelles present. These come out to be about 6×10^{-8} cm^2/sec, decreasing with increasing ionic strength (48, 49).

Micelles are charged if the monomer unit is an electrolyte, since it is the long chain ion that aggregates while those of opposite charge, the counterions, remain unaggregated. The presence of a large charge on micelles is apparent from their electrophoretic mobility, for example (50). The net charge is less than the degree of aggregation, however, since some of the counterions remain associated with the micelle, presumably as part of a Stern layer (see Section V-5) (51). By combining self-diffusion and electrophoretic mobility measurements, the indication is that a typical micelle is made up of about 100 monomer units, but with a net charge of 50 to 70.

The structure of micelles has been a matter of much discussion and dispute (see Ref. 43). Hartley (52) proposed a spherical shape, whereas McBain (53) believed that a lamellar form also existed. The two structures are illustrated in Fig. XIII-14. A roughly spherical shape such as illustrated in Fig. XIII-

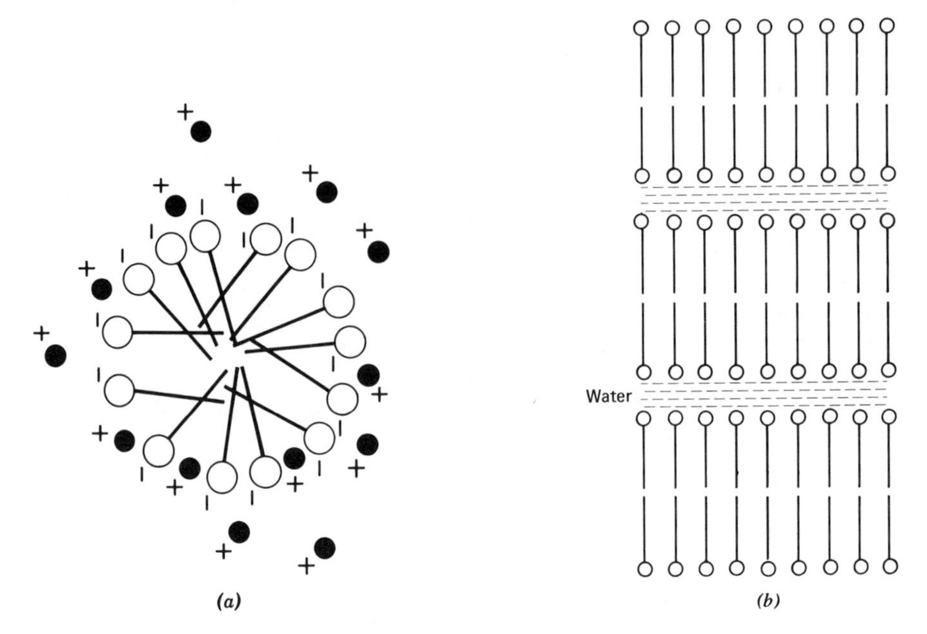

Fig. XIII-14. Schematic representation of (*a*) a spherical micelle and (*b*) a lamellar micelle. (From Ref. 40.)

15 seems likely (43). Fluorescence depolarization studies, for example, indicate that solubilized molecules are in a relatively fluid medium (but see Ref. 54), although neither uv nor nmr spectroscopic data have been unambiguous in deciding whether such molecules are in a fully hydrocarbon medium or, perhaps, in one of the crevices shown in the model of Fig. XIII-15 and thus partly exposed to solvent water. The shape need not be spherical; Tanford (55), for example, proposes an ellipsoidal one.

The actual variation in concentration of the various species present, as a function of the stoichiometric concentration of the detergent, is illustrated in Fig. XIII-16 for the case of an anionic detergent. Beyond the cmc, the concentration of free R^- ions decreases with increasing overall concentration, whereas that of the counterions, for example, Na^+, increases. The activity of the electrolytes, given by the product of the Na^+ and R^- activities, increases slightly and, finally, the concentration of micelles rises nearly linearly with overall composition, starting from the cmc. Fig. XIII-16 is oversimplified in that above the cmc there will be a distribution of micellar sizes, and that small aggregates (dimers, trimers) build up below the cmc as well as being present above it. Details of such distributions have been investigated by means of various mass action law models (56–58).

The detergent–water system is considered from another viewpoint in Fig. XIII-17 (40), which amounts to a qualitative phase diagram or "phase map." The temperature T_k, corresponding to point B, denotes the Kraft temperature, above which the solubility of the colloidal electrolyte increases very rapidly. Note that below T_k

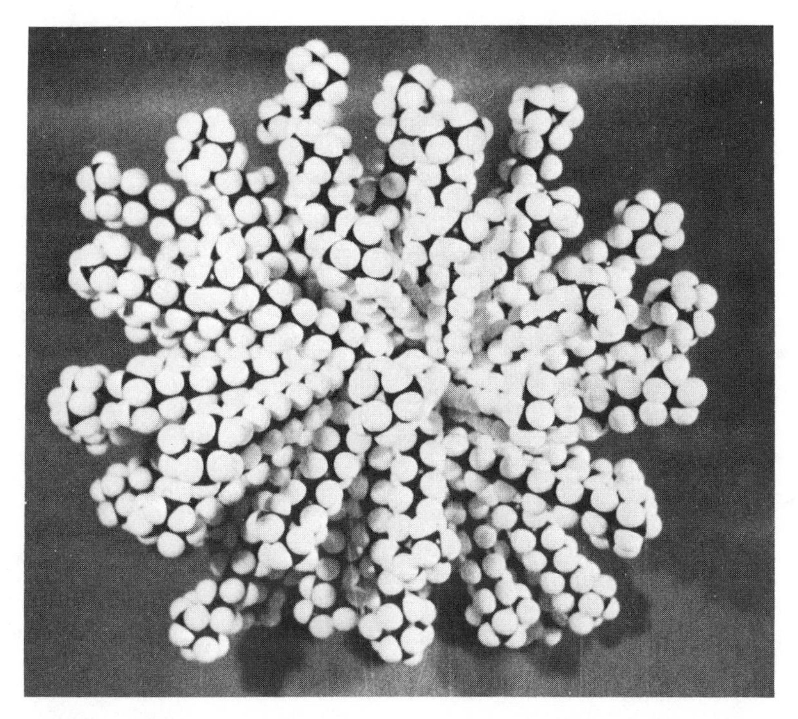

Fig. XIII-15. Model of a dodecyltrimethylammonium ion micelle with an aggregation number of 58. A pyrene molecule is situated among the chains close to the micelle surface. (Reprinted with permission from Ref. 43. Copyright 1979 American Chemical Society.)

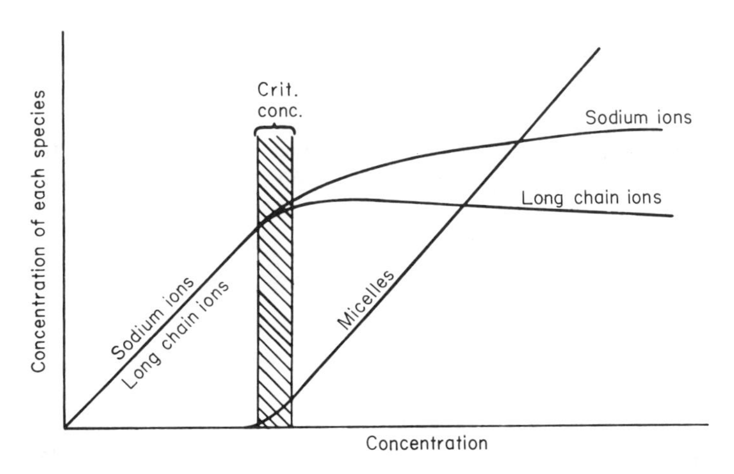

Fig. XIII-16. Concentrations of individual species present in colloidal electrolyte solutions. (From Ref. 40.)

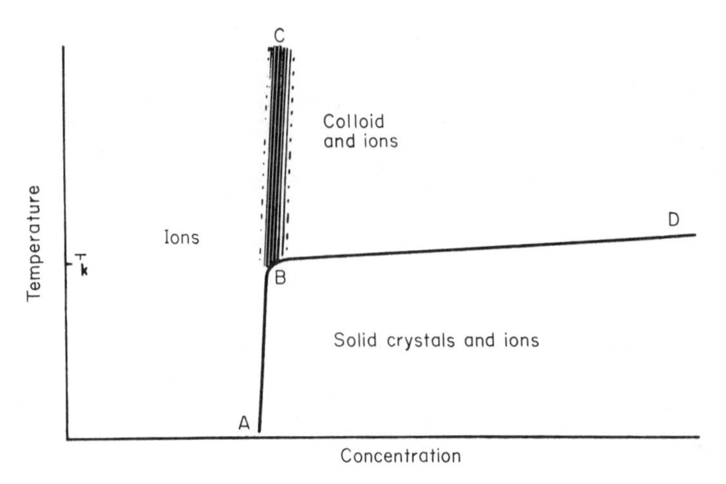

Fig. XIII-17. Phase map for colloidal electrolyte. (From Ref. 40.)

increasing the composition of the system in colloidal electrolyte leads to precipitation of solid soap rather than to micelle formation. It is significant that soaps do not function well below their Kraft temperature.

Other properties of colloidal electrolyte solutions that have been studied include calorimetric measurements of the heat of micelle formation (about 6 kcal/mole for a nonionic species, see Ref. 59) and the effect of high pressure (which decreases the aggregation number, see Ref. 60). Fast reaction relaxation methods (rapid flow mixing, pressure-jump, temperature-jump) tend to reveal two relaxation times t_1 and t_2, the interpretation of which has been subject to much disagreement—see Ref. 61. A "fast" process of $t_1 \sim 1$ msec may represent the rate of addition to or dissociation from a micelle of individual monomer units, and a "slow" process of $t_2 < 100$ msec may represent the rate of total dissociation of a micelle (61; see also Refs. 63–64a).

The traditional colloidal electrolyte is of the M^+R^- type, where R^- is the surfactant ion, studied in aqueous solution. Such salts also form micelles in nonaqueous (and nonpolar) solvents. The structure appears to be one having the polar groups inward if some water is available; very complex structures are present in nearly anhydrous media (see Ref. 65 and under micellar emulsions, Chapter XV-5).

There are many nonionic detergents–often low molecular weight polymers having polar and nonpolar portions. These also form micelles and exhibit solubilization phenomena (see Refs. 66 and 67).

C. Factors in Detergent Action

The fact that successful detergents seem always to show the colloidal properties discussed has led to the thought that micelles must be directly involved in detergent action. McBain (68), for example, proposed that solubilization was one factor in detergent action. Since micelles are able to solubilize dyes and other organic molecules, the suggestion was that oily soil might similarly be incorporated into detergent micelles. As illustrated in Fig. XIII-13, however, detergent ability rises before the cmc is reached and remains practically constant thereafter. Since the concentration of micelles rises steadily from the cmc on, there is thus no direct correlation between their concentration and detergent action. It therefore seems necessary to conclude that detergent action is associated with the long chain monomer ion or molecule and that very likely the properties that make for good detergent action also lead to micelle formation as a *competing* rather than as a contributing process.

In considering possible mechanisms of detergency, it can be said first of all that the type of action illustrated in Figs. XIII-11 and XIII-12 undoubtedly is of importance. Adam and Stevenson (69) show beautiful photographs illustrating how lanolin films adhering to wood fibers are "rolled up" into easily detachable spheres when sodium cetyl sulfate is added to the aqueous medium. This is precisely the kind of effect shown schematically in Fig. XIII-12, with the oil–water–solid contact angle steadily decreasing as the soap concentration is increased. As noted before, this change in contact angle must be due to a decrease in either γ_{WO} or γ_{SW}, or both, and, by the Gibbs equation (see Section III-5), this implies a corresponding adsorption of detergent. It is thus reassuring that soap is known to adsorb on fabrics (see Refs. 70–73), so that a lowering of γ_{SW} must indeed occur (see Section XIII-5D). Also, it is quite evident that detergent-type (polar–nonpolar) molecules will adsorb at an oil–water interface with consequent reduction in γ_{WO}. Thus for fluid soils the expected changes in interfacial tension brought about by the detergent are perhaps sufficient to account for the observed great reduction in adhesion between the soil and the fabric. It might be noted that not only are oil drops detached from a fiber by detergent action, but also they may undergo spontaneous emulsification (see Section XIV-5) and so disappear as a distinct phase.

Much of ordinary soil involves particulate, more or less greasy matter, and an important attribute of detergents is their ability to keep such material suspended in solution once it is detached from the fabric and thus to prevent its redeposition. Were this type of action not present, washing would involve a redistribution rather than a removal of dirt. Detergents thus do possess *suspending power*. For example, carbon suspensions that otherwise would settle rapidly are stable indefinitely if detergent is present, and similarly such other solids as manganese dioxide (74, 75). Evidently, detergent is adsorbed at the particle–solution interface, and suspending action apparently results partly through a resulting change in the charge of the particle and perhaps mainly through a deflocculation of particles that originally were agglomerates. Apart from charge repulsion, the presence of an adsorbed long chain molecule prevents the coalescence of particles in much the same manner as it prevents the contact of surfaces in boundary lubrication (see Section XII-7B). Carboxymethylcellulose is often added to detergent preparations to aid in suspending soil (76); its action is probably similar to that of detergent adsorption but perhaps is especially effective because of the polymeric nature of the material that should give thick and difficultly detachable adsorbed films (see Section XI-2). A related action, and one that is also important to good detergency, is that called *protective action*. This refers to the prevention of the particles of solid from adhering to the fabric.

Another component of detergent formulations is that known as a *builder*—a substance with no detergent properties of its own but one that enhances the performance of a detergent (see Ref. 77). A typical builder would be pyrophosphate, $Na_4P_2O_7$; its effect may be partly as a sequestering agent for the metal ions that contribute to hardness (76). It may play some role in affecting surface charges (78). Recent concerns about excessive phosphate discharge into natural waters have led to a search for alternative builders. There has been some success with mixtures of electrolytes and sequestering agents (79).

The general picture of detergent action that has emerged is that of a balance of opposing forces (see Ref. 39). The soil tends to remain on the fabric either through surface tensional adhesion or mechanical entrapment and, on the other hand, tends to remain in suspension as a result of the suspending power and protective action of the detergent. It is possible to reach a steady state condition, approachable from either side.

D. Adsorption of Detergents on Fabrics

The adsorption of detergent type molecules on fabrics and at the solid–solution interface in general shows a complexity that might be mentioned briefly. Some fairly characteristic data are shown in Fig. XIII-18 (80). There is a break at point *A*,

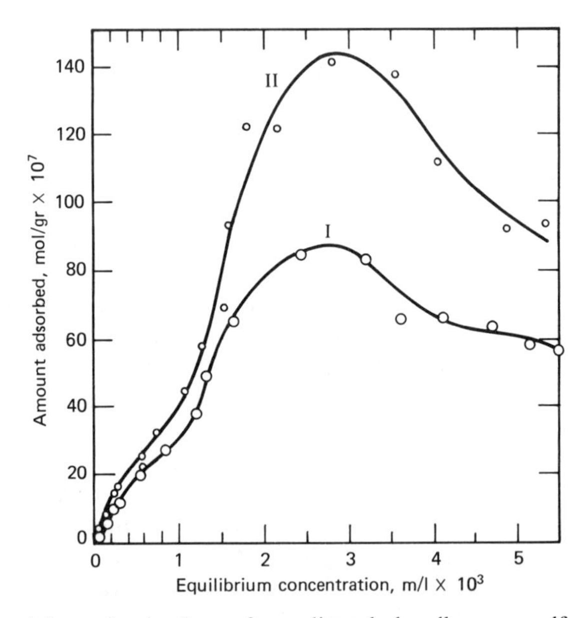

Fig. XIII-18. Adsorption isotherm for sodium dodecylbenzenesulfonate on cotton. Curve I, 30°C; curve II, 0°C. (From Ref. 80.)

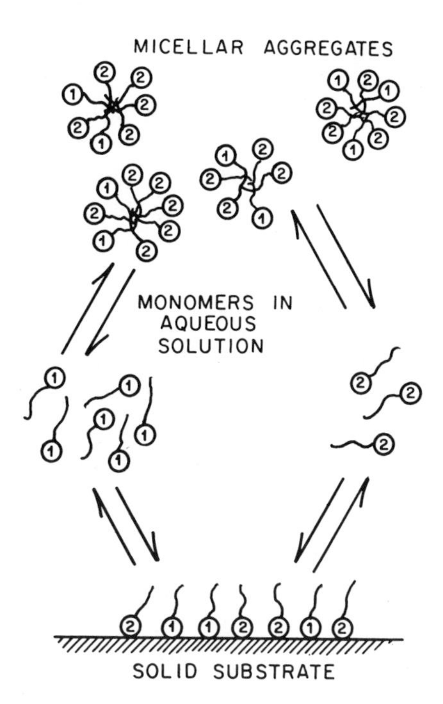

Fig. XIII-19. Schematic distribution of a binary surfactant mixture between aqueous solution and solid surface. (From Ref. 81.)

marking a sudden increase in slope, followed by a *maximum* in the amount adsorbed. Citations to numerous other examples are given by Trogus, Schechter, and Wade (81). The problem is that if such data represent true equilibrium, and no component fractionation is occurring (see Section III-6B), it is possible to argue a second law violation.

The phenomenon occurs not only in fabric–aqueous detergent systems, but also in the currently important problem of tertiary oil recovery. One means of displacing residual oil from a formation involves the use of detergent mixtures (usually petroleum sulfonates), and the degree of loss of detergent through adsorption by the oil-bearing formation is a matter of great importance. The presence of absorption maxima (and even minima) in laboratory determined isotherms is, to say the least, distressing! After reviewing various earlier explanations for an absorption maximum, Trogus, Schechter, and Wade (81) propose perhaps the most satisfactory one so far. Qualitatively, an adsorption maximum can occur if the surfactant consists of at least *two* species (which can be closely related). As illustrated in Fig. XIII-19, what is necessary is that species 2 (say) preferentially forms micelles (has a lower cmc) relative to species 1 and also adsorbs more strongly (has a higher b value in Eq. XI-5, for example).

E. Detergents in Commercial Use

The reader is referred to monographs in the General References section for details but, very briefly, there are three main classes of detergents: anionic, cationic, and nonionic. The first group may be designated as MR, where R is the detergent anion, and includes the traditional soaps (fatty acid salts) and a large number of synthetic detergents such as the sulfates and sulfonates (e.g., sodium dodecylbenzenesulfonate and sodium lauryl sulfate). Cationic detergents, RX, where R is now a detergent cation, include a variety of long chain quaternary amines and amine salts such as Sapamines (e.g., $[(CH_3)_3CCH{=}CH_2NHCOC_{16}H_{33}]_2^+ SO_4^{-2}$). Surfactants of this type find uses in connection with wetting, waterproofing, emulsion formation or breaking, dispersants for inks, and so on.

Nonionic detergents, as the name implies, are not electrolytes, although they do possess the general polar–nonpolar character typical of surfactants. Examples of common types would include polyether esters, for example, $RCO(OCH_2CH_2)_xCH_2CH_2OH$; alkyl-arylpolyether alcohols, for example, $RC_6H_5(OCH_2CH_2)_xCH_2CH_2OH$; and amides, for example, $RCON[(OCH_2CH_2)_xOH]_2$. Detergents of this type compare favorably with soaps and synthetic anionic detergents and find a considerable use in household products such as window- and car-washing preparations (so that incomplete rinsing, which allows for smooth draining, is possible without leaving a powdery residue on drying), insecticides, and detergents for automatic washers.

A problem that became frightening in the early stages of the use of synthetic household detergents was the appearance of mountains of foam in sewage disposal ponds and rivers into which discharge took place. The foam problem is largely under control, through the use of detergents with less foaming tendency and foam dispersants, as is a related problem, that of the biodegradability of detergents. Generally speaking, soaps and straight chain sulfates degrade easily so that rivers can purify themselves with a minimum of harm to fish and animal life, while aromatic and branched chain detergents are to be avoided.

6. Problems

1. The following table lists some of the types of interfacial tensions that occur in systems of practical importance. Each row corresponds to a different system; not all of the types of interfacial tensions are necessarily present in a given system, and of those present there may be some that, from the nature of the situation, are not under control (i.e., from a practical point of view cannot be modified).

For each system certain changes are indicated that constitute a desired goal for that situation, for example, wetting, detergency, and so on. Thus "inc" ("dec") means that it is desirable for good performance to introduce a surfactant that will increase (decrease) the particular surface tension involved.

For each case state which practical situation is involved and discuss briefly why the indicated modifications in surface tension should be desired.

	γ_{SA}	γ_{SW}	γ_{SO}	γ_{OW}	γ_{WA}
(*a*)		dec	inc	dec	
(*b*)	dec	inc			
(*c*)		dec			dec
(*d*)		dec			inc

(S = solid, A = air, W = water or aqueous phase, O = oily or organic water-insoluble phase.)

2. A fabric is made of wool fibers of individual diameter 20 μ and density 1.3 g/cm^3. The advancing angle for water on a single fiber is 120°. Calculate (*a*) the contact angle on fabric so woven that its bulk density is 0.8 g/cc and (*b*) the depth of a water layer that could rest on the fabric without running through. Make (and state) necessary simplifying assumptions.

3. Templeton obtained data of the following type for the rate of displacement of water in a 30-μm capillary by oil (*n*-cetane) (the capillary having previously been wet by water). The capillary was 10 cm long, and the driving pressure was 45 cm of water. When the meniscus was 2 cm from the oil end of the capillary, the velocity of motion of the meniscus was 3.6×10^{-2} cm/sec and when the meniscus was 8 cm from the oil end, its velocity was 1×10^{-2} cm/sec. Water wet the capillary, and the water–oil interfacial tension was 30 dyne/cm.

Calculate the apparent viscosities of the oil and the water. Assuming that both come out to be 0.9 of the actual bulk viscosities, calculate the thickness of the stagnant annular film of liquid in the capillary.

4. Show that for the case of a liquid–air interface Eq. XIII-6 predicts that the distance a liquid has penetrated into a capillary increases with the square root of the time.

5. Calculate ΔG_1, ΔG_2, and W_{pract} (Section XIII-4A) for a bubble in a flotation system for which w_{SL} is 20 erg/cm^2 and γ_L is 30 erg/cm^2. Calculate also ΔG_{1a}, the free energy of adhesion of the bubble to a hemispherical cup in the surface. Take the bubble radius to be 0.15 cm.

6. Fuerstenau and co-workers observed in the adsorption of a long chain ammonium ion RNH_3^+ on quartz that at a concentration of 10^{-3} *M* there was six-tenths of a monolayer adsorbed and the ζ potential was zero. At 10^{-5} *M* RNH_3^+, however, the ζ potential was -60 mV. Calculate what fraction of a monolayer should be adsorbed in equilibrium with the 10^{-5} *M* solution. Assume a simple Stern model.

7. It has been postulated that the tendency for dirt particles to readsorb on fabric is partly dependent on the state of charge of the two solid–liquid interfaces. Describe briefly some experiments that should help to test this postulation.

8. A surfactant is known to lower the surface tension of water and also is known to adsorb at the water–oil interface but not to adsorb appreciably at the water–fabric interface. Explain briefly whether this detergent should be useful in (1) water proofing of fabrics or (2) in detergency or the washing of fabrics.

9. Contact angle is proportional to ($\gamma_{SV} - \gamma_{SL}$), therefore addition of a surfactant that adsorbs at the solid–solution interface should decrease γ_{SL} and therefore increase the quantity above and make θ smaller. Yet such addition in flotation systems increases θ. Discuss what is incorrect or misleading about the opening statement.

10. Where contact angle hysteresis is present, which do you think should be more critical as to bubble adhesion in flotation, the advancing or the receding angle? Explain.

11. It was stated in Section XIII-5D that an adsorption maximum, as illustrated in Fig. XIII-18, implies a second law violation. Demonstrate this. Describe a specific set of operations or a "machine" that would put this violation into practice.

12. Micelle formation can be treated as a mass action equilibrium, for example,

$$40Na^+ + 80R^- = (Na_{40}R_{80})^{-40}$$

The cmc for sodium dodecylbenzenesulfonate is about 10^{-3} *M* at 25°C. Calculate

K for the preceding reaction, assuming that it is the only process that occurs in micelle formation. Calculate enough points to make your own quantitative plot corresponding to Fig. XIII-16. Include in your graph a plot of $(Na^+)(R^-)$. *Note*. It is worthwhile to invest the time for a little reflection on how to proceed before launching into your calculation!

General References

F. M. Fowkes, *Solvent Properties of Surfactant Molecules*, K. Shinoda, Ed., Marcel Dekker, New York, 1967.

D. W. Fuerstenau, *Pure Appl. Chem.,* **24,** 135 (1970).

A. M. Gaudin, *Flotation*, McGraw-Hill, New York, 1957.

J. C. Harris, *Detergency Evaluation and Testing*, Interscience, New York, 1954.

E. Jungermann, *Cationic Surfactants*, Marcel Dekker, New York, 1970.

J. L. Moilliet, B. Collie, and W. Black, *Surface Activity*, E. & F. N. Spon, London, 1961.

Waterproofing and Water Repellency, J. L. Moilliet, Ed., Elsevier, New York, 1963.

S. R. Morrison, *The Chemical Physics of Surfaces*, Plenum, London, 1977.

K. J. Mysels, *Introduction to Colloid Chemistry*, Interscience, New York, 1959.

L. I. Osipow, *Surface Chemistry; Theory and Industrial Applications*, Krieger, New York, 1977.

M. J. Rosen and H. A. Goldsmith, *Systematic Analysis of Surface-Active Agents*, Wiley-Interscience, New York, 1972.

A. M. Schwartz, J. W. Perry, and J. Berch, *Surface Active Agents*, Vol. 2, Interscience, New York, 1958.

D. J. Shaw, *Introduction to Colloid and Surface Chemistry*, Butterworths, London, 1966.

K. L. Sutherland and I. W. Wark, *Principles of Flotation*, Australasian Institute of Mining and Metallurgy, Melbourne, 1955.

Textual References

1. M. C. Wilkinson, A. C. Zettlemoyer, M. P. Aronson, and J. W. Vanderhoff, *J. Colloid Interface Sci.,* **68,** 508 (1979) and preceding papers.
2. T. D. Blake and K. J. Ruschak, *Nature,* **282,** 489 (1979).
3. H. M. Princen, *J. Colloid Interface Sci.,* **34,** 171 (1970).
4. E. W. Washburn, *Phys. Rev*. Ser. 2, **17,** 273 (1921).
5. L. R. Fisher and P. D. Lark, *J. Colloid Interface Sci.,* **69,** 486 (1979).
6. G. E. P. Elliott and A. C. Riddiford, *Nature,* **195,** 795 (1962); also M. Haynes and T. Blake, private communication.
7. R. J. Good and N. J. Lin, *J. Colloid Interface Sci.,* **54,** 52 (1976).
8. J. C. Melrose and C. F. Brandner, *J. Can. Pet. Technol.,* Oct.–Dec., 1 (1974).
9. J. L. Salager, J. C. Morgan, R. S. Schechter, W. H. Wade, and E. Vasquez, *Soc. Pet. Eng. J.,* April, 107 (1979).
10. J. L. Moilliet, Ed., *Water Proofing and Water Repellency*, Elsevier, New York, 1963.
11. R. P. Dron, *Adv. Chem. Ser. No. 43*, American Chemical Society, Washington, D.C., 1964, p. 310.
12. M. G. Flemming and J. A. Kitchener, *Endeavor,* **24,** 101 (1965).
13. J. Leja, *Chem. Can.,* April 1966, p. 2; *J. Chem. Ed.,* **49,** 157 (1972).

14. I. W. Wark, *Chem. Aust.*, **46,** 511 (1979).
15. A. V. Rapacchietta and A. W. Neumann, *J. Colloid Interface Sci.*, **59,** 555 (1977).
16. J. F. Oliver, C. Huh, and S. G. Mason, *J. Colloid Interface Sci.*, **59,** 568 (1977).
17. J. Leja and G. W. Poling, *Preprint, International Mineral Processing Congress*, London, April 1960.
18. V. I. Klassen and V. A. Mokrousov, Eds., *An Introduction to the Theory of Flotation*, Butterworths, London, 1963.
19. A. M. Gaudin, *Flotation*, McGraw-Hill, New York, 1957.
20. S. P. Frankel and K. J. Mysels, *J. Phys. Chem.*, **66,** 190 (1962).
21. A. Vrij, *Discuss. Faraday Soc.*, **42,** 23 (1966).
22. I. B. Ivanov, B. Radoev, E. Manev, and A. Scheludko, *Discuss. Faraday Soc.*, **66,** 1262 (1970).
23. A. F. Taggart, *Handbook of Ore Dressing*, Wiley, New York, 1945.
24. A. M. Gaudin, *Flotation*, McGraw-Hill, New York, 1932.
25. G. W. Poling and J. Leja, *J. Phys. Chem.*, **67,** 2121 (1963).
26. C. Guarnaschelli and J. Leja, *Sep. Sci.*, **1** (4), 413 (1966).
27. N. P. Finkelstein and G. W. Poling, *Miner. Sci. Eng.*, **4,** 177 (1977).
28. R. N. Tipman and J. Leja, *Colloid Polym. Sci.*, **253,** 4 (1975).
29. A. M. Gaudin, *J. Colloid Interface Sci.*, **47,** 309 (1974).
30. A. M. Gaudin and S. C. Sun, *AIME*, Tech. Pub. 2005, May 1946.
31. J. H. Schulman and J. Leja, *Kolloid-Z.*, **136,** 107 (1954).
32. D. W. Fuerstenau and M. C. Fuerstenau, *Min. Eng.*, March 1956.
33. V. A. Arsentiev and J. Leja, *Colloid and Interface Science*, Vol. 5, Academic, 1976.
34. D. J. Johnston and J. Leja, *Miner. Process. Extr. Metall.*, **87,** C237 (1978).
35. F. F. Aplan and D. W. Fuerstenau, *Froth Flotation*, 50th Anniversary Volume, D. W. Fuerstenau, Ed., American Institute of Mining and Metallurgical Engineering, New York, 1962.
36. D. W. Fuerstenau and T. W. Healy, *Adsorptive Bubble Separation Techniques*, Academic Press, 1971, p. 92; D. W. Fuerstenau, *Pure Appl. Chem.*, **24,** 135 (1970).
37. A. M. Gaudin and D. W. Fuerstenau, *Min. Eng.*, October 1955.
38. F. D. Snell, C. T. Snell, and I. Reich, *J. Am. Oil Chem. Soc.*, **27,** 1 (1950).
39. A. M. Schwartz, *Surface and Colloid Science*, E. Matijevic, Ed., Wiley-Interscience, 1972, p. 195.
40. W. C. Preston, *J. Phys. Colloid Chem.*, **52,** 84 (1948).
41. R. J. Williams, J. N. Phillips, and K. J. Mysels, *Trans. Faraday Soc.*, **51,** 728 (1955).
42. J. W. McBain, *Trans. Faraday Soc.*, **9,** 99 (1913).
43. R. M. Menger, *Acc. Chem. Res.*, **12,** 111 (1979).
44. M. J. Schick, S. M. Atlas, and F. R. Eirich, *J. Phys. Chem.*, **66,** 1326 (1962).
45. P. Debye, *J. Phys. Colloid Chem.*, **53,** 1 (1949).
46. H. F. Huisman, *K. Ned. Akad. Van Wetenschappen-Ams.*, Proceedings Series B, **67,** 367 (1964).
47. H. W. Offen, D. R. Dawson, and D. F. Nicoli, *J. Colloid Interface Sci.*, **80,** 118 (1981).
47a. M. Almgren and J. E. Lofroth, *J. Colloid Interface Sci.*, **81,** 486 (1981).
48. D. Stigter, R. J. Williams, and K. J. Mysels, *J. Phys. Chem.*, **59,** 330 (1955).
49. J. Clifford and B. A. Pethica, *J. Phys. Chem.*, **70,** 3345 (1966).

50. D. Stigter, *Rec. Trav. Chim.*, **73**, 593 (1954).
51. D. Stigter, *J. Phys. Chem.*, **68**, 3603 (1964).
52. G. S. Hartley, *Aqueous Solutions of Paraffin-Chain Salts*, Hermann et Cie., Paris, 1936.
53. See *Colloid Chemistry, Theoretical and Applied*, Reinhold, New York, 1944; also see Ref. 40.
54. M. J. Povich, J. A. Mann, and A. Kawamoto, *J. Colloid Interface Sci.*, **41**, 145 (1972).
55. C. Tanford, *J. Phys. Chem.*, **78**, 2468 (1974).
56. R. Nagarajan and E. Ruckenstein, *J. Colloid Interface Sci.*, **60**, 221 (1977).
57. G. Kegeles, *J. Phys. Chem.*, **83**, 1728 (1979).
58. A. Ben-Nalm and F. H. Stilliinger, *J. Phys. Chem.*, **84**, 2872 (1980).
59. J. L. Woodhead, J. A. Lewis, G. N. Malcolm, and I. D. Watson, *J. Colloid Interface Sci.*, **79**, 454 (1981).
60. N. Nishikido, M. Shinozaki, G. Sugihara, M. Tanaka, and S. Kaneshina, *J. Colloid Interface Sci.*, **74**, 474 (1980).
61. G. Kegeles, *Arch. Biochem. Biophy.*, **200**, 279 (1980).
62. E. A. G. Aniansson, S. N. Wall, M. Almgren, H. Hoffmann, I. Kielmann, W. Ulbricht, B. Zana, J. Lang, and C. Tondre, *J. Phys. Chem.*, **80**, 905 (1976).
63. N. Muller, *J. Phys. Chem.*, **76**, 3017 (1972).
64. A. H. Colen, *J. Phys. Chem.*, **78**, 1676 (1974).
64a. M. Grubic and R. Strey, quoted in M. Teubner, *J. Colloid Interface Sci.*, **80**, 453 (1981).
65. N. Muller, *J. Phys. Chem.*, **79**, 287 (1975).
66. M. J. Schwuger, *Kolloid-Z. Z.-Polym.*, **240**, 872 (1970).
67. F. M. Fowkes, *Solvent Properties of Surfactant Solutions*, K. Shinoda, Ed., Marcel Dekker, New York, 1967.
68. *Advances in Colloid Science*, Vol. I, Interscience, New York, 1942.
69. N. K. Adam and D. G. Stevenson, *Endeavour*, **12**, 25 (1953).
70. F. R. M. McDonnell, *Proc. 3rd Int. Congr. Surf. Act., Vol. 3*, 1960, p. 251.
71. J. G. Griffith, *Proc. 3rd Int. Congr. Surf. Act.*, Vol. 4, 1960, p. 28.
72. W. J. Schwartz, A. R. Martin, B. J. Rutkowski, and R. C. Davis, *Proc. 3rd Int. Congr. Surf. Act.*, Vol. 4, 1960, p. 37.
73. A. L. Meader, Jr. and B. A. Fries, *Ind. Eng. Chem.*, **44**, 1636 (1952).
74. J. W. McBain, R. S. Harborne, and A. M. King, *J. Soc. Chem. Ind.*, **42**, 373T (1923).
75. L. Greiner and R. D. Vold, *J. Phys. Colloid Chem.*, **53**, 67 (1949).
76. J. L. Moilliet, B. Collie, and W. Black, *Surface Activity*, E. & F. N. Spon, London, 1961.
77. K. Durham, Ed., *Surface Activity and Detergency*, Macmillan, New York, 1961.
78. K. Durham, *Proc. 2nd Int. Congr. Surf. Act.*, Vol. 4, 1957, p. 60.
79. P. Krings, M. J. Schwuger, and C. H. Krauch, *Naturwissenschaften*, **61**, 75 (1974); P. Berth, G. Jakobi, E. Schmadel, M. J. Schwuger, and C. H. Krauch, *Angew. Chem.*, **87**, 115 (1975).
80. A. Fava and H. Eyring, *J. Phys. Chem.*, **60**, 890 (1956).
81. F. J. Trogus, R. S. Schechter, and W. H. Wade, *J. Colloid Interface Sci.*, **70**, 293 (1979).

CHAPTER XIV

Emulsions and Foams

1. Introduction

This chapter concludes the digression into certain areas of special applicability of surface chemistry, with the possible exception of some of the material on contact catalysis in Chapter XVII. The subjects touched on here are again of widespread technological importance and are contiguous to very large bodies of industrial and semiempirical literature. Again, too, only certain of the more fundamental aspects, as related to surface chemistry, are considered.

Emulsions and foams are grouped in this chapter since in both cases one is dealing with two partially miscible fluids and, almost invariably, also with a surfactant. Both cases involve a dispersion of one of the phases in the other; if both fluids are liquid, the system is called an emulsion, whereas if one fluid is a gas, the system may constitute a foam (or an aerosol). In both cases, also, the dispersed system usually consists of relatively large units, that is, the size of the droplets or gas bubbles will ordinarily range upward from a few tenths of a micron. The systems are generally unstable with respect to separation of the two fluid phases, that is, toward breaking in the case of emulsions and collapse in the case of foams, and their degree of practical stability is largely determined by the charges and surface films at the interfaces.

Although it is hard to draw a sharp distinction, emulsions and foams are somewhat different from systems normally referred to as colloidal. Thus whereas ordinary cream is an oil-in-water emulsion, the very fine aqueous suspension of oil droplets that results from the condensation of oily steam is essentially colloidal and is called an oil hydrosol. In this case the oil occupies only a small fraction of the volume of the system, and the particles of oil are small enough that their natural sedimentation rate is so slow that even small thermal convection currents suffice to keep them suspended; for a cream, on the other hand, as also is the case for foams, the inner phase constitutes a sizable fraction of the total volume, and the system consists of a network of interfaces that are prevented from collapsing or coalescing by virtue of adsorbed films, or electrical repulsions. Exceptions to these situations are considered, however, as in the case of spontaneous emulsification and of emulsions whose droplet size has become comparable to that of the micelles discussed in connection with detergency.

2. Emulsions—General Properties

An emulsion may be defined as a mixture of particles of one liquid with some second liquid and, since almost invariably one of them is aqueous in nature, the two common types of emulsions are oil-in-water (O/W) and water-in-oil (W/O); the term "oil" is used as a general word denoting the water-insoluble fluid.

These two types are illustrated in Fig. XIV-1, and it is clear that one phase (the "outer" phase) is continuous, whereas the other (the "inner" phase) is not. Usually it is not too difficult to decide which is which—experts can do so merely by the feel of the emulsion. A more objective method consists of adding some of the one liquid or the other; the emulsion should be readily diluted if it is the outer phase liquid that has been added. Alternatively, one may add to the system a dye that is soluble in only one of the phases; the dye will disperse readily to give a general color if it is the outer phase in which it is soluble. Finally, O/W emulsions have much higher electrical conductivities than do W/O emulsions.

Apart from chemical composition, an important variable in the description of emulsions is the ratio ϕ of the volume of inner to outer phase. As is discussed further in Section XIV-4B, a natural value for ϕ is 0.74, representing the close packing of spheres. In more dilute emulsions, the inner phase does exist as spheres, and ϕ is therefore the appropriate variable to treat viscosity. For rigid spheres, the Einstein limiting law is (1)

$$\eta = \eta_0(1 + 2.5\phi) \qquad \text{(XIV-1)}$$

Beyond ϕ equal to about 0.02 deviations set in, which can be handled semiempirically by replacing the term in parentheses by a power series in ϕ. Becher (2) and Sherman (3) give the generalized equation

$$\eta_{sp} = \frac{\eta}{\eta_0} - 1 = a\phi + b\phi^2 + c\phi^3 + \cdots \qquad \text{(XIV-2)}$$

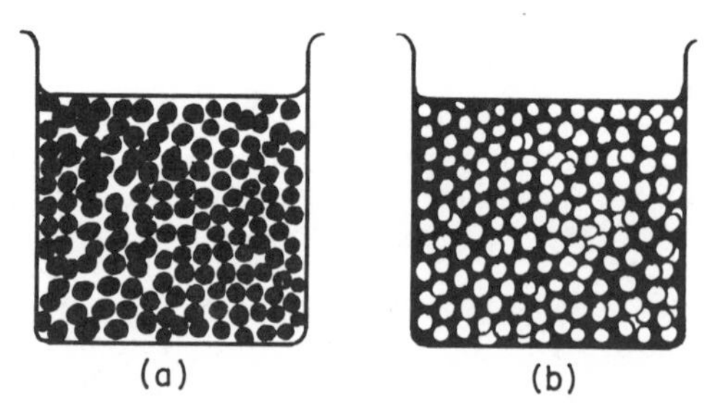

Fig. XIV-1. The two types of emulsion; (*a*) oil in water, O/W; (*b*) water in oil, W/O.

where η_{sp} is called the specific viscosity. Sherman tabulates empirical values for the constants *a*, *b*, and *c* for a number of systems. Theoretical values for *b* are in the range of about 3 to 6 (see Problem 3). Emulsion droplets are not really rigid, and approximate theoretical recognition of this is given by an equation due to Taylor (4):

$$\eta = \eta_0 \left[1 + 2.5\phi \left(\frac{p + \frac{2}{5}}{p + 1} \right) \right] \tag{XIV-3}$$

where *p* is the ratio of inner to outer phase viscosity. In addition, the more concentrated emulsions are not Newtonian in their viscosity, so that η depends on shear rate. Furthermore, η depends not only on ϕ but also on droplet size and size distribution (5). This last is difficult to determine with any accuracy from the usual method of microscopic examination, and often just a mean size or number concentration *n* of droplets is given.

Other physical properties include electrical conductivity, dielectric constant (of W/O emulsions), and electrophoretic mobility measurements. Time dependent effects include rate of shear on viscosity (thixotropy and rheopexy), rate of decrease of *n* with time (coalescence or aggregation), and rate of *creaming*. This last refers to the tendency (as with milk) for an emulsion to separate into a more concentrated and a more dilute emulsion phase.

Local physical properties can be approached through dielectric relaxation, nuclear magnetic resonance (6), and infrared absorption studies. Here some indication is provided of the chemical states in the interfacial versus bulk regions. Where the droplets are very small or even micellar in nature, light scattering (7) or low angle x-ray diffraction (8) can provide some information about dimensions and distribution. The SEM technique (scanning electron microscope—see Section VIII-2B) can provide accurate pictures of the distribution of sizes and shapes of emulsion droplets (9).

The size distribution of emulsion droplets is often determined by means of a *Coulter counter*. As the emulsion flows through a small hole separating two compartments, the electrical resistance between them changes as an emulsion drop passes through the hole. The amount of change is a measure of the drop size, and modern equipment automatically accumulates the data to report a size distribution. Emulsion particle size distributions are often log normal in type, that is, the equation

$$p = \frac{1}{\sigma\sqrt{2\pi}} \exp\left[-\frac{(\ln x - \ln x_m)^2}{2\sigma^2} \right] \tag{XIV-4}$$

is obeyed. Here *p* is the probability of finding an emulsion drop of size *x* if x_m is the mean size and σ is the logarithmic standard deviation. Qualitatively, a log normal distribution differs from the usual Gaussian error curve in giving greater probability to the extremes in *x*. See Ref. 10.

The foregoing survey gives an indication of the complexity of emulsion systems and the wealth of experimental approaches available. We are limited here, however, to some selected aspects of a fairly straightforward nature.

3. Factors Determining Emulsion Stability

If two pure, immiscible liquids, such as benzene and water, are vigorously shaken together, they will form a dispersion, but it is doubtful if one phase or the other is uniquely outer or inner in nature. On stopping the agitation, moreover, phase separation occurs so quickly that it is questionable whether the term emulsion really should be applied to the system. It seems possible to obtain a relatively stable suspension of one pure liquid in another only in the case of oil hydrosols and the like, and it was pointed out before that such systems are more properly considered colloidal suspensions than emulsions.

With these possible types of exceptions, it appears that a surfactant component is always needed to obtain a stable or reasonably stable emulsion. Thus if a little soap is added to the benzene–water system, the result on shaking is a true emulsion, and one that separates out only very slowly. Theories of emulsion stability have therefore concerned themselves with the nature of the interfacial films that are evidently important and with the mechanism of their action in preventing droplet coalescence.

A. Macroscopic Theories of Emulsion Stabilization

It is quite clear, first of all, that since emulsions present a large interfacial area, any reduction in interfacial tension must reduce the driving force toward coalescence and should promote stability. We have here, then, a simple thermodynamic basis for the role of emulsifying agents. Harkins (11) mentions, as an example, the case of the system paraffin oil–water. With pure liquids, the interfacial tension was 41 dyne/cm, and this was reduced to 31 dyne/cm on making the aqueous phase $0.001M$ in oleic acid, under which conditions a reasonably stable emulsion could be formed. On neutralization by $0.001M$ sodium hydroxide, the interfacial tension fell to 7.2 dyne/cm, and if also made $0.001M$ in sodium chloride it became less than 0.01 dyne/cm. With olive oil in place of the paraffin oil, the final interfacial tension was 0.002. These last systems emulsified spontaneously—that is, on combining the oil and water phases no agitation was needed for emulsification to occur.

The surface tension criterion indicates that a stable emulsion should not exist with an inner phase volume fraction exceeding 0.74 (see Problem 2). Actually, emulsions of some stability have been prepared with an inner phase volume fraction as high as 99%, see Lissant (12). Two possible explanations are the following. If the emulsion is very heterogeneous in particle size (note Ref. 10), ϕ may exceed 0.74 by virtue of smaller drops occupying the spaces between larger ones, and so on with successively smaller droplets. Such a system would not be an equilibrium one, of course. It appears, however, that in some cases the thin film separating two emulsion drops may have a *lower* surface tension than that of the bulk interfacial tension (13). As a consequence, two approaching drops will spontaneously deform to give a flat drop–drop contact area, with the drops taking on a polyhedral shape. ϕ can be quite large for a mass of such deformed drops. The appearance of a drop cluster is illustrated in Fig. XIV-2. The triangular space between drops is known as the *Plateau border,* and in this case, the surface of the border does not merge smoothly with the flat drop–drop interface but forms some definite contact angle with it (see Ref. 14). Conversely, it is the existence of such an angle that leads to the formation of dense emulsions of high ϕ value. Emulsions of this type resemble biliquid "foams"; Sebba (14a) has termed the unit cells of such emulsions as "aphrons."

One may rationalize emulsion type in terms of interfacial tensions. Bancroft (15) and later Clowes (16) proposed that the interfacial film of emulsion stabilizing surfactant be regarded as duplex in nature, so that an inner and an outer interfacial tension could be discussed. On this basis, the type of emulsion formed (W/O versus O/W) should be such that the inner surface is the one of higher surface tension. Thus sodium and other alkali metal soaps tend to stabilize O/W emulsions, and the explanation would be that, being more water- than oil-soluble, the film–water interfacial tension should be lower than the film–oil one. Conversely, with the relatively more oil-soluble metal soaps, the reverse should be true, and they should stabilize W/O emulsions, as in fact they do.

The duplex film model contains within it an earlier, more mechanical picture, known as the *oriented wedge* theory (17). It was assumed, of course, that to be stable at all the films must be oriented so that the polar group was in the aqueous phase. It was then argued that surface molecules could be regarded as wedge or cone shaped and should stabilize the type of emulsion that would place the larger end outward. Thus with sodium soaps, the ionic carboxylate group, being larger than the hydrocarbon chain, should face outward and hence stabilize O/W emulsions. The polar end of a zinc soap, on the other hand, binds two carboxyl groups and should be smaller than the *two* hydrocarbon chains forming the other end of the unit. Thus W/O emulsions should be stabilized.

While the oriented wedge theory is clearly too primitive to be of much use (see Ref. 18), the duplex film model, if not taken too seriously, is a helpful one. With simple surfactants, the interfacial film is monomolecular,

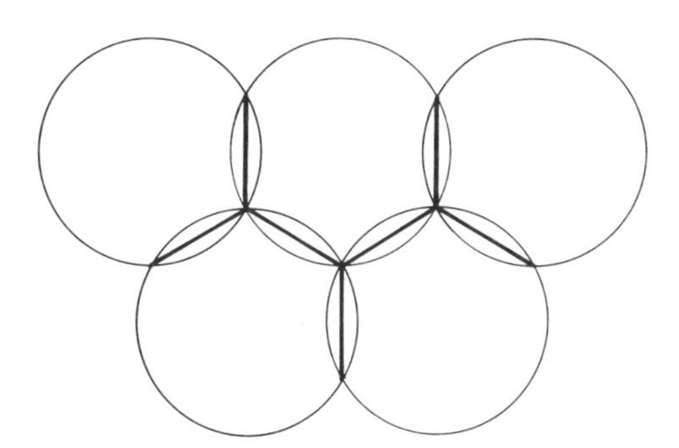

Fig. XIV-2. Spontaneously deformed cluster of emulsion droplets.

not duplex, so the model has only about the same qualitative standing as does Langmuir's principle of independent surface action (Section III-3). On the other hand, it does rationalize the separate consideration of the electrical double layers on each size of the interface, and the development of empirical additivity rules, such as the HLB system (Section XIV-6).

B. Specific Chemical and Structural Effects

The energetics and kinetics of film formation appear to be especially important when two or more solutes are present, since now the matter of monolayer penetration or complex formation enters the picture (see Section IV-8). Schulman and co-workers, in particular, have noted that especially stable emulsions result when the adsorbed film of surfactant material forms strong penetration complexes with a species present in the oil phase. The stabilizing effect of such mixed films may lie in the very low rate at which they desorb (note also Section XIII-4B), although it is pointed out later that the effect also correlates with surface viscosities.

The importance of steric factors in the formation of penetration complexes is made evident by the observation that although sodium cetyl sulfate plus cetyl alcohol gives an excellent emulsion, the use of oleyl alcohol instead of cetyl alcohol leads to very poor emulsions. As illustrated in Fig. XIV-3, the explanation may lie in the difficulty in accommodating the kinked oleyl alcohol chain in the film.

An important aspect of the stabilization of emulsions by adsorbed films is that of the role played by the film in resisting the coalescence of two droplets of inner phase. Such coalescence involves a local mechanical compression at the point of encounter that would be resisted (much as in the approach of two boundary lubricated surfaces discussed in Section XII-7B) and then, if coalescence is to occur, the discharge from the surface region of some of the surfactant material. Alexander (19) pointed out that

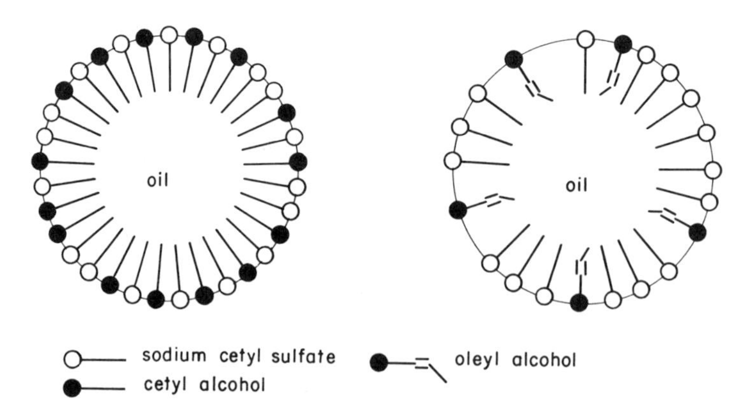

Fig. XIV-3. Steric effects in the penetration of sodium cetyl sulfate monolayers by cetyl alcohol and oleyl alcohol.

desorption may be a hindered process; it thus can provide a barrier to coalescence. Another consequence is that ejected surface material may come out as a solid phase of fibers or crystals. This may be the case with films of polyvalent metal soaps and with protein stabilized emulsions.

With respect to this last type of emulsion, proteins, glucosides, lipids, sterols, and so on, although generally very water soluble, are nonetheless frequently able to impart considerable stability to emulsions (and foams). Agar-agar, saponin, albumin, pectin, gelatin, lecithin, and casein are among the natural substances possessing emulsion-stabilizing properties—these are cited by Berkman and Egloff (20) as being in the order of decreasing effectiveness in the case of benzene–water emulsions.

As discussed in Section IV-11, proteins do spread at the water–air and the water–oil interfaces to give coherent but rather amorphous and viscous films whose structure is not exactly that of the natural protein but involves a partial denaturation through a spreading out of the various side chains. Once spread, these and other materials do not readily return to bulk solution. If a drop is distorted, as in the mechanical working of an emulsion, the surface area is increased and more interfacial film forms, but irreversibly so that if the drop returns to a spherical shape, film material does not go back into solution but may wrinkle the interface or thicken it as expelled curd collects; the same would be true if two drops coalesce. Such processes may account for the rather thick films or membranes that are visible in emulsions stabilized by biological substances (e.g., see Ref. 21). In addition, Cockbain and McRoberts (22) comment that the preferential wettability by one phase or the other of the ejected film fragments may determine the emulsion type.

C. Long-Range Forces as a Factor in Emulsion Stability

There appear to be two stages in the collapse of emulsions: flocculation, in which some clustering of emulsion droplets takes place, and coalescence, in which the number of distinct droplets decreases (see Refs. 23–25). Coalescence rates very likely depend primarily on the film–film surface chemical repulsion and on the degree of irreversibility of film desorption, as discussed. However, if emulsions are centrifuged, a compressed polyhedral structure similar to that of foams results (24–26)—see Section XIV-8—and coalescence may now take on mechanisms more related to those operative in the thinning of foams.

Flocculation, on the other hand, should be sensitive to long-range forces as it involves the first stage of the approach of droplets. First, the oil droplets in O/W emulsions are generally negatively charged; some mobility versus surfactant concentration results are shown in Fig. XIV-4 for the Nujol water system (27) by way of illustration. The manner in which potential should vary across an oil–water interface is shown in Fig. XIV-5, after van den Tempel (28). Here ΔV denotes the surface potential difference between the two phases, and χ is the surface potential jump (see Section V-11). If some electrolyte is present, the solubility of the cations and the anions in general will be different in the two phases. Usually the anions will be somewhat

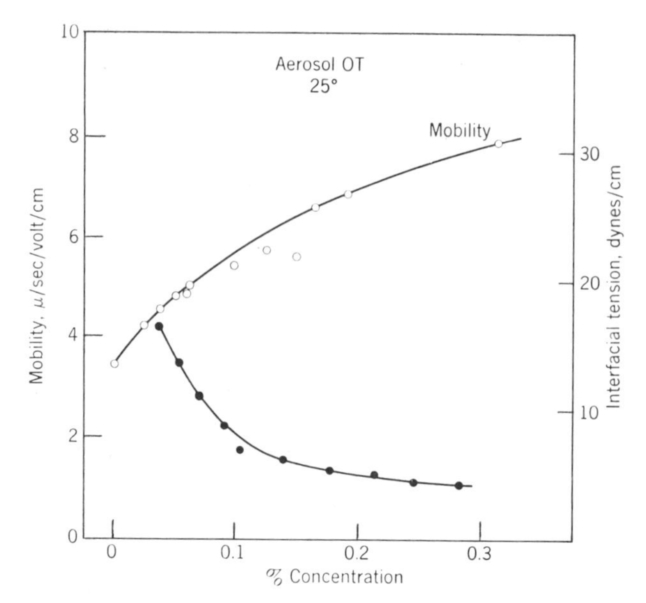

Fig. XIV-4. Mobility of Nujol droplets in solutions of Aerosol OT; interfacial tension of aqueous solutions of Aerosol OT against Nujol. (From Ref. 27.)

more oil soluble than the cations, so that, as illustrated in Figure XIV-5*b*, there should be a net negative charge on the oil droplets.

The repulsion between oil droplets will be more effective in preventing flocculation the greater the thickness of the diffuse layer and the greater the value of ψ_0, the surface potential. These two quantities depend oppositely on the electrolyte concentration, however. The total surface potential should increase with electrolyte concentration, since the absolute excess of anions

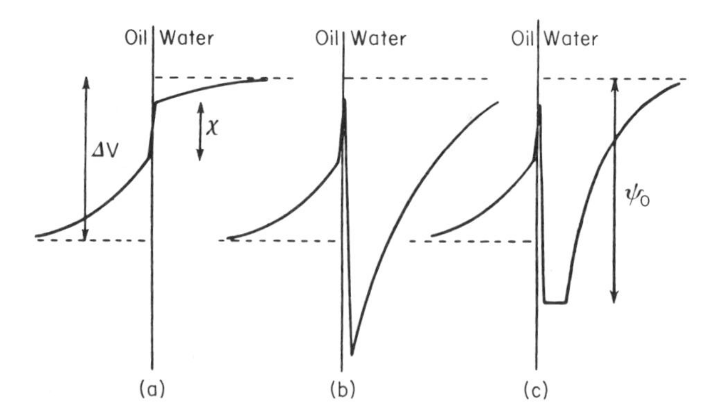

Fig. XIV-5. Variation in potential across an oil–water interface: (*a*) in the absence of electrolyte, (*b*) with electrolyte present, and (*c*) in the presence of soap ions and a large amount of salt. (From Ref. 28.)

over cations in the oil phase should increase. On the other hand, the half-thickness of the double layer decreases with increasing electrolyte concentration. The plot of emulsion stability versus electrolyte concentration may thus go through a maximum.

If an ionic surfactant is present, the potentials should vary as shown in Fig. XIV-5*c*, or similarly to the case with nonsurfactant electrolytes. In addition, however, surfactant adsorption decreases the interfacial tension and thus contributes to the stability of the emulsion. As discussed in connection with charged monolayers (see Section IV-14), the mutual repulsion of the charged polar groups tends to make such films expanded and hence of relatively low π value. Added electrolyte reduces such repulsion by increasing the counterion concentration; the film becomes more condensed and its film pressure increases. It thus is possible to explain qualitatively the role of added electrolyte in reducing the interfacial tension and thereby stabilizing emulsions.

Returning to the general situation, not only is an electrical double layer barrier to coalescence likely to be present, but also the DLVO theory (see Section VII-4B) predicts that a shallow minimum in net interaction potential can exist. A calculated example (29) is shown in Fig. XIV-6 (see also Refs. 30, 31). The effect of such a minimum is that emulsion droplets may flocculate to the distance apart of this secondary minimum without any further tendency to approach each other.

Some studies have been made of W/O emulsions; the droplets are now aqueous and positively charged (32, 33). There is now very little correlation between ξ potential and stability, and Albers and Overbeek (32) suggest that this may result from a rather low electrolyte concentration of the oil external phase, so that the double layer half-thickness is as much as several microns. As a consequence, emulsion droplets lie within each other's double layer and experience relatively little electrical repulsion. In their third paper, these authors also estimated the magnitude of the van der Waals long-range at-

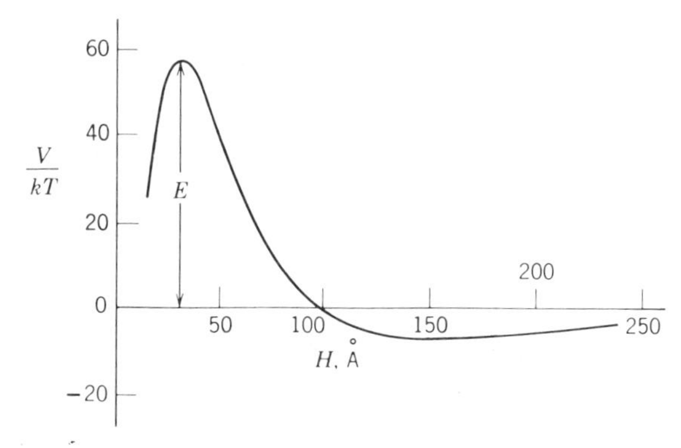

Fig. XIV-6. Calculated interaction energy curve for paraffin oil droplets stabilized by bovine serum albumin. (From Ref. 29.)

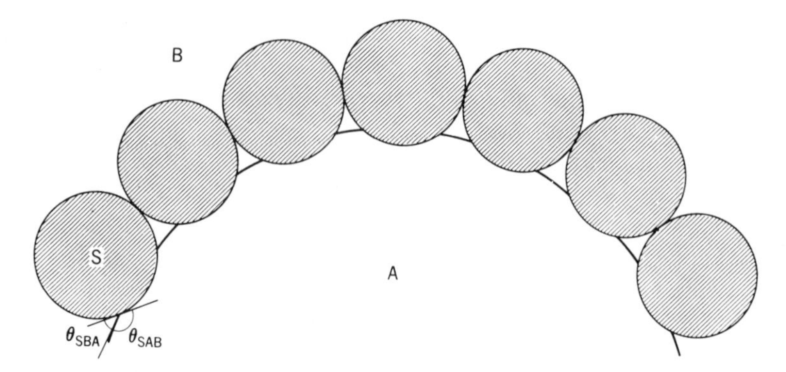

Fig. XIV-7. Stabilization of an emulsion by small particles.

traction from the shear gradient sufficient to detach flocculated droplets (see also Ref. 34).

D. Stabilization of Emulsions by Solid Particles

Powders constitute an interesting type of emulsion-stabilizing agent. For example, benzene–water emulsions are stabilized by calcium carbonate, the solid particles collecting at the oil–water interface and armoring the benzene droplets. Similarly, Scarlett and co-workers (35) report the stabilizing of toluene–water emulsions by pyrite as well as that of water–benzene emulsions by charcoal and by mercuric iodide; glycerol tristearate crystals stabilize water–paraffin oil emulsions (36).

It was pointed out in Section XIII-4A that if the contact angle between a solid particle and two liquid phases is finite, a stable position for the particle is at the liquid–liquid interface. Coalescence is inhibited because it takes work to displace the particle from the interface. In addition, one can account for the type of emulsion that is formed, O/W or W/O, simply in terms of the contact angle value. As illustrated in Fig. XIV-7, the bulk of the particle will lie in that liquid that most nearly wets it, and, by what seems to be a correct application of the oriented wedge principle, this liquid should then constitute the outer phase. Furthermore, the action of surfactants should be predictable in terms of their effect on the contact angle. This was indeed found to be the case in a study by Schulman and Leja (37) on the stabilization of emulsions by barium sulfate.

Stabilization of emulsions by solids at the liquid–liquid interface may be involved in cases where no solid is knowingly added. Whenever the surface film is sufficiently rigid, or in sufficiently slow equilibrium with the bulk phases, then, as discussed, the film material may be forced out as a solid or gel phase as a result of the coalescence of droplets or of mechanically induced distortions in droplet shapes.

4. The Aging and Inversion of Emulsions

As illustrated in Fig. XIV-8, at least three types of aging processes occur for emulsions. The inner phase droplets may undergo flocculation, that is, clustering together without losing their identity; if, as part of or subsequent to flocculation, the flocks undergo a gravity separation, the entire process is called creaming. If coalescence occurs, then eventual breaking of the emulsion must follow, giving two liquid layers. Inversion can be a very dynamic and complex process; it is discussed further below.

A. Flocculation and Coagulation Kinetics

There are two bases on which the kinetics of the flocculation and coagulation of droplets has been approached. The first stems from a relationship due to Smoluchowski (38) for the rate of diffusional encounters of spherical particles:

$$R = 16\pi \mathcal{D} r n^2 \qquad \text{(XIV-5)}$$

where n is the number of particles per cubic centimeter of radius r and $\mathcal{D}$ is their diffusion coefficient. For spheres, the Stokes–Einstein expression for $\mathcal{D}$ is $\mathcal{D} = kT/6\pi\eta r$. If there is an energy barrier to coagulation E^*, then the rate of effective encounters, dn/dt, becomes

$$\frac{dn}{dt} = -kn^2 \qquad \text{(XIV-6)}$$

where

$$k = \frac{8kT}{3\eta} e^{-E^*/kT}$$

The quantity dn/dt is essentially the rate of disappearance of primary particles, and hence the rate of decrease in the total number of particles. Integration gives

$$\frac{1}{n} = \frac{1}{n_0} + kt \qquad \text{(XIV-7)}$$

Equation XIV-7 is rather approximate, and more detailed treatments are given by Kruyt (39) and Becher (2); the problem is that of handling the formation of aggregates of more than two particles. As a result, the form

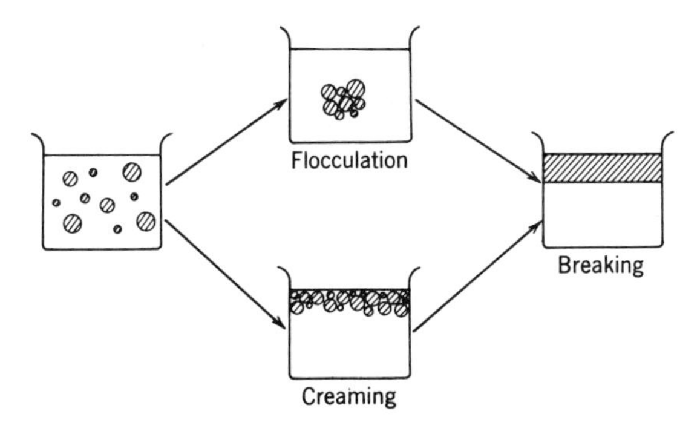

Fig. XIV-8. Types of emulsion instability.

of Eq. XIV-7 is not well obeyed, but $d(1/n)/dt$ is used as a measure of flocculation or coagulation rate.

For example, van den Tempel (28) reports the results shown in Fig. XIV-9 on the effect of electrolyte concentration on flocculation rates of an O/W emulsion. Note that $d(1/n)/dt$ (equal to k in the simple theory) increases rapidly with ionic strength, presumably due to the decrease in double layer half-thickness and perhaps also due to some Stern layer adsorption of positive ions. The preexponential factor in Eq. XIV-6, $k_0 = (8kT/3\eta)$, should have the value of about 10^{-11} cm^3, but at low electrolyte concentration, the values in the figure are smaller by tenfold or a hundredfold. This reduction may be qualitatively ascribed to charge repulsion, and Davies (40) makes use of a limiting form of Eq. V-29 whereby $E^* = \frac{1}{4}\psi_0^2$, where $\frac{1}{4}$ is an empirical factor giving E^* in calories from ψ_0 in millivolts.

The preceding type of treatment relates primarily to flocculation rates, while the irreversible aging of emulsions involves the coalescence of droplets, the prelude to which is the thinning of the liquid film separating the droplets. Similar theories were developed by Spielman (41) and by Honig and co-workers (42), which added hydrodynamic considerations to basic DLVO theory. A successful experimental test of these equations was made by Bernstein and co-workers (43).

There have been some studies of the equilibrium shape of two droplets pressed against each other (see Ref. 44) and of the rate of film thinning (45–47), but these are based on hydrodynamic equations and do not take into account film–film barriers to final rupture. It is at this point, surely, that the chemistry of emulsion stabilization plays an important role.

B. Inversion and Breaking of Emulsions

An interesting effect is that in which an A/B type of emulsion inverts to a B/A type. Generally speaking, the methods whereby inversion may be

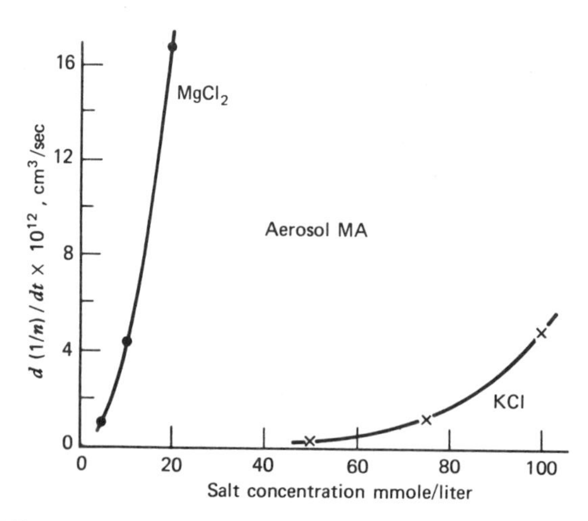

Fig. XIV-9. Effects of electrolyte on the rate of flocculation of Aerosol MA stabilized emulsions. (From Ref. 28.)

caused to take place involve introducing a condition such that the opposite type of emulsion would normally be the stable one. First, an emulsion would *have* to invert if ϕ exceeded 0.74, if the inner phase consisted of uniform rigid spheres; as noted in Section XIV-3, this value of ϕ represents the point of close packing. Actual emulsion droplets are deformable, of course, and not monodisperse, but the preceding criterion does have a limited applicability. Wellman and Tartar (18) observed that benzene–water emulsions stabilized by sodium stearate inverted from O/W to W/O on increasing ϕ, but well before the ideal value of 0.74; Salisbury et al. (48) found that in cold cream preparations, inversion occurred at $\phi = 0.45$. Robertson (49) reported a critical ϕ value of 0.9, and Pickering (50) has reported an O/W emulsion, prepared with soap as the emulsifying agent, that contained 99% oil by volume. Thus while continued addition of inner phase may result in inversion, the effect is not assured and certainly will not be controlled by the theoretical ϕ value of 0.74. As an extreme, Sebba (51) has produced "biliquid foams," that is, emulsions with polyhedral cells of inner liquid and thin film outer liquid looking much like a foam.

Second, it will be recalled that soaps with monovalent ions tend to stabilize O/W type emulsions, whereas those with polyvalent ions stabilize W/O emulsions. It therefore is not surprising that addition of a calcium salt to an O/W emulsion stabilized by a sodium soap can result in inversion. This type of effect was reported by Clowes (16) and is discussed further in Section XIV-6. It is not always necessary to use polyvalent electrolytes; Kremnev and Kuibina (52) report that emulsions of benzene in fairly concentrated (0.3*M*) sodium oleate were broken and then inverted by making the aqueous phase 0.25 to 0.3*M* in sodium chloride or other 1:1 electrolytes. Finally, in addition to inversion by changing the phase volume and by the use of antagonistic agents, inversion may be brought about by temperature changes. Thus Wellman and Tartar (18) found that sodium stearate-stabilized O/W emulsions inverted on cooling.

The general impression is that where the inner phase is not too dilute, the emulsion type is determined by some dynamic balance of various factors and responds fairly readily to a change in conditions. Clowes (16), in particular, made some striking observations on the appearance of emulsions undergoing inversion. On the addition of a calcium salt to a sodium soap-stabilized O/W emulsion, he comments that the oil globules first distorted, then elongated as the critical point was approached, with very marked "Brownian" movement. The elongated sections of aqueous phase then necked in to give a W/O system. The agitated appearance and marked streaming of the two phases at the critical inversion point was probably due to local concentration fluctuations as the added calcium salt mixed with the system with resulting Marangoni effects (Section IV-2B).

Deemulsification or the breaking of an emulsion can be accomplished by the judicious use of one of the preceding methods of emulsion inversion or by methods that accelerate the coalescence rate of droplets. Also, a *phase change* in one of the two liquid phases may be helpful; thus emulsions may be broken by heating to near the boiling point of the inner phase or by freezing and then rewarming. Absorption chromatography has been used as a means of removing the emulsifying agent and thus breaking the emulsion (53).

Shinoda and co-workers have found the phase inversion temperature (PIT) to have very useful correlating properties (see Ref. 54).

5. Spontaneous Emulsification—Micellar or Microemulsions

It was mentioned in Section XIV-3 that some systems undergo spontaneous emulsification, that is, the most gentle addition of one liquid (plus surfactant) to the other is followed by spontaneous kicking or streaming out of fronds of fine emulsified droplets. The effect is not in doubt (see Ref. 2 for illustrations), but it may be just a consequence of local concentration and diffusion gradients and associated Marangoni action. The question involved is whether occurrence of the phenomenon means that the resulting emulsion is positively stable or whether it would still eventually coalesce into bulk phases but now phases mutually saturated with respect to each other and to surfactant.

A more rigorous test of true stability would require that the droplet size respond reversibly to changes in composition or condition, and certain systems do appear to meet this criterion. Bowcott and Schulman (55) found that starting with a coarse emulsion of an O/W type, for example, benzene in water stabilized by a soap such as potassium oleate, addition of long chain alcohols such as hexanol led to a progressive decrease in droplet size until a point was reached such that the mixture was transparent and appeared homogeneous. Actually, the oil droplets had merely become very small, 100 to 500 Å in diameter; they resembled swollen micelles rather than normal oil droplets and were invisible because of their small size as compared to the wavelength of visible light. The presence of a second phase was indicated, however, by light-scattering and x-ray diffraction studies (see Ref. 56). Winsor (57) observed similar behavior and also that such micellar emulsions could be in equilibrium with either oil or aqueous bulk phase.

Schulman and co-workers have called these systems *microemulsions*; but the term used here, *micellar emulsions*, seems to stress better several important special characteristics. Micellar emulsions resemble ordinary emulsions in that ϕ can be around 0.5, that is, the inner phase is not dilute. This inner phase may be aqueous, but highly structured, requiring as it does both detergent and alcoholic constituents, and the small size of the units suggests that even their centers do not reach normal bulk phase properties. For example, while such systems will take up water indefinitely, eventually inverting to a probably normal O/W emulsion, if an electrolyte is added, then the micellar emulsion system can be present as a second phase in equilibrium with bulk aqueous electrolyte solution. These observations imply that the water activity in the units is low or at a higher than normal osmotic pressure, so that spontaneous entry of water can be restrained only by a matching decrease in its external osmotic pressure through the addition of electrolyte. The behavior is that of a system of swollen micelles in equilibrium with their environment and responding reversibly to changes in it (note Ref. 58).

Micellar emulsions have received increasing attention in recent years; they are important, for example, in tertiary oil recovery (59, 60). A model has been proposed

(61) that accounts for the ability of a micellar emulsion phase to be in equilibrium with an aqueous electrolyte phase, and distribution data for the partition of electrolyte, water, and surfactant between these two phases have been examined in terms of the model (62). See Refs. 63 and 64 for recent alternative approaches. Use of D_2O rather than H_2O noticeably affects micellar emulsion properties (65). Other studies include those of viscosity (66), nmr (67, 68), conductivity (69, 70), and both light (71) and neutron scattering (72, 73), and dielectric relaxation (74). The hydrodynamic size of droplets has been estimated by a membrane diffusion method (75). There have been numerous recent studies of reactions, including photochemical reactions and photophysical processes in micellar emulsion systems (see Refs. 76–79). Finally, see Prince (80) for a discussion of the distinction between microemulsions and micelles, and Rosano and co-workers for an earlier review (81) and for a discussion of the variables affecting micellar emulsion formation (81a).

6. The Hydrophile–Lipophile Balance

There is a very large technology that makes use of emulsions, and, somewhat as in flotation, empirical observation still leads theory, in this case with respect to the prediction of the type and stability of emulsion that a given set of constituents will produce. A rough rule, for example, is that the phase in which the stabilizing agent is more soluble will be the external phase (82)—note Table XIV-1. Some elaborations of this type of rule are discussed briefly here.

A numerical rating scheme, known as the Hydrophile–Lipophile Balance (HLB) number, was introduced by Griffin (83) and has come into fairly wide use. First, numbers are assigned on a one-dimensional scale of surfactant action, as given in Table XIV-1; note the correlation with the solubility rule cited. Each surfactant is then rated according to this scale (see Ref. 2 for

TABLE XIV-1
The HLB Scale

Surfactant solubility behavior in water	HLB number		Application
No dispersibility in water	0		
	2		W/O emulsifier
	4		
Poor dispersibility	6		
Milky dispersion; unstable	8		Wetting agent
Milky dispersion; stable	10		
Translucent to clear solution	12	Detergent	O/W emulsifier
	14		
Clear solution	16	Solubilizer	
	18		

a detailed listing), and it is assumed that surfactant mixtures can be assigned an HLB number on a weight prorated basis.

The central assumption of the HLB system can be illustrated as follows. Suppose a certain O/W emulsion is desired. The oil and water phases are emulsified using, say, various proportions of Span 65 (sorbitol tristearate, HLB = 2.1) and Tween 60 (polyoxyethylene sorbitan monostearate, HLB = 14.9). It is found that the optimum emulsion is obtained with 80% Tween 60 and 20% Span 65 (HLB = 12.3). The assumption is then that with any other mixture of surfactants, optimum performance for the particular system will again be at HLB = 12.3 as, for example, if mixtures of Span 85 (sorbitan trioleate, HLB = 1.8) and Tween 20 (polyoxyethylene sorbitan monolaurate, HLB = 16.7) were used in the required proportion, or 70% Tween 20. The *absolute* performance of the two mixtures might differ, but each should be at *its optimum*. The next step, in practice, would be to make up a number of such optimum mixtures and find the one whose absolute performance was best.

Davies (40) carried the additivity principle further by developing a list of HLB functional group numbers, given in Table XIV-2. The empirical HLB number for a given surfactant is computed by adding 7 to the algebraic sum of the group numbers. Thus the calculated HLB number for cetyl alcohol, $C_{16}H_{33}OH$, would be $7 + 1.9 + 16(-0.475) = 1.3$. HLB numbers have also been related to phase inversion temperature (84).

The HLB system has made it possible to organize a great deal of rather messy information and to plan fairly efficient systematic approaches to the optimization of emulsion preparation. If pursued too far, however, the system tends to lose itself in complexities. These can be appreciated by examining Fig. XIV-10 illustrating the possible molecular appearance of the interfacial films as emulsion droplets stabilized by a mixture of Tween 40 and Span 80 approach. It is not surprising that HLB numbers are not really

TABLE XIV-2
Group HLB Numbers

Hydrophilic groups	HLB	Lipophilic groups	HLB
—SO_4Na	38.7	—CH—	−0.475
—COOK	21.1	—CH_2—	−0.475
—COONa	19.1	—CH_3—	−0.475
Sulfonate	about 11.0	—CH=	−0.475
—N (tertiary amine)	9.4	—(CH_2—CH_2—CH_2—O—)	−0.15
Ester (sorbitan ring)	6.8		
Ester (free)	2.4		
—COOH	2.1		
—OH (free)	1.9		
—O—	1.3		
—OH (sorbitan ring)	0.5		

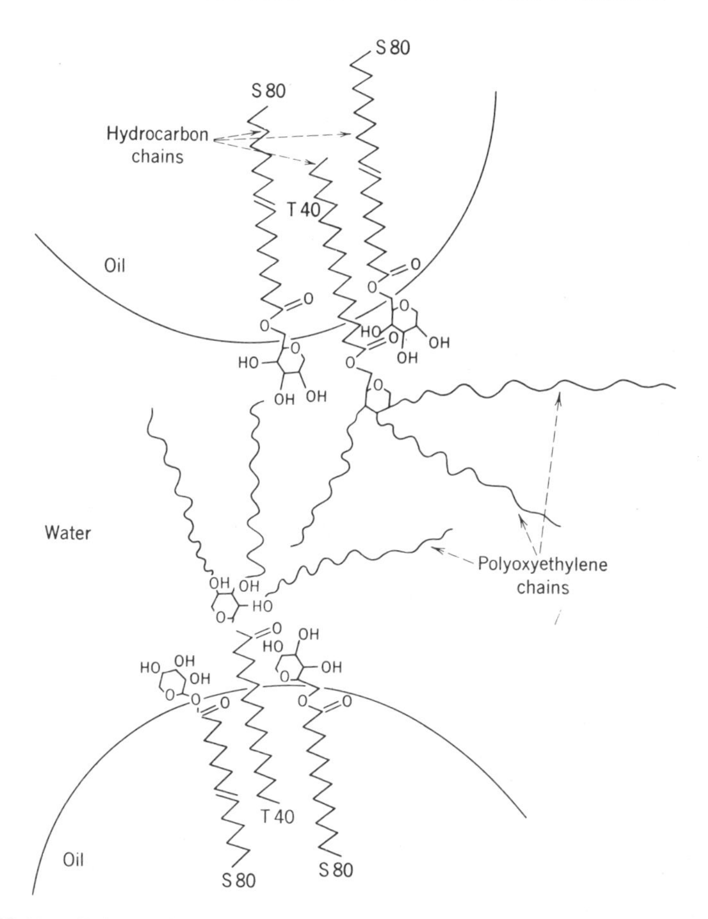

Fig. XIV-10. Schematic representation of orientation of Tween 40 and Span 80 molecules in mixed films adsorbed at the oil–water interface. (From Ref. 86.)

additive; their effective value depends on what particular oil phase is involved, and so on. The question of whether the emulsion will be O/W or W/O depends also on the ϕ value, as noted earlier, and the host of detailed physical characteristics that must be specified in a complete description of an emulsion cannot be encapsulated by a single HLB number.

Shinoda and co-workers (85, 85a) have studied the effect of temperature on HLB numbers, and, in addition, related the HLB system to that of using PIT (phase inversion temperature).

7. Practical Aspects of Emulsions

Spontaneously forming emulsions are the exception in industrial preparations, and a large variety of beating, grinding, mixing, and homogenizing devices are in use. The basic action appears to be that illustrated in Fig.

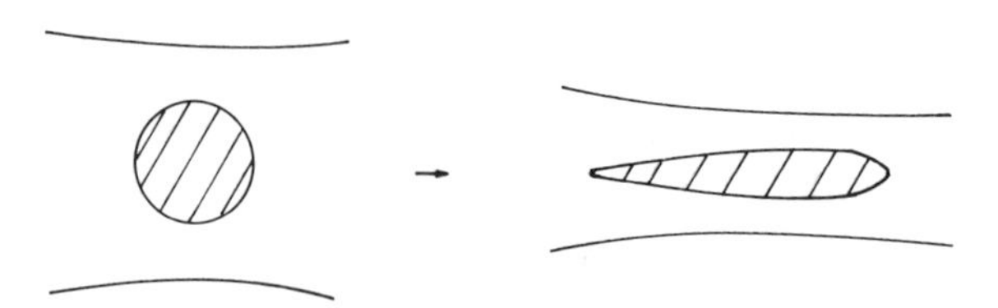

Fig. XIV-11. Drop distortion during emulsification.

XIV-11. That is, a relatively large drop is subjected to fluid or mechanical forces that cause it to elongate; in the case illustrated it does so in the form of a cylinder. In Section II-3 it was pointed out that a cylinder of liquid whose length exceeds its circumference is unstable toward breaking up into a smaller and a larger portion and that a long column or jet of liquid will therefore break up into a sequence of droplets. In emulsification, then, a large drop is drawn out, and surface tensional forces (as well as eddy currents) cause it to break up into a series of much smaller drops; these in turn may suffer further degradation. The process is obviously especially complex in the case of phase volume ratios around unity, since now both liquid phases are being pulverized, and the nascent emulsion presumably is a mixture of both types. The droplets of liquid destined to be the outer phase must coalesce preferentially over those of inner phase material. This picture is the basis for Davies' kinetic treatment, described in the preceding section.

With respect to practical applications of emulsions, a partial listing would include bituminous emulsions, used in road surfacing and in roofing, gasoline emulsions (e.g., for flame throwers), paint emulsions (including latex paints), agricultural sprays, various cleaners and soap preparations, most cosmetic creams and lotions, and various food emulsions such as margarines, ice creams, and salad dressings. Also, many medicinal preparations are in the form of emulsions as, for example, burn ointments of the O/W type and salves of the W/O type for the relief of dry skin. The breaking of emulsions is often of practical importance. Examples would be the breaking of the milk O/W emulsion to obtain butter and the breaking of annoying emulsions of salt water in crude oils.

8. Foams—The Structure of Foams

A foam can be considered as a type of emulsion in which the inner phase is a gas and, as with emulsions, it seems necessary to have some surfactant component present to give stability. The resemblance is particularly close in the case of foams consisting of nearly spherical bubbles separated by rather thick liquid films; such foams have been given the name *kugelschaum* by Manegold (87).

The second type of foam contains mostly gas phase, separated by thin films or laminas. The cells are polyhedral in shape, and the foam can be

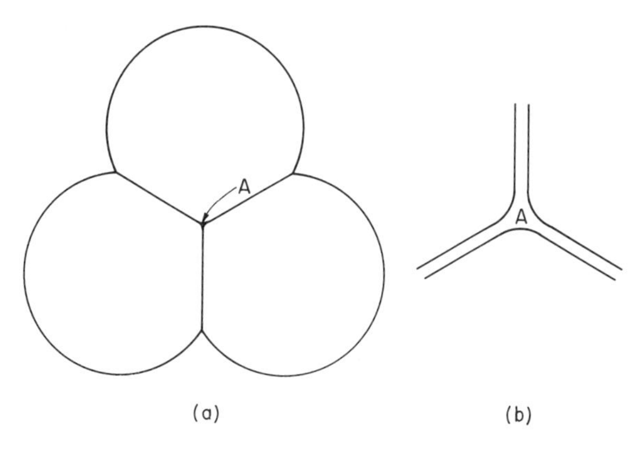

Fig. XIV-12. Plateau's border.

thought of as a space-filling packing of more or less distorted polyhedra; such foams have been called *polyederschaum*. They may result from a sufficient drainage of a kugelschaum or be formed directly, in the case of a liquid of low viscosity. Again, there is a parallel with emulsions—centrifuged emulsions can consist of polyhedral cells of inner liquid separated by thin films of outer liquid. Sebba (87a) reports on foams that range between both extremes, which he calls *microgas emulsions,* that is, clusters of encapsulated gas bubbles.

The discussion here is confined to the more common type of foam, the polyederschaum, as being more relevant to surface chemistry. First, there are some interesting geometric aspects. If three bubbles are joined the appearance will be as in Fig. XIV-12; the three separating film or *septums* meet to give a small triangular column of liquid (perpendicular to the paper in the figure) known as *Plateau's border*. This equilibrium between three fluid laminas was studied in some detail by Plateau† (88) and by Gibbs (89), and the channel known as Plateau's border plays an important role in the mechanism of film drainage. An enlarged drawing of it is shown in Fig. XIV-12*b*, and it is seen, first of all, that the high curvature of the boundaries of the area *A* must mean that a considerable pressure drop occurs between the gas and liquid phases. The resulting tendency for liquid to be sucked from the film into the border plays an important role in foam drainage, as will be seen.

With three bubbles, the septums must meet at 120° if the system is to be mechanically stable. A fourth bubble could now be added as shown in Fig. XIV-13, but this would not be stable. The slightest imbalance or disturbance would suffice to move the septums around until an arrangement such as in Fig. XIV-13*b* resulted. Thus a two-dimensional foam consists of a more or less uniform hexagonal type of network.

† This ancient treatise makes fascinating reading!

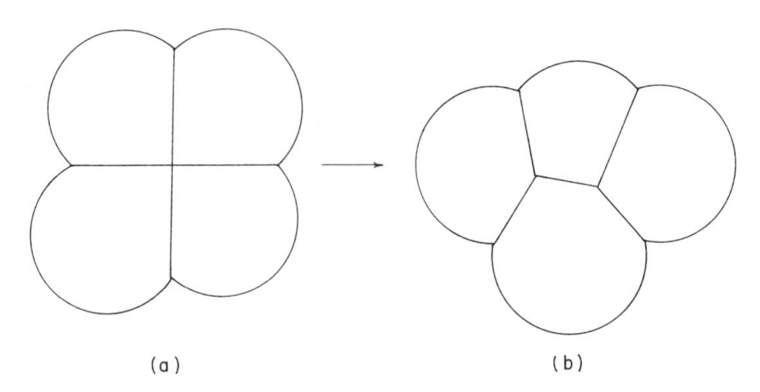

Fig. XIV-13

The situation becomes more complex in the case of a three-dimensional foam. Since the septums should all be identical, again three should meet at 120° angles to form borders or lines, and four lines should meet at a point, at the tetrahedral angle of 109°28′. This was observed to be the case by Matzke (90) in his extensive statistical study of the geometric features of actual foams.

Intuitively, one would like the cells to consist of some single type of regular polyhedron, and one that closely meets the foregoing requirements is the pentagonal dodecahedron (a figure having 12 pentagonal sides). Matzke did find more than half of the faces in actual foams were five-sided, and about 10% of the polyhedra were pentagonal dodecahedrons. Earlier, Desch (91) reached similar conclusions.

However, the pentagonal dodecahedron gives angles that are slightly off (116°33′ between faces and 108° between lines) and, moreover, cannot fill space exactly. This problem of space filling has been discussed by Gibbs (89) and Lord Kelvin (92); the latter showed that a truncated octahedron (a figure having six square and eight hexagonal sides) would fill space exactly and, by suitable curving of the faces, give the required angles. However, experimentally, only 10% of the sides in Matzke's foams were four-sided.

The foregoing discussion leads to the question of whether actual foams do in fact satisfy the conditions of zero resultant force on each side, border, and corner without developing local variations in pressure in the liquid interiors of the laminas. Such pressure variations would affect the nature of foam drainage (see below) and might also have the consequence that films within a foam structure would, on draining, more quickly reach a point of instability than do isolated plane films.

9. Foam Drainage

A. *Drainage of Single Films*

Consider first the simple case of a plane soap film stretched across a rectangular frame, as illustrated in Fig. XIV-14. When first formed, by dipping the frame into a soap solution and raising it carefully (in a closed system to avoid evaporation), the initial film is relatively thick. Drainage begins immediately, however, and with soaps forming liquid (as opposed to solid) interfacial films, a pattern of interference fringes develops rather rapidly. This pattern, as indicated by the shadings in the figure, consists of repeated bands of rainbowlike colors resulting from the interference between light reflected from the front and the back surfaces of the film (see Section IV-3D). As a result of draining, the film thickness decreases from bottom to top and, correspondingly, the fringes are widely spaced at the top but rapidly approach each other as the thickness becomes large in comparison with the wavelength of visible light. The process of thinning is beautiful to watch; as it occurs, each colored band moves downward, and the spacings between them increase. Eventually, the top region takes on a silvery appearance; then, in turn, a black film develops, and the boundary between it and the silver one moves down until, finally, all but the foot of the film is of the black type. The theoretical treatment of these interference effects is somewhat complicated, and the reader is referred to Rayleigh (93), Reinhold and Rucker (94), and Bikerman (95).

The black films are of some special interest; according to Perrin (96), as well as from recent observations by Mysels and co-workers (97) and by Overbeek (98), more than one black film may simultaneously be present, with apparently a stepwise transition in thickness from patches of one type to those of the other. The thinnest of these black films (in the case of soap films) is about 45 Å thick, or not much more

Fig. XIV-14. Drainage of soap films. (Courtesy of Professor K. J. Mysels.)

than twice the thickness of a soap monolayer; the black appearance results from the fact that they are thin enough so that the interference between light reflected from the front and back surfaces is destructive for all visible wavelengths. Horizontal nondraining films have a thickness determined by the degree of suction at the borders, and reflecting a balance between forces, presumably mainly the electrostatic repulsion of the two double layers and the long-range van der Waals attraction (see Section VI-4B). In fact, their study provides one means of investigating such forces (98, 99). Closely related to this subject are studies on the *contact angle* between a plane soap film and the Plateau border (see Ref. 100). A theory of the rupture of black films has been presented by Derjaguin and Prokhorov (101a).

The preceding discussion may have suggested that liquid soap films present a rather quiescent appearance as they thin. Quite the contrary is the case. On looking at such films, one sees a multicolored activity, particularly along the border between the film and its supporting frame. The appearance is one of rapid, fluid motion of a very complex character, but consisting, roughly, of swirls of thinner (and hence differently colored) film rising from the edge and moving inward as well as upward. Examination of these *peacock tails* shows that sizable patches of thinner film form, rise to a level such that their thickness corresponds to the general film thickness, and then fade out. In addition, patches of black film form in the silver film all along the width of the film and rise to a boundary between the silver and black films, throwing up small geysers of silver film as they reach the boundary. A listing of the colors for various film thicknesses is given by Princen and Mason (101).

This brief description of film drainage is given to emphasize the fact that the general thinning of soap films occurs by a mechanism that involves the formation of patches of thinner film at the border, the excess liquid presumably being discharged into the border channel. Drainage is thus determined by an edge effect, and its rate is governed by what happens at the film boundaries; as a consequence, the general rate of thinning varies inversely with the width of the film (102). As discussed earlier, there must be a high curvature at the film edges so that the liquid in the border channel must be at a lower pressure than in the body of the film; it is presumably by virtue of this pressure difference that the thin patches are formed (see Ref. 98).

It might be thought that films should drain by means of a gradual descent of liquid through their interiors, but closer consideration shows that this mechanism actually should not be important except for quite thick films. Such drainage would correspond to a viscous flow of liquid between parallel plates, and Gibbs (89) derived for this case the equation

$$v_{\mathrm{av}} = \frac{\rho g \, \delta^2}{8\eta} \qquad \text{(XIV-8)}$$

whereby the average velocity is given as a function of film thickness δ, liquid viscosity η, and density ρ. For an aqueous film 1 μ thick, Eq. XIV-8 gives a drainage velocity of only 10^{-4} cm/sec. Furthermore, one might expect that the viscosity in the interior of a film should be higher than for bulk liquid, although analysis of some experimental results has suggested that no more than a 10 Å thick layer of altered viscosity can be present (99). This conclusion is supported by some recent esr studies (103), yet is surprising in view of the discussion of Section VI-8.

B. Drainage of Foams

The very thick or kugelschaum type of foam probably does drain by a hydrodynamic mechanism, but the thin film polyederschaum foam has septums similar to the individual soap films discussed and probably drains by the same border mechanism. As a result of such drainage, the film laminas become increasingly thin, and rupture begins to occur here and there. In some cases the uppermost films rupture first, so that the volume of the foam decreases steadily with time, whereas in other cases it is mostly interior laminas that rupture, so that the gas cells become increasingly large and the

Fig. XIV-15. Spontaneous bursting of a vertical plane soap film supported by a 2-cm-wide glass frame. (From Ref. 102.)

foam decreasingly dense. In addition, cell breakage as well as intercell gas diffusion changes the distribution of cell sizes and shapes with time.

It was remarked in the second edition of this book that the mechanism of the rupture of a soap film was not known. Much new information has been added through ingenious experiments. For example, in the case of a soap film spanning a frame such as in Fig. XIV-14, it is now known that rupture originates at the margin, as shown in Fig. XIV-15 (104).

10. The Stability of Foams

A notable characteristic of stable films is their resistance to mechanical disturbance. Gibbs (89) considered the important property to be the elasticity of the film **E**,

$$\mathbf{E} = \frac{2d\gamma}{d \ln \mathcal{A}} \tag{XIV-9}$$

where $\mathcal{A}$ is the area of the film (see Section IV-3C). For the case of a two-component system, Eq. XIV-9 can be put in the form

$$\mathbf{E} = 4(\Gamma_2^1)^2 \frac{d\mu_2}{dm_2} \tag{XIV-10}$$

Here Γ_2^1 is the surface excess of component 2, μ_2 is the chemical potential of that component, and m_2 is its amount per unit area of film. Qualitatively, **E** gives a measure of the ability of a film to adjust its surface tension in an instant of stress. If the surface should be extended, the surface concentration of the surfactant drops, and the local surface tension rises accordingly; the film is thus protected against rupture. For pure liquids, **E** as given would be zero, and this is in accord with the observation that pure liquids do not give a stable foam. Conversely, Eq. XIV-10 indicates that in order for **E** to be large, both Γ_2^1 and $d\mu_2/dm_2$ should be large. In effect, this means that the surfactant concentration should be high, but not too high. This conclusion was confirmed by Bartsch (105), who found that maximum foam stability occurred at fatty acid and alcohol concentrations well below those giving the minimum surface tension. Some measurements of **E** give values in the range of 10 to 40 dyne/cm (106).

The foregoing is an equilibrium or reversible concept, and some transient effects have also been suggested as important to film resilience. Rayleigh (93) noted that surface freshly formed by some insult to the film would have a greater than equilibrium surface tension value (note Fig. II-18). Ross and Haak (107) pointed out that if surfactant from the interior of the film can diffuse rapidly to the surface, a transient thin spot might be restored to its original surface tension by this mechanism before any thickening of the film occurred. The surface tension change would thus be "healed" without the film being restored to its original thickness, and the region would remain mechanically weak for a while. According to this analysis, it is important that a foam stabilizing agent be surface adsorbed only slowly (on a millisecond time scale). Prins and van den Tempel (108) have made a detailed study of the response of a film to external disturbances.

In addition to high elasticity and resilience as properties giving stability to foams, high surface viscosity appears also to be important. Brown et al. (109) found some correlation between the stability of foams containing lauryl alcohol, plus various other surfactants, and film viscosity, and so did Davies (110), some of whose results are shown in Fig. XIV-16. It may be that all that is involved is a decrease in the rate of the border film drainage process with increase in film viscosity. In studies with single soap films, it is quite apparent that border drainage is very much slower in those systems for which the adsorbed film of surfactant is solid in type.

In conclusion, there does not seem to be any very rigorous analysis possible of the interrelation of factors determining foam stability. Qualitatively, foam lifetimes depend on the drainage rate, and then on the stability of the thinned films to evaporation and to mechanical shock, including the

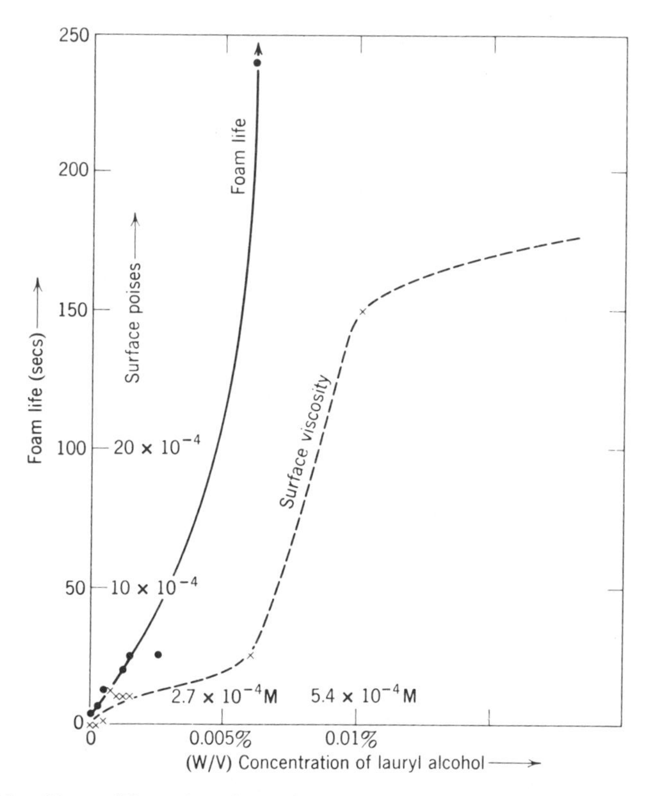

Fig. XIV-16. Foam life and surface viscosity in 0.1% sodium laurate at pH 10 with added lauryl alcohol. The foam life rises to 6.5 min with 0.01% lauryl alcohol and to 30 min with 0.025%. (From Ref. 110.)

shocks transmitted through the mass of the foam as a septum breaks and sudden shifts in adjacent cell walls occur. The general subject has been discussed by Burcik (111) and by Camp and Durham (112).

The action of antifoam agents in preventing foaming can be analyzed qualitatively along lines similar to the preceding material. First, of course, the agent should be able to displace the foam producing surfactant from the interface. The agent should then insert its own poor foam stabilizing characteristics into the system. The subject has been discussed by Ross and coworkers (113).

11. Foaming Agents and Foams of Practical Importance

It has been made clear that foam-stabilizing agents will in general be surfactants but, more specifically, the better foaming agents seem to involve the colloidal electrolyte surfactant discussed in connection with detergency (Section XIII-5B) and biological substances such as proteins, which give films not in rapid equilibrium with the substrate. In addition, as was true for emulsions, solid particles can serve as foam stabilizing agents.

Important types of foam include those used in fire fighting (saponin or hydrolyzed proteins may be used as surfactants, to give rather rigid foams), in froth flotation (see Section XIII-4), and in many household products. Common examples of these last include whipped cream, beaten egg whites, shaving lather, and various cleansing foams.

The prevention of foaming is sometimes quite important, of course. Antifoam agents for water boilers include polyamides (e.g., the triamide of stearic acid with diethylenetriamine). General antifoam agents such as alkyl polysiloxanes and tributylphosphate are in wide use. A number of details are given in the monograph by Bikerman (114).

12. Problems

1. Discuss briefly at least two reasons why two pure immiscible liquids do not form a stable emulsion.

2. Show that the maximum possible value for ϕ is 0.74 in the case of an emulsion consisting of uniform, rigid spheres.

3. Show what the theoretical value of a should be in Eq. XIV-2.

4. Show what the maximum possible value of ϕ is for the case of a two-dimensional emulsion consisting of uniform, rigid circles (or, alternatively, of a stacking of right circular cylinders).

5. Emulsion A has a droplet size distribution that obeys the ordinary Gaussian error curve. The most probable droplet size is 5 μ. Make a plot of $p/p(\text{max})$, where $p(\text{max})$ is the maximum probability, versus size if the width at $p/p(\text{max}) = \frac{1}{2}$ corresponds to a standard deviation of 1.02 μm. Emulsion B has a droplet size distribution that obeys the log normal distribution, Eq. XIV-4, with the same most probable droplet size as for emulsion A and with a logarithmic standard deviation of 1.02. Plot the distribution for emulsion B on the same plot as that for emulsion A.

6. According to the oriented wedge theory of emulsion stabilization, sodium oleate stabilizes O/W emulsions by virtue of being wedge shaped. If, in a particular emulsion, the droplets are 1 μ in diameter, according to this theory what should be the area per hydrocarbon chain if that of the polar group is 45 $Å^2$?

7. Referring to Problem 6, the duplex film theory is an alternative for explaining emulsion stabilization. Consider an oil–water–surfactant system that could exist in O/W or in W/O form, with 1 μ diameter droplets in either case. Calculate the difference between the film–water and the film–oil interfacial tensions if the free energy for the preceding hypothetical inversion process is to be 200 cal/cm^3 of emulsion. Assume ϕ to be 0.5.

8. Consider the case of an emulsion of 1 liter of oil in 1 liter of water having oil droplets of 0.5 μm diameter. If the oil–water interface contains a close-packed monolayer of surfactant of 20 $Å^2$ per molecule, calculate how many moles of surfactant are present.

9. A mixture of 70% Tween 60 and 30% Span 65 gives optimum behavior in a given emulsion system. What composition mixture of sodium lauryl sulfate and cetyl alcohol should also give optimum behavior in the same system?

10. A surfactant mixture having an HLB number of 8 should give a good W/O emulsion in which the oil phase is lanolin. Suggest two possible surfactant mixtures that the perspiring cosmetic chemist might use; he has been told that his formulations must contain 10% cetyl alcohol.

10. Consider the case of two soap bubbles having a common septum. The bubbles have radii of curvature R_1 and R_2, and the radius of curvature of the common septum is R. Show under what conditions R would be zero and under what conditions it would be equal to R_2.

11. Derive Eq. XIV-10 from Eq. XIV-9. State the approximations involved. Explain whether the surface elasticity should be small or large for a surfactant film if the bulk surfactant concentration is about its cmc.

12. Make an estimate of the hydrostatic pressure that might be present in the Plateau border formed by the meeting of three thin black films. Make the assumptions of your calculation clear.

General References

P. Becher, Ed., *Encyclopedia of Emulsion Technology,* Marcel Dekker, New York, 1981.

H. Bennett, J. L. Bishop, Jr., and M. F. Wulfinghoff, *Emulsions,* Chemical Publishing Co., New York, 1968.

S. Berkman and G. Egloff, *Emulsions and Foams,* Reinhold, New York, 1961.

J. J. Bikerman, *Foams,* Springer-Verlag, New York, 1973.

J. T. Davies and E. K. Rideal, *Interfacial Phenomena,* Academic, New York, 1961.

E. Manegold, *Schaum, Strassenbau, Chemie und Technik,* Heidelberg, 1953.

J. L. Moilliet, B. Collie, and W. Black, *Surface Activity,* E. & F. N. Spon, London, 1961.

L. I. Osipow, *Surface Chemistry,* Reinhold, New York, 1962.

L. H. Princen, *Treatise on Coatings,* R. R. Myers and J. S. Long, Eds., Vol. 1, Part 3, Chapter 2, Marcel Dekker, New York, 1972.

P. Sherman, *Emulsion Science,* Academic, New York, 1968.

K. Shinoda, *Principles of Solution and Solubility,* Marcel Dekker, New York, 1978.

Textual References

1. A. Einstein, *Ann. Phys.* **19,** 289 (1906); **34,** 591 (1911).
2. P. Becher, *Emulsions,* Reinhold, New York, 1965.
3. P. Sherman, *Emulsion Science,* Academic, New York, 1968, p. 131.
4. G. I. Taylor, *Proc. Roy. Soc. (London),* **A138,** 41 (1932).
5. P. Sherman, *J. Phys. Chem.,* **67,** 2531 (1963).
6. P. D. Cratin and B. K. Robertson, *J. Phys. Chem.,* **69,** 1087 (1965).
7. A. F. Stevenson, W. Heller, and M. L. Wallach, *J. Chem. Phys.,* **34,** 1789, 1796 (1961).
8. See O. Kratky, *Kolloid-Z.,* **182,** 7 (1962).
9. K. J. Lissant and K. G. Mayhan, *J. Colloid Interface Sci.,* **42,** 201 (1973).
10. L. H. Princen, *Appl. Polym. Symp.,* No. 10, 159 (1969).
11. W. D. Harkins, *The Physical Chemistry of Surface Active Films,* Reinhold, New York, 1952, p. 90.
12. K. J. Lissant, *J. Soc. Cosmet. Chem.,* **21,** 141 (1970).
13. H. M. Princen, M. P. Aronson, and J. C. Moser, *J. Colloid Interface Sci.,* **75,** 246 (1980).
14. H. M. Princen, *J. Colloid Interface Sci.,* **71,** 55 (1979).

14a. F. Sebba, *J. Theor. Biol.,* **78,** 375 (1979).

15. W. D. Bancroft, *J. Phys. Chem.,* **17,** 501 (1913).
16. G. H. A. Clowes, *J. Phys. Chem.,* **20,** 407 (1916).
17. See W. D. Harkins and N. Beeman, *J. Am. Chem. Soc.,* **51,** 1674 (1929).
18. V. E. Wellman and H. V. Tartar, *J. Phys. Chem.,* **34,** 379 (1930).
19. A. E. Alexander, "Surface Chemistry and Colloids," in H. Mark and E. J. W. Verwey, Eds., *Advances in Colloid Sciences,* Vol. 3, Interscience, New York, 1950.
20. S. Berkman and G. Egloff, *Emulsions and Foams,* Reinhold, New York, 1941, pp. 93f, 175.
21. K. J. Packer and C. Rees, *J. Colloid Interface Sci.,* **40,** 206 (1972).
22. E. G. Cockbain and T. S. McRoberts, *J. Colloid Sci.,* **8,** 440 (1953).
23. R. D. Vold and R. C. Groot, *J. Soc. Cosmet. Chem.,* **14,** 233 (1963).
24. S. J. Rehfeld, *J. Phys. Chem.,* **66,** 1966 (1962).
25. S. R. Reddy and H. S. Fogler, *J. Colloid Interface Sci.,* **79,** 105 (1981).
26. R. D. Vold and R. C. Groot, *J. Colloid Sci.,* **19,** 384 (1964).
27. B. D. Powell and A. E. Alexander, *Can. J. Chem.,* **30,** 1044 (1952).
28. M. van den Tempel, *Rec. Trav. Chim.,* **72,** 419 (1953).
29. J. A. Kitchener and P. R. Musselwhite, *Emulsion Science,* P. Sherman, Ed., Academic, New York, 1968, p. 104.
30. F. Huisman and K. J. Mysels, *J. Phys. Chem.,* **73,** 489 (1969).
31. L. H. Princen, *Treatise on Coatings,* R. R. Myers and J. S. Long, Eds., Vol. 1, Part 3, Chapter 2, Marcel Dekker, New York, 1972.
32. W. Albers and J. Th. G. Overbeek, *J. Colloid Sci.,* **14,** 501, 510 (1959); *ibid.,* **15,** 489 (1960).
33. W. Rigole and P. Van der Wee, *J. Colloid Sci.,* **20,** 145 (1965).
34. T. Gillespie and R. M. Wiley, *J. Phys. Chem.,* **66,** 1077 (1962).
35. A. J. Scarlett, W. L. Morgan, and J. H. Hildebrand, *J. Phys. Chem.,* **31,** 1566 (1927).

36. E. H. Lucassen-Reynders and M. van den Tempel, *J. Phys. Chem.*, **67**, 731 (1963).
37. J. H. Schulman and J. Leja, *Trans. Faraday Soc.*, **50**, 598 (1954).
38. M. von Smoluchowski, *Phys. Z.*, **17**, 557, 585 (1916); *Z. Phys. Chem.*, **92**, 129 (1917).
39. H. R. Kruyt, *Colloid Science*, Elsevier, Amsterdam, 1952, p. 378.
40. J. T. Davies, *Proc. 2nd Int. Congr. Surf. Act., London*, Vol. 1, p. 426.
41. L. A. Spielman, *J. Colloid Interface Sci.*, **33**, 562 (1970).
42. E. P. Honig, P. H. Wiersma, and G. J. Roeberson, *J. Colloid Interface Sci.*, **36**, 97 (1971).
43. D. F. Bernstein, W. I. Higuchi, and N. F. H. Ho, *J. Colloid Interface Sci.*, **39**, 439 (1972).
44. H. M. Princen, *J. Colloid Sci.*, **18**, 178 (1963).
45. T. Gillespie and E. K. Rideal, *Trans. Faraday Soc.*, **52**, 173 (1956).
46. S. P. Frankel and K. J. Mysels, *J. Phys. Chem.*, **66**, 190 (1962).
47. S. Hartland and D. K. Vohra, *J. Colloid Interface Sci.*, **77**, 295 (1980).
48. R. Salisbury, E. E. Leuallen, and L. T. Chavkin, *J. Am. Pharm. Assoc.* (*Sci. Ed.*), **43**, 117 (1954).
49. T. B. Robertson, *Kolloid-Z.*, **7**, 7 (1910).
50. S. U. Pickering, *J. Chem. Soc.*, **1907**, 2002; see also E. L. Smith, *J. Phys. Chem.*, **36**, 1401 (1932).
51. F. Sebba, *J. Colloid Interface Sci.*, **40**, 468 (1972).
52. L. Ya. Kremnev and N. I. Kuibina, *Colloid J.* (*USSR*), **17**, 31 (1955).
53. T. Green, R. P. Harker, and F. O. Howitt, *Nature*, **174**, 659 (1954).
54. K. Shinoda and H. Saito, *J. Colloid Interface Sci.*, **26**, 70 (1968).
55. J. E. Bowcott and J. H. Schulman, *Z. Elektrochem.*, **59**, 283 (1955).
56. J. H. Schulman, W. Stoeckenius, and L. M. Prince, *J. Phys. Chem.*, **63**, 1677 (1959).
57. P. A. Winsor, *Trans. Faraday Soc.*, **44**, 376 (1948).
58. L. M. Prince, *J. Colloid Interface Sci.*, **52**, 182 (1975).
59. W. B. Gogarty and H. Surkalo, *J. Pet. Tech.*, **24**, 1161 (1972) and references therein.
60. V. K. Bansal and D. O. Shah, *Micellization, Solubilization, and Microemulsions*, Vol. 1, K. L. Mittal, Ed., Plenum, 1977.
61. A. W. Adamson, *J. Colloid Interface Sci.*, **29**, 261 (1969).
62. W. C. Tosch, S. C. Jones, and A. W. Adamson, *J. Colloid Interface Sci.*, **31**, 297 (1969).
63. J. Biais, P. Bothorel, B. Clin, and P. Lalanne, *J. Colloid Interface Sci.*, **80**, 136 (1981).
64. E. Ruckenstein and R. Krishnan, *J. Colloid Interface Sci.*, **76**, 188 (1980).
65. S. I. Chou and D. O. Shah, *J. Colloid Interface Sci.*, **80**, 49 (1981).
66. J. W. Falco, R. D. Walker Jr., and D. O. Shah, *A.I.Ch.E.J.*, **20**, 510 (1974).
67. D. O. Shah, *Ann. N.Y. Acad. Sci.*, **204**, 125 (1973).
68. C. Kunar and D. Balasubramanian, *J. Phys. Chem.*, **84**, 1895 (1980).
69. R. A. Mackay and R. Agarwal, *J. Colloid Interface Sci.*, **65**, 225 (1978).
70. M. Laguës and C. Sauterey, *J. Phys. Chem.*, **84**, 3503 (1980).
71. C. Hermansky and R. A. Mackay, *J. Colloid Interface Sci.*, **73**, 324 (1980).
72. R. Ober and C. Taupin, *J. Phys. Chem.*, **84**, 2418 (1980).

73. M. Dvolaitzky, M. Laguës, J. P. Le Pesant, R. Ober, C. Sauterey, and C. Taupin, *J. Phys. Chem.*, **84,** 1532 (1980).
74. S. I. Chou and D. O. Shah, *J. Phys. Chem.*, **85,** 1480 (1981).
75. S. I. Chou and D. O. Shah, *J. Colloid Interface Sci.*, **78,** 249 (1980).
76. M. Almgren, F. Grieser, and J. K. Thomas, *J. Am. Chem. Soc.*, **102,** 3188 (1980).
77. J. H. Fendler, *J. Phys. Chem.*, **84,** 1485 (1980).
78. R. A. Mackay, K. Letts, and C. Jones, *Micellization, Solubilization, and Microemulsions,* K. L. Mittal, Ed., Vol. 2, Plenum, 1977.
79. C. A. Jones, L. E. Weaner, and R. A. Mackay, *J. Phys. Chem.*, **84,** 1495 (1980).
80. L. M. Prince, *J. Colloid Interface Sci.*, **52,** 182 (1975).
81. H. L. Rosano and R. C. Peiser, *Rev. Fr. Corps Gras,* April 1969, p. 249.
81a. H. L. Rosano, T. Lan, A. Weiss, W. E. F. Gerbacia, and J. H. Whittam, *J. Colloid Interface Sci.*, **72,** 233 (1979).
82. W. D. Bancroft, *J. Phys. Chem.*, **17,** 501 (1913); *ibid.*, **19,** 275 (1915).
83. W. C. Griffin, *J. Soc. Cosmet. Chem.*, **1,** 311 (1949); *ibid.*, **5,** 249 (1954).
84. C. Parkinson and P. Sherman, *J. Colloid Interface Sci.*, **41,** 328 (1972).
85. K. Shinoda and H. Kunieda, *Encyclopedia of Emulsion Technology,* P. Becher, Ed., Marcell Dekker, New York, 1981.
85a. K. Shinoda and H. Sagitani, *J. Colloid Interface Sci.*, **64,** 68 (1978).
86. J. Boyd, C. Parkinson, and P. Sherman, *J. Colloid Interface Sci.*, **41,** 359 (1972).
87. E. Manegold, *Schaum, Strassenbau, Chemie und Technik,* Heidelberg, 1953, p. 83.
87a. F. Sebba, *J. Colloid Interface Sci.*, **35,** 643 (1971); ACS Symposium Series No. 9, 1975.
88. J. Plateau, *Mem. Acad. Roy. Soc. Belg.*, **33,** (1861), sixth series and preceding papers.
89. J. W. Gibbs, *Collected Works,* Vol. 1, Longmans, Green, New York, 1931, pp. 287, 301, 307.
90. E. B. Matzke, *Am. J. Bot.*, **33,** 58 (1946).
91. C. H. Desch, *Rec. Trav. Chim.*, **42,** 822 (1923).
92. See A. F. Wells, *The Third Dimension in Chemistry,* The Clarendon Press, Oxford, 1956, p. 57; and D. W. Thompson, *Growth and Crystal Form,* Cambridge University Press, Cambridge, England, 1943, p. 551.
93. J. W. S. Rayleigh, *Scientific Papers,* Vol. 2, Cambridge University Press, Cambridge, England, p. 498.
94. A. W. Reinhold and A. W. Rucker, *Trans. Roy. Soc. (London)*, **172,** 447 (1881).
95. J. J. Bikerman, *Foams,* Reinhold, New York, 1953, p. 137.
96. J. Perrin, *Ann. Phys.*, **10,** No. 9, 160 (1918).
97. M. N. Jones, K. J. Mysels, and P. C. Scholten, *Trans. Faraday. Soc.*, **62,** 1336 (1966).
98. J. Th. G. Overbeek, *J. Phys. Chem.*, **64,** 1178 (1960).
99. K. J. Mysels, *J. Phys. Chem.*, **68,** 3441 (1964).
100. K. J. Mysels, *Electroanal. Chem. Interfacial Electrochem.*, **37,** 23 (1972).
101. H. M. Princen and S. G. Mason, *J. Colloid Sci.*, **20,** 453 (1965).
101a. B. V. Derjaguin and A. V. Prokhorov, *J. Colloid Interface Sci.*, **81,** 108 (1981).

102. See K. J. Mysels, K. Shinoda, and S. Frankel, *Soap Films, Studies of Their Thinning and a Bibliography,* Pergamon, New York, 1959.
103. M. J. Povich and J. A. Mann, *J. Phys. Chem.,* **77,** 3020 (1973).
104. K. J. Mysels, *J. Colloid Interface Sci.,* **35,** 159 (1971).
105. O. Bartsch, *Kolloid Chem. Bei.,* **20,** 1 (1925).
106. A. Prins and M. van den Tempel, *J. Phys. Chem.,* **73,** 2828 (1969).
107. S. Ross and R. M. Haak, *J. Phys. Chem.,* **62,** 1260 (1958).
108. A. Prins and M. van den Tempel, *Special Discuss. Faraday. Soc.,* No. 1, 1970, p. 20.
109. A. G. Brown, W. C. Thuman, and J. W. McBain, *J. Colloid Sci.,* **8,** 491 (1953).
110. J. T. Davies, *Proc. 2nd Int. Congr. Surf. Act.,* Vol. 1, p. 220.
111. E. J. Burcik, *J. Colloid Sci.,* **8,** 520 (1953).
112. M. Camp and K. Durham, *J. Phys. Chem.,* **59,** 993 (1955).
113. S. Ross, A. F. Hughes, M. L. Kennedy, and A. R. Mardoian, *J. Phys. Chem.,* **57,** 684 (1953).
114. J. J. Bikerman, *Foams,* Springer-Verlag, New York, 1973.

CHAPTER XV

The Solid–Gas Interface—General Considerations

1. Introduction

These concluding chapters deal with various aspects of a very important type of situation, namely, that in which some adsorbate species is distributed between a solid phase and a gaseous one. From the phenomenological point of view, one observes, on mechanically separating the solid and gas phases, that there is a certain distribution of the adsorbate between them. This may be expressed, for example, as n, the moles adsorbed per gram of solid versus the pressure P. The distribution, in general, is temperature dependent, so the complete empirical description would be in terms of an adsorption function $n = f(P, T)$.

While a *thermodynamic* treatment can be developed entirely in terms of $f(P, T)$, to apply adsorption *models*, it is highly desirable to know n on a per square centimeter basis rather than a per gram basis or, alternatively, to know θ, the fraction of surface covered. In both the physical chemistry and the applied chemistry of the solid–gas interface, the specific surface area Σ is thus of extreme importance.

There has been a great interest in recent years in going beyond macroscopic descriptions of adsorption to obtain detailed information about the structural and chemical state of the solid surface and of the solid–adsorbate complex. All gases below their critical temperature tend to adsorb as a result of general van der Waals interactions with the solid surface. In this case of *physical adsorption,* as it is called, interest centers on the size and nature of adsorbent–adsorbate interactions and on those between adsorbate molecules. There is concern about the degree of heterogeneity of the surface and with the extent to which adsorbed molecules possess translational and internal degrees of freedom.

It has become increasingly appreciated in recent years that the surface structure of the adsorbent may be altered in the adsorption process. Qualitatively, such structural perturbation is apt to occur if the adsorption energy is comparable to the surface energy of the adsorbent (on a per molecular unit basis). As summarized in Table XV-1, physical adsorption (sometimes called *physisorption*) will likely alter the surface structure of a molecular solid adsorbent (such as ice, paraffin, and polymers), but not that of high surface energy, refractory solids (such as the usual metals and metal oxides,

TABLE XV-1
Types of Adsorption Systems

Type of adsorbate–adsorbent interaction	Type of adsorbent	
	Molecular	Refractory
van der Waals (physical adsorption)	Surface restructuring on adsorption	No change in surface structure on adsorption
Chemical (chemisorption)	Chemical reaction modifying the adsorbent	Surface restructuring occurs on adsorption

and carbon black). Chemisorption may alter the surface structure of refractory solids.

The function of this chapter is to summarize some of the general approaches to the determination of the physical and chemical state in both of the types of adsorption systems described.

2. The Surface Area of Solids

The specific surface area of a solid is one of the first things that must be determined if any detailed physical chemical interpretation of its behavior as an adsorbent is to be possible. Such a determination can be made through adsorption studies themselves, and this aspect is taken up in the next chapter; there are a number of other methods, however, that are summarized in the following material. Space does not permit a full discussion, and, in particular, the methods that really amount to a *particle* or *pore size* determination, such as optical and electron microscopy, x-ray diffraction, and permeability studies, are largely omitted.

A. The Meaning of Surface Area

It will be seen that each method for surface area determination involves the measurement of some property that is observed qualitatively to depend on the extent of surface development and that can be related by means of theory to the actual surface area. It is important to realize that the results obtained by different methods differ, and that one should in general *expect* them to differ. The problem is that the concept of surface area turns out to be a rather elusive one as soon as it is examined in detail.

The difficulty in stating just what is meant by the "surface area" of a solid, in one aspect, can be illustrated by considering the somewhat analogous question of what is meant by the "length" of a particular section of coastline. Superficially, it seems easy to say that the length is simply the distance along the shore, between two points. Geography books, however, are prone to make such statements as "because of its many indentations, the coast of Maine is actually some 2000 miles long." It seems reasonable that the coast of Maine should be considered as being greater than that of a straight line drawn from border to border, and that the added length due to harbors, bays, points, and so on should be included. However, once one begins to consider irregularities, the problem is when to stop. Figure XV-1 shows how successive magnifications of a hypothetical stretch of coastline might look. Quite obviously there are indentations on indentations, and with each successive magnification one increases the computed coastline by a more or less constant factor.

There seems to be no logical stopping point to the sequence of successive enlargements if one is after some "absolute" value for the length of the shoreline. One thus arrives at the point of considering irregularities due to individual stones on the beach, due to the individual grains of sand, and due to the surface roughness of each grain, and, finally, of considering the surface area of individual atoms and molecules. In our particular example, the tides and waves would prevent the exact carrying out of the sequence, but, apart from this, the Heisenberg principle comes in eventually to inform us that there is considerable uncertainty in the location of electrons and that the "surface area" of an atom is a somewhat philosophical concept.

Quite obviously, the geographer solves this logical difficulty by arbitrarily deciding that he will not consider irregularities smaller than some specified size. A similar and equally arbitrary choice is frequently made in the estimation of surface areas. The minimum size of irregularities to be considered may be defined in terms of the resolving power of a microscope or the size of adsorbate molecules.

Another approach is the iterative one. In the case of the Maine coastline, one might count the number of coastal villages or landing jetties and take this number as a measure of the length of a stretch of coast. In the case of a solid surface, one might apply a method that attempts to count the number of atoms or molecules in the surface. The result obtained is undoubtedly

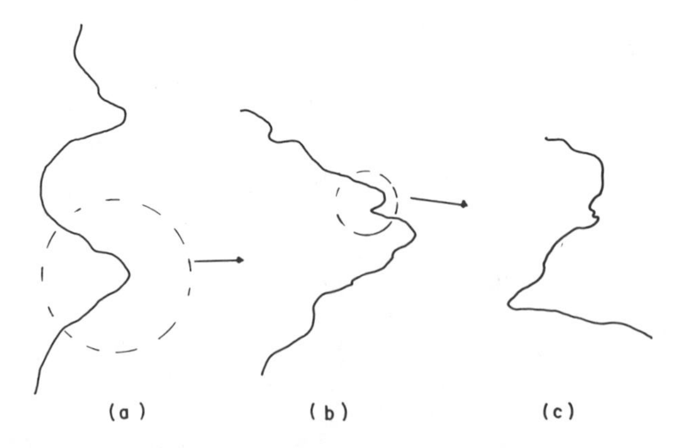

Fig. XV-1. Successive enlargements of hypothetical coast line.

useful, but, again, it does not have the quality of being a unique measure of absolute surface area. Perhaps the closest approach to such a measure of surface area is the geometric area of a plane liquid surface.

Finally, in the case of solids, there is the difficulty that surface atoms and molecules differ in their properties from one location to another. The discussion in Section VII-4 made clear the variety of surface heterogeneity possible in the case of a solid. Those measurements that depend on the state of surface atoms or molecules will generally be influenced differently by such heterogeneities. Thus since the energy and the entropy of a molecule in the surface region are not necessarily proportional to each other, the presence of unstable configurations in the surface may affect the surface free energy and the total surface energy differently. In cases where adsorption depends on surface energy quantities or on interatomic spacings, the amounts of material adsorbed on a series of hypothetical samples of solid of the same total area but differing in surface energy or exposing different crystal planes may be quite different. In such a situation, even the *relative* amounts adsorbed on the different samples would not be a unique measure of the *relative* areas.

The preceding discussion serves to emphasize the point that, although the various topics to be discussed in this and succeeding chapters relate to the common property, surface area, the actual meaning of the answer arrived at is generally characteristic of the method and associated theory. Fortunately, one is very commonly more interested in relative surface areas than in accurate "absolute" values. Any given method will usually be valuable in determining ratios of areas between various materials (subject to the reservation made in the preceding paragraph). The comparisons, moreover, will be particularly valid when they are to be applied to the prediction or correlation of properties closely related to the one used in the measurement. Thus surface areas estimated by adsorption studies are more likely to lead to valid predictions of how additional, similar, adsorbates will behave than to accurate predictions of, say, how the relative heats of immersion or the relative microscopic areas should vary.

B. Methods Requiring Knowledge of the Surface Free Energy or Total Energy

A number of methods have been described in earlier sections whereby the surface free energy or total energy could be estimated. Generally, it was necessary to assume that the surface area was known by some other means; conversely if some estimate of the specific thermodynamic quantity is available, the application may be reversed to give a surface area determination. This is true if the heat of solution of a powder (Section VII-5B), its heat of immersion (Section X-3A), or its solubility increase (Section X-2) is known.

Two approaches of this type, purporting to give absolute surface areas, might be mentioned. Bartell and Fu (1) proposed that the heat of immersion of a powder in a given liquid,

$$q_{\text{imm}} = \Sigma\,(E_{\text{SV}} - E_{\text{SL}}) \tag{XV-1}$$

be combined with data on the temperature dependence of the contact angle for that liquid to obtain the area Σ of the sample. Thus from Eq. X-38,

$$q_{\text{imm}} = \Sigma\left(E_{\text{L}}\cos\theta - \frac{T\gamma_{\text{L}}\,d\cos\theta}{dT}\right) \tag{XV-2}$$

where q_{imm} is on a per gram basis so that Σ is area per gram. The actual use of Eq. XV-2 to obtain Σ values is far from easy (see Ref. 2).

The general type of approach, that is, the comparison of an experimental heat of immersion with the expected value per square centimeter, has been discussed and implemented by numerous authors (3, 4). It is possible, for example, to estimate $(E_{\text{SV}} - E_{\text{SL}})$ from adsorption data or from the so-called isosteric heat of adsorption (see Section XVI-12B). In many cases where approximate relative areas only are desired, as with coals or other natural products, the heat of immersion method has much to recommend it. Pethica and co-workers (5) have stressed, however, that with microporous adsorbents nominal surface areas from heats of immersion can be quite different from those from adsorption studies.

A second method that has been suggested as giving absolute surface areas involves the following procedure. If the solid is first equilibrated with saturated vapor, the molar free energy of the adsorbed material must be equal to that of the pure liquid, and it is a reasonable (but nonthermodynamic) argument that therefore the interfacial energy E_{L}, for the adsorbed film–vapor interface will be the same as for the liquid–vapor interface. Note that the adsorbed film is thought of as being duplex in nature in that the film–vapor interface is regarded as distinct from the film–solid interface. If the solid is now immersed in pure liquid adsorbate, the film–vapor interface is destroyed, and the heat liberated should correspond to $\mathcal{A}E_{\text{L}}$, thus giving the area $\mathcal{A}$.

This assumption was invoked by Harkins and Jura (3) in applying the method, with some success, to a nonporous powder. A concern, however, is that if the adsorbed film is thick, interparticle and capillary condensation will occur, to give too low an apparent area. Recently, however, Rouquerol and co-workers (6) have shown that the heat of immersion becomes constant at relatively low P/P^0 values (P^0 being the saturation pressure) corresponding to only about 1.5 statistical monolayers. Their data for the ground quartz–water system is shown in Fig. XV-2.

C. Rate of Dissolving

A rather different method from the preceding is that based on the rate of dissolving of a soluble material. At any given temperature, one expects the initial dissolving rate to be proportional to the surface area, and an experimental verification of this expectation has been made in the case of rock salt (see Refs. 7, 8). Here, both

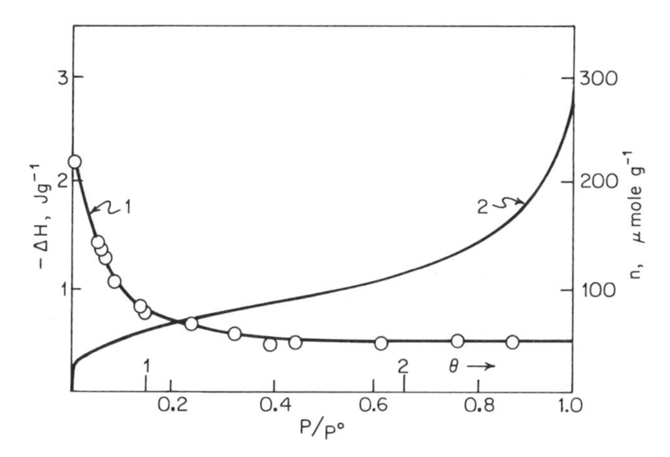

Fig. XV-2. Enthalpy of immersion versus precoverage pressure (curve 1, left ordinate, and the corresponding adsorption isotherm, curve 2, left ordinate) for the ground quartz–water system at 31°C. (From Ref. 6.)

forward and reverse rates are important, and the rate expressions are

$$R_f = -\frac{d(\text{NaCl})}{dt} = k_1 \mathcal{A} \qquad \text{(XV-3)}$$

$$R_b = \frac{d(\text{NaCl})}{dt} = k_2 \mathcal{A} a_{\text{NaCl}} \qquad \text{(XV-4)}$$

where a_{NaCl} denotes the activity of the dissolved salt. By employing a single cubic piece of rock salt, the area $\mathcal{A}$ could be related to W, the mass of salt remaining, since it turned out that the cubic shape was retained approximately even after extensive dissolving had taken place. The expression for R_f was found to hold closely when a steady stream of water flowed past the piece of rock salt, and the net rate law, that is, the sum of R_f and R_b, was obeyed when the salt dissolved in a constant volume of water. The area of a powdered sample could then have been estimated.

In such cases, the rate constants k_1 and k_2 are found to depend on the degree of stirring, although their ratio at equilibrium, of course, does not. The rate controlling step is therefore probably one of diffusion through a boundary liquid film, since the liquid layers next to the solid surface will possess very little velocity normal to the surface. In dissolving, a concentration C_0, equal to that of the saturated solution, is built up in the layer immediately adjacent to the solid, whereas at some distance τ away the concentration is equal to that of the bulk solution C. The dissolving rate is determined by the rate diffusion of solute across this concentration gradient.

Although the rate of dissolving measurements do thus give a quantity identified as the total surface area, this area must include that of a film whose thickness is of the order of a few microns but basically is rather indeterminate. Areas determined by this procedure thus will not include microscopic roughness.

The slow step in dissolving, alternatively, may be that of the chemical process of dissolving itself. This apparently was the case in a study (9) of the rate of dissolving of silica by hydrofluoric acid. The method was calibrated by determining the rate of dissolving of silica rods of geometrically determined area and was then applied to samples of powdered material. The rate was not affected by stirring rate and, moreover, the specific surface area of the silica powder so obtained, about 0.5 m^2/g, was found to agree well with that estimated from adsorption isotherms. It thus appears that where chemical attack on the surface is the rate determining step, dissolving rates may give areas reflecting surface irregularities down to a molecular scale. Clearly, also, the nature of the rate controlling step would have to be established for each particular system studied to know which type of process was rate controlling.

There may be complications. In the case of a semiconductor oxide, NiO, Smart and co-workers (10) found the dissolving rate in acid to vary with the potential at the interface, as determined by the potential determining ion, H^+. As a consequence, the derivative of log (rate) with pH was constant. Further, in the case of MgO, the surface area changed rapidly at first, due to initial attack at defect sites (11).

D. The Mercury Porosimeter

The method to be described does not determine surface area directly but rather the pore size distribution in a porous material or compacted powder. The method is widely used and, moreover, does make use of one of the fundamental equations of surface chemistry, the Laplace equation (Eq. II-7). It will be recalled that Bartell and co-workers (12, 13) determined an average pore radius from the entry pressure of a liquid into a porous plug (Section X-5A).

A procedure that is more suitable for obtaining the actual distribution of pore sizes involves the use of a nonwetting liquid such as mercury (14)—contact angle on glass about 140°. If all pores are equally accessible, only those will be filled for which

$$r > \frac{2\gamma \mid \cos \theta \mid}{\mathbf{P}} \tag{XV-5}$$

so that each increment of applied pressure causes the next smaller group of pores to be filled, with a concomitant increase in the total volume of mercury penetrated into the solid.

Studies of this type were carried out by Ritter and Drake (14, 15), using what is called a mercury porosimeter. The porous solid was placed in a long-necked sample bulb, and the entire dilatometer was then filled with mercury and placed in a protective bomb to which nitrogen pressure up to 60,000 psi could be applied. As pressure was applied, mercury penetrated increasingly into the pores of the solid, and an increasing length of resistance wire was exposed in the neck of the dilatometer; by measuring the resistance of the wire, the volume penetrated could then be calculated. Winslow and Shapiro have given a description of such equipment (16), and a more recent study is that of de Wit and Scholten (17).

The analysis of the direct data, namely, volume penetrated versus pressure, is as follows. Let dV be the volume of pores of radii between r and

$r - dr$; dV will be related to r by some distribution function $D(r)$:

$$dV = -D(r)\,dr \tag{XV-6}$$

For constant θ and γ (the contact angle was found not to be very dependent on pressure), one obtains from the Laplace equation,

$$\mathbf{P}\,dr + r\,d\mathbf{P} = 0 \tag{XV-7}$$

which, in combination with Eq. XV-6, gives

$$dV = D(r)\frac{r}{\mathbf{P}}\,d\mathbf{P} = D(r)\frac{2\gamma\cos\theta}{\mathbf{P}^2}\,d\mathbf{P} \tag{XV-8}$$

or

$$D(r) = \frac{(\mathbf{P}/r)\,dV}{d\mathbf{P}} \tag{XV-9}$$

Thus $D(r)$ is given by the slope of the V versus $\mathbf{P}$ plot. The same distribution function can be calculated from an analysis of vapor adsorption data showing hysteresis due to capillary condensation (see Section XVI-16), and Joyner and co-workers (19) found that the two methods gave very similar results in the case of charcoal, as illustrated in Fig. XV-3. Good and Mikhail (18) have suggested that because of surface roughness, the correct practical contact angle should be taken to be 180°. They find some empirical support in the data of de Wit and Scholten.

The foregoing analysis regards the porous solid as equivalent to a bundle

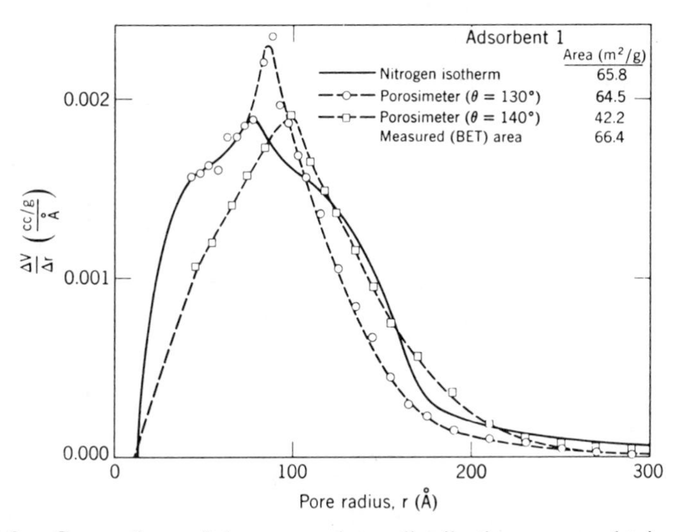

Fig. XV-3. Comparison of the pore volume distribution curves obtained from porosimeter data assuming contact angles of 140° and 130° with the distribution curve obtained by the isotherm method for a charcoal. (From Ref. 19.)

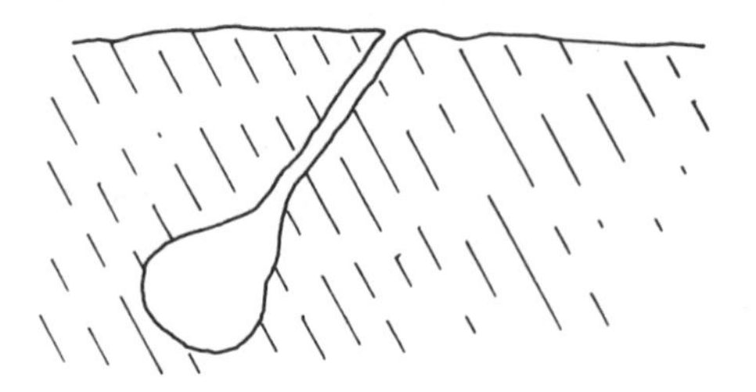

Fig. XV-4. An "ink bottle" pore.

of capillaries of various size, but apparently not very much error is introduced by the fact that such solids in reality consist of interconnected channels, provided that all pores are equally accessible (see Ref. 20); by this it is meant that access to a given sized pore must always be possible through pores that are as large or larger. An extreme example of the *reverse* situation occurs in an "ink bottle" pore, as illustrated in Fig. XV-4. These are pores that are wider in the interior than at the exit, so that mercury cannot enter until the pressure has risen to the value corresponding to the radius of the entrance capillary. Once this pressure is realized, however, the entire space fills, thus giving an erroneously high apparent pore volume for capillaries of that size. Such a situation should also lead to a hysteresis effect, that is, on reducing the pressure, mercury would leave the entrance capillary at the appropriate pressure, but the mercury in the "ink bottle" part would be trapped. Drake and Ritter (21) did find that the plot of V versus $\mathbf{P}$ was in general not retraced on depressurizing, and a typical plot of the hysteresis effect is shown in Fig. XV-5 for activated carbon. The total content of such

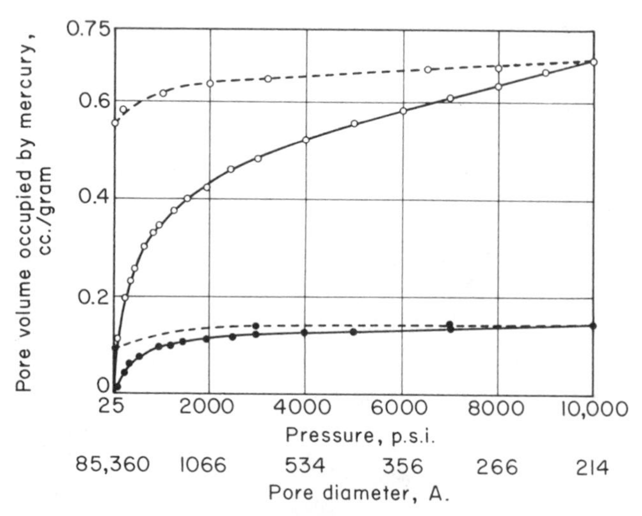

Fig. XV-5. Hysteresis in the pressurization–depressurization cycle for activated carbon and silica alumina gel: ○, activated; ●, silica alumina gel; ———, pressurizing; – –, depressurizing. (From Ref. 21.)

"ink bottles" could be estimated from the end point of the depressurization curve, and this varied from a negligible fraction of the total pore volume in the case of silica–alumina gel to over 80% in that of an activated carbon.

F. Other Methods of Surface Area Estimation

There are a number of other ways of obtaining an estimate of surface area, including such obvious ones as direct microscopic or electron microscopic examination. The rate of charging of a polarized electrode surface can give relative areas. Bowden and Rideal (22) found, by this method, that the area of a platinized platinum electrode was some 1800 times the geometrical or apparent area. Joncich and Hackerman (23) obtained areas for platinized platinum very close to those given by the BET gas adsorption method (see Section XVI-5). The diffuseness of x-ray diffraction patterns can be used to estimate the degree of crystallinity and hence particle size (24). One important general approach, useful for porous media, is that of permeability determination; although somewhat beyond the scope of this book, it deserves at least a brief mention.

A simple law, known as Darcy's law (1956), states that the volume flow rate per unit area is proportional to the pressure gradient, if applied to the case of viscous flow through a porous medium treated as a bundle of capillaries,

$$Q = \frac{r^2 A}{8\eta}\frac{\Delta \mathbf{P}}{l} = K\frac{A\,\Delta \mathbf{P}}{\eta l} \qquad \text{(XV-10)}$$

where Q is the volume flow rate, r is the radius of the capillaries, A is the total cross-sectional area of the porous medium, l is its length, $\Delta\mathbf{P}$ is the pressure drop and η is the viscosity of the fluid. K is called the permeability of the medium.

In applying Eq. XV-10 to an actual porous bed, r is taken to be proportional to the volume of void space $A l \epsilon$, where ϵ is the porosity, divided by the amount of surface; alternatively, then

$$r = \frac{k\epsilon}{A_0(1 - \epsilon)} \qquad \text{(XV-11)}$$

where A_0 is the specific surface area of the particles that make up the porous bed, that is, the area divided by the volume of the particles $Al(1 - \epsilon)$. With the further assumption that the effective or open area is ϵA, K in Eq. XV-10 becomes

$$K = \left(\frac{1}{8}\right)\left(\frac{k}{A_0}\right)^2 \frac{\epsilon^3}{(1 - \epsilon)^2} \qquad \text{(XV-12)}$$

The detailed consideration of these equations is due largely to Kozeny (25); the reader is also referred to Collins (26). However, it is apparent that, subject to assumptions concerning the topology of the porous system, the determination of K provides an estimate of A_0. Somewhat similar equations apply in the case of gas flow; the reader is referred to Barrer (27) and Kraus and co-workers (28).

Although serious discrepancies often develop between the surface areas obtained by different methods, it would be unfair to present an entirely negative picture. Some data given by Emmett (29) for zinc oxide pigments are collected in Table XV-2, and it is seen that agreement between various methods can be quite respectable.

TABLE XV-2
Particle Size Measurements of Zinc Oxide Pigments (29)

	Sample		
	F-1601	K-1602	G-1603
Area by nitrogen absorption at −195°C, m^2/g	9.48	8.80	3.88
Average particle size, μm			
By microscopic count	0.21	0.25	0.49
By adsorption of methyl stearate	0.19	0.24	0.55
By ultramicroscopic count	0.135	0.16	0.26
By permeability	0.12	0.15	0.25
By adsorption of nitrogen	0.115	0.124	0.28

3. The Structural and Chemical Nature of Solid Surfaces

We have considered briefly the important macroscopic description of a solid adsorbent, namely, its specific surface area and, if porous, its pore size distribution. In addition, it is important to know as much as possible about the microscopic structure of the surface, and contemporary surface spectroscopic and diffraction techniques, discussed in Chapter VIII, provide a great deal of such information. On a less informative and more statistical basis are site energy distributions (Sections VII-4D and XVI-14). Also discussed in Section VII-4 is the somewhat large scale type of structure due to surface imperfections and dislocations.

The composition and chemical state of the surface atoms or molecules is very important, especially in the field of contact catalysis, in which mixed surface compositions are common. This aspect is discussed in more detail in Chapter XVII. Since transition metals are widely used in catalysis, the determination of their valence state is of some interest. Indications of the probable chemical state can be obtained from magnetic susceptibility measurements (30), electron spin resonance (31), and the x-ray absorption spectrum whose fine structure reflects the chemical environment of the metal (32, 33). Transition metal ions have low lying electronic energy states due to a splitting of *d*-electron levels whose degree and nature depend on the valence and on the symmetry of the perturbing field of the nearest neighbors. As a consequence, their visible absorption spectrum can give useful chemical information (see Refs. 34, 35, and 36).

Many solids have foreign atoms or molecular groupings on their surfaces that are so tightly held that they do not really enter into adsorption–desorption equilibrium and so can be regarded as part of the surface structure. The partial surface oxidation of carbon blacks has been mentioned as having an important influence on their adsorptive behavior (Section X-3A); depending on conditions, the oxidized surface may be acidic or basic (see Ref. 37), and the surface pattern of the carbon rings may be affected (40). As one other example, the chemical nature of the acidic sites of silica–alumina catalysts has been a subject of much discussion. The main question has been whether the sites represented Brønsted (proton donor) or Lewis (electron acceptor) acids. The evidence is mixed, although infrared spectroscopy has indicated the presence of surface hydroxyl groups (see Refs. 39, 40).

4. The Nature of the Solid–Adsorbate Complex

Before entering the detailed discussion of physical and chemical adsorption in the next two chapters, it is worthwhile to consider briefly and in relatively general terms what type of information can be obtained about the chemical and structural state of the solid–adsorbate complex. The term complex is used to avoid the common practice of discussing adsorption as though it occurred on an inert surface. Three types of effects are actually involved: (1) the effect of the adsorbent on the molecular structure of the adsorbate, (2) the effect of the adsorbate on the structure of the adsorbent, and (3) the character of the direct bond or local interaction between an adsorption site and the adsorbate.

A. Effect of Adsorption on Adsorbate Properties

First, from a thermodynamic or statistical mechanical point of view, the internal energy and entropy of a molecule should be different in the adsorbed state from that in the gaseous state. This is quite apart from the energy of the adsorption bond itself or the entropy associated with confining a molecule to the interfacial region. It is clear, for example, that the adsorbed molecule may lose part or all of its freedom to rotate.

It is of interest in the present context (and is useful later) to outline the statistical mechanical basis for calculating the energy and entropy that are associated with rotation. According to the Boltzmann principle, the time average energy of a molecule is given by

$$\bar{\epsilon} = \frac{\sum N_j \epsilon_j}{\sum N_j} = \frac{\sum g_j \epsilon_j e^{-\epsilon_j/kT}}{\sum g_j e^{-\epsilon_j/kT}} \tag{XV-13}$$

where the summations are over the energy states ϵ_j, and g_j is the statistical weight of the *j*th state. It is the assumption of statistical thermodynamics that this time average energy for a molecule is the same as the instantaneous average energy for a system of such molecules. The average molar energy for the system E is taken to be $N_0\bar{\epsilon}$. Equation XV-13 may be written in the form

$$E = \frac{RT^2 \partial \ln \mathbf{Q}}{\partial T} \tag{XV-14}$$

where $\mathbf{Q}$ is the partition function and is defined as the term in the denominator of Eq. XV-13. The heat capacity is then

$$C_v = \left(\frac{\partial E}{\partial T}\right)_V = \frac{R}{T^2}\frac{\partial^2 \ln \mathbf{Q}}{\partial(1/T)^2} \tag{XV-15}$$

from which the entropy is found to be

$$S = \int_0^T C_v \, d \ln T = \frac{E}{T} + R \ln \mathbf{Q} \tag{XV-16}$$

if the entropy at 0 K is taken to be zero. The Helmholtz free energy A is just $E - TS$ so

$$A = -RT \ln \mathbf{Q} \tag{XV-17}$$

The rotational energy of a rigid molecule is given by $J(J + 1)h^2/8\pi^2 IkT$ where J is the quantum number and I is the moment of inertia, but if the energy level spacing is small compared to kT, integration can replace summation in the evaluation of $\mathbf{Q}_{\text{rot}}$, which becomes

$$\mathbf{Q}_{\text{rot}} = \frac{1}{\pi\sigma}\left(\frac{8\pi^3 IkT}{h^2}\right)^{n/2} \tag{XV-18}$$

Equation XV-18 provides for the general case of a molecule having n independent ways of rotation and a moment of inertia I that, for an asymmetric molecule, is the (geometric) mean of the principal moments. The quantity σ is the symmetry number, or the number of indistinguishable positions into which the molecule can be turned by rotations. The rotational energy and entropy are (41)

$$E_{\text{rot}} = \frac{nRT}{2}$$

$$S_{\text{rot}} = R\left\{\ln\left[\frac{1}{\pi\sigma}\left(\frac{8\pi^3 IkT}{h^2}\right)^{n/2}\right] + \frac{n}{2}\right\} \tag{XV-19}$$

Molecular moments of inertia are about 10^{-39} g/cm^2; thus I values for benzene N_2, and NH_3 are 18, 1.4, and 0.28, respectively, in those units. For the case of benzene gas, $\sigma = 6$ and $n = 3$, and S_{rot} is about 21 cal/°mole at 25°C. On adsorption, all of this entropy would be lost if the benzene were unable to rotate, and part of it if, say, rotation about only one axis were possible (as might be the situation if the benzene was subject only to the constraint of lying flat on the surface). Similarly, all or part of the rotational energy of 0.88 kcal/mole would be liberated on adsorption, or perhaps converted to vibrational energy.

Vibrational energy states are too well separated to contribute much to the entropy or the energy of small molecules at ordinary temperatures, but for higher temperatures this may not be so and, both internal entropy and energy changes may occur due to changes in vibrational levels on adsorption. From a somewhat different point of view, it is clear that even in physical adsorption, adsorbate molecules should be polarized on the surface (see Section VI-7), and in chemisorption more drastic perturbations should occur. Thus internal bond energies of adsorbed molecules may be affected.

Infrared absorption spectroscopy has contributed a great deal of information on this general subject; see Refs. 42–44. As a specific illustration, Blyholder and Richardson (45) in a study of the adsorption of ammonia on an activated iron oxide found that the bands at 3.0 and 6.1 μm shifted only slightly on adsorption, to 3.1 and 6.3 μm, although the intensity ratio changed. Had ammonium ion formation occurred, absorption at 3.4 and 6.9 μm would have been expected, and the conclusion was that ammonia was present as such, although with perhaps some hydrogen bonding. It also appeared, from the shape of the bands, that rotational degrees of freedom had been lost. On coadsorption of some water ammonium ion bands appeared. The infrared spectra of NH_3 and ND_3 adsorbed on ZnO surfaces indicated that both dissociative adsorption and coordinative bonding occurred (46). Fig. XV-6 illustrates the use of infrared absorption spectra in the study of the surface of SiO_2. Dried SiO_2 shows absorption at 3700 cm^{-1}, attributed to isolated surface hydroxyl groups. On hydration of the surface, a broad band at 3400 cm^{-1} grows in, due to adsorbed molecular water.

An interesting point is that infrared absorptions that are symmetry forbidden and hence that do not appear in the spectrum of the gaseous molecule may appear when that molecule is adsorbed. Thus Sheppard and Yates (48) found that normally forbidden bands could be detected in the case of methane and hydrogen adsorbed on glass; this meant that some reduction in molecular symmetry had occurred. In the case of the methane, it appeared from the band shapes that some reduction in rotational degrees of freedom had oc-

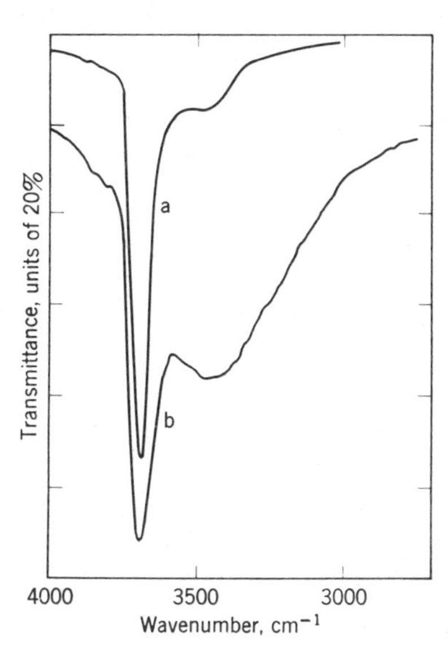

Fig. XV-6. Spectra of SiO_2 suspended in CCl_4: (*a*) predried at 750°C; (*b*) rehydrated. Reprinted by permission from Ref. 47. Copyright by the American Chemical Society.

curred. Figure XV-7 shows the absorption of the normally forbidden N–N stretch when nitrogen is adsorbed on nickel. The appearance of this band and the isotopic components, confirmed that no dissociation into nitrogen atoms occurred.

A few further illustrations of the use of infrared spectra are the following. Yates (49) was able to follow the oxidation of CO chemisorbed on gold by the progressive disappearance of the infrared band of the CO and appearance of that of adsorbed CO_2. Quirk and co-workers (50) have reported on the infrared spectra of pyridine type species complexed with clays. The various types of adsorbed water on partially hydrophobic silicas have been studied by Zettlemoyer and co-workers (51). Boerio and Chen have reported on the infrared spectra of films of dodecanoic acid and 1-octadecanol on metal surfaces (52); spectra of variously deuterated alcohols on alumina as also been reported (56a).

Raman spectroscopy is also possible for surface adsorbed species and in some cases shows less interference from the adsorbent than does conventional infrared spectroscopy (see Refs. 53, 54). It has not so far been possible to obtain microwave spectra of adsorbed molecules, probably because even minor hindrances to rotation would cause excessive line broadening.

Resonance Raman spectra have been obtained for adsorbed films. Examples include various dyes at the glass–water interface (55) and aniline and azo derivatives adsorbed on silica, alumina, and semiconductor surfaces (56). In the case of pyridine adsorbed on surface-roughened silver surfaces, an amazing enhancement of resonance Raman intensity has been reported (57–59). The effect has been given the acronym SERS (surface enhanced Raman scattering). There is a possibility, however, that it is due more to the actual surface area being larger than thought, possibly related to carbon deposits (see Ref. 60).

An indication that the energy states, and hence internal energy content of adsorbed molecules, are different from those for the gaseous state is that the electronic absorption spectrum may be altered. Thus Leftin and Hall (61) observed a new absorption band at 400 mμ on adsorption of α-methylstyrene on a silica–alumina adsorbent. The interpretation in this case was the chemical one that carbonium ion

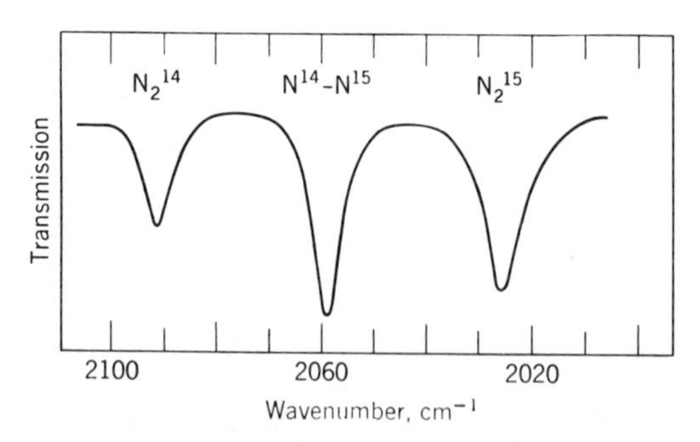

Fig. XV-7. Spectrum of mixture of $^{14}N_2$, $^{15}N_2$, and $^{14}N\,^{15}N$ chemisorbed on nickel. Reprinted by permission from Ref. 44. Copyright by the American Chemical Society.

formation had occurred. Smaller effects take place if the interaction is less specific; Kiselev and co-workers (62) observed shifts of 200 to 300 cm^{-1} (about 700 cal/mole) to longer wavelengths in the ultraviolet absorption spectra of benzene and other aromatic species on adsorption on Aerosil (a silica). Leermakers and co-workers (63) made extensive studies of the spectra and photochemistry of adsorbed organic species. The absorption spectrum of $Ru(bipyridine)_3^{2+}$ is shifted slightly on adsorption on silica gel (64). Absorption peak positions give only the difference in energy between the ground and excited states, so the shifts, while relatively small, could reflect larger but parallel changes in the absolute energies of the two states. In other words, the shifts suggest that appreciable changes in the ground state energies occurred on adsorption. As with the lost rotational energy, these would appear as contributions to the heat of adsorption.

Several electron spin resonance (esr) studies have been reported for adsorption systems (65–68). Esr signals are strong enough to allow the detection of quite small amounts of unpaired electrons, and the shape of the signal can, in the case of adsorbed transition metal ions, give an indication of the geometry of the adsorption site.

The intramolecular perturbations on adsorption discussed so far will be dominant if the surface is sparsely covered. At higher coverages, and certainly if the film is multimolecular, interadsorbate interactions will also be important. These may be treated empirically in terms of an interaction energy, or semiempirically, in terms of assumed intermolecular potential functions. Some aspects of such approaches are considered in the next chapter (Section XVI-10). For the present, the discussion is limited to experimental methods for characterizing the state of the adsorbed film.

One technique is that of nuclear magnetic resonance (NMR). The theory and experimental equipment and procedures have been reviewed by O'Reilly (69). Resing has reviewed the types of results obtainable in some detail (70). NMR studies are limited to atoms having a nonzero nuclear magnetic moment, which from the point of view of adsorption studies limits the scope mainly to hydrogen, fluorine, sodium, aluminum, and silicon. In addition, line widths tend to be such as to obscure chemical shifts. As a consequence, the most interesting work has been done on proton NMR, mainly of water, from the point of view of relaxation times and line widths. Brey and Lawson (71) studied the variation of proton NMR line widths and relaxation times as a function of surface coverage and temperature for water, methanol, and so on, adsorbed on thorium oxide. In the case of *n*-heptane adsorbed on Spheron or diamond, the line width for the first adsorbed layer was very large, but with three or more adsorbed layers, a line structure developed (72). Again, a narrow line is associated with sufficient mobility that local magnetic fields are averaged, and the interpretation was that adsorbed layers beyond the second were mobile. The longitudinal and transverse relaxation times T_1 and T_2 can be interpreted in terms of a correlation time t (73), which is about the time required for a molecule to turn through a radian or to move a distance comparable with its dimensions. Resing and co-workers (74) found that for water adsorbed on a hardwood charcoal t was given approximately by $t_0 e^{Q/RT}$ where t_0 was 3×10^{-23} sec and Q was 12 kcal/mole, but there was an indication that more than one kind of water was present. Similarly, Zimmerman and Lasater (75) studying water adsorbed on a silica gel, found evidence for two kinds of water for surface coverages greater than 0.5 and, as illustrated in Fig. XV-8, a variation of calculated correlation time with coverage, which suggested that surface mobility was fully developed at about the expected monolayer point. A tentative phase identification can be made on the basis that t for water is about 10^{-11} sec at 25°C, while for ice (or "rigid" water) it is about 10^{-5} sec. That is, inspection of Fig. XV-8 would suggest that below $\theta = 0.5$, the adsorbed water was solidlike, while above $\theta = 1$ it was liquidlike. An interesting application is in the use of the variation of NMR line width with equilibrium adsorbate pressure to determine the heat of physical adsorption (76).

A more chemical application of NMR results was made in the case of ethylene adsorbed on a zeolite. NMR shifts were correlated with shifts in

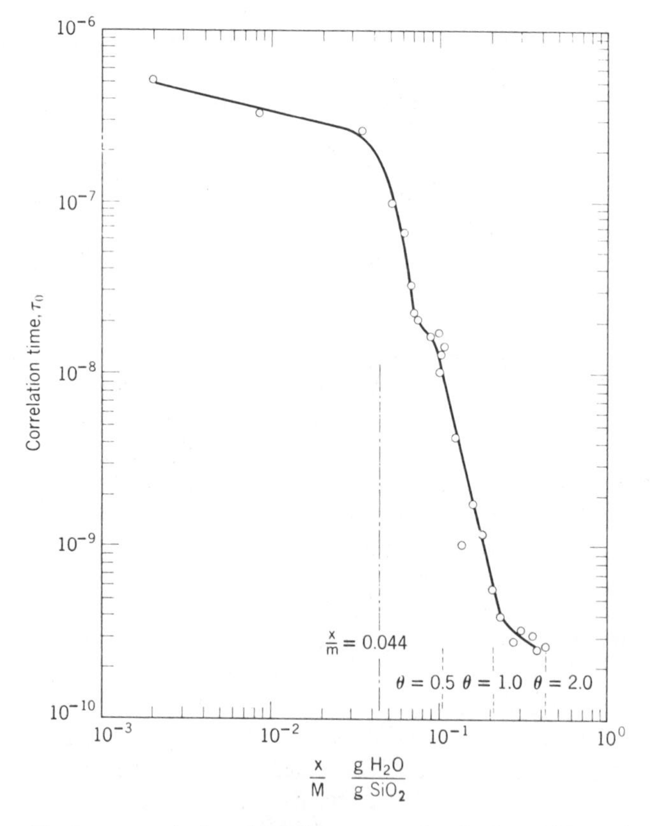

Fig. XV-8. Nuclear correlation times for water adsorbed on silica gel. (From Ref. 75.)

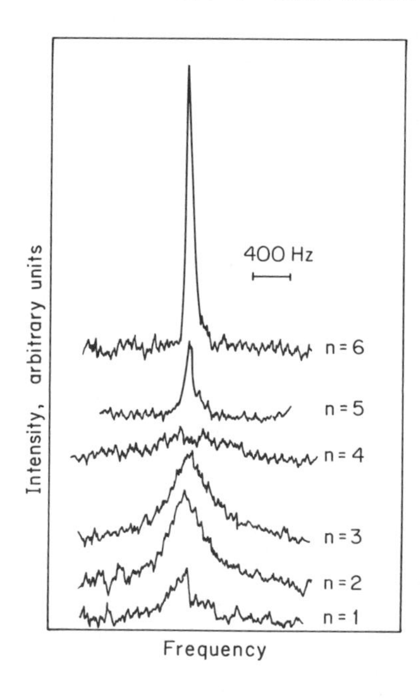

Fig. XV-9. Carbon 13 NMR spectra of benzene sorbed in NA^+ zeolite; n is the number of benzene molecules per cage (see Section XVI-17). (From Ref. 78.)

infrared absorption bands and ascribed to structural deformations (77). Carbon 13 NMR has also been used. Figure XV-9 shows the behavior accompanying the adsorption (or "sorption") of benzene by a zeolite that contains cages capable of holding up to four benzene molecules. The line broadening (and hence correlation time) increased with n, the number of molecules per cage, up to $n = 4$. Beyond this point there was a rapid line narrowing, indicating that a new, highly mobile phase was present (78). In a study of chemisorbed methanol on MgO, ^{13}C NMR indicated molecules to be rigidly bound below a half monolayer coverage, but that higher coverages produced isotropically rotating molecules (79).

A rather different method, but one that yields conclusions similar to those from NMR studies, is that of the determination of the dielectric absorption by adsorbed films. Figure XV-10 illustrates how the dielectric constant for adsorbed water varies with the frequency used as well as with the degree of surface coverage. A characteristic relaxation time τ can be estimated from the frequency dependence of the dielectric constant, especially in combination with that of the dielectric loss. This relaxation time is essentially that for the molecule to move or to reorient so as to follow a changing electric field. For water on α-Fe_2O_3, τ varied from about 1 sec at monolayer coverage, to 10^{-4} sec when several adsorbed layers were present (80). Since the characteristic time was far larger than for liquid bulk water, 10^{-10} sec, the conclusion was that the adsorbed water was present in an icelike, hydrogen bonded structure.

Similar, very detailed studies were made by Ebert (81) on water adsorbed on alumina, with similar conclusions. Water adsorbed on zeolites showed

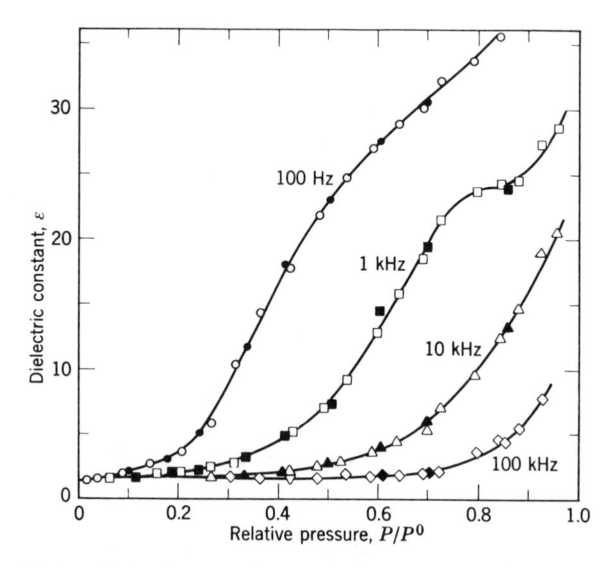

Fig. XV-10. Dielectric isotherms of water vapor at 15°C adsorbed on α-Fe_2O_3 (solid points indicate desorption). A complete monolayer was present at $P/P^\circ = 0.1$ and by $P/P^\circ = 0.8$ several layers of adsorbed water were present. (From Ref. 80.)

a dielectric constant of only 14 to 21, indicating greatly reduced mobility of the water dipoles (82). Similar results were found for ammonia adsorbed in Vycor glass (83). Klier and Zettlemoyer (84) have reviewed a number of aspects of the molecular structure and dynamics of water at the surface of an inorganic material.

The state of an adsorbate is often described as "mobile" or "localized," usually in connection with adsorption models and analyses of adsorption entropies (see Section XVI-3B). A more direct criterion is, in analogy to that of the fluidity of a bulk phase, the degree of mobility as reflected by the surface diffusion coefficient. This may be estimated from the dielectric relaxation time; Resing (70) gives values of the diffusion coefficient for adsorbed water ranging from near bulk liquid values (10^{-5} cm²/sec) to as low as 10^{-9} cm²/sec.

As noted in Chapter VIII, the structure of adsorbed films can sometimes be estimated from diffraction and spectroscopic results. As further examples, a LEED-AES study of tetracyanoethylene adsorbed on KCl suggests that the CN groups are located directly over K^+ ions (85). In a review of the use of angle-resolved photoemission spectroscopy, it is noted that CO adsorbed on Ni (100) surfaces is oriented perpendicular to the surface (86). X-ray scattering studies indicate that Xe adsorbed on spheron (a carbon black) is in a liquidlike state (87).

B. Effect of the Adsorbate on the Adsorbent

It is evident from the preceding material that a great deal of interest has centered on the chemical and physical state of the adsorbate. There is no reason not to expect the adsorbate to affect the surface structure of the adsorbent; in fact, some of the properties described, especially the thermodynamic ones, should properly be assigned to the adsorbate–adsorbent system rather than just to the adsorbate. The same may be true to a lesser extent of dielectric measurements. The subject has been a somewhat neglected one, but there are now many indications that adsorbents are far from inert in their interaction with adsorbates.

First, it is entirely possible that surface heterogeneities and imperfections undergo some reversible redistribution during adsorption. As noted in Section VII-4B, Dunning has considered that since the presence of an adsorbed molecule should alter the energy of special sites (such as illustrated in Fig. VII-6), above some critical temperature for surface mobility, their distribution should depend on the extent of adsorption. Also, the first few layers of the crystalline surface of a solid are distorted (see Fig. VII-5), and this distortion should certainly be altered if an adsorbed layer is present; here no mass transport of atoms of the solid is involved. There is some evidence for this type of effect. Lander and Morrison in their LEED study of germanium surfaces, concluded that a considerable surface rearrangement took place on adsorption of iodine. Ehrlich and co-workers (see Ref. 88) have observed, in field emission microscopy, that surface rearrangement of tungsten can accompany the adsorption or desorption of nitrogen. The subject has been reviewed by Somorjai (89) in connection with chemisorption systems. In physical absorption on molecular solids, surface restructuring is again indicated (90). In the case of *n*-hexane adsorbing on ice, surface restructuring was indicated for temperatures above $-35°C$ (91).

As discussed in connection with Table XV-1, it seems reasonable to expect important surface structural changes to accompany adsorption whenever the adsorption energy is significant in comparison to the bond energy, or molar surface energy, of the solid. Referring to Table III-1, this means that chemisorption on solids such as metals and ionic crystals should affect surface rearrangement. After all, chemisorption energies range from 10 to 50 kcal/mole and so do the values of $E^{s\prime}$ in Table III-1 (92). In the case of physical adsorption, the adsorption energies are generally around that of condensation of the adsorbate, or less than 10 kcal/mole. This means that high surface energy solids such as carbon or the various metals or metal oxides should be relatively inert except for small changes in the pattern of surface distortion. However, this is not so for the adsorption of vapors on low energy solids, such as plastics or molecular solids, as in the case cited of *n*-hexane on ice.

Wu and Copeland (93) point out that the surface thermodynamic properties of the *adsorbent* must change on adsorption in a manner calculable from adsorption data. Thus the change in enthalpy of $BaSO_4$ on adsorption of water was comparable to the heat of adsorption of the water itself, see Section XVI-13A. In summary, with the probable exception of physical adsorption on high energy, rigid solids, adsorbate induced perturbations in the adsorbent appear to be potentially quite important and deserving of much more attention than has been given so far.

C. The Adsorbate–Adsorbent Bond

The immediate site of the adsorbent–adsorbate interaction is presumably that between adjacent atoms of the respective species. This is certainly true in chemisorption, where actual chemical bond formation is the rule, and is largely true in the case of physical adsorption, with the possible exception of multilayer formation, which can be viewed as a consequence of weak, long-range force fields. Another possible exception would be the case of molecules where some electron delocalization is present, as with aromatic ring systems.

It has been customary to assign all of the observed heat of adsorption to the adsorption bond, although the preceding discussion has made it clear that other, less localized contributions can be important. There has been some discussion of the quantum mechanics of bond formation with a surface (see Refs. 94–96), but the subject has been hampered by the nearly total lack of any direct experimental information. Thus no bond lengths are available, or have any infrared absorptions assignable to an adsorption bond been observed. It is generally supposed that the reason for this last situation is that the bond is either too weak or too multicentered for the absorption to be in the usually accessible wavelength region. A weak band around 4.7 μ has been considered as possibly due to the H–Pt bond; this occurs on adsorption of hydrogen on platinum (97).

The difficulty in observing infrared bands due to an adsorption bond is even greater in the case of physical adsorption systems where the interaction is much weaker. Thus from estimates based on adsorption entropies for such systems as ammonia on charcoal, Everett (98) places bands of this type at around 300 μ (corresponding to about 1 kcal/mole). It is possible to make calculations on the adsorption bond in physical adsorption based on an assumed potential function. This has been done in some detail (99, 100), but the approach is modelistic and is more appropriately discussed in the next chapter (Section XVI-10).

5. Problems

1. The contact angle for water on single crystal naphthalene is 89° at 25°C, and $d\theta/dT$ is -0.13 deg/K. Using data from Table III-1 as necessary, calculate the heat of immersion of naphthalene in water in cal/g if a sample of powdered naphthalene of 10 m^2/g is used for the immersion study. (Note Ref. 101).

2. Bartell and Fu (1) were able to determine the adhesion tension, that is, $\gamma_{SV} - \gamma_{SL}$, for the water-silica interface to be 82.8 erg/cm^2 at 20°C and its temperature change to be -0.173 $erg/(cm^2)(K)$. The heat of immersion of the silica sample in water was 15.9 cal/g. Calculate the surface area of the sample, in square centimeters per gram.

3. Harkins and Jura (3) found that a sample of TiO_2 having a thick adsorbed layer

of water on it gave a heat of immersion in water of 0.409 cal/g. Calculate the specific surface area of the TiO_2 in square centimeters per gram.

4. Use the data shown in Fig. XV-2 to estimate the specific surface area of the ground quartz used.

5. The rate of dissolving of a solid is determined by the rate of diffusion through a boundary layer of solution. Derive the equation for the net rate of dissolving. Take C_0 to be the saturation concentration and d to be the effective thickness of the diffusion layer; denote diffusion coefficient by $\mathcal{D}$.

6. Using the curve given by the square points in Fig. XV-3, make a qualitative reconstruction of the original data plot of volume of mercury penetrated per gram versus applied pressure.

7. Optical microscopic examination of a finely divided silver powder shows what appears to be spheres 1 μ in diameter. The density of the material is 3 g/cc. The surface is actually rather rough on a submicroscopic scale, and suppose that a cross section through the surface actually has the appearance of the coastline shown in Fig. XV-1, where the left arrow represents a distance of 0.1 μ and the successive enlargements are each by tenfold.

Make a numerical estimate, with an explanation of the assumptions involved, of the specific surface area that would be found by (*a*) a rate of dissolving study, (*b*) Harkins and Jura who find that at P^0 the adsorption of water vapor is 8.5 cc STP/g (and then proceed with a heat of immersion measurement), and (*c*) a measurement of the permeability to liquid flow through a compacted plug of the powder.

8. Calculate the rotational contribution to the entropy of adsorption of benzene on carbon at 25°C, assuming that the adsorbed benzene has one degree of rotational freedom.

9. Calculate the rotational contribution to the entropy of adsorption of ammonia on silica at −35°C, assuming (*a*) that the adsorbed ammonia retains one degree of rotational freedom and (*b*) that it retains none. In case (*a*) assume that the nitrogen is bonded to the surface.

General References

Advances in Catalysis, Vol. 10, Academic, New York, 1958; Vol. 12, 1960; Vol. 16, 1966.

R. E. Collins, *Flow of Fluids Through Porous Materials,* Reinhold, New York, 1961.

A. E. Flood, Ed., *The Solid-Gas Interface,* Vol. 1, Marcel Dekker, New York, 1967.

S. J. Gregg, *The Surface Chemistry of Solids,* Reinhold, New York, 1951.

M. L. Hair, *Infrared Spectroscopy in Surface Chemistry,* Marcel Dekker, New York, 1967.

T. L. Hill, *Introduction to Statistical Thermodynamics,* Addison-Wesley, Reading, Mass., 1962.

R. R. Irani and C. F. Callis, *Particle Size,* Wiley, New York, 1963.

I. Prigogine, Ed., *Advances in Chemical Physics,* Vol. 9, Interscience, New York, 1965.

S. Ross and J. P. Olivier, *On Physical Adsorption,* Interscience, New York, 1964.

H. Saltsburg, J. N. Smith, Jr., and M. Rogers, Eds., *Fundamentals of Gas-Surface Interactions,* Academic, New York, 1967.

G. A. Somorjai, *Principles of Surface Chemistry,* Prentice-Hall, New York, 1972.

Textual References

1. F. E. Bartell and Y. Fu, "The Specific Surface Area of Activated Carbon and Silica," in *Colloid Symposium Annual,* Vol. 7, H. B. Weiser, Ed., Wiley, New York, 1930.
2. J. W. Whalen and W. H. Wade, *J. Colloid Interface Sci.,* **24,** 372 (1967).
3. See G. Jura, "The Determination of the Area of the Surfaces of Solids" in *Physical Methods in Chemical Analysis,* Vol. 2, W. G. Berl, Ed., Academic, New York, 1951.
4. S. Ergun, *Anal. Chem.,* **24,** 388 (1952).
5. A. J. Tyler, J. A. G. Taylor, B. A. Pethica, and J. A. Hockey, *Trans. Faraday Soc.,* **67,** 483 (1971).
6. S. Partyea, F. Rouquerol, and J. Rouquerol, *J. Colloid Interface Sci.,* **68,** 21 (1979).
7. A. B. Zdanovskii, *Zh. Fiz. Khim.,* **25,** 170 (1951); *Chem. Abstr.,* **48,** 4291c (1964).
8. A. R. Cooper, Jr., *Trans. Faraday Soc.,* **58,** 2468 (1962).
9. W. G. Palmer and R. E. D. Clark, *Proc. Roy. Soc.* (*London*), **A149,** 360 (1935); see also E. Darmois, *Chim. Ind.*-(*Paris*), **59,** 466 (1948) and I. Bergman, J. Cartwright, and R. A. Bentley, *Nature,* **196,** 248 (1962).
10. C. F. Jones, R. L. Segall, R. St.C. Smart, and P. S. Turner, *J. Chem. Soc., Faraday Trans.,* **74,** 1615 (1978).
11. R. L. Segall, R. St.C. Smart, and P. S. Turner, *J. Chem. Soc., Faraday Trans.,* **74,** 2907 (1978).
12. F. E. Bartell and C. W. Walton, Jr., *J. Phys. Chem.,* **38,** 503 (1934).
13. See F. E. Bartell and L. S. Bartell, *J. Am. Chem. Soc.,* **56,** 2202 (1934).
14. H. L. Ritter and L. C. Drake, *Ind. Eng. Chem., Anal. Ed.,* **17,** 782 (1945).
15. L. C. Drake, *Ind. Eng. Chem.,* **41,** 780 (1949).
16. N. M. Winslow and J. J. Shapiro, *ASTM Bull.,* No. 236, p. 39 (1959).
17. L. A. de Wit and J. J. F. Scholten, *J. Catal.,* **36,** 36 (1975).
18. R. J. Good and R. Sh. Mikhail, *Powder Technol.,* **29,** 53 (1981).
19. L. G. Joyner, E. P. Barrett, and R. Skold, *J. Am. Chem. Soc.,* **73,** 3155 (1951).
20. D. H. Everett, in *The Solid–Gas Interface,* E. A. Flood, Ed., Marcel Dekker, New York, 1967.
21. L. C. Drake and H. L. Ritter, *Ind. Eng. Chem., Anal. Ed.,* **17,** 787 (1945).
22. F. P. Bowden and E. K. Rideal, *Proc. Roy. Soc.* (*London*), **A120,** 59 (1928).
23. M. J. Joncich and N. Hackerman, *J. Electrochem. Soc.,* **111,** 1286 (1964).
24. D. J. C. Yates, *Can. J. Chem.,* **46,** 1695 (1968).
25. J. Kozeny, *Śitzber. Akad. Wiss. Wien, Wasserwirtsch. Math. Nat., Kl.* **IIa,** 136, 271 (1927); *Wasserwirtschaft,* **22,** 67, 86 (1927).
26. R. E. Collins, *Flow of Fluids Through Porous Materials,* Reinhold, New York, 1961.
27. R. M. Barrer, *Discuss. Faraday Soc.,* **3,** 61 (1948); see also R. M. Barrer and D. M. Grove, *Trans. Faraday Soc.,* **47,** 826 (1951).
28. G. Kraus, J. W. Ross, and L. A. Girifalco, *J. Phys. Chem.,* **57,** 330 (1953); see also G. Kraus and J. W. Ross, *J. Phys. Chem.,* **57,** 334 (1953).
29. P. H. Emmett, *Catalysis,* Vol. 1, Reinhold, New York, 1954, p. 60.
30. See P. W. Selwood, *Magnetochemistry,* Interscience, New York, 1956.
31. D. E. O'Reilly and D. S. MacIver, *J. Phys. Chem.,* **66,** 276 (1962).

32. Boon-Teng Teo, *Acc. Chem. Res.*, **13**, 412 (4980).
33. P. H. Lewis, *J. Phys. Chem.*, **66**, 105 (1962).
34. B. N. Figgis, *Introduction to Ligand Fields*, Interscience, New York, 1968.
35. See F. S. Stone, *Chem. Ind.*, **41**, 1810 (1963).
36. D. A. Dowden, *Endeavour*, **24**, 69 (1965).
37. H. P. Boehm, *Adv. Catal.*, **16**, 179 (1966).
38. G. R. Henning, *Z. Elektrochem.*, **66**, 629 (1962).
39. F. Boccuzzi, S. Coluccia, G. Ghiotti, C. Morterra, and A. Zecchina, *J. Phys. Chem.*, **82**, 1298 (1978).
40. B. A. Morrow, I. A. Cody, and L. S. M. Lee, *J. Phys. Chem.*, **80**, 2761 (1976).
41. C. Kemball, *Proc. Roy. Soc.* (*London*), **A187**, 73 (1946).
42. L. H. Little, *Infrared Spectra of Adsorbed Molecules*, Academic, New York, 1966.
43. M. L. Hair, *Infrared Spectroscopy in Surface Chemistry*, Marcel Dekker, New York, 1967.
44. R. P. Eischens, *Acc. Chem. Res.*, **5**, 74 (1972).
45. G. Blyholder and E. A. Richardson, *J. Phys. Chem.*, **66**, 2597 (1962); G. D. Parfitt, J. Ramsbotham, and C. H. Rochester, *Trans. Faraday Soc.*, **67**, 841 (1971).
46. T. Morimoto, H. Yanai, and M. Nagao, *J. Phys. Chem.*, **80**, 471 (1976).
47. W. D. Bascom, *J. Phys. Chem.*, **76**, 3188 (1972).
48. N. Sheppard and D. J. C. Yates, *Proc. Roy. Soc.* (*London*), **A238**, 69 (1956).
49. D. J. C. Yates, *J. Colloid Interface Sci.*, **29**, 194 (1969).
50. S. Olejnik, A. M. Posner, and J. P. Quirk, *Spectrochim. Acta*, **27A**, 2005 (1971).
51. K. Klier, J. H. Shen, and A. C. Zettlemoyer, *J. Phys. Chem.*, **77**, 1458 (1973).
52. F. J. Boerio and S. L. Chen, *J. Colloid Interface Sci.*, **73**, 176 (1980).
53. P. J. Hendra, I. D. M. Turner, E. J. Loader, and M. Stacey, *J. Phys. Chem.*, **78**, 300 (1974).
54. J. F. Rabolt, R. Santo, and J. D. Swalen, *Appl. Spectrosc.*, **13**, 549 (1979).
55. T. Takenaka and K. Yamasaki, *J. Colloid Interface Sci.*, **78**, 37 (1980).
56. J. F. Brazdll and E. B. Yeager, *J. Phys. Chem.*, **85**, 995, 1005 (1981).
56a. J. Lavalley, J. Calliod, and J. Travert, *J. Phys. Chem.*, **84**, 2083 (1980).
57. J. A. Creighton, C. G. Blatchford, and M. G. Albrecht, *J. Chem. Soc., Faraday Trans. II*, **75**, 790 (1979).
58. C. S. Allen and R. P. van Duyne, *Chem. Phys. Lett.*, **63**, 455 (1979).
59. A. Otto, *Surf. Sci.*, **75**, L392 (1978).
60. M. R. Mahoney, M. W. Howard, and R. P. Cooney, *Chem. Phys. Lett.*, **71**, 59 (1980); M. W. Howard, R. P. Cooney, and A. J. McQuillan, *J. Raman Spectrosc.*, **9**, 273 (1980).
61. H. P. Leftin and W. K. Hall, *J. Phys. Chem.*, **66**, 1457 (1962).
62. V. N. Abramov, A. V. Kiselev, and V. I. Lygin, *Russ. J. Phys. Chem.*, **37**, 1507 (1963).
63. See L. D. Weis, T. R. Evans, and P. A. Leermakers, *J. Am. Chem. Soc.*, **90**, 6109 (1968) and references therein.
64. J. Namnath and A. W. Adamson, unpublished work.
65. G. P. Lozos and B. M. Hoffman, *J. Phys. Chem.*, **78**, 200 (1974).
66. M. B. McBride, T. J. Pinnavala, and M. M. Mortland, *J. Phys. Chem.*, **79**, 2430 (1975).

67. S. Abdo, R. B. Clarkson, and W. K. Hall, *J. Phys. Chem.,* **80,** 2431 (1976).
68. M. F. Ottaviani and G. Martini, *J. Phys. Chem.,* **84,** 2310 (1980).
69. D. E. O'Reilly, *Adv. Cataly.,* **12,** 31 (1960).
70. H. A. Resing, *Adv. Mol. Relaxation Processes,* **1,** 109 (1967–68); *ibid.,* **3,** 199 (1972).
71. W. S. Brey, Jr. and K. D. Lawson, *J. Phys. Chem.,* **68,** 1474 (1964).
72. D. Graham and W. D. Phillips, *Proc. 2nd Int. Congr. Surf. Act., London,* Vol. II, p. 22.
73. N. Bloembergen, E. M. Purcell, and R. V. Pound, *Phys. Rev.,* **73,** 679 (1948).
74. H. A. Resing, J. K. Thompson, and J. J. Krebs, *J. Phys. Chem.,* **68,** 1621 (1964).
75. J. R. Zimmerman and J. A. Lasater, *J. Phys. Chem.,* **62,** 1157 (1958).
76. C. L. Kibby, V. Yu. Borovkov, and V. B. Kazansky, *J. Catal.,* **46,** 275 (1977).
77. G. M. Muha and D. J. C. Yates, *J. Chem. Phys.,* **49,** 5073 (1968).
78. V. Yu. Borovkov, W. K. Hall, and V. B. Kazanski, *J. Catal.,* **51,** 437 (1978).
79. I. D. Gay, *J. Phys. Chem.,* **84,** 3230 (1980).
80. E. McCafferty and A. C. Zettlemoyer, *Discuss. Faraday Soc.,* 239 (1971).
81. G. Ebert, *Kolloid-Z.,* **174,** 5 (1961).
82. K. R. Foster and H. A. Resing, *J. Phys. Chem.,* **80,** 1390 (1976).
83. I. Lubezky, U. Feldman, and M. Folman, *Trans. Faraday Soc.,* **61,** 1 (1965).
84. K. Klier and A. C. Zettlemoyer, *J. Colloid Interface Sci..* **58,** 216 (1977).
85. H. Saijo, N. Uyeda, and E. Suito, *J. Chem. Soc., Faraday Trans. II,* **73,** 517 (1977).
86. E. W. Plummer and T. Gustafsson, *Science,* **198,** 165 (1977).
87. G. W. Brady, D. B. Fein, and W. A. Steele, *Phys. Rev. B,* **15,** 1120 (1977).
88. G. Ehrlich and F. G. Hudda, *J. Chem. Phys.,* **35,** 1421 (1961).
89. G. A. Somorjai, *Principles of Surface Chemistry,* Prentice-Hall, New York, 1972, p. 239ff.
90. A. W. Adamson and M. W. Orem, *Progress in Surface and Membrane Science,* D. A. Cadenhead, J. F. Danielli, and M. D. Rosenberg, Eds., Academic, New York, 1974, Vol. 8.
91. A. W. Adamson, L. M. Dormant, and M. W. Orem, *J. Colloid Interface Sci.,* **25,** 206 (1966); M. W. Orem and A. W. Adamson, *ibid.,* **31,** 278 (1969).
92. G. Ehrlich, *Br. J. Appl. Phys.,* **15,** 349 (1964).
93. Y. C. Wu and L. E. Copeland, *Adv. Chem. Ser.* No. 33, 348 (1961).
94. T. B. Grimley, *Adv. Catal.,* **12,** 1 (1960).
95. W. J. Dunning, *The Solid–Gas Interface,* E. A. Flood, Ed., Marcel Dekker, New York, 1966.
96. J. Koutecky, *Adv. Chem. Phys.,* **9,** 85 (1965).
97. W. A. Pliskin and R. P. Eischens, *Abstracts of the April 1959 Meeting, American Chemical Society, Boston.*
98. D. H. Everett, *Proc. Chem. Soc.* (*London*), 1957, p. 38.
99. W. A. Steele and G. D. Halsey, Jr., *J. Phys. Chem.,* **59,** 57 (1955).
100. J. A. Barker and D. H. Everett, *Trans. Faraday Soc.,* **58,** 1608 (1962).
101. J. B. Jones and A. W. Adamson, *J. Phys. Chem.,* **72,** 646 (1968).

CHAPTER XVI

Adsorption of Gases and Vapors on Solids

1. Introduction

The subject of gas adsorption is indeed a very broad one, and no attempt is made to give complete coverage to the voluminous literature on it. Instead, as in past chapters, the principal models or theories are taken up partly for their own sake and partly as a means of introducing characteristic data.

As stated in the introduction to the previous chapter, adsorption is described phenomenologically in terms of an empirical adsorption function $n = f(P, T)$ where n is the amount adsorbed. As a matter of experimental convenience, one usually determines the adsorption *isotherm* $n = f_T(P)$; in a detailed study, this is done for several temperatures. Figure XVI-1 displays some of the extensive data of Drain and Morrison (1). It is fairly common in physical adsorption systems for the low pressure data to suggest that a limiting adsorption is being reached, as in Fig. XVI-1*a*, but for continued further adsorption to occur at pressures approaching the saturation or condensation pressure P^0 (which would be close to 1 atm for N_2 at 75°K), as in Fig. XVI-1*b*.

Alternatively, data may be plotted as n versus T at constant pressure, or as P versus T at constant n. One thus has adsorption *isobars* and *isosteres* (note Problem 1).

As also noted in the preceding chapter, it is customary to divide adsorption into two broad classes, namely, *physical adsorption* and *chemisorption*. Physical adsorption equilibrium is very rapid in attainment (except when limited by mass transport rates in the gas phase or within a porous adsorbent) and is reversible, the adsorbate being removable without change by lowering the pressure. It is supposed that this type of adsorption occurs as a result of the same type of relatively nonspecific intermolecular forces that are responsible for the condensation of a vapor to a liquid, and in physical adsorption the heat of adsorption should be in the range of heats of condensation. Physical adsorption is usually important only for gases below their critical temperature, that is, for vapors.

Chemisorption may be rapid or slow and may occur above or below the critical temperature of the adsorbate. It is distinguishable, qualitatively, from physical adsorption in that chemical specificity is higher and that the energy of adsorption is large enough to suggest that full chemical bonding has occurred. Gas that is chemisorbed may be difficult to remove, and desorption may be accompanied by chemical changes. For example, oxygen adsorbed on carbon is held very strongly; on heating it comes off as a mixture of CO and CO_2 (2). Because of its nature, chemisorption is expected to be limited to a monolayer; as suggested in connection with Fig. XVI-1, physical adsorption is not so limited and, in fact, may occur on top of a chemisorbed

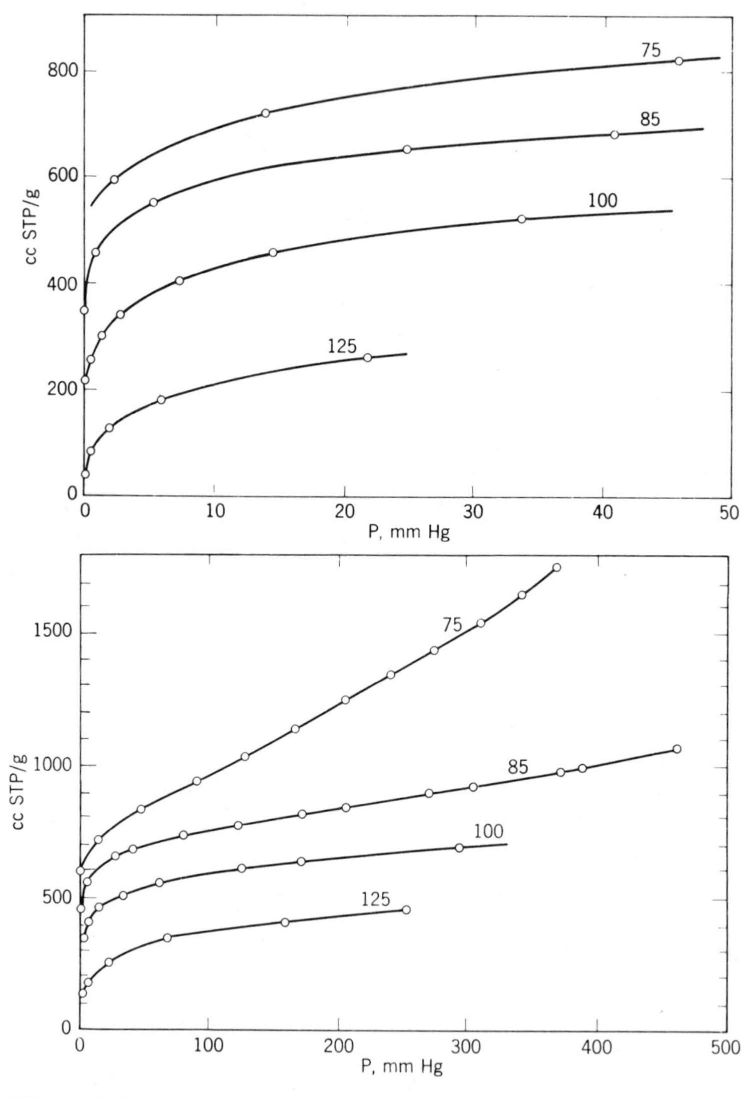

Fig. XVI-1. Adsorption of N_2 on rutile; temperatures indicated are in °K. (*a*) Low pressure region; (*b*) high pressure region.

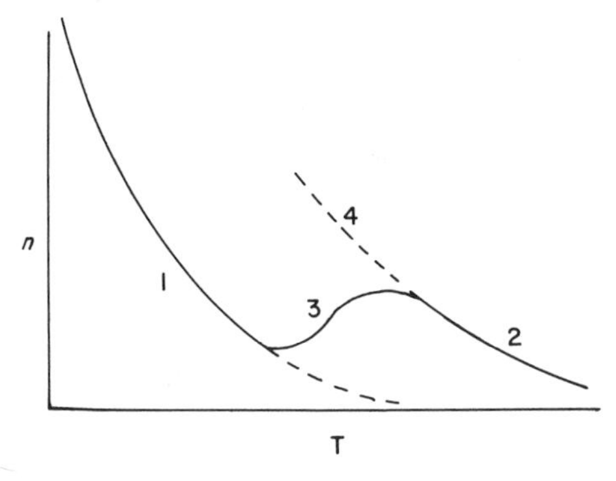

Fig. XVI-2. Transition between physical and chemical adsorption.

layer (note Section XI-1C) as well as alongside it. The infrared spectrum of CO_2 adsorbed on γ-alumina suggests the presence of both physically and chemically adsorbed molecules (3).

Chemisorption may be slow and the rate behavior indicative of the presence of an activation energy; it may, in fact, be possible for a gas to be physically adsorbed at first, and then, more slowly, to enter into some chemical reaction with the surface of the solid. At low temperatures, chemisorption may be so slow that for practical purposes only physical adsorption is observed, whereas, at high temperatures, physical adsorption is small (because of the low adsorption energy), and only chemisorption occurs. An example is that of hydrogen on nickel, for which a schematic isobar is shown in Fig. XVI-2. Curve 1 shows the normal decrease in physical adsorption with temperature, and curve 2, that for chemisorption. In the transition region, curve 3, the rate of chemisorption is slow, but not negligible, so the location of the points depends on the equilibration time allowed; curve 3 is therefore not an equilibrium one and is not retraced on cooling, but rather some curve between 3 and 4, depending on the rate.

As is made evident in the next section, there is no sharp dividing line between these two types of adsorption, although the extremes are easily distinguishable. It is true that most of the experimental work has tended to cluster at these extremes, but this is more a reflection of practical interests and of human nature than of anything else. At any rate, although this chapter is ostensibly devoted to physical adsorption, much of the material can be applied to chemisorption as well. For the moment, we do assume that the adsorption process is reversible in the sense that equilibrium is reached and that on desorption the adsorbate is recovered unchanged.

2. The Adsorption Time

A useful approach to the phenomenon of adsorption is from the point of view of the adsorption time, as discussed by de Boer (4). Consider a molecule in the gas phase that is approaching the surface of the solid. If there were no attractive forces at all between the molecule and the solid, then the time of stay of the molecule in the vicinity of the surface would be of the order of a molecular vibration time, or about 10^{-13} sec, and its accommodation coefficient would be zero. By this last it is meant that the molecule retains its original energy. A "hot" molecule striking a cold surface should then rebound with its original energy, and its reflection from the surface would be specular.

If attractive forces are present, then according to an equation by Frenkel (see Ref. 2), the average time of stay τ of the molecule on the surface will be

$$\tau = \tau_0 e^{Q/RT} \tag{XVI-1}$$

where τ_0 is 10^{-12} to 10^{-13} sec and Q is the interaction energy, that is, the energy of adsorption. If τ is as large as several vibration periods, it becomes reasonable to consider that adsorption has occurred; temperature equilibration between the molecule and the surface is approached and, on desorption the molecule leaves the surface in a direction that is independent of that of its arrival; the accommodation coefficient is now said to be unity.

In addition to Q and τ, a quantity of interest is the surface concentration Γ, where

$$\Gamma = Z\tau \tag{XVI-2}$$

Here, if Z is expressed in moles of collisions per square centimeter per second, Γ is in moles per square centimeter. We assume the condensation coefficient to be unity, that is, that all molecules that hit the surface stick to it. At very low Q values, Γ as given by Eq. XVI-2 is of the order expected just on the basis that the gas phase continues uniformly up to the surface so that the net surface concentration (e.g., Γ_2^s in Eq. XI-22) is essentially zero. This is the situation prevailing in the first two rows of Table XVI-1. The table, which summarizes the spectrum of adsorption behavior, shows that with intermediate Q values of the order of a few kilocalories, Γ rises

TABLE XVI-1
The Adsorption Spectrum

Q, kcal/mole	τ, sec (25°C)	Γ, net moles/cm^2	Comments
0.1	10^{-13}	0	Adsorption nil; specular reflection; accommodation coefficient zero
1.5	10^{-12}	0	Region of physical adsorption;
3.5	4×10^{-11}	10^{-12}	accommodation coefficient
9.0	4×10^{-7}	10^{-8}	unity
20.0	100		Region of chemisorption
40.0	10^{17}		

to a level comparable to that for a complete monolayer. This intermediate region corresponds to one of physical adsorption.

The third region is one for which the Q values are of the order of chemical bond energies; the τ values become quite large, indicating that desorption may be slow, and Γ as computed by Eq. XVI-2 becomes preposterously large. Such values are evidently meaningless, and the difficulty lies in the assumption embodied in Eq. XVI-2 that the collision frequency gives the number of molecules hitting and *sticking* to the surface. As monolayer coverage is approached, it is to be expected that more and more impinging molecules will hit occupied areas and rebound without experiencing the full Q value. One way of correcting for this effect is taken up in the next section, which deals with the Langmuir adsorption equation.

3. The Langmuir Adsorption Isotherm

The following several sections deal with various theories or models for adsorption. It turns out that not only is the adsorption isotherm the most convenient form in which to obtain and plot experimental data, but it is also the form in which theoretical treatments are most easily developed. One of the first demands of a theory for adsorption then, is that it give an experimentally correct adsorption isotherm. Later it is shown that this test is insufficient and that a more sensitive test of the various models requires a consideration of how the energy and entropy of adsorption vary with the amount adsorbed.

A. Kinetic Derivation

The derivation that follows is essentially that given by Langmuir (5) in 1918, in which one writes separately the rates of evaporation and of condensation. The surface is assumed to consist of a certain number of sites S of which S_1 are occupied and $S_0 = S - S_1$ are free. The rate of evaporation is taken to be proportional to S_1, or equal to k_1S_1, and the rate of condensation proportional to the *bare surface* S_0 and to the gas pressure, or equal to k_2PS_0. At equilibrium,

$$k_1S_1 = k_2PS_0 = k_2P(S - S_1) \qquad \text{(XVI-3)}$$

Since S_1/S equals θ, the fraction of surface covered, Eq. XVI-3, can be written in the form

$$\theta = \frac{bP}{1 + bP} \qquad \text{(XVI-4)}$$

where

$$b = \frac{k_2}{k_1} \qquad \text{(XVI-5)}$$

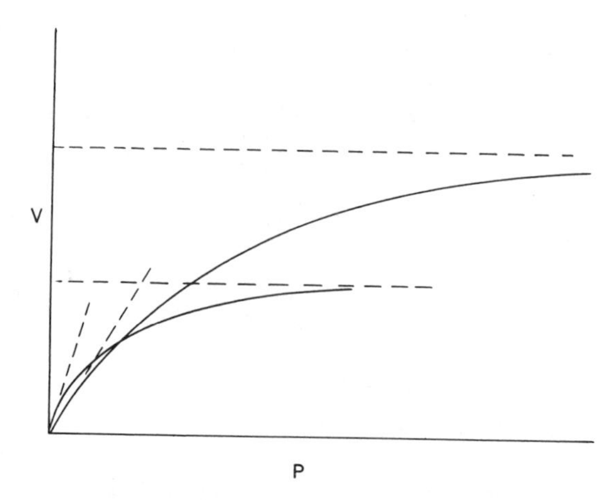

Fig. XVI-3. Langmuir isotherms.

Alternatively, θ can be replaced by n/n_m, where n_m denotes the moles per g adsorbed at the monolayer point. Thus

$$n = \frac{n_m bP}{1 + bP} \tag{XVI-6}$$

It is of interest to examine the algebraic behavior of Eq. XVI-6. At low pressure, the amount adsorbed becomes proportional to the pressure

$$n = n_m bP \tag{XVI-7}$$

whereas at high pressure, n approaches the limiting value n_m. Some typical shapes are illustrated in Fig. XVI-3. For convenience in testing data, Eq. XVI-6 may be put in the linear form

$$\frac{P}{n} = \frac{1}{bn_m} + \frac{P}{n_m} \tag{XVI-8}$$

A plot of P/n versus P should give a straight line, and the two constants n_m and b may be evaluated from the slope and intercept. In turn, n_m may be related to the area of the solid:

$$n_m = \frac{\Sigma}{N_0\sigma^0} \tag{XVI-9}$$

where Σ denotes the specific surface area of the solid, and σ^0 is the area of a site. The amount of a gas or a vapor that is adsorbed is often measured volumetrically, and may be expressed in terms of v, the cubic centimeters STP adsorbed per gram. Substitution of v for n and of v_m for n_m leaves Eqs. XVI-6 through XVI-8 unchanged algebraically. Since $n = v/v_0$, where v_0 is the STP number, 22,400 cm^3/mole, Eq. XVI-9 becomes

$$v_m = \frac{\Sigma v_0}{N_0\sigma^0} \tag{XVI-10}$$

If σ^0 can be estimated, Σ can be calculated from the experimental n_m or v_m.

If there are several competing adsorbates, a derivation analogous to the foregoing gives

$$n_i = \frac{n_{mi} b_i P_i}{1 + \sum_j b_j P_j} \tag{XVI-11}$$

The rate constants k_1 and k_2 may be related to the concepts of the preceding section as follows. First, k_1 is simply the reciprocal of the adsorption time, that is,

$$k_1 = \left(\frac{1}{\tau_0}\right) e^{-Q/RT} \tag{XVI-12}$$

In evaluating k_2, if a site can be regarded as a two-dimensional potential box, then the rate of adsorption will be given by the rate of molecules impinging on the site area σ_0. From gas kinetic theory,

$$k_2 = \frac{N_0 \sigma^0}{(2\pi MRT)^{1/2}} \tag{XVI-13}$$

and the Langmuir constant b becomes

$$b = \frac{N_0 \sigma^0 \tau_0 e^{Q/RT}}{(2\pi MRT)^{1/2}} \tag{XVI-14}$$

It will be convenient to write b as

$$b = b_0 e^{Q/RT} \qquad b_0 = \frac{N_0 \sigma^0 \tau_0}{(2\pi MRT)^{1/2}} \tag{XVI-15}$$

where b_0 is of the nature of a frequency factor. Thus for nitrogen at its normal boiling point of 77°K, b_0 is 9.2×10^{-4}, with pressure in atmospheres, σ^0 taken to be 16.2 Å^2 per site (actually, this is the estimated molecular area of nitrogen), and τ_0, as 10^{-12} sec.

B. Statistical Thermodynamic Derivation

The preceding derivation, being based on a definite mechanical picture, is easy to follow intuitively; kinetic derivations of an *equilibrium* relationship suffer from a common disadvantage, namely, that they usually assume more than is necessary. It is quite possible to obtain the Langmuir equation (as well as other adsorption isotherm equations) from examination of the statistical thermodynamics of the two states involved.

The following derivation is modified from that of Fowler and Guggenheim (6, 7). The adsorbed molecules are considered to differ from gaseous ones in that their potential energy and local partition function (see Section XV-4A) have been modified and that, instead of possessing normal translational motion, they are confined to *localized* sites without any interactions between adjacent molecules but with an adsorption energy Q.

Since translational and internal energy (of rotation and vibration) are independent, the partition function for the gas can be written

$$\mathbf{Q}^g = \mathbf{Q}^g_{\text{trans}} \mathbf{Q}^g_{\text{int}} \tag{XVI-16}$$

We write for the adsorbed or surface state

$$\mathbf{Q}^s = \mathbf{Q}^s_{\text{site}} \mathbf{Q}^s_{\text{int}} e^{Q/RT} \tag{XVI-17}$$

where the significance of the site partition function $\mathbf{Q}^s_{\text{site}}$ is explained later and the inclusion of the term $e^{Q/RT}$ means that $\mathbf{Q}^s$ is referred to the gaseous state. Furthermore, $\mathbf{Q}^s$ is a function of temperature only and not of the degree of occupancy of the sites. The complete partition function is obtained by multiplying $\mathbf{Q}^s$ by the number of distinguishable ways of placing N molecules on S sites. This number is obtained as follows: there are S ways of placing the first molecule, $(S - 1)$ for the second, and so on; and for N molecules, the number of ways is

$$S(S - 1) \cdots [S - (N - 1)] \qquad \text{or} \qquad S!/(S - N)!$$

Of these, $N!$ are indistinguishable since the molecules are not labeled, and the complete partition function for N molecules becomes

$$\mathbf{Q}^s_{\text{tot}} = \frac{S!}{(S - N)!N!} (\mathbf{Q}^s)^N \tag{XVI-18}$$

The Helmholtz free energy of the adsorbed layer is given by $-kT \ln \mathbf{Q}^s_{\text{tot}}$ (Eq. XV-20), and with the use of Sterling's approximation for factorials $x! = (x/e)^x$ one obtains

$$A^s = kT[-S \ln S + N \ln N + (S - N) \ln (S - N) - N \ln \mathbf{Q}^s] \tag{XVI-19}$$

The chemical potential μ^s is given by $(\partial A^s/\partial N)_T$, so that

$$\mu^s = kT \ln \frac{N}{S - N} - kT \ln \mathbf{Q}^s \tag{XVI-20}$$

For the gas phase,

$$\mu^g = -kT \ln \mathbf{Q}^g \tag{XVI-21}$$

and on equating the two chemical potentials (remembering that $\theta = N/S$) one obtains

$$\frac{\theta}{1 - \theta} = \frac{\mathbf{Q}^s}{\mathbf{Q}^g} \tag{XVI-22}$$

It is now necessary to examine the partition functions in more detail. The energy states for translation are assumed to be given by the quantum mechanical picture of a particle in a box. For a one-dimensional box of length a,

$$\epsilon_n = \frac{n^2h^2}{8a^2m} \tag{XVI-23}$$

so that the one-dimensional translational partition function is

$$\begin{aligned} \mathbf{Q}_{\substack{\text{trans} \\ \text{1 dim}}} &= \sum_n \exp\left(-\frac{\epsilon_n}{kT}\right) \simeq \int_0^\infty \exp\left(-\frac{n^2h^2}{8a^2mkT}\right) dn \\ &= \left(\frac{2\pi mkT}{h^2}\right)^{1/2} a \end{aligned} \tag{XVI-24}$$

where ordinarily the states are so close together that the sum over the quantum numbers n may be replaced by the integral (perhaps contrary to intuition, this is true even if a is of molecular magnitude). For a two-dimensional box,

$$\mathbf{Q}_{\substack{\text{trans} \\ \text{2 dim}}} = \left(\frac{2\pi mkT}{h^2}\right) a^2 \tag{XVI-25}$$

and in three dimensions,

$$\mathbf{Q}_{\text{trans}} = \left(\frac{2\pi mkT}{h^2}\right)^{3/2} \frac{kT}{P} \tag{XVI-26}$$

where a^3 is now replaced by the volume and, in turn, by kT/P. Substitution of Eqs. XVI-26 and XVI-17 into Eq. XVI-22 gives

$$\frac{\theta}{1 - \theta} = \frac{(h^2/2\pi mkT)^{3/2}}{kT} \frac{\mathbf{Q}^s_{\text{site}}\mathbf{Q}^s_{\text{int}}}{\mathbf{Q}^g_{\text{int}}} e^{Q/RT}P \tag{XVI-27}$$

or

$$\frac{\theta}{(1 - \theta)} = bP \tag{XVI-28}$$

which is the same as the Langmuir equation, Eq. XVI-4, but with b_0 of Eq. XVI-16 given by

$$b_0' = \frac{(h^2/2\pi mkT)^{3/2}}{kT} \frac{\mathbf{Q}^s_{\text{site}}\mathbf{Q}^s_{\text{int}}}{\mathbf{Q}^g_{\text{int}}} \tag{XVI-29}$$

The two expressions for b_0 may be brought into formal identity as follows. On adsorption, the three degrees of translational freedom can be supposed to appear as two degrees of translational motion within the confines of a two-dimensional box of area $a^2 = \sigma_0$, plus one degree of vibration in the adsorption bond, normal to the surface. The partition function for the first is given by Eq. XVI-25. For one degree of vibrational freedom the energy states are

$$\epsilon_n = (n + \tfrac{1}{2})\, h\nu^0 \tag{XVI-30}$$

in the case of a harmonic oscillator. The partition function is

$$\mathbf{Q}_{\mathrm{vib}} = \sum_n \exp(-\epsilon_n/kT) = \frac{e^{h\nu^0/2kT}}{e^{h\nu^0/kT} - 1} \qquad \text{(XVI-31)}$$

If the adsorption bond is weak so that $h\nu^0/kT \ll 1$, expansion of Eq. XVI-31 gives $\mathbf{Q}_{\mathrm{vib}} \simeq kT/h\nu^0$. If we now make the following identification:

$$\mathbf{Q}^s_{\mathrm{site}} = \underset{\sigma^0\ \mathrm{box}}{\mathbf{Q}^s_{\mathrm{trans}}} \times \underset{\mathrm{ads\ bond}}{\mathbf{Q}^s_{\mathrm{vib}}} \qquad \text{(XVI-32)}$$

then

$$\mathbf{Q}^s_{\mathrm{site}} = \frac{2\pi mkT}{h^2}\sigma^0\frac{kT}{h\nu^0} \qquad \text{(XVI-33)}$$

Since ν^0 corresponds to $1/\tau_0$, we have

$$b_0' = \frac{N\sigma^0\tau_0}{(2\pi MRT)^{1/2}}\left(\frac{\mathbf{Q}^s_{\mathrm{int}}}{\mathbf{Q}^g_{\mathrm{int}}}\right) = b_0\frac{\mathbf{Q}^s_{\mathrm{int}}}{\mathbf{Q}^g_{\mathrm{int}}} \qquad \text{(XVI-34)}$$

Thus the kinetic and statistical mechanical derivations may be brought into identity by means of a specific series of assumptions, including the assumption that the internal partition functions are the same for the two states (see Ref. 8). As discussed in Section XV-4A, this last is almost certainly not the case because as a minimum effect some loss of rotational degrees of freedom should occur on adsorption.

C. *Adsorption Entropies*

1. Configurational Entropies. The factorial expression in Eq. XVI-18 may be called the configurational partition function; it is that part of the partition function that has to do with the ways of arranging a given state. Thus

$$\mathbf{Q}^s_{\mathrm{config}} = \frac{S!}{(S-N)!N!} \qquad \text{(XVI-35)}$$

Since $\mathbf{Q}^s_{\mathrm{config}}$ has no temperature dependence, we have from Eq. XVI-20 that $\mathbf{S}^s_{\mathrm{config}} = k \ln \mathbf{Q}^s_{\mathrm{config}}$. On applying Sterling's approximation and dividing through by N, to obtain S^s_{config} on a per molecule basis, the result is

$$S^s_{\mathrm{config}} = -k\left[\frac{1-\theta}{\theta}\ln(1-\theta) + \ln\theta\right] \qquad \text{(XVI-36)}$$

This is an *integral* entropy; the differential entropy is obtained by the operation $\bar{S} = \partial(NS)/\partial N = S + N(\partial S/\partial N)$, which yields

$$\bar{S}^s_{\mathrm{config}} = -k\ln\frac{\theta}{1-\theta} \qquad \text{(XVI-37)}$$

(The same result can be obtained from Eq. XVI-20 since $\bar{S} = -\partial\mu/\partial T$, considering only the configurational term in that equation.)

Thus the thermodynamic description of the Langmuir model is that the energy of adsorption Q is constant and that the entropy of adsorption varies with θ according to Eq. XVI-37.

To further emphasize the special nature of the configurational entropy assumption embodied in the Langmuir model, let us repeat the derivation assuming instead that the surface molecules are mobile. In terms of the kinetic derivation, this amounts to setting the rate of condensation equal to k_2PS rather than to k_2PS_0, with the result that $\theta = bP$. The effect on the statistical thermodynamic derivation is first that the factorial grouping in Eq. XVI-18 becomes just $1/N!$ multiplied by S^N (since there are now S ways of placing each of the N molecules), with the result that Eq. XVI-22 becomes

$$\theta = \frac{\mathbf{Q}^s}{\mathbf{Q}^g} \tag{XVI-38}$$

and second that in obtaining $\mathbf{Q}^s_{\text{site}}$, a^2 in Eq. XVI-25 is replaced by the total area $\mathcal{A} = S\sigma^0$ rather than just by the site area σ^0. On making these substitutions, the result is

$$\theta = bP \tag{XVI-39}$$

$$b = b_0' e^{Q/RT}$$

The configurational entropies are now

$$S^s_{\substack{\text{config}\\ \text{mobile}}} = -k \ln \theta + k \tag{XVI-40}$$

$$\bar{S}^s_{\substack{\text{config}\\ \text{mobile}}} = -k \ln \theta \tag{XVI-41}$$

All of these entropies may be put on a per mole basis by replacing k by R.

The case of a vapor adsorbing on its own liquid surface should certainly correspond to mobile adsorption. Here, θ is unity and $P = P^0$, the vapor pressure. The energy of adsorption is now that of condensation Q_v, and it will be convenient to define the Langmuir constant for this case as b^0; thus, from Eq. XVI-39,

$$1 = b^0P^0 = b_0P^0 \exp\left(\frac{Q_v}{RT}\right) \tag{XVI-42}$$

If, furthermore, we write $c = b/b^0$ and $x = P/P^0$, the Langmuir equation can be put in the form

$$\theta = \frac{cx}{1 + cx} \tag{XVI-43}$$

2. *Entropies of Adsorption.* The Langmuir model is not usually considered to imply any particular value for the *total* entropy change on adsorption, but the statistical thermodynamic approach makes it easy to postulate various possible values. In considering, for example, the differential entropy change that should occur when 1 mole of gas at 1 atm pressure adsorbs at surface coverage θ, the matter becomes one of assembling the various possible contributions.

First, the total translational partition function for 1 mole of gas is

$$\mathbf{Q}^{g}_{\text{tot}} = \frac{1}{N!}\left[\left(\frac{2\pi mkT}{h^2}\right)^{3/2} V\right]^N \tag{XVI-44}$$

On applying Sterling's approximation, replacing V/N by kT/P, and using Eq. XVI-20 and setting $P = 1$ atm, the final result, known as the Sackur–Tetrode equation, is

$$\bar{S}^{0,g}_{\text{trans}} = R\ln(T^{5/2}M^{3/2}) - 2.30 \tag{XVI-45}$$

For nitrogen at 77°K, $\bar{S}^{0,g}_{\text{trans}}$ is 29.2 EU(cal/K mole).

For localized adsorption, we need the contribution from the adsorption bond. Per degree of vibrational freedom, we obtain on applying Eq. XVI-20 to Eq. XVI-31,

$$S^s_{\text{vib}} = R\left\{\frac{h\nu^0/kT}{e^{h\nu^0/kT} - 1} - \ln(1 - e^{-h\nu^0/kT})\right\} \tag{XVI-46}$$

which approximates to $R[1 - \ln(h\nu^0/kT)]$. If T is 77 K and ν^0 is taken to be 10^{12} sec^{-1} (see Ref. 9), then S^s_{vib} is 2.9 EU per degree of freedom. Second, if the adsorption site is regarded as a two-dimensional potential box, as implied by the kinetic derivation given above, the corresponding translational entropy must be evaluated. The total partition function for N molecules is just $[(2\pi mkT/h^2)\sigma^0]^N$, and the usual operation yields

$$\bar{S}^s_{\substack{\text{trans}\\ \sigma^0\text{ box}}} = R\ln(MT\sigma^0) + 63.8 \tag{XVI-47}$$

For nitrogen at 77°K, with $\sigma^0 = 16.2$ Å^2, this entropy becomes 11.4 EU.

If the adsorbed gas is mobile, the total partition function is $(S^N/N!)$ $[(2\pi mkT/h^2)\mathcal{A}]^N$, which gives (see Ref. 7)

$$S^s_{\substack{\text{trans}\\ \text{2 dim}}} = -R\ln\theta + R\ln(MT\sigma^0) + 65.8 \tag{XVI-48}$$

remembering that $\mathcal{A}/N = \sigma^0/\theta$. Alternatively,

$$\bar{S}^s_{\substack{\text{trans}\\ \text{2 dim}}} = -R\ln\theta + R\ln(MT\sigma^0) + 63.8 \tag{XVI-49}$$

Notice that the nonconfigurational part of Eq. XVI-49 is just the entropy given by Eq. XVI-47.

We can now proceed with various estimates of the entropy of adsorption $\Delta\bar{S}^0_{\text{ads}}$. Two extreme positions are sometimes taken (see Ref. 10). First, one assumes that for localized adsorption the only contribution is the configurational entropy. Thus

$$\begin{aligned}\Delta\bar{S}^0_{\substack{\text{ads}\\ \text{local}}} &= -R\ln\left(\frac{\theta}{1-\theta}\right) - \bar{S}^{0,g}_{\text{trans}}\\ &= \qquad 0 \qquad - 29.2\\ &= -29.2\end{aligned} \tag{XVI-50}$$

where the numbers are for nitrogen at 77 K and $\theta = 0.5$. This contrasts with the value for a mobile film

$$\begin{aligned}\Delta \bar{S}^0_{\substack{\text{ads}\\ \text{mobile}}} &= -R \ln \theta + S^s_{\substack{\text{trans}\\ \sigma^0 \text{ box}}} - \bar{S}^{0,g}_{\text{trans}} \\ &= 1.4 \quad + 11.4 \quad - 29.2 \\ &= -16.4\end{aligned} \tag{XVI-51}$$

This difference looks large enough to be diagnostic of the state of the adsorbed film. However, to be consistent with the kinetic derivation of the Langmuir equation it was necessary to suppose that the site acted as a potential box and, furthermore, that a weak adsorption bond of ν^0 corresponding to $1/\tau_0$ was present. With these provisions we obtain

$$\begin{aligned}\Delta \bar{S}^0_{\substack{\text{ads}\\ \text{local}}} &= -R \ln \left(\frac{\theta}{1-\theta}\right) + S^s_{\substack{\text{trans}\\ \sigma^0 \text{ box}}} + S^s_{\substack{\text{vib}\\ \text{ads bond}}} - \bar{S}^{0,g}_{\text{trans}} \\ &= 0 \quad + 11.4 \quad + 2.9 \quad - 29.2 \\ &= -14.9\end{aligned} \tag{XVI-52}$$

Thus the entropy of "localized" adsorption can range widely, depending on whether the site is viewed as equivalent to a strong adsorption bond of negligible entropy or as a potential box plus a weak bond (see Ref. 8). In addition, estimates of $\Delta \bar{S}^0_{\text{ads}}$ should include possible surface vibrational contributions in the case of mobile adsorption, and all calculations are faced with possible contributions from a loss in rotational entropy on adsorption as well as from change in the adsorbent structure following adsorption (see Section XV-4B). These uncertainties make it virtually impossible to affirm what the state of an adsorbed film is from entropy measurements alone; for this, additional independent information about surface mobility and vibrational surface states is needed. By way of illustration, Ross (11) found that $\Delta \bar{S}^0_{\text{ads}}$ for n-butane on Spheron 6 carbon was about the value given by Eq. XVI-51 for a mobile film; yet the surface diffusion coefficient showed an activation energy of around 6 kcal, so the adsorbed butane could not actually have been in a mobile state.

D. Lateral Interaction

It is assumed in the Langmuir model that while the adsorbed molecules occupy sites of energy Q they do not interact with each other. An approach due to Fowler and Guggenheim (6) allows provision for such interaction. The probability of a given site being occupied is N/S, and if each site has z neighbors, the probability of a neighbor site being occupied is zN/S, so the fraction of adsorbed molecules involved is $z\theta/2$, the factor one-half correcting for double counting. If the lateral interaction energy is ω, the added energy of adsorption is $z\omega\theta/2$, and the added differential energy of adsorption is just $z\omega\theta$.

The modified Langmuir equation becomes

$$\frac{\theta}{1-\theta} = b'P$$

$$b' = b_0 \exp\left(\frac{Q + z\omega\theta}{RT}\right) = b \exp\left(\frac{z\omega\theta}{RT}\right) \tag{XVI-53}$$

It is convenient to illustrate the lateral interaction effect by plotting θ versus the Langmuir variable bP for various values of $\beta = z\omega/RT$, as shown in Fig. XVI-4. For $\beta < 4$, the isotherms are merely steepened versions of the Langmuir equation, but if $\beta > 4$, a maximum and a minimum in bP appear so that a two-phase equilibrium is implied, with phases of θ values lying at the ends of the dotted line.

The quantity $z\omega$ will depend very much on whether adsorption sites are close enough for neighboring adsorbate molecules to develop their normal van der Waals attraction; if, for example, $z\omega$ is taken to be about one-fourth of the energy of vaporization (12), β would be 2.5 for a liquid obeying Trouton's rule and at its normal boiling point. The critical pressure P_c, that is, the pressure corresponding to $\theta = 0.5$ with $\beta = 4$, will depend on both Q and T. A way of expressing this follows, with the use of the definitions of Eqs. XVI-42 and XVI-43 (13):

$$\frac{P_c}{P^0} = \frac{1}{e^2 c} \tag{XVI-54}$$

This is useful since c can be estimated by means of the BET equation (see Section XVI-5). A number of more or less elaborate variants of the preceding treatment of lateral interaction have been proposed. Thus, Kiselev and co-workers, in their very extensive studies of physical adsorption, have proposed an equation of the form

$$K_1 x = \frac{\theta}{1-\theta} - K_1 K_n \theta x \tag{XVI-55}$$

where K_n reflects the degree of adsorbate–adsorbate interaction and $c = K_1 + K_1 K_n$ so that Eq. XVI-55 reduces to Eq. XVI-43 if K_n is zero (14). Misra (15) has summarized other (and semiempirical) Langmuir-like adsorption isotherms. A fundamental approach by Steele (16) treats monolayer adsorption in terms of interatomic

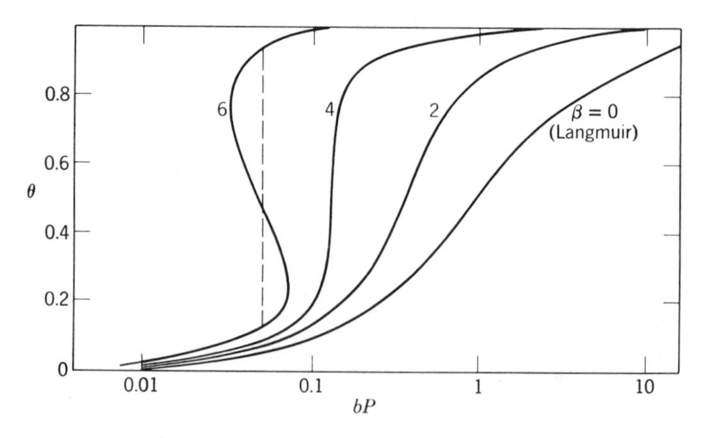

Fig. XVI-4. Langmuir plus lateral interaction isotherms.

potential functions, and includes pair and higher order interactions. Young and Crowell (17) and Honig (18) give additional details on the general subject.

E. Experimental Applications of the Langmuir Equation

A variety of experimental adsorption data has been found to fit the Langmuir equation fairly well, and many representative examples are given by Brunauer (19) and Young and Crowell (17). Data are generally plotted according to the linear form, Eq. XVI-8, and the constants b and n_m are calculated from the best fitting straight line. The specific surface area can be obtained from Eq. XVI-9, provided σ^0 is known. There is some ambiguity at this point. A widely used practice is to take σ^0 to be the molecular area of the adsorbate, estimated from liquid or solid densities; this was done by Brunauer and co-workers in their study of the adsorption of various gases on charcoal (20). The same practice is conventional in the case of adsorption from solution (see Section XI-3). On the other hand, the Langmuir model is cast around the concept of adsorption sites, whose spacing one would suppose to be characteristic of the adsorbent. However, to obtain site spacings requires much more knowledge about the surface structure than normally has been available, and actually the widespread use of an adsorbate-based value of σ^0 has mainly been *faute de mieux*. See Section XVI-5B for an additional discussion of σ^0 values.

A true fit to the Langmuir equation implies that n_m and Q are independent of temperature, and systems obeying the form of Eq. XVI-8 often fail this more severe test. In some cases, multilayer formation, discussed in the next section, is a source of trouble, but in general the Langmuir model is too simple for really detailed agreement to be expected with experimental systems. However, more sophisticated models also introduce more semiempirical parameters, so the net gain, apart from data fitting, has not been great. The simple Langmuir model has therefore retained great general utility as well as providing the point of departure for many of the proposed refinements (as for example, Eq. XVI-53).

4. Experimental Procedures

The remainder of the chapter is concerned with increasingly specialized developments in the study of gas adsorption, and before proceeding to this material, it seems desirable to consider briefly some of the experimental techniques that are important in obtaining gas adsorption data. A detailed general review of these has been given by Ross and Oliver (21).

Perhaps the most widely used technique is that which makes use of pressure–volume measurements to determine the amount of adsorbate gas before and after exposure to the adsorbent. An elementary type of vacuum line suitable for this purpose is shown in Fig. XVI-5. The adsorbent, usually a powdered solid, is placed in bulb A, which is kept at the desired temperature of adsorption T_1; the vacuum line also contains a manometer B and a gas buret C. One first determines the "dead space,"

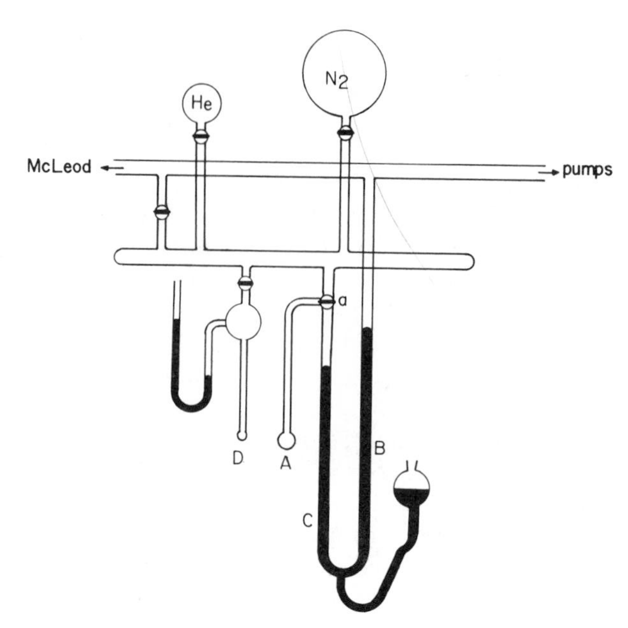

Fig. XVI-5. Vacuum line for gas adsorption measurements. *A*, sample bulb; *B* and *C*, gas buret and manometer; *D*, vapor pressure thermometer; *a*, three-way stopcock.

or the gas volume in bulb *A*, up to the three-way stopcock *a*, and this is done by evacuating the line and then admitting some nonadsorbed gas such as helium to some pressure and volume reading on the manometer and buret. Stopcock *a* is then turned so as to connect *A* and *C*, and the change in pressure and volume readings is noted. Since the dead space includes some volume at T_1 and some at room temperature T_2, it is necessary to make two independent measurements of the above type if each volume is to be determined. The helium is then removed, and the adsorbate gas is admitted while stopcock *a* is closed. The amount of gas is determined from the manometer and buret readings. The stopcock is then turned to connect *A* and *C* and, from the new readings taken after a suitable equilibration time, the amount of adsorbed gas can be calculated, *D* is vapor pressure thermometer, immersed in the same bath as the sample bulb. Some special aspects related to very low pressure work are discussed by Hobson (22). Davis *et al.* (23) describe an undergraduate laboratory procedure. Automated apparatus of the type described is available commercially.

A second general type of procedure, due to McBain (24), is to determine n by a direct weighing of the amount of adsorption. McBain used a delicate quartz spiral spring, but modern equipment generally makes use of a microbalance or a transducer (e.g., see Ref. 25). Again, largely automated equipment for gravimetric adsorption analysis is available commercially.

If the total surface area is small (say a few hundred square centimeters), the amount adsorbed becomes so little that measurements are difficult by normal procedures. Thus the change in pressure–volume product on admitting gas to the adsorbent becomes so small that precision is impaired.

One way of avoiding this problem is to set T_1 so that the vapor pressure of the

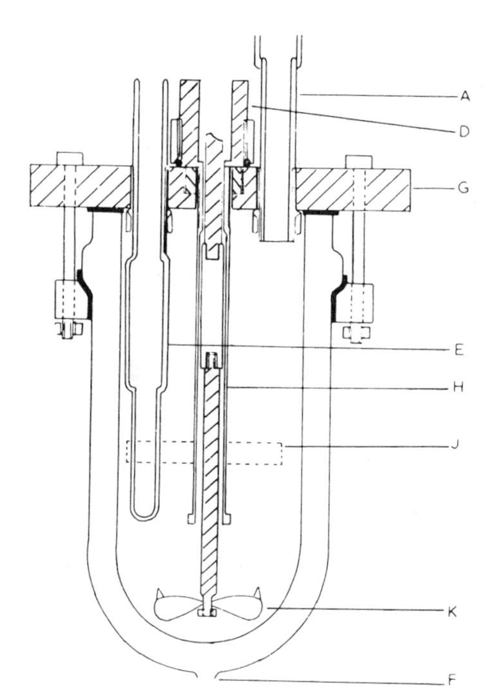

Fig. XVI-6. Heat of immersion calorimeter. *A*, solvent and dry nitrogen inlet system; *B*, holding brackets for fixing the calorimeter in the thermostat bath; *C*, four-lead heater supply and measurement system; *D*, stirrer stuffing box; *E*, thermistor pockets; *F*, connections for evacuating the calorimeter jackets; *G*, nylon lid; *H*, heater support and stirrer guide; *J*, sample holder; *K*, stirrer blade. (From Ref. 29.)

liquid adsorbate is very low, (e.g., krypton at −195°C) (26, 27). Monolayer formation usually occurs when a few tenths of saturation pressure is reached, regardless of its *absolute* value; and if P^0 is very small, even small amounts of adsorption will cause relatively large changes in the pressure–volume product of the gas in the vacuum line.

Ultrahigh vacuum techniques have become common, especially in connection with surface spectroscopic and diffraction studies, but also in adsorption on very clean surfaces. The techniques have become rather specialized and the reader is referred to Ref. 22 and citations therein.

The heat of adsorption is an important experimental quantity. The heat evolution with each of successive admissions of adsorbate vapor may be measured directly by means of a calorimeter described by Beebe and coworkers (28). Alternatively, the heat of immersion in liquid adsorbate of adsorbent having various amounts preadsorbed on it may be determined. The difference between the two values is related to the integral heat of adsorption (see Section X-3B) between the two degrees of coverage. An example of a contemporary calorimeter is shown in Fig. XVI-6 (29); see also Ref. 30. A quartz crystal oscillator may be used as a temperature sensor (31). See also Section XV-2B for additional references on calorimetry. Finally, adsorption can be measured by chromatographic procedures (17), as was true in the case of adsorption from solution.

5. The BET and Related Isotherms

Adsorption isotherms are by no means all of the Langmuir type as to shape, and Brunauer (19) considered that there are five principal forms, as illustrated in Fig. XVI-7. Type I is the Langmuir type, roughly characterized by a monotonic approach to a limiting adsorption that presumably corresponds to a complete monolayer. Type II is very common in the case of physical adsorption and undoubtedly corresponds to multilayer formation. For many years it was the practice to take point B, at the knee of the curve, as the point of completion of a monolayer, and surface areas obtained by this method are fairly consistent with those found using adsorbates that give type I isotherms. Type III is relatively rare—a recent example is that of the adsorption of nitrogen on ice (32)—and seems to be characterized by a heat of adsorption equal to or less than the heat of liquefication of the adsorbate. Types IV and V are considered to reflect capillary condensation phenomena in that they level off before the saturation pressure is reached and may show hysteresis effects.

This description is traditional, and some further comment is in order. The flat region of the type I isotherm has never been observed up to pressures approaching P^0; this type typically is observed in chemisorption, at pressures far below P^0. Types II and III approach the P^0 line asymptotically; experimentally, such behavior is observed for adsorption on powdered samples, and the approach toward infinite film thickness is actually due to interparticle condensation (33) (see Section X-5C) although such behavior is expected even for adsorption on a flat surface if bulk liquid adsorbate wets the adsorbent. Types IV and V specifically refer to porous solids.

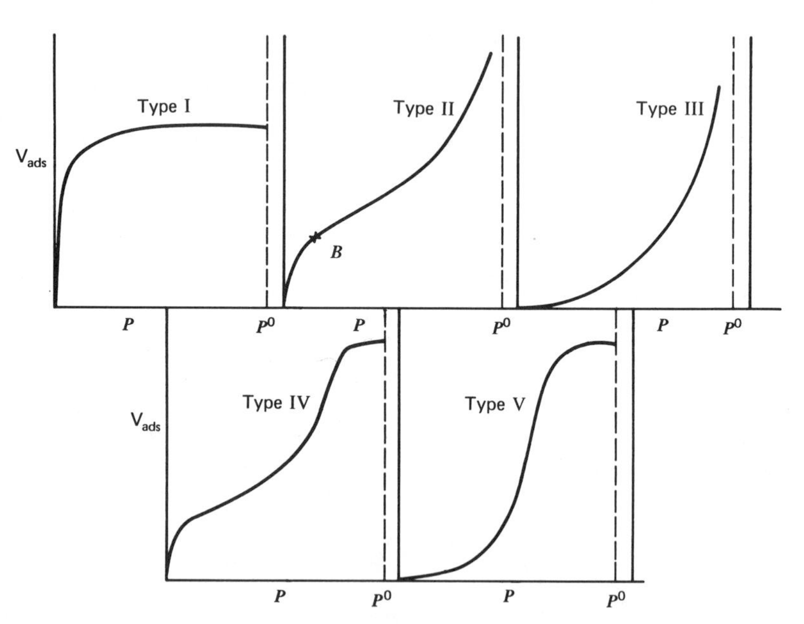

Fig. XVI-7. Brunauer's five types of adsorption isotherms. (From Ref. 19.)

There is a need to recognize at least the two additional isotherm types shown in Fig. XVI-8. These are two simple types possible for adsorption on a flat surface for the case where bulk liquid adsorbate rests on the adsorbent with a finite contact angle (34, 35).

A. Derivation of the BET Equation

Because of their prevalence in physical adsorption studies on high energy, powdered solids, type II isotherms are of considerable practical importance. Brunauer, Emmett, and Teller (20) showed how to extend Langmuir's approach to multilayer adsorption, and their equation has come to be known as the BET equation. The derivation that follows is the traditional one, based on a detailed balancing of forward and reverse rates.

The basic assumption is that the Langmuir equation applies to each layer, with the added postulate that for the first layer the heat of adsorption Q may

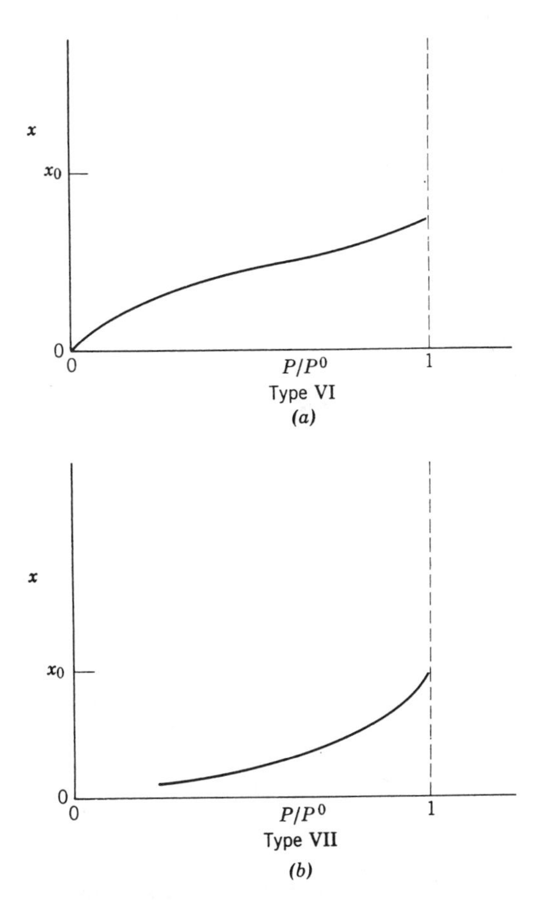

Fig. XVI-8. Two additional types of adsorption isotherms expected for nonwetting adsorbate–adsorbent systems. (From Ref. 34.)

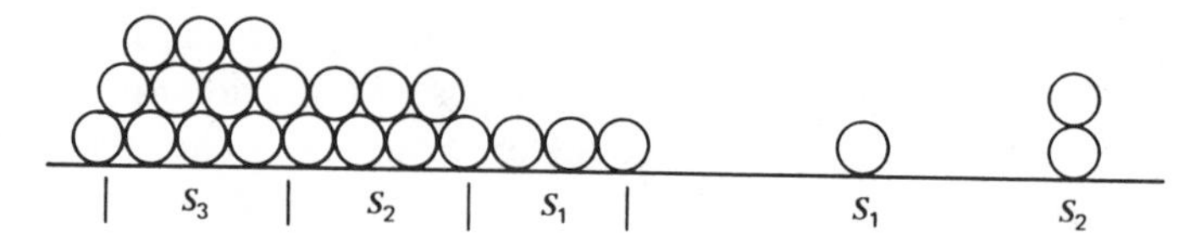

Fig. XVI-9. The BET model.

have some special value, whereas for all succeeding layers, it is equal to Q_v, the heat of condensation of the liquid adsorbate. A further assumption is that evaporation and condensation can occur only from or on exposed surfaces. As illustrated in Fig. XVI-9, the picture is one of portions of uncovered surface S_0, of surface covered by a single layer S_1, by a double layer S_2, and so on. The condition for equilibrium is taken to be that the amount of each type of surface reaches a steady state value with respect to the next deeper one. Thus for S_0

$$a_1PS_0 = b_1S_1e^{-Q_1/RT} \tag{XVI-56}$$

and for all succeeding surfaces,

$$a_iPS_{i-1} = b_iS_ie^{-Q_v/RT} \tag{XVI-57}$$

It then follows that

$$S_1 = yS_0, \qquad S_2 = xS_1$$

and

$$S_i = x^{i-1}S_1 = yx^{i-1}S_0 = cx^iS_0$$

where

$$y = \frac{a_1}{b_1}Pe^{Q_1/RT}$$

$$x = \frac{a_i}{b_i}Pe^{Q_v/RT} \tag{XVI-58}$$

and

$$c = \frac{y}{x} = \frac{a_1b_i}{b_1a_i}e^{(Q_1-Q_v)/RT} \simeq e^{(Q_1-Q_v)/RT} \tag{XVI-59}$$

Then

$$\frac{n}{n_m} = \frac{\sum_{i=1}^{\infty} iS_i}{\sum_{i=0}^{\infty} S_i} = cS_0\frac{\sum_{i=1}^{\infty} ix^i}{S_0 + S_0c\sum_{i=1}^{\infty} x^i} \tag{XVI-60}$$

Insertion of the algebraic equivalents to the sums yields

$$\frac{n}{n_m} = \frac{cx/(1-x)^2}{1 + cx/(1-x)} \tag{XVI-61}$$

which rearranges to

$$\frac{n}{n_m} = \frac{cx}{(1 - x)[1 + (c - 1)x]} \qquad x = \frac{P}{P^0} \qquad \text{(XVI-62)}$$

The essential next step is to take the ratio of frequency terms (a_i/b_i) to be the same as for liquid adsorbate–vapor equilibrium so that combination of Eqs. XVI-58 and XVI-47 (where $b_0^0 = a_i/b_i$) identifies x as equal to P/P^0. Usually, also, it is assumed that $a_1/b_1 = a_i/b_i$ in that the approximate form of Eq. XVI-59 is used in interpreting the constant c.

Although the preceding derivation is the easier to follow, the BET equation also may be derived from statistical mechanics by a procedure similar to that described in the case of the Langmuir equation (36, 37).

B. Properties of the BET Equation

The BET equation filled an annoying gap in the interpretation of adsorption isotherms, and at the time of its appearance in 1938 it was also hailed as a general method for obtaining surface areas from adsorption data. The equation can be put in the form

$$\frac{x}{n(1 - x)} = \frac{1}{cn_m} + \frac{(c - 1)x}{cn_m} \qquad \text{(XVI-63)}$$

so that n_m and c can be obtained from the slope and intercept of the straight line best fitting the plot of $x/n(1 - x)$ versus x. The specific surface area can then be obtained through Eq. XVI-9 if σ^0 is known. In the case of multilayer adsorption it seems reasonable to take σ^0 as an adsorbate (as opposed to an adsorbent site) area, based on either the solid or the liquid density, depending on the temperature. Values that give reasonably self-consistent areas are (in square angstroms per molecule): N_2, 16.2; O_2, 14.1; Ar, 13.8; Kr, 19.5; n-C_4H_{10}, 18.1. These and values for other adsorbates are reviewed critically by McClellan and Harnsberger (38). The preceding values are close to the calculated ones from the liquid densities at the boiling points and are, in this respect, reasonable for a multilayer adsorption situation. There may be special cases. Pierce and Ewing (39) suggest that for graphite surfaces it is the lattice spacing of the adsorbent that controls and that the effective molecular area for N_2 becomes 20 Å^2 rather than the usual 16.2 Å^2.

From the experimental point of view, the BET equation is easy to apply, and the surface areas so obtained are reasonably consistent (see Section XVI-8 for further discussion). The equation in fact has become the standard one for practical surface area determinations, usually with nitrogen at 77 K as the adsorbate, but in general with any system giving type II isotherms. On the other hand, the region of fit usually is not very great—the linear region of a plot according to Eq. XVI-64 typically lies between a P/P^0 of 0.05 and 0.3, as illustrated in Fig. XVI-17. The typical deviation is such that the best-fitting BET equation predicts too little adsorption at low pressures and too much at high pressures.

The BET equation also seems to cover three of the five isotherm types described in Fig. XVI-7. Thus for c large, that is, $Q_1 \gg Q_v$, it reduces to the Langmuir equation, Eq. XVI-43, and for small c values, type III isotherms result, as illustrated in Fig. XVI-10. However, the adsorption of relatively inert gases such as nitrogen and argon on polar surfaces generally gives c values around 100, corresponding to type II isotherms. For such systems the approximate form of Eq. XVI-62,

$$\frac{n}{n_m} = \frac{1}{1 - x} \tag{XVI-64}$$

works quite well in the usual region of fit of the BET equation, and a "one point" method of surface area estimation thus follows (see Ref. 40). The method has been incorporated in commercial rapid surface area determination equipment.

C. Modifications of the BET Equation

The very considerable success of the BET equation stimulated various investigators to consider modifications of it that would correct certain approximations and give a better fit to type II isotherms. Thus if it is assumed

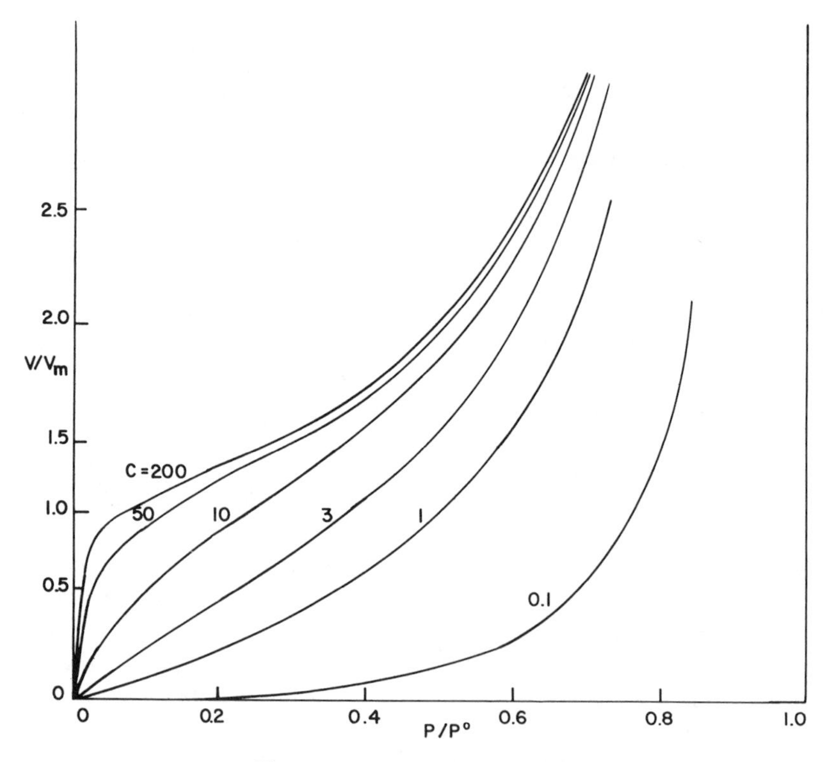

Fig. XVI-10. BET isotherms.

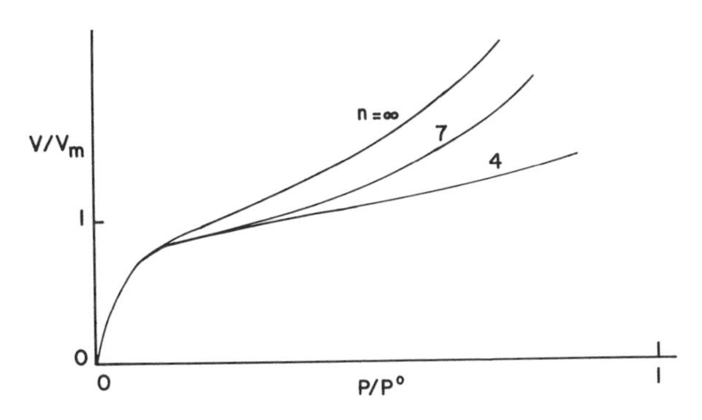

Fig. XVI-11. BET isotherms with adsorption limited to n layers.

that multilayer formation is limited to n layers, perhaps because of the opposing walls of a capillary being involved, one obtains (1)

$$v = \frac{[v_m cx/(1 - x)][1 - (n + 1)x^n + nx^{n+1}]}{1 + (c - 1)x - cx^{n+1}} \qquad \text{(XVI-65)}$$

As can be seen in Fig. XVI-11, by choosing the appropriate n value, the amount of adsorption predicted at large P/P^0 values is reduced, and a better fit to data can usually be obtained. Kiselev and co-workers have proposed an equation paralleling Eq. XVI-55, but which reduces to the BET equation if K_n, the lateral interaction parameter, is zero (14). There are other modifications that also give better fits to data but also introduce one or more additional parameters. Young and Crowell (17) and Gregg and Sing (41) may be referred to for a more detailed summary of such modifications, and one further example will suffice here. If the adsorbed film is not quite liquidlike so that $a_i/b_i \neq b_0^0$, its effective vapor pressure will be $P^{0\prime}$ rather than P^0 (note Section X-6C), and the effect is to multiply x by a factor k (42).

6. Isotherms Based on the Equation of State of the Adsorbed Film

It was shown in Section X-3B that a formal calculation can be made of the film pressure, or reduction in the surface free energy of the solid, by evaluating the integral of $nd \ln P$. Conversely, if it is assumed that the adsorbed material can be treated as a two-dimensional film of one of the types found with monolayers on liquid substrates, one has a large new source of isotherm equations.

A. *Film Pressure–Area Diagrams from Adsorption Isotherms*

According to Eq. X-12, straightforward application of the Gibbs equation gives

$$\pi_{\text{at } P_1} = \frac{RT}{\Sigma} \int_0^{P_1} nd \ln P \qquad \text{(XVI-66)}$$

Graphical integration of adsorption data is often more conveniently carried out by using Eq. XVI-66 in the form

$$\pi_{\text{at } P_1} = \frac{RT}{\Sigma}\left(\int_{P/v \text{ as } v\to 0}^{P_1/v_1} n d \ln \frac{P}{n} + n_1\right) \qquad \text{(XVI-67)}$$

At low pressures a linear or Henry's law region may be reached such that the ratio P/n is constant; the plot of n versus $\ln (P/n)$ will thus have zero area except insofar as departure from linearity occurs. If the isotherm is linear up to n_1, Eq. XVI-67 reduces to

$$\pi = \frac{RTn_1}{\Sigma} = \Gamma RT \qquad \text{or} \qquad \pi\sigma = RT \qquad \text{(XVI-68)}$$

Gregg (43; see also Ref. 41) has surveyed the types of force–area diagrams obtainable from adsorption isotherms. Qualitatively, there is enough similarity between films adsorbed on solids and those on liquids to suggest that two-dimensional equations of state might be useful.

B. Adsorption Isotherms from Two-Dimensional Equations of State

There are a number of types of equations of state that apply to solids, liquids, or gases and to their two-dimensional analogs; a few of these are considered in the following sections.

Ideal Gas Law. Here Eq. XVI-68 applies, and on reversing the procedure that led to it one finds

$$kP = \theta \qquad \text{(XVI-69)}$$

The ideal gas law equation of state thus leads to a linear or Henry's law isotherm.

A natural modification of the ideal gas law would include a covolume term

$$\pi(\sigma - \sigma^0) = RT \qquad \text{(XVI-70)}$$

Then

$$\ln P = \frac{1}{RT}\int \sigma \, d\pi = \int \frac{d\theta}{\theta(1-\theta)^2} \qquad \text{(XVI-71)}$$

(remembering that $\theta = \sigma^0/\sigma$), from which

$$kP = \frac{\theta}{1-\theta} e^{\theta/(1-\theta)} \qquad \text{(XVI-72)}$$

As illustrated in Fig. XVI-12, Eq. XVI-72 bears a strong resemblance to the Langmuir equation—see also de Boer (4)—to the point that it is doubtful whether the two could always be distinguished experimentally. An equivalent form, obtained by Volmer (44) is

$$\ln kP = \ln \pi + \frac{b\pi}{RT} \qquad \text{(XVI-73)}$$

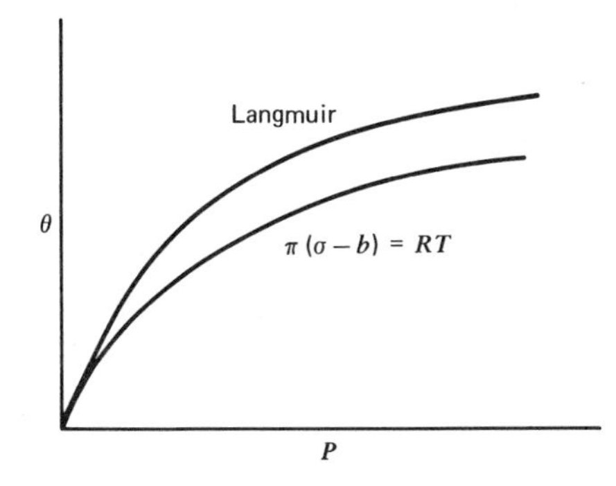

Fig. XVI-12. Comparison of Langmuir and modified two-dimensional gas law isotherms.

and Kemball and Rideal (45) found this form to fit fairly well their data on the adsorption of various organic vapors on mercury.

van der Waals Equations of State. A logical step to take next is to consider equations of state that contain both a covolume term and an attractive force term, such as the van der Waals equation. De Boer (4) and, more recently, Ross and Oliver (21) have given this type of equation much emphasis. This writer, also, has explored in some detail the types of adsorption isotherms that can be obtained (46).

It must be remembered that, in general, the constants a and b of the van der Waals equation depend on volume and on temperature. Thus a number of variants are possible, and some of these and the corresponding adsorption isotherms are given in Table XVI-2. All of them lead to rather complex adsorption equations, but the general appearance of the family of isotherms from any one of them is as illustrated in Fig. XVI-13. The dotted line in the figure represents the presumed actual course of that particular isotherm and

TABLE XVI-2
Two-Dimensional Equations of State and Corresponding Isotherms

Equation of state	Corresponding isotherm
Ideal gas type	
$\pi\sigma = RT$	$\ln kP = \ln \theta$
$\pi(\sigma - \sigma^0) = RT$	$\ln kP = \theta/(1 - \theta) + \ln [\theta/(1 - \theta)]$
van der Waals type	
$(\pi + a/\sigma^2)(\sigma - \sigma^0) = RT$	$\ln kP = \theta/(1 - \theta) + \ln [\theta/(1 - \theta)] - c\theta$
$(\pi + a/\sigma^3)(\sigma - \sigma^0) = RT$	$\ln kP = \theta/(1 - \theta) + \ln [\theta/(1 - \theta)] - c\theta^2$
$(\pi + a/\sigma^3)(\sigma - \sigma^0/\sigma) = RT$	$\ln kP = 1/(1 - \theta) +$ $1/2 \ln [\theta/(1 - \theta)] - c\theta$ $(c = 2a/\sigma^0RT)$
Viral type	
$\pi\sigma = RT + \alpha\pi - \beta\pi^2$	$\ln kP = (\phi^2/2\omega) + (1/2\omega)(\phi + 1)$ $[(\phi - 1)^2 + 2\omega)]^{1/2}$ $-\ln \{(\phi - 1) + [(\phi - 1)^2 + 2\omega]^{1/2}\}$ $(\phi = 1/\theta,\ \omega = 2\beta RT/\alpha^2)$

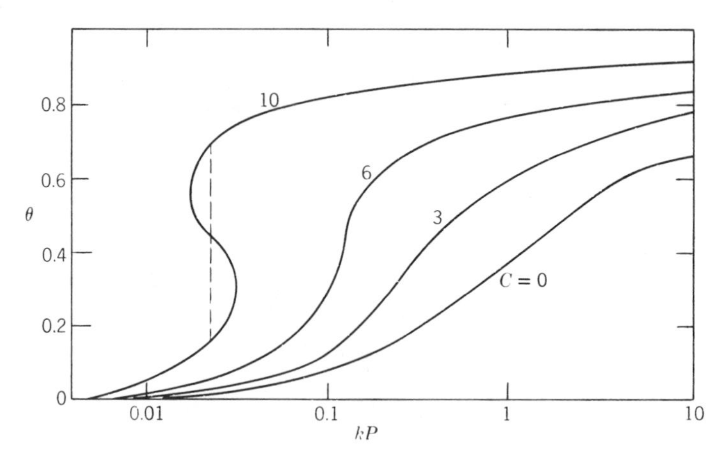

Fig. XVI-13. The van der Waals equation of state isotherm.

corresponds to a two-dimensional condensation from gas to liquid. Notice the general similarity to the plots of the Langmuir plus the lateral interaction equation shown in Fig. XVI-4.

Equations of this type may well apply to selected cases of adsorption in the submonolayer region. Thus Ross and Winkler (47) found that the adsorption of nitrogen on a carbon black at 77.8 K obeyed the third equation of Table XVI-2 for θ values between 0.1 and 0.5. The difficulty is that, as with the various modified BET equations, there are enough possibilities for algebraic variations and enough parameters (n_m, c, and k), that it is difficult to know how much significance should be attached to ability to fix data. The matter of possible phase changes is discussed in Section XVI-11 and that of the complicating effect of surface heterogeneity, in Section XVI-14.

Condensed Films. Harkins and Jura (48) proposed that type II isotherms be represented by Eq. IV-44 in the form

$$\pi = b - a\sigma \qquad \text{(XVI-74)}$$

On applying the Gibbs transformation, one obtains

$$\ln \frac{P}{P^0} = B - \frac{A}{n^2} \qquad \text{(XVI-75)}$$

where

$$A = \frac{a\Sigma^2}{2RT} \quad \text{and} \quad \Sigma = kA^{1/2} \qquad \text{(XVI-76)}$$

According to Eq. XIV-75, a plot of ln (P/P^0) versus $1/n^2$ should give a straight line, and, indeed, many type II isotherms do, for example, nitrogen on titanium dioxide. As an empirical observation, k in Eq. XVI-76 was independent of the nature of the solid for a given adsorbate; its value for nitrogen at 77 K was 4.06 for Σ in square meters per gram.

7. The Potential Theory

A. The Polanyi Treatment

A still different approach to multilayer adsorption is that which considers that there is a potential field at the surface of a solid into which adsorbate molecules "fall." The adsorbed layer thus resembles the atmosphere of a planet—it is most compressed at the surface of the solid and decreases in density outward. The general idea is quite old, but was first formalized by Polanyi in about 1914—see Brunauer (19). As illustrated in Fig. XVI-14, one can draw surfaces of equipotential that appear as lines in cross-sectional view of the surface region. The space between each set of equipotential surfaces corresponds to a definte volume, and there will thus be a relationship between potential ϵ and volume ϕ.

If we consider the case of a gas in adsorption equilibrium with a surface, there must be no net free energy change on transporting a small amount from one region to the other. Therefore, since the potential ϵ_x represents the work done by the adsorption forces when adsorbate is brought up to a distance x from the surface, there must be a compensating compressional increase in the free energy of the adsorbate. Thus

$$\epsilon_x = \int_{\mathbf{P}_g}^{\mathbf{P}^x} V\, d\mathbf{P}. \tag{XVI-77}$$

The mass of adsorbate present is given by

$$w = \sum \int_0^\infty (\rho_x - \rho_g)\, dx \tag{XVI-78}$$

and the equation of state of the adsorbate gives the necessary relationship

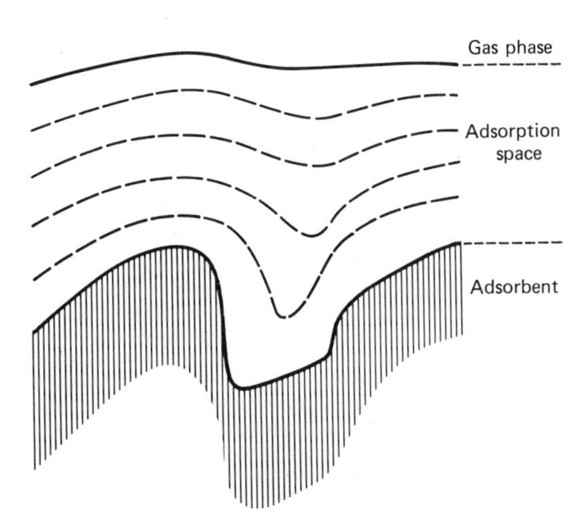

Fig. XVI-14. Isopotential contours (From Ref. 19.)

between the density ρ and the pressure. These equations allow any adsorption isotherm to be converted to an ϵ versus ϕ plot, called the *characteristic curve* or, alternatively, to a ρ versus ϕ plot. A schematic example of the latter is given in Figure XVI-15 to illustrate that at high temperatures the adsorbate remains gaseous and is merely compressed in the potential field, but at some lower temperature the pressure of the adsorbate will reach P^0 and condensation to liquid will occur.

The treatment at this point is esentially thermodynamic—it does not imply any particular function for $\epsilon(\phi)$, neither does it provide a determination of Σ. However, the function $\epsilon(\phi)$ should be temperature independent, that is, for a given system all isotherms should give the same characteristic function. In many cases this seems to be true. Thus McGavack and Patrick (49) found that for sulfur dioxide on silica gel the same characteristic curve was obtained for eight temperatures between 373 and 193 K. The treatment can be applied to adsorption in the submonolayer region, in which case ϵ is really a function of θ rather than of ϕ, that is, one is dealing with a heterogeneous surface. It turns out that the calculation of a site energy distribution for a heterogeneous surface amounts to obtaining a characteristic curve for the surface (see Section XVI-14).

In the case of multilayer adsorption it seems reasonable to suppose that condensation to a liquid film occurs (as in curves 1 or 2 of Fig. XVI-15). If one now assumes that the amount adsorbed can be attributed entirely to such a film, and that the liquid is negligibly compressible, the thickness x of the film is related to n by

$$n = \frac{\Sigma x}{V_l} \tag{XVI-79}$$

where V_l is the molar volume of the liquid, and the potential at x, being just

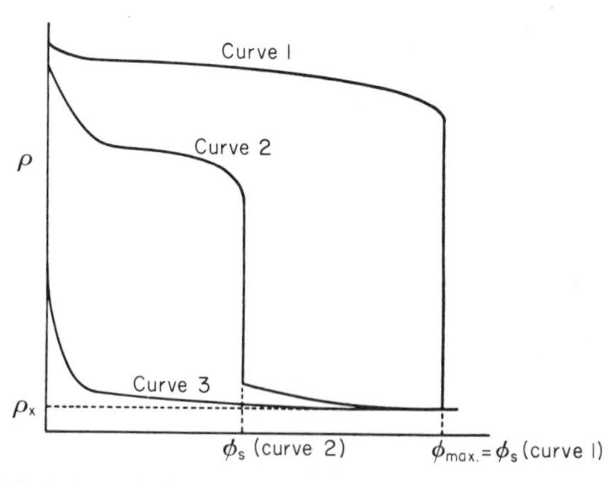

Fig. XVI-15. Variation of the density with the adsorbed phase according to the potential theory (From Ref. 19.)

sufficient to cause condensation, is given by

$$\epsilon_x = RT \ln \frac{P^0}{P_g} \tag{XVI-80}$$

Where P_g is the pressure of vapor in equilibrium with the adsorbed film. The characteristic curve is now just $RT \ln P_0/P$ versus x (or against n, if Σ is not known). Dubinin and co-workers (see Ref. 50) have made much use of a semiempirical relation between x and ϵ_x:

$$x = x_0 \exp(-b\epsilon^2) \tag{XVI-81}$$

where, for porous solids, x_0 is taken to be the pore volume, and x, the volume adsorbed at a given P^0/P_g value. See Ref. 51 for a variant of Eq. XIV-81 known as the Dubinin–Radushkevich equation.

B. Correspondence between the Potential Theory and That of a Two-Dimensional Film

There is an interesting correspondence between the potential treatment and the concept of a surface or film pressure. According to Eq. XVI-78, the three-dimensional pressure P varies with the distance from the surface in some manner depending on $\epsilon(x)$ and on the three-dimensional equation of state of the adsorbate. On the other hand, Eq. III-42 becomes, in the present context,

$$\pi = \int_0^\infty (\mathbf{P} - P_g)\, dx \tag{XVI-82}$$

Thus since $\mathbf{P}(x)$ is known, then for every P_g a corresponding π value can be obtained.

One special case is of some significance, namely, that of a square-well potential in which ϵ has the value ϵ_0 out to x_0 and is zero thereafter. The pressure is $\mathbf{P}^0$ inside the well and P_g outside, so that

$$\pi = (\mathbf{P}^0 - P_g)x_0 \cong \mathbf{P}^0 x_0 \tag{XVI-83}$$

where $\mathbf{P}^0$ is determined by Eq. XVI-78 and the equation of state of the adsorbate. Also, σ is equal to V/x_0. If, then, the three-dimensional equation of state is given by some function $V = f(P)$, the two-dimensional equation of state will be that same function, that is, $x_0\sigma = f(\pi/x_0)$. Thus if $f(P)$ is given by the van der Waals equation.

$$\left(P + \frac{a}{V^2}\right)(V - b) = RT \tag{XVI-84}$$

the film pressure as calculated from the Gibbs equation will vary with σ according to the relationship

$$\left(\pi + \frac{a}{x_0^2\sigma^2}\right)\left(\sigma - \frac{b}{x_0}\right) = RT \tag{XVI-85}$$

One would thus expect the critical temperature of the adsorbed film to be the same as that for the bulk state. However, if the adsorbate molecules really are constrained to move on a surface rather than in a thin but still basically three-dimensional layer, an analysis by de Boer (4) indicates that the critical temperature of a two-dimensional van der Waals gas should be half of the normal value. Furthermore, of course, the square-well treatment does not take into account surface orientation and polarization effects that would make the equation of state of the adsorbed material differ from that for the bulk adsorbate.

C. Isotherms Based on an Assumed Variation of Potential with Distance

As demonstrated the monolayer approach can be regarded as a square-well case of the potential theory, and the logical next step is to consider other shapes for the potential well. Of particular interest now is the case of multilayer adsorption, and a reasonable assumption is that the principal interaction between the solid and the adsorbate is of the dispersion type, so that for a plane solid surface the potential should decrease with the inverse cube of the distance (see Section VI-4). To avoid having an infinite potential at the surface, the potential function may be written

$$\epsilon(x) = \frac{\epsilon_0}{(a + x)^3} \qquad \text{(XVI-86)}$$

where a is a distance of the order of a molecular radius.†

On combining Eqs. XVI-81 and XVI-86, one obtains

$$RT \ln \frac{P^0}{P} = \frac{\epsilon_0}{(a + x)^3} \qquad \text{(XVI-87)}$$

where P is the gas pressure, the subscript g no longer being needed for clarity. On solving for x and substituting into Eq. XVI-80,

$$n = -\alpha + \beta\omega^{-1/3} \qquad \text{(XVI-88)}$$

where

$$\alpha = \frac{a\,\Sigma}{V_l} \qquad \beta = \left(\frac{\Sigma}{V_l}\right)\left(\frac{\epsilon_0}{RT}\right)^{1/3} \qquad \omega = \ln \frac{P^0}{P}$$

The physical model is thus that of a liquid film, condensed in the inverse cube potential field, whose thickness increases to infinity as P approaches P^0.

Equation XVI-88 turns out to fit type II adsorption isotherms quite well—generally better than does the BET equation. Furthermore, the exact form of the potential function is not very critical; if an inverse square dependence is used, the fit tends to be about as good as with the inverse cube law, and the equation now resembles Eq. XVI-76. Here again, quite similar equations have resulted from deductions based on rather different models.

† Equation XV-86 is based on the assumption that the solid is continuous and actually is a rather poor approximation for the case of a molecule in the first few layers of a multilayer on a crystalline surface. Also, the constant ϵ_0 would better be written $(\epsilon_{0,SL} - \epsilon_{0,LL})$ (52), where each interaction constant is of the form given by Eq. VI-25, that is, $\epsilon_{0,ST} = (\pi/6)n_S A_{SL}, \epsilon_{0,LL} = (\pi/6)n_L A_{LL}$. Blake (53) has investigated both experimentally and theoretically the variation of disjoining pressure (Section VI-6) with film thickness.

The general approach goes back to Frenkel (54) and has been elaborated on by Halsey (55), Hill (56), and McMillan and Teller (57). A form of Eq. XVI-88, with $a = 0$,

$$\left(\frac{n}{n_m}\right)^n = \frac{A}{\ln (P^0/P)} \qquad A = \frac{\epsilon_0}{x_m^n RT} \tag{XVI-89}$$

where x_m is the film thickness at the monolayer point, is frequently referred to as the *Frenkel–Halsey–Hill* equation. In this form, the power n in $\epsilon = \epsilon_0/r^n$ may be left as an empirical parameter, and in Halsey's tabulation of n values for various systems (55), values between 2 and 3 are fairly common. Pierce (58) finds $n = 2.75$ for nitrogen essentially independent of the solid.

As with the BET equation, a number of modifications of Eqs. XVI-88 or XVI-89 have been proposed, adding complexity and empirical parameters. Greenlief and Halsey (59) proposed

$$P = bn \exp\left[\frac{c}{b}n\right] + P^0 \exp[-a(v/v_m)^{-3}] \tag{XVI-90}$$

where c is a negative and a and b are positive constants, n is the amount of gas adsorbed, and v and v_m have the usual meanings. Equation XVI-90 approaches the desired Henry's law form (see Section XVI-10) at low pressures and avoids the BET catastrophe (see Section XVI-13A) at the high pressure limit. Interestingly, it can give isotherms of the type of Fig. X-15. Adamson has proposed (60)

$$kT \ln \frac{P^0}{P} = \frac{g}{x^3} + \epsilon_0 e^{-ax} - \beta e^{-\alpha x} \tag{XVI-91}$$

The first term on the right is the common inverse cube law, the second is taken to be the empirically more important form for moderate film thicknesses (and also conforms to the polarization model, Section XVI-7D), and the last term allows for structural perturbation in the adsorbed film relative to bulk liquid adsorbate. The equation has been useful in relating adsorption isotherms to contact angle behavior (see Section X-6). Roy and Halsey (61) have used a similar equation; earlier, Halsey (62) allowed for surface heterogeneity by assuming a distribution of ϵ_0 values in Eq. XVI-89. Dubinin's equation (Eq. XVI-82) has been mentioned; another variant has been used by Bonnetain and co-workers (63).

D. The Polarization Model

An interesting alternative method for formulating $\epsilon(x)$ was proposed in 1929 by de Boer and Zwikker (64), who suggested that the adsorption of nonpolar molecules be explained by assuming that the polar adsorbent surface induces dipoles in the first adsorbed layer and that these in turn induce dipoles in the next layer, and so on. As shown in Section VI-7, this approach leads to

$$\epsilon(x) = \epsilon_0 e^{-ax} \tag{XVI-92}$$

which, in combination with Eqs. XVI-80 and XVI-81, gives

$$RT \ln \frac{P^0}{P} = \epsilon_0 e^{-ax} \tag{XVI-93}$$

or

$$\ln \ln \frac{P^0}{P} = \ln \frac{\epsilon_0}{RT} - \frac{aV_l}{\Sigma} n \qquad \text{(XVI-94)}$$

Thus a plot of log log (P^0/P) versus n should give a straight line, and, indeed, Eq. XVI-94 is quite successful.

The polarization theory was severely criticized by Brunauer (19) on the grounds that the effect was not large enough, and the polarization theory has been largely ignored. Some recent indications suggest that this neglect may be mistaken. Bewig and Zisman (65) found that adsorption of n-hexane on various metals produced large changes in surface potential difference ΔV, which meant that large induced dipole moments were present, of the order of 0.3 debye (D). From an estimated polarizability, the necessary surface fields amounted to about 10^7 V/cm. Similarly, rare gases adsorbed on metals can produce sizable changes in ΔV (see Ref. 66). In particular, Pritchard (67) found ΔV values of 0.2 to 0.8 V for xenon adsorbed at $-183°$C on copper, nickel, gold, and platinum; furthermore, a sharp decrease in the slope of the ΔV versus n plot appeared to coincide with completion of the monolayer. Benson and King (68) have suggested that the adsorption of inert gases on alumina is largely due to the polarization of the adsorbate by strong local electrostatic fields. Also, graphitic surfaces appear to have strong surface fields due to the separation of charge between the negative cloud of π electrons and the positive carbon atoms. Some further, more recent, evidence has accumulated. The various spectroscopic evidence noted in Section XV-4 indicates appreciable polarization of adsorbed species. Material in Section XVI-10 supports the conclusion that even on molecular solids only a fraction of the adsorption energy of an adsorbent is due to dispersion forces; moreover, the potential field of the adsorbent decays nearly exponentially as it is coated with intervening layers of inert preadsorbed film. It may thus prove to be fairly general that adsorption in the first few layers is due as much to an electrostatic polarization interaction (Eq. VI-40) as to dispersion forces.

The term multilayer adsorption, as used in this chapter, has meant adsorption in roughly the BET range of P/P^0 values or, at most, n/n_m values of 3 (the problem of deep multilayer formation as $P/P^0 \rightarrow 1$ was considered in Section X-6). The theoretical problem is mainly that of estimating the induced dipole interaction in the second and third layer, and it is not really fair to consider the molecules as point spheres in making the calculation. The approach taken in Section VI-7 made empirical allowance for this difficulty by writing a in Eq. XVI-92 as

$$a = -\frac{1}{d_0} \ln \left(\frac{\alpha}{d^3}\right)^2 \qquad \text{(XVI-95)}$$

where d might be smaller than the molecular diameter d_0, in which case Eq. XVI-94 takes the form

$$\ln \ln \frac{P^0}{P} = \ln \frac{\epsilon_0}{RT} - \left[\ln \left(\frac{d^3}{\alpha}\right)^2\right]\left(\frac{n}{n_m}\right) \qquad \text{(XVI-96)}$$

If Eq. XVI-96 is applied to typical type II isotherms, such as CO or N_2 on silica "C" (16) or N_2 on KCl (65), α/d^3 comes out to be about 0.4, from which d is about an atomic radius (69).

The polarization treatment of $\epsilon(x)$ suggests much more strongly than does the dispersion one that strong orientational effects should be present in multilayers and that their structure could be considerably perturbed from that of the normal liquid. As was seen in Section X-6, this perturbation is essential to the explanation of contact angle phenomena. The problem is, of course, extremely difficult from a rigorous point of view. Hill noted in 1952 (56) that there was no satisfactory theory of the liquid state, even for monatomic liquids, and that the detailed treatment of a liquid in a combination electrical and dispersion potential was still far away. While some progress has been made, the subject remains a very difficult one (see Ref. 70).

8. Comparison of the Surface Areas from the Various Multilayer Models

It has been emphasized repeatedly that the fact that an isotherm equation fits data is an insufficient test of its validity (although not of its practical usefulness). To further stress this point, the data for a particular isotherm for the adsorption of nitrogen on powdered potassium chloride at 78 K are plotted in Fig. XVI-16 according to four equations; in each case there is a satisfactory fit. For simple gases such as nitrogen, oxygen, and argon, the BET equation generally fits over the range of P/P^0 values of 0.05 to 0.3; the potential theory (Eq. XVI-88), over the range 0.1 to 0.8; and similarly for the polarization equation, Eq. XVI-96. There is thus little to choose from between the various models, but partly because of tradition and familiarity,

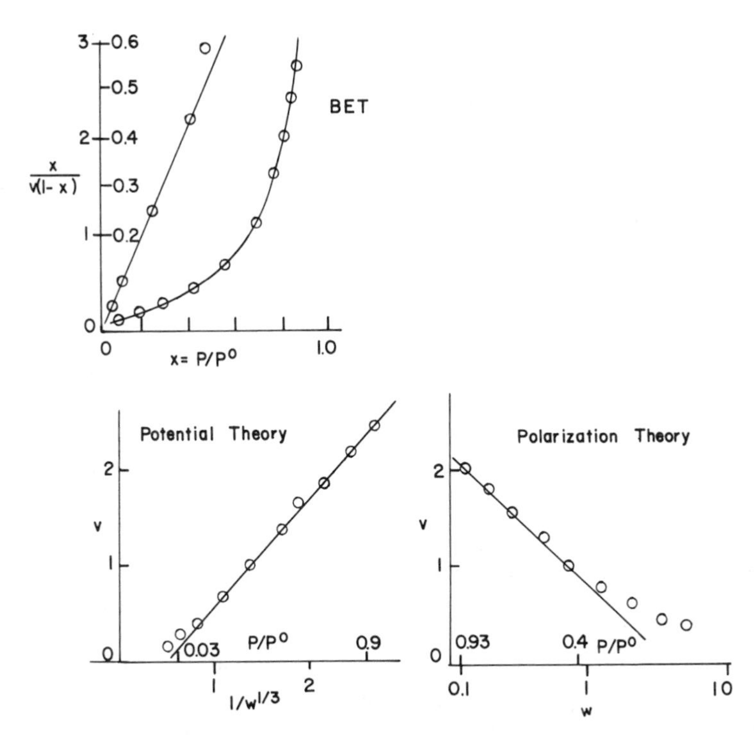

Fig. XVI-16. Adsorption of nitrogen on potassium chloride at 79 K, plotted according to various equations. (Data from Ref. 71.)

TABLE XVI-3

Gas/K	BET	HJ	Ratio of values[a] Potential theory[b]	Polarization theory[c]	Characteristic isotherm
			Σ of egg albumin[d]/Σ of KCl 2		
O_2/90	4.0	5.2	5.5	5.3	5.2
Ar/90	4.0	7.0	5.7	5.6	6.7
N_2/78	5.3	5.6	5.7	4.4	5.4
	4.4	5.9	5.6	5.1	5.8
	±15%	±12%	±2%	±6%	±10%
			Σ of TiO_2/Σ of KCl 2		
O_2/78	5.2	8.8	8.1	6.7	8.1
N_2/78	6.4	6.8	7.9	6.7	7.1
	5.8	7.8	8.0	6.7	7.6
	±10%	±13%	±1%	±0%	±7%
			Σ of Si(C)/Σ Sterling S450		
CO/78	3.0	3.1	2.9	2.7	3.0
N_2/78	2.9	3.2	2.6	2.8	3.0
C_2H_5Cl/195	2.7	2.4	3.2	2.5	2.4
	2.9	2.9	2.9	2.7	2.8
	±4%	±11%	±7%	±4%	±9%

[a] Data are from the following sources: egg albumin, Ref. 72, KCL, Ref. 71; TiO_2, Ref. 72; Si(C) and Sterling S450, Ref. 74.
[b] Eq. XVI-88.
[c] Eq. XVI-96.
[d] This was a lyophilized, dry protein.

and partly because n_m enters in it so explicitly, the BET equation is in fact almost exclusively used.

A seemingly more stringent test would be to determine whether the ratios of areas for various solids as obtained by means of a given isotherm equation are independent of the nature of the adsorbate. The data summarized in Table XVI-3 were selected from the literature mainly because each author had obtained areas for two or more solids using two or more adsorbates. It is seen that the BET equation gives ratios of areas that may vary by 10 to 15% from one adsorbate to another; the performance of the Harkins–Jura equation is somewhat better, as is that of the potential theory in its various forms, that is, using either an inverse cube or a polarization potential, or as a characteristic isotherm. Note that the BET analysis gives noticeably different ratios than do the other four; this is probably a reflection of the fact that the region of fit of the BET equation is over a lower P/P^0 range than that for the other models. The relative merit of various models is discussed in more detail in Section XVI-13.

9. The Characteristic Isotherm and Related Concepts

The relatively uniform success of these various plots suggests that, except as modified by changes in Σ, the shape of the isotherm in the multilayer region tends to be characteristic of the adsorbate and independent of the nature of the solid. The test of this is that on plotting log n versus some arbitrary function of P/P^0, the isotherms for a given adsorbate and various solids should be identical except for a vertical displacement. As noted in the first edition (1960; see Ref. 75), this prediction turns out amazingly well. Figure XVI-17 shows the superposition of a number of nitrogen adsorption isotherms. Each adsorbate gives the same isotherm shape, to within experimental error, and if each is adjusted vertically by choosing the best value of n_m, all of the data fall on a single curve over the range of 0.3 to 0.95 in P/P^0. Below P/P^0 about 0.3 there are increasing differences between the curves, presumably as the individual nature of the vaious solids make themselves felt. Somewhat similar early observations have been made by Pierce (58) and by Halász and Schay (76), and by Dubinin (50).

The curve of Fig. XVI-17 is essentially a characteristic curve of the Polanyi theory, but in the form plotted it might better be called a *characteristic isotherm*. Furthermore, as would be expected from the Polanyi theory, if the data for a given adsorbate are plotted with $RT \ln (P/P^0)$ as the abscissa instead of just $\ln (P/P^0)$, then a nearly invariant shape is obtained for different temperatures. The plot might then be called the *characteristic adsorption curve*.

The existence of this situation (for nonporous solids) explains why the

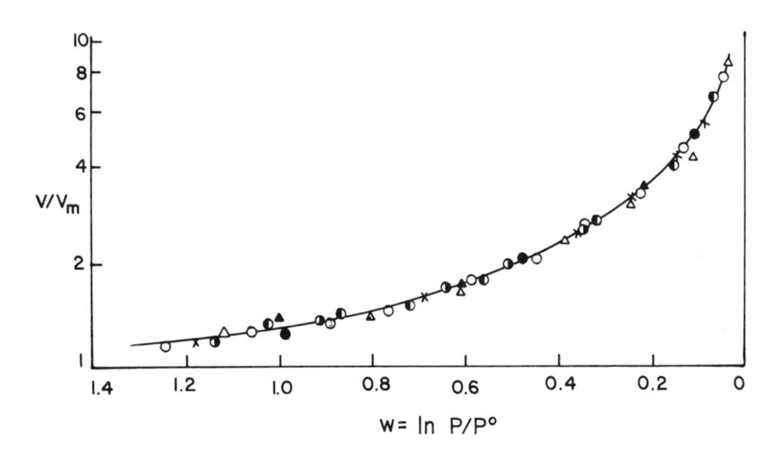

Fig. XVI-17. Characteristic isotherm for nitrogen at 78 K on various solids: ○, KCl-1 (Ref. 71); △, egg albumin 61 (Ref. 71); ◐, bovine albumin 68 (Ref. 72); ×, titanium dioxide [From W. D. Harkins and G. Jura, *J. Am. Chem. Soc.*, **66**, 919 (1944)]; ◑, Graphon (Ref. 77); ▲, egg albumin 59 (Ref. 72); ●, polyethylene. [From A. C. Zettlemoyer, A. Chand, and E. Gamble, *J. Am. Chem. Soc.*, **72**, 2752 (1950).]

ratio test discussed above and exemplified by the data in Table XVI-3 works so well. Esssentially, *any* isotherm fitting data in the multilayer region *must* contain a parameter that will be found to be proportional to surface area. In fact, this observation explains the success of the point "B" method (Section XVI-5) and other single point methods, since for *any* P/P^0 value in the characteristic isotherm region, the measured n is related to the surface area of the solid by a proportionality constant that is independent of the nature of the solid.

The characteristic isotherm concept was elaborated by de Boer and co-workers (78, 79). By accepting a reference n_m from a BET fit to a standard system and assuming a density for the adsorbed film, one may convert n/n_m to film thickness t. The characteristic isotherm for a given adsorbate may then be plotted as t versus P/P^0. For any new system, one reads t from the standard t-curve and v from the new isotherm, for various P/P^0 values. De Boer and co-workers' t values are given in Table XVI-4. A plot of t versus n should be linear if the experimental isotherm has the same shape as the reference characteristic isotherm, and the slope gives Σ:

$$\Sigma = 15.47\, v/t = \frac{3.26 \times 10^5\, n}{t} \tag{XVI-97}$$

where Σ is in square meters per gram, v in cubic centimeters STP per gram, and t, in angstrom units. Sing (79) (see also Ref. 41) has reviewed much of the literature on such uses of the characteristic isotherm. To avoid referencing to the BET equation, he proposes using n/n_x instead of t, where n_x is the amount adsorbed at $P/P^0 = 0.4$.

TABLE XVI-4
The t Plot for N_2 at 78 K[a]

P/P^0	t Å	P/P^0	t Å	P/P^0	t Å	P/P^0	t Å
0.08	3.51	0.32	5.14	0.56	6.99	0.80	10.57
0.10	3.68	0.34	5.27	0.58	7.17	0.82	11.17
0.12	3.83	0.36	5.41	0.60	7.36	0.84	11.89
0.14	3.97	0.38	5.56	0.62	7.56	0.86	12.75
0.16	4.10	0.40	5.71	0.64	7.77	0.88	13.82
0.18	4.23	0.42	5.86	0.66	8.02	0.90	14.94
0.20	4.36	0.44	6.02	0.68	8.26	0.92	16.0[b]
0.22	4.49	0.46	6.18	0.70	8.57	0.94	17.5[b]
0.24	4.62	0.48	6.34	0.72	8.91	0.96	19.8[b]
0.26	4.75	0.50	6.50	0.74	9.27	0.98	22.9[b]
0.28	4.88	0.52	6.66	0.76	9.65		
0.30	5.01	0.54	6.82	0.78	10.07		

[a] From Ref. 78.

[b] These are extrapolated values and undoubtedly contain an important contribution from interparticle condensation.

To return to the discussion of Section XVI-9, the existence of a characteristic isotherm for each adsorbate is encouraging in the sense that it appears that investigators insterested primarily in relative surface area values can choose their isotherm equation on the basis of convenience and with little concern as to whether it is a fundamentally correct one. On the other hand, one would like to know if there are any points of experimental distinction that are diagnostic of the correctness of an isotherm model. Such an evaluation is only possible to some extent and is discussed in Section XVI-13.

10. Potential Theory as Applied to Submonolayer Adsorption

A major field of development has been that of applying explicit expressions for the short-range adsorbent–adsorbate interaction potential to the calculation of the adsorption energy and free energy. The potential functions may be of the type of Eq. VII-8, that is

$$\epsilon(r) = -\frac{A}{r^6} + \frac{B}{r^{12}} \tag{XVI-98}$$

Alternatively, an exponential repulsion term may be used

$$\epsilon(r) = -\frac{A}{r^6} + Be^{-Cr} \tag{XVI-99}$$

The constants may be evaluated from cohesion energy and compressibility data, for a liquid or a solid, or from gas nonideality, for vapors. Figure XVI-18 shows some typical plots (80).

One procedure is to compute

$$n_a = \mathcal{A}\int_a^\infty (C - C_g)\, dr \tag{XVI-100}$$

where

$$C = C_g e^{-\epsilon/kT}$$

The concentration in the gas phase C_g is given by P/RT, and we have

$$k\mathcal{A} = \frac{n_a}{P} = \mathcal{A}\frac{1}{kT}\left[\int_a^\infty (e^{-\epsilon/kT} - 1)\, dr\right] \tag{XVI-101}$$

where k is the Henry's law constant, that is, the limiting slope of the adsorption isotherm. It is assumed that adsorption is occurring at sufficiently low θ that θ is proportional to pressure, and, of course, that the surface is uniform. The distance a is taken to be that at which ϵ is zero; that is, van der Waals attraction and the short-range repulsion have become equal.

If the coefficients of Eq. XVI-98 or XVI-99 are obtained from other data (compressibility, gas nonideality, etc.), the integral of Eq. XVI-101 can be evaluated absolutely. Comparison of the calculated Henry's law constant with an experimental n_a/P then gives the surface area of the adsorbent. The calculation is sensitive to the value chosen for a, and this parameter can be evaluated empirically as that which gives the best fit to the temperature dependence of the Henry's law constant k. In

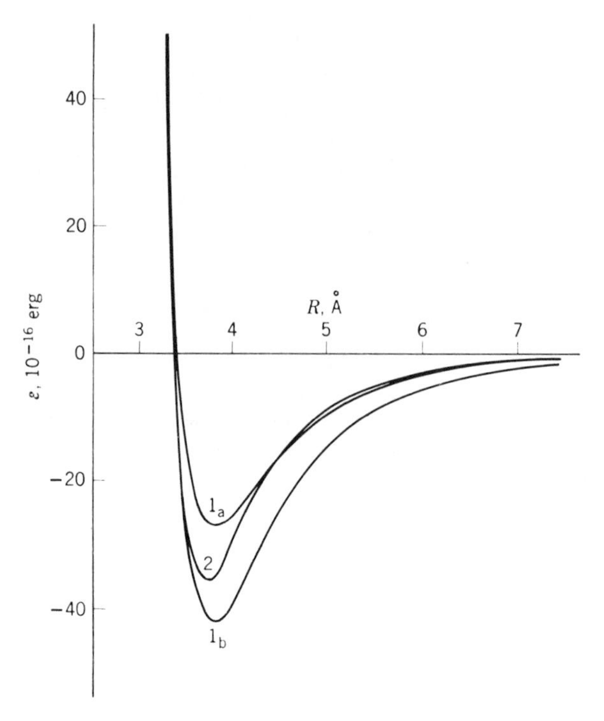

Fig. XVI-18. Potential laws for the van der Waals interaction between carbon atoms. Curves 1a and 1b are according to Eq. XVI–98, with constants evaluated from cohesion energy and compressibility, respectively (of graphite), and curve 2 is according to Eq. XVI-99. Reprinted with permission of *J. Phys. Chem.* (Ref. 80). Copyright by the American Chemical Society.

this manner $\mathscr{A}$ can be determined without making any specific assumption as to the cross-sectional area of the adsorbate.

The preceding approach is essentially that of Steele and Halsey (81); much work has been done by Everett and co-workers (see Ref. 82), and the general subject has been reviewed by Steele (70). One problem is that the integral is very sensitive to the function chosen to represent $\epsilon(r)$, and there is not a rigorous, independent way of evaluating the constants for the adsorbent–adsorbate potential from the properties of the bulk adsorbent and of the adsorbate liquid or vapor. The problem of recognizing an adsorbent surface as a periodic array of atoms (rather than a smooth surface) has been examined in some detail by Steele (83) and by Ricca and co-workers (84).

A further development of this general approach has been to consider the region of slightly higher θ values such that n_a/P begins to deviate from constancy and to attribute the deviation to lateral interactions. These again may be calcuated from one or another assumed potential function if the surface concentration $n_a/\mathscr{A}$ is known and, conversely, the magnitude of the deviation from Henry's law allows a calculation of $\mathscr{A}$. There is again a matter of choice of a molecular size, now in the form of a two-dimensional collision cross section; this may either be taken from data on the deviation from ideality of the gaseous adsorbate, or by use of the preceding analysis of k itself. An added complication is that the adsorbed molecules do not really move in a plane but have vibrational motions perpendicular to the surface, so that $n_a/\mathcal{A}$ does not give the true average distance apart. The procedure has now become very complex, with several cycles of approximations needed to find that set of parameters giving the best fit to k, to the deviation from k, and to their temperature dependencies. Depending on whether departure from planarity of the molecular motion in the adsorbed film was allowed for, Barker and Everett (82) obtained values of 86 and 128 m^2/g for a carbon sample. The effect of different choices of the form of the interaction potential between adsorbate molecules is discussed further by Johnson and Klein (85), using data for argon on graphitized carbon black (P-33); the interpretation of the data was found not to be very sensitive to the form of the *repulsion* part of this potential. Everett, in reviewing this approach

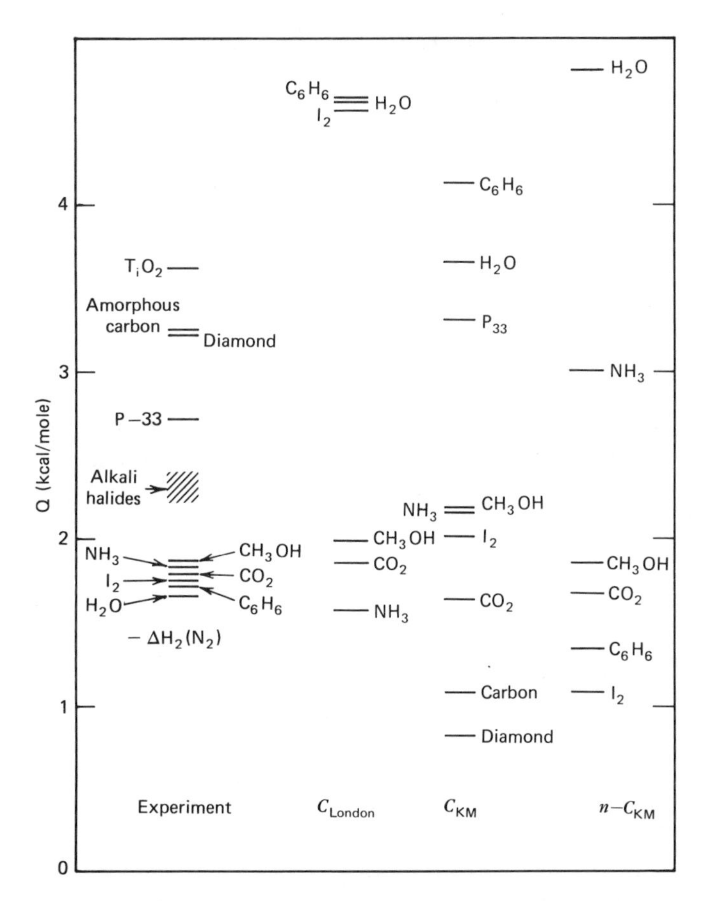

Fig. XVI-19. Experimental versus calculated interaction energies. The constant C is the coefficient in Eq. VI-25 as valuated according to Eq. VI-19 (London) or Eq. VI-21 (KM). The last column (density times the Kirkwood–Müller C) is drawn such that the methanol is correct; TiO_2, diamond, P-33 carbon, and amorphous carbon are off scale. (From Ref. 88.)

(86), concludes that it should be less sensitive to surface heterogeneity as well as less dependent on absolute theoretical calculations than is the method based on direct evaluation of the Henry's law constant itself. Additional references in this area are Steele (87) and Halsey (88).

Most of the foregoing calculations have been made for graphite (or pyrolytic carbon) as the adsorbent, and there has been criticism of the assumption that dispersion only forces are involved in adsorption on such adsorbents, and also more specificially of the assumption that adsorption of N_2 on solids generally is due only to dispersion forces (as implied, e.g., by equations such as Eq. XVI-87). Molecular solids tend to give rather uniform surfaces when used as an adsorbent (89), and a series of studies was made of N_2 adsorption on solids such as ice, NH_3, CO_2, CH_3OH, I_2, and benzene (90). The adsorption interaction must be entirely van der Waals, since there are no broken chemical bonds at the surface of a molecular solid. The experimental heats of adsorption are shown in Fig. XVI-19 (along with literature values for various carbons and alkali halides), and the results of calculations using various forms of the dispersion potential, Eqs. VI-19 and VI-21. Not only are the absolute values incorrect, but also, much more important, the *ordering* of the various adsorbents does not agree with experimental observation. The conclusion was that polarization interactions were at least as important as dispersion ones. The subject is discussed further in Section XVI-13.

11. Adsorption Steps and Phase Transformations

The isotherms so far illustrated have all been of a continuous appearance, and for many years this was the only type to which much attention was paid. It is now recognized that smooth isotherms are very often a consequence of surface heterogeneity and that various types of adsorbate (and perhaps adsorbate–adsorbent complex) phase transformations probably do occur, but are visible only with very uniform surfaces. Halsey in his 1965 Kendall Award paper assembled in one diagram all of the various types of transformations that might occur. This is shown in Fig. XVI-20.

First, what appears to be a two-dimensional condensation from dilute to condensed film may occur in the submonolayer region. Note that such behavior is predicted both for a localized film with lateral interaction (Eq. XVI-53) and for nonideal mobile films, such as ones obeying the two-dimensional van der Waals and related equations, as summarized in Table XVI-2. Many examples are now known, an early one being that of Kr on NaBr (91). A more recent study is that of Kr on the 0001 face of graphite, shown in Fig. XVI-21 (92). The dashed lines sketch a probable phase diagram; below about 85 K there is a two-phase region of gas–solid film equilibrium and above this temperature, a region of gas–liquid film equilibrium. The triple point for two-dimensional gas–liquid–solid equilibrium is thus at 85 K. Similar phase transformations have been reported for other vapors adsorbed on carbon (93) and on other adsorbents (94, 95). In the case of ethanol adsorbed on the (001) surface of LiF, a transition from a disordered to an ordered film was observed at around 150 K, from diffraction measurements using a beam of He atoms (96).

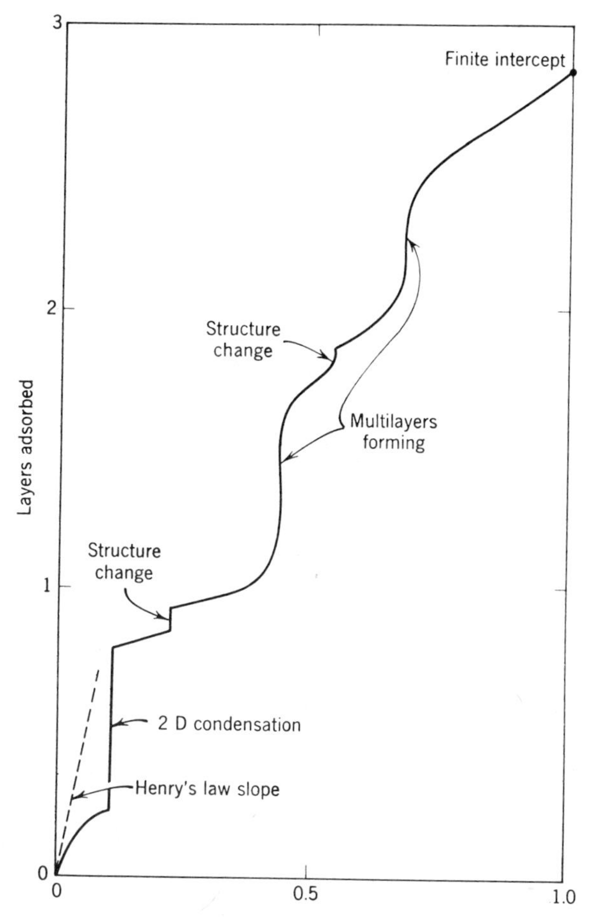

Fig. XVI-20. An exemplary adsorption isotherm (After Halsey.)

Figure XVI-22 shows combined isotherm and calorimetric heat of adsorption data for argon on boron nitride. Note the resemblance to the hypothetical case of Fig. XVI-21. The feature at B_1 is thought to represent a transition from a liquid- to a solidlike film. The transition B_2 occurs between the second and third adsorption layer. A phase transition in the multilayer region has also been observed for ethyl chloride on Sterling MT graphite (98).

One precondition for a vertical step is presumably that the surface be sufficiently uniform that the transition does not occur at different pressures on different portions, with a resulting smearing out of the step feature. It is partly on this basis that graphitized carbon and certain other adsorbents have been considered to have rather uniform surfaces. The argument is a reasonable one but has not been given a general proof, and, certainly, the converse conclusion that a surface giving a smooth isotherm must necessarily be heterogeneous is less tenable.

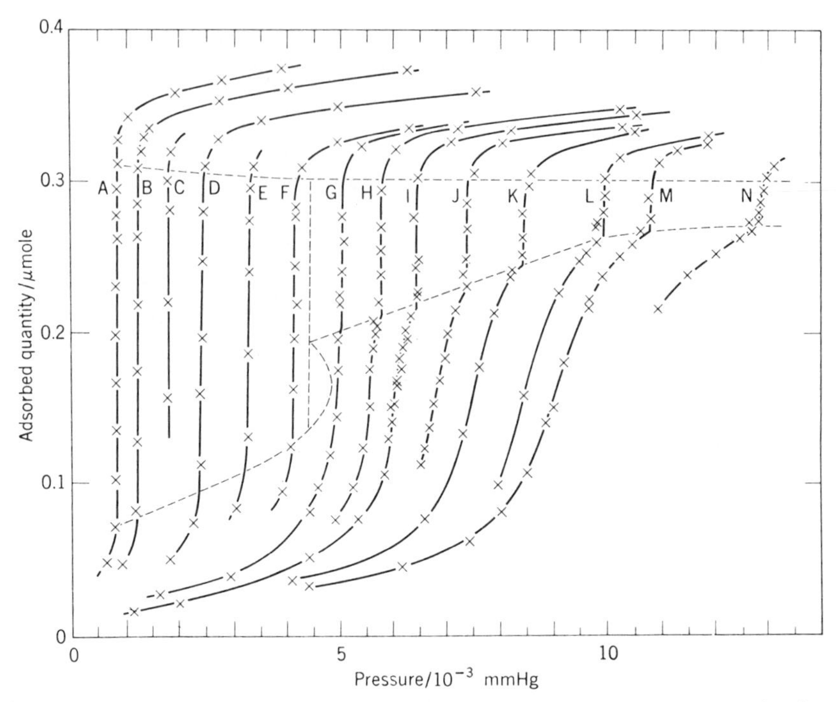

Fig. XVI-21. Adsorption isotherms of Kr on the (0001) face of graphite in the monolayer domain. *A*, 79.24; *B*, 80.54; *C*, 81.77; *D*, 82.83; *E*, 83.84; *F*, 84.69; *G*, 85.33; *H*, 85.74; *I*, 86.12; *J*, 86.58; *K*, 87.08; *L*, 87.61; *M*, 87.81; *N*, 88.46 K. (From Ref. 92.)

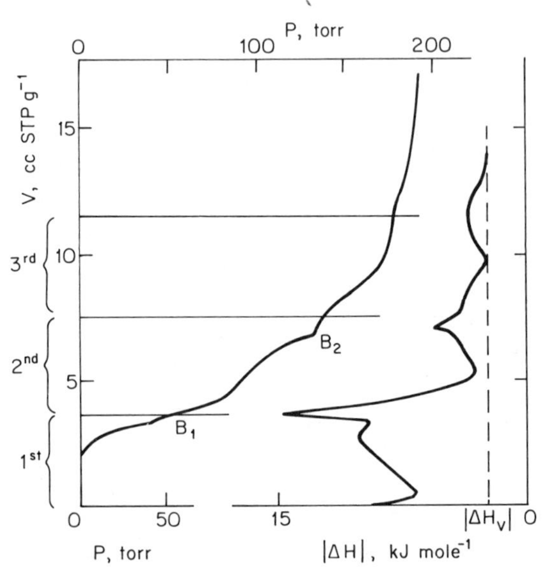

Fig. XVI-22. Adsorption of argon on boron nitride at 77.6 K. Left: adsorption isotherm showing transitions at B_1 and B_2. Right: calorimetric isosteric heat of adsorption. (From Ref. 97.)

12. Adsorption in the Case of a Nonwetting System

As noted in connection with Fig. X-15, if the liquid adsorbate does not wet the solid adsorbent, that is, rests on it with a finite contact angle, then the adsorption isotherm for the vapor *must* cut the $P = P^0$ point at a finite film thickness. Examples of such isotherms are discussed in Ref. 98a for the cases of water and various organic vapors on polymer surfaces. The adsorption of ammonia on carbon black appears also to give this type of isotherm at sufficiently low temperatures (98b).

13. Thermodynamics of Adsorption

A. Theoretical Considerations

We take up here some aspects of the thermodynamics of adsorption that are of special relevance to gas adsorption. Two types of processes are of interest: the first may be called integral adsorption and may be written, for an adsorbent X_1 and an adsorbate X_2,

$$n_1^s X_1 \text{ (adsorbent at } T) + n_2^g X_2 \text{ (gaseous adsorbate at } P, T) = n_2^s X_2 \text{ adsorbed on } n_1^s X_1 \qquad \text{(XVI-102)}$$

If the process is carried at constant volume, the heat evolved Q_i will be equal to an energy change ΔE_2 or, per mole of adsorbate, $q_i = \Delta \mathrm{E}_2$ (small capital letters will be used to denote mean molar quantities). Alternatively, the process may be

$$n_2^g X_2 \text{ (gas, } P, T) = n_2^s \text{ (adsorbed at composition } n_2'^s/n_1'^s \text{ constant)} \qquad \text{(XVI-103)}$$

The heat evolved will now be a *differential heat* of adsorption, equal, at constant volume to Q_d or per mole, to $q_d = \Delta \bar{\mathrm{E}}_2$, where $\Delta \bar{\mathrm{E}}_2$ is the change in partial molar energy. It follows that

$$q_d = \left(\frac{\partial Q_i}{\partial n_2^s}\right)_{T,V} \qquad \text{(XVI-104)}$$

1. Adsorption Heats and Entropies. It is not necessary, phenomenologically, to state whether the process is adsorption, absorption, or solution, and for the adsorbent–adsorbate complex formal equations can be written, such as

$$dG^s = \bar{\mathrm{G}}_1^s \, dn_1^s + \bar{\mathrm{G}}_2^s \, dn_2^s \qquad \text{(XVI-105)}$$

and

$$G^s = n_1^s \bar{\mathrm{G}}_1^s + n_2^s \bar{\mathrm{G}}_2^s \qquad \text{(XVI-106)}$$

However, a body of thermodynamic treatment has been developed on the basis that the adsorbent is inert and with attention focused entirely on the adsorbate. The abbreviated presentation given here is based on that of Hill (see Refs. 56 and 99) and of Everett (100). First, we have the defining relationships:

$$dE^s = T\,dS^s - \pi\,d\mathcal{A} + \mu_2\,dn_2^s \qquad E^s = TS^s - \pi\mathcal{A} + \mu_2 n_2^s \qquad \text{(XVI-107)}$$

$$dH^s = T\,dS^s - \pi\,d\mathcal{A} + \mu_2\,dn_2^s \qquad H^s = TS^s - \pi\mathcal{A} + \mu_2 n_2^s \qquad \text{(XVI-108)}$$

$$dA^s = -S^s\,dT - \pi\,d\mathcal{A} + \mu_2\,dn_2^s \qquad A^s = -\pi\mathcal{A} + \mu_2 n_2^s \qquad \text{(XVI-109)}$$

$$dG^s = -S^s\,dT - \pi\,d\mathcal{A} + \mu_2\,dn_2^s \qquad G^s = -\pi\mathcal{A} + \mu_2 n_2^s \qquad \text{(XVI-110)}$$

The second set of equations is obtained from the first set by the Gibbs integration at constant intensive variables, as was done in obtaining Eq. III-74. It is convenient, in dealing with a surface species, to introduce some special definitions, two of which are:

$$d\mathcal{H}^s = T\,dS^s + \mathcal{A}\,d\pi + \mu_2\,dn_2^s \qquad \mathcal{H}^s = TS^s + \mu_2 n_2^s \qquad \text{(XVI-111)}$$

$$d\mathcal{G}^s = -S^s\,dT + \mathcal{A}\,d\pi + \mu_2\,dn_2^s \qquad \mathcal{G}^s = \mu_2 n_2^s \qquad \text{(XVI-112)}$$

The expressions for $\mathcal{H}^s$ and $\mathcal{G}^s$ now have the same appearnace as those for H and G in a bulk system.

Now, considering process XVI-103, where P is the equilibrium pressure of the adsorbate, we have

$$\Delta\bar{G}_2 = \bar{G}_2^s - \bar{G}_2^{0,g} = RT \ln P \qquad \text{(XVI-113)}$$

where the quantity $\bar{G}_2^s$ is defined as $(\partial G_2^s/\partial n_2^s)_{T,A} = \mu_2$. A standard thermodynamic relation is that

$$\left[\frac{\partial(\bar{G}_2^{0,g}/T)}{\partial T}\right]_{P,n_2^s} = -\frac{\bar{H}_2^{0,g}}{T^2} \qquad \text{(XVI-114)}$$

where the superscript zero means that a standard state (gas at 1 atm) is referred to. It follows from the preceding definitions that

$$\left[\frac{\partial(\bar{G}_2^s/T)}{\partial T}\right]_{A,n_2^s} = -\frac{\bar{H}_2^s}{T^2} \qquad \text{(XVI-115)}$$

In combination with Eq. XVI-113 we obtain

$$\left(\frac{\partial \ln P}{\partial T}\right)_\Gamma = -\frac{\bar{H}_2^s - \bar{H}_2^g}{RT^2} = -\frac{\Delta\bar{H}_2}{RT^2} = \frac{q_{st}}{RT^2} \qquad \text{(XVI-116)}$$

dropping unnecessary superscripts for the gas phase, assumed to be ideal, and remembering that $n_2^s/\mathcal{A} = \Gamma$, q_{st} is called the *isosteric heat of adsorption*.

Since $\bar{\mathcal{G}}_2^s = (\partial\mathcal{G}_2^s/\partial n_2^s)_\pi = \mu_2$ and $\mathcal{G}_2^s = \bar{\mathcal{G}}_2^s$, an alternative statement to Eq. XVI-113 is

$$\Delta\mathcal{G}_2 = \mathcal{G}_2^s - G_2^{0,g} = RT \ln P \qquad \text{(XVI-117)}$$

and

$$\left(\frac{\partial(\mathcal{G}_2^s/T)}{\partial T}\right)_\pi = -\frac{\mathcal{H}_2^s}{T^2} \tag{XVI-118}$$

it follows that

$$\left(\frac{\partial \ln P}{\partial T}\right)_\pi = -\frac{\mathcal{H}_2^s - \mathrm{H}_2^g}{RT^2} = -\frac{\Delta\mathcal{H}_2}{RT^2} = \frac{q_\pi}{RT^2} \tag{XVI-119}$$

The quantity $\Delta\mathcal{H}_2$ has been called (by Hill) the equilibrium heat of adsorption. It follows from the foregoing definitions that

$$\mathcal{H}_2^s = \mathrm{H}_2^s + \frac{\pi}{\Gamma} \tag{XVI-120}$$

also, it can be shown (101) that

$$\Delta\mathcal{H}_2 = \Delta\bar{\mathrm{H}}_2 + \frac{T}{\Gamma}\left(\frac{\partial \pi}{\partial T}\right)_{P,\Gamma} \tag{XVI-121}$$

To summarize, the four common heat quantities are

1. Integral calorimetric heat

$$q_i = \left(\frac{Q_i}{n_2^s}\right)_V \tag{XVI-122}$$

2. Differential calorimetric heat

$$q_d = \left(\frac{\partial Q_i}{\partial n_2^s}\right)_{T,V} \tag{XVI-104}$$

3. Isosteric or differential thermodynamic heat

$$q_{\mathrm{st}} = -\Delta\bar{\mathrm{H}}_2 = q_d + RT \tag{XVI-123}$$

4. Integral thermodynamic heat

$$q_\pi = -\Delta\mathcal{H}_2 = q_i + RT - \frac{\pi}{\Gamma} \tag{XVI-124}$$

It follows from the defining relationships that

$$\bar{\mathrm{H}}_2^s = T\bar{\mathrm{S}}_2^s + \bar{\mathrm{G}}_2^s \tag{XVI-125}$$

and

$$\mathcal{H}_2^s = T\mathrm{S}_2^s + \mathcal{G}_2^s \tag{XVI-126}$$

so that

$$\Delta\bar{\mathrm{S}}_2 = \bar{\mathrm{S}}_2^s - \mathrm{S}_2^{0,g} = \frac{\Delta\mathrm{H}_2 - \Delta\bar{\mathrm{G}}_2}{T} \tag{XVI-127}$$

and

$$\Delta \mathrm{s}_2 = \mathrm{s}_2^s - \bar{\mathrm{s}}_2^{0,g} = \frac{\Delta\mathcal{H}_2 - \Delta\mathcal{G}_2}{T} \tag{XVI-128}$$

where

$$\Delta\bar{\mathrm{G}}_2 = \Delta\mathcal{G}_2 = RT \ln P \tag{XVI-129}$$

Thus from an adsorption isotherm and its temperature variation, one can calculate either the differential or the integral entropy of adsorption as a function of surface coverage. The former probably has the greater direct physical meaning, but the latter is the quantity usually first obtained in a statistical thermodynamic adsorption model.

The adsorbed state often seems to resemble liquid adsorbate, as in the approach of the heat of adsorption to the heat of condensation in the multilayer region. For this reason, a common choice for the standard state of free adsorbate is the pure liquid. We now have

$$\Delta\bar{\mathrm{G}}_{2(l)} = \Delta\mathcal{G}_{2(l)} = RT \ln x \tag{XVI-130}$$

and

$$\Delta\bar{\mathrm{H}}_{2(l)} = \bar{\mathrm{H}}_2^s - \bar{\mathrm{H}}_2^{0,l} \qquad \Delta\mathcal{H}^0_{2(l)} = \mathcal{H}_2^s - \bar{\mathrm{H}}_2^{0,l} \tag{XVI-131}$$

$$\Delta\bar{\mathrm{S}}_{2(l)} = \bar{\mathrm{S}}_2^s - \bar{\mathrm{S}}_2^{0,l} \qquad \Delta\mathrm{s}_{2(l)} = \bar{\mathrm{S}}_2^s - \mathrm{s}_2^{0,l} \tag{XVI-132}$$

Also

$$q_{\mathrm{st}(l)} = RT^2 \left(\frac{\partial \ln x}{\partial T}\right)_\Gamma = q_{\mathrm{st}} - \Delta\mathrm{H}_v \tag{XVI-133}$$

and

$$q_{\pi(l)} = RT^2 \left(\frac{\partial \ln x}{\partial T}\right)_\pi = q_\pi - \Delta\mathrm{H}_v \tag{XVI-134}$$

Thus the new thermodynamic heats and entropies of adsorption differ from the preceding ones by the heats and entropies of vaporization of liquid adsorbate.

There are alternative ways of defining the various thermodynamic quantities. One may, for example, treat the adsorbed film as a phase having volume, so that P, V terms enter into the definitions. A systematic treatment of this type has been given by Honig (102), who also points out some additional types of heat of adsorption.

Finally, it is perfectly possible to choose a standard state for the surface phase. De Boer (10) makes a plea for taking that value of π^0 such that the average distance apart of the molecules is the same as in the gas phase at STP. This is a hypothetical standard state in that π^0 for an ideal two-dimensional gas with this molecular separation would be 0.338 dyne/cm at 0°C. The standard molecular area is then $4.08 \times 10^{-16}\,T$. The main advantage of this choice is that it simplifies the relationship between translational entropies of the two- and the three-dimensional standard states.

2. *Thermodynamic Quantities for the Adsorbent.* It is also possible to calculate the change in thermodynamic quantities for the *adsorbent*, in the adsorption process, and this has been discussed by Copeland and Young (103). One problem, however, is how to define quantities such as $\bar{G}_1^s$ (in Eq. XVI-105) when the system consists of adsorbent particles so that there is no way of making a minute change dn_1^s without changing the specific surface area. This is handled by taking the change Δn_1^s, corresponding to the addition of one particle, but treating the thermodynamic quantities as continuous functions by basing them on the locus through the points representing successive increments of Δn_1^s. On this basis, one can proceed to apply ordinary two-component thermodynamics.

Thus for the differential process Eq. XVI-103, dn_1^s is zero, and it follows from Eq. XVI-105 that

$$\Delta G = \int_0^{n_2^s} \bar{G}_2 \, dn_2^s = RT\int_0^{n_2^s} \ln P \, dn_2^s \qquad \text{(XVI-135)}$$

Also, from the Gibbs equation,

$$d\pi = \frac{n_2^s}{\Sigma' \, n_1^s} RT \, d \ln P$$

where Σ' is now the area per *mole* of adsorbent. Since n_1^s is constant, we have, in combination with Eq. XVI-113,

$$\Sigma' \, \pi = \frac{RT}{n_1^s} \int_0^{n_2^s} n_2^s \, d \ln P = \frac{1}{n_1^s}\left(n_2^s \, \Delta\bar{G}_2 - \int_0^{n_2^s} \Delta\bar{G}_2 \, dn_2^s\right) \qquad \text{(XVI-136)}$$

or

$$\Sigma' \, \pi = \frac{1}{n_1^s} \int_0^{n_2^s} n_2^s \, d \, \Delta\bar{G}_2 = -\Delta\bar{G}_1 \qquad \text{(XVI-137)}$$

since by Eqs. XVI-105 and XVI-106, $0 = n_1^s \, d\bar{G}_1^s + n_2^s \, d\bar{G}_2^s$. The differential heats of adsorption are then

$$\Delta\bar{H}_1 = \left[\partial \frac{(\Delta\bar{G}_1/T)}{\partial(1/T)}\right]_{n_1^s, n_2^s, \Sigma'} \qquad \text{(XVI-138)}$$

and

$$\Delta\bar{H}_2 = \left[\frac{\partial(\Delta\bar{G}_2/T)}{\partial(1/T)}\right]_{n_1^s, n_2^s, \Sigma'} = -RT^2\left(\frac{\partial \ln P}{\partial T}\right)_\Gamma \qquad \text{(XVI-139)}$$

(or the same as Eq. XVI-116). $\Delta\bar{H}_1$ is probably most conveniently obtained as

$$\Delta\bar{H}_1 = \frac{\Delta H}{n_1^s} - \frac{n_2^s}{n_1^s} \Delta\bar{H}_2 \qquad \text{(XVI-140)}$$

It can be shown that

$$\Delta H = -\pi\mathcal{A} + n_2^s RT^2 \left(\frac{\partial \ln P}{\partial T}\right)_\pi \tag{XVI-141}$$

(so that, on dividing through by n_2^s, it is seen that $\Delta\mathcal{H}_2 = \Delta H/n_2^s + \pi/\Gamma$). The entropies may be obtained using relationships paralleling Eq. XVI-125. Wu and Copeland (104) applied this analysis to data on the adsorption on $BaSO_4$. A more detailed analysis of the extensive data of Drain and Morrison (1) on the adsorption of N_2 on rutile is shown in Fig. XVI-23. Here, the quantity $[\Delta H_1]$ is defined as $n_1\Delta\bar{H}_1/n_2$ and corresponds to the total enthalpy change for the adsorbent *per mole of adsorbate*. The values of $[\Delta H_1]$ are far from negligible and indicate that, in the symmetric approach, both adsorbent and adsorbate partial molal quantities are significant. The same is true for free energy quantities, suggesting that if a detectably volatile adsorbent were used, it is to be expected that its vapor pressure would be changed, following adsorption. Experiments of this type should be very interesting to carry out, for example, studies on the adsorption of vapors on ice have been reported (89, 106), and it should be possible to determine whether the adsorption of a hydrocarbon vapor on ice between, say, 0°C and −50°C, affects the ice vapor pressure. From the extent to which this happens, an estimate can be made of the depth of ice surface that is in equilibrium with the vapor phase.

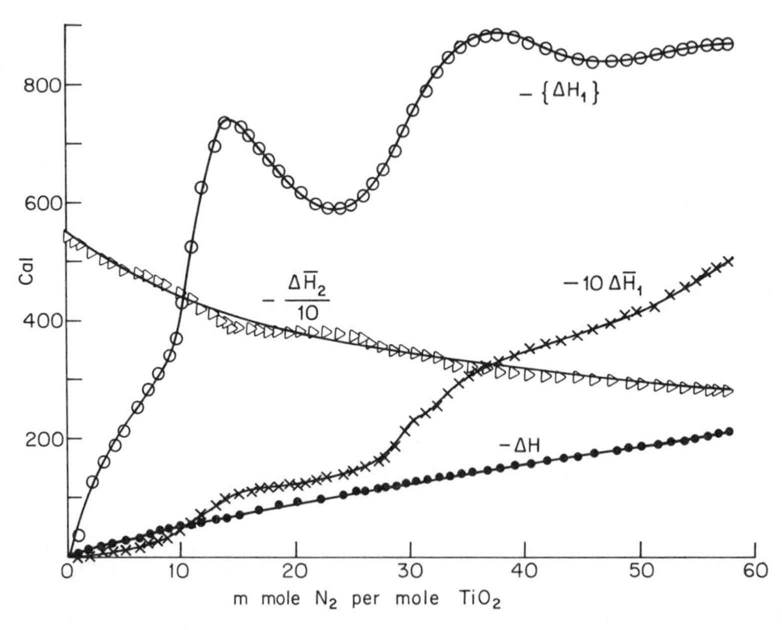

Fig. XVI-23. Variation with surface occupancy of the various enthalpy quantities for the case of nitrogen on rutile. (From Ref. 105.)

In the case of refractory solids, structural changes might be detectable by other means. As a minimum effect, some strain relief in the solid surface should occur on adsorption, and an interesting attempt to calculate this contribution to adsorption energetics was made by Cook et al. (107).

B. Experimental Heats and Entropies of Adsorption

Before taking up the results of measurements of heats and entropies of adsorption, it is perhaps worthwhile to review briefly the various alternative procedures for obtaining these quantities.

The *integral heat* of adsorption Q_i may be measured calorimetrically by determining directly the heat evolution when the desired amount of adsorbate is admitted to the clean solid surface. Alternatively, it may be more convenient to measure the heat of immersion of the solid in pure liquid adsorbate. Immersion of clean solid gives the integral heat of adsorption at $P = P^0$, that is, $Q_i(P^0)$ or $q_i(P^0)$, whereas immersion of solid previously equilibrated with adsorbate at pressure P gives the difference $[q_i(P^0) - q_i(P)]$, from which $q_i(P)$ can be found (108, 109). The *differential heat* of adsorption q_d may be obtained from the slope of the Q_i versus n_2^s plot, or by measuring the heat evolved as small increments of adsorbate are added (110).

Alternatively, q_{st} may be obtained from the application of Eq. XVI-116 to adsorption data at two or more temperatures (see Ref. 77). Similarly, q_i is obtainable from isotherm data by means of Eq. XVI-124, but now only provided that isotherms down to low pressures are available so that Gibbs integrations to obtain π values are possible.

The *partial molar* entropy of adsorption $\Delta\bar{s}_2$ may be determined from q_d or q_{st} through Eq. XVI-127, and hence is obtainable either from calorimetric heats plus an adsorption isotherm or from adsorption isotherms at more than one temperature. The *integral entropy* of adsorption can be obtained from isotherm data at more than one temperature, through Eqs. XVI-119 and XVI-128, in which case complete isotherms are needed. Alternatively, ΔS_2 can be obtained from the calorimetric q_i plus a single complete adsorption isotherm, using Eq. XVI-124. This last approach has been recommended by Jura and Hill (108) as giving more accurate integral entropy values.

Turning now to the results of such measurements, perhaps the first point of interest is whether the calorimetric and thermodynamic heats of adsorption do in fact agree according to Eqs. XVI-123 and XVI-124. This appears to be the case. Brunauer (19) gives several examples involving vapors such as carbon dioxide, methanol, nitrogen, and water, adsorbed on charcoal, in which q_d and q_{st} agree within experimental error. Greyson and Aston (111) found that the detailed and rather complex variation of q_d with amount adsorbed, in the case of neon on graphitized carbon, agreed very closely with their q_{st} values. Note in Fig. XVI-24*c* that both isosteric and calorimetric values are shown. The question is not trivial; such agreement is not assured

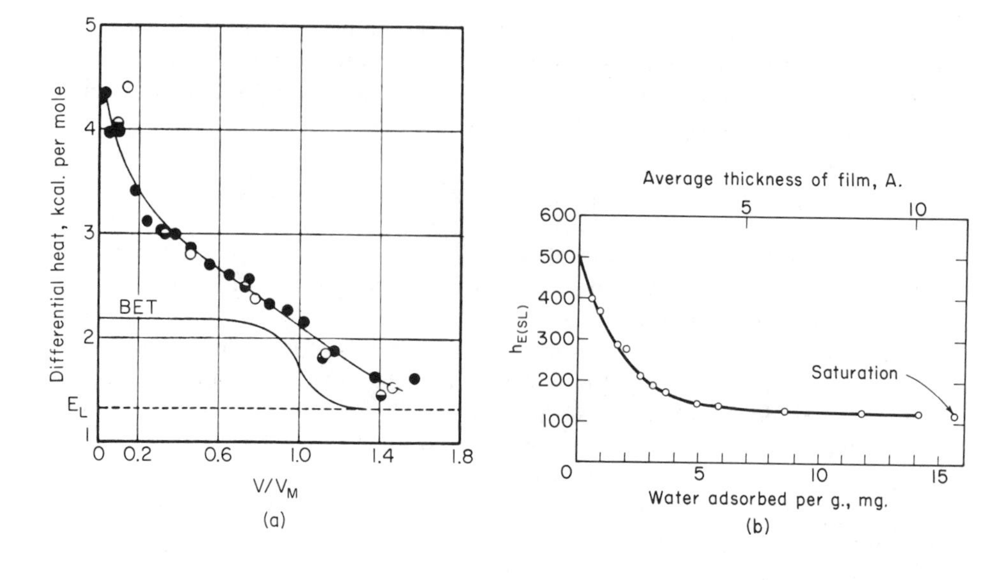

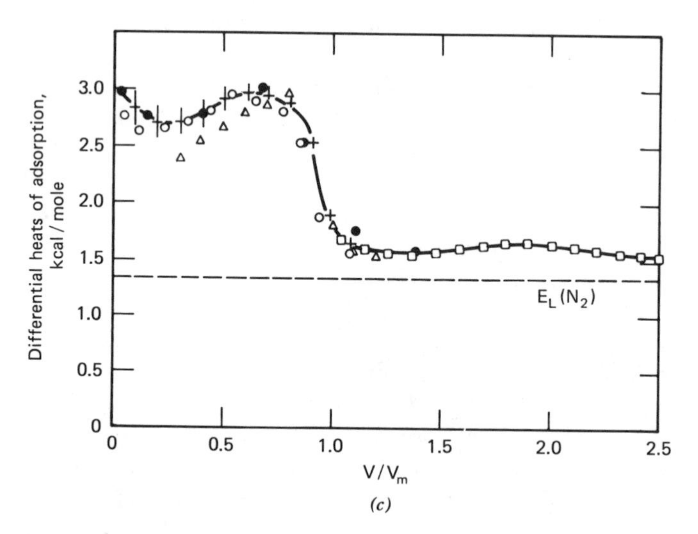

Fig. XVI-24. (*a*) Differential heat of adsorption of nitrogen on carbon black (Spheron 6) at 78.5 K. (From Ref. 110.) (*b*) Thickness of the water film and the energy of emersion per square centimeter of the surface of titanium dioxide. (From Ref. 112.) (*c*) Differential heat of adsorption of nitrogen of Graphon, except for ○ and ●, which were determined calorimetrically. (From Ref. 77.) (*d*) Isosteric heats of adsorption cf *n*-hexane on ice; v_m = 0.073 cc STP. (From Ref. 113.) (*e*) Isosteric heats of adsorption of Ar on graphitized carbon black having the indicated number of preformed layers of ethylene. (From Ref. 114.)

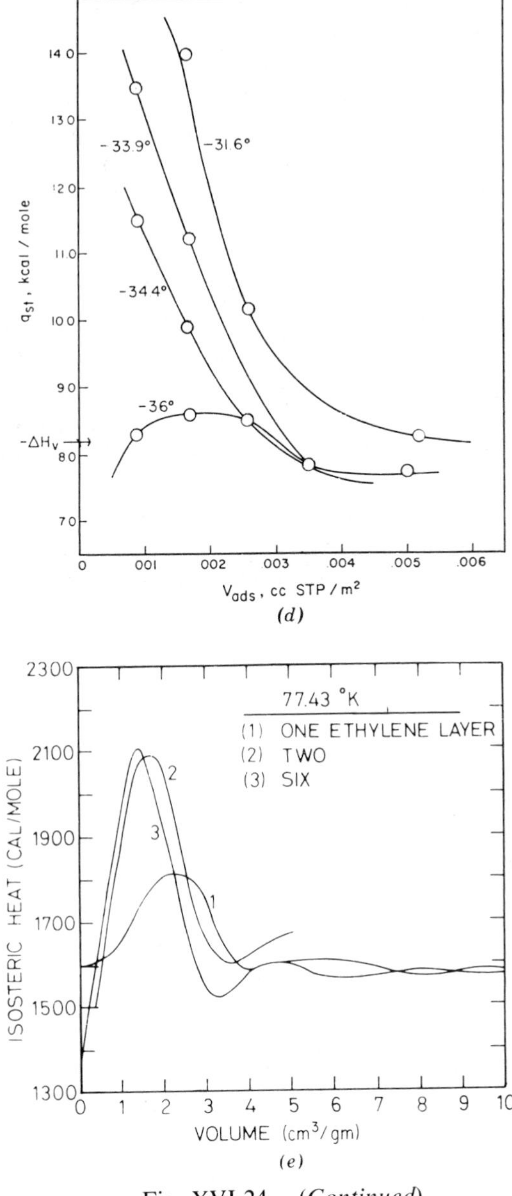

(d)

(e)

Fig. XVI-24. *(Continued)*

in the case of systems showing hysteresis (see Section XVI-16), and it has been difficult to affirm it on rigorous thermodynamic grounds in the case of a heterogeneous surface.

Differential heats of adsorption generally decrease steadily with increasing amount adsorbed and, in the case of physical adsorption tend to approach the heat of liquefaction of the adsorbate as P approaches P^0. Some illustrative data are shown in Fig. XVI-24. The presumed monolayer point may be marked by a sharp decrease in q_{st}, as in Fig. XVI-24*c*; a more steady decrease, as in Figs. XVI-24*a*,*b*, probably indicates surface heterogeneity. The dramatic change in behavior around $-35°C$ for *n*-hexane on ice is attributed to surface clatherate formation at the higher temperatures. Figure XVI-24*e* shows that only two monolayers of ethylene were sufficient to shield Ar from graphitized carbon black since no further change in the q_{st} behavior occurred. Gregg and Sing (41) give other examples.

Some representative plots of entropies of adsorption are shown in Fig. XVI-25; in general, $T\,\Delta\bar{S}_2$ is comparable to $\Delta\bar{H}_2$, so that the entropy contribution to the free energy of adsorption is important. Notice in Figs. XVI-25*a* and *b* how nearly the entropy plot is mirror image of the enthalpy plot. As a consequence, the maxima and minima in the separate plots tend to cancel to give a smoothly varying free energy plot, that is, adsorption isotherm.

As with enthalpies of adsorption, the entropies tend to approach the entropy of condensation as P approaches P^0, in further support of the conclusion that the nature of the adsorbate is approaching that of the liquid state.

Finally, one frequently observes that $\Delta\bar{S}_2$ goes through a minimum at or

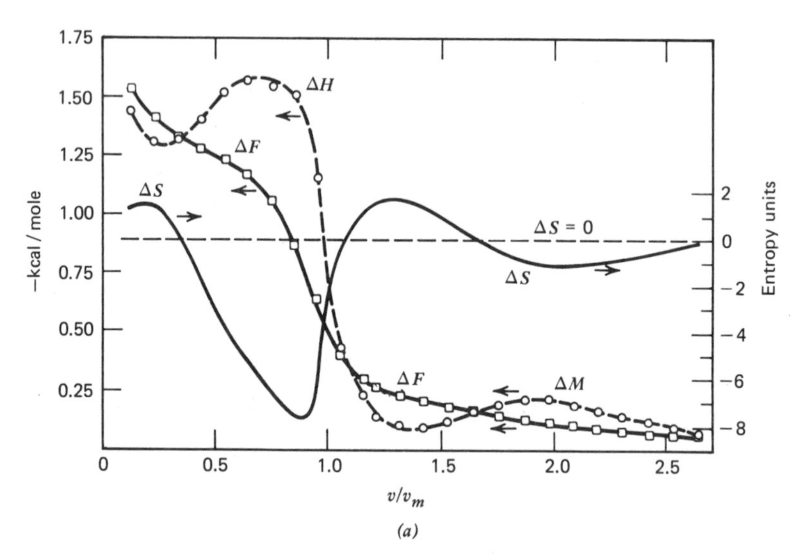

Fig. XVI-25. (*a*) Entropy, enthalpy, and free energy of adsorption relative to the liquid state of nitrogen on Graphon at 78.3°K. (From Ref. 77.) (*b*) Isosteric heats and partial molar entropies of adsorption of Ar, N_2, and CO on ice at 77°C. (From Ref. 115.) (*c*) Differential entropies of adsorption of *n*-hexane on (1) 1700°C heat treated Spheron 6, (2) 2800°C heat treated (3) 3000°C heat treated, and (4) Sterling MT-1, 3100°C heat treated. (From Ref. 14.)

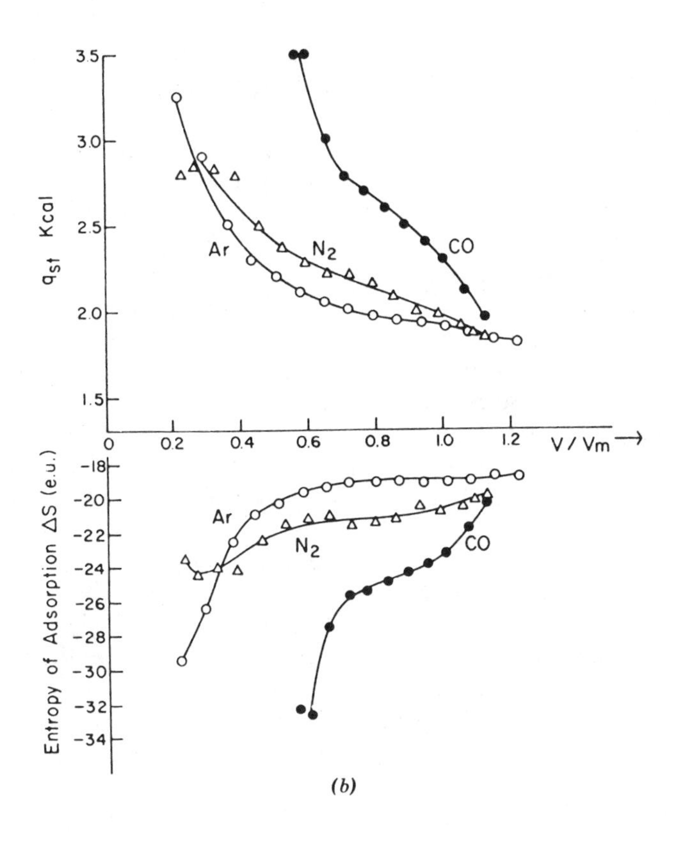

(b)

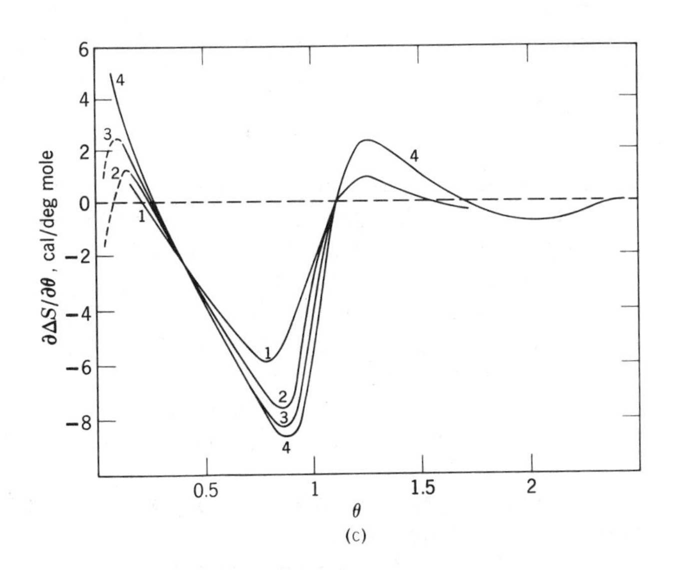

(c)

Fig. XVI-25. (*Continued*)

near $n = n_m$. An interesting example of this effect is provided by some data of Isirikyan and Kiselev (14) for the adsorption of *n*-hexane on various carbons, illustrated in Fig. XVI-25*c*. The minimum deepens with degree of graphitization and hence with presumed increasing degree of surface uniformity. Hill et al. (108) report a very clear-cut minimum in $\Delta \bar{S}_2$ in the case of nitrogen or Graphon—another carbon that appears to have a very uniform surface.

The accepted explanation for the minimum is that it represents the point of complete coverage of the surface by a monolayer; according to Eq. XVI-37, $\bar{S}_{config}$ should go to minus infinity at this point, but in real systems, an onset of multilayer adsorption occurs, and this provides a counteracting positive contribution. Some further discussion of the behavior of adsorption entropies in the case of heterogeneous adsorbents is given in Section XVI-14.

14. Critical Comparison of the Various Models for Adsorption

As pointed out in Section XVI-8, agreement of a theoretical isotherm equation with data at one temperature is a necessary but quite insufficient test of the validity of the premises on which it was derived. Quite differently based models may yield equations that are experimentally indistinguishable and even algebraically identical. In the multilayer region, it turns out that in a number of cases the isotherm shape is relatively independent of the nature of the solid and that any equation fitting it can be used to obtain essentially the same relative surface areas for different solids, so that consistency of surface area determination does not provide a sensitive criterion either.

The data on heats and entropies of adsorption do allow a more discriminating test of an adsorption model, although even so only some rather qualitative conclusions can be reached. The discussion of these follows.

A. The Langmuir–BET Model

This model calls for localized adsorption in all layers, with $\Delta \bar{H}_2$ a constant for the first layer and equal to the heat of condensation for all succeeding ones. The Langmuir model is perfectly acceptable for adsorption at low P/P^0 values, but as was seen in Section XVI-3C, it is not easily distinguished from mobile adsorption in terms of $\Delta \bar{S}_2$ and, in Section XVI-6B, it was noted that the distinction is also difficult in terms of algebraic form. It appears that independent data on the surface diffusion coefficient or on the energy states of the adsorbent–adsorbate complex are essential to any firm diagnosis (see also Section XV-4). The complicating effect of surface heterogeneity is discussed in Section XVI-14.

Turning to the multilayer region, the actual assumptions of the BET model are not realistic. As illustrated in Fig. XVI-24*a*, it does not give the correct variation of q_d with θ (but partly because of surface heterogeneity) or the correct value of $\Delta\bar{s}_2$. Basically, the available evidence suggests that the adsorbed film approaches bulk liquid in properties as P approaches P^0, and while the BET assumption as to the adsorption energy correctly reflects this behavior, the assumption of localized multilayers is not consistent with it and gives an erroneous configurational entropy. Related to this is a catastrophe that Cassel (117) has pointed out, namely, that the integral I of Eq. X-40 is infinity in the BET model. Some further evidence of the liquidlike state of multilayer films was provided by Arnold (118), who found that his data on the desorption of oxygen–nitrogen mixtures on titanium dioxide could be accounted for by assuming the second and third layers to possess the molar entropy of normal liquid and that Raoult's law applied, whereas the BET treatment gave much poorer agreement. As other examples, Whalen and co-workers found highly nonlinear BET plots for N_2 and Ar on Teflon 6 (119), N_2 isotherms for variously dehydrated TiO_2 (rutile) gave erroneous BET areas (120), and, finally, BET areas will be in error if micropores are present (see Ref. 121).

Brunauer (see Refs. 122 and 123) defends these defects as deliberate approximations needed to obtain a practical two-constant equation. The assumption of a constant heat of adsorption in the first layer represents a balance between the effects of surface heterogeneity and of lateral interaction, and the assumption of a constant instead of a decreasing heat of adsorption for the succeeding layers balances the overestimate of the entropy of adsorption. These comments do help to explain why the model works as well as it does. However, since these approximations are inherent in the treatment, one can see why the BET model does not lend itself readily to any detailed insight into the real physical nature of multilayers. In summary, the BET equation will undoubtedly maintain its usefulness in surface area determinations, and it does provide some physical information about the nature of the adsorbed film, but only at the level of approximation inherent in the model. Mainly, the c value provides an estimate of the first layer heat of adsorption, averaged over the region of fit.

B. Two-Dimensional Equation of State Treatments

There is little doubt that, at least with type II isotherms, we can tell the approximate point at which multilayer adsorption sets in. The concept of a two-dimensional phase seems relatively sterile as applied to multilayer adsorption, except insofar as such isotherm equations may be used as empirically convenient, since the thickness of the adsorbed film is not easily allowed to become variable.

On the other hand, as applied to the submonolayer region, the same comment can be made as for the localized model. That is, the two-dimensional nonideal gas equation of state is a perfectly acceptable concept, but one that in practice is remarkably difficult to distinguish from the localized adsorption picture. If there can be even a small amount of surface heterogeneity the distinction becomes virtually impossible (see Section XVI-14). Even the cases of phase change are susceptible to explanation on either basis. Ross and Olivier (21), in their extensive development of the van der Waals equation of state model have, however, provided a needed balance to the Langmuir picture. This writer anticipates a gradual merging of the two approaches as submonolayer adsorption comes to be viewed in terms of the energy states of the adsorbent–adsorbate complex, of adsorbate–adsorbate interactions, and of their distribution if the surface is heterogeneous. This sort of information needs to be obtained from independent measurements such as described in Chapter XV, however, rather than from values of parameters obtained empirically by the fitting of adsorption equations to data.

C. The Potential Model

Returning to multilayer adsorption, the potential model appears to be fundamentally correct. It accounts for the empirical fact that systems at the same value of $RT \ln (P/P^0)$ are in essentially corresponding states, and that the multilayer approaches bulk liquid in properties as P approaches P^0. However, the specific treatments must be regarded as still somewhat primitive. The various proposed functions for $\epsilon(r)$ can only be rather approximate. Even the general appearing Eq. XVI-81 cannot be correct, since it does not allow for structural perturbations that make the film different from bulk liquid. Such perturbations should in general be present and *must* be present in the case of liquids that do not spread on the adsorbent (Section X-5). The last term of Eq. XVI-91, while reasonable, represents at best a semiempirical attempt to take structural perturbation into account.

The existence of a characteristic isotherm (or of a t-plot) gives a very important piece of information about the adsorption potential, at least for polar solids for which the observation holds. The direct implication is that film thickness t, or alternatively v/v_m, is determined by P/P^0 *independent of the nature of the adsorbent*. We can thus write

$$\frac{n}{n_m} = f\left(\frac{P}{P^0}\right) \tag{XVI-142}$$

where the function f (and any associated constants) is independent of the nature of the solid (but may vary with that of the adsorbate).

Unfortunately none of the various proposed forms of the potential theory satisfy this criterion! Equation XVI-88 clearly does not; Eq. XVI-89 would except that f includes the constant A, which contains the dispersion energy ϵ_0, which in turn depends on the nature of the adsorbent. Equation XVI-94 fares no better if, according to its derivation, ϵ_0 reflects the surface polarity of the adsorbent (note Eq. VI-40). It would seem that after one or at most two layers of coverage, the adsorbate film is effectively insulated from the adsorbent.

This insulation does in fact seem to occur. Figure XVI-26 shows some results of Dubinin and co-workers (124). Nitrogen adsorption isotherms at $-185°C$ were de-

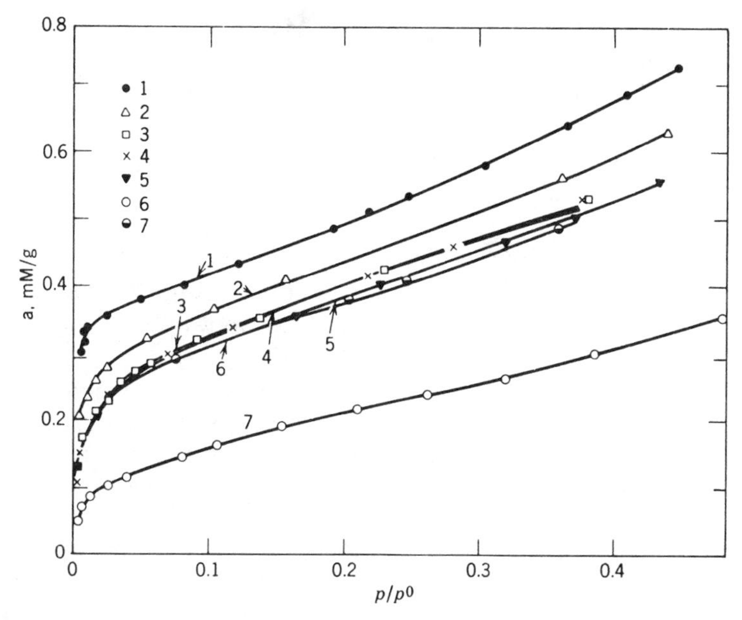

Fig. XVI-26. Nitrogen adsorption isotherms at −195°C on pure carbon black: (1) on uncoated surface; (2) to (6) with 0.6, 1.1, 1.5, 4.5, and 6.2 statistical monolayers of benzene, respectively; (7) powdered benzene. (From Ref. 124.)

termined for a carbon black having various amounts of preadsorbed benzene. After only about 1.5 statistical monolayers of benzene no further change took place in the nitrogen isotherms. The isotherms are all of the same shape, furthermore, so that they fit essentially the same characteristic isotherm; the only effect would be a decrease in the calculated n_m between 0 and 1.5 layers of precoating. A similar behavior was reported by Halsey and co-workers (125) for Ar and N_2 adsorption in TiO_2 having increasing amounts of preadsorbed water. In this case, a limiting isotherm was reached with about four statistical layers of water, the approach to the limiting form being approximately exponential. Interestingly, the final isotherm, in the case of N_2 as adsorbate, was quite similar to that for N_2 on a directly prepared and probably amorphous ice powder (32, 106). On the other hand, N_2 adsorption on carbon with increasing thicknesses of preadsorbed methanol decreased steadily—no limiting isotherm was reached (124).

Clearly, it is more desirable somehow to obtain detailed structural information on multilayer films so as perhaps to settle the problem of how properly to construct the potential function. Some attempts have been made to develop statistical mechanical treatments of molecules in a potential field, but with limited success—the problem is a most difficult one (see Refs. 11, 56, 126).

15. Adsorption on Heterogeneous Surfaces

The discussion on adsorption models in the preceding section was rather restrained because it turns out that for nearly all systems studied an overriding effect makes it virtually impossible to make an experimental verification of the validity of the model or to set up any but remotely austere fundamental theoretical treatments. The effect is that of surface heterogeneity. It has been made very clear that solid surfaces are not in general uniform (e.g., Section VII-4) and the data of Figs. XVI-24 and XVI-25 provide a direct indication of nonuniformity of heats and entropies of adsorption.

A. Site Energy Distributions

To keep the situation manageable, we confine ourselves to adsorption in the submonolayer region, and the problem of adsorption on a heterogeneous surface can then be formulated in a general way by noting that, regardless of model, the fraction of surface covered should be some function θ of Q, P, and T, where Q is an adsorption energy. If the surface is heterogeneous, the probability of there being an adsorption energy between Q and $Q + dQ$ can be described by a distribution function $f(Q)\, dQ$. The experimentally observed adsorption will be the sum of all the adsorptions on the different kinds of surface, and so will be a function Θ of P and T. Thus

$$\Theta(P, T) = \int_0^\infty \theta(Q, P, T) f(Q)\, dQ \qquad \text{(XVI-143)}$$

Alternatively, an integral distribution function F may be defined as giving the fraction of surface for which the adsorption energy is greater than or equal to a given Q,

$$f(Q) = \frac{dF}{dQ} \qquad \text{(XVI-144)}$$

whence

$$\Theta(P, T) = \int_0^1 \theta(Q, P, T)\, dF \qquad \text{(XVI-145)}$$

The variation in q_d with Θ will not in general be the same as $F(\Theta)$, since some adsorption will be occurring on all portions of the surface so that the heat liberated on adsorption of dn_2^s moles will be a weighted average. There is one exception, however, namely adsorption at 0 K; the adsorption will occur sequentially on portions of increasing Q value so that $q_d(\Theta)$ now gives $F(\Theta)$. This circumstance was made use of by Drain and Morrison (1), who determined $q_d(\Theta)$ for argon, nitrogen, and oxygen on titanium dioxide at a series of temperatures and extrapolated to 0 K. The procedure is a difficult one and not without some approximations in the extrapolation. Clearly, it would be very desirable to find a way of solving the integral equation so that site or adsorption energy distributions could be obtained from data at customary temperatures.

One approach is to assume functions for both $\theta(Q, P, T)$ and $f(Q)$ such that integration is possible. Thus if $f(Q) = \alpha e^{-Q/nRT}$ and $\theta(Q, R, T)$ is the Langmuir equation, then the Freundlich equation results (see Section XI-1A). See Ref. 126a for the case of a multicomponent system.

Ross and Olivier (21) took $f(Q)$ to be Gaussian and $\theta(Q, P, T)$ to be the two-dimensional van der Waals equation and have provided extensive tabulations of the solutions to Eq. XVI-145 for various choices of the parameters.

The whole approach can be bypassed; Temkin (127) took all sites to vary in Q according to the equation $Q = Q_0(1 - \alpha\Theta)$, and direct substitution into the expression for b in the Langmuir equation gives

$$\ln P = -\ln(b_0 e^{Q_0/RT}) + \ln \frac{\Theta}{1 - \Theta} + \frac{Q_0 \alpha \Theta}{RT} \qquad \text{(XVI-146)}$$

However, returning to the integral equation, a second type of procedure is to choose a function for $\theta(Q, P, T)$ such that the integral equation can be inverted to give $f(Q)$ from the observed isotherm. Hobson (128) chose a local isotherm function that was essentially a stylized van der Waals form with a linear low pressure region followed by a vertical step to $\theta = 1$. Sips (129) showed that Eq. XVI-143 could be converted to a standard transform if the Langmuir adsorption model was used. One writes

$$\Theta(P, T) = \int_0^\infty \frac{f(Q) bP}{1 + bP} \, dQ = \int_0^\infty \frac{e^{Q/RT} f(Q)}{e^{Q/RT} + 1/b_0 P} \, dQ \qquad \text{(XVI-147)}$$

or

$$\frac{1}{RT} \Theta \frac{1}{b_0 y} = \int_1^\infty \left[\frac{f(RT \ln x)}{x + y} \right] dx \qquad \text{(XVI-148)}$$

where $y = 1/b_0 P$ and $x = e^{Q/RT}$. The equation is now in the form

$$\chi(y) = \int_1^\infty \frac{\phi(x)\, dx}{(x + y)} \qquad \text{(XVI-149)}$$

for which the solution is

$$\phi(x) = \frac{\chi(xe^{-\pi i}) - \chi(xe^{\pi i})}{2\pi i} \qquad \text{(XVI-150)}$$

It is necessary, for this procedure, to express Θ as an analytical function of P. Thus, if $\Theta(P, T) = AP^c$,

$$\chi(y) = \frac{A}{RTb_0^c} y^{-c} \qquad \text{(XVI-151)}$$

and by Eq. XVI-150,

$$\phi(x) = \frac{(A/RTb_0^c)[(xe^{-\pi i})^{-c} - (xe^{\pi i})^{-c}]}{2\pi i} = \frac{A}{RTb_0^c} \sin \pi c \frac{x^{-c}}{\pi} \qquad \text{(XVI-152)}$$

or

$$f(Q) = \frac{A}{RTb_0^c} \sin \pi c \frac{1}{\pi} \exp\left(-\frac{cQ}{RT}\right) \qquad \text{(XVI-153)}$$

which is a more exact derivation of Eq. XI-13. If one chooses $\Theta = AP^c/(1 + AP^c)$ so as to have a form that gives a limiting Θ, the resulting $f(Q)$ is Gaussianlike. Honig and Reyerson (130) used $\Theta = [P/(A + P)]^c$, which fit their data for nitrogen on titanium dioxide over the range of Θ from 0.5 to 0.9 and gave an $f(Q)$ that steadily decreased with increasing Q. The data would also fit the equation $\Theta = A \ln [(kP + C)/(P + C)]$ but now the $f(Q)$ obtained from Eq. XVI-150 was a constant over a certain range of Q values and zero outside that range.

At this point it is evident that the solution to the general integral equation, Eq. XVI-143, can be sensitive to the choice of analytical expressions, especially for $\Theta(P, T)$, and when the choice is dictated by the mathematical limitations as to what forms allow a transform to be found, the fit to data may not be very good. As a consequence it is very difficult to know how significant the resulting $f(Q)$ really is.

An alternative procedure is to solve Eq. XVI-143 by successive approximations (131). This may be done as follows. First, a relative $\theta(bP)$ is chosen; this may be *any* function in which b is proportional to $e^{Q/RT}$ so that Langmuir type equations as well as those of Table XVI-2 may be used. The function is then tabulated or plotted, as illustrated in Fig. XVI-27*b*, in which the Langmuir equation is chosen, but merely as the simplest example. As a *first* approximation $\theta(bP)$ is replaced by a step function; in the illustration, the

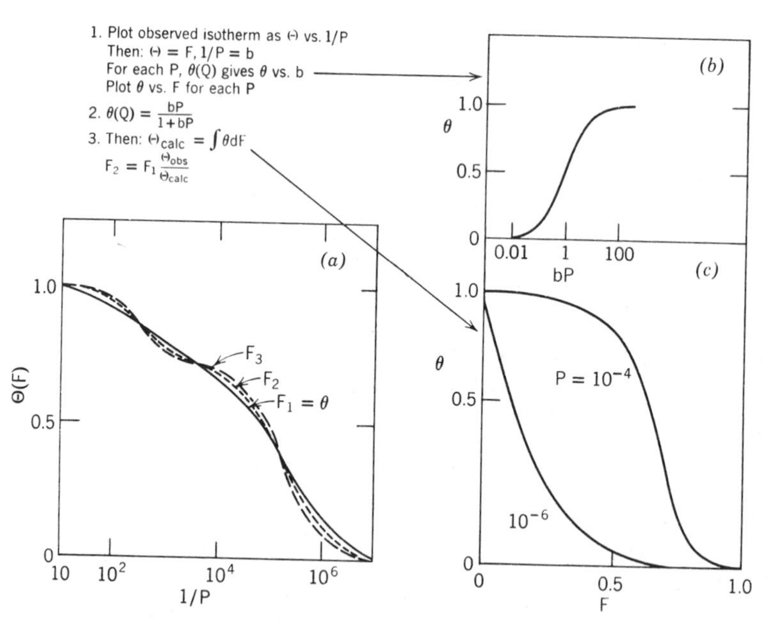

Fig. XVI-27. Outline of procedure for obtaining site energy distribution. (*a*) ———, adsorption isotherm plotted as θ versus $1/P$; – – –, successive approximations to F versus b. (*b*) $\theta(Q)$ (in this case, the Langmuir equation). (*c*) θ versus F plots for two different pressures. (From Ref. 131.)

step would be at $bP = 1$ so that for $bP < 1$, $\theta = 0$, and for $bP > 1$, $\theta = 1$. For a step function the situation is similar to that at 0 K, that is, adsorption sites fill in strict order, a site of given b (and hence Q) value filling at $P = 1/b$ or, conversely, for pressures up to a given value of $1/P$, all sites b value equal to or less than that are filled, and no others. Consequently, in this first approximation, the plot of Θ versus $1/P$ is also the plot of F versus b, Fig. XVI-27*a*.

The second approximation is made by noting that for a *given* value of P, for each b a value of θ follows from Fig. XVI-27*b* and of F from Fig. XVI-27*a*. By taking a series of b values, an auxiliary plot of θ versus F may be constructed, as shown in Fig. XVI-27*c*. Now the overall surface occupancy Θ must be given by $\int \theta \, dF$, so the area under the auxiliary plot gives a Θ_{calc} for that P. However, for that P the experimental Θ in general will be somewhat different, and the adjustment is made that the second approximation to F, F_2, is given by $F_2 = F_1 \Theta_{obs}/\Theta_{calc}$. This F_2 is then entered on the F versus b plot at the b value equal to $1/P$. The whole procedure is now repeated for a second choice of pressure, yielding an F_2 for another b value, that is, for b equal to one over this second choice of pressure. In this way a series of points is obtained through which the second approximation to the integral distribution plot of F versus b is drawn.

A third round of approximations may be needed, as illustrated in Fig. XVI-27*a*. The procedure is the same as for the second round. A value of P is chosen, then by taking a series of b values an auxiliary plot is constructed whose area gives a Θ_{calc}. From this the third approximation to F, F_3 is obtained for the b value equal to one over the chosen P. On repeating with a series of chosen P values, the F_3 versus b plot is constructed.

It turns out that in general there is *no exact* solution to Eq. XVI-143; this is simply because the experimental data have error and because the assumed θ function will not be exactly correct for that system. The consequence is that successive approximations to the F versus b plot do not converge exactly, especially for nearly homogeneous surfaces. A point is reached, often fairly quickly, where successive approximations do not differ by much and, moreover, nonsystematically so that no continuing trends are apparent. The whole procedure, of course, can be carried out by means of a digital computer.

The final F versus b plot may be converted to one of F versus $Q(b) = b_0 e^{Q/RT}$ in the case of the Langmuir equation and thence to a plot of $f(Q)$ versus Q. An example of such plots is obtained by applying the preceding procedure to the data of Drain and Morrison (1), using the BET equation as the local isotherm function (132), as shown in Fig. XVI-28. Figure XVI-28 illustrates how closely the procedure can approximate the true site energy distribution. The site energy distribution will depend on the *adsorbate* as well as on the adsorbent; it is a property of a given system. The effect of changing adsorbate, however, is usually more one of shifting the energy scale than of making major shape changes in the distribution.

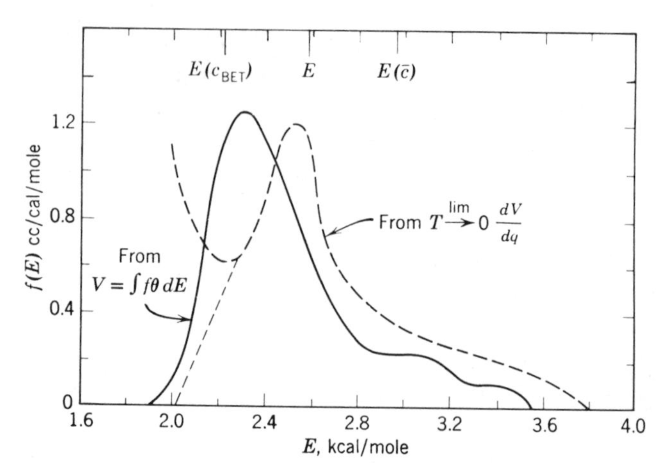

Fig. XVI-28. Site energy distributions for Ar on rutile from the data of Drain and Morrison (1). Solid line: calculated from the 85°K isotherm. Dashed line: obtained by Drain and Morrison by extrapolation to 0°K; the light dashed line shows their assumed closure at low energies.

In the case of a very heterogeneous surface, the only significant information obtainable is that of the site energy distribution, and this is essentially independent of any assumptions about the adsorption model. A gross error in the assumed entropy of adsorption (e.g., b_0) can be detected in that the distribution will fail to be temperature independent, and if a model providing for lateral interactions is used, the distribution will be shifted along the Q scale by an amount corresponding to the average lateral interaction.

The analysis is thus relatively exact for heterogeneous surfaces and is especially valuable for analyzing changes in an adsorbent following one or another treatment. An example is shown in Fig. XVI-29 (133). This type of application has also been made to nuclear irradiated titanium dioxide (134) and to carbon blacks and silica–alumina catalysts (135). House and Jaycock (136) compared the Ross and Olivier (21) and Adamson (131) procedures as applied to Kr adsorption on anatase (TiO_2); each could represent the data, but the site energy distribution depended on the method and choice of local isotherm function (Langmuir versus two-dimensional van der Waals).

One can write Eq. XVI-143 as a set of n equations for n data points, to be solved simultaneously. A computer program for this approach has been developed by House and Jaycock (137) for several specific assumed local isotherm functions, and has been applied to adsorbed argon on NaCl and nitrogen on silica surfaces (138). Dormant (see Ref. 105) pointed out that, given some experimental error, a relatively small set of different energy patches of adsorption sites will suffice—the addition of more patches will not improve the fit. Thus the extensive data of Drain and Morrison (1), *including their calorimetric data*, could be fit using the Langmuir local isotherm function and just *eight* patches. Some 70 data points were fit.

On the more theoretical side, Oh and Kim (139) considered the statistical mechanics of adsorption on a heterogeneous surface. Cerefolini and co-workers (140, 141) have investigated useful specific forms for the site energy distribution function.

Turning to the case of relatively homogeneous surfaces, the use of the site energy distribution analysis illuminates an awkward situation. As before, various adsorption models may be used to obtain an $f(Q)$, but since $f(Q)$ now represents a fairly narrow site energy distribution, different adsorption models give noticeably different distributions. This is illustrated in Fig. XVI-30 for the case of nitrogen on BN (142) (data from Ref. 143); the three distributions and associated choices for $\theta(Q, P, T)$ gave entirely comparable agreements with the experimental results. The somewhat paradoxical situation is thus that in the case of a nearly homogeneous surface, *neither* the adsorption model *nor* the site energy distribution can be affirmed with any great assurance.

The difficulties do not stop at this point. The application of the preceding site energy distribution analysis with models assuming lateral interaction implicitly assumes the heterogeneities to be present in patches. This need not be the case and, in fact, a complete description of the statistics of heterogeneities would have to

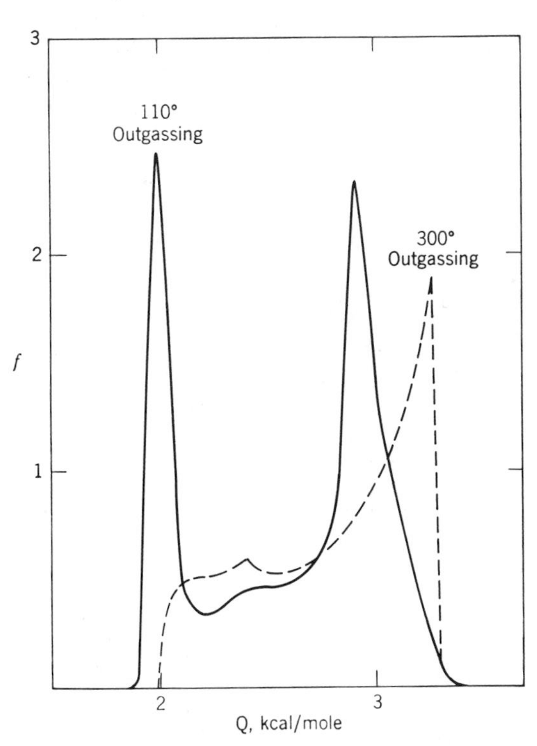

Fig. XVI-29. Site energy distribution for nitrogen adsorbed on Silica SB. (From Ref. 133.) Reprinted with permission from *J. Phys. Chem.* Copyright by the American Chemical Society.

include the distribution of site energies adjacent to a site of given energy. At this point there are probably too many variables to be extracted from adsorption data alone, although the comparison of isotherms for adsorbates of varying size may help (e.g., nitrogen versus butane—see Ref. 142). A formal statistical mechanical approach to the problem was made by Steel (144).

B. *Thermodynamics of Adsorption on Heterogeneous Surfaces*

It is generally assumed that isosteric thermodynamic heats obtained for a heterogeneous surface retain their simple relationship to calorimetric heats (Eq. XVI-132), although it may be necessary in a thermodynamic proof of this to assume that the chemical potential of the adsorbate does not show discontinuities as Θ is varied (145). An analytical proof for the special case in which $\theta(Q, P, T)$ is the Langmuir equation has been given (146).

There is a drastic effect of surface heterogeneity on the adsorption entropy, as noted by Everett (9). This is illustrated in Fig. XVI-31, in which the variation of Δs_2 with Θ is shown, first for various adsorption models assuming a uniform surface, and then the values calculated for a heterogeneous system (using the site energy distribution for nitrogen on titanium dioxide from Ref. 131), again assuming various models. The concern here is with how the configurational entropy varies with Θ rather than with absolute adsorption entropies, and the point is that $\bar{S}_{config}$ is so altered on a heterogeneous surface that a distinction between models is not possible. The physical explanation is that in the case of a heterogeneous surface, each added

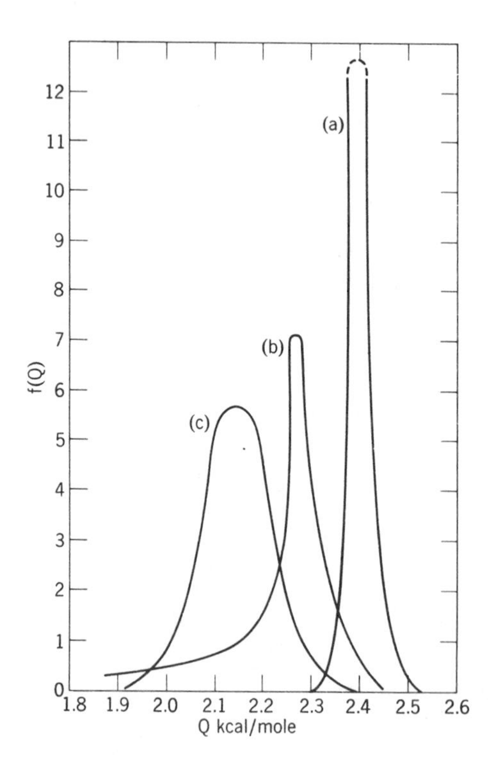

Fig. XVI-30. Interaction energy distributions for N_2 on BN. (*a*) Langmuir. (*b*) Langmuir plus lateral interaction. (*c*) Van der Waals. (From Ref. 142.)

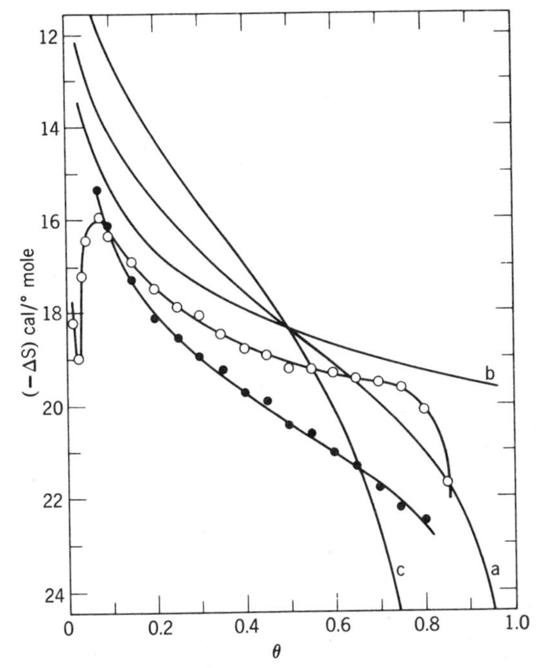

Fig. XVI-31. Ideal and calculated adsorption entropies for N_2 on TiO_2. Solid lines: ideal entropy from (*a*) Langmuir, (*b*) ideal gas, and (*c*) van der Waals model, uniform surface. These have been made to coincide at $\theta = 0.5$; ○, calculated, using $f(Q)$ of Ref. 134 and Langmuir model (from Ref. 142); ●, calculated similarly, using van der Waals model.

increment of adsorbate is confined to the small portion of the surface that is being filled. For an extremely heterogeneous surface, $\bar{S}_{config}$ would be zero except at the extremes of $\Theta \rightarrow 0$ and $\Theta \rightarrow 1$ (except as complicated by multilayer adsorption), irrespective of model.

There is positive aspect to the situation. Since for a heterogeneous surface neither the site energy distribution nor the configurational entropy are very model sensitive, the former gives relatively unambiguous information about the energetics of the system, and the entropy of adsorption is now assignable to the entropy change in forming the adsorbent–adsorbate complex without it being necessary to estimate configurational entropy corrections.

16. Rate of Adsorption

There have been indications that an alternative method of surface area determination is that of measuring the rate of adsorption and of noting that point at which a break occurs. Thus Jura and Powell (147) found kinetic areas that agreed fairly well with BET values, in the case of ammonia on a cracking catalyst and nitrogen on titanium dioxide. Similarly, Calvet (148) found that a break in heat evolution occurred at the monolayer point in the case of water on titanium dioxide.

Such effects may not be chemical kinetic ones. Benson and co-workers (72), in a study of the rate of adsorption of water on lyophilized proteins, comment that the empirical rates of adsorption were very markedly complicated by the fact that the samples were appreciably heated by the heat evolved on adsorption. In fact, it appeared that the actual adsorption rates were very fast and that the time dependence of the adsorbate pressure above the adsorbent was simply due to the time variation of the temperature of the sample as it cooled after the initial heating when adsorbate was first introduced.

Deitz and Carpenter (149) found that argon and nitrogen adsorbed only slowly on diamond under conditions such that the final θ was only a few percent. They were able to rule out heat transfer as rate controlling and concluded that while the adsorption process per se was very rapid, the nature of the diamond surface changed slowly to a more active one on the cooling of the sample prior to an adsorption run. The change appeared to be reversible and may constitute a type of illustration of the Dunning effect discussed in Section XV-4B. A related explanation was invoked by Good and co-workers (150). In studies of the heat of immersion of Al_2O_3 and SiO_2 in water they noticed a slow residual heat evolution, which they attributed to slow surface hydration.

In conclusion, any observation of slowness in attainment of physical adsorption equilibrium should be analyzed with caution and in detail. When this has been done, the phenomenon has either been found to be due to trivial causes or else some unsuspected and interesting other effects were operative.

17. Adsorption on Porous Solids—Hysteresis

As a general rule, adsorbates above their critical temperatures do not give multilayer type isotherms. In such a situation, a porous absorbent behaves like any other, unless the pores are of molecular size, and at this point the distinction between adsorption and absorption dims. Below the critical temperature, multilayer formation is possible and capillary condensation can occur. These two aspects of the behavior of porous solids are discussed briefly in this section.

A. Molecular Sieves

McBain coined the term "persorption" to apply to the situation where the pores were small enough to act as molecular sieves, so that different apparent surface areas would be obtained according to the size of the adsorbate molecules (151). Zeolites have been of much interest in this connection because the open way in which the (Al, Si)O_4 tetrahedra join gives rise to large cavities and large windows into the cavities. This is illustrated in Fig. XVI-32 (152). As a specific example, chabasite ($CaAl_2Si_4O_{12}$) has cages about 10 Å in diameter, with six openings into it or windows of about 4 Å diameter. Monatomic and diatomic gases, water, and *n*-alkanes can enter into such cavities, but larger molecules do not. Thus isobutane can

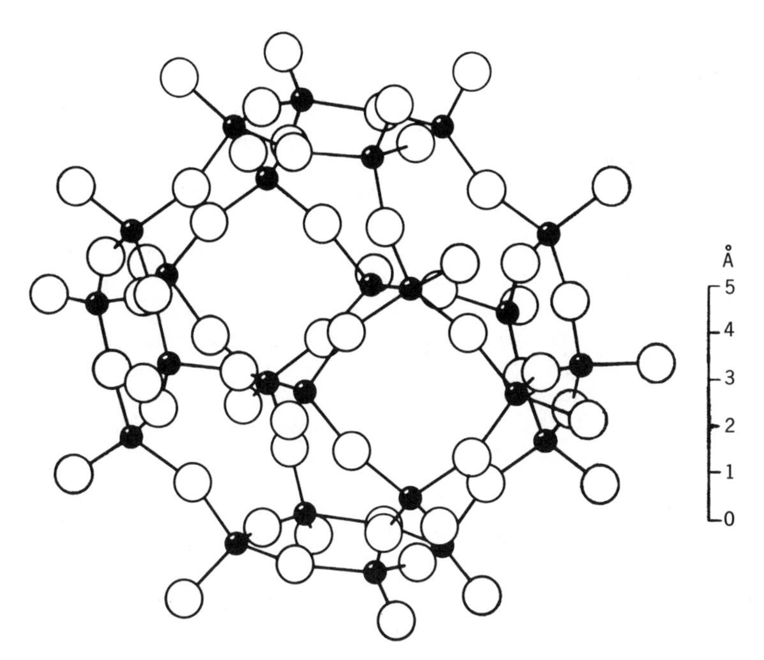

Fig. XVI-32. The arrangement of (Al, Si)O_4 tetrahedra that gives the cubo-octahederal cavity found in some felspathoids and zeolites. (From Ref. 152.)

be separated from *n*-alkanes and, on the basis of rates, even propane from ethane (153). In addition, there appear to be pores of smaller size that can distinguish between hydrogen and nitrogen (154). The replacement of the calcium by other ions (zeolites have ion exchange properties—note Section XI-5C) considerably affects the relative adsorption behavior, and various synthetic zeolites having various window diameters in the range of 4 to 10 Å are available under the name of Linde Molecular Sieves (see Ref. 152).

The adsorption isotherms are often approximately Langmuir in type (under conditions such that multilayer formation is not likely), and both n_m and b vary with the cation present. Isotherms of this appearance are reported, for example, by Yates (155) for ethylene on various faujasite type zeolites (two types of adsorption sites were postulated). The filling of the zeolitic cavities appears to be progressive, and an isotherm similar to Eq. XVI-82, but with ϵ^n rather than ϵ^2 has been proposed (156).

A special case of persorption might be considered that of clathrate compounds. Here, cages are present, but without access windows, so for "adsorption" to occur the solid usually must be crystallized in the presence of the "adsorbate." Thus quinol crystallizes in such a manner that holes several angstroms in diameter occur and, if crystallization takes place in the presence of solvent or gas molecules of small enough size, one or more such molecules may be incorporated into each hole. The union may thus be stoichiometric, but topological rather than specific chemical factors are involved; in the case of quinol, such diverse species as sulfur dioxide, methanol, formic acid, and nitrogen may form clathrate compounds (157).

A particularly interesting clathrating substance is ice; the cavities consist of six cages of 5.9 Å diameter and two of 5.2 Å per unit cell (158), and a variety of molecules are clathrated, ranging from rare gases to halogens and hydrocarbons. The ice system provides an illustration that it is not always necessary to carry out an *in situ* crystallization to obtain a clathrate. Barrer and Ruzicka (159) report a spontaneous clathrate formation between ice powder and xenon and krypton at −78°C, and in an attempted adsorption study of ethane on ice at −96°C it was found that, again, spontaneous ethane hydrate formation occurred (106); similar behavior was found for the adsorption of CO_2 by ice (160). These cases could be regarded as extreme examples of an adsorbate induced surface rearrangement!

B. Capillary Condensation

Below the critical temperature of the adsorbate, adsorption is generally multilayer in type, and the presence of pores may have the effect not only of limiting the possible number of layers of adsorbate (see Eq. XVI-65) but also of introducing capillary condensation phenomena. A wide range of porous adsorbents is now involved and usually having a broad distribution of pore sizes and shapes, unlike the zeolites. The most general characteristic of such adsorption systems is that of hysteresis; as illustrated in Fig. XVI-33, the desorption branch lies to the left of the adsorption branch; in addition, the isotherms may tend to flatten as P/P^0 approaches unity (note Fig. XVI-7). We are concerned in this section with loops of the type shown in Fig. XVI-33*b* and *c*. The open loop of Fig. XVI-33*a* has at least two types of explanations; one is that in connection with Fig. XV-4, namely ink-bottle pores that can trap adsorbate. A perhaps more plausible explanation in the case of vapors (as contrasted to liquid mercury) is that irreversible change

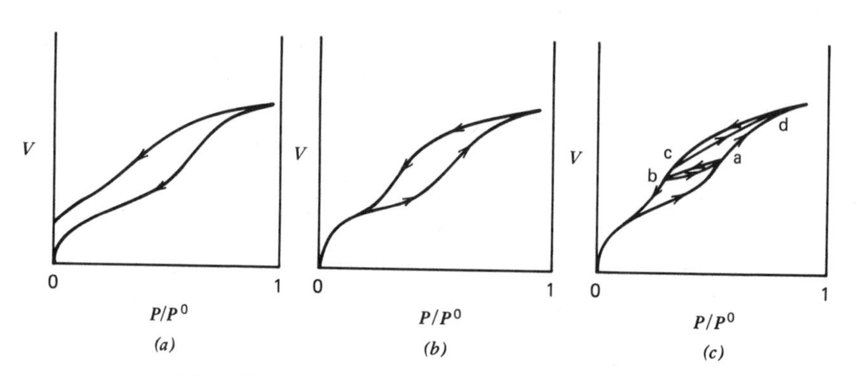

Fig. XVI-33. Hysteresis loops in adsorption.

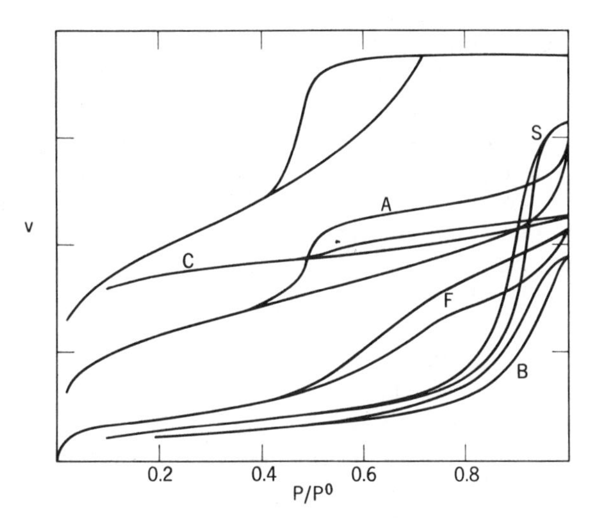

Fig. XVI-34. Nitrogen isotherms; the volume adsorbed is plotted on an arbitrary scale. The upper scale shows pore radii corresponding to various relative pressures. Samples: A, Oulton catalyst; B, bone char number 452; C, activated charcoal; F, Alumina catalyst Fl2; G, porous glass; S, silica aerogel. (From Ref. 162.)

may occur in the pore structure on adsorption so that the desorption situation is truly different from the adsorption one (161).

The variety of shapes of hysteresis loops of the closed variety that may be observed in practice is illustrated in Fig. XVI-34 for nitrogen adsorbed on various solids (see also Ref. 163). Figure XVI-35 shows some scanning curves obtained for the *n*-decane–porous Vicor system, from Ref. 164.

The basic explanation, in terms of capillary condensation, is attributed to Zsigmondy (165) who attributed hysteresis to contact angle hysteresis

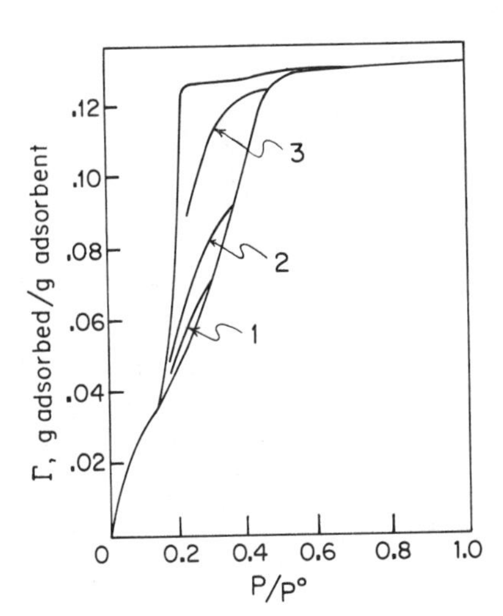

Fig. XVI-35. Equilibrium adsorption and desorption isotherms; *n*-decane on porous Vycor. Curves 1, 2, and 3 show the desorption behavior for successively higher degrees of adsorption, short of saturation. (From Ref. 164.)

due to impurities; this might account for behavior of the type shown in Fig. XVI-33*a*, but not in general for the many systems having retraceable closed hysteresis loops. Most of the early analyses and many current ones are in terms of a model representing the adsorbent as a bundle of various-sized capillaries. Cohan (166) suggested that the adsorption branch—curve *abc* in Fig. XVI-33*b*—represented increasingly thick film formation whose radius of curvature would be that of the capillary r, so that at each stage the radius of capillaries just filling would be given by the corresponding form of the Kelvin equation (Eq. III-20):

$$x_a = e^{-\gamma V/rRT} \tag{XVI-154}$$

At c all such capillaries would be filled and on desorption would empty by retreat of a meniscus of curvature $2/r$, so that at each stage of the desorption branch *dea* the radius of the capillaries emptying would be

$$x_d = e^{-2\gamma V/rRT} \tag{XVI-155}$$

The section *cd* can be regarded as due to relatively large cone-shaped pores that would fill and empty without hysteresis. Since P_a should be complicated by an adsorption potential (i.e., ordinary multilayer) as well as a radius of curvature contribution, the effect of this general analysis is to stress the use of the desorption branch to obtain a pore-size distribution. The basic procedure stems from those of Barrett et al. (167) and Pierce (162; see also Ref. 50) wherein the effective meniscus curvature is regarded as given by the capillary radius minus the thickness of ordinary multilayer adsorption expected at that P/P^0. This last can be estimated from adsorption data on similar but nonporous material, and for this de Boer's t-curve (see Table XVI-4) is widely used. The general calculation is as follows. After each stepwise decrease in P_d (pressure on desorption), an effective capillary radius is calculated from Eq. XVI-155, and the true radius is obtained by adding the estimated multilayer thickness. The exposed pore volume and pore area can then be calculated from the volume desorbed in that step. For all steps after the first, the desorbed volume is first corrected for that from multilayer thinning on the sum of the areas of previously exposed pores. In this way a tabulation of cumulative pore volume of pores of radius greater than a given r is obtained and, from the slopes of the corresponding plot, a pore-size distribution results. Such a distribution was compared in Fig. XV-3 to the one from mercury porosimetry, with excellent agreement. See Dubinin (169) for a more elaborate treatment of the de Boer type of approach.

Sing (see Ref. 170 and earlier papers) has developed an important modification of the de Boer t-plot idea. The latter rests on the observation of a characteristic isotherm (Section XVI-9), that is, on the conclusion that the adsorption isotherm is independent of the adsorbent in the multilayer region. Sing recognized that there were differences for different adsorbents, and used an appropriate "standard isotherm" for each system, the standard isotherm being for a nonporous adsorbent of composition similar to that of the porous one being studied. He then defined a quantity $\alpha_s = (n/n_x)_s$ where n_x is the amount adsorbed by the nonporous reference material at the selected P/P^0. The α_s values are used to correct pore radii for multilayer adsorbed in much the same manner as with de Boer. Lecloux and Pirard (171) have discussed further the use of standard isotherms.

Everett (168) has pointed out that the bundle of capillaries model can be outrageously wrong for real systems so that the results of the preceding type of analysis, while internally consistent, may not give more than the roughest kind of information about the real pore structure. One problem is that of "ink bottle" pores (Fig. XV-4), which empty at the capillary vapor pressure of the access channel but then discharge the contents of a larger cavity. Barrer et al. (172) have also discussed a variety of geometric situations in which filling and emptying paths are different.

Brunauer and co-workers (173, 174) have proposed a "modelless" method for obtaining pore size distributions; no specific capillary shape is assumed. Use is made of the general thermodynamic relationship due to Kiselev (175)

$$\gamma \, d\mathcal{A} = \Delta\mu \, dn \qquad \text{(XVI-156)}$$

where $d\mathcal{A}$ is the surface that disappears when a pore is filled by capillary condensation, $\Delta\mu$ is the change in chemical potential, equal to $RT \ln P/P^0$, and dn is the number of moles of liquid taken up by the pore. It can be shown that the Kelvin equation is a special case of Eq. XVI-156 (Problem 33). The integral form of Eq. XVI-156 is

$$\mathcal{A} = -\frac{RT}{\gamma}\int_{n_h}^{n_s} \ln\frac{P}{P^0}\, dn \qquad \text{(XVI-157)}$$

One now defines a hydraulic pore radius r_h as

$$r_h = \frac{V}{\mathcal{A}} \qquad \text{(XVI-158)}$$

where V is the volume of a set of pores and $\mathcal{A}$, their surface area. The remaining procedure is similar to that described. For each of successive steps along the desorption branch, the volume of that group of pores that empty is given by the change in n, and their area by Eq. XVI-157. Their hydraulic radius follows from Eq. XVI-158. The total pore surface area can be estimated from application of the BET equation to the section of the isotherm before the hysteresis loop, and a check on the pore distribution analysis is that the total pore area calculated from the distribution agrees with the BET area. Figure XVI-36 shows the pore volume distribution of a hardened Portland cement paste as calculated by the Brunauer–Mikhail–Bodor (173) modelless method and by the older method using the Kelvin equation and assuming cylindrical pores. The two distributions are comparable, recognizing that the hydraulic radius of a cylinder is one fourth of its diameter.

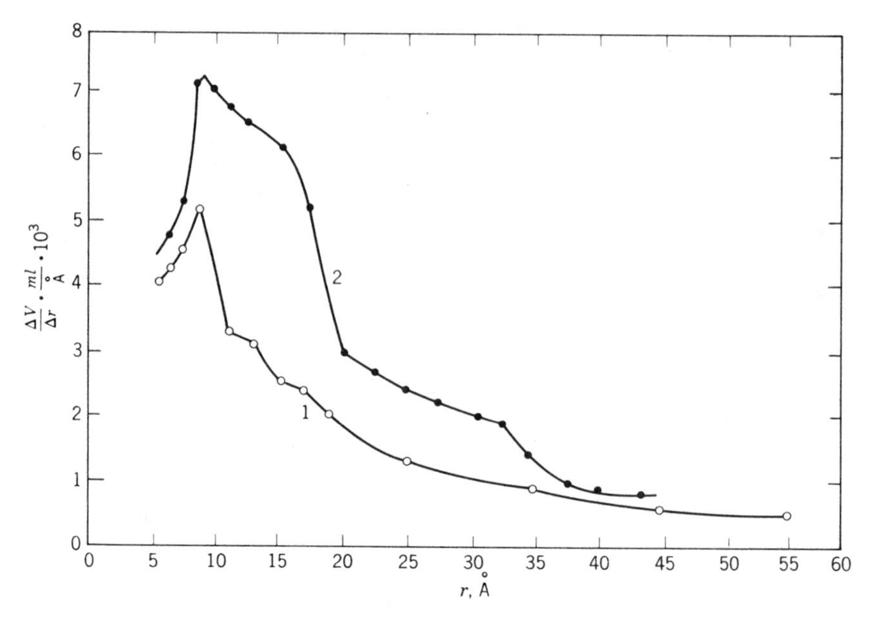

Fig. XVI-36. Pore volume distribution of a hardened Portland cement paste. Curve 1: modelless method (see text); curve 2: assuming cylindrical pores. (From Ref. 173.)

More detailed information about the pore system can be obtained from "scanning curves," illustrated in Fig. XIV-33*c* and Fig. XVI-35. Thus if adsorption is carried only up to point *a* and then desorption is started, the lower curve *ab* will be traced; if at *b* absorption is resumed, the upper curve *ab* is followed, and soon. Any complete model should account in detail for such scanning curves and, conversely, through their complete mapping much more information can be obtained about the nature of the pores. Rao (176) and Emmett (177) have summarized a great deal of such behavior.

A potentially powerful approach is that of Everett (178), who treats the pore system as a set of domains, independently acting in a first approximation. Each domain consists of those elements of the adsorbent that fill at a particular $x_{a(j)}$ and empty at a particular other relative pressure $x_{d(j)}$, the associated volume being V_j. Each domain is thus characterized by these three variables, and a plot of the function $V(x_a, x_d)$ would produce a surface in three dimensions something like a relief map. This is illustrated in Fig. XVI-37, the surface topography being shown by the periodic sectionings. Consider what should happen on increasing pressure x_a to $x_a + dx_a$: all domains of filling pressure in this interval should fill, but these domains can and in general would have a range of emptying pressure x_d ranging from $x_d = 0$ to $x_d = x_a$. The section at this x_a gives this x_d distribution, and its area times dx_a gives the total volume of such domains and hence the volume increment dV_a on the adsorption isotherm. Since the x_d of any domain cannot exceed its x_a (it cannot empty at a higher pressure than it fills!), the base of the topological map must be a 45° triangle.

On adsorption, one sweeps through the series of cross sections for successive

x_a's; on desorption, however, a step dx_d empties all domains of that emptying pressure, but these will now have a distribution of filling pressures, that can range from $x_a = x_d$ up to $x_a = 1$. The distribution is thus given by the indicated section parallel to the x_a axis, and the desorption volume increment dV_d, by its area times dx_d. Thus the detailed map predicts the adsorption and desorption branches, as well as, of course, the complete detail of all scanning loops. The problem of deducing such a map *from* adsorption data is similar to that discussed in Section XVI-15A in connection with site energy distributions, only now the much more difficult task of obtaining a two-dimensional distribution is involved.

A concluding comment might be made on the temperature dependence of adsorption in such systems. One can show by setting up a piston and cylinder experiment that mechanical work must be lost (i.e., converted to heat) on carrying a hysteresis system through a cycle. An irreversible process is thus involved, and the entropy change in a small step will not in general be equal to $\delta q/T$. As was pointed out by LaMer (179), this means that second-law equations such as Eq. XVI-116 no longer have a simple meaning. In hysteresis systems, of course, two sets of q_{st} values can be obtained, from the adsorption and from the desorption branches. These usually are not equal and neither of them in general can be expected to equal the calorimetric heat. Another way of stating the problem is that the system is not locally reversible. The adsorption following an *increase* of x by δx is not retraced on *decreasing* the pressure by δx. This means that extreme caution should be exercised in treating q_{st} values as though they represented physical heat quantities, although it is certainly possible that in individual cases or in terms of particular models the discrepancy between q_{st} and a calorimetric heat may not be serious (see Ref. 180).

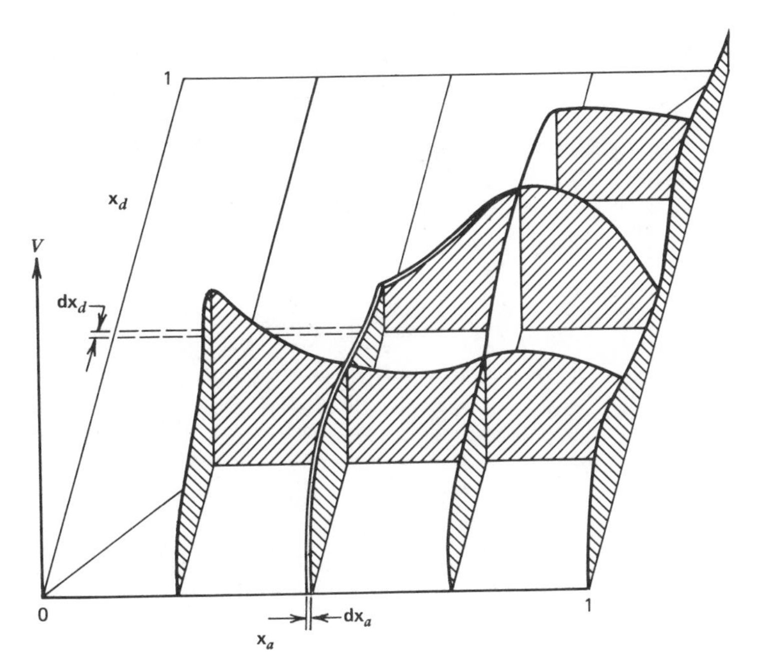

Fig. XVI-37. Perspective view of cross sections of a hypothetical domain map.

C. *Micropore Analysis*

Adsorbents such as some silica gels and types of carbons and zeolites have pores of the order of molecular dimensions, that is, from several up to 10 to 15 Å in diameter. Adsorption in such pores is not readily treated as a capillary condensation phenomenon—in fact, there is typically no hysteresis loop. What happens physically is that as multilayer adsorption develops, the pore becomes filled by a meeting of the adsorbed films from opposing walls. The adsorption isotherm may start off looking like one of the high BET *c*-value curves of Fig. XVI-10, but will then level off much like a Langmuir isotherm (Fig. XVI-3) as the pores fill and the surface area available for further adsorption greatly diminishes. The BET-type equation for adsorption limited to *n* layers (Eq. XVI-65) will sometimes fit this type of behavior.

A method for obtaining a micropore size distribution has been proposed by Mikhail, Brunauer, and Bodor (181), which is an extension of the *t*-curve method for obtaining surface areas (Section XVI-9). In this method, a plot of cubic centimeters STP adsorbed per gram v versus the value of t for the corresponding P/P^0 (as given, for example, by Table XVI-4) should, according to Eq. XVI-97, give a straight line of slope proportional to the specific surface area Σ. As illustrated in Fig. XVI-38. such plots may bend

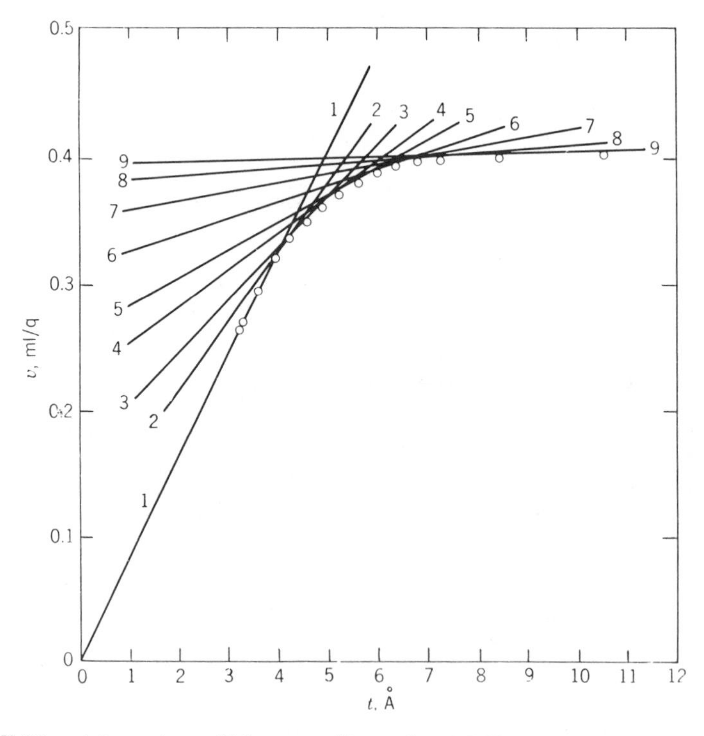

Fig. XVI-38. Adsorption of N_2 on a silica gel at 77.3°K, expressed as a v versus t plot, illustrating a method for micropore analysis. (From Ref. 181.)

over. This is now interpreted not as a deviation from the characteristic isotherm principle but rather as an indication that progressive reduction in surface area is occurring as micropores fill. The proposal of Mikhail et al. was that the slope at each *point* gave a correct surface area for that P/P^0 and v value. The drop in surface area between successive points then gives the volume of micropores that filled at the average P/P^0 of the two points, and the average t value, the size of the pores that filled. In this way a pore size distribution can be obtained. The method has given reasonable results (see also Ref. 174), but some legitimate criticisms of its claims have been made by Dubinin (182). The writer would add that the assumption of the validity of the t-curve method is least tenable just in the relatively low P/P^0 region where micropore filling should occur.

18. Problems

1. Read off points from Fig. XVI-1 and plot a set of corresponding isosteres and isobars.

2. Derive the general form of the Langmuir equation, Eq. XVI-11.

3. The separate adsorption isotherms for gases A and B on a certain solid obey the Langmuir equation, and it may be assumed that the mixed or competitive adsorption obeys the corresponding form of the equation.

Gas A, by itself, adsorbs to a θ of 0.01 at $P = 100$ mm Hg, and gas B, by itself, adsorbs to $\theta = 0.01$ at $P = 10$ mm Hg; T is 77 K in both cases. (*a*) Calculate the difference between Q_A and Q_B, the two heats of adsorption. Explain briefly any assumptions or approximations made. (*b*) Calculate the value for θ_A when the solid, at 77 K, is equilibrated with a mixture of A and B such that the final pressures are 200 mm Hg each. (*c*) Explain whether the answer in *b* would be raised, lowered, or affected in an unpredictable way if all of the preceding data were the same but the surface was known to be heterogeneous. The local isotherm function can still be assumed to be the Langmuir equation.

4. Dye adsorption from solution may be used to estimate the surface area of a powdered solid. Suppose that if 1 g of a bone charcoal is equilibrated with 100 cm^3 of initially $10^{-4}M$ methylene blue, the final dye concentration is $0.6 \times 10^{-4}M$, while if a 2-g amount had been used, the final concentration would have been $0.4 \times 10^{-4}M$.

Assuming that the Langmuir equation is obeyed, calculate the specific surface area of the bone charcoal in square meters per gram. The molecular area of methylene blue in a monolayer may be taken to be 65 Å^2.

5. Calculate the value of the first three energy levels according to the wave mechanical picture of a particle in a one-dimensional box. Take the case of nitrogen in a 4 Å box. Calculate $\mathbf{Q}_{\substack{\text{trans}\\ \text{(1 dim)}}}$ at 78 K using the integration approximation and by directly evaluating the sum $\sum_n \exp(-\epsilon_n/kT)$.

6. Calculate $\Delta\bar{S}_2$ at $\theta = 0.1$ for argon at 77 K that forms a weak adsorption bond with the adsorbent, having three vibrational degrees of freedom.

7. Discuss the physical implications of the conditions under which b_0 as given by the kinetic derivation of the Langmuir equation is the same as b_0', the constant as given by the statistical thermodynamic derivation.

8. The standard entropy of adsorption $\Delta\bar{S}_2$ of benzene on a certain surface was found to be -25.2 EU at 323.1°K; the standard states being the vapor at 1 atm and

the film at an area of $22.5 \times T$ Å^2 per molecule. Discuss, with appropriate calculations, what the state of the adsorbed film might be, particularly as to whether it is mobile or localized.

9. Derive Eq. XVI-54.

10. Show that the critical value of β in Eq. XVI-53 is indeed 4, that is, the value of β above which a maximum and a minimum in bP appear. What is the critical value of θ?

11. Drain and Morrison (1) report the following data for the adsorption of N_2 on rutile at 75°K, where P is in millimeters of mercury and v in cubic centimeters STP per gram.

P	v	P	v	P	v	P	v
1.17	600.06	275.0	1441.14	455.2	2418.34	498.6	3499.13
14.00	719.54	310.2	1547.37	464.0	2561.64	501.8	3628.63
45.82	821.77	341.2	1654.15	471.2	2694.67		
87.53	934.68	368.2	1766.89	477.1	2825.39		
127.7	1045.75	393.3	1890.11	482.6	2962.94		
164.4	1146.39	413.0	2018.18	487.3	3107.06		
204.7	1254.14	429.4	2144.98	491.1	3241.28		
239.0	1343.74	443.5	2279.15	495.0	3370.38		

Plot the data according to the BET equation and calculate v_m and c, and the specific surface area in square meters per gram.

The saturation vapor pressure P^0 of N_2 is given by

$$\log P^0 = -\frac{339.8}{T} + 7.71057 - 0.0056286$$

(Ref. 41), in mm Hg.

12. When plotted according to the linear form of the BET equation, data for the adsorption of N_2 on Graphon at 77 K give an intercept of 0.005 and a slope of 1.5 (both in cubic centimeters STP per gram). Calculate Σ assuming a molecular area of 16 Å^2 for N_2. Calculate also the heat of adsorption for the first layer (the heat of condensation of N_2 is 1.3 kcal/mole). Would your answer for v_m be much different if the intercept were taken to be zero (and the slope the same)? Comment briefly on the practical significance of your conclusion.

13. Consider the case of the BET equation with $c = 1$. Calculate for this case the heat of adsorption for the process:

A (liquid adsorbate at T) = A (adsorbed, in equilibrium with pressure P, at T)

for θ values of 0.1 and of 1.5. Calculate also the entropies of adsorption for the same θ values. Finally, derive the corresponding two-dimensional equation of state of the adsorbed film.

14. Hüttig (183) proposed the equation

$$\frac{x(1 + x)}{v} = \frac{1}{cv_m} + \frac{x}{v_m} \qquad \text{(XVI-159)}$$

This may be derived on the BET approach, but assuming that *each layer is an independently acting* Langmuir film whose maximum coverage is equal to the actual extent of the layer beneath it. Make this derivation.

15. An equation very similar to the BET equation can be derived by assuming a multilayer structure, with the Langmuir equation applying to the first layer, and Raoult's law to succeeding layers, supposing that the escaping tendency of a molecule is unaffected by whether it is covered by others or not. Make this derivation, adding suitable assumptions as may be needed.

16. Calculate and plot an isotherm according to Eq. XVI-65. Assume $c = 200$ and $n = 4$, and plot n/n_m versus P/P^0.

17. Plot the data of Problem 11 according to Eq. XVI-75; calculate n_m assuming k to be 4.06 as given in the text.

18. Show that $\bar{S}_{\text{config}} = 0$ for adsorption obeying the Dubinin equation (Eq. XVI-81).

19. Plot the data of Problem 11 according to Eq. XVI-88 and according to Eq. XVI-89. Comment.

20. Plot the data of Problem 11 according to Eq. XVI-94. Comment.

21. Construct a v versus t plot for the data of Problem 11 (assume that Table XVI-4 can be used) and calculate the specific surface area of the rutile.

22. An absorption system follows Eq. XVI-89 in the form $\ln v = B - (1/n) \ln \ln (P^0/P)$ with $n = 2.75$ and $B = 3.2$. Assuming now that you are presented with data that fall on the curve defined by this equation, calculate the corresponding BET v_m and c values.

23. Plot the data of Table XVI-4 as v/v_m versus P/P^0, and plot according to (*a*) the BET equation, (*b*) Eq. XVI-89, and (*c*) Eq. XVI-94.

24. Use the data of Fig. XVI-1 to calculate q_{st} for a range of v values and, in conjunction with Problem 11, plot q_{st} versus n/n_m.

25. As a simple model of a heterogeneous surface, assume that 15% of it consists of sites of $Q = 2$ kcal/mole; 50% of sites $Q = 3$ kcal/mole; and the remainder, of sites of $Q = 4$ kcal/mole. Calculate Θ (P, T) for nitrogen at 77 K and at 90 K, assuming the adsorption to follow the Langmuir equation with b_0 given by Eq. XVI-15. Calculate q_{st} for several Θ values and compare the result with the assumed integral distribution function.

26. If Θ (P, T) is $\Theta = bP/(1 + bP)$, show what result for $\phi(x)$ follows from Eq. XVI-150.

27. Sketch what the domain plot of Fig. XVI-37 might look like if there were no hysteresis loop, that is, if the adsorption were reversible.

28. The projection of a domain plot onto its base makes a convenient two-dimensional graphical representation for describing adsorption–desorption operations. Here, the domain region that is filled can be indicated by shading the appropriate portion of the 45° base triangle. Indicate the appropriate shading for (*a*) adsorption up to $x_a = 0.7$; (*b*) such adsorption followed by desorption to $x_d = 0.4$; and (*c*) followed by readsorption from $x_d = 0.4$ to $x_a = 0.6$.

29. Table XVI-5 shows the map of domain volumes (in arbitrary units). For example, the number 0.15 centered at $x_a = 0.5$ and $x_d = 0.4$ means that pores having filling x values between 0.45 and 0.55 and emptying x values between 0.45 and 0.35 contribute a volume of 0.15 units. (*a*) Calculate and plot the adsorption and desorption isotherms for this system. (*b*) Show the scanning curve for desorption following adsorption up to $x = 0.8$. (*c*) Show the scanning curve for adsorption up to $x = 0.9$, desorption down to $x = 0.4$, followed by adsorption. In each case give the total volume contained by the system at the end of each step.

30. The nitrogen adsorption isotherm is determined for a finely divided, nonporous solid. It is found that at $\theta = 0.5$, P/P^0 is 0.05 at 77 K, and P/P^0 is 0.2 at 90 K.

TABLE XVI-5

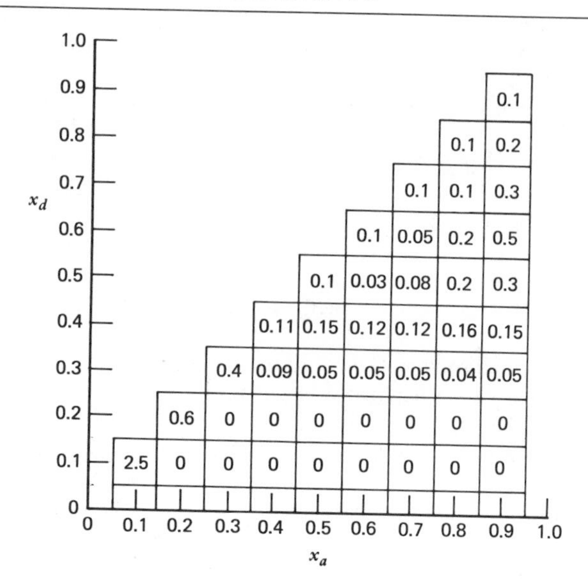

x_d \ x_a	0.1	0.2	0.3	0.4	0.5	0.6	0.7	0.8	0.9
0.9									0.1
0.8								0.1	0.2
0.7							0.1	0.1	0.3
0.6						0.1	0.05	0.2	0.5
0.5					0.1	0.03	0.08	0.2	0.3
0.4				0.11	0.15	0.12	0.12	0.16	0.15
0.3			0.4	0.09	0.05	0.05	0.05	0.04	0.05
0.2		0.6	0	0	0	0	0	0	0
0.1	2.5	0	0	0	0	0	0	0	0

Calculate the isosteric heat of adsorption, and $\Delta\bar{S}^0$ and $\Delta\bar{G}^0$ for adsorption at 77 K. Write the statement of the process to which your calculated quantities correspond. Explain whether the state of the adsorbed N_2 appears to be more nearly gaslike or liquidlike. The normal boiling point of N_2 is 77 K, and its heat of vaporization is 1.35 kcal/mole.

31. Discuss physical situations in which it might be possible to observe a vertical step in the adsorption isotherm of a gas on a heterogeneous surface.

32. Derive the expressions for the partition function for two-dimensional translation in the case of an adsorbate that is in the surface standard state recommended by de Boer. Express your result further as $\bar{S}^{0,s}_{\text{trans}}$ and then express this standard state entropy in terms of $\bar{S}^{0,g}_{\text{trans}}$ (the answer should be in the form $\bar{S}^{0,s}_{\text{trans}} = \bar{S}^{0,g}_{\text{trans}} + b \ln T + c$).

33. Derive Eq. XVI-156. Derive from it the Kelvin equation (Eq. III-20).

General References

J. H. de Boer, *The Dynamical Character of Adsorption,* The Clarendon Press, Oxford, 1953.

E. A. Flood, Ed., *The Solid-Gas Interface,* Vols. 1 and 2, Marcel Dekker, New York, 1967.

R. H. Fowler and E. A. Guggenheim, *Statistical Thermodynamics,* Cambridge University Press, Cambridge, England, 1952.

S. J. Gregg and K. S. Sing, *Adsorption, Surface Area and Porosity,* Academic, New York, 1967.

S. Ross and J. P. Olivier, *On Physical Adsorption,* Interscience, New York, 1964.

H. Saltzburg, J. N. Smith, Jr., and M. Rogers, Eds., *Fundamentals of Gas–Surface Interactions,* Academic, New York, 1967.

W. A. Steele, *The Interaction of Gases with Solid Surfaces*, Pergamon, New York, 1974.

D. M. Young and A. D. Crowell, *Physical Adsorption of Gases*, Butterworths, London, 1962.

Textual References

1. L. E. Drain and J. A. Morrison, *Trans. Faraday Soc.*, **49,** 654 (1953).
2. M. T. Coltharp and N. Hackerman, *J. Phys. Chem.*, **72,** 1171 (1968).
3. S. J. Gregg and J. D. F. Ramsay, *J. Phys. Chem.*, **73,** 1243 (1969).
4. J. H. de Boer, *The Dynamical Character of Adsorption*, The Clarendon Press, Oxford, 1953.
5. I. Langmuir, *J. Am. Chem. Soc.*, **40,** 1361 (1918).
6. R. H. Fowler and E. A. Guggenheim, *Statistical Thermodynamics*, Cambridge University Press, Cambridge, England, 1952.
7. D. M. Young and A. D. Crowell, *Physical Adsorption of Gases*, Butterworths, London, 1962.
8. J. E. Lennard-Jones and A. F. Devonshire, *Proc. Roy. Soc.*, **A156,** 6, 29 (1936).
9. D. H. Everett, *Proc. Chem. Soc. (London)*, **1957,** 38.
10. J. H. de Boer and S. Kruyer, *K. Ned. Akad. Wet. Proc.*, **55B,** 451 (1952).
11. J. W. Ross and R. J. Good, *J. Phys. Chem.*, **60,** 1167 (1956).
12. S. Ross and J. P. Olivier, *The Adsorption Isotherm*, Rensselaer Polytechnic Institute, Troy, New York, 1959, p. 39f.
13. T. L. Hill, *J. Chem. Phys.*, **15,** 767 (1947).
14. A. A. Isirikyan and A. V. Kiselev, *J. Phys. Chem.*, **66,** 205, 210 (1962).
15. D. N. Misra, *J. Colloid Interface Sci.*, **77,** 543 (1980).
16. W. A. Steele, *J. Chem. Phys.*, **65,** 5256 (1976).
17. D. M. Young and A. D. Crowell, *Physical Adsorption of Gases*, Butterworths, London, 1962.
18. J. M. Honig, *The Solid–Gas Interface*, E. A. Flood, Ed., Marcel Dekker, New York, 1967.
19. S. Brunauer, *The Adsorption of Gases and Vapors*, Vol. 1, Princeton University Press, Princeton, New Jersey, 1945.
20. S. Brunauer, P. H. Emmett, and E. Teller, *J. Am. Chem. Soc.*, **60,** 309 (1938).
21. S. Ross and J. P. Olivier, *On Physical Adsorption*, Interscience, New York, 1964.
22. J. P. Hobson, *A.I.Ch.E. Symp. Ser.*, **68,** No. 125, 16 (1972).
23. B. W. Davis, G. H. Saban, and T. F. Moran, *J. Chem. Ed.*, **50,** 219 (1973).
24. J. W. McBain and A. M. Bakr, *J. Am. Chem. Soc.*, **48,** 690 (1926).
25. T. D. Blake and W. H. Wade, *J. Phys. Chem.*, **75,** 1887 (1971).
26. R. A. Beebe, J. B. Beckwith, and J. M. Honig, *J. Am. Chem. Soc.*, **67,** 1554 (1945).
27. A. J. Rosenberg, *J. Am. Chem. Soc.*, **78,** 2929 (1956).
28. G. H. Amberg, W. B. Spencer, and R. A. Beebe, *Can. J. Chem.*, **33,** 305 (1955).
29. A. J. Tyler, J. A. G. Taylor, B. A. Pethica, and J. A. Hockey, *Trans. Faraday Soc.*, **67,** 483 (1971).

30. See J. J. Chessick and A. C. Zettlemoyer, in *Advances in Catalysis,* Vol. 11, Academic, New York, 1959, and W. H. Wade, M. L. Deviney, Jr., W. A. Brown, M. H. Hnoosh, and D. R. Wallace, *Rubber Chem. Technol.,* **45,** 117 (1972).
31. J. E. Gardner, J. S. Riney, and W. H. Wade, *Rev. Sci. Inst.,* **38,** 652 (1967).
32. A. W. Adamson and L. Dormant, *J. Am. Chem. Soc.,* **88,** 2055 (1966).
33. W. H. Wade and J. W. Whalen, *J. Phys. Chem.,* **72,** 2898 (1968).
34. A. W. Adamson, *J. Colloid Interface Sci.,* **27,** 180 (1968).
35. M. E. Tadros, P. Hu, and A. W. Adamson, *J. Colloid Interface Sci.,* **49,** 184 (1974).
36. A. B. D. Cassie, *Trans. Faraday Soc.,* **41,** 450 (1945).
37. T. Hill, *J. Chem. Phys.,* **14,** 263 (1946), and succeeding papers.
38. A. L. McClellan and H. F. Harnsberger, *J. Colloid Interface Sci.,* **23,** 577 (1967).
39. C. Pierce and B. Ewing, *J. Phys. Chem.,* **68,** 2562 (1964).
40. M. J. Katz, *Anal. Chem.,* **26,** 734 (1954).
41. S. J. Gregg and K. S. W. Sing, *Adsorption, Surface Area, and Porosity,* Academic, New York, 1967.
42. R. B. Anderson, *J. Am. Chem. Soc.,* **68,** 686 (1946).
43. S. J. Gregg, *J. Chem. Soc.,* **1942,** 696.
44. M. Volmer, *Z. Phys. Chem.,* **115,** 253 (1925).
45. C. Kemball and E. K. Rideal, *Proc. Roy. Soc. (London),* **A187,** 53 (1946).
46. A. W. Adamson, unpublished work (1954).
47. S. Ross and W. Winkler, *J. Colloid Sci.,* **10,** 319 (1955); see also *J. Am. Chem. Soc.,* **76,** 2637 (1954).
48. W. D. Harkins and G. Jura, *J. Am. Chem. Soc.,* **66,** 1366 (1944).
49. J. McGavack, Jr., and W. A. Patrick, *J. Am. Chem. Soc.,* **42,** 946 (1920).
50. M. M. Dubinin, *Russ. J. Phys. Chem. (Eng. Transl.),* **39,** 697 (1965); *Q. Rev.,* **9,** 101 (1955).
51. See M. J. Sparnaay, *Surf. Sci.,* **9,** 100 (1968).
52. A. W. Adamson and I. Ling, *Advances in Chemistry No. 43,* American Chemical Society, Washington, D.C., 1964.
53. T. D. Blake, *J. Chem. Soc., Faraday Trans. I,* **71,** 192 (1975).
54. Y. L. Frenkel, *Kinetic Theory of Liquids,* The Clarendon Press, Oxford, 1946. (Reprinted by Dover Publications, 1955.)
55. G. D. Halsey, Jr., *J. Chem. Phys.,* **16,** 931 (1948).
56. T. L. Hill, *Adv. Catal.,* **4,** 211 (1952).
57. W. G. McMillan and E. Teller, *J. Chem. Phys.,* **19,** 25 (1951).
58. C. Pierce, *J. Phys. Chem.,* **63,** 1076 (1959).
59. C. M. Greenlief and G. D. Halsey, *J. Phys. Chem.,* **74,** 677 (1970).
60. See A. W. Adamson, *J. Colloid Interface Sci.,* **44,** 273 (1973); A. W. Adamson and I. Ling in *Adv. Chem.,* **43,** 57 (1964).
61. N. N. Roy and G. D. Halsey, Jr., *J. Chem. Phys.,* **53,** 798 (1970).
62. G. D. Halsey, Jr., *J. Am. Chem. Soc.,* **73,** 2693 (1951).
63. J. Ginous and L. Bonnetain, *CR,* **272,** 879 (1971).
64. J. H. de Boer and C. Zwikker, *Z. Phys. Chem.,* **B3,** 407 (1929).
65. K. W. Bewig and W. A. Zisman, *J. Phys. Chem.,* **68,** 1804 (1964).
66. A. Eberhagen, *Forstchr. Phys.,* **8,** 245 (1960).
67. J. Pritchard, *Nature,* **194,** 38 (1962).

68. S. W. Benson and J. W. King, Jr., *Science,* **150,** 1710 (1965).
69. A. W. Adamson, unpublished work.
70. W. A. Steele, *The Solid–Gas Interface,* E. A. Flood, Ed., Marcel Dekker, New York, 1967.
71. A. G. Keenan and J. M. Holmes, *J. Phys. Colloid Chem.,* **53,** 1309 (1949).
72. S. W. Benson and D. A. Ellis, *J. Am. Chem. Soc.,* **72,** 2095 (1950).
73. J. R. Arnold, *J. Am. Chem. Soc.,* **71,** 2104 (1949).
74. C. Pierce and B. Ewing, *J. Phys. Chem.,* **68,** 2562 (1964).
75. A. W. Adamson, *The Physical Chemistry of Surfaces,* Interscience, New York, 1960, and unpublished work, 1954.
76. I. Halász and G. Schay, *Acta Chim. Acad. Sci. Hung.,* **14,** 315 (1956).
77. L. G. Joyner and P. H. Emmett, *J. Am. Chem. Soc.,* **70,** 2353 (1948).
78. B. C. Lippens, B. G. Linsen, and J. H. de Boer, *J. Catal.,* **3,** 32 (1964); J. H. de Boer, B. C. Lippens, B. G. Linsen, J. C. P. Broekhoff, A. van den Heuvel, and Th. J. Osinga, *J. Colloid Interface Sci.,* **21,** 405 (1956). See also R. W. Cranston and F. A. Inkley, *Adv. Catal.,* **9,** 143 (1957).
79. M. R. Bhambhani, P. A. Cutting, K. S. W. Sing, and D. H. Turk, *J. Colloid Interface Sci.,* **38,** 109 (1972).
80. C. Pisani, F. Ricca, and C. Roetti, *J. Phys. Chem.,* **77,** 657 (1973).
81. W. A. Steele and G. D. Halsey, Jr., *J. Chem. Phys.,* **22,** 979 (1954).
82. J. A. Barker and D. H. Everett, *Trans. Faraday Soc.,* **58,** 1608 (1962).
83. W. A. Steel, *Surf. Sci.,* **36,** 317 (1973).
84. L. Battezzati, C. Pisani, and F. Ricca, *J. Chem. Soc., Far. Trans. II,* **71,** 1629 (1975). L. Battezzati, C. Pisani, and F. Ricca, *J. Chem. Soc., Faraday Trans. II,* **71,** 1629 (1975).
85. J. D. Johnson and M. L. Klein, *Trans. Faraday Soc.,* **60,** 1964 (1964).
86. D. H. Everett in *Surface Area Determination, Prox. Int. Symp.,* Bristol, 1969, Butterworths, London.
87. W. A. Steele, *J. Phys.,* **38,** C4-61 (1977).
88. G. D. Halsey, *Surf. Sci.,* **72,** 1 (1978).
89. A. W. Adamson and M. W. Orem, *Progr. Surf. Membrane Sci.,* **8,** 285 (1974).
90. L. M. Dormant and A. W. Adamson, *J. Colloid Interface Sci.,* **28,** 459 (1968).
91. B. B. Fisher and W. G. McMillan, *J. Am. Chem. Soc.,* **79,** 2969 (1957).
92. Y. Larher, *J. Chem. Soc., Faraday Trans. I,* **70,** 320 (1974).
93. W. A. Steele and R. Karl, *J. Colloid Interface Sci.,* **28,** 397 (1968).
94. Y. Nardon and Y. Larher, *Surf. Sci.,* **42,** 299 (1974).
95. A. Enault and Y. Larher, *Surf. Sci.,* **62,** 233 (1977).
96. B. F. Mason and B. R. Williams, *J. Chem. Phys.,* **56,** 1895 (1972).
97. Y. Grillet and J. Rouquerol, *J. Colloid Interface Sci.,* **77,** 580 (1980).
98. B. W. Davis and C. Pierce, *J. Phys. Chem.,* **70,** 1051 (1966).
98a. J. Tse and A. W. Adamson, *J. Colloid Interface Sci.,* **72,** 515 (1979) and preceding papers.
98b. G. Bomchi, N. Harris, M. Leslie, and J. Tabony; quoted in J. W. White, *J. Chem. Soc. Faraday Trans. I,* **75,** 1535 (1979) and following papers.
99. T. L. Hill, *J. Chem. Phys.,* **17,** 520 (1949).
100. D. H. Everett, *Trans. Faraday Soc.,* **46,** 453 (1950).
101. R. N. Smith, *J. Am. Chem. Soc.,* **74,** 3477 (1952).
102. J. M. Honig, *J. Colloid Interface Sci.,* **70,** 83 (1979).
103. L. E. Copeland and T. F. Young, *Adv. Chem.,* **33,** 348 (1961).

104. Y. C. Wu and L. E. Copeland, *Adv. Chem.*, **33**, 357 (1961).
105. L. M. Dormant and A. W. Adamson, *J. Colloid Interface Sci.*, **75**, 23 (1980).
106. A. W. Adamson, L. Dormant, and M. Orem, *J. Colloid Interface Sci.*, **25**, 206 (1967).
107. M. A. Cook, D. H. Pack, and A. G. Oblad, *J. Chem. Phys.*, **19**, 367 (1951).
108. G. Jura and T. L. Hill, *J. Am. Chem. Soc.*, **74**, 1598 (1952).
109. A. C. Zettlemoyer, G. J. Young, J. J. Chessick, and F. H. Healey, *J. Phys. Chem.*, **57**, 649 (1953).
110. G. L. Kington, R. A. Beebe, M. H. Polley, and W. R. Smith, *J. Am. Chem. Soc.*, **72**, 1775 (1950).
111. J. Greyson and J. G. Aston, *J. Phys. Chem.*, **61**, 610 (1957).
112. W. D. Harkins and G. Jura, *J. Am. Chem. Soc.*, **66**, 919 (1944).
113. M. W. Orem and A. W. Adamson, *J. Colloid Interface Sci.*, **31**, 278 (1969).
114. C. Prenzlow, *J. Colloid Interface Sci.*, **37**, 849 (1971).
115. N. K. Nair and A. W. Adamson, *J. Phys. Chem.*, **74**, 2229 (1970).
116. T. L. Hill, P. H. Emmett, and L. G. Joyner, *J. Am. Chem. Soc.*, **73**, 5102 (1951).
117. H. M. Cassel, *J. Chem. Phys.*, **12**, 115 (1944); *J. Phys. Chem.*, **48**, 195 (1944).
118. J. R. Arnold, *J. Am. Chem. Soc.*, **71**, 104 (1949).
119. J. W. Whalen, W. H. Wade, and J. J. Porter, *J. Colloid Interface Sci.*, **24**, 379 (1967).
120. R. E. Day and G. D. Parfitt, *Trans. Faraday Soc.*, **63**, 708 (1967).
121. S. J. Gregg and J. F. Langford, *Trans. Faraday Soc.*, **65**, 1394 (1969).
122. S. Brunauer, L. E. Copeland, and D. L. Kantro, *The Solid-Gas Interface*, E. A. Flood, Ed., Marcel Dekker, New York, 1966.
123. S. Brunauer, in *Surface Area Determination, Proc. Int. Symp.*, Bristol, 1969, Butterworths, London.
124. A. I. Sarakhov, M. M. Dubinin, and Yu. F. Bereskina, *Izv. Akad. Nauk. SSSR, Ser. Khim.*, 1165 (July 1963).
125. F. E. Karasz, W. M. Champion, and G. D. Halsey, Jr., *J. Phys. Chem.*, **60**, 376 (1956).
126. W. A. Steele, *The Solid–Gas Interface*, E. A. Flood, Ed., Marcel Dekker, New York, 1966.
126a. Ch. Sheindorf, M. Rebhun, and M. Sheintuch, *J. Colloid Interface Sci.*, **79**, 136 (1981).
127. See B. M. W. Trapnell, *Chemisorption*, Academic, New York, 1955, p. 124.
128. J. P. Hobson, *Can. J. Phys.*, **43**, 1934 (1965).
129. R. Sips, *J. Chem. Phys.*, **16**. 490 (1948).
130. J. M. Honig and L. H. Reyerson, *J. Phys. Chem.*, **56**, 140 (1952); J. M. Honig and P. C. Rosenbloom, *J. Chem. Phys.*, **23**, 2179 (1955).
131. A. W. Adamson and I. Ling, *Adv. Chem.*, **33**, 51 (1961).
132. L. M. Dormant and A. W. Adamson, *J. Colloid Interface Sci.*, **38**, 285 (1972).
133. J. W. Whalen, *J. Phys. Chem.*, **71**, 1557 (1967).
134. A. W. Adamson, I. Ling, and S. K. Datta, *Adv. Chem.*, **33**, 62 (1961).
135. P. Y. Hsieh, *J. Phys. Chem.*, **68**, 1068 (1964); *J. Catal.*, **2**, 211 (1963).
136. W. A. House and M. J. Jaycock, *J. Colloid Interface Sci.*, **47**, 50 (1974).
137. W. A. House and M. J. Jaycock, *Colloid and Polymer Sci.*, **256**, 52 (1978).
138. W. A. House, *J. Chem. Soc. Faraday Trans. I*, **74**, 1045 (1978).
139. B. K. Oh and S. K. Kim, *J. Chem. Phys.*, **67**, 3416 (1977).

140. G. F. Cerofolini, *Z. Phys. Chemie, Leipzig,* **258,** 937 (1977).
141. G. F. Cerofolini, M. Jaroniec, and S. Sokolowski, *Colloid Poly. Sci.,* **256,** 471 (1978).
142. A. W. Adamson, I. Ling, L. Dormant, and M. Orem, *J. Colloid Interface Sci.,* **21,** 445 (1966).
143. S. Ross and W. W. Pultz, *J. Colloid Sci.,* **13,** 397 (1958).
144. W. A. Steel, *J. Phys. Chem.,* **67,** 2016 (1963); see also J. M. Honig, *Adv. Chem.,* **33,** 239 (1961).
145. D. H. Everett, private communication.
146. L. G. Helper, *J. Chem. Phys.,* **16,** 2110 (1955).
147. G. Jura and R. E. Powell, *J. Chem. Phys.,* **19,** 251 (1951).
148. J. Calvet, *CR,* **232,** 964 (1951).
149. V. R. Deitz and F. G. Carpenter, in *Adv. Chem.,* **33,** 146 (1961).
150. C. A. Guderjahn, D. A. Paynter, P. E. Berghausen, and R. J. Good, *J. Phys. Chem.,* **63,** 2066 (1959).
151. J. W. McBain, *Colloid Symp. Monograph,* **4,** 1 (1926).
152. R. M. Barrer, *Proc. 10th Colston Symp.,* Butterworths, London, 1958, p. 6.
153. R. M. Barrer, *Discuss. Faraday Soc.,* **7,** 135 (1949); *Q. Rev.,* **3,** 293 (1949).
154. E. Rabinowitsch and W. C. Wood, *Trans. Faraday Soc.,* **32,** 947 (1936).
155. D. J. C. Yates, *J. Phys. Chem.,* **70,** 3693 (1968).
156. M. M. Dubinin and V. A. Astakhov, *Adv. Chem.,* **102,** (1971).
157. H. M. Powell, *J. Chem. Soc.,* **1954,** 2658.
158. R. M. Barrer and W. I. Stuart, *Proc. Roy. Soc. (London),* **A243,** 172 (1957).
159. R. M. Barrer and D. J. Ruzicka, *Trans. Faraday Soc.,* **58,** 2262 (1962).
160. A. W. Adamson and B. R. Jones, *J. Colloid Interface Sci.,* **37,** 831 (1971).
161. A. Bailey, D. A. Cadenhead, D. H. Davies, D. H. Everett, and A. J. Miles, *Trans. Faraday Soc.,* **67,** 231 (1971).
162. C. Pierce, *J. Phys. Chem.,* **57,** 149 (1953).
163. R. I. Razouk, Sh. Nashed, and F. N. Antonious, *Can. J. Chem.,* **44,** 877 (1966).
164. R. S. Schechter, W. H. Wade, and J. A. Wingrave, *J. Colloid Interface Sci.,* **59,** 7 (1977).
165. R. Zsigmondy, *Z. Anorg. Chem.,* **71,** 356 (1911).
166. L. H. Cohan, *J. Am. Chem. Soc.,* **60,** 433 (1938); see also *ibid.,* **66,** 98 (1944).
167. E. P. Barrett, L. G. Joyner, and P. P. Halenda, *J. Am. Chem. Soc.,* **73,** 373 (1951).
168. D. H. Everett, *Proc. 10th Colston Symp,* Butterworths, London, 1958, p. 95.
169. M. M. Dubinin, *J. Colloid Interface Sci.,* **77,** 84 (1980).
170. G. D. Parfitt, K. S. W. Sing, and D. Urwin, *J. Colloid Interface Sci.,* **53,** 187 (1975).
171. A. Lecloux and J. P. Pirard, *J. Colloid Interface Sci.,* **70,** 265 (1979).
172. R. M. Barrer, N. McKenzie, and J. S. S. Reay, *J. Colloid Sci.,* **11,** 479 (1956).
173. S. Brunauer, R. Sh. Mikhail, and E. E. Bodor, *J. Colloid Interface Sci.,* **24,** 451 (1967).
174. J. Hagymassy, Jr., I. Odler, M. Yudenfreund, J. Skalny, and S. Brunauer, *J. Colloid Interface Sci.,* **38,** 20 (1972).
175. A. V. Kiselev, *Usp. Khim.,* **14,** 367 (1945).
176. K. S. Rao, *J. Phys. Chem.,* **45,** 517 (1941).
177. P. H. Emmett, *Chem. Rev.,* **43,** 69 (1948).

178. D. H. Everett, *The Solid-Gas Interface,* Vol. 2, E. A. Flood, Ed., Marcel Dekker, New York, 1966; D. H. Everett, *Trans. Faraday Soc.,* **51,** 1551 (1955).
179. V. K. LaMer, *J. Colloid Interface Sci.,* **23,** 297 (1967) (posthumous paper).
180. G. L. Kington and P. S. Smith, *Trans. Faraday Soc.,* **60,** 705 (1964).
181. R. Sh. Mikhail, S. Brunauer, and E. E. Bodor, *J. Colloid Interface Sci.,* **26,** 45 (1968).
182. M. M. Dubinin, *J. Colloid Interface Sci.,* **46,** 351 (1974).
183. G. F. Hüttig, *Monatsh. Chem.,* **78,** 177 (1948).

CHAPTER XVII

Chemisorption and Catalysis

1. Introduction

In this concluding chapter we take up some of those aspects of the adsorption of gases on solids in which the adsorbent–adsorbate bond approaches an ordinary bond in strength and in which the chemical nature of the adsorbate may be significantly different in the adsorbed state. Such adsorption is generally called chemisorption, although as was pointed out in the introduction to Chapter XVI the distinction between physical adsorption and chemisorption is sometimes blurred, and many of the principles of physical adsorption apply to both kinds of adsorption. One experimental distinction is that we now deal almost entirely with submonolayer adsorption, since in chemisorption systems the heat of adsorption in the first layer ordinarily is much greater than that in succeeding layers. In fact, most chemisorption systems involve temperatures above the critical temperature of the adsorbate, so that the usual treatments of multilayer adsorption do not apply.

At one time the twin subjects of chemisorption and catalysis were so closely intertwined as to be virtually indistinguishable. Chemisorption was the mode of adsorption, and heterogeneous or contact catalysis the interesting consequence. The industrial importance of catalytic systems tended to bias the research toward those systems of special catalytic relevence. The massive development in recent years of high vacuum technology and associated spectroscopic and diffraction techniques (see Chapter VIII) has brought the field of chemisorption to maturity as a distinct field of surface chemistry with research interests undirected toward catalysis (although often relatable to it). The molecular emphasis of modern chemisorption has benefited the field of catalysis by giving depth and scope to the understanding of the surface chemistry of catalytic processes. There is also a drawing together of catalysis involving metal surfaces and organometallic chemistry; metal cluster compounds now being studied are approaching indistinguishability from polyatomic metal patches on a supporting substrate.

The plan of this chapter is as follows. We discuss chemisorption as a distinct topic, first from the molecular and then from the phenomenological points of view. Heterogeneous catalysis is then taken up, but now first from the phenomenological (and technologically important) viewpoint, and then in terms of current knowledge about surface structures at the molecular level. Section XVII-6 takes note of the current interest in *photo-driven* surface reactions.

As on previous occasions, the reader is reminded that no very extensive coverage of the literature is possible in a textbook such as this one, and that the emphasis is primarily on principles and their illustration. Several monographs are available for more detailed information (see General References). Refs. 1 to 5 list some useful review articles.

2. Chemisorption—The Molecular View

A. LEED Structures

The technique of low energy electron diffraction (LEED, Section VIII-3) has provided a considerable amount of information about the manner in which a chemisorbed layer arranges itself. Somorjai (5) summarizes the LEED results for a number of systems. Just as one example, oxygen chemisorbed on (110) planes of nickel gives a (2 × 1) lattice or mesh of oxygen atoms; CO gives a (1 × 1) mesh of chemisorbed CO molecules. In the case of chemisorbed hydrogen, the pattern is (1 × 2), but is now of the surface nickel atoms, since LEED does not detect hydrogen atoms because of their low electron density.

The organization of chemisorbed molecules may change with degree of coverage. Carbon monoxide adsorbed on (100) faces of Pd is bonded to specific, random sites up to a coverage of about half a monolayer; after this point there is a switch to an adsorption layer that is out of registry with the crystal lattice (6). Other cases, reviewed by Somorjai (5), of disordered or essentially random adsorption at low coverages giving way to an ordered structure (usually in register with the substrate, however) at higher coverages, include Xe on Cu (100) and Ir (100) surfaces and benzene on Pt (111) surfaces. As illustrated in Fig. XVII-1, low temperature chemisorption of CO on (110) planes of tungsten shows a succession of LEED patterns (7). This is presumably a matter of surface packing, the binding always being W—CO. On heating, EID and thermal desorption show that several chemically different new states can form.

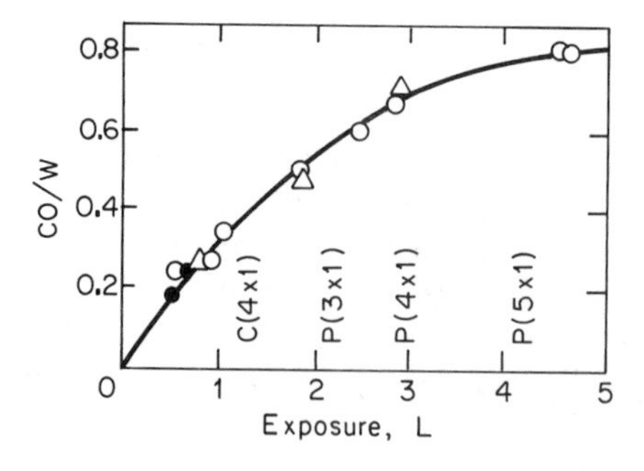

Fig. XVII-1. Surface coverage and symmetry of LEED patterns versus exposure of W (110) planes to CO. Exposure is in units of Langmuirs, (L), or 1 × 10^{-6} torr sec. (From Ref. 7.)

B. Spectroscopy of Chemisorbed Species

The spectroscopic methods described in Chapter VIII for the characterization of the surface composition and state of solids are generally also applicable to chemisorbed species. A widely used tool is that of infrared spectroscopy (see also Section XV-4). A few examples are as follows. Eischens and Jacknow (8) observed a band at 2128 cm^{-1} when $^{15}N_2$ was adsorbed on a silica supported nickel surface, and a shift to 2160 if $^{14}N—^{15}N$ was used. The isotopic shift identified the dinitrogen molecule, establishing that it was not largely dissociated into atoms on the surface. The proposed surface structure was $Ni—N{\equiv}N^+$, and force constants of 3×10^5 dyne/cm and 19.1 $\times 10^5$ dyne/cm were calculated for the Ni—N and N—N bonds, respectively. Changes in the N_2 frequency could be followed as reactive second gases were added. In another case, adsorption of mixtures of CO and NO on various metal surfaces led to a band at 2260 cm^{-1}, attributed to the formation of chemisorbed isocyanate, M—NCO (9).

Originally, infrared spectra were obtained as transmission spectra using high surface area powders. However, sufficient sensitivity can be obtained by using a high angle of incidence and samples mounted on a reflective metal surface to permit infrared spectra on single crystal adsorbents to be obtained (10). It is possible to connect LEED and other studies made with single crystal surfaces to the results of phenomenological measurements on the complex surfaces of actual catalysts.

Other techniques, such as electron spectroscopy for chemical analysis (ESCA), ultraviolet photoelectron spectroscopy (UPS), and electron stimulated desorption (ESD) have found use (see Ref. 5). Figure XVII-2 shows an ESCA spectrum of CO chemisorbed on tungsten, clearly indicating the two types of binding, so-called α-CO and β-CO.

Characteristic vibrational frequencies of surface adsorbed species may also be determined by high resolution electron energy loss spectroscopy (EELS). Figure XVII-3 shows the sequence of spectra as ethanol reacts with preadsorbed oxygen on Cu (100) planes (12). There have been numerous applications electron paramagnetic resonance (EPR) spectroscopy, as another type of example. Hall and co-workers (13) found, by this means, evidence for O^- when molecular oxygen is adsorbed on a reduced molybdena–alumina catalyst. One may use a radical trapping molecule to stabilize radicals that are formed on surfaces, as in the case of hydrogen and propylene adsorption on ZnO (14).

C. Work Function and Related Measurements

The work function across a phase boundary, discussed in Sections V-11 and VIII-2C, is strongly affected by the presence of adsorbed species. Conversely, work function changes can be diagnostic of changes in types of surface binding. Somorjai (5) summarizes a number of results of measurements of $\Delta\Phi$ due to chemisorption—values range from -1.5 V for CO on

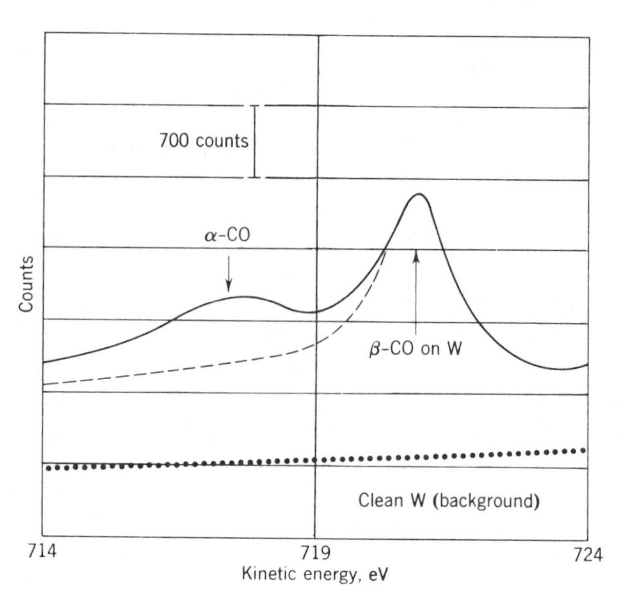

Fig. XVII-2. ESCA spectrum for CO chemisorbed on tungsten. Solid upper curve shows the α- and β-types of adsorbed CO; dashed line shows the spectrum after selective desorption of the α-type; dotted line gives the clean tungsten surface. (From Ref. 1, as adopted from Ref. 11.)

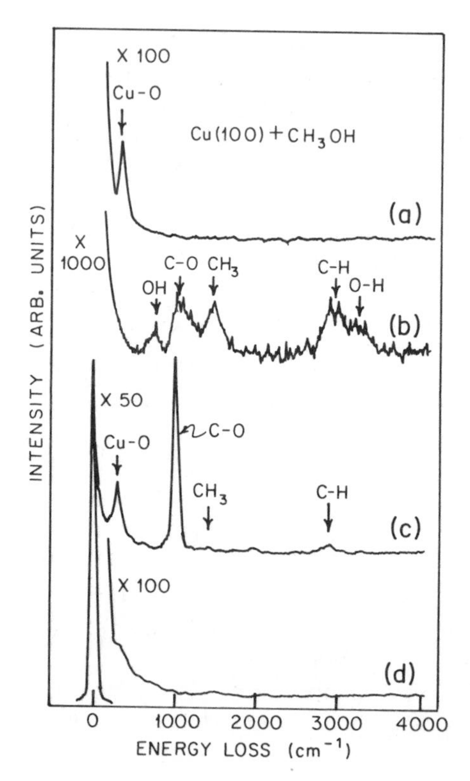

Fig. XVII-3. Reaction of methanol with preadsorbed oxygen on Cu (100) planes. (*a*) 50 L of O_2 exposure at 470 K. (*b*) Multilayer of methanol ice condensed on (*a*) at 100 K. (*c*) Warmed to 370 K briefly with resulting methoxide formation. (*d*) Warmed to 420 K with resulting methoxide decomposition. (From Ref. 12.)

iron to 1.6 V for oxygen on nickel. Changes in surface state may be observed. Oxygen chemisorbed on clean aluminum may either increase or decrease the work function; depending on conditions, a chemisorbed surface layer may form or the oxygen may be incorporated into the subsurface region (15a).

Different types of chemisorption sites may be observed, each with a characteristic $\Delta\Phi$ value. Several adsorbed states appear to exist for CO chemisorbed on tungsten, as noted. These states of chemisorption probably have to do with different types of chemisorption bonding, maybe involving different types of surface sites. Much of the evidence has come initially from flash desorption studies, discussed immediately following.

D. Flash Desorption

A very powerful technique in studying both adsorption and desorption rates is that of flash desorption. The general procedure (see Refs. 2, 15b, 16a and citations following) is to expose a clean filament or surface (usually of a metal) to a known low pressure of gas that flows steadily over it. The pressure may be quite low, for example, 10^{-7} mm Hg or less, so that even nonactivated adsorption can take some minutes for complete monolayer coverage to be achieved. While the technique is difficult, it does allow a calculation of how much adsorption should have occurred in the time allowed if the sticking probability were unity. The surface is then heated, electrically if a filament, and by an incident light beam if a flat single crystal surface, so that the adsorbed gas is evolved, and the increase in gas pressure in the system allows a calculation to be made of how much actually was adsorbed and hence of the sticking probability.

If the heating is quickly to a high temperature, or a "flashing," all adsorbed gas is removed indiscriminately. If, however, the heating is gradual, then separate, successive desorptions may be observed. Thus in the case of CO on W (100), a single peak α appears at around 550 K, and a set of incompletely resolved desorption peaks, β_1, β_2, β_3, in the region 1100 to 1700 K (16b). The α state is thought to be a metal carbonyl-like W—CO binding; the C—O bond is much weakened in the strongly bound β-states, and there may be partial dissociation into C and O. Eirich (17) reports several adsorbed states for N_2 on tungsten, weakly held α- and γ-states that may involve molecular nitrogen, and a strongly (81 kcal/mole) held β-state that probably consists of atomically bound N atoms.

One may also examine the rate of appearance of reaction *products*. Figure XVII-4 shows a set of desorption peaks for the reaction products of HCOOD on Ni (110) planes (2).

The types of chemisorption states may be quite different on different crystal faces of the same metal. Figure XVII-5 shows the desorption spectra for H_2 chemisorbed on 100 and 111 faces of tungsten (18). The γ-state is thought to represent molecular hydrogen, while the β-states involve H atoms adsorbed at various types of sites. If the heating is at a constant rate b,

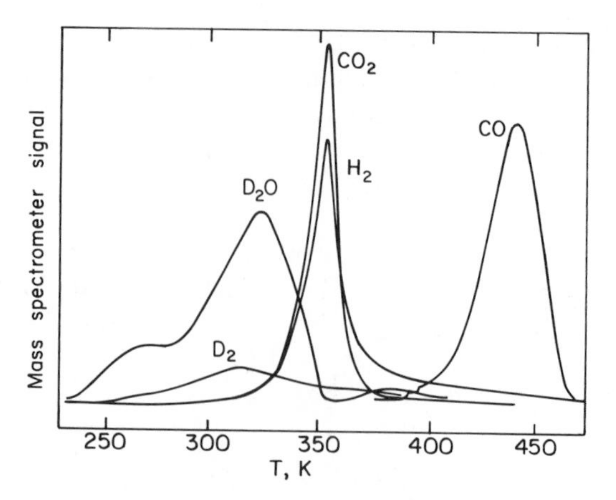

Fig. XVII-4. The temperature-programmed reaction spectrum for products resulting from HCOOD adsorption and decomposition on clean Ni (110) planes. (Reprinted with permission from Ref. 2. Copyright 1979 American Chemical Society.)

analysis (19) gives

$$-b\frac{dn}{dT} = -\frac{dn}{dt} = An^m e^{-(E_d^*/RT)} \tag{XVII-1}$$

where n is the number of adsorbed atoms, A is the frequency factor, E_d^* the activation energy for desorption, and m is the order of the desorption kinetics. The exponential makes the rate slow at low temperatures; by the time high temperatures are reached, the rate is again small because n is small. Alternatively, if T_p is the desorption temperature, a plot of ln b/T_p^2 versus $1/T_p$ should be linear with a slope of E^*/R (see Ref. 2).

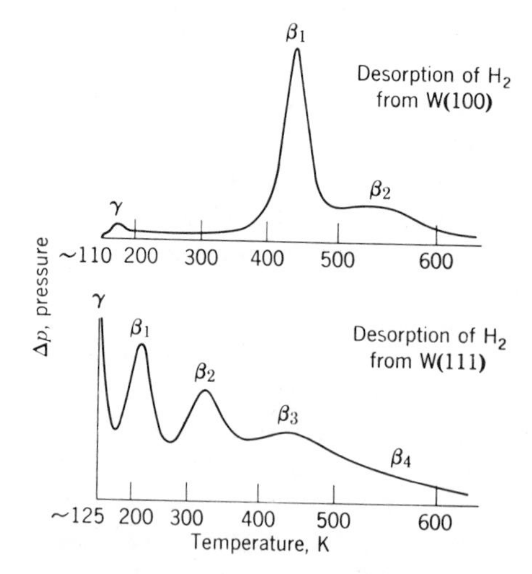

Fig. XVII-5. Molecular hydrogen desorption spectra from monolayer covered tungsten (100) and (111) faces. (From Ref. 18.)

A great deal of information of this type is now available, and the foregoing represent samplings to illustrate the complexity of many chemisorption systems—especially those of catalytic activity.

3. Chemisorption Isotherms

In considering isotherm models for chemisorption, it is important to remember the types of systems that are involved. As pointed out, conditions are generally such that physical adsorption is not important, nor is multilayer adsorption, in determining the equilibrium state, although the former especially can play a role in the kinetics of chemisorption.

Because of the relatively strong adsorption bond supposed to be present in chemisorption, the fundamental adsorption model has been that of Langmuir (as opposed to that of a two-dimensional nonideal gas). The Langmuir model is therefore basic to the present discussion, but for economy in presentation, the reader is referred to Section XVI-3 as prerequisite material. However, the Langmuir equation (Eq. XVI-4) as such,

$$\theta = \frac{bP}{1 + bP} \qquad \text{(XVII-2)}$$

is not often obeyed in chemisorption systems. Ordinarily, complications appear, and the following material is largely concerned with the necessary specializations of the Langmuir model that are needed. These involve a review of ways of treating surface heterogeneity and lateral interactions and the new isotherm forms that arise if adsorption requires the presence of two adjacent sites or if dissociation occurs on adsorption.

A. Variable Heat of Adsorption

It is not surprising, in view of the material of the preceding section, that the heat of chemisorption often varies with the degree of surface coverage. It is convenient to consider two types of explanation (actual systems involving some combination of the two). First, the surface may be heterogeneous, so that a site energy distribution is involved (Section XVI-14).

If the differential distribution function is exponential in Q (Eq. XVI-153), the resulting $\Theta(P, T)$ is that known as the Freundlich isotherm

$$\Theta(P, T) = AP^c \qquad \text{(XVII-3)}$$

where c is generally less than unity and is therefore often written as $1/n$. (In discussing situations involving surface heterogeneity, we use Θ to denote the average surface coverage, and θ, that of a given patch of uniform Q). The linear form of Eq. XVII-3 is

$$\log v = \log (v_m A) - n \log P \qquad \text{(XVII-4)}$$

and quite a few experimental isotherms obey this form over some region of

Θ values. A recent example is illustrated in Fig. XVII-6. Notice that D_2 is significantly more strongly adsorbed than H_2.

The Freundlich equation is clearly defective as a model because it predicts infinite Θ at infinite pressure or, alternatively, because the $f(Q)$ involved requires infinite Q at zero coverage and zero Q at infinite coverage. The equation is therefore useful at all only in the middle range of an adsorption isotherm. The difficulty can be patched by supposing that

$$\Theta = \frac{AP^c}{1 + AP^c} \tag{XVII-5}$$

for which $f(Q)$ is nearly Gaussian in shape and, moreover, now can be normalized.

A second supposition is that

$$Q = Q_0(1 - \alpha\Theta) \qquad \text{or} \qquad \Theta_Q = \frac{1}{\alpha}\left(\frac{Q_0 - Q}{Q_0}\right) \tag{XVII-6}$$

This leads to an equation that, in the middle range of Θ values, approximates to

$$\Theta = \frac{RT}{Q_0\alpha} \ln P + \text{constant} \tag{XVII-7}$$

a form known as the Temkin isotherm (21); see also Eq. XVI-146). Again, this is an equation that may be useful for fitting the middle region of an adsorption isotherm. However, it is often quite difficult to distinguish between these various equations. Thus the middle range of the adsorption of nitrogen on iron powder could be fitted to either a Langmuir form (Eq. XVI-2) or to the Freundlich or the Temkin equations (13).

It would seem better to transform chemisorption isotherms into corresponding site energy distributions in the manner reviewed in Section XVI-14 than to make choices of analytical convenience regarding the $f(Q)$ function. The second procedure tends to give equations whose fit to data is empirical and deductions from which can be spurious. An alternative ap-

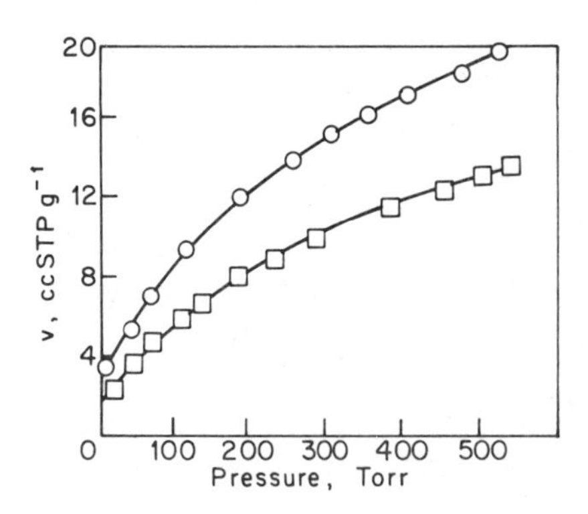

Fig. XVII-6. Adsorption isotherms for D_2, O, and for H_2, □, at 77°K for a reduced supported Mo catalyst. (Reprinted with permission from Ref. 20. Copyright 1979 American Chemical Society.)

proach, possible in chemisorption, is to study separately the rates of adsorption and desorption; this is discussed in Section XVII-4.

Surface heterogeneity may merely be a reflection of different types of chemisorption, as is probably the case for the examples of Figs. XVII-2 and XVII-5. The presence of various crystal planes, as in powders leads to heterogeneous adsorption behavior. Heterogeneity may be adventitious, resulting from surface imperfections or from impurities. It may be deliberate; many catalysts make use of alloys and other combinations of active surfaces. In the case of alloys, it should be noted that the surface composition is not necessarily the same as the bulk composition—one or another component may be in surface excess just as with liquid solutions. For example, copper–nickel alloys appear to have a highly copper-rich surface (22).

The second general cause of a variable heat of adsorption is that of adsorbate–adsorbate interaction. In physical adsorption, the effect usually appears as a lateral *attraction*, ascribable to van der Waals forces acting between adsorbate molecules. A simple treatment led to Eq. XVI-53).

Such attractive forces are relatively weak in comparison to chemisorption energies, and it appears that in chemisorption, *repulsion* effects may be more important. These can be of two kinds. First, there may be a short-range repulsion affecting nearest neighbor molecules only, as if the spacing between sites is uncomfortably small for the adsorbate species. A repulsion between the electron clouds of adjacent adsorbed molecules would then give rise to a short-range repulsion, usually represented by an exponential term of the type employed in Eq. VII-17. In the treatment of lateral interaction given in Section XVI-3D, the increment to the differential heat of adsorption was $z\omega\theta$, where z is the number of nearest neighbors and ω is the interaction energy, negative in the case of repulsion. As discussed by Fowler and Guggenheim (23), a more elaborate approach allows for the fact that, at equilibrium, neighboring sites will be occupied more often than the statistical expectation if the situation is energetically favored (ω positive) and, conversely, if there is lateral repulsion (ω negative), nearest neighbor sites will be less frequently occupied than otherwise expected.

A second type of repulsion could be long range. If adsorption bond formation polarizes the adsorbate, or strongly orients an existing dipole, the adsorbate film will consist of similarly aligned dipoles that will have a mutual electrostatic repulsion. The presence of such dipoles can be inferred from the change in surface potential difference ΔV on adsorption, as mentioned in Section V-6C. The resulting interaction should show both a short-range Coulomb repulsion due to adjacent dipoles whose positive and negative charges would repel each other, and which could be considered as part of ω, and a long-range repulsion due to the dipole field, which is inverse cube in distance.

The short-range repulsion is very difficult to estimate quantitatively, and comparisons of predicted versus observed shapes of Q versus θ plots (e.g., Ref. 12) encounter the difficulty that a treatment in terms of ω alone is relatively meaningless; some knowledge is needed of surface heterogeneity. Note, for example, the formal resemblance of the dependence of Q on Θ in Eq. XVII-6, supposing surface heterogeneity, and Eq. XVI-53, supposing lateral interaction.

B. Effect of Site and Adsorbate Coordination Number

Since in chemisorption systems it is reasonable to suppose that the strong adsorbent–adsorbate interaction is associated with specific adsorption sites, a situation that may arise is that the adsorbate molecule occupies or blocks the occupancy of a second adjacent site. This means that each molecule effectively requires two adjacent sites. An analysis (12) suggests that, in terms of the kinetic derivation of the Langmuir equation, the rate of adsorption should now be

$$\text{rate of adsorption} = k_2P\left(\frac{z}{z-\theta}\right)(1-\theta)^2 \tag{XVII-8}$$

while the rate of desorption is still

$$\text{rate of desorption} = k_1\theta \tag{XVII-9}$$

where z is again the number of nearest neighbors to a site. On equating the two rates, one obtains a quadratic form of the Langmuir equation,

$$bP = \left(\frac{z-\theta}{z}\right)\left[\frac{\theta}{(1-\theta)^2}\right] \tag{XVII-10}$$

If the adsorbed molecule occupies two sites because it dissociates, the desorption rate takes on the form

$$\text{rate of desorption} = k_1\left[\frac{(z-1)^2}{z(z-\theta)}\right]\theta^2 \tag{XVII-11}$$

so that the isotherm becomes

$$(b'P)^{1/2} = \frac{\theta}{1-\theta} \tag{XVII-12}$$

where

$$b' = \frac{k_2}{k_1}\left(\frac{z}{z-1}\right)^2$$

It should be cautioned that the correctness of the factors involving z has not really been verified experimentally and that the algebraic forms involved are modelistic.

Halsey and Yeates (24) add an interaction term, writing Eq. XVII-12 in the form

$$\left(\frac{P}{P_0}\right)^{1/2} = \left(\frac{\theta}{1-\theta}\right)\exp\left[\left(\frac{2w}{kT}\right)\left(\frac{\theta-1}{2}\right)\right] \tag{XVII-13}$$

where w is an interaction energy. Eq. XVII-13 was fit by data for the adsorption of H_2 on Ni (110) and Pt (111), for example, with respective w/kT values of -1.5 and 2.3 (note the observation of both positive and negative w values).

The preceding treatments are based on the concept of localized rather than mobile adsorption. The distinction may be difficult experimentally; note Ref. 23 and the discussion in connection with Fig. XVI-30. There are also conceptual subtleties; see Section XVII-5.

C. Adsorption Thermodynamics

The thermodynamic treatment that was developed for physical adsorption applies, of course, to chemisorption, and the reader is therefore referred to Section XVI-12. As in physical adsorption the chief use that is made of adsorption thermodynamics is in the calculation of heats of adsorption from temperature dependence data, that is, the obtaining of q_{st} values. As in physical adsorption, these should be the same as the calorimetric differential heats of adsorption (except for the small difference RT), probably even for heterogeneous surfaces. There is, however, much more danger in chemisorption work that the data do not represent an equilibrium adsorption; the Beeck criterion for surface mobility discussed in Section XVII-5 represents an extreme case where only a portion of the surface is supposed to be in equilibrium with the gas phase. It is well to remember that in such situations q_{st} values need have no simple physical meaning.

Entropies of adsorption are likewise obtainable in the same manner as discussed in Chapter XVI.

4. Kinetics of Chemisorption

A. Activation Energies

It was noted in Section XVI-1 that chemisorption may become slow at low temperatures so that even though it is favored thermodynamically the only process actually observed may be that of physical adsorption. Such slowness implies an activation energy for chemisorption, and the nature of this effect has been much discussed.

The classic explanation for the presence of an activation energy in the case where dissociation occurs on chemisorption is that of Lennard-Jones (23), and is illustrated in Fig. XVII-7. The curve labeled $M + X_2$ represents the variation of potential energy as the molecule X_2 approaches the surface; there is a shallow minimum corresponding to the energy of physical adsorption and located at the sum of the van der Waals radii for the surface atom and the adsorbate molecule. The curve labeled $M + 2X$ represents the potential energy for two atoms X; far from the surface, it is separated from the first curve by the dissociation energy of X_2. If an atom is brought up to the surface and can engage in chemical bond formation with it, then

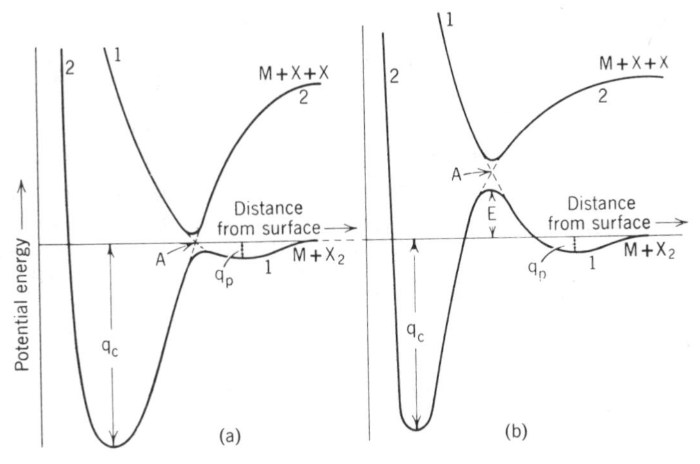

Fig. XVII-7. Potential energy curves for (*a*) physical adsorption and (*b*) chemisorption. (From Ref. 13.)

the curve $M + 2X$ (i.e., twice that for one atom) follows. The minimum is now deeper and, as shown in the figure in somewhat exaggerated fashion, at a smaller distance than the one for physical adsorption. The two curves cross, and the adsorbate therefore can pass from the first to the second; the actual potential curves mix and become rounded at the crossing region because of a wave mechanical effect. Thus in case (*a*) in the figure, chemisorption should occur easily; a physically adsorbed molecule might pass directly over to a chemisorbed state before losing its energy of physical adsorption or, at the worst, only a small activation energy would be required to surmount the barrier. In case (*b*), however, an activation energy equal to a sizable fraction of the dissociation energy of X_2 is required to pass from the physically adsorbed to the chemisorbed state, and chemisorption should be quite noticeably activated.

Figure XVII-7 simplifies the situation in that the interatomic distance $X—X$ is not indicated although it also affects the energy, that is, the barrier to chemisorption may involve an energy of stretching the $X—X$ bond to match the distance between sites. Thus for the case of the adsorption of hydrogen on various carbon surfaces, the picture can be taken to be that of a hydrogen molecule approaching a pair of surface carbon atoms, with simultaneous H—H bond stretching and C—H bond formation as the final state of chemisorbed hydrogen atoms is attained:

```
    H—H        H— —H
   /   \       |    |
  /     \      |    |
—C-——C—     C——C
```

An early calculation by Sherman and Eyring (25) led to a theoretical variation of the activation energy with the C—C distance which showed a minimum (of 7 kcal/mole) at a spacing of 3.5 Å. The qualitative explanation for the minimum was that if the C—C distance is too large, then the H—H bond must be stretched considerably before much gain due to incipient C—H bond formation can occur. If the C—C distance is too short, H—H repulsion again raises the activation energy. In the case of graphite and diamond, the C—C spacings are 1.42 and 1.54 Å, respectively, leading to calculated activation energies of about 45 kcal/mole, or much higher than the experimental values of 22 and 14 (26). However, if one disregards nearest neighbor distances (a perhaps questionable action), spacings of about 2.8 Å can be found between certain carbon atom pairs, and the calculated activation energies thus can be made to agree with the experimental ones (13). Incidentally, the presence of such alternative choices of site pairs constitutes another source of difficulty in applying equations such as Eq. XVII-10, based on a single assumed z value.

Another kind of problem, explored in calculations of H atom adsorption on graphite, is that surface relaxation can significantly alter the calculated energies (27).

A more elaborate theoretical approach develops the concept of surface molecular orbitals and proceeds to evaluate various overlap integrals (28). Calculations for hydrogen on Pt 111 planes were consistent with flash desorption and LEED data. In general, the greatly increased availability of LEED structures for chemisorbed films has allowed correspondingly detailed theoretical interpretations, as, for example of the commonly observed (C2 × 2) structure (29; note also Ref. 3).

B. Rates of Adsorption

Mention was made in Section XVII-2D of flash desorption; this modern technique gives specific information about both the adsorption and the desorption of specific molecular states, at least when applied to single crystal surfaces. The kinetic theory involved is essentially that used in Section XVI-3A. It will be recalled that the adsorption rate was there taken to be simply the rate at which molecules from the gas phase would strike a site area σ^0 times the fraction θ of unoccupied sites. For greater generality, it is necessary to include an additional factor, the *condensation coefficient* c, which gives the fraction of molecules hitting area σ^0 that stick. Also, if the adsorption is activated, the fraction of molecules hitting and sticking that can proceed to a chemisorbed state is given by $\exp(-E_a^*/RT)$. The adsorption rate constant of Eq. XVI-13 becomes

$$k_2 = \frac{Nc\sigma^0 \exp(-E_a^*/RT)}{(2\pi MRT)^{1/2}} = A \exp\left(\frac{-E_a^*}{RT}\right) \qquad \text{(XVII-14)}$$

The rate of adsorption is then

$$R_a = \frac{d\theta}{dt} = k_2 f(\theta) P \qquad \text{(XVII-15)}$$

where $f(\theta)$ is the fraction of available surface, taken to be $(1 - \theta)$ in the simple Langmuir derivation, but capable of taking on other forms as, for example, if the adsorbing molecule must find two adjacent unoccupied sites. See Ref. 30 for a calculation of the probability of this. A second quantity, the *sticking* coefficient, is defined as $s = cf(\theta)\exp(-E_a^*/RT)$ and represents the practical efficiency of collisions of gas molecules with the surface.

Alternatively, the treatment can be put in the framework of absolute rate theory in which the equilibrium constant for forming the activated or transition state is invoked. This transition state would have the configuration of the system at the potential maximum in Fig. XVII-7*b*, and in the formal development it turns out that the condensation coefficient c is replaced by an expression involving the partition function of the transition state (13, 31).

Where E_a^* is appreciable, adsorption rates may be followed by ordinary means. Scholten and co-workers (32) were able to follow the adsorption of nitrogen on an iron catalyst of 6 m^2/g specific surface area by means of a vacuum balance; the weight of the catalyst was 87 g and the weight of adsorbed nitrogen at $\theta = 1$ was 39 mg. The rate decreased rapidly with increasing θ, so the data were reported in terms of $(1/P)(d\theta/dt)$ versus θ, and from the change in rate with temperature at a given θ, the plot of E_a^* versus θ shown in Fig. XVII-8 was obtained. Note the break at about the same point as in the plot of the variation of Q with θ.

A variation of E_a^* with θ is not uncommon, and if the empirical relation $E_a^* = E_0^* + \alpha\theta$ is used, Eq. XVII-15 becomes

$$\frac{d\theta}{dt} = Af(\theta)P\exp\left(-\frac{E_0^* + \alpha\theta}{RT}\right) \qquad \text{(XVII-16)}$$

so that at a given temperature, the rate should vary according to $f(\theta)e^{-\alpha\theta}$. Equation XVII-16 is a form of what is known as the *Elovich* equation (33).

In the preceding example the rate, in the range $\theta = 0.07$ to 0.22, could be expressed by the equation

$$\frac{d\theta}{dt} = 21.9P_{N_2}e^{(132.4\theta/R)}e^{-(5250+77{,}500\theta)/RT} \qquad \text{(XVII-17)}$$

with pressure in centimeters of mercury and time in minutes. Equation XVII-17 can be put in the form of Eq. XVII-16; $f(\theta)$ is presumably $(1 - \theta)^2$ since adsorption on two sites is presumed but, over the small range of θ values involved, this in turn can be approximated by $e^{-2.22\theta}$, which amounts to a negligible amendment to the other exponential terms. In this same range of θ values, the frequency factor (A in Eq. XVII-14) increased by 10^5-fold, indicating that some type of progressive change in the condensation coefficient or, alternatively, in the partition function for the transition state, was occurring.

The general picture presented by the preceding example is fairly representative, that is, both the energy of activation and the frequency factor tend to increase as

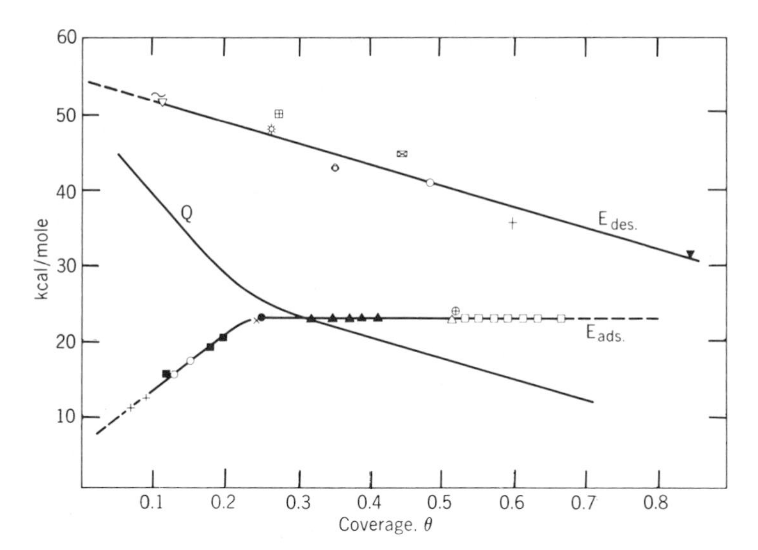

Fig. XVII-8. Activation energies of adsorption and desorption, and heat of chemisorption for nitrogen on a single promoted, intensively reduced iron catalyst. O is calculated from $Q = E_{\text{des}} - E_{\text{ads}}$. (From Ref. 32.)

Q decreases or θ increases. The physical explanation of this correlation would seem to be that active sites have a relatively negative entropy of adsorption, perhaps because the adsorbate is very highly localized, but the high energy of adsorption assists in the bond deformations needed for chemisorption to occur. Thus in Fig. XVII-7*b*, a potential energy curve for $M + 2X$ having a deeper minimum and hence larger Q value would intersect the $M + X_2$ curve lower to give a lower activation energy for adsorption.

Returning to the system of Fig. XVII-8, it was evident that beyond θ = 0.22 a change in the nature of the surface state must have occurred; the activation energies and heats of adsorption were more nearly constant, and the frequency factor now decreased with increasing θ. Scholten et al. (32) interpreted this region as being one of mobile adsorption. It is possible, however, that nitrogen was now ceasing to undergo dissociation, or was now adsorbing on crystal planes that allowed more vibrational and rotational degrees of freedom.

C. Rates of Desorption

Desorption will always be activated, since the minimum E_d^* is that equal to the energy of adsorption Q. Thus

$$E_d^* = E_a^* + Q \tag{XVII-18}$$

This means that desorption activation en^rgies can be much larger than those for adsorption and very dependent on θ, since the variation of Q with

θ now contributes directly. The rate of desorption may be written, following the kinetic treatment of the Langmuir model,

$$R_d = -\frac{d\theta}{dt} = \frac{1}{\tau_0} \exp\left(-\frac{Q}{RT}\right) \exp\left(-\frac{E_a^*}{RT}\right) f'(\theta) \qquad \text{(XVII-19)}$$

where $f'(\theta)$ is simply θ in the Langmuir derivation (Section XVI-3), but may take on other forms in chemisorption systems as, for example, θ^2 if two surface atoms must associate to desorb.

In the case of nitrogen on iron, the experimental desorption activation energies are also shown in Fig. XVII-8; the desorption rate was given by the empirical expression

$$-\frac{d\theta}{dt} = 4.8 \times 10^{14}\theta^2 e^{-10.64\theta} e^{[-55{,}000+29{,}200\theta]/RT} \qquad \text{(XVII-20)}$$

again with time in minutes. Note the presence of the term in θ^2; that in $e^{-10.64\theta}$ could represent an empirical compensation to θ^2 for the statistics of finding two adjacent sites. The general picture, however, is that of an adsorbed state consisting of nitrogen atoms, which associate to desorb as N_2. A second point of interest is that the plot of E_d^* versus θ did not show the break at $\theta = 0.22$ that was found for E_a^* (and by inference, for Q). The interpretation, in terms of transition state theory, would be that the activated state was very similar in nature to the adsorbed state, so that while its variation with θ would affect the adsorption kinetics, it would not affect that for desorption.

One might expect the frequency factor A for desorption to be around 10^{13}/sec (note Eq. XVI-1). Much smaller values are sometimes found, as in the case of the desorption of Cs from Ni surfaces (34), for which the adsorption lifetime obeyed the equation $\tau = 1.7 \times 10^{-3} \exp(3300/RT)$ sec (R in calories per mole per degree Kelvin). A suggested explanation was that surface diffusion must occur to desorption sites for desorption to occur.

5. Surface Mobility

The matter of surface mobility has come up at several points in the preceding material. The subject has been a source of confusion—see Ref. 24. Actually, two kinds of concepts seem to have been invoked. The first is that invoked in the discussion of physical adsorption, which has to do with whether the adsorbate can move on the surface so freely that its state is essentially that of a two-dimensional nonideal gas. For an adsorbate to be mobile in this sense surface barriers must be small compared to kT. This type of mobile adsorbed layer seems unlikely to be involved in chemisorption.

In general it seems more reasonable to suppose that in chemisorption specific sites are involved and that therefore definite potential barriers to lateral motion should be present. The adsorption should therefore obey the statistical thermodynamics of a localized state. On the other hand, the kinetics of adsorption and of catalytic processes will depend greatly on the frequency and nature of such surface jumps as do occur. A film can be fairly mobile in this kinetic sense and yet not be expected to show any significant deviation from the configurational entropy of a localized state.

Field emission studies have provided very detailed information about surface mobility in this second sense, although only for a rather limited number of types of systems. It is possible to deposit a few adsorbed molecules on one side of a tip, and to observe, through the changing pattern of emission intensity, the rate and manner in which they disperse on the surface on heating to various temperatures (see Fig. VIII-4). Thus nitrogen atoms begin to migrate easily around 111 planes of tungsten at about 400 K (35). Chen and Gomer (36) find $D = 7 \times 10^{-8} \exp(550/T)$ cm²/sec for the diffusion of Xe on W (110) and comment that the activation energy for diffusion is one-fourth of that for desorption, a ratio comparable to that found in chemisorption. Interestingly, the low temperature diffusion of hydrogen atoms on W (110) occurs through a tunneling mechanism (37). Hayward and Trapnell (15b) comment that surface migration tends to become important at about half the temperature at which evaporation is observable; this conclusion, while derived from observations on tungsten, may well be a fairly general one. Moreover, it is possible, with the field ion emission technique, to observe individual atoms and hence to follow individual atomic migrations. With sufficient observations, rudimentary statistics on the mean surface displacement per unit time, and hence a rough value of the surface diffusion coefficient, can be obtained.

Macroscopic mobility measurements are difficult to carry out and are not common. As an early example, Bosworth (38) was able to follow the spreading of a spot of potassium adsorbed on tungsten by means of the enhanced ultraviolet photoelectric effect wherever potassium was present. A diffusion coefficient of 6×10^{-6} cm²/sec was estimated at 480 K, with an activation energy of 15 kcal/mole. In a rather different study, the diffusional migration of a long chain fatty acid adsorbed on silica gel was interpreted as occurring through evaporative hopping (39).

The surface *self-diffusion* of Ni atoms on a nickel 111 surface was followed by forming a spot of radioactive nickel and observing the spreading in successive radioautographs (40). The results obeyed the equation

$$\mathcal{D}_s = 300e^{-E^*/RT} \text{ cm}^2/\text{sec} \qquad \text{(XVII-21)}$$

where E^* was 38 ± 4 kcal/mole.

A much more indirect approach is that of comparing adsorption entropies with theoretical values. The reader is referred to Hayward and Trapnell for details (15b), but the principles are essentially those discussed in Sections XVI-3C and XVI-13, and the difficulties are also much the same. It seems doubtful whether either the absolute value or the variation with θ of adsorption entropies has any firm diagnostic utility.

To summarize, possibly three different definitions of surface mobility are in use. The statistical thermodynamic criterion states that the entropy of the adsorbed film should contain a contribution ascribable to free translation in two dimensions (a rather unlikely situation in the writer's opinion). The equilibrium thermodynamic criterion is that through surface diffusion (or any other means) adsorbed atoms find the lowest free energy arrangement on the surface on a time scale short compared to the adsorption experiment. The third definition of mobility is that the adsorbed film should exhibit macroscopic surface migration or flow.

The first definition has not in fact been a very fruitful one; so many factors can affect the adsorption entropy that the experimental value for a particular system has not been diagnostic of surface mobility of the statistical thermodynamic type. The second definition is a useful one, especially in chemisorption where, unlike in physical adsorption, the adsorption bond can be strong enough that nonmobility can indeed mean that adsorbed species have a random rather than an equilibrium surface distribution.

The third definition is essentially a rheological one and is capable of unambiguous experimental application; it also represents the extension to films of the criteria we use for bulk phases and, of course, it is the basis for distinguishing states of films on liquid substrates. Thus as discussed in Chapter IV, solid films should be ordered and should show elastic and yield point behavior; liquid films should be coherent and show viscous flow; gaseous films should be in rapid equilibrium with all parts of the surface. While less fine distinctions can be made in the case of films on solid substrates, the approach is the same. In the writer's opinion, only the third definition of mobility should be used, as a matter both of practicality and of consistency.

6. The Chemisorption Bond

The chemisorption bond can be approached from at least two viewpoints. The extensive development of surface diffraction and spectroscopic studies has yielded a wealth of detail about the adsorbed state, as indicated in Section XVII-2. Analysis of LEED intensity data has also permitted the estimation of adsorbate–substrate bond lengths, summarized in Table XVII-1. Bond energies can be obtained from flash or temperature programmed desorption data, if coupled with knowledge of the activation energy for adsorption, as seen in Section XVII-4. The older approach to obtaining bond energies is, of course, through the calculation of the isosteric heat of adsorption q_{st} from adsorption isotherms obtained for more than one temperature. Most of the values reported in this section have been so obtained; they are approximate in the sense that q_{st} usually varies with surface coverage so that no one value exists for a given system.

The chemisorption bond is now approached from several theoretical points of view, summarized in what follows. The subject has been reviewed by Gomer (4).

A. *The Localized Bond Approach*

A quite fruitful approach has been to regard chemisorption as simply a bond formation between an atom of the adsorbate molecule and one of the

TABLE XVII-1
Adsorbate–Substrate Bond Lengths from LEED[a]

Substrate	Adsorbate	Bond length, Å	Reference	Predicted,[b] Å
Ni (001)	O	1.97	41	1.90
	S	2.18	41	2.28
	Se	2.27	41	2.41
	Na	3.37	42	3.10
Ni (110)	O	1.91	19	1.90
	S	2.17	14	2.28
Ni (111)	S	2.62	44	2.28
Ag (001)	Se	2.80	45	2.61
Ag (111)	I	2.75	46	2.77
Al (100)	Na	3.52	47	3.32
Mo (001)	N	2.02	48	2.08
W (110)	O	2.08	49	2.05

[a] From J. C. Buchholz and G. A. Somorjai, *Acc. Chem. Res.*, **9,** 333 (1976).

[b] See L. Pauling, *The Chemical Bond*, Cornell University Press, Ithaca, New York, 1966.

adsorbent, obeying the same energetics as if the process were one of formation of a diatomic molecule. Thus for the adsorption of alkali metal atoms on metal surfaces, the following steps may be written

$$\begin{aligned} &M(g) = M^+(g) + e^-(g) && Q_1 = -I \text{ (ionization potential)} \\ &e^-(g) = e^- \text{ (in metal)} && Q_2 = W \text{ (work function)} \qquad \text{(XVII-22)} \\ &M^+(g) = M^+ \text{ (adsorbed)} && Q_3 = e^2/4r_0 \end{aligned}$$

The quantity Q_3 corresponds to the potential energy of the ion at a distance r_0 from the surface, due to the electrostatic image force (i.e., as a property of the metal as a conductor, the ion is attracted as though by an equal and opposite charge positioned a distance r_0 below the surface). Calculated interaction energies for various alkali metals on tungsten agree moderately well with the experimental Q values (see Ref. 13).

Where covalent bonding is assumed, a possible procedure is to use the electronegativity system (see Ref. 50) for estimating the bond energy. For example, in the chemisorption (with dissociation) of hydrogen on tungsten,

$$2W + H_2 = 2W\text{—}H \qquad \text{(XVII-23)}$$

and

$$Q = 2E_{W\text{—}H} - E_{H\text{—}H} \qquad \text{(XVII-24)}$$

assuming that no W—W bonds need be broken. The H—H bond energy is known, and the value of $E_{W—H}$ can be obtained from the relationship

$$E_{W—H} = \tfrac{1}{2}(E_{W—W} - E_{H—H}) + 23(X_W - X_H)^2 \qquad \text{(XVII-25)}$$

where the X's are the respective electronegativities. The energy of adsorption is then

$$Q = E_{W—W} + 46(X_W - X_H)^2 \qquad \text{(XVII-26)}$$

The W—W bond energy should be about one-sixth of the sublimation energy (note Section III-1B), and there are various schemes for estimating electronegativities, of which Mulliken's (51) is perhaps the most fundamental. As an approximation, since $(X_W - X_H)$ relates to the ionic character of the bond, this difference can be estimated from the surface dipole moment as obtained from surface potential difference measurements (52, 53).

Alternatively, $E_{W—H}$ can be prorated between the value for a purely ionic bond (using Eqs. XVII-22 but with $Q_3 = e^2/r_0$) and that for a purely covalent bond [Eq. XVII-25 with $(X_W - X_H) = 0$], by the wave-mechanical approximation (54)

$$E_{W—H} = \frac{fE_i - (1 - f)E_c}{2f - 1} \qquad \text{(XVII-27)}$$

where the subscripts i and c are the two extreme estimates, and f is the fractional ionic character given by $f = \mu/er_0$, μ being the dipole moment from surface potential difference measurements. Some values given by Hayward and Trapnell (15b) using this procedure are summarized in Table XVII-2. Some additional experimental values are given by Somorjai (5).

Since one recipe for getting electronegativities involves combining selected bond energies (56), it might be guessed that the whole preceding approach could be replaced by taking the surface bond energy as simply that in a representative compound. Thus per mole of oxygen, the heat of formation of $WO_3(g)$ from the gaseous elements is 193 kcal, which compares well with the Q of 194 kcal/mole for the chemisorption of oxygen on tungsten; similarly, per CO, the heat of formation of $Ni(CO)_4(g)$ from Ni(g) and CO(g) is 35 kcal/mole, while the chemisorption Q is 42 kcal/mole (see Ref. 15b).

The same observation seems to apply to adsorption on carbon. Thus the chemisorption of hydrogen can be written as

$$2C + H_2 = 2CH \qquad \text{(XVII-28)}$$

If a normal C—H bond energy value is assumed (e.g., one-quarter of the total bond energy in methane), then Q can be calculated as

$$Q = 2E_{C—H} - E_{H—H} \qquad \text{(XVII-29)}$$

The approach is similar, but more complicated, in the case of the adsorption of more complex molecules. In all these instances, of course, the calculated Q depends on the assumption made as to the nature of the bonds formed.

If, for example, hydrogen adsorption on a particular surface (e.g., copper) actually is occurring on oxide-coated portions rather than on free metal, this fact must be recognized in any calculation of Q.

The Q values tabulated in Table XVII-2 indicate a moderate degree of success in the localized bond approach, and to this extent it does appear that the nature of the surface chemical bond is not very different from the ordinary chemical bond in simple compounds. Within this framework, however, there is room for sizable specific effects, as indicated by the more extreme discrepancies between the calculated and observed Q values. Some of these specific aspects are discussed further in the following section.

B. Metals

In general, the initial heats of adsorption on various metals do follow a common pattern, irrespective of the adsorbate (at least for the common ones such as hydrogen, nitrogen, ammonia, carbon monoxide, ethylene, and acetylene). The usual order of decreasing Q values is: Ta > W > Cr > Fe > Ni > Rh > Cu > Au; this is illustrated in Fig. XVII-9 (15b). It is evident, first, that transition metals are the most active elements in chemisorption and, second, that the activity decreases in proceeding from left to right in a given transition row. This pattern of behavior strongly suggests that the ability of a metal to use d orbitals in forming an adsorption bond is involved, and attempts have been made to find a correlation with the degree of d character in the metal–metal bonding, but with only partial success (15b, 57). A correlation with catalytic activity toward ethylene hydrogenation, is illustrated in Fig. XVII-10. Further evidence comes from magnetic susceptibility data. Selwood and co-workers (see Ref. 58) found that the saturation magnetization of nickel powder decreased linearly with the amount of chemisorbed hydrogen and estimated that 0.7 unpaired d electrons were used per

TABLE XVII-2
Calculated and Experimental Heats of Chemisorption (from Refs. 13 and 55)

System		Q, kcal/mole	
Adsorbate	Solid	Calculated	Observed
H_2	W	37	45
H_2	Fe	20	32
H_2	Ni	18	30
O_2	W	74	194
O_2	Pt	54	70
N_2	W	106	95
CO	Ni	35	42
C_2H_4	W	73	102
NH_3	C	32	17

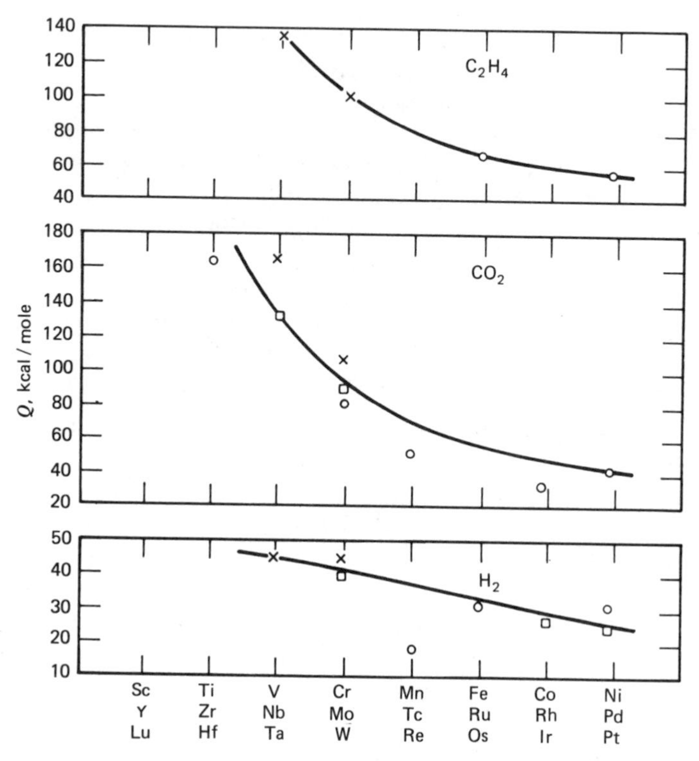

Fig. XVII-9. Heats of chemisorption for the transition metals, plotted according to periodic classification. ○, □, and × denote first, second, and third periods, respectively. (From Ref. 13.)

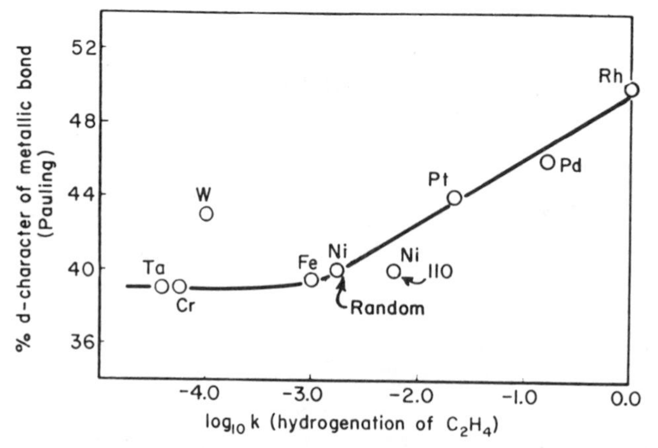

Fig. XVII-10. Correlation of catalytic activity toward ethylene dehydrogenation and percent d character of the metallic bond in the metal catalyst. (From Ref. 59.)

hydrogen atom adsorbed. Also, while hydrogen physically adsorbed at −196°C did not affect the magnetization, if measured at −196°C, a decrease occurred as the system was allowed to warm up and the hydrogen became chemisorbed (58, 59). Similar effects on the specific magnetization were found with other adsorbates such as benzene, cyclohexane, ethylene, and ethane.

Several of the adsorbates mentioned, such as carbon monoxide and ethylene, are capable of forming very stable coordination compounds with transition metals (see Ref. 60), and it may be that correlations with this type of chemistry should be sought rather than with the character of the metal–metal bonding in the metal. Low-lying, empty *d* orbitals are involved in such coordinative bonding, and their availability does vary in accord with the general trends in chemisorption Q values. The success of the calculation of Q for carbon monoxide on nickel from the heat of formation of $Ni(CO)_4$ gives support to this view. A complicating aspect is that *d* orbitals are not equivalent if the environment is not symmetrical, and they generally are split into two or more groups. Thus in octahedral complexes, the d_{z^2} and $d_{x^2-y^2}$ orbitals form one group, used in bonding, and the d_{xy}, d_{xz}, and d_{yz} orbitals form another, occupied by metal electrons. The coordinate bond energy is affected by this splitting, and it may be that the ligand field effect, as such removal of degeneracy is called, is important in chemisorption (61, 62) as well as in some cases of photodesorption (62). The development of metal cluster compounds, that is, of coordination compounds having three or more metals bonded to each other as a central unit, has stimulated some analogies to chemisorption of the same ligands. CNDO (complete neglect of differential overlap—a very useful wave-mechanical approach) calculations have been made, for example, on the interaction of carbon monoxide with clusters of Ni atoms on 100 and 111 faces (63, 64; see Ref. 65 for a review). Interestingly, it appears that the results are dependent on how many Ni atoms are included as well as on the configuration of the CO molecule. A quite different approach to chemisorption on metals is to consider that adsorbate interactions are due to indirect interactions by way of the conduction electrons (67).

C. Semiconductors

Some aspects of adsorption on oxides and other semiconductors can be treated in terms of the electrical properties of the solid, and these are reviewed briefly here. More details can be found in Refs. 15b and 6.

In many crystals there is sufficient overlap of atomic orbitals of adjacent atoms so that each group of a given quantum state can be treated as a crystal orbital or band. Such crystals will be electrically conducting if they have a partly filled band; but if the bands are all either full or empty, the conductivity will be small. Metal oxides constitute an example of this type of crystal; if exactly stoichiometric, all bands are either full or empty, and there is little electrical conductivity. If, however, some excess metal is present in an oxide, it will furnish electrons to an empty band formed of the 3*s* or 3*p* orbitals of the oxygen ions, thus giving electrical conductivity. An example is ZnO, which ordinarily has excess zinc in it.

If adsorption of oxygen on such an oxide involves the process

$$O_2 + 4e^- = 2O^{-2}(\text{ads}) \quad \text{(XII-30)}$$

adsorption will tend to be limited to the extent that excess zinc is present, that is, it will be small and moreover will reduce the conductivity by removing electrons from the conduction band; both predictions are confirmed experimentally. This type of adsorption has been called depletive. The situation is illustrated qualitatively in Fig. XVII-11 for the case where a surface electron acceptor state or adsorbate is present. Since the system remains electrically neutral, positive donor ions accumulate near the surface to complete an electrical double layer.

On the other hand, an oxide such as NiO is oxygen rich, in the sense that occasional Ni^{2+} ions are missing, electroneutrality being preserved by some of the nickel being in the plus three valence state. These Ni^{3+} ions take electrons from the otherwise filled conduction bands, thus again providing the condition needed for electrical conductivity. Oxygen adsorption according to Eq. XVII-30 can draw on the electrons in the slightly depleted band (or, alternatively, can produce unlimited additional Ni^{3+}) and so should be able to proceed to monolayer formation. Furthermore, since adsorption will make for more vacancies in a nearly filled band, electrical conductivity should rise. Again, the predictions are borne out experimentally (68).

In general, then, anion forming absorbates should find *p*-type semiconductors (such as NiO) more active than insulating materials, and these, in turn, more active than *n*-type semiconductors (such as ZnO). It is not necessary that the semiconductor type be determined by an excess or deficiency of a native ion; impurities, often deliberately added, can play the same role. Thus if Li^+ ions are present in NiO, in lattice positions, additional Ni^{3+}

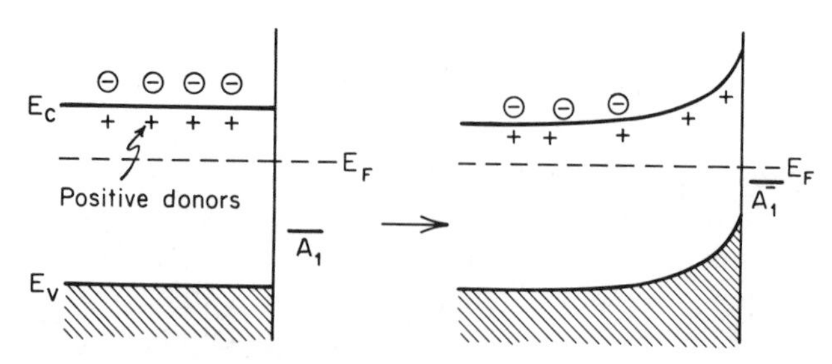

Fig. XVII-11. Band bending with a negative charge on the surface states. E_v, E_f, and E_c are the energies of the valance band, the Fermi level, and the conduction level, respectively. (From Ref. 66.)

ions must also be present to maintain electroneutrality; these now compete for electrons with oxygen and reduce the activity toward oxygen adsorption.

A quantitative treatment for the depletive adsorption of iogenic species on semiconductors is that known as the boundary layer theory (15b, 69), in which it is assumed that, as a result of adsorption, a charged layer of depth l is formed. It is further assumed that all defect-produced charged centers are neutralized within this depth, so that, if the defect density is n_0 (e.g., extra zinc atoms per cubic centimeter in ZnO), $n_0 l$ equals the adsorbate surface concentration, or

$$\theta = n_0 l \sigma^0 \tag{XVII-31}$$

where σ^0 is the site area, about 10^{-15} cm^2. The charge density ρ in this layer is taken to be constant, and thus simply en_0, so that integration of the Poisson equation (Eq. V-5) yields

$$V_s = \frac{2\pi\rho l^2}{D} = \frac{2\pi e n_0 l^2}{D} \tag{XVII-32}$$

Elimination of l between Eqs. XVII-31 and XVII-32 gives

$$\theta = \sigma^0 \left(\frac{DV_s n_0}{2\pi e}\right)^{1/2} \tag{XVII-33}$$

where V_s is the change in surface potential due to adsorption. Equation XVII-33 approximates to

$$\theta = 3 \times 10^{-2}(n_0 V_s)^{1/2} \tag{XVII-34}$$

with V_s now in practical volts, D assumed to be 10, and σ^0 being 10^{-15}. If defects amount to, say, 10^{18}/cc and V_s is taken to be 1 V, in accord with commonly observed values of the surface potential change on adsorption, θ would be about 0.003. Thus the amount of adsorption from this mechanism is quite small, although the effect on the electrical state of the surface is very large. Also, there can be photo effects; illumination of the surface can change the electron density in the boundary layer and lead to either increased or decreased adsorption, depending on the particular type of semiconductor.

Adsorption on semiconductor surfaces constitutes an important phenomenon since it can affect the performance of the semiconductor as an electrical component, usually adversely (see Ref. 70). Thus water, oxygen, and carbon dioxide adsorb on germanium and silicon surfaces to form semiconductor oxide, and such adsorption can produce a p-type layer on an n-type surface. Conversely, such effects can allow the detection of quite small amounts of adsorbed material.

Some complicated, yet understandable, effects occur if a semiconductor is coated with a thin film of a catalytic metal. For example, the dehydrogenation of formic acid over silver deposited on a semiconductor shows a decreasing activation energy with decreasing n-type conductivity of the support (71).

```
   |      ..              ..   |
 —Si  :O:  ←—Al—→  :O:  Si— (Lewis acid)
   |      ..    ↓     ..   |
               ..
              :O:
               ..
             —Si—
               |

              ..
          H  :O:  H⁺
   |      ..  ..                  ..   |
 —Si  :O:  ←——Al——→  :O:  Si— (Brönsted acid)
   |      ..    ↓             ..   |
               ..
              :O:
               ..
             —Si—
               |
```

Fig. XVII-12

D. Acid–Base Systems

Still another type of adsorption system is that in which either a proton transfer occurs between the adsorbent site and the adsorbate, or in which a Lewis acid–base type of reaction occurs. An important group of solids having acid sites is that of the various silica–aluminas, widely used as cracking catalysts. The sites center on surface aluminum ions, but could be either proton donor (Brønsted acid) or Lewis acid in type, as illustrated in Fig. XVII-12. The type of site can be distinguished by infrared spectroscopy, since an adsorbed base, such as ammonia or pyridine, should be either in the ammonium or pyridinium ion form or in coordinated form. The type of data obtainable is illustrated in Fig. XVII-13, which shows a portion of the infrared spectrum of pyridine adsorbed on a Mo(IV)-Al_2O_3 catalyst. In the presence of some surface water both Lewis and Brønsted types of adsorbed pyridine are seen, as marked in the figure. Thus the features 1450 and 1620

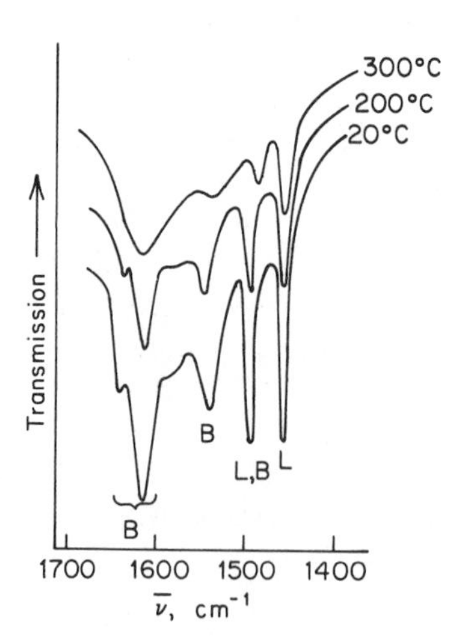

Fig. XVII-13. Spectra of pyridine adsorbed on a water containing molybdenum oxide (IV)-Al_2O_3 catalyst. L and B indicate features attributed to pyridine adsorbed on Lewis and Brønsted acid sites, respectively. (Reprinted with permission from Ref. 72. Copyright 1976 American Chemical Society.)

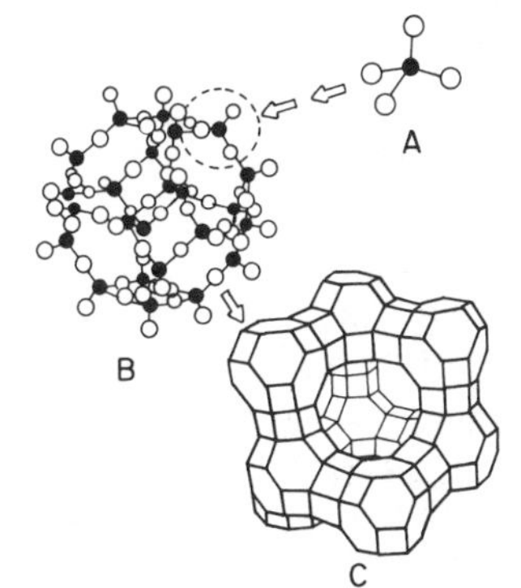

Fig. XVII-14. The framework structure of faujasite. (*a*) Tetrahedral arrangement of silican (or aluminum) atoms sharing oxygen atoms. (*b*) Sodalite unit consisting of 24 SiO_4^- and AlO_4^- tetrahedral. (*c*) Zeolite superstructure consisting of tetrahedral arrangement of sodalite units connected by oxygen bridges forming hexagonal prisms. (From Ref. 1.)

cm^{-1} are attributed to pyridine bound to Lewis acid sites, while those at 1540 and above 1600 cm^{-1} are attributed to pyH^+. The proportion of Brønsted sites increased with increasing surface water. Some further examples and discussion may be found in Ref. 73.

As another aspect, coordinate bond formation can be regarded as a reaction between a Lewis acid (the metal ion) and a Lewis base (the ligand), and, for example, the very strong surface acidity developed on partially dehydrating nickel sulfate has been attributed to the presence of a vacant nickel *d* orbital on the surface (74). This acidity may be titrated, using bases in nonaqueous solvents or by the adsorption of basic gases; such centers may be catalytically active (as in paraldehyde depolymerization in this particular case).

A new dimension to acid–base systems has been developed with the use of zeolites. As illustrated in Fig. XVII-14, the alumino–silicate faujasite has an open structure of interconnected cavities. By exchanging alkali metal for H^+ (or for NH_4^+ and then driving off ammonia) acid zeolites can be obtained whose acidity is comparable to that of sulfuric acid, and having excellent catalytic properties (see Section XVII-7). An important added feature is that the size of the channels and cavities, which can be controlled, gives selectivity in that only reactants or products below certain dimensions can get in or out. See Refs. 1 and 75 for additional discussion.

7. Mechanisms of Heterogeneous Catalysis

The sequence of events in a surface catalyzed reaction comprises: (1) diffusion of reactants to the surface (usually considered to be fast); (2) adsorption of the reactants on the surface (slow if activated); (3) surface diffusion of reactants to active sites (if the adsorption is mobile); (4) reaction of the adsorbed species (often rate-determining); (5) desorption of the reaction products (often slow); and (6) diffusion of the products away from the surface. Processes 1 and 6 may be rate-determining where one is dealing with a porous catalyst (64). For a more detailed discussion see Ref. 78.

A. *Adsorption or Desorption as the Rate-Determining Step*

Process 2, the adsorption of the reactant or reactants, is often quite rapid, as noted in Section XVI-16; on the other hand, process 4, the desorption of the products, must always be activated at least by Q, the heat of adsorption, and is much more apt to be slow. In fact, because of this expectation, certainly seemingly paradoxical situations have arisen. For example, the catalyzed exchange between hydrogen and deuterium on metal surfaces may be quite rapid at temperatures well below room temperature and under circumstances such that the rate of desorption of the product HD appeared to be slow. It was therefore suggested by Rideal (79) that an alternative mechanism was involved.

The earlier supposition, due to Bonhoeffer and Farkas (80) was, as illustrated in Fig. XVII-15*a*, that hydrogen and deuterium each chemisorbed as atoms, and that exchange took place through the random recombination of H and D, with desorption of the resulting HD. Rideal's suggestion was that the reaction took place between chemisorbed atoms and a colliding or physically adsorbed molecule, as illustrated in Fig. XVII-15*b*. Similar alternative mechanisms could be written for the catalysis of ortho–para hydrogen conversion by metal surfaces. Another example would be the Fischer–Tropsch synthesis, for which the net reaction is

$$n\mathrm{CO} + (2n + 1)\mathrm{H_2} = n\mathrm{H_2O} + \mathrm{C}_n\mathrm{H}_{2n+2} \qquad \text{(XVII-35)}$$

The Rideal mechanism for this reaction is illustrated in Fig. XVII-15*c*, whereby hydrogen reacts with chemisorbed CO, C, CH_2, and so on.

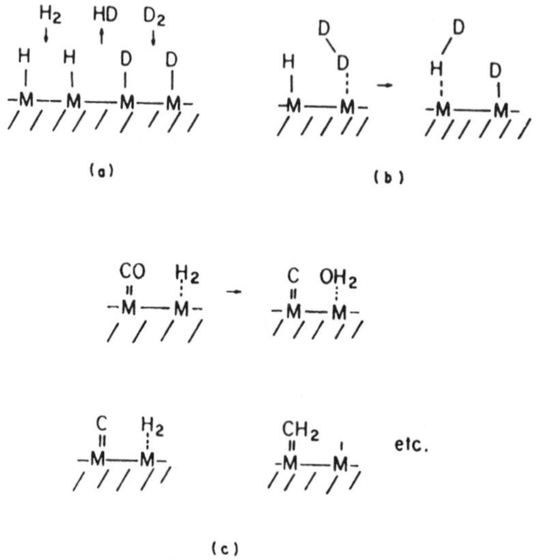

Fig. XVII-15. (*a* and *b*) Mechanisms for the catlyzed exchange of hydrogen with deuterium. (*c*) Rideal mechanism for the Fischer–Tropsch reaction.

Some detailed calculations have been made by Tully (81) on the trajectories for Rideal type processes. Thus the collision of an oxygen atom with a carbon atom bound to Pt results in a CO that departs with essentially all of the reaction energy as vibrational energy.

The decision between such seemingly clear-cut mechanistic alternatives turns out to be rather difficult (see Ref. 82). Often more than one kind of adsorption site is present; examples given in Section XVII-2 suggest strongly that physically or at least molecularly adsorbed species are often present, along with *several* strongly bound and possibly dissociated forms of the adsorbate. The current trend toward the study of catalytic reactions using well-characterized single-crystal type surfaces is establishing new and more definitive answers to questions of the type posed in Fig. XVII-15. The writing of M—H, M=C, M=CO, and so on, for the chemisorbed species is giving way to more accurate bonding representations as studies of metal cluster catalysts continue (note Refs. 78 and 83–86).

B. Reaction Within the Adsorbed Film as the Rate-Determining Step

Reactions within adsorbed films are either unimolecular or bimolecular or the result of some sequence of such steps, as is the case for reactions in bulk phases; termolecular processes can be ignored. The rate law for the surface reaction itself is generally written by analogy to the usual Mass Action law; that is, for a bimolecular process, the rate is taken to be proportional to the product of the two surface concentrations. The apparent rate law, that is, the rate in terms of gas pressures, depends on the form of the adsorption isotherm, as discussed later. Alternatively, surface reactions can be treated in terms of absolute rate theory (87).

It is of interest at this point to consider briefly why it is that a contact catalyst is able to serve as such, that is, why it is able to provide a reaction path that is faster than the homogeneous one. One reason, in the case of bimolecular reactions, is simply that the concentration of the reacting species may be much higher in the surface film than it is in the gas phase. A catalyst may thus be effective purely because of the concentration factor, and it is not necessary that the surface reaction itself be any different in character than the homogeneous one.

This type of explanation may apply to the case of the alumina catalyzed dehydration of ethanol to ethylene, for which an acid–base mechanism has been proposed (88):

$$\begin{array}{ccc} C_2H_5OH + H^+ \text{ (site)} & \rightleftharpoons & C_2H_5\overset{+}{O}H_2 \text{ (ads)} \\ & & \Big\Updownarrow {\scriptstyle -H_2O \text{ (ads)}} \\ H_2C{=}CH_2 & \underset{}{\overset{-H^+ \text{ (site)}}{\rightleftharpoons}} & \overset{+}{C}H_3CH_2 \text{ (ads)} \end{array} \qquad \text{(XVII-36)}$$

This is essentially the same mechanism as is written for the homogeneous reaction in solution.

In many cases, however, well-designed catalysts provide intrinsically different reaction paths, and the specific nature of the catalyst surface can be quite important. This is clearly the case with unimolecular reactions for which the surface concentrations effect is not applicable.

One specific factor that appears to be important in catalytic activity is the precise nature of the atomic spacings of the catalyst sites. This point was mentioned in Section XVII-4A in connection with the effect of the C—C distance in diamond, graphite, and so on on the activation energy for hydrogen adsorption. Beeck and co-workers (89) found that nickel films condensed on a glass plate presented mainly (110) planes and were much more active than unoriented films. As another example, the rate of the copper catalyzed reaction between hydrogen and oxygen was found to be twice as fast on (111) planes as on (110) planes.

Another and still important type of emphasis was given by Balandin (90). The essential feature was that the geometry of the catalyst surface should in some appropriate way match that of the adsorbate. For example, in the hydrogenation of benzene, in order for the molecule to lie flat, the catalyst surface should have hexagonal symmetry, as would be the case for (111) planes of face-centered cubic lattices and of hexagonal close-packed lattices. There are a number of examples where body-centered materials are much less active than face-centered ones, or where (111) planes are more active than (110) or (100) planes, with much of such evidence now coming from field and field ion microscopy and from LEED and related studies. For example, as illustrated in Fig. XVII-16, the surface of a Pt single crystal cut so as to have (111) *steps* was more catalytically active (toward dehydrocyclization of *n*-heptane) than were pure (111) surfaces (91). It thus appears that not only is the "primary" surface structure of a catalyst important, that is, the symmetry and spacing of surface atoms, but also the "secondary" surface structure—steps, edges, and other microtopological features.

The importance of the morphology of the catalyst surface was perhaps first suggested by Taylor (92) and remains a lively topic today. *Specific*

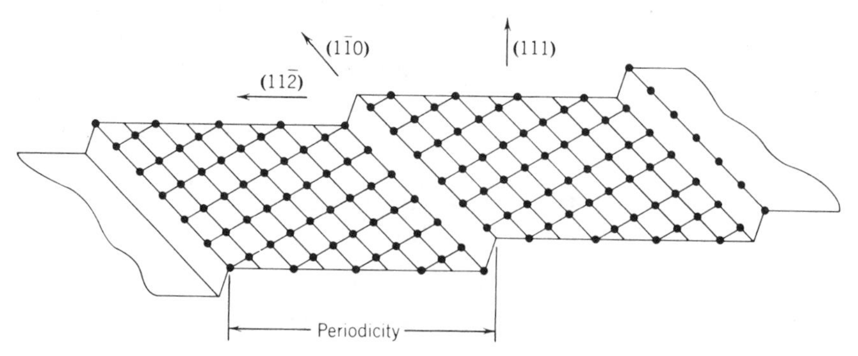

Fig. XVII-16. Single crystal platinum cut so as to have short (111) steps. (From Ref. 91.)

catalytic activity may depend on the particle size of the catalyst, and this aspect has been discussed by Boudart (93).

In summary, the surface geometry of a catalyst appears to affect the activation energies and strengths of adsorption and desorption, and thus to control the nature, the concentration, and the mobility of surface species. If these are radicals, reaction occurs through various recombination reactions and with a rate enhanced because of the surface concentration effect. Alternatively, the catalyst may act as an acid (or a base), promoting reactions through carbonium or oxonium ion intermediates. It should, in fact, be possible to develop physical adsorption catalysts that act through the concentrating and orienting effect of multilayer adsorption.

8. Influence of the Adsorption Isotherm on the Kinetics of Heterogeneous Catalysis

One of the reasons why the kinetics of heterogeneous catalysis is a more difficult field of study than that of the kinetics of homogeneous systems is that the experimental observable is the concentration of reactants and products in the gas phase rather than in the actual surface phase in which reaction occurs. The rate law in terms of surface concentrations might be called the true rate law, and the one analogous to that for a homogeneous system. What is observed, however, is an apparent rate law giving the dependence of the rate on the various gas pressures. The true and the apparent rate laws are related by the adsorption equilibrium, since this connects surface with gas phase concentrations. It is the nature of this relationship that is under discussion here.

The problem of treating surface encounters where the surface is heterogeneous or where complex lateral enteraction effects are present is virtually insurmountable, and as a consequence some very drastic simplifying assumptions have to be made. In extreme form, these amount to assuming that the film is sufficiently mobile that adsorbed molecules undergo many encounters before desorbing, so that mass action rate expressions can be written; and the simple Langmuir equation is obeyed. This is known as the *Langmuir–Hinshelwood* model, and some of its elementary applications are outlined in the following. See Ref. 94 for a more detailed discussion.

It will be recalled (Eq. XVI-11) that in the case of competitive adsorption the Langmuir equation takes the form

$$S_i = S\left(\frac{b_i P_i}{1 + \sum b_j P_j}\right) \qquad \text{(XVII-37)}$$

where

$$b_j = b_{0j} e^{Q_j/RT} \qquad \text{(XVII-38)}$$

If a surface reaction is bimolecular in species A and B, the assumption is that the rate is proportional to $S_A \times S_B$. We now proceed to apply this interpretation to a few special cases.

A. *Unimolecular Surface Reactions*

We suppose the type reaction to be

$$A \rightarrow C + D$$

and that the surface reaction proceeds according to the rate law

$$\frac{dn_A}{dt} = -kS_A \qquad \text{(XVII-39)}$$

or

$$\frac{dn_A}{dt} = -kS\left(\frac{b_A P_A}{1 + b_A P_A + b_C P_C + b_D P_D}\right) \qquad \text{(XVII-40)}$$

If the products C and D are weakly adsorbed, Eq. XVII-40) reduces to

$$\frac{dn_A}{dt} = kS\left(\frac{b_A P_A}{1 + b_A P_A}\right) = \frac{k' P_A}{1 + b_A P_A} \qquad \text{(XVII-41)}$$

which means that the apparent rate law should show a behavior similar to that of the Langmuir equation, that is, at low P_A the rate should be proportional to P_A but should reach a limiting rate kS at high P_A.

If one or more of the products is strongly adsorbed, Eq. XVII-40 takes on another limiting form, of the type

$$\frac{dn_A}{dt} = -kS\left(\frac{b_A P_A}{1 + b_C P_C}\right) = -\frac{k' P_A}{1 + b_C P_C} \qquad \text{(XVII-42)}$$

(where product C is more strongly adsorbed than A or D; in the limit of very strong adsorption of C, the right-hand side of Eq. XVII-41 becomes $-k' P_A / b_C P_C$).

Just as the surface and apparent kinetics are related through the adsorption isotherm, the surface or true activation energy and the apparent activation energy are related through the heat of adsorption. The apparent rate constant k' in these equations contains two temperature dependent quantities, the true rate constant k and the parameter b_A. Thus

$$k' = k b_A S_A = k b_{0A} S_A e^{Q_A} \qquad \text{(XVII-43)}$$

If the slight temperature dependencies of S_A and of b_{0A} are neglected, then

$$\frac{d \ln k'}{dt} = \frac{E_{app}}{RT^2} = \frac{E_{true} - Q_A}{RT^2} \qquad \text{(XVII-44)}$$

or

$$E_{app} = E_{true} - Q_A$$

The apparent activation energy is then less than the actual one for the surface reaction per se by the heat of adsorption. Most of the algebraic forms cited are complicated by having a composite denominator, itself temperature dependent, which must be allowed for in obtaining k' from the experimental data. However, Eq. XVII-44 would apply directly to the low pressure limiting form of Eq. XVII-41. Another limiting form of interest results if one product dominates the adsorption so that the rate law becomes

$$\frac{dn_A}{dt} = -\frac{k'P_A}{P_C} \tag{XVII-45}$$

It follows that

$$E_{app} = E_{true} - Q_A + Q_C \tag{XVII-46}$$

It should not be inferred from the foregoing that the heat of adsorption effect is the only one modifying the activation energy of a catalyzed reaction from that for the homogeneous one. The true or surface activation energy may itself be quite different from that for the homogeneous reaction. As an example, the true activation energy for the tungsten catalyzed decomposition of ammonia is only 39 kcal/mole, as compared to the value of about 90 kcal/mole for the gas phase reaction.

B. Bimolecular Surface Reactions

Continuing the formal development of the influence of the adsorption isotherm on the apparent reaction kinetics, we next consider the case of a reaction that is bimolecular on the surface,

$$A + B \rightarrow C + D$$

and whose surface reaction rate law is

$$\frac{dn_A}{dt} = -kS_A S_B \tag{XVII-47}$$

The general expression for the apparent rate law is now

$$\frac{dn_A}{dt} = -kS\left[\frac{b_A b_B P_A P_B}{(1 + b_A P_A + b_B P_B + \sum b_{prod} P_{prod})^2}\right] \tag{XV-47}$$

Only two of the many possible special cases need be considered. Thus if the products and reactants are weakly adsorbed,

$$\frac{dn_A}{dt} = -k'P_A P_B \tag{XVII-48}$$

A is weakly adsorbed as well as the products but B is strongly adsorbed, one finds

$$\frac{dn_A}{dt} = -\frac{k'P_A}{P_B} \tag{XVII-49}$$

so that retardation by a *reactant* is possible. The hydrogenation of pyridine on metal oxide catalysts shows retardation both by pyridine and by the reaction products, for example, (95).

Rate laws have also been observed that correspond to there being two kinds of surface, one adsorbing reactant A and the other, reactant B, and with the rate proportional to $S_A \times S_B$. For traditional discussions of Langmuir–Hinshelwood rate laws see Refs. 87, 96, and 97.

C. *Effect of Isotherm Complexities*

Equations XVII-17 and XVII-19 provide a good illustration of the type of complexity that can occur in both the entropy and the energy of adsorption; it is apparent that any attempt to combine such forms with mass action rate expressions would lead to equations difficult to use and even more difficult to verify. Also mentioned in Section XVII-4B, however, was the qualitative observation that where Q varies with θ, there often tends to be a compensating variation in the frequency factor of a catalyzed reaction. Thus on an empirical level, the effect of surface heterogeneity on the variation of k values with θ is often not as serious as might be expected.

The difficulty of establishing a mechanism by fitting mass action rate expressions to an experimental rate law under conditions such that surface coverage is varying widely has tended to limit the usefulness of Langmuir–Hinshelwood-type equations. It is better to work under conditions such that limiting forms can be expected to hold, and in complex catalytic systems emphasis is more on the identification of reaction intermediates by means of isotopic exchange studies, infrared spectroscopy, surface acidity studies, and so on, than on attempts to fit theoretical rate laws in detail.

9. Mechanisms of a Few Catalyzed Reactions

A short discussion of a few selected catalyzed reactions follows because of their importance and partly because of the great deal of study that has gone into them and the consequent appreciation of the complexities involved. The reactions chosen, the ammonia synthesis, Fischer–Tropsch type reactions, ethylene dehydrogenation, and the catalytic cracking of hydrocarbons, represent merely the writer's choice of a balanced group of interesting systems. The section concludes with a brief discussion of photoassisted surface reactions, a topic of some current interest.

A. *Ammonia Synthesis*

A discussion of commercial catalysts that are effective has been given by Brunauer and Emmett (98) and by Frankenburg (99). The general type is an iron–iron oxide catalyst with added aluminum and potassium oxides as promoters, the latter being helpful only in the presence of aluminum oxide. Poisons include carbon dioxide and monoxide, the former probably absorbing on the potassium oxide and the latter, on iron sites, and hydrogen and oxygen. More recently, much work has been done using single-crystal tungsten surfaces—the ease with which LEED, work function, flash desorption, and so on, studies can be made with single-crystal tungsten has provided much information about the detailed chemisorption chemistry of N_2 and H_2 separately, as well as of NH_3 and of its surface catalyzed decomposition. A paper by McAllister and Hansen (100) provides a good example.

The study of the mechanism of the ammonia synthesis is made complex by the wide variety of catalysts involved and the need to include an explanation of such effects as promotion; it is also made complicated by the sheer mass of data available. However, the course of the reaction appears to involve the following types of steps:

1. $N_2(g) = N_2(ads)$
2. $N_2(ads) = 2\ N(ads)$
3. $H_2(g) = H_2(ads)$
4. $H_2(ads) = 2\ H(ads)$
5. $N + H = NH$ (imide formation)
6. $NH + H = NH_2$ (amide formation)
7. $NH_2 + H = NH_3(ads)$
8. $NH_3(ads) = NH_3(g)$

There is a general consensus that the dominant requirement for an ammonia catalyst is that it be able to accomplish the difficult task of activating the N—N bond, but yet not form so tight a surface binding (e.g., as in actual nitride formation) that the nitrogen is unable to react further. Activation of the hydrogen is important, but secondary, as evidenced by the general inactivity of hydrogenation catalysts in ammonia synthesis. Reaction 2 may be complex since, although exchange with nitrogen-15 does occur (101), it is slow under conditions of rapid ammonia synthesis but is highly accelerated in the presence of some hydrogen. The sequence 3 and 4 is again much simplified. There appear to be at least two types of chemisorption of hydrogen on singly or doubly promoted iron catalysts, in addition to the low temperature physical adsorption; each type shows its own behavior with respect to H–D exchange or ortho–para hydrogen conversion.

The observed rate law depends on the type of catalyst involved; with promoted iron catalysts a rather complex dependence on nitrogen, hydrogen, and ammonia pressures is observed, and it has been difficult to obtain any definitive form from experimental data. A mechanism proposed by Temkin and Pyzhev (102) and developed further by Love and Emmett (99, 103) has been rather successful, however. The principal assumption is that the rate-determining step is the adsorption of nitrogen, so that the rate of ammonia formation is proportional to the nitrogen pressure and to the fraction of surface free of ammonia and hydrogen, assumed to be in equilibrium with the intermediates M—H, M—NH and M—NH_2, where M is the catalyst surface.

An alternative approach has been to study the kinetics of ammonia *decomposition*. McAllister and Hansen (100) found

$$-\frac{d(\mathrm{NH_3})}{dt} = a + bP_{\mathrm{NH_3}}^{2/3} \qquad \text{(XVII-51)}$$

the values of a and b being different for (111), (100), and (110) faces of their catalyst, single-crystal tungsten. They concluded that most of the surface was covered by the species W—N, the a term in Eq. XVII-51 being due to the slow step: $2W\text{—}N \rightarrow W_2N + \frac{1}{2}N_2$. The second term in the rate law could not be fit by the Temkin–Pyzhev mechanism, and instead one involving equilibrium between surface species $W_2N_3H_2$ and WNH and gaseous ammonia was proposed. Hansen and co-workers have also made theoretical calculations of the bonding of nitrogen and of hydrogen atoms to hypothetical W (100) surfaces; see Ref. 104.

B. Fischer–Tropsch Type Reactions

The Fischer–Tropsch synthesis was mentioned earlier (Eq. XVII-35) in connection with the Rideal mechanism and involves the production of hydrocarbons (plus water) from carbon monoxide and hydrogen using, mainly, cobalt catalysts. The process constitutes an important method for the production of gasolines, diesel fuels, and hydrocarbons in general. It was discovered (or rather disclosed) in 1913 and was extensively developed around 1925 by Fischer and Tropsch (see Ref. 105) and was of special importance in Germany because the necessary 2:1 ratio of H_2 to CO could be obtained readily from coal-produced water gas.

The catalysts in use include Co–ThO_2–MgO mixtures supported on kieselguhr, variations of these with the thorium dioxide replaced by manganese and other metals, or with nickel, copper, or zinc in place of the cobalt, so that quite a variety is involved. With the cobalt catalysts, typical operating conditions involve 1 to 10 atm pressure and around 200°C; under these conditions, some 50% of the liquid product is in the C_4 to C_{10} range of hydrocarbons.

Fischer and Tropsch proposed a mechanism similar to that shown in Fig. XVII-15*c*, except that all the species were thought to be chemisorbed. The essential feature was that metal carbide was supposed to be the intermediate. Although patents have been issued on the basis of this carbide mechanism, there has been a continuing thread of skepticism concerning it. Thus Weller et al. (106) found that precarbiding of the cobalt catalyst *inhibited* the synthesis; used samples of catalyst showed no carbide by x-ray analysis, although the precarbided ones did. Krummer and co-workers (107) found that ^{14}C-containing carbon deposited on catalysts by pretreatment with carbon monoxide did not appear in the hydrocarbon products of the synthesis.

A more generally accepted mechanism at present is that suggested by Elvins and Nash (108) and elaborated by Storch and co-workers (109), which does not involve carbide intermediates. As shown in Fig. XVII-17, oxygen-

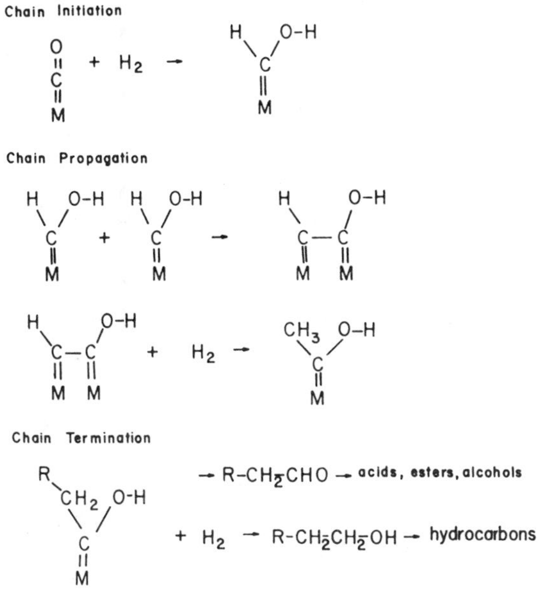

Chain Initiation

$$\underset{\underset{M}{\|}}{\overset{O}{\underset{\|}{\overset{\|}{C}}}} + H_2 \rightarrow \underset{\underset{M}{\|}}{H\diagdown C \diagup O\text{-}H}$$

Chain Propagation

$$\underset{\underset{M}{\|}}{H\diagdown C \diagup O\text{-}H} + \underset{\underset{M}{\|}}{H\diagdown C \diagup O\text{-}H} \rightarrow \underset{\underset{M\quad M}{\|\quad \|}}{H\diagdown C-C \diagup O\text{-}H}$$

$$\underset{\underset{M\quad M}{\|\quad \|}}{H\diagdown C-C \diagup O\text{-}H} + H_2 \rightarrow \underset{\underset{M}{\|}}{CH_3\diagdown C \diagup O\text{-}H}$$

Chain Termination

$$\underset{\underset{M}{\|}}{R\text{-}CH_2\diagdown C \diagup O\text{-}H} \rightarrow R\text{-}CH_2\text{-}CHO \rightarrow \text{acids, esters, alcohols}$$

$$+ H_2 \rightarrow R\text{-}CH_2\text{-}CH_2\text{-}OH \rightarrow \text{hydrocarbons}$$

Fig. XVII-17. Mechanism of the Fischer–Tropsch reaction.

ated or alcoholic type intermediates are proposed instead. Kummer and co-workers (110) have provided some direct evidence for the correctness of this type of mechanism by showing that ^{14}C-labeled alcohols, introduced during the course of a Fischer–Tropsch reaction led to the production of labeled hydrocarbons.

As also suggested by the sequence in Fig. XVII-17, this general catalytic system is important in alcohol production (see Ref. 97), especially methanol. The effective catalyst is now zinc oxide, promoted by admixture with other oxides, of a high melting and not easily reduced type such as chromic, ferrous, and magnesium. The kinetics have been studied especially by Natta and co-workers (see Ref. 111), with resulting rate expressions of complexity comparable to those for the ammonia synthesis.

Still another variation is that known as the oxosynthesis, or the hydro-formulation of a double bond:

$$R\text{—}CH\text{=}CH_2 + CO + H_2 \rightarrow R\text{—}CH_2\text{—}CH_2\text{—}CHO \quad \text{or } R\text{—}CH(CHO)\text{—}CH_3 \qquad \text{(XVII-52)}$$

Catalysts similar to those in the Fischer–Tropsch synthesis are used.

C. *Hydrogenation of Ethylene*

The catalytic hydrogenation of ethylene occurs on various metal catalysts, such as nickel, including active or skeletal forms produced by dissolving out silicon or aluminum from $NiSi_2$ or NiAl, $NiAl_2$ alloys, reduced copper, platinum, rhodium, iron, and chromium. A review of the literature has been given by Eley (112; see also Ref. 97) and, as with the preceding systems, only a brief discussion is attempted here.

Depending on the catalyst and the relative hydrogen and ethylene pressures, the kinetics may correspond to Eq. XVII-50 with reactant B, ethylene, strongly adsorbed, so that the rate law Eq. XVII-50 results. Alternatively, the rate law may be

$$\frac{dP_{H_2}}{dt} = -kP_{H_2} \tag{XVII-53}$$

Prior to about 1955 most of the studies of ethylene chemisorption and hydrogenation were carried out by gas analysis and gas volumetric adsorption techniques; since then there has been increasing use of a variety of physical methods including magnetic susceptibility, infrared absorption, and others.

Early studies, such as that of Jenkins and Rideal (113), made it clear that carbon–hydrogen bonds were broken in the chemisorption of ethylene on nickel. Work such as that of McKee (114) established that breaking of the C–C bond could also occur; methane production was observed on desorption of chemisorbed ethylene on nickel. Infrared spectra of chemisorbed ethylene on nickel suggested the presence of the species

```
H       H
 \     /
  C==C
  |   |
  M   M
```

and, on admission of hydrogen, of chemisorbed radicals such as $CH_3CH_2\cdot$ (115).

Ethylene hydrogenation has, of course, been studied with other catalysts than Ni. In the case of evaporated Ir films, the rate law was found to be $R = kP_{C_2H_4}P_{H_2}^{1/2}$, the mechanism being essentially that described above (116). Also, ZnO and oxides of La, Ce, and Th prepared *in situ* are hydrogenation catalysts (117). A summary of other hydrogenation reactions may be found in Ref. 78.

The hydrogenolysis, that is, conversion of ethane to methane over nickel and other catalysis is now well known (see Ref. 118). The rate obeys the expression

$$\text{rate} = kP_E^n P_H^m \tag{XVII-54}$$

where P_E and P_H are the ethane and hydrogen partial pressures (118, 119); the exponents n and m are about 1 and -1.3, respectively. Interestingly, catalytic activity plummeted as copper was alloyed with the nickel (note Section XVII-3 and Ref. 21) although there was very little effect on the activity for a dehydrogenation reaction, that of conversion of cyclohexane to benzene. There is clearly a complete surface *chemistry* in hydrogenation–dehydrogenation systems.

D. Catalytic Cracking of Hydrocarbons and Related Reactions

A number of related reactions of hydrocarbons are catalyzed by acidic oxide type of materials. These include the *cracking* of high molecular weight hydrocarbons to lower molecular weight ones to produce gasolines from oils; *reforming*, which involves isomerizations and molecular weight redistributions through hydrogenation-dehydrogenation steps; and *alkylation*, which may amount to the reverse of cracking. The Houdry process, an early cracking procedure, was announced in 1933 (120) and made use of activated bentonite clay as the catalyst. Current catalysts also include synthetic aluminosilicates prepared by precipitation of alumina in the presence of silica gel, followed by filtration, washing, drying, and calcining. The activity is quite definitely associated with the presence of acid sites (121, 122), although in the discussion relating to Fig. XVII-13 it was pointed out that there is still some question as to whether these are Brønsted or Lewis in type.

As mentioned in Section XVII-6D, zeolites, including those in the acid form, have become important industrial catalysts for cracking and reforming. The special feature here is the added ability to select reactants and products on the basis of size and shape (note Fig. XVII-14).

The general features of the cracking mechanism involve carbonium ion formation by a reaction of the type

$$RH + H^+(\text{acid site}) \rightarrow R^+(\text{ads}) + H_2 \qquad \text{(XVII-55)}$$

where the acid site might, for example, be in the Brønsted form. The resulting carbonium ions then undergo various rearrangement and clevage reactions such as

$$CH_3{-}CH_2{-}\overset{+}{C}H_2 \rightarrow \underset{\underset{H^+}{|}}{CH_3}{-}CH{=}CH_2 \rightarrow CH_3{-}\overset{+}{C}H{-}CH_3$$

$$R\overset{+}{C}H{-}CH_2{-}CH_3 \rightarrow R{-}CH{=}CH_2 + \overset{+}{C}H_3 \qquad \text{(XVII-56)}$$

$$R{-}CH{=}CH_2 + \overset{+}{C}H_3 \rightarrow R{-}CH(CH_3){-}\overset{+}{C}H_2$$

and terminating by the reverse of reaction XVII-55.

Because of the great industrial importance of these processes, a great deal of performance information is now available, and rather precise control of the degree of unsaturation, isomerization, and aromatic character of the product can be achieved; far more so than in the early, noncatalytic thermal cracking procedures. However, as can be imagined, the detailed kinetics are quite complicated, and much remains to be done in the way of fundamental research.

E. Photochemical and Photoassisted Processes at Surfaces

Some mention should be made of a topic of current interest, that of photoassisted and photochemical processes at surfaces. By photoassisted is meant a process that can occur in the dark but occurs with less energy consumption if the surface is irradiated. An example is the electrolytic decomposition of water using TiO_2 as one electrode; the requisite voltage is much reduced on illumination (see Ref. 123). The photodecomposition can be induced without applied voltage if short enough wavelength light is used, and works even better if platinized Pt has been deposited on the TiO_2 (124; see also 125).

Photohydrogenation of acetylene and ethylene occurs on irradiation of TiO_2 exposed to the gases, but only if TiOH surface groups are present as a source of hydrogen (126). Platinized TiO_2 powder shows, in the presence of water, photochemical oxidation of hydrocarbons (127). Some of the postulated reactions are:

$$
\begin{aligned}
&(TiO_2) + h\nu \rightarrow e^- + \text{hole}^+ \\
&H_2O + \text{hole}^+ \rightarrow \cdot OH + H^+ \\
&H^+ + e^- \rightarrow H\cdot \\
&O_2 + e^- \rightarrow \cdot O_2^- \xrightarrow{H\cdot} \cdot HO_2^- \\
&HO_2^- + \text{hole}^+ \rightarrow \cdot HO_2 \\
&2\,\cdot HO_2^- \rightarrow O_2 + H_2O_2 \xrightarrow{\cdot O_2^-} \cdot OH + OH^- + O_2 \\
&RH + \cdot OH\ (\text{or}\ \cdot HO_2) \rightarrow ROH + \cdot H \\
&RH + \text{hole}^+ \rightarrow RH^+ \xrightarrow{\text{hole}^+} RH^{2+}
\end{aligned}
\qquad \text{(XVII-57)}
$$

Again with platinized TiO_2, ultraviolet irradiation can lead to oxidation of aqueous CN^- (128) and to the water–gas shift reaction, $CO + H_2O = H_2 + CO_2$ (129).

Other surfaces have been used. Irradiation with visible light of rhodamine B adsorbed on CdS can lead to *N*-dealkylation reactions (130). Neither is it necessary that semiconductor surfaces be used. Alkyl ketones adsorbed on Vycor glass show a different stereochemistry of photoproducts from that observed in bulk solution (131). An interesting use of the adsorbed state to confine nascent radical pairs has recently been described (132).

While hopes are high, photochemical systems of the type described have not yet found practical application. The photovoltaic cell or "solar panel" is at present the only system with important (although limited) commercial use (see Ref. 133).

10. Problems

1. Give four specific experimental tests, measurements, or criteria that would be considered good evidence for characterizing adsorption in a given system as either physical adsorption or chemisorption.

2. The Langmuir as a unit is defined in the legend to Fig. XVII-1. Considering the specific case of CO at 200°C how many Langmuir units (L) exposure would be required to give 10% surface coverage on, say, W, if every molecule hitting the surface was chemisorbed? Take the molecular area of CO to be 16 Å^2.

3. In a flash desorption experiment for CO from Ni (110), using a constant rate of heating, b/T_p^2 was 1×10^{-4} and 5×10^{-6} (arbitrary units) at T_p values of 450 K and 420 K, respectively. Calculate the activation energy for the desorption process.

4. Discuss what conclusions can be drawn from the spectra shown in Fig. XVII-3.

5. It was pointed out that a bimolecular reaction can be accelerated by a catalyst just from a concentration effect. As an illustrative calculation, assume that A and B react in the gas phase with 1:1 stoichiometry and according to a bimolecular rate law, with the second-order rate constant k equal to 10^{-5} l-mole/sec at 0°C. If now an equimolar mixture of the gases is condensed to a liquid film on a catalyst surface and the rate constant in the condensed liquid solution is taken to be the same as for the gas phase reaction, calculate the ratio of times for half reaction in the gas phase and on the catalyst surface at 0°C. Assume further that the density of the liquid phase is 1000 times that of the gas phase.

6. Calculate the surface self-diffusion coefficient for Ni atoms on a Ni (111) surface at 600 K. Calculate also how far, on the average, a given surface atom will diffuse at this temperature in 1 hr.

7. Comment, with reference to Fig. XVII-6, why D_2 should adsorb more strongly than H_2.

8. What is the physical meaning of P_0 in Eq. XVII-13?

9. Calculate the entropy of adsorption $\Delta\bar{S}_2$ for several values of θ for the case of nitrogen on an iron catalyst. Use the data of Scholten and co-workers given in Section XVII-4B.

10. The quantity Θ_Q in Eq. XVII-6 corresponds to $F(Q)$ of Eq. XVI-145. Derive $\Theta(P, T)$ using these relationships and show how the result can be reduced to Eq. XVII-7; also compare your result with Eq. XVI-146, and discuss the reasons for the differences.

11. Assuming that I_2 adsorbs on copper as atoms, estimate the heat of adsorption in two independent ways.

12. A rate of chemisorption obeys the rate law

$$\frac{d\theta}{dt} = Ae^{-b\theta}$$

Show whether a plot of θ versus $\log t$ should be linear (a form of the Elovich equation).

13. Calculate the depth of the boundary layer in the case of adsorption on a semiconductor of $n_0 = 10^{18}$ defects per cubic centimeter, $D = 10$, $V_s = 1$ (practical volt), and $\sigma^0 = 10^{-15}$ cm^2.

14. Discuss whether the iogenic chemisorption of hydrogen on (a) ZnO and (b) NiO should be limited to a very small θ value.

15. The Pt catalyzed decomposition of NO (into N_2 and O_2) is found to obey the experimental rate law

$$\frac{dP_{\mathrm{NO}}}{dt} = -\frac{kP_{\mathrm{NO}}}{P_{\mathrm{O}_2}}$$

Assuming adsorbed gases obey the Langmuir equation, derive this rate law starting with some reasonable assumed mechanism for the surface reaction.

If the heat of adsorption of NO is 20 kcal/mole and that of O_2 is 25 kcal/mole, show what the actual activation energy for the surface reaction should be, given that the apparent activation energy (i.e., from the temperature dependence of k above) is 15 kcal/mole.

16. Some early observations on the catalytic oxidation of SO_2 to SO_3 on platinized asbestos catalysts led to the following observations: (1) the rate was proportional to the SO_2 pressure and was inversely proportional to the SO_3 pressure; (2) the apparent activation energy was 30 kcal/mole; (3) the heats of adsorption for SO_2, SO_3 and O_2 were 20, 25, and 30 kcal/mole, respectively.

By using appropriate Langmuir equations, show that a possible explanation of the rate data is that there are two kinds of surface present, S_1 and S_2, and that the rate determining step is

$$\mathrm{SO_2(ads\ on\ }S_1) + \mathrm{O_2(ads\ on\ }S_2) \rightarrow \text{intermediates}$$

On this basis, what would you expect the dependence of rate on pressure to be (*a*) at low oxygen pressure and (*b*) during the initial stages of reaction when negligible SO_3 is present? Finally, calculate the true activation energy, assuming the preceding rate determining step.

17. According to Schwab, the kinetics of the decomposition of ammonia on Pt are:

(*a*) At low N_2 pressure: $$\frac{dP_{\mathrm{NH}_3}}{dt} = -\frac{kP_{\mathrm{NH}_3}}{P_{\mathrm{H}_2}}$$

(*b*) At low H_2 pressure: $$\frac{dP_{\mathrm{NH}_3}}{dt} = -\frac{k'P_{\mathrm{NH}_3}}{P_{\mathrm{N}_2}}$$

where P denotes partial pressure. Write a simple Langmuir–Hinshelwood mechanism that gives these limiting rate laws.

18. Write a possible reaction sequence for the photochemical oxidation of aqueous CN^- ion on TiO_2.

General References

P. G. Ashmore, *Catalysis and Inhibition of Chemical Reactions,* Butterworths, London, 1963.

M. Boudart, *Adv. Catal.,* **20,** 153 (1969).

E. Draublis, R. D. Gretz, and R. I. Jaffee, Eds., *Molecular Processes on Solid Surfaces,* McGraw-Hill, New York, 1968.

P. H. Emmett, Ed., *Catalysis,* Reinhold, New York, 1956.

M. Green, Ed., *Solid State Surface Science,* Marcel Dekker, New York, 1969.

N. B. Hannay, Ed., *Treatise on Solid State Chemistry,* Vol. 6A, *Surfaces I* and Vol. 6B, *Surfaces II,* Plenum, New York, 1976.

D. O. Hayward and B. M. W, Trapnell, *Chemisorption,* Butterworths, London, 1964.

H. Saltzburg, J. N. Smith, Jr., and M. Rogers, Eds. *Fundamentals of Gas-Surface Interactions,* Academic, New York, 1967.

G. A. Somorjai, *Principles of Surface Chemistry,* Prentice-Hall, Englewood Cliffs, New Jersey, 1972.

G. A. Somorjai, Ed., *The Structure and Chemistry of Solid Surfaces,* Wiley, New York, 1969.

H. H. Storch, N. Golumbic, and R. B. Anderson, *The Fischer-Tropsch and Related Systems,* Wiley, New York, 1951.

C. L. Thomas, *Catalytic Processes and Proven Catalysts,* Academic, New York, 1970.

Textual References

1. A. W. Sleight, *Science,* **208,** 895 (1980).
2. R. J. Madix, *Acc. Chem. Res.,* **12,** 265 (1979).
3. J. C. Buchholz and G. A. Somorjai, *Acc. Chem. Res.,* **9,** 333 (1976).
4. R. Gomer, *Acc. Chem. Res.,* **8,** 420 (1975).
5. G. A. Somorjai, *Principles of Surface Chemistry,* Prentice-Hall, Englewood Cliffs, New Jersey, 1972; G. A. Somorjai and L. L. Kesmodel, *MTP International Review of Science,* Butterworths, London, 1975.
6. J. C. Tracy and P. W. Palmberg, *J. Chem. Phys.,* **51,** 4852 (1969).
7. Ch. Steinbrüchel and R. Gomer, *Surf. Sci.,* **67,** 21 (4977).
8. R. P. Eischens and J. Jacknow, *Proc. 3rd Int. Congr. Catal.,* Amsterdam 1964, North-Holland Publishing Co., Amsterdam, 1965.
9. M. L. Unland, *Science,* **179,** 567 (1973).
10. M. A. Chesters, J. Pritchard, and M. L. Sims, *Chem. Commun.,* 1454 (1970); J. Pritchard and M. L. Sims, *Trans. Faraday Soc.,* **66,** 427 (1970).
11. J. T. Yates, Jr., T. E. Madey, and N. E. Erickson, *Surf. Sci.,* **43,** 257 (1974).
12. B. A. Sexton, *Surf. Sci.,* **88,** 299 (1979).
13. S. Abdo, R. F. Howe, and W. K. Hall, *J. Phys. Chem.,* **82,** 969 (1978).
14. T. Matsuzaki, T. Uda, A. Kazusaka, G. W. Keulks, and R. F. Howe, *J. Am. Chem. Soc.,* **102,** 7511 (1980).

15a. R. L. Wells and T. Fort, Jr., *Surf. Sci.,* **33,** 172 (1972).

15b. D. O. Hayward and B. M. W. Trapnell, *Chemisorption,* Butterworths, London, 1964.

16a. G. Eirich, *J. Phys. Chem.,* **60,** 1388 (1956); *J. Appl. Phys.,* **32,** 4 (1961).

16b. Y. Viswanath and L. D. Schmidt, *J. Chem. Phys.,* **59,** 4184 (1973).

17. G. Eirich, *Proc. 3rd Int. Congr. Catal.,* North-Holland, Amsterdam, 1965, p. 113.
18. T. E. Madey, *Surf. Sci.,* **29,** 571 (1972).
19. P. W. Tamm and L. D. Schmidt, *J. Chem. Phys.,* **54,** 4775 (1971).
20. W. S. Millman, F. H. Van Cauwelaert, and W. K. Hall, *J. Phys. Chem.,* **83,** 2764 (1979).
21. See B. M. W. Trapnell, *Chemisorption,* Academic, New York, 1955, p. 124.
22. D. A. Cadenhead and N. J. Wagner, *J. Catal.,* **27,** 475 (1972).
23. R. Fowler and E. A. Guggenheim, *Statistical Thermodynamics,* Cambridge University Press, Cambridge, England, 1952, p. 437.

24. G. D. Halsey and A. T. Yeates, *J. Phys. Chem.*, **83,** 3236 (1979).
25. A. Sherman and H. Eyring, *J. Am. Chem. Soc.*, **54,** 2661 (1932).
26. R. M. Barrer, *J. Chem. Soc.*, **1936,** 1256, and succeeding papers.
27. R. Dovesi, C. Pisani, F. Ricca, and C. Roetti, *Chem. Phys. Lett.*, **44,** 104 (1976).
28. W. H. Weinberg and R. P. Merrill, *Surf. Sci.*, **33,** 493 (1972).
29. T. L. Einstein, and J. R. Schrieffer, *Phys. Rev. B*, **7,** 3629 (1973); *idem, J. Vac. Sci. Technol.*, **9,** 956 (1972).
30. K. J. Vette, T. W. Orent, D. K. Hoffman, and R. S. Hansen, *J. Chem. Phys.*, **60,** 4854 (1974).
31. S. Glasstone, K. J. Laidler, and H. Eyring, *The Theory of Rate Processes*, McGraw-Hill, New York, 1941.
32. J. J. F. Scholten, P. Zwietering, J. A. Konvalinka, and J. H. de Boer, *Trans. Faraday Soc.*, **55,** 2166 (1959).
33. S. Yu. Elovich and G. M. Zhabrova, *Zh. Fiz. Khim.*, **13,** 1761 (1939).
34. M. B. Liu and P. G. Wahlbeck, *J. Phys. Chem.*, **80,** 1484 (1976).
35. J. Eisinger and J. T. Law, *J. Chem. Phys.*, **30,** 410 (1959).
36. J. R. Chen and R. Gomer, *Surf. Sci.*, **94,** 456 (1980).
37. R. DiFoggio and R. Gomer, *Phys. Rev. Lett.*, **44,** 1258 (1980).
38. R. C. L. Bosworth, *Proc. Roy. Soc. (London)*, **A154,** 112 (1936).
39. V. Slawson, J. Mead, and A. W. Adamson, *J. Phys. Chem.*, **85,** 116 (1981).
40. J. R. Wolfe and H. W. Weart, *The Structure and Chemistry of Solid Surfaces*, G. A. Somorjai, Ed., Wiley, New York, 1969.
41. J. E. Demuth, D. W. Jepsen, and P. M. Marcus, *Phys. Rev. Lett.*, **32,** 1182 (1973).
42. S. Anderson and J. B. Pendry; *J. Phys. Chem.*, **5,** L41 (1972).
43. J. E. Demuth, D. W. Jepsen, and P. M. Marcus, presented at the Physical Electronics Conference, State College, Pennsylvania, 1975.
44. J. E. Demuth, D. W. Jepsen, and P. M. Marcus, *Phys. Rev. Lett.*, **32,** 1182 (1974).
45. A. Ignatiev, F. Jona, D. W. Jepsen, and P. M. Marcus, *Surf. Sci.*, **40,** 439 (1973).
46. F. Forstmann, W. Berndt, and P. Buttner, *Phys. Rev. Lett.*, **30,** 17 (1973).
47. B. M. Hutchins, T. N. Rhodin, and J. E. Demuth, presented at the American Physical Society Meeting, Denver, Colorado, 1975.
48. A. Ignatiev, F. Jona, D. W. Jepsen, and P. M. Marcus, *Surf. Sci.*, **49,** 189 (1975).
49. M. Van Howe and S. Y. Tong, presented at the Physical Electronics Conference, State College, Pennsylvania, 1975; *Phys. Rev. Lett.*, **35,** 1092 (1975).
50. See M. C. Day, Jr., and J. Selbin, *Theoretical Inorganic Chemistry*, Reinhold, New York, 1962, p. 112.
51. R. S. Mulliken, *J. Chem. Phys.*, **2,** 782 (1934); **3,** 573 (1935).
52. D. D. Eley, *Discuss. Faraday Soc.*, **8,** 34 (1950).
53. R. V. Culver and F. C. Tompkins, *Adv. Catal.*, **11,** 67 (1959).
54. I. Higuchi, T. Ree, and H. Eyring, *J. Am. Chem. Soc.*, **79,** 1330 (1957).
55. See B. M. W. Trapnell, *Chemisorption*, Academic, New York, 1955.
56. L. Pauling, *The Nature of the Chemical Bond*, Cornell University Press, Ithaca, 1960.
57. W. A. Hickmont and G. Ehrlich, *J. Phys. Chem. Solids*, **5,** 47 (1958).

58. R. E. Dietz and P. W. Selwood, *J. Chem. Phys.,* **35,** 270 (1961).
59. P. W. Selwood, *J. Am. Chem. Soc.,* **78,** 3893 (1956).
60. F. A. Cotton and G. Wilkinson, *Advanced Inorganic Chemistry,* 2nd ed., Interscience, New York, 1966.
61. W. J. Dunning, *The Gas–Solid Interface,* Vol. 1, E. A. Flood, Ed., Marcel Dekker, New York, 1966.
62. F. S. Stone, *An. R. Soc. Esp. Fis. Quim.,* **61B,** 109 (1965).
63. G. Blyholder, *J. Phys. Chem.,* **79,** 756 (1975).
64. P. Politzer and S. D. Kasten, *J. Phys. Chem.,* **80,** 385 (1976).
65. T. N. Rhodin and D. L. Adams, *Treatise on Solid State Chemistry,* Vol. 6A, *Surfaces I,* N. B. Hannay, Ed., Plenum, New York, 1976.
66. S. R. Morrison, *Treatise on Solid State Chemistry,* Vol. 6B, *Surfaces II,* N. B. Hannay, Ed., Plenum, New York, 1976.
67. T. B. Grimley, *J. Am. Chem. Soc.,* **90,** 3016 (1968).
68. W. E. Garner, F. S. Stone, and P. F. Tiley, *Proc. Roy. Soc.* (*London*), **A211,** 472 (1952).
69. F. S. Stone, *Adv. Catal.,* **13,** 1 (1962).
70. *The Electrochemistry of Semiconductors,* P. J. Holmes, Ed., Academic, New York, 1962.
71. G. Schwab, *J. Colloid Interface Sci.,* **34,** 237 (1970).
72. T. Fransen, O. van der Meer, and P. Mars, *J. Phys. Chem.,* **80,** 2103 (1976).
73. W. K. Hall, *Acc. Chem. Res.,* **8,** 257 (1975).
74. T. Takeshita, R. Ohnishi, T. Matsui, and K. Tanabe, *J. Phys. Chem.,* **69,** 4077 (1965).
75. L. D. Rollman, *J. Catal.,* **47,** 113 (1977).
76. H. W. Haynes, Jr., *Catal. Rev.,* **17,** 276 (1978).
77. P. B. Weisz and C. D. Prater, *Adv. Catal.,* **6,** 143 (1943).
78. T. E. Mady, J. T. Yates, Jr., D. R. Sandstrom, and R. J. H. Voorhoeve, *Treatise on Solid State Chemistry,* Vol. 6B, *Surfaces II,* N. B. Hannay, Ed., Plenum, New York, 1976.
79. E. K. Rideal, *Proc. Cambridge Phil. Soc.,* **35,** 130 (1938).
80. K. F. Bonhoeffer and A. Farkas, *Z. Phys. Chem.,* **B12,** 231 (1931).
81. J. C. Tully, *Acc. Chem. Res.,* **14,** 188 (1981).
82. D. D. Eley and P. R. Norton, *Discuss. Faraday Soc.,* No. 41, 135 (1966).
83. J. C. Hemminger, E. I. Muertterties, and G. A. Somerjai, *J. Am. Chem. Soc.,* **101,** 62 (1979).
84. R. Dovesi, C. Pisani, F. Ricca, and C. Roetti, *J. Chem. Phys.,* **65,** 4116 (1976).
85. L. W. Anders and R. S. Hansen, *J. Chem. Phys.,* **59,** 5277 (1973).
86. G. C. Smith, T. P. Chojnacki, S. R. Dasgupta, K. Iwatate, and K. L. Watters, *Inorg. Chem.,* **14,** 1419 (1973); A. L. Robinson, *Science,* **185,** 772 (1974).
87. P. H. Emmett, *Catalysis,* Reinhold, New York, 1954.
88. H. Pines and J. Manassen, *Adv. Catal.,* **16,** 49 (1966).
89. O. Beeck, A. Wheeler, and A. E. Smith, *Phys. Rev.,* **55,** 601 (1939).
90. A. A. Balandin, *Z. Phys. Chem.,* **B3,** 167 (1929).
91. G. A. Somorjai, *Catal. Rev.,* **7,** 87 (1972).
92. H. S. Taylor, *Proc. Roy. Soc.,* **A108,** 105 (1925).
93. M. Boudart, *Adv. Catal.,* **20,** 153 (1969).
94. R. J. Madix, *The Chemical Physics of Solid Surfaces and Heterogeneous Catalysis,* Vol. 4, Elsevier, Amsterdam, 1981.

95. J. Sonnemans, J. M. Janus, and P. Mars, *J. Phys. Chem.,* 2107 (1976).
96. G. Schwab, H. S. Taylor, and R. Spence, *Catalysis,* Van Nostrand, New York, 1937.
97. P. G. Ashmore, *Catalysis and Inhibition of Chemical Reactions,* Butterworths, London, 1963.
98. P. H. Emmett, *J. Chem. Educ.,* **7,** 2571 (1930); S. Brunauer and P. H. Emmett, *J. Am. Chem. Soc.,* **62,** 1732 (1940).
99. W. G. Frankenburg, *Catalysis,* Vol. 3, P. H. Emmett, Ed., Reinhold, New York, 1955, p. 171.
100. J. McAllister and R. S. Hansen, *J. Chem. Phys.,* **59,** 414 (1973).
101. G. G. Joris and H. S. Taylor, *J. Chem. Phys.,* **7,** 893 (1939).
102. M. Temkin and V. Pyzhev, *Acta Physicochim.* (*USSR*), **12,** 327 (1940).
103. K. S. Love and P. H. Emmett, *J. Am. Chem. Soc.,* **63,** 3297 (1941).
104. L. W. Anders, R. S. Hansen, and L. S. Bartell, *J. Chem. Phys.,* **62,** 1641 (1975).
105. R. B. Anderson, *Catalysis,* Vol. 4, Reinhold, New York, 1956, pp. 1, 29; see also H. H. Storch, *Adv. Catal.,* **1,** 115 (1948).
106. S. Weller, L. J. E. Hofer, and R. B. Anderson, *J. Am. Chem. Soc.,* **70,** 799 (1948).
107. J. T. Kummer, T. W. DeWitt, and P. H. Emmett, *J. Am. Chem. Soc.,* **70,** 3632 (1948).
108. O. C. Elvins and A. W. Nash, *Nature,* **118,** 154 (1926).
109. H. H. Storch, N. Golumbic, and R. B. Anderson, *The Fischer-Tropsch and Related Synthesis,* Wiley, New York, 1951. See also R. B. Anderson, *Catalysis,* Vol. 4, P. H. Emmett, Ed., Reinhold Publishing, New York, 1956.
110. J. T. Kummer, H. H. Podgurski, W. B. Spencer, and P. H. Emmett, *J. Am. Chem. Soc.,* **73,** 564 (1951).
111. G. Natta, *Catalysis,* Vol. 3, P. H. Emmett, Ed., Reinhold, New York, 1955.
112. D. D. Eley, *Catalysis,* Vol. 3, P. H. Emmett, Ed., Reinhold, New York, 1955, p. 49.
113. G. I. Jenkins and E. K. Rideal, *J. Chem. Soc.,* 2490 (1955).
114. D. W. McKee, *J. Am. Chem. Soc.,* **84,** 1109 (1962).
115. R. P. Eischens and W. A. Pliskin, *Adv. Catal.,* **10,** 2 (1958).
116. P. Mahaffy, P. B. Masterson, and R. S. Hansen, *J. Chem. Phys.,* **64,** 3911 (1976).
117. H. Imamura and W. E. Wallace, *J. Phys. Chem.,* **84,** 3145 (1980).
118. J. H. Sinfelt, J. L. Carter, and D. J. C. Yates, *J. Catal.,* **24,** 283 (1972).
119. J. H. Sinfelt, *Catal. Rev.,* **9,** 147 (1974).
120. R. V. Shankland, *Adv. Catal.,* **6,** 271 (1954).
121. L. B. Roland, M. W. Tamele, and J. N. Wilson, *Catalysis,* Vol. 7, P. H. Emmett, Ed., Reinhold, New York, 1960, p. 1.
122. C. L. Thomas, *Catalytic Processes and Proven Catalysts,* Academic, New York, 1970.
123. M. S. Wrighton, A. B. Ellis, P. T. Wolczanski, D. I. Morse, H. B. Abrahamson, and D. S. Ginley, *J. Am. Chem. Soc.,* **98,** 2774 (1976).
124. S. Sato and J. M. White, *Chem. Phys. Lett.,* **72,** 83 (1980).
125. T. Yamase and T. Ikawa, *Inorg. Chim. Acta,* **45,** L55 (1980).
126. A. H. Boonstra and C. A. H. A. Mutsaers, *J. Phys. Chem.,* **79,** 2025 (1975).

127. I. Izumi, W. W. Dunn, K. O. Wilbourn, F. F. Fan, and A. J. Bard, *J. Phys. Chem.*, **84,** 3207 (1980).
128. K. Kogo, H. Yoneyama, and H. Tamura, *J. Phys. Chem.*, **84,** 1705 (1980).
129. S. Sato and J. M. White, *J. Am. Chem. Soc.*, **102,** 7206 (1980).
130. T. Takizawa, T. Watanabe, and K. Honda, *J. Phys. Chem.*, **82,** 1391 (1978).
131. Y. Kubokawa and M. Anpo, *J. Phys. Chem.*, **79,** 2225 (1975).
132. G. A. Epling and E. Florio, *J. Am. Chem. Soc.*, **103,** 1237 (1981).
133. E. A. Perez-Albuerne and Y. Tyan, *Science*, **208,** 902 (1980).

Index

Abhesion, 428
Abrasion, surface effect of, 262
Accommodation coefficient, 520
Acid-base sites: in adsorption, 626
  in cracking reactions, 639
Acids, carboxylic, adsorption of: on carbon, 376, 381
  on metals, 378
  on Spheron 6, 376
  surface areas from, 381
Acids, carboxylic, films of: on solids, 173
  monolayers of, 124
    fluorinated, 138
    two-dimensional vapor pressures, 140
Activation energy in chemisorption, 611
Activators, in flotation, 438
Adhesion, 424–428
  and Antonow's rule, 424
  of bubbles in flotation, 442
  contact angle and, 425
  in detergency, 447
  energy versus free energy of, 427
  and friction, 414
  ideal, 424
  meniscus pressure and, 425
  peel test of, 425
  practical, 425
  roughness and, 428
  shear strength, 425
  weak boundary layer in, 426
  work of, 64, 424
Adhesion tension, 339
Adhesive failure, 427
Adsorbate: diffusion of, 510
  effect of, on adsorbent, 510
  effect of adsorption on, 503
  mobile *vs.* localized, 510
  see also Adsorption; Films; Monolayers
Adsorbate-adsorbent bond, 512
Adsorbates, photochemistry of, 507, 640
Adsorbed films: in boundary lubrication, 415
  structural perturbation in, 539, 545, 573
Adsorbent: effect of adsorbate on, 510
  thermodynamic quantities for, 563
Adsorbent-adsorbate bond, 512
Adsorption: adsorbent thermodynamic quantities in, 563
  apparent, 383
  from binary liquid systems, 382
  bond in, 512, 553, 618
  and capillary condensation, 584
  characteristic curve for, 544
  characteristic isotherm in, 551
  and contact angle, 356
  definition of, 492
  desorption rate in, 615
  of detergents, 454
  domain model for, 588
  effect of: on adsorbate, 503
    on adsorbent, 510, 564, 602
  of electrolytes, 388
  entropy of, 526, 561, 569, 581
    as criterion for monolayer point, 568
    on heterogeneous surfaces, 581
  expansion on solids due to, 284
  experimental procedures in, 531
  film pressure in, 336, 356, 539
  heat of, 561, 565
  Henry's law region of, 540, 553
  on heterogeneous surfaces, 372, 573, 607
    thermodynamics of, 580
  hysteresis in, 584
  at interfaces, interrelations between, 384
  isotherms: BET, 533, 570
    BET modifications, 538
    characteristic, 551

comparison of, 549
composite, 383
of composition change, 383
condensed film type, 542
dielectric, 510
Dubinin, 545
equation of state type, 539, 571
Fowler-Guggenheim, 529
Frenkel-Halsey-Hill, 547
Freundlich, 373, 574, 607
Harkins-Jura, 542
ideal two-dimensional gas type, 540
Kiselev, 530
Langmuir, 387, 521–531, 570, 607
Polanyi type, 543
polarization model, 547
potential theory type, 543, 572
stepped, 556
Temkin, 575, 608
t-plots of, 552, 572
types of, 534–535
van der Waals two-dimensional gas, 541
kinetics of, 611
lateral interaction in, 529, 609
localized *vs.* mobile, 528, 616
micropores and, 590
models, comparison of, 570
multilayer, 377, 533
negative, 393
of nitrogen on rutile, 518
phase transformations in, 556
physical, 335, 517
vs. chemisorption, 519
on porous solids, 582
rate of physical, 581
site energy distribution in, 574
from solution, 369
capillary condensation in, 377
constitutive effects in, 376
of dyes, 382
of electrolytes, 388
Freundlich equation for, 373
heat of, 388
Langmuir equation for, 371
multilayer, 377
of polymers, 378
reciprocity relations in, 387
and solubility, 375
surface area from, 381, 393
and surface heterogeneity, 372
standard states for, 562
statistical thermodynamics of, 503, 523
steps in, 556
submonolayer, potential theory and, 553
surface free energy change from, 335, 539
surface heterogeneity and, 372, 573, 607, 614
surface mobility in, 616
surface restructuring on, 511
thermodynamics of, 559, 580
time, 519
t-plot for, 552
see also Catalysis; Chemisorption
Adsorption isobars, 517
Adsorption isosteres, 517
Adsorption isotherm, 336, 370, 517
characteristic curve in, 544
characteristic isotherm in, 551
composite, 383
of composition change, 383
and contact angle, 356
exemplary, 557
isotherm function, 370
Adsorption time, 519
AEAPS, 297
Aerosol OT, emulsions with, 468
AES, 310, 296
Aging, of precipitates, 389
Albumin: bovine, adsorption on, 551
egg, adsorption on, 551
monolayers of, 154, 156
Alcohols: evaporation through films of, 144
lubrication by, 416
in micellar emulsions, 474
monolayers of, 130, 137
Alkali metal halides, surface energy of, 270
Alkali metals, adsorption of, on tungsten, 618
Alkanes contact angles of, on solids, 351
solid liquid interfacial tensions of, 327
Alkylammonium ions, adsorption of, 391–392
Alumina: adsorption on, 393
heat of immersion of, 335
Amagat equation, *see* Equation
Amalgams, electrocapillary effect with, 217
Amides, monolayers of, 130
Ammonia: decomposition of, 635
synthesis of, 634
Amontons' law, 402
n-Amyl alcohol, solutions of, 85
Angle of contact, *see* Contact angle
Antifoam agents, 486
Anomalous water, *see* Water
Antonow's rule, *see* Equation
APD, 295
Aphrons, 465
APS, 298
Area, surface, *see* Surface area
Argon: adsorption of, 566, 568, 578

molecular area of, 537
surface energy of, 270
Autophobic, *see* Films
Autophobic systems, 354

Balance, *see* Film, balance for
Barium stearate, built-up films of, 172
Barium sulfate: adsorption of ions by, 389
flotation of, 445
heat of immersion of, 335
stabilization of emulsions by, 470
Barytes, *see* Barium sulfate
Becker and Doring, *see* Equation
Beilby layer, 262
Benefaction, in flotation, 438
Benzene: adsorption of, on Graphon and Spheron, 385
from solution, 384
spreading coefficient of, on water, 106
solutions of, 67
BET, *see* Adsorption isotherm
Biological substances: as emulsifiers, 467
and long range forces, 247
at oil-water interfaces, 158
monolayers of, 153
Blacktop roads: stripping of, 428
water repellency of, 437
Blodgett, *see* Films
Boltzmann, *see* Equation
Bond: adsorbate-adsorbent, 512
chemisorption, 618
Bonding, hydrophobic, 245
Boron nitride, nitrogen adsorption on, 558
Boundary layer, weak, 426
Boundary lubrication, 415
mechanism of, 417
pressurized film model for, 421
Breaking of emulsions, 472
Bubbles, adhesion of, 442
Builders, in detergency, 454
Built-up films, 172
Burgers vector, 278
Butane, molecular area of, 537

Calcite, flotation of, 445
Calcium carbonate, decomposition of, 286
Calorimeter: for adsorption, 533
for heats of solution, 282
Camphor, movement of, on water, 100, 111
Capacity, of electrical double layer, 192, 209, 217
Capillaries, 10
rate of entry into, 436
vapor pressure in, 55
Capillary action: and adhesion, 425
electric field and, 35
and wetting, 436
Capillary adsorption, from solution, 377
Capillary condensation, 584
Capillary constant, 12
Capillary depression, 12
Capillary electrometer, 211
Capillary method, for electroosmosis, 201
Capillary rise: exact solutions for, 12
experimental aspects of, 17
weight of liquid column in, 13
and wetting, 436
Capillary waves, 168
damping of, 171
see also Surface tension
Captive bubble method, 342
Carbon: adsorption on, from solution, 385
argon adsorption on, 566
hydrogen adsorption on, 612
nitrogen adsorption on, 566, 573, 585
effect of benzene precoating, 523
pore size in, 499–500
stearic acid adsorption on, 382
see also Charcoal; Graphite; Graphon; Sterling carbon
Carbon dioxide, friction on solid, 410
Carbon monoxide: adsorption of, on nickel, 623
desorption of, from tungsten, 605
Carbon tetrachloride, surface tension and energy of, 51
Catalase, monolayers of, 158
Catalysis: kinetics of, 627
and adsorption isotherm, 631
Langmuir-Hinshelwood mechanisms, 631
mechanisms of, 627
Rideal mechanism for, 628
by semiconductors, 623
and surface LEED structures, 602
see also Adsorption; Chemisorption
Catalysts: acid-base sites of, 626
d-electrons, role of, 622
nitrogen adsorption on, 585
CELS, *see* Spectroscopy
Cetyl alcohol, 146
Chabasite, adsorption of gases by, 582
Chalk, squeaking of, 408
Characteristic adsorption curve, *see* Adsorption
Characteristic curve, *see* Adsorption
Charcoal, adsorption on, 385. *See also* Carbon
Charge: of electrical double layer, 186
at mercury-aqueous solution interface, 209
Charged monolayers, *see* Monolayers

Chemical potential, *see* Potential
Chemiluminescence of fatty acid films, 153
Chemisorption: acid-base sites in, 626
  activation energies in, 611
  bond in, 618
  boundary layer model for, 623
  and catalytic activity of metals, 621
  desorption rate in, 615
  and ESCA analysis, 604
  flash desorption in, 605
  heats of, 607, 618
  isotherms for, 607
  kinetics of, 611
  lateral interactions in, 609
  LEED structures in, 602
  localized bonds in, 618
  on metals, 621
  nature of, 519
  vs. physical adsorption, 493, 612
  rate of adsorption in, 613
  on semiconductors, 623
  site coordination number and, 610
  spectroscopic techniques in, 603
  surface heterogeneity and, 607
  thermodynamics of, 616
  work function changes in, 605
  see also Adsorption; Catalysis
Chlorophyll monolayers, 123, 153
Cholesterol, monolayers of, 138, 140, 157
Chromatography: and adsorption, 377
  of emulsions, 474
Clapeyron equation, *see* Equation
Clatherates, 583
Clays, structures of, 390
Cleaning, *see* Detergency
Cleaving of crystals, 281
Cloud chamber, 325
Clusters: critical, 321
  in nucleation, 320
Cohesion, work of, 64
Cold-working, 263
Collectors, in flotation, 438
Colloidal electrolytes, 448
  Kraft temperature of, 450
  micelle formation by, 449
  penetration of, into monolayers, 139
  phase map for, 452
  properties of, 448
  surface tension of aqueous, 26, 67
  see also Detergents; Micelles; Soaps
Colloids: flocculation of, 246
  stability of, 204
Condensation coefficient, 613. *See also* Nucleation
Condensors, steam, and water repellency, 437
Conductance: of electrolyte solutions, 199
  surface, 200
Contact angle, 338
  and adhesion, 425
  and adsorption behavior, 356
  apparent, 355
  of composite surface, 340
  and critical surface tension, 350
  in detergency, 447
  and film pressure, 356
  in flotation, 439
  and gravitational field, 348
  and heat of immersion, 351, 496
  hysteresis in, 344
  measurement of, 341
  microscopic, 355
  models for, 357
  of moving meniscus, 436
  nonuniform surfaces, and, 340
  on porous surface, 340
  and surface roughness, 340
  table of values of, 349
  temperature dependence of, 344, 349, 351
  thermodynamic, 355
  and water repellency of fabrics, 340, 437
  and wetting, 433
  Young's (or Young-Dupré) equation for, 338
Contact catalysis, *see* Catalysis
Contact potential, *see* Surface potential
Copper: adsorption of xanthates on, 443
  friction of, 411
  lifetime of surface atoms of, 262
  reaction of methanol on, 604
  surface tension of, 281
Copper oxide, reduction of, 287
Copper sulfate, hydration of, 285
Correlation times, *see* Nuclear magnetic resonance
Corrosion, 287
Cotton, adsorption on, 454
Coulomb's law, in friction, 407
Coulter counter, for emulsions, 463
Cracking, of hydrocarbons, 639
Creaming, of emulsions, 463, 471
Critical micelle concentration, 448
Critical surface tension, *see* Surface tension
Crystals: covalently bonded, 266
  defects in, 277
  dislocations in, 277
  edge energy of, 272
  equilibrium shape of, 264
  equilibrium surface of, 275

factors affecting surfaces of, 274
friction between, 412
growth of, 327
habit of, 266, 276, 328
heat capacity of, 283
ionic, 270
heats of solution of, 282
liquid, 251
molecular, 273
rare gas, 267
screw dislocations in, 278
slip plane in, 278
stresses in, 263
work of cleaving of, 281
see also Metals; Solids
Curvature: effect of, on surface tension, 55
radii of, 7
Cyclohexane: adsorption of, 386
solutions of, 67
Cylinders, instability of, 9

Darcy's law, 501
Debye, *see* Equation
Debye-Huckel theory, 187
Defects, *see* Crystals
Deformed interfaces, 34
Deposited films, *see* Built-up Films; Films
Depressants, in flotation, 438, 444
Desorption: in emulsion stability, 467
rates of, 615
Detergency, 446–456
contact angle and, 447
factors in, 452
and surface tension, 447
Detergents: adsorption of, 454
commercial, 455
see also Soaps
Diamond, 267
Dielectric constant: of adsorbates, 510
of thin films, 252
Dielectric loss, in adsorbed films, 509
Differential heat of adsorption, *see* Adsorption
Diffraction, electron: of films, 124
low energy electron (LEED), 305
in chemisorption, 602
Diffuse double layer, *see* Double layer
Dilational modulus, of surfaces, 90
Dipole-dipole forces, 234, 236
Dipole-induced dipole potential, 249
Discrete charge effect, 195, 227
Disjoining pressure, 248
Dislocations, *see* Crystals
surface, 277
Dispersion potential, *see* Potential
Dissolving rate, 285, 496
of monolayers, 146
and surface area, 496
Distearoyl-lecithin, monolayers of, 159
Distortion, surface, of crystals, 268, 270, 272, 308
Dividing surface, *see* Surface
Dixanthogen, in flotation, 443
DLVO theory, 244
Docosyl sulface, mixed films of, 141
Dodecahedra, in foams, 480
n-dodecane, solutions of, 67
Dodecyl ammonium chloride, in flotation, 445
Dodecyldimethylammonium chloride, surface tension of, 67
Domain model, in hysteresis, 588
Donnan effect, in reactions of monolayers, 151, 163. *See also* Equation
Double layer, 185
capacity of, 194
in charged films, 162
diffuse, 189, 195
discrete charge effect in, 195, 227
electric moment of, 200
and emulsions, 469
in flotation, 446
free energy of, 195
Helmboltz planes in, 195, 215
in overvoltage, 226
parts of, 193
repulsion between two, 196, 245
and solubility of powders, 333
Stern treatment, 191, 193
and zeta potential, 198
Drop weight method, *see* Surface tension
Dubinin-Radushkevich, *see* Equation
Du Noüy tensiometer, *see* Surface tension
Duplex film, 104, 109, 465
Dust abatement, 435
Dyes, adsorption of, 382

Edge energy, of crystals, 272
EELS, 296
Efflorescence, of solids, 286
Einstein law, for viscosity, 462
Effect: discreteness of charge, 195, 227
electrokinetic, 199
EID, 297
EIS, 296
Elasticity: of films, 484
of monolayers, 117
Electrical, *see* Double Layer; Films; Monolayers; Potential
Electrical potential, and surface area, 501

Electric field, effect of, on capillarity, 35. *See also* Capillary waves; Surface tension
Electrocapillarity, 206–217
Electrochemical, *see* Potential
Electrochemical potential, 220
Electrode phenomena, irreversible, 224
Electrode potentials, 222
Electrokinetic effects, 199. *See also* Zeta potential
Electrolytes: adsorption of, 388
  effect of on electrocapillary curve, 207
  see also Colloidal electrolytes; Salts
Electron diffraction, 262
  high energy (HEED), 295
  low energy (LEED), 305, 602
Electronegativity, and chemisorption bond, 620
Electron microscopy, of monolayers, 124, 134. *See also* Microscopy
Electron paramagnetic resonance, of adsorbates, 124, 507
Electroosmosis, 200
Electrophoresis, 199
Electro-polishing, 263
ElEED, 295
Ellipsometry: in corrosion studies, 287
  of films, 121
  and surface excess, 82
  and thickness of surface region, 57
Emersion, *see* Heat, of immersion
Emission, from monolayers, 135
Emulsification: mechanism of, 478
  spontaneous, 474
Emulsions, 462–478
  aging of, 470
  breaking of, 472
  chromatography of, 474
  coagulation rate of, 471
  creaming of, 463, 471
  and electrical double layer, 469
  electrical potential across interfaces in, 469
  flocculation of, 471
  hydrophile-lipophile balance and, 475
  inversion of, 470
  and long range forces, 467
  and Marangoni effect, 474
  micellar, 474
  micro, 474
  microgas, 479
  mobility of droplets in, 468
  oil-in-water, 462, 465, 470, 475
  phase inversion temperature of, 474
  phase volume ratio in, 462, 464
  practical aspects of, 477
  stabilization of, by solid particles, 470
    stearic and structural effects in, 466, 476
  stabilization theories of, 464
  viscosity of, 462
  water-in-oil, 462, 465, 470, 475
Energy, edge, of crystals, 272. *See also* Surface energy
Engulfment, 337
Enthalpy, *see* Surface enthalpy
Entropy: configurational, 526
  differential, 526, 528
  integral, 526
  see also Surface entropy
Enzymes, monolayers of, 158
Eötvös equation, *see* Equation
Equation: Amagat, 86
  Antonow, 109
  Becker and Doring, 321
  BET, 533, 549
  Boltzmann, 186
  Clapeyron, 130
  Debye-Hückel, 187
  Donnan, 151, 163
  Dubinin-Radushkevich, 545
  Elovich, 614
  Einstein, for viscosity, 462
  Eötvös, 51
  Fick, 146
  Fowler Guggenheim, 529
  Fowler and Nordheim, 302
  Frenkel-Halsey-Hill, 547
  Freundlich, 373, 574, 607
  Frumkin, 216
  Gibbs, 70, 209
    verification of, 78
  Gibbs-Duhem, 73
  Girifalco, Good, and Fowkes, 357
  Girifalco and Good, 107, 357
  Good and Fowkes, 107
  Gouy-Chapman, 187
  Guggenheim, 51
  Harkins Jura, 542
  Helmholtz condensor, 116, 188
  Kelvin, 54, 266, 332, 586
  Kirkwood-Muller, 555
  Kozeny, 501
  Langmuir, 88, 370, 521–531, 570, 607
    kinetic derivation of, 521
    linear form for, 522
    statistical derivation of, 523
  Lippmann, 208
  London, 235, 555
  Navier-Stokes, 168

Poiseuille, 118, 202
Poisson, 186
Polanyi type, 543
Ramsay and Shields, 51
Sackur-Tetrode, 528
Singer, 156
Smoluchowski, 201, 471, 500
Stokes-Einstein, 471
Szyszkowski, 70, 375
Tafel, 225
Tate, 20
Temkin, 575, 608
van der Waals, 86
two-dimensional, 541
Vonnegut, 35
Washburn, 436
Young, 338
thermodynamics of, 354
Young and Dupré, 338
Young and Laplace, 6
see also Law
Equations of state, *see* Adsorption
Equilibrium heat of adsorption, *see* Adsorption
ESCA, 311
Esin and Markov effect, 227
ESR, *see* Electron paramagnetic resonance
Esters, hydrolysis of monolayers of, 150
Ethanol, solutions of, 68, 76
Ethylene hydrogenation, 637
Evaporation, 143
coefficient, 145
EXAFS, 298
Exchange current, 226
Expansion of solids during adsorption, 284
Extinction coefficient, 122

Fabrics, water repellent, 437
Fatty acids, lubrication by, 416, 418
Faujasite, structure of, 627
FDS, 297
Feathers, contact angle on, 340
FEM, 300
Fick's law, 146
Field emission, microscopy, 300
and work function, 302
Field ion microscopy (FIM), 300
Films: autophobic, 173
balance for, 101, 112
black, 159, 482
built up (Blodgett), 172
charge, 162
chemiluminescence of, 153
drainage of, 480
duplex, 104, 109
in emulsions, 465
elasticity of, 484
at liquid-liquid interfaces, 161
osmotic model for, 88
of phospholipids, 158
polymer, 160
pressure of, 83, 112, 356
from adsorption, 539
in contact angle phenomena, 356
correspondence to bulk pressure, 128, 545
in non-wetting systems, 353
soap, 8, 480
spontaneous bursting of, 483
transfer of, 173
viscosity of, anomalous, 252
see also Adsorbates; Monolayers
FIM, 301
Fischer-Tropsch synthesis, 628, 636
Flash desorption, *see* Chemisorption
Flocculation: of emulsions, 471
of lyophobic colloids, 245
value, 205
Flotation, 438–446
bubble adhesion in, 442
contact angle and, 439
of metallic minerals, 443
of non-metallic minerals, 444
parlor demonstration of, 441
penetration effects in, 444
pH effect on, 445
Fluorescence, in monolayers, 123, 153
Foams, 478–487
drainage of, 483
elasticity of films in, 484
kugelschaum type, 478
microgas type, 479
polyederschaum type, 479
polyhedra present in, 480
practical aspects, 486
stability of, 484
structure of, 478
Forces, long range, 232–254
evidence for, 251
in solution, 244
verification of, 241
van der Waals, 233, 236
see also Potential
Formic acid, decomposition of, on nickel, 606
Fowkes, *see* Equation
Fowler and Nordheim, *see* Equation
Free energy, *see* Surface free energy; Surface tension

Freezing point, surface tension effect on, 333
Frenkel-Halsey-Hill, *see* Equation
Freundlich, *see* Adsorption; Equation
Friction: and adhesion, 402–428
area of contact in, 406
and boundary lubrication, 415
coefficient of, 415, 417
Coulomb's law in, 407, 415
of ice and snow, 410
interfacial potential and, 218
and load, 418
between nonmetals, 412
of oxide coated metals, 411
plastic deformation in, 406
on plastics, 413
rolling, 408
skid marks and, 409
and sliding speed, 415
static, 407
stick-slip, 407
temperature, local, during, 404
temperature variation of, 417
between unlubricated surfaces, 402
see also Boundary lubrications; Lubrication
Frothing agents, in flotation, 438

Galvani, *see* Potential
Gangue, in flotation, 438
Gaseous films, *see* Monolayers
Gibbs-Duhem, *see* Equation
Gibbs equation, *see* Equation
Gibbs monolayers, 83
Girifalco, *see* Equation
Glass: boundary lubrication of, 416
contact angle of alkanes on, 344
dispersion attraction of, 247
titania coated, contact angle on, 346
zeta potential of, 204
Glasses, formation of, 324
Gliadin, 154
Goethite, flotation of, 445
Gold: friction on, 411
surface tension of, 281
wetting of, by water, 348
Good, *see* Equation
Gouy, *see* Equation; Potential
Gouy-Chapman theory, 187
Graphite: adsorption of Kr on, 559
friction on, 412
stearic acid adsorption on, 382
see also Carbon; Charcoal; Graphon; Spheron
Graphon adsorption on, 385
fatty acid adsorption on, 382
heat of immersion of, 335
nitrogen adsorption on, 566, 568
see also Carbon; Charcoal; Graphite; Spheron
Guggenheim equation, *see* Equation

Hamaker constant, 240
Harkins and Jura, *see* Adsorption
Heat: of adsorption: in chemisorption, 607, 618, 621
differential, 560, 565
equilibrium, 561
from heats of immersion, 334, 565
and hysteresis, 589
integral, 559, 565
isosteric, 560
on metals, 621
rate effects due to, 582
from solution, 388
and surface heterogeneity, 580
of immersion, 334
and surface area, 495
of solution, *see* Surface energy
Heat development of nuclei, 327
HEED, 295
Helmholtz, *see* Double layer; Equation
Helmholtz condensor formula, *see* Equation
Hemi-micelle, 391
Henry's law, in adsorption, 540, 553
n-Heptane, solutions of, 87
n-Heptyl alcohol, solutions of, 85
Heterogeneity, *see* Surface
Hexafluoropropylene, friction on, 413
n-Hexane, adsorption of, 87
on Spheron, 568
Hexyl alcohol: adsorption of, on silica gel, 378
solutions of, 85
HLB, *see* Hydrophile-lipophile balance
Hofmeister series, 205, 214
Houdry process, 639
HRELS, 296
Hydrocarbons: adsorption of, on water, 87
cracking of, 639
lubrication by, 416
Hydrodynamic lubrication, 415
Hydroformulation reaction, 637
Hydrogen: adsorption bond with platinum, 512
chemisorption on Mo catalyst, 608
heats of adsorption on metals, 622
Hydrogenation of ethylene, 637
Hydrophile-lipophile, index number, 476
Hydrophile-lipophile balance, and emulsions, 475
Hydrophobic interactions, 244–245

Hydroxydecanoic acid, monolayer of, 134
α-Hydroxystearic acid, 150
Hysteresis: in adsorption, 584
  in contact angle, 344

Ice, abhesion of, 428
  argon, nitrogen, and carbon monoxide adsorption on, 568
  as clatherating solid, 584
  hexane adsorption on, 566
  and snow, friction on, 410
  water interface, interface, interfacial tension of, 333, 327
  see also Water
Ideal gas, two-dimensional, 84
Immersion, *see* Heat
Independent surface action, principle of, 63
Index of refraction, complex, 122
Infra-red, *see* Spectroscopy
Ink bottle pores, 500
Interactions, molecular, 233
Interfaces: liquid: lifetime of molecule at, 57
    treatment of, 56
  liquid-liquid, films at, 161
  mercury-aqueous solution, 212
  oil-water, protein films at, 158
  orientation at, 63
  silver halide, 218
  solid-gas, 260, 359, 492
    mobility of, 260
  solid-liquid, 332
  solid-melt, 333
  solid-vapor, 359, 517
  solution-solid, 369
  see also Surface
Interfacial tension, values of, 40. *See also* Surface tension
Interferometry, of films, 123
Intermediate films, *see* Monolayers
Inversion, of emulsions, 472
Ion atmosphere, 187
Ion dipole force, 234
Ion exchange, 394
IRE, 296
Iron catalyst, nitrogen adsorption on, 615
Irreversible electrode phenomena, 224
Isobar, *see* Adsorption
Isooctane, solutions of, 66
Isostere, *see* Adsorption
Isosteric heat of adsorption, 560
Isotherms, *see* Adsorption
ISS, 297

Jet, oscillating, *see* Surface tension

Kelvin, *see* Equation
Kiselev isotherm, 530
Kozeny, *see* Equation
Kraft temperature, 450
Krypton, friction on, 410
  molecular area of, 537
  in surface area determinations, 533
Kugelschaum, 478

Langmuir, *see* Adsorption; Equation
Langmuir-Hinshelwood mechanisms, 631
Langmuir-Schaeffer method for contact angle, 343
Laplace, *see* Equation
Lateral interaction, in adsorption, 391
Latex, adsorption on, 391
Lattice model, 158
Launderometer, 446
Lauric acid, boundary lubrication by, 418
Law: Amontons, 402
  Coulomb, 233, 237
  Coulomb's, in friction, 407
  Darcy's, 501
  Lennard-Jones potential, 268, 270
  parabolic, in corrosion, 286
  von Weimarn, 328
  see also Equation
LEED, adsorbate-substrate bond lengths from, 619. *See also* Electron diffraction
LEIS, 297
Lenses, 106, 111
Linear tension, 111, 348
Lipids, as emulsion stabilizer, 467
  films of, 158
Lippmann, *see* Electrocapillarity; Equation
Liquid crystals, 138, 173, 251
Liquid films, *see* Monolayers
Liquids: nature of interface of, 57
  supercooling of, 323, 326
  surface tensions of (Table), 40
London, *see* Potential
Long chain compounds, names for, 102
Long range forces, 232
Low energy electron diffraction, *see* Electron diffraction
Lubrication, boundary, 415
  hydrodynamic, 415
  see also Friction
Lyophobic cooloids, *see* Colloids

McBain balance, in adsorption, 532
Macroscopic objects, interaction between, 238
Magnetic field, *see* Surface tension
Marangoni effect, 110, 474

Maximum bubble pressure, *see* Surface tension
MBRS, 297
Melts, supercooling of, 323, 326
Membrane potential, *see* Donnan effect
Meniscus profile, 360
Mercury: adsorption of water vapor on, 96
  aqueous solution interface, 212
  films at surface of, 161
  supercooling of, 327
Mercury porosimeter, 498
Metals: boundary lubrication of, 411, 415
  chemisorption on, 622
  corrosion of, 286
  ethylene catalyzed reaction on, 622
  friction between, 411
  surface energy of, 273
  surface free energy of, 281
  tarnishing of, 286
  see also Crystals; Solids
Methanol, reaction of, on copper, 604
Mica: structure of, 390
  dispersion attraction of, 244
Micellar emulsions, 474
Micellar models for monolayers, 132
Micelles, 448
  diffusion of, 449
  solubilization by, 449
Microemulsions, 474
Microgas emulsions, *see* Foams
Micropores, 590
Microscopy, 294
  electron, 294
    of films, 134
  field emission (FEM), 300
  field ion (FIM), 300
  optical, 294
  scanning electron (SEM), 299
    of emulsions, 463
Microtome method, 79
Minerals, flotation of, 438
Mixing, excess free energy of, 141
Mobility, in chemisorption, 616
  of deposited films, 173
  diffusional, 423
  in physical adsorption, 528
  of solids, 260
Model, Good, and Fowkes, 357
  potential-distortion, 359
Molecular crystals, surface energy of, 273
Molecular sieves, 377, 583
Molybdenum disulfide, friction on, 412
Moment of inertia, 504
Monolayers: of biological substances, 153
  charged, 162–168
  compressibility of, 117
  diffraction by, 124
  dilational modulus of, 139, 141, 142
  effect of metal ions on, 135
  elasticity of, 170
  of esters, hydrolysis of, 150
  evaporation through, 143
  experimental methods for, 111
  free energy of mixing in, 139
  gaseous state of, 125, 128
  Gibbs, 83, 89
  intermediate state of, 132
  liquid states of, 125, 129–134
  on liquid substrates, 100
  microscopy of, 124
  mixed, 139
  molecular structure and type of, 136, 153
  nuclear magnetic resonance of, 124
  optical properties of, 121
  osmotic model for, 129
  penetration of, 139, 141, 142
  photochemistry of, 153
  polymerization in, 153
  of polymers, 160
  pressurized, in boundary lubrication, 421
  protein, 153–160
    at oil-water interface, 158, 161
    penetration of, 157
    reactions in, 157
    spreading of, 155
  rate of dissolving of, 146
  reactions in, 148–153
  rheology of, 135, 143
  at solid-gas interface, *see* Adsorption
  solid state of, 126, 133
  states of, 124
  substrate, effect of, on, 134
  surface potentials of, 114
  transfer of, 173
  vapor pressures (two-dimensional) of, 129, 140
  vapors adsorption by, 143
  viscosity of, 117–121, 135
  see also Adsorbate; Films
Monomolecular film, *see* Films; Monolayers
Multilayer adsorption, *see* Adsorption
Myristic acid, films of, 127, 140

Navier-Stokes, *see* Equation
Necking, of liquid column, 10
Negative adsorption, *see* Adsorption
Newmann's method for contact angle measurement, 342
Nickel, heat of chemisorption on, 619
  on carbon, 566, 567, 573

formic acid decomposition on, 606
on iron catalyst, 615
nitrogen adsorption on, 506
Nitriles, monolayers of, 130
Nitrogen, adsorption of: on boron nitride, 580
on carbon, 566, 567, 573
on iron catalyst, 615
molecular area of, 537
on nickel, 506
on silica, 579
on synthetic ammonia catalysts, 635
on titanium dioxide, 518, 581
t-plot for, 552
on various solids, 551
NMR, *see* Nuclear magnetic resonance
Nonelectrolytes, adsorption of from solution, 369, 382
Nonionic detergents, 456
Nonmetals, friction between, 412
NPD, 295
Nuclear magnetic resonance: of adsorbates, 507
correlation times of adsorbates, 507
of films, 124
of thin films, 252
Nucleation, 55, 319–328
by SO_3, 329

Octacosanoic acid, lubrication by, 418
Octadecanol: evaporation through films of, 145
evaporation resistance of, 145
mixed films of, 141
Octadecyltrimethylamine, 165
Octahedra, truncated, in foams, 480
n-Octane, adsorption of, 87, 88
n-Octyl alcohol, solutions of, 85
Oil, recovery of, 436
Oil slicks, 438
Oil-water interfaces: films at, 158, 161
potential at, 167
Oleic acid, mixed films of, 140
Oleyl alcohol, in emulsions, 466
Optical properties of films, 121
Orange OT, solubilization of, 449
Ores, flotation of, 443
Orientation, *see* Interfaces
Orientation force, 63, 234
Oriented wedge theory, 465
Oscillating jet method, 37
OSEE, 296
Osmotic pressure, 88
Osmotic pressure model for films, 129
Oulton catalyst, adsorption of nitrogen on, 585
Over-voltage, 224
Oxidation reactions in monolayers, 152
Oxide films, effect of, on friction, 411
Oxides, surface energy of, 273
Oxo-synthesis, 637
Oxygen, adsorption: on nickel oxide, 624
on titanium dioxide, 336

Palmitic acid, mixed films of, 140
Palmitoleic acid, mixed films of, 140
Paper, wetting of, 434
Parabolic law in corrosion, 286
Parachor, 52
Paraffin, surface tension of, 281
Paraffin oil: contact angle of, 347
emulsions of, 464
suspension of, 469
Particle size, 275. *See also* Surface area
Partition function, 503, 524
and Helmholtz free energy, 504
for rotation, 504
for translation, 525, 528
for vibration, 526
Pavement, friction of, 409
Peacock tails, in draining soap films, 482
Pendant drop, *see* Surface tension
Penetration, 139
in flotation, 444
see also Monolayers
Pentadecylic acid, films of, 131
n-Pentane, adsorption of, 87, 88
Pepsin, mixed films of, 158
Permeability and surface area, 501
Permittivity of vacuum, 190, 238
Persorption, 582
PES, *see* Spectroscopy
Phase changes: in breaking emulsions, 473
in monolayers, 128
in vapor adsorbed films, 556
Phase map for colloidal electrolytes, 452
Phase rule, two-dimensional, 127
Phase transformations in adsorption, 556
Phase volume ratio, in emulsions, 462, 473
PhD, 295
Phospholipids, 141, 158
Photochemical processes, 640
Photochemistry, of adsorbates, 507. *See also* Monolayers
Photoelectron, *see* Spectroscopy
Physical Adsorption, *see* Adsorption
Physisorption, *see* Adsorption
Plastics, friction of, 413
Plateau's border, 465, 479

PLAWM trough, 83
Plowing, in friction, 407
Pockels point, 101
Point of zero charge, 391
Poiseuille equation, *see* Equation
Poisson, *see* Equation
Polanyi adsorption model, *see* Adsorption
Polarization: in chemisorption, 609
  in electrocapillarity, 211
  of electrodes, 224
  see also Adsorption; Potential
Polishing, 262
Polonium electrode method, 115
Polydimethylsiloxane, viscosity of thin films of, 252
Polyederschaum, definition of, 479
Polyhexafluoropropylene, friction on, 413
Polymers: adsorption of, 378
  atactic, 160
  isotactic, 160
  Langmuir equation for adsorption of, 380
  monolayers of, 154
  syndiotactic, 160
  see also Films; Monolayers
Polytetrafluoroethylene: contact angle on, 348, 351
  heat of immersion of, 335
Polywater, *see* Water
Pore, ink bottle, 500
  size distribution of, 498, 587
Porosimeter, mercury, 498
Porous solids, adsorption by, 582
Porphyrin esters, monolayers of, 138
Portland cement, pore size distributions in, 588
Potassium chloride: adsorption of nitrogen on, 549, 551
  flotation of, 445
Potential: absolute difference between phases, 218, 223
  chemical, 193, 220
  contact, 114, 221
  Debye, 235
  determining ion for, 193, 392
  difference of between phases, 219
  dipole-induced dipole, 234, 249
  dispersion, 235, 239
    calculated and observed, 241, 555
    retarded, 240
  Donnan, 163
  electrochemical, 220
  electrode, 222
    irreversible, 224
  Galvani, 168, 219
    across interface, 167
  Gouy, 164
  interfacial, 168
  Keesom, 234
  London, 235
  orientation, 234
  polarization, 249–251
  rational, 212
  real, 220
  retarded, 240
  sedimentation, 203
  streaming, 202
  surface, 114, 161, 221
  surface jump, 219
  Volta, 219
  Zeta, 198
    in flotation, 446
Potential determining ion, 193
Potential-distortion model, 359
Potential theory, and two-dimensional equations of state, 545. *See also* Adsorption
Powders: contact angles for, 344
  emulsion stabilization by, 470
  heat capacity of, 283
  solubility of, 332
Pressure, effect of, on surface tension, 56
  disjoining, 248
  film, *see* Films, pressure of
Pressurized film model, 421
Protective action, in detergency, 454
Proteins, as emulsion stabilizers, 467
  molecular weight of, 153
  see also Monolayers
Proton magnetic resonance, *see* Nuclear magnetic resonance
Pyrex, adsorption of polymers on, 380
Pyridine, infra-red spectrum of adsorbed, 626

Quartz: attraction between plates of, 241
  heat of immersion of, 497
  spiral method for adsorption, 532
  zeta potential of, 392
  see also Silica

Radial distribution function, 62
Radioactive tracers, in monolayer studies, 81, 124
Radius of curvature, principle, 7
Raman spectroscopy, *see* Spectroscopy
Ramsay and Shields equation, *see* Equation
Rare gas crystals, surface energies of, 267
Rate of adsorption, 581, 611
Rate of spreading, 109
Reactions, *see* Monolayers

Real potential, *see* Potential
Rehbinder effect, 285
Resilience of films, and foam stability, 484
Retardation effect, in electrophoresis, 200
Retarded dispersion potential, 240
RHEED, 295
Rheology, *see* Monolayers
Rideal mechanism in catalysis, 628
Ring method, *see* Surface tension
Rotating drop, *see* Surface tension
Rotation effect, *see* Surface tension
Roughness, and adhesion, 428. *See also* Contact angle
Rule: Schulze-Hardy, 205, 246, 390
  Traube's, 92, 373
Ruthenium trisbipyridine: adsorption spectrum of adsorbed, 507
  hydrolysis of films of, 151
  monolayers of, 135
Rutile, nitrogen adsorption on, 518

Salts: friction between, 412
  powdered, solubility of, 332
  supersaturation of solutions of, 327
  see also Electrolytes
Scanning curves in adsorption, 585
Scanning electron microscopy, *see* Microscopy
Scattering, inelastic, of ions, 297
Schiller layers, 247
Schulze-Hardy, 205, 246, 390
Screw dislocation, *see* Crystals
Sedimentation, *see* Potential
Seizure, during friction, 408, 414
SEM, 299
Semiconductors, in catalysis, 623
Septum, in foams, 480
SERS, 506
Sessile bubble or drop, *see* Surface tension
Sessile drop method, 342
SEXAFS, 297
Shearing, in friction, 405, 414
SHEED, 295
Silica: dissolving rate of, 497
  heat of immersion of, 335
  nitrogen adsorption on, 579
  see also Quartz
Silica-alumina gel: acidity of surface sites of, 626
  pore size in, 500
Silica gel: adsorption on, 374, 376, 378
  contact angles on, 347
  hexyl alcohol adsorption on, 378
  nitrogen adsorption on, micropore analysis from, 588
  sulfur dioxide adsorption on, 544
Siloxane, polydimethyl, 171
Silver halide interface, 218
SIMS, 297
Singer, *see* Equation
Sintering, 261
SI system of units, 189, 237
Site energy distribution in adsorption, 574
  for rutile, 578
  for silica, 579
Skid marks, 409
Sliding speed, and friction, 405, 423
Smoluchowski, *see* Equation
Snow, friction on, 410
Soap bubble, pressure difference in, 6
Soap films, experiments with, 6. *See also* Emulsions; Films
Soaps, phase map for, 452. *See also* Colloidal electrolytes; Detergents
Sodium cetylsulfate, as emulsion stabilizer, 466
Sodium chloride: dissolving rate of, 497
  flotation of, 445
  solutions of, 68
  surface distortion of, 272
  surface energy of 273
Sodium dodecylbenzenesulfonate, adsorption of, on cotton, 454
Sodim dodecyl sulfate: adsorption of, on alumina, 393
  in flotation, 445
  penetration by, 141
  properties of aqueous, 448
  solutions of, 81
Sodium dodecyl sulfonate, in flotation, 445
Sodium laurate, foam stabilization by, 485
Sodium stearate, surface tension of aqueous, 33
Soil: removal of, 446
  standard, indetergency, 446
Solid-adsorbate complex, nature of, 503
Solid films, *see* Monolayers
Solid-melt interfacial tension, 327
Solids: area of contact between, in friction, 404
  effect of adsorbed gases on, 335
  expansion of, due to adsorption, 284
  friction between, 412
  heat of immersion of, *see* Heat
  nature of surfaces of, 274, 403, 502
  polishing of, 262
  rate of dissolving, 496
  reactions at surfaces of, 285

sintering of, 261
surface area of, 493
surface composition of, 310
surface defects of, 277
surface diffusion of, 261
surface free energy of, 263
surface mobility of, 261
surface spectroscopy of, 296
surface structure of, 275, 305
surface tension of, and rigidity modulus, 279
tensile strength of, 284
see also Crystals; Interfaces; Metals
Solid state, *see* Monolayers
Sols, flocculation rate of, 205
lyophobic, 245
Solubility, and surface free energy, 332
Solubilization, 449
Solutions: adsorption from, 369–396
regular, 65
supersaturation of, 328
surface tension of, 65–70
see also Surface tension
Specific heat, *see* Surface heat capacity
Specific surface area, *see* Surface Area
Spectroscopy: absorption: infra-red, 505
ultraviolet, 506
visible, 506
appearance potential (APS), 298
auger electron (AES), 310
characteristic energy loss (CELS), 296
electron, of chemisorbed species, for chemical analysis (ESCA), 311, 604
electron impact (EIS), 296
of films, infra-red, 123, 505
infra-red, of adsorbates, 505
of chemisorbed species, 603, 626
of silica, 505
ion neutralization (INS), 296
photoelectron, 297
Raman, of adsorbates, 506
secondary ion mass (SIMS), 297
soft x-ray (SXS), *see* SXAPS
soft x-ray adsorption (SXAS), 298
soft x-ray emission (SXES), *see* XES
ultraviolet photoelectron (UPS), 297
visible, of adsorbates, 507
x-ray photoelectron (XPS), 297
Spheron: adsorption on, from solution, 376
benzene adsorption on, 386
hexane adsorption on, 568
see also Carbon; Charcoal; Graphon
Spinning drop, *see* Surface tension
Spreading: and adhesion, 427
coefficient of, 104, 109, 339, 341
and wetting, 435
kinetics of, 109
of liquid on liquid, 104
of oleic acid, 110
prevention of, 110
Standard state, *see* Monolayers
Static friction, 407
Statistical thermodynamics, *see* Adsorption
Stearic acid: adsorption on graphitized surface, 382
films of, on mercury, 123
monolayers of, 136
solubility of monolayers of, 147
transfer ratio of, 124
Stearyltrimethylammonium ion, monolayers of, 165
Steel, friction on, 405, 413
Sterling carbon, *n*-hexane adsorption on, 568. *See also* Carbon; Charcoal; Graphite; Graphon
Stern, *see* Double layer
Stick-slip friction, 408
Stokes-Einstein, *see* Equation
Strain, *see* Crystal
Streaming, *see* Potential
Stress, surface, 272
Stribeck curve, 415
Structure, of solvent near surface, 133
Sugden's Parachor, *see* Parachor
Sulfur, monodisperse sols of, 328
Sulfur dioxide, adsorption of, on silica gel, 544
Supersaturation, 55, 328
critical pressures for, 326
see also Nucleation
Surface: composition determination of, 310–315
dividing, 59, 70
structure determination of, 305
Surface area: absolute, 493, 553
from adsorption models, 549
adsorption from solution and, 381, 393
from BET equation, 537
determination of, 381
from dissolving rate, 496
by dye adsorption, 382
from electrical potential, 501
from heat of immersion, 496
from Langmuir equation, 523
meaning of, 493
from mercury porosimeter, 498
from multilayer adsorption, 549, 551
from negative adsorption, 393
from permeability, 493
of solids, 493
from stearic acid adsorption, 382

from submonolayer adsorption, 553
see also Porous solids
Surface charge: and double layer theory, 188
effect of, on flotation, 445
in electrocapillarity, 209
and repulsion of spheres, 198, 245
Surface diffusion, of nickel, 617
Surface distortion of solids, by adsorption, 493, 511
surface tensional, 266, 272
Surface elasticity, *see* Elasticity
Surface energy, 50, 53, 61
from heats of solution, 282
of ionic crystals, 273
of rare gas crystals, 270
of solids, estimation of, 280
Surface enthalpy, 50
Surface entropy, 50
Surface excess: definition of, 59, 70
measurement of, 78
specific reduced, 370
Surface films, *see* Adsorption; Deposited films; Films; Monolayers
Surface free energy, 1, 49
and adsorption, 335
calculation of, for liquids, 61
of crystals, from work of cleaving of, 281
dispersion component of, table of values for, 108
from equilibrium crystal shape, 283
from solubility, 332
from strain rate of metal, 281
vs. surface tension, 58, 263
units of, 4
see also Surface tension
Surface heat capacity, 50
Surface heterogeneity, 267
and adsorption, 372, 573, 607
Surface mobility, in chemisorption, 616
Surface phase, thermodynamics of, 76, 141
Surface potential, 127
measurement of, 114
see also Potential
Surface region: pressure variation through, 60, 545
thickness of, 57
see also Interfaces
Surfaces: autophobic, 354
ellipticity of light reflected from, 57, 121
reactions at solid, 285
Surface stresses, of crystals, 272
Surface tension, 4
of colloidal electrolytes, *see* Colloidal electrolytes
and compressibility, 62
critical, 350, 352
and detergency, 447
dispersion component of, 108
effect of curvature on, 55
effect of electric field on, 36
effect of pressure on, 56
of fused salts, 69
measurement of: by capillary rise, 10
by capillary wave method, 38, 168
by drop weight method, 20
dynamic methods, 36
by film balance, 112, 161
by maximum bubble pressure, 18
by oscillating jet method, 37
by pendant drop method, 28
by ring method, 23
by rotating drop method, 35
by sessile drop or bubble method, 29, 161
by tensiometer method, 23
by Wilhelmy slide method, 24, 112, 161
mechanical analogy to, 57
role of, in emulsions, 464
of solids, experimental estimation of, 280, 332–337
theoretical estimation of, 266–274
of solutions, 65–70
vs. surface free energy, 4, 58, 263
temperature dependence of, 51, 52–54
units of, 4
values of, 40
see also Surface free energy
Surface to tension, 60
Surface viscosity, *see* Viscosity
Suspending power, in detergency, 453
SXAPS, 298
SXES, 296
Sylvite, *see* Potassium chloride
Szyszkowski, 375. *See also* Equation

Tafel, *see* Equation
Talc, friction on, 412
Tarnishing, 286
Tate, *see* Equation
TDS, 297
Teflon, *see* Polytetrafluoroethylene
Temkin, *see* Equation
Tensile strength, 284
Tensiometer, *see* Surface tension
Tension: linear, 111, 348
surface, 4
Tertiary oil recovery, 436
Thermionic work function, *see* Work function
Thermodynamic quantities for adsorbent, 563

Thermodynamics: of liquid interfaces, 49
statistical, of surfaces, 503
Thermodynamics of adsorption, 559, 580
on heterogeneous surfaces, 581
Thioindigo, monolayers of, 153
Thromboresistant films, 160
Tilting plate method, 343
Titanium dioxide: heat of immersion of, 335
nitrogen adsorption on, 518, 551, 576
oxygen adsorption on, 336
water adsorption on, 566
Tobacco mosaic virus, 247
Topochemical reactions, 286
Topotactic, 286
T-plots, *see* Adsorption
TPRS, 297
Transfer coefficient, 226
Transfer ratio, 124
Trimethylpentane, adsorption of, 87
Triolean, 152
Trypsin, films of, 158

Ultramicroscope, 124
Ultraviolet irradiations of films, 158
Units, SI, 237
UPS, 297

Van der Waals equation, *see* Equation
Van der Waals forces, 233
Vaporization coefficient, 261
Vapor pressure above curved surface, 54
Vapors: adsorption of, by monolayers, 143
supercooling of, 320, 326
Vibrating electrode method, 115
Viscometers, surface, 118–121
Viscosity: dilational, 117
of emulsions, 462
and foam stability, 485
of monolayers, 117, 161
shear, 117
see also Monolayers
Vitamin K_1, interaction with films of, 157
Volta potential, *see* Surface potential
Vonnegut, *see* Equation
von Weimarn's law, 328

Washburn, *see* Equation
Wasserman antibody, monolayers of, 157
Water: adsorption of: on aluminia, 627
on mercury, 96
on titanium dioxide, 566
anomalous, 55, 253
nmr of absorbed, 508
poly, *see* Water, anomalous
supercooling of, 327
supersaturation of vapor of, 326
vapor pressure above curved surface of, 54
vicinal, 252
see also Ice
Water-ethanol: surface excess quantities for, 76
surface tension of solutions of, 68
Water flooding, 436
Water-in-oil emulsions, 462
Water-organic liquids, works of adhesion of, 65
Waterproofing of fabrics, 340, 437
Water repellency, 437
Waves, *see* Capillary waves; Surface tension
Wear, and boundary lubrication, 419
Welding, during friction, 414
Wetting, 433–438
applications of, 434
as capillary phenomenon, 435
and contact angle, 433, 436
speed of, 435
Wilhelmy slide method: in contact angle measurement, 342
for film pressure measurement, 112, 161
for surface tension, 24
Wine, and Marangoini effect, 110
Wires, surface tensional contraction of, 280
Work of adhesion, *see* Adhesion
Work of cohesion, *see* Cohesion
Work function, 221, 302
and chemisorption, 603
Wulff construction, 264, 283

Xanthates, in flotation, 443
XES, 296
XPS, 297

Young, *see* Equation
Young and Dupre, *see* Equation
Young and Laplace, *see* Equation

Zeolites, 394, 583, 627
Zeta, *see* Potential
Zeta potential, and flotation, 445
Zinc oxide, adsorption on, 624
Zisman plot, 351